Manufacturing Planning and Control Systems

Manufacturing Planning and Control Systems

Thomas E. Vollmann
International Institute for Management Development, Switzerland

William L. Berry
University of North Carolina

D. Clay Whybark
University of North Carolina

Third Edition

IRWIN

Homewood, IL 60430
Boston, MA 02116

In memory of Joseph Orlicky and Oliver Wight, pioneers in manufacturing planning and control
and
Don Fogarty, a colleague in manufacturing planning and control, a friend, and a competitor

This symbol indicates that the paper in this book is made from recycled paper. Its fiber content exceeds the recommended minimum of 50% waste paper fibers as specified by the EPA.

Sponsoring editor: Richard T. Hercher, Jr.
Project editor: Jess Ann Ramirez
Production manager: Carma W. Fazio
Cover designer: Stuart Paterson, Image House, Inc.
Interior designer: Larry J. Cope
Compositor: Weimer Typesetting Co., Inc.
Typeface: 10/12 Times Roman
Printer: R. R. Donnelley & Sons Company

Library of Congress Cataloging-in-Publication Data

Vollmann, Thomas E.
    Manufacturing planning and control systems / Thomas E. Vollmann,
  William L. Berry, D. Clay Whybark.—3rd ed.
        p.   cm.
    Includes bibliographical references and index.
    ISBN 0-256-08808-X
    1. Production planning.   2. Production control.   I. Berry,
  William L.   II. Whybark, D. Clay.   III. Title.
  TS176.V63   1991
  658.5—dc20   91–26037

Printed in the United States of America
1 2 3 4 5 6 7 8 9 0 DOC 8 7 6 5 4 3 2 1

Preface

Recently one of the authors visited the coffee area of the DeKalb Market in Atlanta, Georgia. One stand offered coffee beans from several different countries, where they would roast and grind any mixture of beans to customer specifications. If the customer wanted, they would even add a flavoring—all while customers waited or finished their other shopping. After a few minutes' wait, one-half pound of Ethiopian Harrar dark roast, fine-ground coffee with no flavoring was testimony to the claims. When the story was repeated to a group of manufacturing executives later, their response was, "That's what manufacturing of the future will have to do."

Several observations of progressive firms confirm their intentions to become like the DeKalb Market coffee vendor. Major efforts to reduce the time to design, manufacture, and deliver products are under way. The programs have a variety of objectives, from reducing lead time to improving time-based competitive responses. Time compression is also found as part of total quality management (TQM) programs.

TQM programs and the Malcom Baldrige award have focused increased attention on the need for better quality. The processes for improving quality start with customer requirements, just as for the DeKalb Market coffee vendor. Customer focus has, in turn, led to customer-driven manufacturing and incorporating customer requirements more directly into the manufacturing processes.

Forecasting these requirements has not become any easier; in fact, just the contrary. In an informal poll of several hundred executives, not one said forecasting was getting easier. The implications are clear: if the demands can't be forecast, the manufacturing function must be designed to respond to the demands. To do this rapidly, more and more of the manufacturing decisions are being delegated to the factory floor. In essence, the customer is saying what is to be made, the due date is *now,* and the work force is figuring out how to do it—routinely!

As the manufacturing world moves toward the "zero everything" (zero inventory, zero lead time, zero defects, zero waste) vision of the future, fundamental changes will take place in the factory. These changes will necessitate changes in manufacturing planning and control (MPC) systems as well. This edition of the book addresses some of these changes and anticipates others.

COMMENTS ON THE THIRD EDITION

The many changes in competition, market demands and opportunities, global forces, and manufacturing technology mean many firms find themselves needing to change their MPC systems. This requires them to manage the transition from one configuration of manufacturing planning and control to another. Consequently, we have added a new chapter describing how changes in market requirements lead to necessary changes in MPC systems. Increasingly, leading-edge companies improve their competitiveness by integrating their MPC systems with their strategic needs. We also address the integration of the material requirements planning (MRP) and just-in-time (JIT) approaches, which may be required by the transition.

Another part of our response to the changes taking place in the manufacturing world was to move the just-in-time chapter forward in the book and give it a more thorough treatment. We have also increased the integration of the JIT viewpoint in most of the subsequent chapters.

We have also added a new chapter devoted to advanced JIT concepts. The chapter points the way to enhancements of existing MPC systems for JIT manufacturing. It presents a framework for consideration of refinements in practice and needs for research on JIT approaches.

In response to several reviewer suggestions, the chapter on purchasing has been eliminated in this edition. The key material on scheduling the "outside factory" (the vendor) has been included in this edition in a new chapter on production activity control.

All of the other chapters have had revisions and/or additions of new material. To provide reduced lead times and more rapid response to customer demands, many firms have moved much closer to flow-type manufacturing systems. In response, we have included new material on an MPC system developed for the process industries. We have also added some material on the theory of constraints to augment the section on OPT.

Although there have been many changes, it is important to draw attention to the things that have remained constant. Our original model of the manufacturing planning and control system continues to serve us well. It has been broad enough to survive the vast changes currently taking place in manufacturing and to include the new approaches to manufacturing planning and control that are being developed. We have made some minor modifications to the model (and, more important, to the book itself) to overcome what many users thought was an excessive MRP bias. The revised model accommodates the newer approaches; all the functions we describe must be performed. Different circumstances require different emphases or responses for some of the functions, but we have not encountered a company that did not perform all that are described in our model.

Many reviewers and users of the book have suggested that we need to change the chapter sequence. We continue to be surprised by the variety of suggestions that have been made. We believe there are many different customer requirements for chapter order, each of which represents a valid way of teaching a course on manufacturing planning and control. We have tried, therefore, to minimize roadblocks to the way *you* want to go through the material. The chapters are relatively independent, so you can develop your own path through them and not feel obligated to use our sequence.

That said, we have largely stuck with our original design. We have found that starting with MRP record processing has been pedagogically sound. We find that understanding time-phased records is helpful for understanding the entire MPC system. At the same time, we have also positioned the JIT material early so that readers can view MPC systems with the broader perspective provided through understanding JIT, as well as how improvements in manufacturing processes have a profound impact on MPC system design.

The division between the basic material in the first half of the book and the advanced material in the second also has stood up well, and we have maintained that separation. If we have any suggestions for the user, they would be to cover the MRP and JIT chapters first and to do the basic chapter in a subject before the advanced (even though we have seen some course outlines where the ordering of material started with the latter chapters).

APICS CERTIFICATION EXAMINATION PROGRAM

The American Production and Inventory Control Society (APICS) has an ongoing program of certification in production and inventory management (CPIM) based on successful completion of five of the following examinations:

- Capacity management
- Inventory management
- Just-in-time
- Master planning
- Material requirements planning
- Production activity control

The specific content of the examinations is constantly being updated, and it is very important to secure copies of the study guides and sample examination questions from APICS as part of any plan of study. Each study guide has a list of references, as do the chapters in this book. APICS also provides training aids and books of reprinted articles that show the concepts in practice. The *APICS Dictionary* is also an important reference for the certification examinations.

Using the combination of this textbook and the materials provided by APICS is an excellent way to prepare for the certification examinations. The APICS publications tend to focus on case examples rather than on basic methods. The latter type of material is better provided in a text. A similar difference applies to the problems furnished from both sources. The sample tests provided by APICS use the short answer format found in the examinations. In fact, all of the sample questions (40 for each examination) are test items that were on examinations at one time. In this book, we ask more open-ended kinds of problems that require calculation and discussion. In some cases, there are no single right answers; discussion of the alternatives leads to a deeper understanding of manufacturing planning and control systems.

In the balance of this section, we detail the primary and linkage chapters that address the subject matter of each examination. We also provide information on what is not in our book, sources where the gaps can best be filled, and at least one more reference that complements the materials presented here. We do, however, feel the need to add a few words of caution. Manufacturing planning and control cannot be easily partitioned into five or six parts. This is a dynamic, ever-changing field that needs to be understood in its entirety. Moreover, the changes will continue, and professionals in this area need a framework for understanding change and seeing the opportunities for their companies. We feel that the framework will come from reading all of this book, or at least the first 10 chapters. This set of chapters is recommended reading before going into the more detailed study of each area in the remaining chapters.

Master Planning

The primary chapters dealing with the materials in the master planning examination are 6, 7, 8, 9, 14, 15, 16, and 18. Chapters 1, 2, 3, 4, and 19 describe linkages to other MPC systems. Chapter 10 deals with issues of implementation. The primary chapters provide coverage of most of the subject matter in this examination. There are two notable exceptions, however. The examination requires some knowledge of basic statistics including the standard deviation and areas under the normal distribution curve. There is also a need to understand some fundamentals of extrinsic forecasting, such as regression. An article by Chambers, Mullick, and Smith in the July 1971 *Harvard Business Review* provides a good overview of these ideas. For readers interested in more depth on forecasting, we recommend the book *Forecasting Methods for Management,* 5th edition, by Makridakis and Wheelwright (Wiley, 1989).

Material Requirements Planning

The primary chapters addressing the material requirements planning examination are 1, 2, 9, 10, 11, and 18. Chapters 3, 4, 5, 6, 7, 8, and 19 deal with linkages to other MPC systems. In addition, Chapter 17 provides background information on basic inventory models. The primary chapters provide coverage of all of the topical areas of the examination, except for the interfaces with other MPC systems. These are covered in the chapters dealing with linkage. For readers desiring another source of reading, we recommend the book *MRPII: Making It Happen,* 2nd edition, by Tom Wallace (Oliver Wight Limited Publications, 1990).

Inventory Management

The primary chapters that delineate the material for the inventory management examination are 2, 3, 6, 11, 16, 17, and 18. Chapters 4, 5, 7, 8, and 9 provide the linkages to other MPC systems and concepts. Chapter 10 deals with implementation issues. This examination also requires knowledge of some basic concepts in statistics and accounting. Included are statistical concepts like the areas under the normal distribution curve and accounting methods such as LIFO versus FIFO inventory valuation. We recommend the book *Production and Inventory Management,* 2nd edition, by Fogarty, Blackstone, and Hoffmann (South-Western, 1991) as an additional reference.

Capacity Management

The primary chapters dealing with the materials found on the capacity management examination are 1, 3, 4, 7, 8, and 15. The chapters that provide direct linkages to other MPC systems are 2, 5, 6, and 9. Chapter 10 treats questions of implementation. The capacity management examination requires understanding the difference between capacity and load and the difference between nominal, standing, or rated capacity and actual or demonstrated capacity. Also required is awareness of efficiency and utilization. These are explained in the APICS study guide. We also recommend that *Capacity Management Reprints* (APICS, 1987) be studied for this examination.

Production Activity Control

The primary set of chapters for preparation for the production activity control examination are 1, 2, 3, 4, 5, and 13. Chapters providing linkages to production activity control are 6, 7, 8, and 9. Chapter 10 is concerned with implementation issues. These chapters deal with essentially all of the materials outlined in the study guide, with the exception of some basics in cost accounting such as the costing of work-in-process inventory, shrinkage, and scrap, as well as standard costing techniques and variances. A good additional reading is the book *Production Activity Control* by Carter and Melnick (Dow Jones-Irwin, 1987).

Just-in-Time

The primary set of chapters for the just-in-time (JIT) examination module includes 3, 5, 9, 12, and 19. Chapters 1, 4, 7, 8, 10, and 17 all provide linkages or supportive theory. In fact, JIT has been integrated into most chapters of this new edition. Chapter 9 deals with how market requirements dictate MPC system choices and the integration of MRP and JIT approaches to MPC system design. We note in Chapter 3 (which focuses most directly on JIT) that JIT is much more than MPC systems. The examination reflects this wider scope and includes topics such as total quality control, process analysis and layout, JIT philosophy, and implementation issues. Good additional sources for the JIT examination are *World Class Manufacturing* by R. J. Schonberger (Free Press, 1986) and *Attaining Manufacturing Excellence: Just-in-Time Manufacturing, Total Quality, Total People Involvement* by R. W. Hall (Dow Jones-Irwin, 1987).

ACKNOWLEDGMENTS

All editions of *Manufacturing Planning and Control Systems* have benefited from the comments of reviewers. In the first edition, we had:

Gene Groff, Georgia State University,
Robert Millen, Northeastern University, and
Richard Penlesky, Marquette University.

Reviewers for the second edition were:

Jeff Miller, Boston University,
William Sherrard, San Diego State University, and
Urban Wemmerlov, University of Wisconsin.

In this edition, we were helped by the reviews of:

Stanley Brooking, University of Southern Mississippi,
Henry Crouch, Pittsburgh State University,
Marilyn Helms, University of Tennessee,
Robert Johnson, Pennsylvania State University,

Ted Lloyd, Eastern Kentucky University,
Dan Reid, University of New Hampshire,
Dwight Smith-Daniels, Arizona State University,
Herman Stein, Bellarmine College, and
Glen Wilson, Middle Tennessee State University.

As before, our ideas have been greatly shaped by the manufacturing practitioners and college students whom we teach and the colleagues at our schools. The deans who have supported us in this edition are Paul Rizzo, University of North Carolina; George McGurn, Boston University; and Juan Rada, IMD. We thank all of these friends and associates for their support and help.

Bob Fetter, our consulting editor for the first and second editions, was a champion as well as a critical reviewer. In addition, special thanks go to Herman Stein, who double-checked all of the problem material in this edition, and to Cecil Bozarth and Rick Metters for their work in completely rewriting the instructor's manual.

As in the past, if there are any errors, each of us intends to blame them on his coauthors.

Finally, we want to express our appreciation to our wives for their support during all those days spent working on the book: Tani, Jane, and Neva.

Thomas E. Vollmann
William L. Berry
D. Clay Whybark

Contents

Chapter 1

Manufacturing Planning and Control

In this chapter we describe key functions of the manufacturing planning and control (MPC) system. The MPC system concerns planning and controlling the *manufacturing process* (including materials, machines, people, and suppliers). Both the MPC system and the manufacturing process are designed to meet the dictates of the marketplace and to support overall company strategy. An effective MPC system can provide substantial competitive advantage for a company in its markets. However, what's effective today may not be tomorrow. Markets, technologies, and competitive pressures all change constantly. Changes in the company and manufacturing strategies may be required as a result. These in turn may require changes in the processes and MPC system. This chapter presents a framework for thinking about functions performed by the MPC system, their design, and appropriate responses to changes in markets, technologies, and competitive pressures. Specifically, this chapter is organized around three managerial concerns:

- The MPC system defined: What are the typical tasks performed and what can they do for a company?
- An MPC system framework: What are the key system components and how do they respond to the company's needs?
- Evolution of the MPC system: What forces drive changes in the MPC system and how do companies respond to the forces?

This chapter also provides the model for later chapters in the book. Each will start with an introduction describing the chapter's content and highlighting managerial questions addressed. We also direct you to related parts of the book in this introductory section. You won't see, within the chapters, footnotes or references to other pages or chapters of the book. Each chapter concludes with a set of principles, references, and problems.

THE MPC SYSTEM DEFINED

Here we define the MPC system, describe benefits it provides for the firm, and indicate surrounding management issues. Basically the MPC system provides information to efficiently manage the flow of materials, effectively utilize people and equipment, coordinate internal activities with those of suppliers, and communicate with customers about market requirements. A key in this definition is managers' need to use the information to make intelligent decisions. The MPC system does not make decisions or manage the operations—managers perform those activities. The system provides the support for them to do so wisely.

Typical MPC Support Tasks

Typical management activities supported by the MPC system include:

- Plan capacity requirements and availability to meet marketplace needs.
- Plan for materials to arrive on time in the right quantities needed for product production.
- Ensure utilization of capital equipment and other facilities is appropriate.
- Maintain appropriate inventories of raw materials, work in process, and finished goods—in the correct locations.
- Schedule production activities so people and equipment are working on the correct things.
- Track material, people, customers' orders, equipment, and other resources in the factory.
- Communicate with customers and suppliers on specific issues and long-term relationships.
- Meet customer requirements in a dynamic environment that may be difficult to anticipate.
- Respond when things go wrong and unexpected problems arise.
- Provide information to other functions on the physical and financial implications of the manufacturing activities.

Costs and Benefits of MPC Systems

In many firms, accomplishing all these activities requires a large number of MPC professionals. Often the MPC system tasks involve the largest number of indirect persons working in the manufacturing area—sometimes in the whole company. In many companies, manufacturing planning and control has been a big problem. These companies are characterized by poor customer service, excessive inventories, inappropriate equipment or worker utilization, high rates of part obsolescence, and large numbers of expediters dedicated to "fire fighting." These symptoms of an inappropriate or ineffective MPC system are the bane of many managers and their firms. In fact, poor MPC performance has often been a major cause of firm liquidation. For example, one large U.S. producer of farm equipment allowed its inventories of finished goods to get so large and unmatched to the required market mix, that the firm was necessarily acquired by another company—for a fraction of what the company might have been worth.

In contrast, some companies have realized handsome payoffs from their investments in MPC systems. Consider the results reported by the following:

- The Tennant company, in a two-year period with its MPC system in place, achieved the following payoffs:

  Purchased material inventory reduced by 42 percent ($3,129,000).

  Production rate increased by 66 percent.

  Assembly efficiency increased from 45 to 85 percent.

  Delivery promises met on time increased from 60 to 90 percent.

- In Europe, Kumera OY implemented an MPC system in six months during a period of heavy competitive pressure and:

  Tripled its gross margin.

  Increased inventory turnover from 2.5 to 10 times per year.

  Eliminated late delivery penalties.

  Used funds generated from these improvements to install new equipment that provided a distinct competitive advantage.

- In the emerging economies of Eastern Europe, MPC is paying off. In Hungary, for example, the Videoton company:

  Reduced inventories despite high levels of uncertainty.

  Increased flexibility to respond to market changes.

  Improved the utilization of its labor force and equipment.

- Even our friends in the developing world have realized benefits from investments in MPC systems. In the Peoples' Republic of China, for example, the Optical Equipment Company (using a personal computer):

  Reduced work-in-process inventories by 20 percent.

  Improved equipment utilization.

  Increased the profit rate by 5.4 percent.

  Decreased late deliveries.

AN MPC SYSTEM FRAMEWORK

Companies carry out manufacturing planning and control activities in several different areas in different degrees of detail. The needs of different firms call for different activities, as well. In this section we provide an overall framework for viewing these different aspects of the MPC system.

MPC System Activities

Overall direction for manufacturing planning and control is provided by a company game plan linking and coordinating the company's various departments (e.g., engineering, marketing, finance). The game plan is top management's responsibility. It should always be consistent with strategic plans, departmental budgets, and the firm's capabilities. Manufacturing plans, an integral part of the game plan, specify the production output required to achieve the overall objectives.

In any firm, manufacturing planning and control encompasses three distinct aspects or phases. The first phase is *creating the overall manufacturing plan* for the manufacturing part of the company game plan. It must be stated in production terms, such as end items or product options. The second phase is *performing the detailed planning of material and capacity needs* to support the overall plans. The third and final MPC phase is *executing these plans* on the shop floor and in purchasing. Figure 1.1 depicts these three phases.

The System and the Framework

Figure 1.1, a simplified schematic of a modern MPC system, shows the skeletal framework for all the required activities. The full system includes other data inputs, system modules, and feedback connections. Figure 1.1 is divided into three parts. The top third, or **front end,** is the set of activities and systems for overall direction setting. This phase establishes the company objectives for manufacturing planning and control. Demand management encompasses forecasting customer/end product demand, order entry, order promising, accommodating interplant and intercompany demand, and spare parts requirements. In essence, demand management coordinates all activities of the business that place demands on manufacturing capacity. Production planning provides the production input to the company game plan and determines the manufacturing role in this agreed-upon strategic plan. The master production schedule (MPS) is the disaggregated version of the production plan. That is, it states which end items or product options manufacturing

FIGURE 1.1 Manufacturing Planning and Control System (simplified)

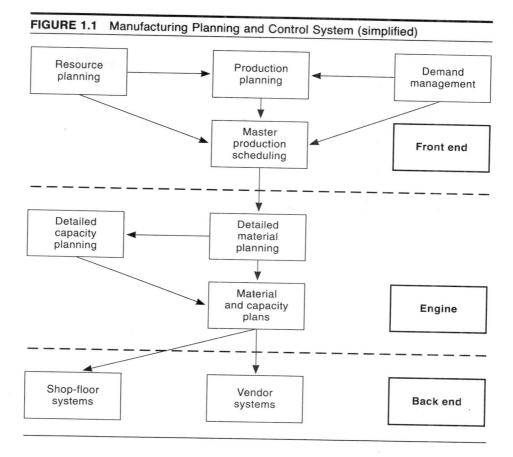

will build in the future. The MPS must sum up to the production plan. Resource planning provides the capacity necessary to produce the required product now and in the future. In the long run this means bricks and mortar, while in the short run it means labor and machine hours. Resource planning provides the basis for managing the match between manufacturing plans and capacity.

The middle third, or **engine,** in Figure 1.1 is the set of systems for accomplishing the detailed material and capacity planning. The master production schedule feeds directly into the detailed material planning module. Firms with a limited product range can specify rates of production for developing these plans. However, for firms producing a wide variety of products with many parts per product, detailed material planning can involve calculating requirements for thousands of parts and components, using a formal logic called **material requirements planning (MRP).** MRP determines (explodes) the period-by-period (time-phased) plans for all component parts and raw materials required to produce all the products in the MPS. This material plan can thereafter be utilized in the detailed capacity planning systems to compute labor or machine center capacity required to manufacture all the component parts.

The bottom third, or **back end,** of Figure 1.1 depicts the execution systems. Here again, the system's configuration depends on the process's needs. For example, firms producing a large variety of products using thousands of parts often group all equipment of a similar type into a single work center. Their shop-floor control systems establish priorities for all shop orders at each work center so the orders can be properly scheduled. Other firms will group mixtures of equipment that produce a similar set of parts into work centers called **production cells.** For them, production rates and **just-in-time (JIT)**–based shop-floor systems for execution are appropriate. Purchasing systems provide detailed planning information for vendor scheduling. This information relates to existing purchase orders as well as to planned purchase orders.

This three-phase framework for manufacturing planning and control is supported by a widely available set of MPC systems and software, from master production scheduling to the back-end systems. Moreover, the software is integrated to follow the framework. That is, the MPS produces the right input for the development of detailed material and capacity plans, which in turn provides the right input to the execution systems.

In firms using MRP systems, execution of the detailed material and capacity plans involves the scheduling of machine and other work centers. In these factories, scheduling must reflect such routine events as starting and completing orders for parts and any problem conditions, such as breakdowns or absenteeism. These schedules must usually be updated at least once per day in factories with complex manufacturing processes for producing parts and components.

Purchased parts require an analogous detailed schedule. In essence, purchasing is the procurement of outside-work-center capacity. It must be planned and scheduled well to minimize final customers' overall cost. Good purchasing systems typically separate the procurement activity from routine order release and order follow-up. Procurement, a highly professional job, involves contracting for vendor capacity and establishing ground rules for order release and order follow-up.

Some firms have used the execution phase for other critical activities. For example, the Elliot company uses a direct analog of shop-floor scheduling to plan and control its engineering and drafting activities.

A final activity tied to execution is the measurement of actual results. As products are manufactured, the rate of production and timing of specific completion can be compared to plans. As shipments are made to customers, measures of actual customer service can be obtained. As capacity is used, it too can be compared to plans.

Matching the MPC System with the Needs of the Firm

The requirements placed on the design of an MPC system will vary with the nature of the production process, customer expectations, and the needs of management. In addition, MPC requirements are not static. As we get some aspect of the firm under routine control, we have an opportunity to tackle new problems. The result is differing emphasis on the various MPC system modules.

MPC technology has changed over time. Material requirements planning, for example, was never a practical approach until random access computers became available. A more recent change is the use of on-line systems. On-line systems allow a fundamental operating difference. Printed paper reports are dramatically reduced, and the planning process can be redone on a daily cycle. This reduces inventories and has other benefits, but it also makes planning and execution of MPC systems much more dynamic. Changes in the way the users deal with the system and do their routine work become necessary.

The process itself is not static, which complicates things even more. In some of the most recent advances in MPC systems, the major breakthroughs have been physical process changes. Job shops have been configured to become lines. Use of small groups of equipment dedicated to production of a group of parts or products (cellular manufacturing) is growing. Fundamental changes are occurring in the relationships between firms and their suppliers. Just-in-time manufacturing as well as just-in-time purchasing are growing in importance.

There's an important distinction to draw in the process of matching MPC system design to firm requirements. The strategy is *not* the system. The system supports the execution of the strategy. Perhaps the best analogy is to a household thermostat. The thermostat measures the temperature and tells the furnace when to add more heat. This is the system. The homeowner sets the desired temperature. That is the strategy. The distinction is important because although the system is quite general, its use is very individual. Only after clearly understanding the strategy, the associated set of tasks the system is to perform, and the elements of the system itself, can the appropriate match be made.

An MPC Classification Schema

Figure 1.2 shows the relationship between MPC system approaches, the complexity of the manufactured product as expressed in the number of subparts, and the repetitive nature of production, expressed as the time between successive units. Figure 1.2 also shows some example products that fit these time and complexity scales.

FIGURE 1.2   MPC Classification Schema

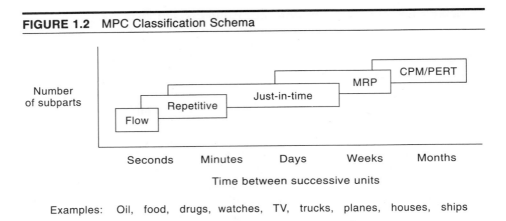

Examples: Oil,   food,   drugs,   watches,   TV,   trucks,   planes,   houses,   ships

Several MPC approaches presented in Figure 1.2 are appropriate for products that fit in various points in the schema. The figure demonstrates that the MPC emphasis changes as the nature of the product, process, or both changes. For example, as a product grows in volume over time, the MPC emphasis might shift from right to left. In any event, it's necessary to perform all the activities depicted in Figure 1.1. How they are performed, however, changes significantly.

The lower left-hand corner of Figure 1.2 shows a flow-oriented manufacturing process typical of many chemical, food, petroleum, and bulk product firms. Since products are produced in streams instead of discrete batches, virtually no time elapses between successive units. The MPC systems primarily concern flow rates that become the master production schedule. Typically, these products have relatively few component parts, so "engine" management is straightforward. Depending on how components are purchased, the back end may involve some complexity. Typically, these firms' major cost is for raw materials; transportation costs can also be significant.

Repetitive manufacturing activities are found in many plants that assemble similar products (e.g., automobiles, watches, microcomputers, pharmaceuticals, and televisions). For such products, component-part management is necessary, but everything is coordinated with the flow or assembly rate for the end items.

In the middle of the figure we show a large application area for just-in-time systems. Many firms today try to move processes from right to left in the figure. That is, they try to make processes more repetitive as opposed to unique, and to achieve the MPC operating conditions of repetitive manufacturing (shorter cycles, reduced lead times, lower inventories, and the like). JIT is shown as spanning a wide variety of products and processes. This MPC approach is increasingly being integrated with more traditional MRP-based systems. The goal is to achieve better MPC system performance and to reduce costs of maintaining the MPC system.

Figure 1.2 also shows material requirements planning as spanning a wide area. MRP is key to any MPC system involving management of a complicated parts situation. The majority of manufacturing firms have this sort of complexity, and MRP-based systems continue to be widely applied. For many firms, successful use of MRP is an important step in evolving their approaches to MPC. Once routine MRP operation is achieved, portions of the product and processes that can be executed with JIT methodologies can be selected.

The last form of MPC depicted in Figure 1.2, the project type, is applied to unique long lead time products, such as ships and highly customized products. Here the primary concern is usually management of the time dimension. Related to time is cost. Project management attempts to continually assess partially completed projects' status in terms of expected completion dates and costs. Some firms have successfully integrated MRP approaches with the problems of project management. This is particularly effective in planning and controlling the combined activities of engineering and manufacturing.

EVOLUTION OF THE MPC SYSTEM

The activities shown in Figure 1.1 are performed in every manufacturing company, whether large or small. MPC system design, however, depends strongly on the company's attributes at a particular point in time. The key to keeping the MPC system matched to evolving company needs ensuring system activities are synchronized and focused on the firm's strategy. This ensures the detailed MPC decision making is in harmony with the company's game plan. But the process is not static—the need for matching is ongoing.

The Changing Competitive World

Figure 1.3 depicts some manufacturing firms' typical responses to changing marketplace dictates. New technology, products, processes, systems, and techniques permit new competitive initiatives; global competition intensifies many of these forces. Marketplace dictates drive revisions in company strategy, which in turn often call for changes in manufacturing strategy, manufacturing processes, and MPC systems.

Shorter product life cycles come about partly because consumers have access to products from all over the world. This has spawned the move to "time-based competition." Who can get to the market quickest? Similarly, today's market insists on ever higher quality requirements, which in turn have led to many changes in manufacturing practices. Cost pressures have trans-

---

FIGURE 1.3 Evolutionary Responses to Forces for Change

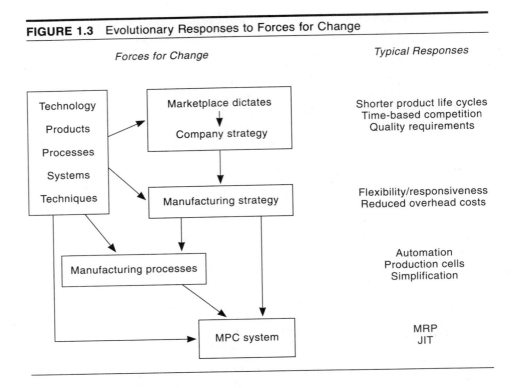

Forces for Change

Typical Responses

Shorter product life cycles
Time-based competition
Quality requirements

Flexibility/responsiveness
Reduced overhead costs

Automation
Production cells
Simplification

MRP
JIT

---

lated into reductions of all manufacturing cost components from material and labor to overhead and energy.

But increasingly, cost and quality are the ante to play the game—winning requires flexibility and responsiveness in dealing with ever more fickle customer demands. Clearly, these pressures and responses require changes in both the MPC system and the underlying manufacturing processes. As Figure 1.3 shows, typical MPC responses are MRP and JIT. Process responses include automation, simplification, and production cells for cellular manufacturing.

Reacting to the Changes

If the MPC system has remained unchanged for a significant length of time, it may no longer be appropriate to the company's needs. The system, as the strategy and processes themselves, must change to meet the dictates of the market. In many instances, this may imply simply a different set of evaluative criteria for the MPC system. In other cases, new modules or information may be required. In yet other cases, entire MPC activities may need to be eliminated. For example, JIT systems frequently move materials so quickly through the factory that shop-floor systems to track them are not needed.

The need for evolution in MPC systems implies the need for periodic auditing that compares system responses to the marketplace's requirements. The audit must address not only the system's focus, but also the concomitant training of people and the match with current objectives. Although the MPC framework in Figure 1.1 is general, its application is specific and evolving. Keeping it on track is an essential feature of MPC itself.

CONCLUDING PRINCIPLES

This chapter lays the groundwork for the rest of the book. Defining and adjusting the MPC system to support the manufacturing activity are a dynamic challenge. We hope that, as you read the rest of the book, you constantly ask how the general framework applies in specific instances, and what is happening to ensure a better match between MPC system design and marketplace dictates. From the chapter we draw the following principles:

- Manufacturing planning and control systems should support the strategy and tactics pursued by the firm in which they are implemented.
- Different manufacturing processes often dictate the need for different designs of the MPC system.
- An effective MPC system can contribute to competitive performance by lowering costs and providing greater responsiveness to the market.
- The framework for MPC is general, and all three phases must be performed; but specific applications necessarily reflect particular company conditions and objectives.
- The MPC system's design will vary, depending on the distinctive needs of the firm and the current state of its MPC system.
- The system should evolve to meet changing requirements in the market, technology, products, and manufacturing processes.

REFERENCES

Blackburn, Joseph D., ed. *Time-Based Competition: The Next Battle Ground in American Manufacturing.* Homewood, Ill.: Business One Irwin, 1991.

Buffa, E. S., and J. G. Miller. *Production-Inventory Systems: Planning and Control.* 3rd ed. Homewood, Ill.: Richard D. Irwin, 1979.

Byrkett, D. L.; M. N. Ozden; and J. M. Patton. "Integrating Flexible Manufacturing Systems with Traditional Manufacturing Planning and Control." *Production and Inventory Management Journal* 29, no. 3 (3rd quarter 1988).

Fogarty, D. W.; J. H. Blackstone, Jr.; and T. R. Hoffmann. *Production and Inventory Management.* 2nd ed. Cincinnati: Southwestern, 1991.

Hall, R. W. *Attaining Manufacturing Excellence.* Homewood, Ill.: Dow-Jones Irwin, 1987.

Harmon, R. L., and L. Anderson. *Reinventing Manufacturing.* New York: Free Press, 1990.

Holstein, W. K. "Production Planning and Control Integrated." *Harvard Business Review,* May/June 1968.

Hyer, N. L. "The Potential of Group Technology for U.S. Manufacturing." *Journal of Operations Management* 4, no. 3 (1984).

Jacobs, F. R., and V. A. Mabert. *Production Planning, Scheduling, and Inventory Control: Concepts, Techniques, and Systems.* 3rd ed. Atlanta: Institute of Industrial Engineering, Monograph Series, 1986.

Johnson, G. A. *APICS Bibliography.* Falls Church, Va.: American Production and Inventory Control Society, 1981.

Krajawski, L. J.; B. E. King; L. P. Ritzman; and D. S. Wong, "Kanban, MRP, and Shaping the Manufacturing Environment," *Management Science* 33, no. 1 (1987) pp. 39–57.

Miller, J. G. "Fit Production Systems to the Task." *Harvard Business Review,* January/February 1981.

Mize, J. H., and D. J. Seifert. "CIM—A Global View of the Factory." IEE Fall Conference, 1985.

Plossl, G. W. *Manufacturing Control—The Last Frontier for Profits.* Reston, Va.: Reston, 1973.

————. *Production and Inventory Control.* 2nd ed. Englewood Cliffs. N. J.: Prentice-Hall, 1985.

Rosenthal, S. R. "Progress toward the 'Factory of the Future.'" *Journal of Operations Management* 4, no. 3 (1984).

Schmenner, R. W., and R. L. Cook, "Explaining Productivity Differences in North Carolina Factories." *Journal of Operations Management* 5, no. 3 (1985).

Schultz, T. "MRP to BRP: The Journey of the 80's." *Production and Inventory Management Review and APICS News,* October 1981, pp. 29–32.

Van Dierdonck, R., and J. G. Miller. "Designing Production Planning and Control Systems." *Journal of Operations Management* 1, no. 1 (1980).

Wallace, T. F. *MRPII: Making it Happen.* Essex Junction, Vt.: Oliver Wight, Limited Publications, 1985.

Wight, O. W. *Production and Inventory Management in the Computer Age.* Boston: Cahners Books International, 1974.

DISCUSSION QUESTIONS

1. A fundamental principle of MPC systems is that management decisions can be improved with better information systems. Provide some examples of this concept from your own experience. For instance, when would you have been able to make better use of your time had you had better information?

2. The discussion of the framework for manufacturing planning and control seems to imply the overall direction setting must be done before detailed material and capacity planning activities can be accomplished. The latter must be done before executing the plans is possible. Do you agree? Give an example supporting your position.

3. Apply the MPC framework to a college setting. In particular, identify the front end, engine, and back end.

4. Assuming college students are "raw materials" processed into "finished goods," what various transactions need to be processed as this occurs? What happens if they aren't done well?

5. Someone asks you to advise her on installing an MPC system. She starts with the process of selecting the type of system. What questions do you pose?

Chapter 2

Material Requirements Planning

This chapter deals with material requirements planning (MRP), a basic tool for performing the detailed material planning function in the manufacture of component parts and their assembly into finished items. MRP is used by many companies that have invested in batch production processes. MRP's managerial objective is to provide "the right part at the right time" to meet the schedules for completed products. To do this, MRP provides *formal* plans for each part number, whether raw material, component, or finished good. Accomplishing these plans without excess inventory, overtime, labor, or other resources is also important.

Chapter 2 is organized around the following seven topics:

- Material requirements planning in manufacturing planning and control: Where does MRP fit in the overall MPC system framework and how is it related to other MPC modules?
- Record processing: What is the basic MRP record and how is it produced?
- Technical issues: What additional technical details and supporting systems should you recognize?
- Using the MRP system: Who uses the system, how is it used, and how is the exact match between MRP records and physical reality maintained?
- System dynamics: How does MRP reflect changing conditions, and why must transactions be processed properly?
- The MRP data base: What computer files support MRP?
- MRP system examples: How does MRP play out in actual firms?

MRP's relationship to other manufacturing planning and control (MPC) concepts is shown in Chapter 1. Many just-in-time (JIT) concepts have emerged as basic approaches for designing MPC systems in some companies. These concepts are discussed in Chapter 3. Advanced MRP techniques are presented in Chapter 11.

MATERIAL REQUIREMENTS PLANNING IN MANUFACTURING PLANNING AND CONTROL

Joe Orlicky, whom many authorities regard as the father of modern MRP, talked of MRP as a "Copernican revolution." MRP is as different from the traditional approaches to manufacturing planning and control as the Copernican model of the earth rotating around the sun was to the older model of the sun rotating around the earth. For companies assembling end items from components produced in batch manufacturing processes, MRP is central to the development of detailed plans for part needs. It is often where companies start in developing their MPC systems. Facility with time-phased planning and the associated time-phased records is basic to understanding many other aspects of the MPC system. Finally, although introduction of JIT and investments in repetitive manufacturing processes have brought about fundamental changes in detailed material planning for some firms, companies continue to adapt the MRP approach or enhance their existing systems. For these reasons, we have chosen to describe MRP early in the book.

For firms using MRP, the general MPC framework depicted in Figure 2.1 shows that detailed requirements planning is characterized by the use of time-phased (period-by-period) requirement records. Several other supporting activities are shown in the front end, engine, and back end of the system as well. The front end of the MPC system produces the master production schedule (MPS). The back end, or execution system, deals with shop-floor scheduling of the factory and with managing materials coming from vendor plants.

The detailed material planning function represents a central system in the engine portion of Figure 2.1. For firms preparing detailed material plans using MRP, this means taking a time-phased set of master production schedule requirements and producing a resultant time-phased set of component part and raw material requirements.

In addition to master production schedule inputs, MRP requires two other basic inputs. A bill of material shows, for each part number, what other part numbers are required as direct components. For example, for a car, it could show five wheels required (four plus the spare). For each wheel, the bill of materials could be a hub, tire, valve stem, and so on. The second basic input to MRP is inventory status. To know how many wheels to make for a given number of cars, we must know how many are on hand, how many of those are already allocated to existing needs, and how many have already been ordered.

The MRP data make it possible to construct a time-phased requirement record for any part number. The data can also be used as input to the detailed capacity planning models. Developing material and capacity plans is an iterative process where the planning is carried out level by level. For example, planning for a car would determine requirements for wheels, which in turn determines requirements for tires, and so on. But planning for tires has to be

FIGURE 2.1   Manufacturing Planning and Control System

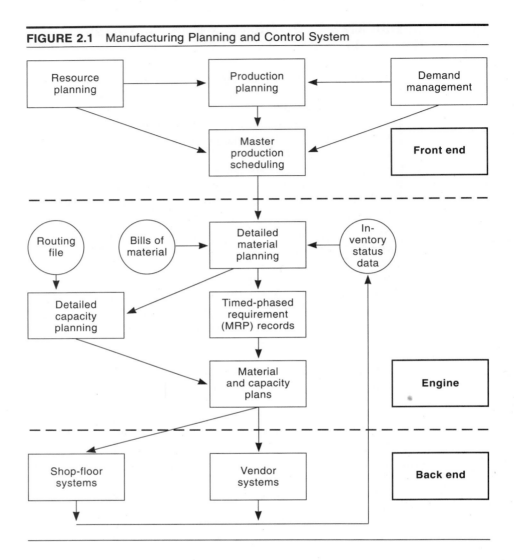

done *after* the planning for wheels; if the company wants to build 10 cars (50 wheels) and has 15 complete wheels on hand, it only needs 35 more—and 35 tires. If 20 wheels have already been ordered, only 15 more must be made to complete the 10 cars.

An MRP system serves a central role in material planning and control. It translates the overall plans for production into the detailed individual steps necessary to accomplish those plans. It provides information for developing capacity plans, and it links to the systems that actually get the production accomplished.

MRP and MRPII

The "engine" section of Figure 2.1 provides disaggregation of a master production schedule into the resultant detailed plans for each manufactured and purchased part number. That is, the MPS is a plan for end items or product options as offered to the customers. MRP is the detailed planning process for components to support the MPS. The box labeled "Detailed material planning" in this section represents the system that does the disaggregation. MRP was a major breakthrough in MPC made feasible by random access computers and data base management systems. Later it was realized that due dates for shop orders could be updated by MRP replanning. This resulted in recasting MRP as more than a disaggregation technique for planning; it was now also seen as a dynamic priority-setting scheme so shop-floor and vendor operations could do a better job of execution.

When execution on the shop floor was improved, attention naturally turned to the front end. The question was how to establish and maintain a viable master production schedule—one that could be executed. When better master production scheduling was incorporated into MRP-based MPC systems, people began to describe them as **closed loop MRP systems.**

Additional enhancements included better capacity planning procedures at the front end, engine, and back end levels. As this occurred, users of these systems began to consider them less as MPC systems and more as company-wide systems. It was now possible to include financial plans based on the detailed MPC planning process. Because execution was improved, the resultant plans became more and more believable. Simulation possibilities were added, along with various ways to examine "what-if" scenarios. This overall vision of an MRP system for planning and controlling company operations was so fundamentally different from the original concepts of MRP that a new term seemed appropriate. Oliver Wight coined the term **MRPII.** In this case, MRP did not stand for material requirements planning; MRPII means **manufacturing resource planning.** The old term MRP is now sometimes referred to as "mrp" or "little MRP." These terms are now widely accepted, and most MPC professionals clearly understand the distinction between MRP, mrp, and MRPII.

In this book we will continue to use the more generic term, manufacturing planning and control (MPC) systems. Our reasoning will become clear as we go along, but the fundamental issue is one of evolution. Just as *little MRP* has evolved into *MRPII,* there have been further improvements (such as just-in-time) and more are coming. The question is what to call all of this with a minimum of confusion. We find it useful to distinguish between the front end or direction-setting phase of MPC, the detailed disaggregation phase, and the execution aspects associated with detailed shop-floor schedules and procurement. We will see that some of the latest MPC enhancements work in more than one of these areas at the same time.

RECORD PROCESSING

In this section, we present the MRP procedures starting with the basic MRP record, its terminology, timing conventions, and construction. We then turn to an example illustrating coordination of planning component parts and end items. We examine several aspects of this coordination and the relationships that must be accounted for. We then look at linking MRP records to reflect all the required relationships. We intend to show clearly how each MRP record can be managed independently while the *system* keeps them coordinated.

The Basic MRP Record

At the heart of the MPC system is a universal representation of the status and plans for any single item (part number), whether raw material, component part, or finished good: the MRP time-phased record. Figure 2.2 displays the following information:

The anticipated future usage of or demand for the item *during* each period (i.e., **gross requirements**).

Existing replenishment orders for the item due in at the *beginning* of each period (i.e., **scheduled receipts**).

The current and projected inventory status for the item at the *end* of each period (i.e. **projected available balance**).

Planned replenishment orders for the item at the *beginning* of each period (i.e., **planned order releases**).

The top row in Figure 2.2 indicates periods that can vary in length from a day to a quarter or even longer. The period is also called a **time bucket.** A

FIGURE 2.2   The Basic MRP Record

Period		1	2	3	4	5
Gross requirements			10		40	10
Scheduled receipts		50				
Projected available balance	4	54	44	44	4	44
Planned order releases					50	
Lead time = 1 period Lot size = 50						

widely used time bucket or period is one week. A timing convention is that the current time is the beginning of the first period. The initial available balance of four units is shown prior to period 1. The number of periods in the record is called the **planning horizon.** In this simplified example, the planning horizon is five periods. The planning horizon indicates the number of future periods for which plans are made.

The second row, "Gross requirements," is the anticipated future usage of (or demand for) the item. The gross requirements are **time phased,** which means they're stated on a unique period-by-period basis, rather than aggregated or averaged; that is, gross requirements are stated as 10 in period 2, 40 in period 4, and 10 in period 5, rather than as a total requirement of 60 or as an average requirement of 12 per period. This method of presentation allows for special orders, seasonality, and periods of no anticipated usage to be explicitly taken into account. A gross requirement in a particular period will be unsatisfied unless the item is **available** during that period. Availability is achieved by having the item in inventory or by receiving either a scheduled receipt or a planned replenishment order in time to satisfy the gross requirement.

Another timing convention comes from the question of availability. The item must be available at the *beginning* of the time bucket in which it's required. This means plans must be so made that any replenishment order will be in inventory at the beginning of the period in which the gross requirement for that order occurs.

The "Scheduled receipts" row describes the status of all open orders (work in process or existing replenishment orders) for the item. This row shows the quantities ordered and when we expect these orders to be completed. Scheduled receipts result from previously made ordering decisions and represent a source of the item to meet gross requirements. For example, the gross requirements of 10 in period 2 cannot be satisfied by the 4 units presently available. The scheduled receipts of 50, due in period 1, will satisfy the gross requirement in period 2 if things go according to plan. Scheduled receipts represent a commitment. For an order in the factory, necessary materials have been committed to the order, and capacity at work centers will be required to complete it. For a purchased item, similar commitments have been made to a vendor. The timing convention used for showing scheduled receipts is also at the *beginning* of the period; that is, the order is shown in the period during which the item will be available to satisfy a gross requirement.

The next row in Figure 2.2 is "Projected available balance." The timing convention in this row is the *end* of the period; that is, the row is the projected balance *after* replenishment orders have been received and gross requirements have been satisfied. For this reason, the "Projected available balance" row has an extra time bucket shown at the beginning. This bucket shows the balance *at the present time;* that is, in Figure 2.2, the beginning available balance is 4 units. The quantity shown in period 1 is the projected balance at the *end* of period 1. The projected available balance shown at the end of a period is available to meet gross requirements in the next (and succeeding) periods. For example, the 54 units shown as the projected available balance at the end of period 1 result from adding the 50 units scheduled to be received to the beginning balance of 4 units. The gross requirement of 10 units in period 2 reduces the projected balance to 44 units at the end of period 2. The term projected *available* balance is used, instead of projected *on-hand* balance, for a very specific reason. Units of the item might be on hand physically but not available to meet gross requirements because they are already promised or allocated for some other purpose.

The "Planned order releases" row is determined directly from the "Projected available balance" row. Whenever the projected available balance shows a quantity insufficient to satisfy gross requirements (a negative quantity), additional material must be planned for. This is done by creating a planned order release in time to keep the projected available balance from becoming negative. For example, in Figure 2.2, the projected available balance at the end of period 4 is 4 units. This is not sufficient to meet the gross requirement of 10 units in period 5. Since the lead time is one week, the MRP system creates a planned order at the beginning of week 4 providing a **lead time offset** of one week. As we have used a lot size of 50 units, the projected available balance at the end of week 5 is 44 units. Another way that this logic is explained is to note that the balance for the end of period 4 (4 units) is the beginning inventory for period 5, during which there's a gross requirement of 10 units. The difference between the available inventory of 4 and the gross requirement of 10 is a **net requirement** of 6 units in period 5. Thus, an order for at least 6 units must be planned for period 4 to avoid a shortage in period 5.

The MRP system produces the planned order release data in response to the gross requirement, scheduled receipt, and projected available data. When a planned order is created for the most immediate or current period, it is in the **action bucket.** A quantity in the action bucket means some action is needed now to avoid a future problem. The action is to release the order, which converts it to a scheduled receipt.

The planned order releases are *not* shown in the scheduled receipt row because they haven't yet been released for production or purchasing. No material has been committed to their manufacture. The planned order is analogous to an entry on a Christmas list, since the list comprises plans. A scheduled receipt is like an order mailed to a catalog firm for a particular Christmas gift, since a commitment has been made. Like Christmas lists versus mailed orders, planned orders are much easier to change than scheduled receipts. Not converting planned orders into scheduled receipts any earlier than necessary has many advantages.

The basic MRP record just described provides the correct information on each part in the system. Linking these single part records together is essential in managing all the parts needed for a complex product or customer order. Key elements for linking the records are the bill of materials, the explosion process (using inventory and scheduled receipt information), and lead time offsetting. We consider each of these before turning to how the records are linked into a system.

An Example Bill of Materials. Figure 2.3 shows a snow shovel, end item part number 1605. The complete snow shovel is assembled (using four rivets and two nails) from the top handle assembly, scoop assembly, scoop-shaft connector, and shaft. The top handle assembly, in turn, is created by combining the welded top handle bracket assembly with the wooden handle using two nails. The welded top handle bracket assembly is created by welding the

FIGURE 2.3 The 1605 Snow Shovel Shown with Component Parts and Assemblies

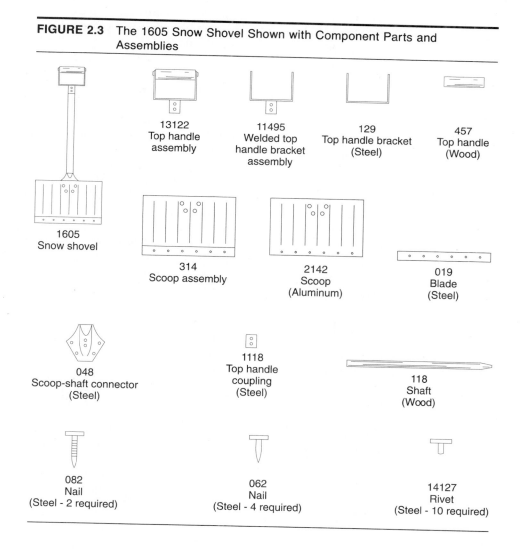

top handle coupling to the top handle bracket. In a similar way, the scoop assembly combines the aluminum scoop with the steel blade using six rivets.

Explaining even this simple assembly process is a cumbersome task. Moreover, such diagrams as Figure 2.3 get more complicated as the number of subassemblies, components, and parts used increases, or as they are used in increasingly more places (e.g., rivets and nails). Two techniques that get at this problem nicely are the **product structure diagram** and the **indented bill of materials (BOM)** shown in Figure 2.4. Both provide the detailed information of Figure 2.3, but the indented BOM has the added advantage of being easily printed by a computer.

Note that both the product structure diagram and the indented BOM show exactly what goes into what instead of being just a parts list. For example, to make one 13122 top handle assembly, we see by the product structure diagram that one 457 top handle, two 082 nails, and one 11495 bracket assembly are needed. The same information is shown in the indented BOM; the three required parts are indented and shown, one level beneath the 13122. Note also that we *don't* need a top handle bracket (129) or a top handle coupling (1118) to produce a top handle assembly (13122). These are only needed to produce a bracket assembly (11495). In essence, the top handle assembly does not care *how* a bracket assembly is made, only that it *is* made. Making the bracket assembly is a separate problem.

Before leaving our brief discussion of bills of material, it is important to stress that the bill of material used to support MRP may differ from other company perceptions of a bill of materials. The BOM to support MRP must be consistent with the way the product is manufactured. For example, if we're making red cars, the part numbers should be for red doors. If green cars are desired, the part numbers must be for green doors. Also, if we change to a different set of subassemblies, indentations on the BOM should change as well. Engineering and accounting may well not care what color the parts are or what the manufacturing sequence is.

Gross to Net Explosion.   **Explosion** is the process of translating product requirements into component part requirements, taking existing inventories and scheduled receipts into account. Thus, explosion may be viewed as the process of determining, for *any* part number, the quantities of *all* components needed to satisfy its requirements, and continuing this process for *every* part number until all purchased and/or raw material requirements are exactly calculated.

As explosion takes place, only the component part requirements net of any inventory or scheduled receipts are considered. In this way, only the *necessary* requirements are linked through the system. Although this may seem like an obvious goal, the product structure can make determination of net requirements more difficult than it seems. To illustrate, let's return to the snow shovel example.

FIGURE 2.4 Parts for Snow Shovel

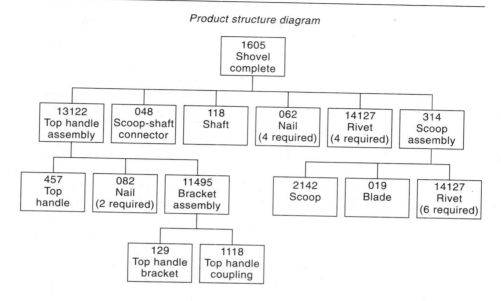

Product structure diagram

Indented bill of materials (BOM)

1605 Snow Shovel

    13122 Top Handle Assembly (1 required)
       457   Top Handle (1 required)
       082   Nail (2 required)
      11495   Bracket Assembly (1 required)
         129   Top Handle Bracket (1 required)
        1118 Top Handle Coupling (1 required)

    048   Scoop-Shaft Connector (1 required)
    118   Shaft (1 required)
    062   Nail (4 required)
    14127   Rivet (4 required)
    314   Scoop Assembly
      2142   Scoop (1 required)
      019   Blade (1 required)
      14127   Rivet (6 required)

Suppose the company wanted to produce 100 snow shovels, and we were responsible for making the 13122 top handle assembly. We are given current inventory and scheduled receipt information from which the gross requirements and net requirements for each component of the top handle can be calculated, as shown in Figure 2.5.

FIGURE 2.5 Gross and Net Requirement Calculations for the Snow Shovel

Part description	Part number	Inventory	Scheduled receipts	Gross requirements	Net requirements
Top handle assembly	13122	25	—	100	75
Top handle	457	22	25	75	28
Nail (2 required)	082	4	50	150	96
Bracket assembly	11495	27	—	75	48
Top handle bracket	129	15	—	48	33
Top handle coupling	1118	39	15	48	—

The gross and net requirements in Figure 2.5 may not correspond to what we feel they should be. It might at the outset seem that since one top handle coupling (1118) is used per shovel, the gross requirements should be 100 and the net requirement 46, instead of the 48 and zero shown. To produce 100 shovels means we need (have a demand for) 100 top handle assemblies (part 13122). Twenty-five of these 100 can come from inventory, resulting in a net requirement of 75. As we need to make only 75 top handle assemblies, we need 75 top handles and bracket assemblies. This 75 is the *gross* requirement for parts 457 and 11495 (as indicated by the circled numbers in Figure 2.5). Since 2 nails (part 082) are used per top handle assembly, the gross requirement for 082 is 150. The 25 units of top handle assembly inventory contain some implicit inventories of handles, brackets, and nails, which the gross to net process takes into account. Looking on down, we see that there are 27 units of the bracket assembly in inventory, so the net requirement is for 48. This becomes the gross requirement for the bracket and coupling. Since there are 39 top handle couplings in inventory and 15 scheduled for receipt, there is *no* net requirement for part 1118.

Gross to net explosion is a key element of MRP systems. It not only provides the basis for calculating the appropriate quantities but also serves as the communication link between part numbers. It's the basis for the concept of **dependent demand;** that is, the "demand" (gross requirements) for top handles depends on the net requirements for top handle assemblies. To correctly do the calculations, the bill of material, inventory, and scheduled receipt data are all necessary. With these data, the dependent demand can be exactly calculated. It need not be forecast. On the other hand, some **independent demand** items, such as the snow shovel, are subject to demand from outside the firm. The need for snow shovels will have to be forecast. The concept of dependent demand is often called the fundamental principle of MRP. It provides the way to remove uncertainty from the requirement calculations.

Lead Time Offsetting. Gross to net explosion tells us how many of each subassembly and component part are needed to support a desired finished product quantity. What it doesn't do, however, is tell us *when* each component and subassembly is needed. Referring back to Figures 2.3 and 2.4, clearly the top handle bracket and top handle coupling need to be welded together before the wooden top handle is attached. These relationships are known as **precedent relationships.** They indicate the order in which things must be done.

In addition to precedent relationships, determining when to schedule each component part also depends on how long it takes to produce the part (that is, the lead time). Perhaps the top handle bracket (129) can be fabricated in one day, while the top handle coupling (1118) takes two weeks. If so, it would be advantageous to start making the coupling before the bracket, since they are both needed at the same time to make a bracket assembly.

Despite the need to take lead time differences into account, many systems for component part manufacturing ignore them. For example, most furniture manufacturers base production on what is called a **cutting.** In the cutting approach, if a lot of 100 chairs were to be assembled, then 100 of each part (with appropriate multiples) are started at the same time. Figure 2.6 is a Gantt chart (time-oriented bar chart) showing how this cutting approach would be applied to the snow shovel example. (Note processing times are shown on the chart.)

Figure 2.6 shows clearly that the cutting approach, which starts all parts as soon as possible, will lead to unnecessary work-in-process inventories. For example, the top handle bracket (129) doesn't need to be started until the end of day 9, since it must wait for the coupling (1118) before it can be put into its assembly (11495), and part 1118 takes 10 days. In the cutting approach, parts are scheduled earlier than need be. This results from using **front schedule** logic (that is, scheduling as early as possible).

What should be done is to **back schedule**—start each item as late as possible. Figure 2.7 provides a back schedule for the snow shovel example. The schedules for parts 1118, 11495, 13122, and 1605 don't change, since they form a critical path. All of the other parts, however, are scheduled later in this approach than in the front scheduling approach. A substantial savings in work-in-process inventory is obtained by this shift of dates.

Back scheduling has several obvious advantages. It will reduce work-in-process, postpone the commitment of raw materials to specific products, and minimize storage time of completed components. Implementing the back schedule approach, however, requires a system. The system must have accurate BOM data and lead time estimates, some way to ensure all component parts are started at the right times, and some means of tracking components and subassemblies to make sure they are all completed according to plans. The cutting approach is much simpler, since all component parts are started at the same time and left in the pipeline until needed.

MRP achieves the benefits of the back scheduling approach *and* performs the gross to net explosion. In fact, the combination of back schedules and gross to net explosion is the heart of MRP.

FIGURE 2.6 Gantt Chart for Cutting Approach to Snow Shovel Problem (front or earliest start schedule)

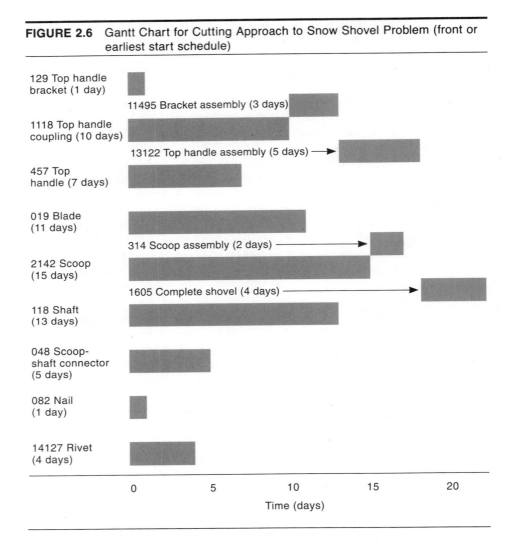

Linking the MRP Records

Figure 2.8 shows the linked set of individual time-phased MRP records for the top handle assembly of the snow shovel. We have already used the first five periods of the 082 nail record shown in Figure 2.8 as the record in Figure 2.2. To see how that record fits into the whole, we start with the snow shovels themselves. We said 100 snow shovels were going to be made, and now we see the timing. That is, the "Gross requirements" row in the MRP record for part number 13122 in Figure 2.8 shows the total need of 100 time phased as

FIGURE 2.7   Gantt Chart Based on Back Schedule (latest start)

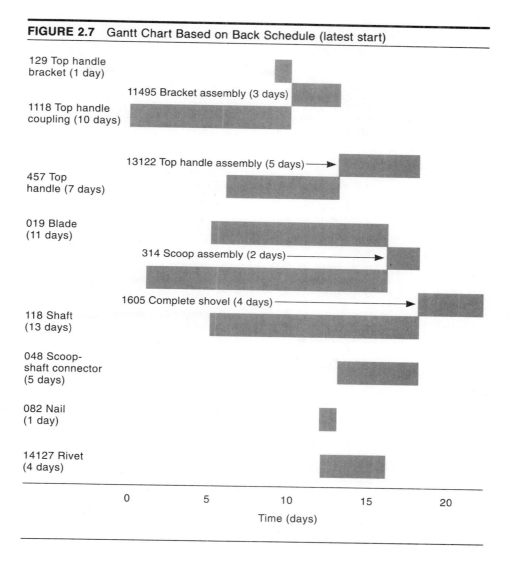

20 in week 2, 10 in week 4, 20 in week 6, 5 in week 7, 35 in week 9, and 10 in week 10. Since each snow shovel takes a top handle assembly, the "Gross requirements" row for the top handle shows when shovel assembly is to begin. Note the total planned orders for the top handle assembly is the net requirement of 75 that we calculated before in the gross to net calculations of Figure 2.5.

The lead time for the top handle assembly is two weeks, calculated as the five days processing time shown in Figure 2.6 plus five days for paperwork. The lead time for each of the other records is similarly calculated;

FIGURE 2.8   MRP Records for the Snow Shovel Top Handle Assembly

13122 Top handle assembly
Lead time = 2

		Week									
		1	2	3	4	5	6	7	8	9	10
Gross requirements			20		10		20	5		35	10
Scheduled receipts											
Projected available balance	25	25	5	5	0	0	0	0	0	0	0
Planned order releases			(5)		20	5		35	10		

457 Top handle
Lead time = 2

		Week									
		1	2	3	4	5	6	7	8	9	10
Gross requirements			(5)		20	5		35	10		
Scheduled receipts				25							
Projected available balance	22	22	17	42	22	17	17	0	0	0	0
Planned order releases						18	10				

082 Nail
(2 required)
Lead time = 1
Lot size = 50

		Week									
		1	2	3	4	5	6	7	8	9	10
Gross requirements			10		40	10		70	20		
Scheduled receipts		50									
Projected available balance	4	54	44	44	4	44	44	24	4	4	4
Planned order releases				50		50					

11495 Bracket assembly
Lead time = 2

		Week									
		1	2	3	4	5	6	7	8	9	10
Gross requirements			5		20	5		35	10		
Scheduled receipts											
Projected available balance	27	27	22	22	2	0	0	0	0	0	0
Planned order releases				3		35	10				

129 Top handle bracket
Lead time = 1

		Week									
		1	2	3	4	5	6	7	8	9	10
Gross requirements				3		35	10				
Scheduled receipts											
Projected available balance	15	15	15	12	12	0	0	0	0	0	0
Planned order releases				23	10						

1118 Top handle coupling
Lead time = 3
Safety stock = 20

		Week									
		1	2	3	4	5	6	7	8	9	10
Gross requirements				3		35	10				
Scheduled receipts			15								
Projected available balance	39	39	54	51	51	20	20	20	20	20	20
Planned order releases				4	10						

one week (five days) of paperwork time is added to the processing time and the total rounded to the nearest five-day week. The current inventories and scheduled receipts for each part are those shown in Figure 2.5. The scheduled receipts are shown in the appropriate periods. Using the two-week lead time and recognizing a net requirement of five units in week 4 for the top handle assembly, we see the need to plan an order for week 2 of five units.

This planned order release of five units in week 2 becomes a gross requirement in week 2 for the top handles as shown by the circles in Figure 2.8. Note also the gross requirements for the nails and brackets in period 2 derive from this same planned order release (with two nails per top handle assembly). Thus, the communication between records is the dependent demand that we saw illustrated before in the gross to net calculations of Figure 2.5.

The remaining planned order releases for the top handle assembly exactly meet the net requirements in the remaining periods, offset for the lead time. The ordering policy used for these items is called **lot-for-lot** (i.e., as required) sizing. An exception to the lot-for-lot procedure is the ordering of nails, which is done in lots of 50. In the case of the nails, the total planned orders will not necessarily add up to the net requirements.

Another part for which there is a discrepancy between the planned orders and the net requirements calculated in Figure 2.5 is the top handle coupling. For this part, a safety stock of 20 units is desired. This means the planned order logic will schedule a planned order release to prevent the projected available balance from going below the safety stock level of 20 units. For the top handle couplings, this means a total of 4 units must be planned for period 2 and 10 for period 3 to maintain the 20-unit safety stock.

The one element we have yet to clearly show is the back scheduling effect. We saw in Figure 2.7 that it would be desirable to delay the start of the top handle bracket (part 129) so that this item is completed at the same time as the top handle coupling (part 1118). The MRP records show that the start of the first planned order for part 129 isn't until week 4, two weeks after the first planned order for part 1118. Both of these planned orders are to satisfy a gross requirement of 35 derived from the planned order for the bracket assembly in week 5. We see then that the orders are back scheduled. This relationship can be more complicated than our example, since the planned order release timing depends on the safety stock and inventory levels, as well as the lead times. The MRP system, however, coordinates all of that information and determines the appropriate planned order release dates, based on back scheduling.

At this point, we see fully the linking of the MRP time-phased records. The "Planned order releases" row for the top handle assembly (13122) becomes (with the appropriate multiplier) the "Gross requirements" row for each of its components (part 457, 082, and 11495), and they are linked together. Once all the gross requirements data are available for a particular record, the individual record processing logic is applied and the planned order releases for the part are passed down as gross requirements to its components, following the product structure (BOM) on a level-by-level basis. In some cases, parts will receive their requirements from more than one source (common parts), as is true for the nails and rivets in the snow shovel. In these cases, gross requirements will reflect needs from more than one planned order release source. Again, the system accounts for this and incorporates it into the gross to net logic.

The MRP records take proper account of gross to netting. They also incorporate back scheduling and allow for explicit timings, desired lot sizing procedures, safety stocks, and part commonality. Even more important, however, is independence of the part number planning. With the MRP approach, the person planning show shovels need not explicitly coordinate his planning with planning of the component parts. The MRP system accomplishes the coordination. Whatever is done to the MRP record for the snow shovels will result in a set of planned orders that the system will correctly pass down as gross requirements to its components. This means plans for each part number can be developed independently of the product structures, and the plans at each level will be communicated correctly to the other levels.

TECHNICAL ISSUES

In this section, we briefly introduce some technical issues to consider in designing MRP systems.

Processing Frequency

Thus far we've looked only at the static construction of the MRP records and how they're linked together. Since conditions change and new information is received, the MRP records must be brought up to date so plans can be adjusted. This means processing the MRP records anew, incorporating current information. Two issues are involved in the processing decision: how frequently the records should be processed and whether all the records should be processed at the same time.

Processing all of the records in one computer run is called **regeneration.** This signifies that *all* part number records are completely reconstructed each time the records are processed. An alternative is **net change** processing, in which only those records affected by the new or changed information are reconstructed. The key issue raised by contrasting regeneration and net change is the frequency of processing.

The appropriate frequency for processing the MRP time-phased records depends on the firm, its products, and its operations. The most common practice is weekly processing using regeneration. But some firms regenerate every two weeks or even monthly, while others process all the MRP time-phased records twice per week or even more often.

The prime motivation for less frequent processing is the computational cost. This can be especially high with regeneration, since a new record is created for every active part number in the product structure file at each processing. But computational time required varies considerably from company to company, depending on the computer approach used, the amount of part numbers, the complexity of product structure, and other factors. For companies using regeneration, typically 8 to 24 hours of central processing unit (CPU) time are required. For example, at one time, at the Hill-Rom Company in Batesville, Indiana, regeneration for approximately 13,000 active part numbers was done over each weekend, requiring approximately 16 hours on an IBM 370–145 computer.

The problem with processing less frequently is that the portrayal of component status and needs expressed in the records becomes increasingly out of date and inaccurate. This decrease in accuracy has both anticipated and unanticipated causes. As the anticipated scheduled receipts are received and requirements satisfied, the inventory balances change. As unanticipated scrap, requirement changes, stock corrections, or other such transactions occur, they cause inaccuracies if not reflected in all the time-phased records influenced by the transactions. Changes in one record are linked to other time-phased records as planned order releases become gross requirements for lower-level components. Thus, some change transactions may cascade throughout the product structure. If these transactions are not reflected in the time-phased records early enough, the result can be poor planning.

More frequent processing of the MRP records increases computer costs but results in fewer unpleasant surprises. When the records reflecting the changes are produced, appropriate actions will be indicated to compensate for the changes.

A logical response to the pressure for more frequent processing is to reduce the required amount of calculation by processing only the records affected by the changes. This net change approach only creates a new part number record when a transaction makes the present component plan inaccurate. Although this could be done as the transaction occurs, typically all transactions are accumulated daily and then processed overnight.

The argument for the net change approach is that it can reduce computer time enough to make daily or more frequent processing possible. Companies where this can be done have the added advantage of smoothing the computational requirements over the week. On the other hand, daily processing of part of the records could lead to even greater overall computational cost than weekly regeneration. Since only some of the records are reviewed at each processing, there's a need for very accurate computer records and transaction processing procedures. Some net change users do an occasional regeneration to clean up all records.

As computers get faster, firms are increasingly adopting daily processing of records, either through regeneration or through net change. The state of the art in hardware and software technology now supports on-line systems with a daily updating cycle.

Bucketless Systems

To some extent, the problems of timing are tied to the use of time buckets. When the buckets are small enough, the problems are reduced significantly. However, smaller buckets mean more buckets, which increases review, storage, and computation costs. A bucketless MRP system specifies the exact release and due dates for each requirement, scheduled receipt, and planned order. The managerial reports are printed out on whatever basis is required, including by exact dates.

Bucketless MRP systems are a better way to use the computer. Above and beyond that, the approach allows better maintenance of lead time offsets and provides more precise time-phased information. The approach is consistent with state-of-the-art software, and many firms now use bucketless systems. The major addition is that the planning cycle itself is bucketless. That is, plans are revised as necessary, not on a periodic schedule, and the entire execution cycle is also shortened.

Lot Sizing

In the snow shovel example of Figure 2.8, we use a fixed lot size (50 units for the nails) and the lot-for-lot procedure. The lot size of 50 for the nails could have been someone's estimate of a good lot size or the result of calculation. The time-phased information can be used in combination with other data to develop lot sizes conforming to organizational needs. We might reach the conclusion, for the top handle (1118) in Figure 2.8, that it's undesirable to set up the equipment for 4 parts in week 2, and again for 10 parts in week 3, so we'd combine the two orders. The time-phased record permits us to develop such **discrete lot sizes** that will exactly satisfy the net requirements for one or more periods.

Several formal procedures have been developed for lot sizing the time-phased requirements. The basic trade-off usually involves elimination of one or more setups at the expense of carrying inventory longer. In many cases, discrete lot sizes possible with MRP are more appealing than fixed lot sizes. Compare the residual inventory of nails in week 10, with that of the bracket assemblies in Figure 2.8, for example.

At first glance the lot-for-lot technique seems a bit too simple-minded since it doesn't consider any of the economic trade-offs or physical factors. However, batching planned orders at one level will increase gross requirements at the next level in the product structure. So larger lot sizing near the end item level of the bill of materials cascades down through all levels. Thus, it turns out that lot-for-lot is better than we might expect in actual practice, particularly at the intermediate levels in the bill of materials. This is especially the case when a product structure has many levels, and the cascading effect becomes greatly magnified. This cascading effect can be mitigated to some extent for components and raw materials that are very common. When this is the case, again lot sizing may be appropriate. As a consequence, many firms employ lot sizing primarily at the end item and basic component levels, while intermediate subassemblies are planned on a lot-for-lot basis.

Safety Stock and Safety Lead Time

Carrying out detailed component plans is sometimes facilitated by including **safety stocks** and/or **safety lead times** in the MRP records. Safety stock is a buffer of stock above and beyond that needed to satisfy the gross requirements. Figure 2.8 illustrates this by incorporating safety stock for the top handle coupling. Safety lead time is a procedure whereby shop orders or purchase orders are released and scheduled to arrive one or more periods before necessary to satisfy the gross requirements.

Safety stocks can be incorporated into MRP time-phased records. The result is that the projected available balance doesn't fall below the safety stock level instead of reaching zero. To incorporate safety lead time, orders are issued (planned) earlier and are scheduled (planned) to be received into inventory before the time that the MRP logic would indicate as necessary. Figure 2.9 shows the top handle bracket from Figure 2.8 being planned with a one-week safety lead time. Notice that both the planned release and planned receipt date are changed. Safety lead time is not just inflated lead time.

Both safety stock and safety lead time are used in practice and can be used simultaneously. However, both are hedges indicating that orders should be released (launched) or that they need to be received when, in fact, this is not strictly true. To use safety stocks and safety lead times effectively, we must understand the techniques' influence on plans. If they aren't well understood, wrong orders can be sent to the factory, meaning workers will try to get out part A because of safety lead time or safety stock when, in fact, part B will be required to meet a customer order.

Safety stock tends to be used in MRP systems where uncertainty about quantities is the problem (e.g., where some small amount of scrap, spare part

FIGURE 2.9   MRP Record with Safety Lead Time

Part 129			1	2	3	4	5	6	7	8	9	10
Part 129	Gross requirements				3		35	10				
Top handle bracket lead time = 1	Scheduled receipts											
Lot-for-lot	Projected available balance	15	15	15	12	35	10	0	0	0	0	0
Safety lead time = 1	Planned order releases				23	10						

demand, or other unplanned usage is a frequent occurrence). Safety lead time, on the other hand, tends to be used when the major uncertainty is the timing rather than the quantity. For example, if a firm buys from a vendor who often misses delivery dates, safety lead time may provide better results than safety stock.

Low-Level Coding

If we refer once again to Figure 2.4, we see that the rivet (part 14127) is a common part. The "Planned order" row for completed shovels will be passed down as gross requirements to the rivet. But there are additional requirements for the rivets (14127) from the scoop assembly (314). If we process the time-phased record for this common part before all of its gross requirements have been accumulated, the computations must be redone.

The way this problem is handled is to assign **low-level code numbers** to each part in the product structure or the indented BOM. By convention, the top final assembly level is denoted as level 0. In our example, the snow shovel would have a low-level code of 0. All immediate component part numbers of this part (13122, 048, 118, 062, 14127, and 314 in Figure 2.4) are given the low-level code number 1. The next level down (part numbers 457, 082, 11495, 2142, 019, and 14127) are low-level coded 2. Note the common part (rivet) has just been recoded as level 2, indicating it is used lower in the product structure. The higher the level codes, the lower in the product structure the part is used. Consequently, the last level code assigned to a part indicates the lowest level of usage and is the level code retained for that part. We finish the example when part numbers 129 and 1118 are coded level 3. The level code assigned to any part number is based on the part's usage in all products manufactured by the organization.

Once low-level codes are established, MRP record processing proceeds from one level code to the next, starting at level code 0. This ensures all gross requirements have been passed down to a part before its MRP record is processed. The result is planning of component parts coordinated with the needs of all higher-level part numbers. Within a level, the MRP record processing is typically done in part number sequence.

Pegging

Pegging relates all the gross requirements for a part to all the planned order releases or other sources of demand that created the requirements. The pegging records contain the specific part number or numbers of the sources of all gross requirements. At level 0, for example, pegging records might contain the specific customer orders to be satisfied by the gross requirements in the end item time-phased records. For lower-level part numbers, the gross requirements are most often pegged to planned orders of higher-level items, but might also be pegged to customer orders if the part is sold as a service part.

Pegging information can be used to go up through the MRP records from a raw material gross requirement to some future customer order. In this sense, it's the reverse of the explosion process. Pegging is sometimes compared to **where-used data.** Where-used data, however, indicate for each part number, the part numbers of all items on which the part is used. Pegging, on the other hand, is a *selective* where-used file. Pegging shows only the specific part numbers that produce the specific gross requirements in each time period. Thus, pegging information can trace the impact of a material problem all the way up to the order it would affect.

Firm Planned Orders

The logic used to illustrate the construction of an MRP record for an individual part number is automatically applied for every processed part number. The result is a series of planned order releases for each part number. If changes have taken place since the last time the record was processed, planned order releases can be very different from one record-processing cycle to the next. Since planned orders are passed down as gross requirements to the next level, the differences can cascade throughout the product structure.

One device for preventing this cascading down through the product structure is to create a **firm planned order (FPO).** FPO, as the name implies, is a planned order that the MRP system *does not* automatically change when conditions change. To change either the quantity or timing of a firm planned order, managerial action is required. This means the trade-offs in making the change can be evaluated before authorization.

The FPO provides a means for temporarily overriding the system to provide stability or to solve problems. For example, if changes are coming about because of scrap losses on open orders, the possibility of absorbing those variations with safety stock can be evaluated. If more rapid delivery of raw material than usual is requested (say by using air freight) to meet a special need, lead time can be reduced for that one order. An FPO means the system won't use the normal lead time offset from the net requirement for that order.

Service Parts

Service part demand must be included in the MRP record if the material requirements are not to be understated. The service part demand is typically based on a forecast and is added directly into the gross requirements for the part. From the MRP system point of view, the service part demand is simply another source of gross requirements for a part, and the sources of all gross requirements are maintained through pegging records. The low-level code for a part used exclusively for service would be zero. If it's used as a component part as well, the low-level code would be determined the same way as for any other part.

As actual service part needs occur, it's to be expected that demand variations will arise. These can be partially buffered with safety stocks (inventories specifically allocated to service part usage) or by creative use of the MRP system. By careful examination of pegging records, expected shortage conditions for manufacturing part requirements can sometimes be satisfied from available service parts. Conversely, critical service part requirements can perhaps be met with orders destined for higher-level items. Only one safety stock inventory is needed to buffer uncertainties from both sources, however.

Planning Horizon

In Figure 2.8, the first planned order for top handle assemblies occurs in week 2 to meet period 4's gross requirement of 10 units. This planned order of 5 units in week 2 results in a corresponding gross requirement in that week for the bracket assembly (part 11495). This gross requirement is satisfied from the existing inventory of part 11495. But a different circumstance occurs if we trace the gross requirements for 35 top handle assemblies in week 9.

The net requirement for 35 units in week 9 becomes a planned order release in week 7. This, in turn, becomes a gross requirement for 35 bracket assemblies (part 11495) in week 7 and a planned order release in week 5. This passes down to the top handle coupling (part 1118), which creates a planned order release for 4 units in week 2. This means the **cumulative lead time** for the top handle assembly is 7 weeks (from release of the coupling order in week 2 to receipt of the top handle assemblies in week 9).

Scheduled Receipts versus Planned Order Releases

A true understanding of MRP requires knowledge of certain key differences between a scheduled receipt and a planned order. We noted one such difference before: the scheduled receipt represents a commitment, whereas the planned order is only a plan—the former is much more difficult to change than the latter. A scheduled receipt for a purchased item means a purchase order, which is a formal commitment, has been prepared. Similarly, a scheduled receipt for a manufactured item means there's an open shop order. Raw materials and component parts have *already* been specifically committed to that order and are no longer available for other needs. One major result of this distinction, which can be seen in Figure 2.8, is that planned order releases explode to gross requirements for components, but scheduled receipts (the open orders) don't.

A related issue is seen from the following question: Where would a scheduled receipt for the top handle assembly (13122) in Figure 2.8 of, say, 20 units in week 2 be reflected in the records for the component parts (457, 082, and 11495)? The answer is nowhere! Scheduled receipts are not reflected in the current records for component parts. For that scheduled receipt to exist, the component parts would have already been assigned to the shop order representing the scheduled receipt for part 13122 and removed from the available balances of the components. As far as MRP is concerned, the 20 part 457s, 40 part 082s, and 20 part 11495s don't exist! They're on their way to becoming 20 part 13122s. The 13122 record controls this process, not the component records.

USING THE MRP SYSTEM

In this section, we discuss critical aspects of using the MRP system to ensure that MRP system records are exactly synchronized with physical flows of material.

The MRP Planner

The persons most directly involved with the MRP system outputs are planners. They are typically in the production planning, inventory control, and purchasing departments. Planners have the responsibility for making detailed decisions that keep the material moving through the plant. Their range of discretion is carefully limited (e.g., without higher authorization, they can't change plans for end items destined for customers). Their actions, however, are reflected in the MRP records. Well-trained MRP planners are essential to effective use of the MRP system.

Computerized MRP systems often encompass tens of thousands of part numbers. To handle this volume, planners are generally organized around logical groupings of parts (such as metal parts, wood parts, purchased electronic parts, or West Coast distribution center). Even so, reviewing each record every time the records are processed wouldn't be an effective use of the planners' time. At any time, many records require no action, so the planner only wants to review and interpret those that do require action.

The primary actions taken by an MRP planner are:

1. Release orders (i.e., launch purchase or shop orders when indicated by the system).
2. Reschedule due dates of existing open orders when desirable.
3. Analyze and update system planning factors for the part numbers under her control. This would involve such things as changing lot sizes, lead times, scrap allowances, or safety stocks.
4. Reconcile errors or inconsistencies and try to eliminate root causes of these errors.

5. Find key problem areas requiring action now to prevent future crises.
6. Use the system to solve critical material shortage problems so actions can be captured in the records for the next processing. This means the planner works *within* formal MRP rules, *not* by informal methods.
7. Indicate where further system enhancements (outputs, diagnostics, etc.) would make the planner's job easier.

Order Launching.   **Order launching** is the process of releasing orders to the shop or to vendors (purchase orders). This process is prompted by MRP when a planned order release is in the current time period, the **action bucket.** Order launching converts the planned order into a scheduled receipt reflecting the lead time offset. Order launching is the opening of shop and purchase orders; closing these orders occurs when scheduled receipts are received into stockrooms. At that time, a transaction must be processed—to increase the on-hand inventory and eliminate the scheduled receipt. Procedures for opening and closing shop orders have to be carefully defined so all transactions are properly processed.

The orders indicated by MRP as ready for launching are a function of lot sizing procedures and safety stock as well as timing. We saw this in Figure 2.8 where we worked with lot-for-lot approaches and fixed lot sizes. A key responsibility of the planner is managing with awareness of the implications of these effects. For example, not *all* of a fixed lot may be necessary to cover a requirement, or a planned order that's solely for replenishment of safety stock may be in the action bucket.

When an order is launched, it's sometimes necessary to include a shrinkage allowance for scrap and other process yield situations. The typical approach allows some percentage for yield losses that will increase the shop order quantity above the net amount required. To effect good control over open orders, the *total* amount, including the allowance, should be shown on the shop order, and the scheduled receipt should be reduced as actual yield losses occur during production.

Allocation and Availability Checking.   A concept closely related to order launching is **allocation**—a step prior to order launching that involves an availability check for the necessary component or components. From the snow shovel example, if we want to assemble 20 of the top handle assembly (13122) in period 4, the availability check would be whether sufficient components (20 of part 457, 40 of part 082, and 20 of part 11495) are available. If not, the shop order for 20 top handle assemblies (13122) should not be launched, because it cannot be executed without component parts. The planner role is key here, as well. The best course of action might be to release a partial order. The planner should evaluate that possibility.

Most MRP systems first check component availability for any order that a planner desires to launch. If sufficient quantities of each component are available, the shop order can be created. If the order is created, then the system allocates the necessary quantities to the particular shop order. (Shop orders are assigned by the computer, in numerical sequence.) The allocation means this amount of a component part is mortgaged to the particular shop order and is, therefore, not available for any other shop order. Thus, the amounts shown in Figure 2.8 as projected available balances may not be the same as the physical inventory balances. The physical inventory balances could be larger, with the differences representing allocations to specific shop orders that have been released, but whose component parts have not been removed from inventory.

After availability checking and allocation, **picking tickets** are typically created and sent to the stockroom. The picking ticket calls for a specified amount of some part number to be removed from some inventory location, on some shop order, to be delivered to a particular department or location. When the picking ticket has been satisfied (inventory moved), the allocation is removed and the on-hand balance is reduced accordingly.

Availability checking, allocation, and physical stock picking are a type of double-entry bookkeeping. The result is that the quantity physically on hand should match what the records indicate is available plus what is allocated. If they don't match, corrective action must be taken. The resulting accuracy facilitates inventory counting and other procedures for maintaining data integrity.

Exception Codes

Exception codes in MRP systems are used "to separate the vital few from the trivial many." If the manufacturing process is under control and the MRP system is functioning correctly, exception coding typically means only 10 to 20 percent of the part numbers will require planner review at each processing cycle. Exception codes are in two general categories. The first, checking the input data accuracy, includes checks for dates beyond the planning horizon, quantities larger or smaller than check figures, nonvalid part numbers, or any other desired check for incongruity. The second category of exception codes directly supports the MRP planning activity. Included are the following kinds of exception (action) messages or diagnostics:

1. Part numbers for which a planned order is now in the most immediate time period (the action bucket). It's also possible to report any planned orders two to three periods out to check lead times, on-hand balances, and other factors while there's some time to respond, if necessary.
2. Open order diagnostics when the present timing and/or amount for a scheduled receipt is not satisfactory. Such a message might indicate that an open order exists that's not necessary to cover any of the requirements in the planning horizon. This message might suggest order cancelation caused by an engineering change that substituted some new part for the

one in question. The most common type of open order diagnostic shows scheduled receipts that are timed to arrive either too late or too early and should, therefore, have their due dates revised to reflect proper factory priorities. An example of this is seen with each of the three scheduled receipts in Figure 2.8. The 457 top handle open order of 25 could be delayed one week. A one-week delay is also indicated for the 082 nail scheduled receipt. For part 1118 (the top handle coupling), scheduled receipt of 15 could be delayed from week 2 until week 5. Another open order exception code is to flag any past-due scheduled receipt (scheduled to have been received in previous periods, but for which no receipt transaction has been processed). MRP systems assume a past-due scheduled receipt will be received in the immediate time bucket.

3. A third general type of exception message indicates problem areas for management; in essence, situations where level-0 quantities can't be satisfied unless the present planning factors used in MRP are changed. One such exception code indicates a requirement has been offset into the past period and subsequently added to any requirement in the first or most immediate time bucket. This condition means an order should have been placed in the past. Since it wasn't, lead times through the various production item levels must be compressed to meet the end item schedule. A similar diagnostic indicates the allocations exceed the on-hand inventory— a condition directly analogous to overdrawing a checking account. Unless more inventory is received soon, the firm won't be able to honor all pick tickets issued, and there will be a material shortage in the factory.

Bottom-Up Replanning

Bottom-up replanning—using pegging data to solve material shortage problems—is best seen through an example. Let's return again to Figure 2.8, concentrating on the top handle assembly and the nails (parts 13122 and 082). Let's suppose the scheduled receipt of 50 nails arrives on Wednesday of week 1. On Thursday, quality control checks them and finds the vendor sent the wrong size. This means only 4 of the 10 gross requirement in week 2 can be satisfied. By pegging this gross requirement up to its parent planned order (5 units of 13122 in week 2), we see that only 7 of the gross requirement for 10 units in week 4 can be satisfied (the 5 on hand plus 2 made from 4 nails). This, in turn, means only 7 snow shovels can be assembled in week 4.

The pegging analysis shows that 3 of the 10 top handle assemblies can't be available without taking some special actions. If none are taken, the planned assembly dates for the snow shovels should reflect only 7 units in week 4, with the additional 3 scheduled for week 5. This should be done if we can't overcome the shortfall in nails. The change is necessary because the 10 snow shovels now scheduled for assembly in week 4 also explode to other parts—parts that won't be needed if only 7 snow shovels are to be assembled.

There may, however, be a critical customer requirement for 10 snow shovels to be assembled during week 4. Solving the problem with bottom-up replanning might involve one of the following alternatives (staying *within* the MRP system, as planners must do):

1. Issue an immediate order to the vendor for six nails (the minimum requirement), securing a promised lead time of two days instead of the usual one week. This will create a scheduled receipt for six in week 2.
2. Order more nails for the beginning of week 3, and negotiate a one-week reduction in lead time (from two weeks to one week) for fabricating this one batch of part 13122. The planned order release for five would be placed in week 3 and converted to a firm planned order, so it wouldn't change when the record is processed again. The negotiation for a one-week lead time might involve letting the people concerned start work earlier than week 3 on the two part 13122s, for which material already exists, and a reduction in the one-week paperwork time included in the lead times.
3. Negotiate a one-week lead time reduction for assembling the snow shovels; place a firm planned order for 10 in week 5, which will result in a gross requirement for 10 top handle assemblies in period 5 instead of period 4.

Thus, we see the solution to a material shortage problem might be made by compressing lead times throughout the product structure using the system and bottom-up replanning. Planners work within the system using firm planned orders and net requirements to develop workable (but not standard) production schedules. The creativity they use in solving problems will be reflected in the part records at the next MRP processing cycle. All implications of planner actions will be correctly coordinated throughout the product structure.

It's important to note that the resolution of problems can't *always* involve reduced lead time and/or partial lots. Further, none of these actions are free. In some cases, customer needs will have to be delayed or partial shipments made. Pegging and bottom-up replanning will provide advance warning of these problems so customers can take appropriate actions.

An MRP System Output

Figure 2.10 is an MRP time-phased record for one part number out of a total of 13,000 at the Batesville, Indiana, facility of the Hill-Rom Company. The header information includes the date the report was run, part number and description, planner code number, buyer code number (for purchased parts), unit of measure for this part number (pieces, pounds, etc.), rejected parts that have yet to receive disposition by quality control, safety stocks, shrinkage allowance for anticipated scrap loss, lead time, family data (what other parts are similar to this one), year-to-date scrap, usage last year, year-to-date usage, and order policy/lot size data. The policy code of 3 for this part means the order policy is a **period order quantity (POQ).** In this case,

FIGURE 2.10 Example MRP Record

```
DATE- 01/21                         MATERIAL STATUS-PRODUCTION SCHEDULE

******PART NUMBER*******
NONJEK OPTY SSV LAM PP UPHL                                                                    FAMILY
                                    DESCRIPTION          SAFETY    SHRINKG   LEAD      MIN ORD  DATA
USTRO40                  3/16x7/8 MR P & C STL STRAP     STOCK     ALLOWNE   TIME      POINT
                                                         497       1         08
****USAGE****
YTD    LAST YR   YTD     ***** ORDER POLICY AND LOT SIZE DATA *******************************
SCRAP                   POLICY     PERIODS   STANDARD  MINIMUM   MAXIMUM   MULTIPLE
                        CODE       TO COMB.  QUANTITY  QTY       QTY       QTY
                        3          04
```

```
                 PAST DUE   563    564    565    566    567    568    569    570    571    572    573
                            01/22  01/29  02/05  02/12  02/19  02/26  03/05  03/12  03/19  03/26  04/02

REQUIREMENTS       495
SCHEDULED RECEIPTS               483
PLANNED RECEIPTS                                              491              516
AVAILABLE ON-HAND 1,500                                       491
PLANNED ORDERS     491

                 574    575    576    577    578    579    580    581    582    583    584    585
                 04/09  04/16  04/23  04/30  05/07  05/14  05/21  05/28  06/04  06/11  06/18  06/25

REQUIREMENTS       337
SCHEDULED RECEIPTS       508    508    25     25     25
PLANNED RECEIPTS                337                          516              334
AVAILABLE                       334    334
PLANNED ORDERS                         334

                 586    587    588    589-592  593-596  597-600  601-604  605-608  609-612
                 07/16  07/23  07/30  08/06    09/03    10/01    10/29    11/26    12/24

                 VACATION

******EXCEPTION MESSAGES*********
PLANNED ORDER OF   491 FOR M-WK  568   OFFSET INTO A PAST PERIOD BY 03 PERIODS

********PEGGING DATA (ALLOC)*********
790116    455 JN25220

********PEGGING DATA (REQMT)*********
790205    483 F 17144     790305   516 F 19938     790409   337 F 17144
790507    334 F 19938
```

"periods to comb. = 04" means each order should combine four periods of net requirements.

The first time bucket is "past due." After that, weekly time buckets are presented for the first 28 weeks of data; thereafter, 24 weeks of data are lumped into 4-week buckets. In the computer itself, a bucketless system is used with all data kept in exact days, with printouts prepared in summary format for one- and four-week buckets. The company maintains a manufacturing calendar; in this example, the first week is 563 (also shown as 1/22), and the last week is 612.

In this report, safety stock is subtracted from the on-hand balance (except in the past-due bucket). Thus, the exception message indicating that a planned order for 491 should have been issued three periods ago creates no major problem, since the planner noted that this amount is less than the safety stock. This report also shows the use of safety lead time. *Planned* receipts are given a specific row in the report and are scheduled one week ahead of the actual need date. For example, the 337-unit planned order of week 565 is a planned receipt in week 573, although it's not needed until week 574.

The final data in the report is the pegging data section tying specific requirements to the part numbers from which those requirements came. For example, in week 565 (shop order no. 790205), the requirement for 483 derives from part number F17144. MRP records are printed at this company only for those part numbers for which exception messages exist.

SYSTEM DYNAMICS

Murphy's law states that if anything can go wrong, it will. Things are constantly going wrong, so it's essential that the MRP system mirror actual shop conditions, that is, both the physical system and the information system have to cope with scrap, incorrect counts, changes in customer needs, incorrect bills of material, engineering design changes, poor vendor performance, and a myriad of other mishaps.

In this section, we look at the need for quick and accurate transaction processing and review the MRP planner's replanning activities in coping with change. We discuss sources of problems occurring as a result of data base changes plus actions to ensure the system is telling the truth, even if the truth hurts.

Transactions during a Period

To illustrate transaction processing issues, we use a simple example for one part. Figure 2.11 shows an MRP record (for part 1234) produced over the weekend preceding week 1. The planner for part 1234 would receive this MRP record on Monday of week 1.

FIGURE 2.11   MRP Record for Part 1234 as of Week 1

		1	2	3	4	5
Gross requirements		30	20	20	0	45
Scheduled receipts		50				
Projected available balance	10	30	10	40	40	45
Planned order releases		50		50		

Lead time = 2
Lot size = 50

The planner's first action would be to try to launch the planned order for 50 units in period 1; that is, the MPC system would first check availability of the raw materials for this part and then issue an order to the shop to make 50, if sufficient raw material is available. Launching would require allocating the necessary raw materials to the shop order, removing the 50 from the "Planned order release" row for part 1234, and creating a scheduled receipt for 50 in week 3, when they're needed. Thereafter, a pick ticket would be sent to the raw material area and work could begin.

Let's assume during week 1 the following changes occurred, and the transactions were processed:

- Actual disbursements from stock for item 1234 during week 1 were only 20 instead of the planned 30.
- The scheduled receipt for 50 due in week 1 was received on Tuesday, but 10 units were rejected, so only 40 were actually received into inventory.
- The inventory was counted on Thursday and 20 additional pieces were found.
- The requirement date for the 45 pieces in week 5 was changed to week 4.
- Marketing requested an additional five pieces for samples in week 2.
- The requirement for week 6 has been set at 25.

The resultant MRP record produced over the weekend preceding week 2 is presented as Figure 2.12.

Rescheduling

The MRP record shown in Figure 2.12 illustrates two important activities for MRP planners: (1) indicating the sources of problems that will occur as a result of data base changes, and (2) suggesting actions to ensure the system is

FIGURE 2.12 MRP Record for Part 1234 as of Week 2

		2	3	4	5	6
Gross requirements		25	20	45	0	25
Scheduled receipts			50			
Projected available balance	50	25	55	10	10	35
Planned order release				50		

Lead time = 2
Lot size = 50

telling the truth. Note the scheduled receipt presently due in week 3 is not needed until week 4. The net result of all the changes to the data base means it's now scheduled with the wrong due date, and the due date should be changed to week 4. If this change is not made, this job may be worked on ahead of some other job that is really needed earlier, thereby causing problems. The condition shown in Figure 2.12 would be highlighted by an MRP exception message, such as "reschedule the receipt currently due in week 3 to week 4."

Complex Transaction Processing

So far, we've illustrated system dynamics by using a single MRP record. However, an action required on the part of an MRP planner may have been caused by a very complex set of data base transactions involving several levels in the bill of materials. As an example, consider the MRP records shown in Figure 2.13, which include three levels in the product structure. Part C is used as a component in both parts A and B as well as being sold as a service part. Part C, in turn, is made from parts X and Y. The arrows in Figure 2.13 depict the pegging data.

The part C MRP record is correctly stated at the beginning of week 1. That is, no exception messages would be produced at this time. In particular, the two scheduled receipts of 95 and 91, respectively, are scheduled correctly, since delaying either by one week would cause a shortage, and neither has to be expedited to cover any projected shortage.

While the two scheduled receipts for part C are currently scheduled correctly, transactions involving parts A and B can have an impact on the proper due dates for these open orders. For example, suppose an inventory count adjustment for part A resulted in a change in the 30-unit planned order release

---

FIGURE 2.13   MRP Record Relationships for Several Parts

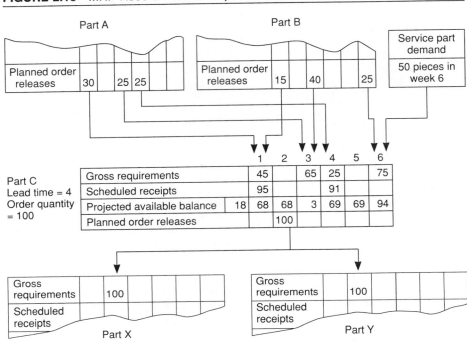

Note: This example is based on one originally developed by Joseph Orlicky, *Material Requirements Planning* (New York: McGraw-Hill, 1975), chap. 3, pp. 44–64.

---

from week 1 to week 3. In this case, the 95 units of part C would not be needed until week 3, necessitating a reschedule. Similarly, any change in timing for the planned order release of 25 units of part A in week 4 would call for a reschedule of the due date for 91 units of part C. Finally, suppose a transaction requiring 75 additional units of part B in week 5 were processed. This would result in an immediate release of an order for 100 units of part C. This might necessitate rescheduling for parts X and Y. The point here is that actions required on the part of an MRP planner can occur because of a complex set of data base transactions involving many different parts. They may not necessarily directly involve the particular part being given attention by the MRP planner.

Procedural Inadequacies

MRP replanning and transaction processing activities are two essential aspects of ensuring the MPC data base remains accurate. However, while these activities are necessary, they aren't sufficient to maintain accurate records.

Some of the procedures used to process transactions simply may be inadequate to the task.

To illustrate inadequate transaction procedures, let's return to the example in Figure 2.13. Note that, if 4 or more pieces are scrapped on the shop order for 95, there will be a shortage in week 3, necessitating rescheduling of the order for 91 one week earlier.

It's even more interesting to see what would happen if 4 pieces were scrapped on the order for 95, and this scrap transaction weren't processed. If the scrap isn't reported, MRP records would appear as shown in Figure 2.13 indicating no required rescheduling—when, in fact, that's not true. *If* the shortage were discovered by the person in charge of the stockroom when he or she puts away this order, then only one week would be lost before the next MRP report shows the problem. If, however, the stockroom person doesn't count, or if the person who made the scrap puts the defective parts at the bottom of the box where they go undetected by quality control, then the problem will only be discovered when the assembly lines are trying to build As and Bs in week 3. Such a discovery comes under the category of unpleasant surprises. An interesting sidelight to this problem is that the cure will be to rush down to the shop to get at least 1 piece from the batch of 91. The very person who failed to report the earlier scrap may well now be screaming, "Why don't those idiots know what they need!"

Still another aspect of the scrap reporting issue can be seen by noting the 95 and 91 were originally issued as lot sizes of 100. This probably means 5 and 9 pieces of scrap have occurred already, and the appropriate adjustments have been made in the scheduled receipt data. Note that, if these adjustments had *not* been made, the two scheduled receipts would show as 100 each. The resultant 14 (or 5 + 9) pieces (that don't, in fact, exist) would be reflected in the MRP arithmetic. Thus, the projected available balance at the end of period 5 would be 83 (or 69 + 14); this is more than enough to cover the gross requirement of 75 in period 6, so the planned order release for 100 in period 2 would not exist and the error would cascade throughout the product structure. Further, even if shop orders are carefully counted as they are put into storage, the five-piece shortage in period 1 is not enough to cause the MRP arithmetic to plan an order. Only after period 4 (the beginning of period 5) will the additional nine pieces of scrap be incorporated in the MRP record showing a projected shortage in period 6. This will result in an immediate order, to be completed in one week instead of four! What may be obvious is that, if accurate counting isn't done, then the shortage is discovered in week 6, when the assembly line goes down. This means procedures for issuing scrap tickets when scrap occurs and procedures for ensuring good parts are accurately counted into inventory must be in place. If not, all the MPC systems will suffer.

The long and the short of all this is that we have to believe the numbers, and an error of as little as *one* piece can cause severe problems. We have to know the truth. We have to tightly control transactions. Moreover, we have to develop iron-clad procedures for processing MPC data base transactions.

THE MRP DATA BASE

To install and derive maximum benefit from a material planning and control system, a large integrated data base is usually required. The computer hardware and software design aspects of common data bases are beyond the scope of this book. Their importance in MPC systems compels us to briefly identify the required primary files and communication links. In this section, we treat the data files that would usually be required to support the engine of the MPC system in Figure 2.1. We make no claim that the following approach is optimal, or that you might not group the data in different ways. Rather, our objective is to illustrate one way the data might be supplied and to identify the elements needed. The enormity of the overall data base even for small firms is awesome, but data needs exist *even if there is no formal system,* as long as people are manufacturing products.

The Item Master File

The data on an individual part are often contained in two files. The information that remains the same (or nearly so) from period to period is found in the **item master file,** while information on part status is found in the subordinate file. The item master file typically contains all the data needed to completely describe each part number. These data are used for MRP, purchasing, cost accounting, and other company functions. The objective is to hold, in one file, all of the static data describing the attributes of individual part numbers. Included are part number, name, low-level code, unit of measure, engineering change number, drawing reference, release date, planner code, order policy code, lead time, safety stock, standard costs, and linkages to other data files, such as routing, where used, and bill of material.

The Subordinate Item Master File

A **subordinate item master file** is often used for changing or dynamic data about individual part numbers. Included are current allocations and the shop order number to which each allocation is tied, time-phased scheduled receipts and associated order numbers, time-phased gross requirements, planned orders, firm planned orders, pegging data, and linkages to the item master file.

The Bill of Material File

The bill of material file is typically established on a **single-level basis,** with each part number linked only to the part numbers of the immediate components required to produce it; that is, the linkages are to one level farther down in the product structure only. By successively linking the part numbers, a full bill of material for each part can be developed from the individual single-level linkings. Data elements held in this file usually include component part numbers required to make each individual part, number of each required, units of measure, engineering change numbers, effectivity dates, active/inactive coding, and where-used information.

The Location File

The **location file** keeps track of the set of exact physical storage locations for each part number. This can be a highly dynamic file, since the data elements usually include departments, rows, bays, tiers, quantities, units of measure, in dates, original quantities, date of last activity, and so on.

The Calendar File

The **calendar file** converts the shop day calendar used by the firm to a day/date/year calendar. The file also provides for phenomena such as annual vacations and holidays.

Open Order Files

An entire set of files is maintained to support the scheduled receipts (open orders) in the MRP system. These involve both purchase orders and shop orders. For the purchase orders, we need open purchase orders, open quotations, a vendor master file, vendor performance data, alternate sources, and price/quantity information. Another set of records needs to be maintained to support shop orders in the factory. Included are data files describing open orders, routings, work centers, employees, shifts, tooling, and labor/performance reporting.

Other File Linkages

In addition to the data files needed for the engine, many other data files are necessary to flesh out the entire MPC system. Among them are files for forecasting, capacity planning, production scheduling, cost accounting, budgeting, order entry, shop-floor control, distribution, invoicing, payroll, job standards, and engineering.

MRP SYSTEM EXAMPLES

Next we have two examples of how MRP systems function in companies. The first example shows that the principles are key, even when the system is manually applied. The second is a fully integrated system illustration.

Ethan Allen Furniture Company—A Manual Application

The Ethan Allen manual method for detailed component scheduling is based on MRP logic, which utilizes product structures to create demand dependency, lead time offsetting, and gross to netting. The Ethan Allen approach establishes the assembly date and then prepares a back schedule or Gantt chart of the form shown in Figure 2.7. The parts are scheduled according to their individual lead times to be ready for final assembly. For MRP planning purposes, Ethan Allen incorporates a one-week safety lead time to allow parts to be exactly counted and prepared for final assembly. When the firm changed to a back schedule–based approach from a cutting approach, an immediate reduction of 15 to 20 percent in the work-in-process inventory and lead times was achieved in the converted plants. (Old ways die hard, however, and though back scheduling's benefits were clearly demonstrated, one otherwise well-run factory took eight years to convert to the new system.)

The manual MRP-like system is satisfactory for Ethan Allen furniture factories for two basic reasons. First, although a bill of materials for most items would show indentations of several levels, in fact, subassemblies tend to be phantoms; that is, they are not stored. For example, drawers are not stored. They are produced when the end item is produced, being assembled at the same time from basic components. Thus, there's little need for gross to netting on a level-by-level explosion basis. End item needs are multiplied by the number of components per end item, and gross to netting is performed only against component inventories. The second reason this approach works well for the company is there's very little commonality among parts and no spare parts requirements. Therefore, component parts generally don't receive gross requirements from more than one source.

Manual MRP systems are still in use at many Ethan Allen factories. However, they have converted several factories to the computer and expect to convert others as their own knowledge increases, as growth dictates, and as the ever-decreasing costs of computers make these conversions more attractive.

Jet Spray—An Integrated On-Line Example

Jet Spray Corporation manufactures and sells dispensers for noncarbonated cold beverages (e.g., lemonade) and hot beverages (e.g., coffee). Jet Spray uses a software package called Data 3 for manufacturing planning and control. The package is an integrated on-line bucketless system encompassing MRP, capacity planning, shop-floor control, master production scheduling, inventory management, and other functions.

One part at Jet Spray, 3273, is a beverage bowl used for the TJ3 model cold drink dispenser. It's also sold as a replacement or service part, as part number S3273. Figure 2.14 is a portion of the inquiry record for the S3273, where the time-phased record is in vertical format. There are 11 on hand, and a series of customers orders marked C/O. The record begins with the oldest date for a past-due order (8/4); all past-due orders are on credit hold. The projected balance goes negative at 9/25. A work order (W/O 78430) for 96 pieces is due on 9/29, but it has a reschedule-in (RSI) message to expedite it to 9/26. Other exception messages are given to help the planner manage the item.

Figure 2.15 is a portion of the report that extracts the exception code information from the individual MRP records like Figure 2.14 into one overall report for items made in the plastic finishing department. This is where plastic parts sold as service parts are packaged. Item S3273 is shown with the information from Figure 2.14. By looking at the overall report shown as Figure 2.15, work can be efficiently released to the plastic finishing department on a daily basis.

Figure 2.16 presents the inquiry record for part 3273 (not the service part S3273). It contains information derived from the service part record (e.g., the 96 units of WOA 78450—shown as 23 and 73 that haven't yet been picked) and from records for the end item on which this part is used.

FIGURE 2.14   Jet Spray MRP System Inquiry (part S3273)

```
   9/29/     8:17:23      MRP INQUIRY FOR PART: S3273          C/N   1  RNMRP001
   DESCRIPTION: BOWL  PINCH TYPE
   QTY ON HAND:          11                              LEADTIME DAYS :   3
   SAFETY-STCK:          0                       D    BUYER/PLANNER : 030

              REQUIRED  RECEIVABLE                 PROJECTED
   TYP CRDER # QUANTITY  QUANTITY  MESSAGE          BALANCE     DUE DATE OPT
   C/O 9007987     1                                   10         8/04/
   C/O 9009239     1                                    9         9/08/
   C/O 99C9314     1                                    8         9/10/
   C/O 9009344     1                                    7         9/11/
   C/O 0201038     3                                    4         9/19/
   C/O 9009811     1                                    3         9/24/
   C/O 9009830     1                                    2         9/24/
   C/O 9009875     2                                              9/25/
   C/O 9009879     5                                    5-        9/25/
   C/O 9009917     1                                    6-        9/25/
   C/O 9009944     2                                    8-        9/26/
   W/O 0078430               96  RSI 09/26/            88         9/29/
   F/C            10                                   78         9/29/
   W/O 0078450               96                       174        10/03/
   C/O 0201118    50                                  124        10/03/
   C/O 0201129     5                                  119        10/03/
   PLO                       96  OPEN SCH REC         215        10/06/
   F/C            65                                  150        10/06/
   F/C            65                                   85        10/13/
   PLO                       96  OPEN SCH REC         181        10/20/
   F/C            65                                  116        10/20/
```

FIGURE 2.15 Jet Spray MRP System Exception Messages (part S3273)

```
001   9/26/    20.56.49             JET SPRAY CORPORATION                              MCSUPVSR       RNMRPP02        PAGE   17
                                    MRP EXCEPTION MESSAGE REPORT                                        SEQUENCED BY MAKE/BUY
                         FOR BUYER/PLANNER: 030 - J.CAPPADONA--PLASTIC FINISHING
                                                                                                                    MAKE/BUY
PART NUMBER --RI DESCRIPTION -----------------* ACTION MESSAGES---------------------------------------------------* CODE

S3170          BOWL GASKET                      OPEN A SCHEDULED RECEIPT DUE  9/26/   FOR QTY OF        1,000                IF
                                                OPEN A SCHEDULED RECEIPT DUE 10/03/   FOR QTY OF        1,000
                                                OPEN A SCHEDULED RECEIPT DUE 10/13/   FOR QTY OF        1,000
                                                OPEN A SCHEDULED RECEIPT DUE 10/27/   FOR QTY OF        1,000

S3273          BOWL  PINCH TYPE                 RESCHEDULE IN  W/O NO. 0078430 TO  9/26/   FROM  9/29/  . QTY IS     96      IF
                                                OPEN A SCHEDULED RECEIPT DUE 10/06/   FOR QTY OF          96
                                                OPEN A SCHEDULED RECEIPT DUE 10/20/   FOR QTY OF          96
                                                OPEN A SCHEDULED RECEIPT DUE 10/27/   FOR QTY OF          96

S3338          BOWL COVER                       RESCHEDULE IN  W/O NO. 0079380 TO  9/26/   FROM 10/03/  . QTY IS     50      IF
                                                OPEN A SCHEDULED RECEIPT DUE 10/06/   FOR QTY OF         100
                                                OPEN A SCHEDULED RECEIPT DUE 10/27/   FOR QTY OF         100
```

FIGURE 2.16 Jet Spray MRP System Inquiry (part 3273)

TYP ORDER #	REQUIRED QUANTITY	MESSAGE	PROJECTED BALANCE	DUE DATE	
9/29/		MRP INQUIRY FOR PART: 3273			
DESCRIPTION: BOWL TJ3					
QTY ON HAND: -1,034			LEADTIME DAYS : 6		
SAFETY-STCK: 0			BUYER/PLANNER : 020		
MISC. SHORT: 0					
WOA 0078450	23		1011	9/23/	
WOA 0078450	73		938	9/23/	
DEP	96		842	9/24/	
DEP	20		822	9/26/	
DEP	126		696	10/01/	
DEP	113		583	10/06/	
DEP	96		487	10/07/	
DEP	96		391	10/15/	
DEP	145		246	10/22/	
DEP	150		96	10/24/	
PLO		500 OPEN SCH REC	596	10/29/	
DEP	120		476	10/29/	
DEP	96		380	10/29/	
DEP	150		230	10/30/	
DEP	96		134	11/04/	+

Figure 2.17 shows the daily exception message report for the plastic molding department. It's used to help schedule the extensive changeovers of the molding machines. The exception message for 3273 indicated in Figure 2.16 can be seen in Figure 2.17.

Figures 2.14 through 2.17 show an integrated set of real-time MRP records. They're used on a daily basis at Jet Spray. Many "standard" MRP reports are rarely, if ever, printed. They're replaced with a "paperless" system. Each day the planners, using video screens, take the actions required for that day. The MRP planning is on a daily cycle. This results in lead time and inventory reductions, but it comes at the cost of all procedures and support activities, such as stockrooms executing instructions in a more timely mode. Again we see pressures for high levels of data integrity and performance. One of Jet Spray's goals is to pick orders from stock on the same day they're created by MRP planners.

CONCLUDING PRINCIPLES

Chapter 2 provides an understanding of the MRP approach to detailed material planning. It describes basic techniques, some technical issues, and how MRP systems are used in practice. MRP, with its time-phased approach to planning, is a basic building-block concept for materials planning and control systems. Moreover, there are many other applications of the time-phased record. We see the most important concepts or principles of this chapter as follows:

- Effective use of an MRP system allows development of a forward-looking (planning) approach to managing material flows.

FIGURE 2.17 Jet Spray MRP System Exception Messages (part 3273)

```
001   9/26/
                              JET SPRAY CORPORATION
                           MRP EXCEPTION MESSAGE REPORT
                  FOR BUYER/PLANNER: 020            PLASTIC MOLDING

PART NUMBER --RI DESCRIPTION -------------*  ACTION MESSAGES-------------------------*  CODE

3223       SPACER/JT JS EVAP        OPEN A SCHEDULED RECEIPT DUE 10/24/   FOR QTY OF    5,000    IF

3273       BOWL TJ3                 OPEN A SCHEDULED RECEIPT DUE 10/29/   FOR QTY OF     500     IF

3715       FUNNEL COVER HC2 HCL     RESCHEDULE IN W/O NO. 0078990 TO  9/26/   FROM 10/02/86,  QTY IS    2,000    IF
```

- The MRP system provides a coordinated set of linked product relationships, thereby permitting decentralized decision making on individual part numbers.
- All decisions made to solve problems must be done within the system, and transactions must be processed to reflect the resultant changes.
- Effective use of exception messages allows focusing attention on the "vital few," not on the "trivial many."
- System records must be accurate and reflect the factory's physical reality if they're to be useful.
- Procedural inadequacies in processing MRP transactions need to be identified and corrected to ensure material plans are accurate.

REFERENCES

Cox, James F., and Richard R. Jesse, Jr. "An Application of MRP to Higher Education." *Decision Sciences* 12, no. 2 (April 1981), pp. 240–60.

Davis, E. W. *Case Studies in Material Requirements Planning.* Falls Church, Va.: American Production and Inventory Control Society, 1978.

Gray, C. D. *The Right Choice.* Essex Junction, Vt.: Oliver Wight Ltd., 1987.

Herrick, Terry L. "End Item Pegging Made Easier." *Production and Inventory Management,* 3rd quarter 1976.

Miller, J. G., and L. G. Sprague. "Behind the Growth in Materials Requirements Planning." *Harvard Business Review,* September/October 1975.

Myers, K. A.; R. J. Schonberger; and A. Amsari. "Requirements Planning for Control of Information Resources." *Decision Sciences* 14, no. 1 (January 1983), pp. 19–33.

New, C. *Requirements Planning.* New York: Halsted Press, 1973.

Orlicky, J. *Material Requirements Planning.* New York: McGraw-Hill, 1975.

Steinberg, E.; W. B. Lee; and B. M. Khumawala. "MRP Applications in the Space Program," *Journal of Operations Management* 1, no. 2 (1981).

Steinberg, E.; B. M. Khumawala; and R. Scarnell. "Requirements Planning Systems in the Health Care Environment." *Journal of Operations Management* 2, no. 4 (August 1982).

Wight, Oliver. *Manufacturing Resource Planning: MRPII.* Essex Junction, Vt.: Oliver Wight Ltd., 1984.

DISCUSSION QUESTIONS

1. Why is the MRP activity in the "engine" part of the MPC system shown in Figure 2.1?
2. What additional information would be helpful to you in using or following the basic MRP record?
3. Compare a bill of material (BOM) and a cookbook recipe.

4. How does the *system* coordinate the individual item records and provide back schedule information?
5. Provide examples of potential differences between the information system and the physical reality for university activities. What are the consequences of some of these mismatches?
6. What are some of the reasons for wanting to process the records in an MRP system frequently? Provide examples and consequences of delaying the processing of the information.
7. The chapter uses a Christmas list and Christmas gift order analogy for planned order releases and scheduled receipts. What are other analogies of these two concepts? Why is it important to keep them separate in the MRP records?
8. What are the implications of *not* allocating material to a shop order after availability checking?
9. Give some examples of transaction processing for individual students at a university. What happens if they are not done well?

PROBLEMS

1. MacRonald's Restaurant sells three kinds of hamburgers—regular, super, and super-duper. The bills of material are:

Regular burger		Super burger		Super-duper burger	
⅛ lb. patty	1.0	¼ lb. patty	1.0	⅛ lb. patty	2.0
Regular bun	1.0	Sesame bun	1.0	Sesame bun	1.0
Pickle slice	1.0	Pickle slices	2.0	Pickle slices	4.0
Catsup	0.1 oz.	Catsup	0.2 oz.	Lettuce	0.3 oz.
		Onion	0.2 oz.	Catsup	0.2 oz.
				Cheese	0.5 oz.
				Onion	0.2 oz.

a. If the product mix is 20 percent regulars, 45 percent supers, and 35 percent super-dupers, and Mac sells 200 burgers a day, how much hamburger meat is used per day?
b. How many pickle slices are needed per day?
c. Suppose buns are delivered every second day. Mac is ready to order. His on-hand balance of regular buns is 25, and his on-hand balance of sesame buns is 20. How many buns should be ordered?
d. Reconsider question c if Mac has 10 regular hamburgers, 5 supers, and 15 super-dupers all made.

2. The following illustration shows how to assemble your P301 computer.

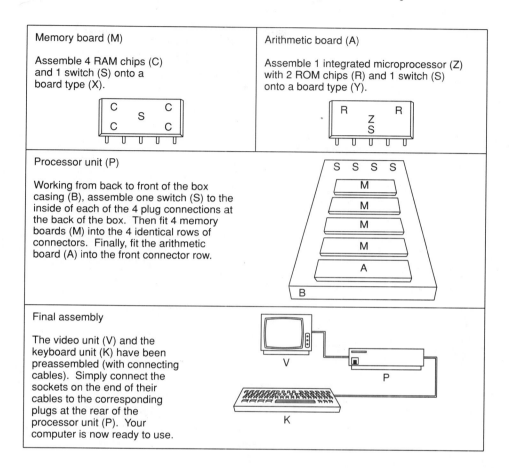

Memory board (M)	Arithmetic board (A)
Assemble 4 RAM chips (C) and 1 switch (S) onto a board type (X).	Assemble 1 integrated microprocessor (Z) with 2 ROM chips (R) and 1 switch (S) onto a board type (Y).

Processor unit (P)

Working from back to front of the box casing (B), assemble one switch (S) to the inside of each of the 4 plug connections at the back of the box. Then fit 4 memory boards (M) into the 4 identical rows of connectors. Finally, fit the arithmetic board (A) into the front connector row.

Final assembly

The video unit (V) and the keyboard unit (K) have been preassembled (with connecting cables). Simply connect the sockets on the end of their cables to the corresponding plugs at the rear of the processor unit (P). Your computer is now ready to use.

a. Draw the product structure tree corresponding to the assembly instructions.
b. Determine low-level codes for the following items: (1) A—Arithmetic board, (2) B—Box casing, (3) C—RAM chip, (4) K—Keyboard unit, (5) M—Memory board, (6) P—Processor unit, (7) R—ROM chip, (8) S—Switch, (9) V—Video unit, (10) X—Board type X, (11) Y—Board type Y, and (12) Z—Integrated microprocessor.
c. Assume no inventory of any item. How many of each part should be available to assemble one completed unit?

3. Given the following information related to the Roxy Renolds chair, draw and label the product structure diagram and determine low-level codes for all items.

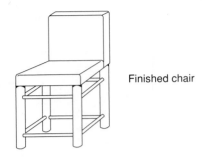

Finished chair

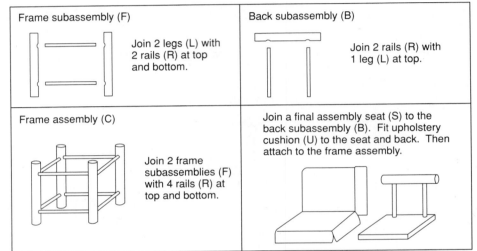

Frame subassembly (F)	Back subassembly (B)
Join 2 legs (L) with 2 rails (R) at top and bottom.	Join 2 rails (R) with 1 leg (L) at top.
Frame assembly (C)	Join a final assembly seat (S) to the back subassembly (B). Fit upholstery cushion (U) to the seat and back. Then attach to the frame assembly.
Join 2 frame subassemblies (F) with 4 rails (R) at top and bottom.	

4. Develop an MRP spreadsheet record for six periods using the following parameters for the items:

Requirements	20 units/period
Lead time	1 period
Lot size	40 units
Safety stock	0 units
Inventory	2 units
Scheduled receipt	40 units in period 1

a. In what periods are there planned order releases?
b. What happens to the timing and number of planned order releases if 10 units of safety stock are required?

5. The recipe for 6 drinks of Little Nellie's Triple Polar Bear Comforter calls for 2 dashes of Angostura, 1½ quarts of Southern Comfort, 1 quart of Polish vodka, ½ quart of Cointreau, and 1 bag of ice (for the head?).

   a. Nellie is planning a party for 12 people and wants to make Polar Bear Comforters for openers. How much Polish vodka does she need for 12 drinks if her vodka supply is totally gone?

   b. How many quarts of Southern Comfort must be bought if Nellie already has half a quart of Comfort on hand?

   c. Nellie is planning her party for November 15. It's now November 11. The liquor store will deliver quart bottles if given one-day notice. Fill in the following MRP record for Cointreau if Nellie has ½ quart on hand and wants the rest delivered. Assume she can mix the drinks the day of the party.

November		11	12	13	14	15
Gross requirements						
Scheduled receipts						
Projected available balance						
Planned order releases						

6. The Big B Bike and Trike Shop produces two basic bikes called A and B. Each week, Paul, the owner, plans to assemble 10 A bikes and 5 B bikes. Given this information and the following product structure diagrams for A and B, fill out the MRP records (inventory status files) for component parts G and Y for the next seven weeks.

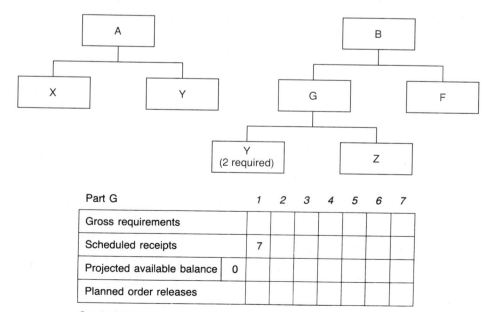

Part G		1	2	3	4	5	6	7
Gross requirements								
Scheduled receipts		7						
Projected available balance	0							
Planned order releases								

Q = lot for lot; LT = 1; SS = 0.

Part Y		1	2	3	4	5	6	7
Gross requirements								
Scheduled receipts		10						
Projected available balance	28							
Planned order releases								

Q = lot-for-lot; LT = 2; SS = 0.

Suppose 10 units of safety stock are required for part Y. What changes would result in the records? Would the MRP system produce any exception messages?

7. Given the following product structure diagram, complete the MRP records for parts A, B, and C.

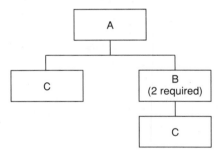

Part A		1	2	3	4	5	6
Gross requirements		5	15	18	8	12	22
Scheduled receipts							
Projected available balance	21						
Planned order releases							

Q = 20; LT = 1; SS = 0.

Part B		1	2	3	4	5	6
Gross requirements							
Scheduled receipts		32					
Projected available balance	20						
Planned order releases							

Q = 40; LT = 2; SS = 0.

Part C		1	2	3	4	5	6
Gross requirements							
Scheduled receipts							
Projected available balance	50						
Planned order releases							

Q = lot-for-lot; LT = 1; SS = 10.

8. Lucy Davis wants to make 20 picture frames per week with the following product structure (indented bill of materials):

> X Picture frame
>   Y Subassembly (2 required)
>   Z Fastener
>   Z Fastener

Other data:

	On hand	L.T.	SS	Q
X	41	1	0	25
Y	52	2	0	50
Z	60	1	10	Lot-for-lot

Construct the MRP records for weeks 1–5 using a spreadsheet.

9. Consider the following product structure and inventory information:

Item	Inventory
A	10
B	40
C	60
D	60

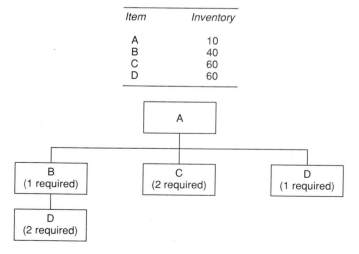

Lead time = 1 week for all items. There are no scheduled receipts for any item. How many units of product A can be delivered to customers at the start of next week under each of the following circumstances? (Treat each independently; that is, only a, only b, or only c.)

a. The bill of materials for B is wrong. It actually takes 2 units of B to make an A.

b. The inventory for D is only 40 units.

c. There was need to scrap 10 units of the inventory for item C.

10. Consider the information contained in the planned order row of the following MRP record:

Period		1	2	3	4	5
Gross requirements		5	30	20	20	0
Scheduled receipts			40			
Projected available balance	10	5	15	35	15	15
Planned order releases			40			

Q = 40; LT = 1; SS = 2.

What transactions would lead to shifting the planned order to period 1 or period 3?

11. The MPC system at the Duckworth Manufacturing Company is run weekly to update the master production schedule (MPS) and MRP records. At the start of week 1, the MPS for end products A and B is:

Master production schedule						
Week number	1	2	3	4	5	6
Product A	10	—	25	5	10	—
Product B	5	20	—	20	—	20

One unit of component C is required to manufacture one unit of either end product A or B. Purchasing lead time for component C is two weeks, an order quantity of 40 units is used, and no (zero) safety stock is maintained for this item. Inventory balance for component C is 5 units at the start of week 1, and there's an open order (scheduled receipt) for 40 units due to be delivered at the beginning of week 1.

a. Complete the MRP record for component C as it would appear at the beginning of week 1:

Week		1	2	3	4	5	6
Gross requirements							
Scheduled receipts							
Projected available balance							
Planned order releases							

b. During week 1, the following transactions occurred for component C:
   1. The open order for 40 units due to be received at the start of week 1 was received on Monday of week 1 with a quantity of 30. (Ten units of component C were scrapped on this order.)
   2. An inventory cycle count during week 1 revealed that five units of component C were missing. An inventory adjustment of -5 was, therefore, processed.
   3. Ten units of component C were actually disbursed (instead of the 15 units planned for disbursement to produce end products A and B). (The MPS quantity of five in week 1 for product B was canceled due to a customer order cancelation.)
   4. The MPS quantities for week 7 include 15 units for product A and zero units for product B.
   5. Due to a change in customer order requirements, marketing has requested that the MPS quantity of 25 units for product A scheduled in week 3 be moved to week 2.
   6. An order for 40 units was released.

Given this information, complete the MRP record for component C as it would appear at the beginning of week 2:

Week		2	3	4	5	6	7
Gross requirements							
Scheduled receipts							
Projected available balance							
Planned order releases							

What action(s) are required by the inventory planner at the start of week 2 as a result of transactions occurring during week 1?

12. Consider the following MRP record for Cactus Cups:

Week		1	2	3	4	5	6
Gross requirements		25	30	5	15	5	10
Scheduled receipts			40		0	15	
Projected available balance	35		20	15		10	
Planned order releases							

Q = lot-for-lot; LT = 5; SS = 0.

Suppose 5 units of the scheduled receipt for 40 units due on Monday of week 2 are scrapped during week 1, and no scrap ticket is issued. Furthermore, assume this lot isn't counted before it's put away in the stockroom on Monday of week 2, but recorded as a receipt of 40 units. What impact will these actions have on factory operations?

13. Ward Manufacturing Company has collected 20 periods of data for two of its products, the Snarf and the Barf. Using the two data sets as gross requirements data, construct MRP time-phased records. Using a spreadsheet program, create four time-phased records, two for each data set. For one case, use a lot size of 150 units, and for the other use an order quantity equal to the total net requirements for the next three periods. In all four records, start with a beginning inventory of 150 units, and compute the average inventory level held over the 20 periods. What do the results mean?

	Demand			Demand	
Period	Snarf	Barf	Period	Snarf	Barf
1	51	77	11	45	73
2	46	83	12	52	88
3	49	90	13	49	15
4	55	22	14	48	21
5	52	10	15	43	85
6	47	80	16	46	22
7	51	16	17	55	88
8	48	19	18	53	75
9	56	27	19	54	14
10	51	79	20	49	16

14. Foremost Furniture has the following BOM for one of its products:

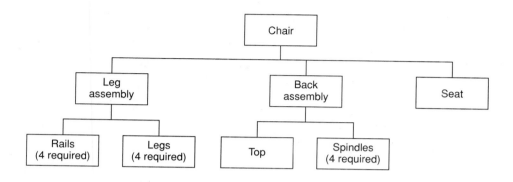

Additional data follow:

	Inventory	Lead time (weeks)
Chairs	100	1
Leg assembly	50	2
Back assembly	25	1
Seats	40	3
Rails	100	1
Legs	150	1
Tops	30	2
Spindles	80	2

The company wants to produce 400 chairs in week 5 and 300 in week 6. Develop a material plan for the parts.

15. Use a spreadsheet program to develop the MRP records for parts A and B from the following product structure. Use the data from the following table:

Part	A	B
Requirements	50/period	—
Initial inventory balance	68	8
Lead time	1	1
Lot size	lot-for-lot	250
Safety stock	10	—
Scheduled receipt	—	250 in period 1

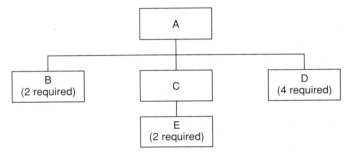

a. What are the planned orders for item B?
b. On an unlucky day (it must have been the 13th), the planner for part A found that the inventory was wrong by 13 units. Instead of 68, there were only 55 on hand. What happens to the planned orders for part B?
c. Use the spreadsheet model to generate a 10-period material plan for parts A and B. Assume that the 68 units of inventory for A was correct. Suppose that in period 1, actual demand for part A was 60 units instead of 50. Regenerate the spreadsheet for periods 2 through 11. What changes occur in the material plans for parts A and B?
d. What is the impact if the 60-unit actual demand for period 1 is repeated for five periods, in each case with a planned demand of 50 units?

Chapter 3

Just-In-Time

This chapter addresses just-in-time (JIT) approaches for manufacturing planning and control. Although a fairly recent development, JIT is one of the basic bodies of knowledge for modern approaches to manufacturing planning and control (MPC). JIT embodies much more than MPC systems. JIT is both a philosophy and a set of techniques. Moreover, the techniques go beyond those traditionally considered part of a manufacturing planning and control system. JIT makes major changes in the actual practice of manufacturing, which in turn affects MPC execution. JIT greatly reduces the complexity of detailed material planning, the need for shop-floor tracking, work-in-process inventories, and the transactions associated with shop-floor and purchasing systems. The chapter concentrates on the MPC aspects of JIT but necessarily touches on broader aspects as well. It's organized around the following seven topics:

- JIT in manufacturing planning and control: What are JIT's key features and how does it impact MPC systems?
- A JIT example: How can the basic principles of JIT be illustrated in one simplified example?
- JIT applications: What are some concrete examples of JIT practice?
- Nonrepetitive JIT: How can JIT concepts be applied to the nonrepetitive manufacturing environment?
- JIT in purchasing: What are the changes required and the payoffs to both vendor and vendee?
- JIT software: What features of computer packages support JIT?
- Managerial implications: What changes are required to fully pursue the benefits of JIT?

JIT is one of two fundamental approaches to material planning and control. Chapter 2 describes the other—material requirements planning (MRP). JIT techniques have the most influence on the "back-end" execution concepts in Chapter 5. In addition, the integration of JIT and MRP, as well as the market requirements that drive MPC choices, are described in Chapter 9. Chapter 12 gives an advanced analysis of JIT, and a Finnish system similar to JIT is described in Chapter 19. More fundamentally, JIT influences will be referred to in many other chapters where the techniques used are essentially different when JIT is in place.

JIT IN MANUFACTURING PLANNING AND CONTROL

Figure 3.1 shows how just-in-time programs relate to our manufacturing planning and control framework. The shaded area indicates the portions of MPC systems that can be affected by implementation of JIT. It affects all areas, but the primary application area is in back-end execution. However, JIT extends beyond manufacturing planning and control. JIT programs raise fundamental questions about what's really important in manufacturing and how

FIGURE 3.1 Manufacturing Planning and Control System and JIT

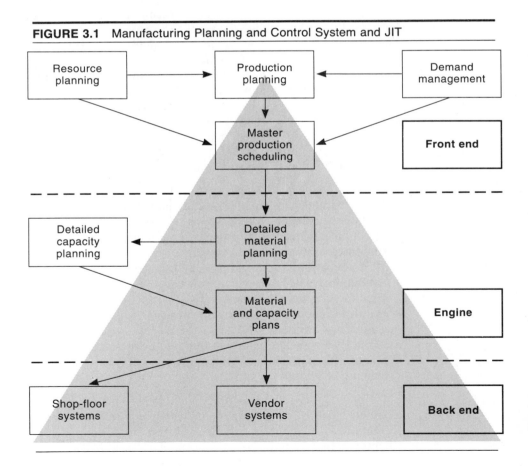

performance should be evaluated. For this reason, we start this section with a discussion of the major elements of an overall JIT program. Thereafter, we turn to the impact of just-in-time on the MPC system and the overhead cost savings from reduced MPC system transaction processing. The section closes by describing cornerstones of an overall JIT program and the linkages to MPC systems.

Major Elements of Just-in-Time

Many definitions have been put forward for just-in-time, and they have been evolving as JIT is being more globally adopted. Several of the more popular current definitions focus on JIT as an approach to minimize waste in manufacturing. This focus is so broad as to be of limited use. It helps to subdivide waste into time, energy, material, and errors. A useful common denominator running through this and other JIT definitions is a broad philosophy of pursuing zero inventories, zero transactions, and zero "disturbances"; that is, *zero disturbances* means routine execution of schedule, day in–day out.

The literature on JIT is largely one of cases. The best-known JIT examples are from firms with high-volume repetitive manufacturing methods, such as the classic case of Toyota. The most important features of these applications have been the elimination of discrete batches in favor of production rates, the reduction of work-in-process inventories, production schedules that level the capacity loads and keep them level, mixed model plans and master production schedules where all products are made more or less all the time instead of changing over from one model to another, visual control systems where workers remember how to build the products and execute the schedule without paperwork or complex overhead support, and direct ties to vendors who deliver high-quality goods frequently. All of these have MPC implications.

Just-in-time objectives are typically achieved through several physical system changes or projects. One of them is setup time reduction and a drive toward constantly smaller lot sizes. This is necessary to make all of the products constantly. It's also consistent with reducing inventory levels. Setup times are typically reduced by applying fairly common industrial engineering techniques for analyzing the setup process itself, often by workers themselves using a video camera. The results of setup time reduction have been impressive indeed. Changeovers of several hours have been reduced to less than 10 minutes. The goal now being achieved by many firms is expressed by Shigeo Shingo: SMED (single-minute exchange of dies—meaning all changeovers are less than 10 minutes).

Another physical program associated with JIT has been pursuit of improved quality through process improvement. Most JIT firms have engaged in programs of quality awareness and statistical process control. In a repetitive manufacturing system, any quality problem will result in a stoppage of the entire flow line, unless undesirable buffer inventories are held.

Quality improvement has taken many forms and is largely beyond our present scope. Two critical aspects for JIT are *TPM* and *poka-yoke*. TPM can stand for both total preventative maintenance and total productive maintenance. The focus is to apply the same diligence of product quality approaches to equipment and process quality. Poka-yoke means foolproof operations. The intent is to guarantee quality by building checking operations into processes so that the quality of every part is evaluated as it's created. This also ensures low cost since the cost of finding defects is lowest when they're found at the same time they're made. All these quality-related concepts have an impact on the MPC system requirements and design.

In a JIT program, most people now include the goal of continual improvement as a maxim for day-to-day operations. Every day, each worker should get better in some dimension, such as fewer defects, more output, or fewer stoppages. Continual improvement, as a cornerstone of JIT, means making thousands of small improvements in methods, processes, and products in a never-ceasing quest for excellence. Most JIT implementations include a strong degree of worker involvement and worker participation. In the words of a union official at the GM/Toyota (NUMMI) plant in Fremont, California, "This is the way work ought to be. [With JIT] this plant employs our hearts and minds, not just our backs."

JIT firms tend to group their equipment for **cellular manufacturing:** a group of machines manufactures a particular set of parts. The equipment layout minimizes both travel distances and inventories between machines. Cells are typically U-shaped to increase worker interactions and reduce material handling. Cross-trained workers can run several of the machines. Cellular manufacturing can make "capacity" quite flexible, so surges or mix changes are more readily handled. An extension of the cellular concept is the plant within a plant, where a portion of a factory focuses solely on one group of products.

In summary, a JIT orientation includes several action programs: (1) reduction of setup times and lot sizes; (2) a "no defects" goal in manufacturing; (3) a focus on continual improvement; and (4) an effort to involve workers and use their knowledge to a greater extent. Figure 3.2 summarizes the benefits of undertaking a comprehensive JIT program. This summary is the aggregate benefits achieved by a group of JIT companies.

JIT's Impact on Manufacturing Planning and Control

As we discussed above, JIT covers more than MPC activities but does impact all three areas of our MPC framework (front end, engine, and back end). JIT's primary contribution is in the back end. JIT provides for greatly streamlined execution on the shop floor and in purchasing. JIT offers the potential for eliminating large portions of standard shop-floor control systems, for sharply reducing costs of detailed shop scheduling, for signifi-

FIGURE 3.2   JIT Benefit Summary

	Improvement	
	Aggregate percentage (3–5 years)	Annual percentage
Manufacturing cycle time reduction	80–90%	30–40%
Inventory reductions:		
Raw materials	35–70	10–30
Work-in-process	70–90	30–50
Finished goods	60–90	25–60
Labor cost reductions:		
Direct	10–50	3–20
Indirect	20–60	3–20
Space requirements reduction	40–80	25–50
Quality cost reduction	25–60	10–30
Material cost reduction	5–25	2–10

Source: George W. Plossl, *Just-in-Time: A Special Roundtable* (Atlanta, Ga.: George Plossl Educational Services, Inc., 1985).

cant reductions in work-in-process and lead times, and for better vendor scheduling.

However, although JIT makes its major impact in the back end, it's not without influence on the front end and engine. In the detailed MRP planning of the engine, JIT plans a strong role in reducing the number of part numbers planned and the number of levels in the bill of materials. Many part numbers formerly planned by MRP analysts can now be treated by different means, such as "phantoms" (i.e., as part numbers still in the bill of materials but not expected to be transacted into and out of inventories). The result is often an order of magnitude reduction in the complexity of detailed material planning, with a concomitant reduction in the planning personnel required.

In the front end, JIT also gives rise to some important changes. Production plans and master production schedules will be required that provide whatever degree of level capacity loading is necessary for smooth shop operations. In many cases, this also requires a rate-based MPS; that is, so many units per hour or day. This drive toward more stable, level, daily-mix schedules dictates many of the required JIT activities, such as setup time reduction. To the extent that lead times are sufficiently reduced, many firms that had to provide inventories in anticipation of customer orders (made-to-stock firms) now find themselves more like make-to-order or assemble-to-order companies able to respond to customer orders. This, in turn, can impact demand management activities.

Execution under JIT is based on the concept that orders will move through the factory so quickly that it's not necessary to track their progress with a complex shop-floor control system. A similar argument holds for purchased items. If they're converted into finished goods within hours or days of receipt, it's unnecessary to put them into stockrooms, pick them, and go through all the details normally associated with receipts from vendors. Instead, the JIT firm can simply pay the vendor for the purchased components in whatever products are completed each time period; there will be so little work-in-process inventory that it's not worth either party keeping track of it for purposes of accrual payments.

The concept of updating component inventory balances when finished items are received into stock is called **backflushing.** Instead of creating detailed work-in-process accounting systems based on shop-order transactions, some JIT firms just reduce component part inventory balances by exploding the bills of material for whatever has been delivered into finished goods. It's worth noting, however, that backflushing implies a very high level of data integrity.

The JIT approach in execution is focused on simplicity. The intent is to design the manufacturing cells, products, and systems so goods flow through very routinely. With problems of quality and disturbances largely eliminated, routine execution becomes just that: routine. Simple systems can be employed by shop people without detailed records or the need for extensive overhead staff support.

The Hidden Factory

A manufacturing firm can be thought of as comprising two "factories." One makes products and the other (the hidden factory) processes transactions on papers and computer systems. Over time, the former factory has been decreasing in relative cost, compared with the latter. The annual survey of manufacturing firms in North America by the Boston University Manufacturing Round Table has consistently found rising overhead costs to be an important concern of manufacturing managers. A major driver for these costs is transactions. **Logistical transactions** include ordering, execution, and confirmation of materials moving from one location to another. Included are the costs of personnel in receiving, shipping, expediting, data entry, data processing, accounting, and error follow-up. Under JIT, the goal is to eliminate the vast majority of this work and the associated costs. Work orders that accompany each batch of material as it moves through the factory are eliminated under JIT. The fundamental concept is that if the flow can be simplified, fast, and guaranteed, then there's no need for paperwork.

Balancing transactions are largely associated with the planning that generates the logistical transactions. Included are production control, purchasing, master scheduling, forecasting, and customer order processing/maintenance. In most companies, balancing transaction costs are 10 to 20 percent of the total manufacturing overhead costs. JIT again offers a significant opportunity to sharply reduce these costs. MRP planning can be cut by perhaps 75 to 90 percent in complexity. The improvements generated by vendor scheduling can be extended. Vendor firms will no longer have to process *their* sets of transactions that are paid for in hidden factory costs.

Quality transactions extend far beyond what one normally thinks of as quality control. Included are transactions associated with identification and communication of specifications, certification that other transactions have indeed taken place, and recording of required backup data. Many of the costs of quality identified by Juran and others are largely associated with transactions. JIT, with closer coupling of production and consumption, has faster quality monitoring and response capability.

Still another category is **change transactions.** Included are engineering changes and all those that update manufacturing planning and control systems, such as routings, bills of material, and material specifications. Engineering change transactions are some of the most expensive of any in the company. A typical engineering change, for example, might require a meeting of people from production control, line management, design engineering, manufacturing engineering, and purchasing. The change has to be approved, scheduled, and monitored for execution.

One way that firms attack the hidden factory is by finding ways to significantly reduce the number of transactions. JIT is a major weapon in this attack. Stability is another attack, and again JIT is important since it is based on stabilized operations. Still another attack on hidden factory transaction costs is through automation of transactions (as with bar coding) and eliminating redundancies in data entry. More integrated systems are also providing payoffs, and so are better data entry methods. But we remain convinced that stability and transaction elimination should be pursued before turning to automation of transactions. JIT is clearly a key in achieving hidden factory cost reductions.

JIT Building Blocks in MPC

As Figure 3.3 shows, JIT links four fundamental building blocks: product design, process design, human/organizational elements, and manufacturing planning and control. JIT provides the connecting link for these four areas.

Critical connecting activities in product design include quality, designing for manufacture in cells, and reducing the number of "real" levels in the bill of materials to as few as possible. Some firms say there should be no more than two or three levels. (Phantom levels not controlled separately aren't counted.) By not having more than three levels in the bill of materials, products only

FIGURE 3.3 Building Blocks for Just-in-Time

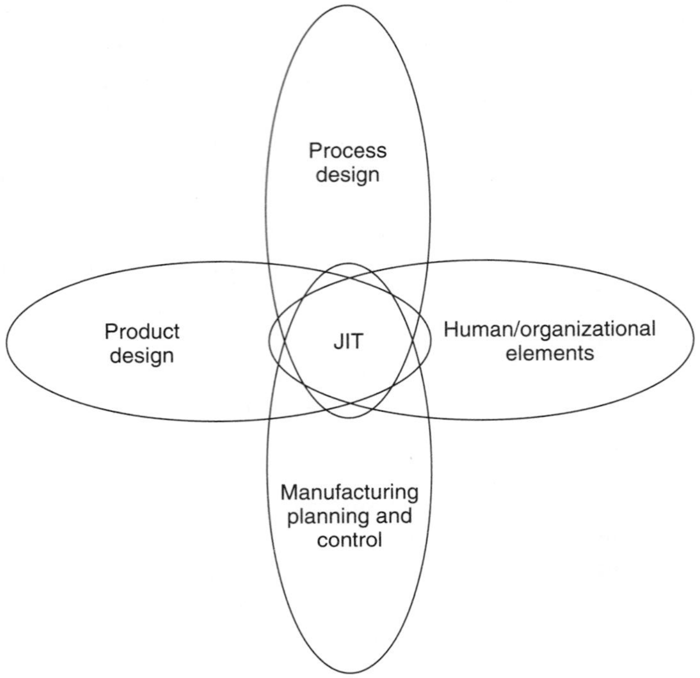

have to go into inventory and out again, with MRP-based planning, once or twice as they are produced.

Reducing the bill of material levels and designing the manufacturing process are closely related. For fewer levels to be practical, the number of product conversion steps included in one routing must be reduced through process design changes. Another key concept in process design is cellular manufacturing. Equipment in cellular manufacturing is positioned (often in a U shape) to achieve rapid flow of production with minimal inventories. The object is to concentrate on material velocity. Jobs must flow through in short cycle times, so detailed tracking is unnecessary.

Band width is another important notion in designing manufacturing processes. A wide band width system has enough surge capacity to take on a fairly mixed set of products, and some variation in demand for the products as well. The impact on MPC system design is through the focus on inventory and throughput time reduction, which means inventory is not built to level out capacity requirements. JIT systems are designed to be responsive to as large a set of demands as possible. Superior manufacturing processes support greater band width. The objective is to be able to make any product, right behind any other, with minimal disruption.

Human/organizational elements are the third building block for JIT. An aspect of this is the **whole person concept,** which will continually apply training, study, process improvement, and whatever else is needed to eliminate recurring problems. The objective is continual learning and improvement. Human/organizational elements recognize that workers' range of capabilities and level of knowledge are often more important assets to the firm than equipment and facilities. Education and cross training are continuing investments in the human asset base. As the asset base's capabilities grow, need for overhead support is reduced and overhead personnel can be redeployed to address other issues.

Linking human/organization elements into the other activities has a significant impact on operation of the production process and MPC system. Band width and not building inventories to utilize direct labor mean surge capacity must be available. Surge capacity in direct labor personnel means these people will not be fully utilized in direct production activities. In fact, the whole person concept is based on the premise of hiring *people,* not just their muscles. As a consequence, direct workers are cross trained to take on many tasks not usually associated with "direct labor." This includes equipment maintenance, education, process improvement, data entry, and scheduling. From a JIT standpoint, the human/organizational elements building block puts a greater emphasis on scheduling by workers and less on scheduling by a centralized staff function. The entire process is fostered by the inherent JIT push toward simplification. With no defects, zero inventories, no disturbances, and fast throughput, detailed scheduling is easier; moreover, any problems tend to be local in nature and amenable to solution on a decentralized basis.

The final building block in Figure 3.3 is the manufacturing planning and control system and its link to JIT. Applying JIT requires most of the critical MPC functions described in this book. It will always be necessary to do master production scheduling, production planning, capacity planning, and material requirements planning based on explosion. If the bill of materials is reduced to two or three levels, the detailed material planning and associated transaction costs can be cut significantly. If all detailed tracking is done by direct laborers under the whole person concept, additional savings can be achieved.

We see then that JIT has the potential for changing the character of production control in a company, since it reduces MPC transactions. JIT can significantly reduce the size of the "hidden factory" that produces papers and computer transactions instead of products. Table 3.1 provides a more detailed listing of JIT's building blocks and objectives. Many of these will be described in the next section, which presents a detailed JIT example.

TABLE 3.1   JIT Objectives and Building Blocks

Ultimate objectives:
- Zero inventory.
- Zero lead time.
- Zero failures.
- Flow process.
- Flexible manufacture.
- Eliminate waste.

Building blocks:
- Product design:
  Few bill of material levels.
  Manufacturability in production cells.
  Achievable quality.
  Appropriate quality.
  Standard parts.
  Modular design.
- Process design:
  Setup/lot size reduction.
  Quality improvement.
  Manufacturing cells.
  Limited work-in-process.
  Production band width.
  No stockrooms.
  Service enhancements.
- Human/organizational elements:
  Whole person.
  Cross training/job rotation.
  Flexible labor.
  Continual improvement.
  Limited direct/indirect distinction.
  Cost accounting/performance measurement.
  Information system changes.
  Leadership/project management.
- Manufacturing planning and control:
  Pull systems.
  Rapid flow times.
  Small container sizes.
  Paperless systems.
  Visual systems.
  Level loading.
  MRP interface.
  Close purchasing/vendor relationships.
  JIT software.
  Reduced production reporting/inventory transaction processing.
  Hidden factory cost reductions.

A JIT EXAMPLE

In this section we develop a detailed but simple example to illustrate the basic concepts of JIT. The intent is to show how MPC approaches based on MRP would be modified to implement JIT and to show the necessary building blocks (Table 3.1) to achieve this. The product for our example is a one-quart saucepan produced in four models by the Muth Pots and Pans Company. (See Figure 3.4.) The product's brochure sums up its importance: "If you ain't got a Muth, you ain't got a pot." We'll look at elements of a JIT program for the saucepan that range from leveling production to redesigning the product. Some of these elements have a direct MPC relevance; others will affect MPC only indirectly.

Leveling the Production

We start the saucepan's JIT program by considering how to "level and stabilize" production. This means not only planning a level output of one-quart saucepans but planning to produce the full mix of models each day (or week or some other short interval, if volumes are not sufficiently high to warrant daily production). Full-mix production in a short interval provides for less inventory buildup in each model. Moreover, the schedule can respond to actual custom order conditions more quickly. Level output also implies "freezing" a plan for a time to stabilize the production and related activities on the

FIGURE 3.4   The 151 One-Quart Saucepan Line

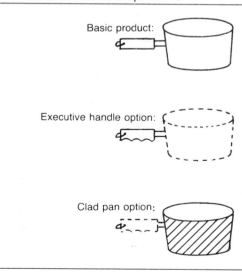

Basic product:

Executive handle option:

Clad pan option:

floor. Before looking at how this might be done specifically, let's consider Muth's manufacturing situation with traditional MRP-based approaches.

At the present, Muth uses production planning to set the overall production rate. In setting this rate, it's necessary to build inventories in anticipation of the Christmas season peak in demand. The annual forecast for each of the four models is given in Figure 3.5. A product structure and part listing is provided in Figure 3.6. A master production schedule, for each of the four models, is exploded to produce a material requirements planning record for each of the 14 component part numbers shown on the part listing in Figure 3.6. Safety stock is carried for all components, and production is in the lot sizes indicated in Figure 3.6. Figure 3.7 gives lead times and routing data; lead times are computed on the basis of two days per operation, rounded up to the next whole week using five-day weeks. A typical MRP record is shown as Figure 3.8.

To plan for level production, the first step is converting the forecasts to the daily requirements for each model. Using a 250-day year, this conversion is shown in Figure 3.9. Note the difference between the current lot sizes and the daily requirements. Daily production will put pressure on process design to reduce setup times. Two other possible mixed model master production schedules are shown in Figure 3.9, in addition to the one based on daily production batch sizes. The first shows quantities to be produced if hourly batches are to be made. The second shows an MPS with the minimum batch size of one for model 151B.

Pull System Introduction

A "pull" system of material flow control occurs when a work center is only authorized to produce when it has been signaled that there's a need for more parts in a downstream (user) department. In general, this means no work center is allowed to produce parts just to keep workers or equipment busy. It also means no work center is allowed to "push" material to a downstream work center. All movements and production are authorized by a signal from a downstream work center when it has a need for component parts.

FIGURE 3.5   Annual Forecast Data

| *Completed pan model number* | *Description of model* | | *Annual forecast* |
	Handle	*Metal*	
151A	Basic	Sheet	200,000
151B	Basic	Clad	2,500
151C	Executive	Sheet	25,000
151D	Executive	Clad	100,000

FIGURE 3.6   Product Structure and Parts List

| | | | Finished item number | | | |
Models (end items)	Lot size	Safety stock	151A	151B	151C	151D
Complete pan 151A	8,000	5,000	X			
Complete pan 151B	900	1,000		X		
Complete pan 151C	3,000	3,000			X	
Complete pan 151D	6,000	5,000				X

Component part

	Lot size	Safety stock	151A	151B	151C	151D
Regular pan 1936	14,000	10,000	X		X	
Clad pan 1937	8,000	6,000		X		X
Basic han. ass. 137	14,000	8,000	X	X		
Exec. han. ass. 138	8,000	5,000			X	X
Basic han. set 244	9,000	8,000	X	X		
Exec. han. set 245	9,000	8,000			X	X
Bas. han. base 7731	14,000	8,000	X	X		
Ex. han. base 7735	12,000	5,000			X	X
Ring 353	24,000	15,000	X	X	X	X
Rivets 4164	100,000	50,000	X	X	X	X
Sheet metal 621	1 coil	1 coil	X		X	
Clad sheet 624	1 coil	1 coil		X		X
Handle sheet 685	1 coil	1 coil	X	X	X	X
Plastic beads 211	5 tons	1 ton	X	X	X	X

FIGURE 3.7 Routing and Lead Time Data

Department	Item	Routing	Lead time
Final assembly	Complete pan	1. Spot weld	2 days
		2. Inspect	2 days
		3. Package	2 days
		Total = 6 days = 2 weeks	
Punch press	Pan	1. Blank and form	2 days
		2. Roll lip	2 days
		3. Test for flat	2 days
		4. Straighten	2 days
		5. Inspect	2 days
		Total = 10 days = 2 weeks	
Handle base	Handle base	1. Blank and form	2 days
		2. Inspect	2 days
		Total = 4 days = 1 week	
Handle assembly	Handle assembly	1. Rivet	2 days
		2. Inspect	2 days
		Total = 4 days = 1 week	
Injection molding	Plastic handle set	1. Mold	2 days
		2. Deburr	2 days
		3. Inspect	2 days
		Total = 6 days = 2 weeks	

Purchased items

Purchasing	Sheet metal		Purchased lead time one week for all items
	Clad sheet metal		
	Plastic beads		
	Ring		
	Rivets		

Figure 3.8 MRP Record for Basic Handle Assembly (Part 137)

		Week									
		1	2	3	4	5	6	7	8	9	10
Gross requirements			8		8	3	8		8		8
Scheduled receipts											
Proj. avail. bal.	10	10	16	16	8	19	11	11	17	17	9
Planned order rel.		14			14			14			

Q = 14; LT = 1; SS = 8. (All quantities are in thousands.)

FIGURE 3.9 Master Production Schedule Data*

	Model			
Option configurations:	*151A*	*151B*	*151C*	*151D*
Handle	Basic	Basic	Executive	Executive
Pan	Sheet	Clad	Sheet	Clad
Annual forecast (units)	200,000	2,500	25,000	100,000
Possible mixed model master production schedules:				
Daily batch MPS	800	10	100	400
Hourly batch MPS	100	1.25	12.5	50
Minimum batch MPS	80	1	10	40

*Data are based on a 250-day year and an eight-hour work day.

Frequently, it's believed that the pull system creates the benefits in JIT. In part, primary payoffs come from the discipline required to make the system work. Included are lot size reductions, limited work-in-process, fast throughput, and guaranteed quality.

The signals for communicating downstream work center demand vary widely. They include rolling a colored golf ball from a downstream work center to its supplying work center when it wants parts; yelling "Hey, we need some more"; sending an empty container back to be filled; and using cards (kanbans) to say more components are needed. A widely used technique is to paint a space on the floor that holds a specified number of parts. When the space is empty, the producing department is authorized to produce material to fill it. The consuming or using department takes material out of the space as they need it; typically, this only occurs when the space authorizing *their* output is empty. For the Muth example, we'll use an empty container as the signal for more production; that is, whenever a using department empties a container, it sends the container back to the producing department. An empty container represents authorization to fill it up.

Given that Muth has committed to a level schedule where all models are made every day, the firm is almost ready to move into a pull mode of operation. Two additional issues need to be faced. First, there's the question of stability. For most pull-type systems, it's necessary to keep the schedule firm (frozen) for some reasonable time. This provides stability to the upstream work centers, as well as overall balance of the work flow. For Muth, assume the schedule is frozen for one month, with the daily batch quantities given in Figure 3.9 (1,310 pots per day).

The second issue is determining the container sizes to transport materials between work centers—a fairly complicated issue. It involves material handling considerations, container size commonality, congestion in the shop, proximity of work centers, and, of course, setup costs. For example, consider the container used between handle assembly and final assembly for the basic handle, part 137 (810 being used per day). The center is currently producing in lots of 14,000. We'll choose a container size that holds 100 pieces representing just under an eighth of a day's requirements. Note this choice puts a great deal of pressure on the handle assembly work center to reduce setup times.

Figure 3.10 shows the flow of work in Muth's new system for handle assembly to the final assembly line. Only two containers are used for part 137; while one is being used at the final assembly line, the other is being filled at handle assembly. This approach is very simple and is facilitated by the two departments being in close proximity. Figure 3.11 shows the factory layout. A worker from the final assembly line or a material handler can return empty containers. Any empty container is a signal to make a new batch of handles (i.e., fill it up). It's interesting to note the difference in average inventory that will be held in this system, compared with the former MRP methods and the lot size of 14,000. The system with a small container approaches "zero inventory," with an average inventory of about 100 units. Compare this to the inventories shown in the MRP record of Figure 3.8 (average inventory = 14,400).

This pull system example has no buffer at either work center. It would be possible to add another container, which would allow greater flexibility in handle assembly, at the cost of extra inventory in the system. As it is, the final assembly area would use up a container in just under one hour. This means the system has to be responsive enough for the empty container to be returned to handle assembly and a batch made in this time frame. An extra container allows more time for responding to a make signal (an empty container) and also allows more flexibility in the supplying department. The extra inventory helps resolve problems—for example, when several production requests for different parts (containers) arrive at the same time.

Product Design

To illustrate the implications for product design, consider the basic and executive handles for Muth's one-quart saucepan shown in Figure 3.4. There are two differences between the handles: the grips and the ring placement. With some redesign of the plastic parts on the executive handle, the handle base becomes a common part and the ring placement is common between the two handle models; the methods for handle assembly could also be standardized. The only difference would be the choice of plastic handle parts. Such a redesigned handle base is shown in Figure 3.12.

In addition to the improvements this design change makes in handle subassembly, there are potential impacts in other areas as well. For example, handle bases would have one combined lot of production instead of two, with attendant reductions in inventory. It might now be possible to run the handle

FIGURE 3.10 Pull System for Muth Pots and Pans

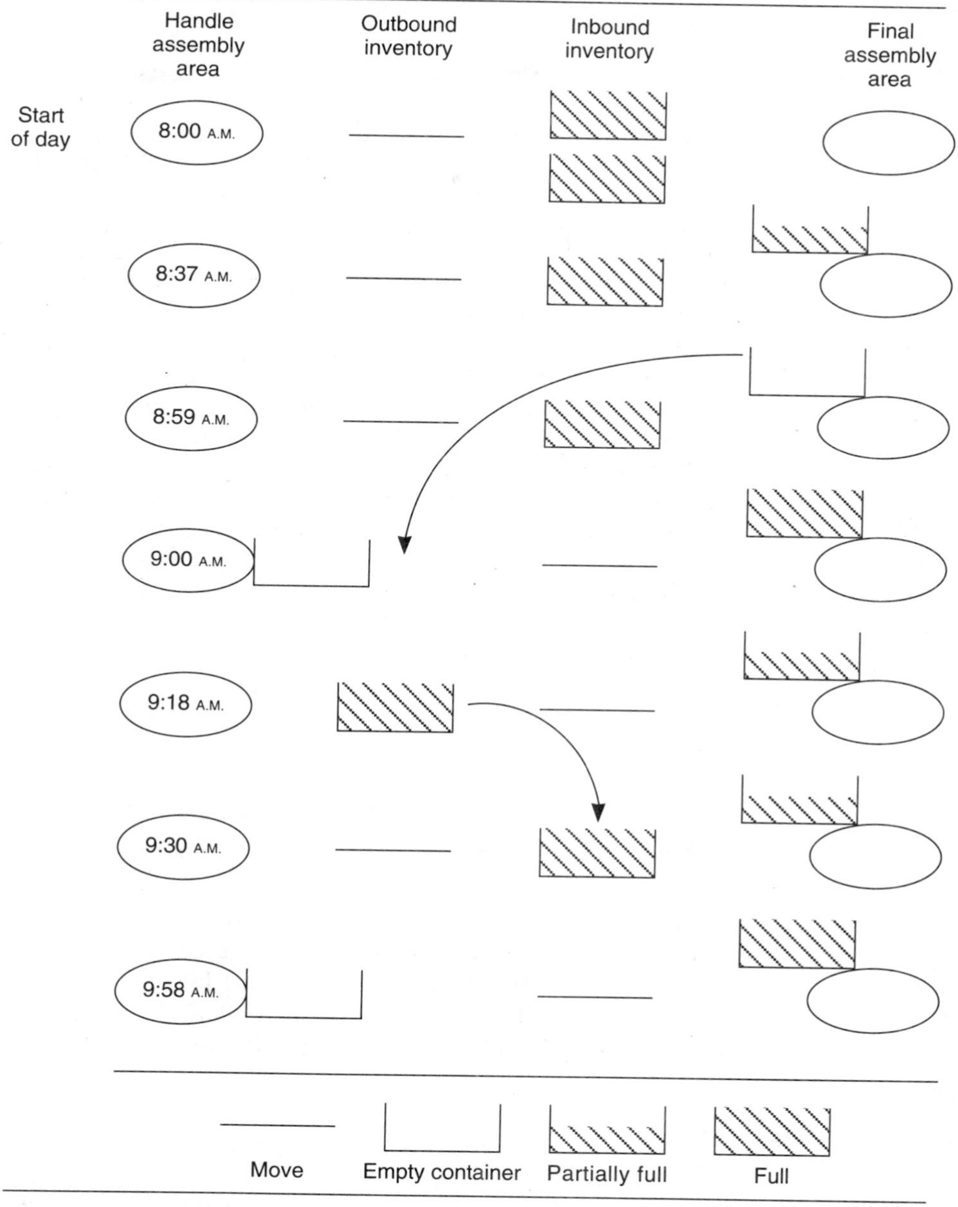

	Handle assembly area	Outbound inventory	Inbound inventory	Final assembly area

Start of day — 8:00 A.M.

8:37 A.M.

8:59 A.M.

9:00 A.M.

9:18 A.M.

9:30 A.M.

9:58 A.M.

Move   Empty container   Partially full   Full

FIGURE 3.11   Factory Layout

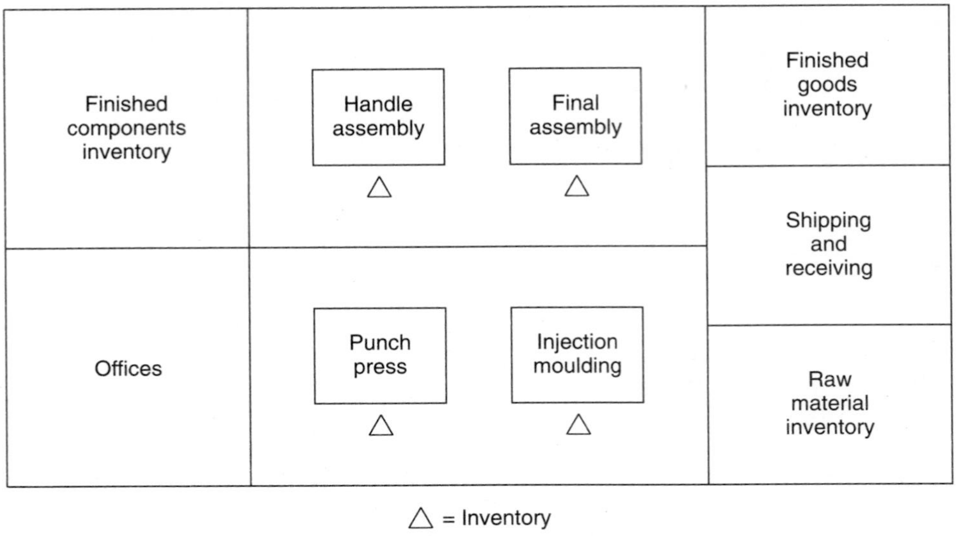

$\triangle$ = Inventory

FIGURE 3.12   Redesigned Handle Base

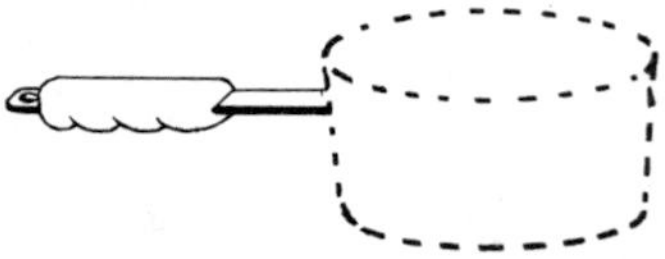

base area on a pull system as well, with containers passing between the handle base area and the handle subassembly area. Another advantage is a simplification in the bill of materials, a reduction in the number of parts that must be planned and controlled with MRP, and a concomitant reduction in the number of transactions that have to be processed.

Process Design

The product redesign, in turn, opens opportunities for process improvement. For example, it may now be possible to use the same equipment to attach both kinds of plastic handles to the handle base. Perhaps a cell can be formed, where handle bases are made and assembled as a unit. Figure 3.13 shows one way this might be accomplished, including an integration of the handle assembly cell and the final assembly line in a U-shaped layout. Note in this exam-

FIGURE 3.13 Cellular Manufacturing of Handle Assembly and Final Assembly

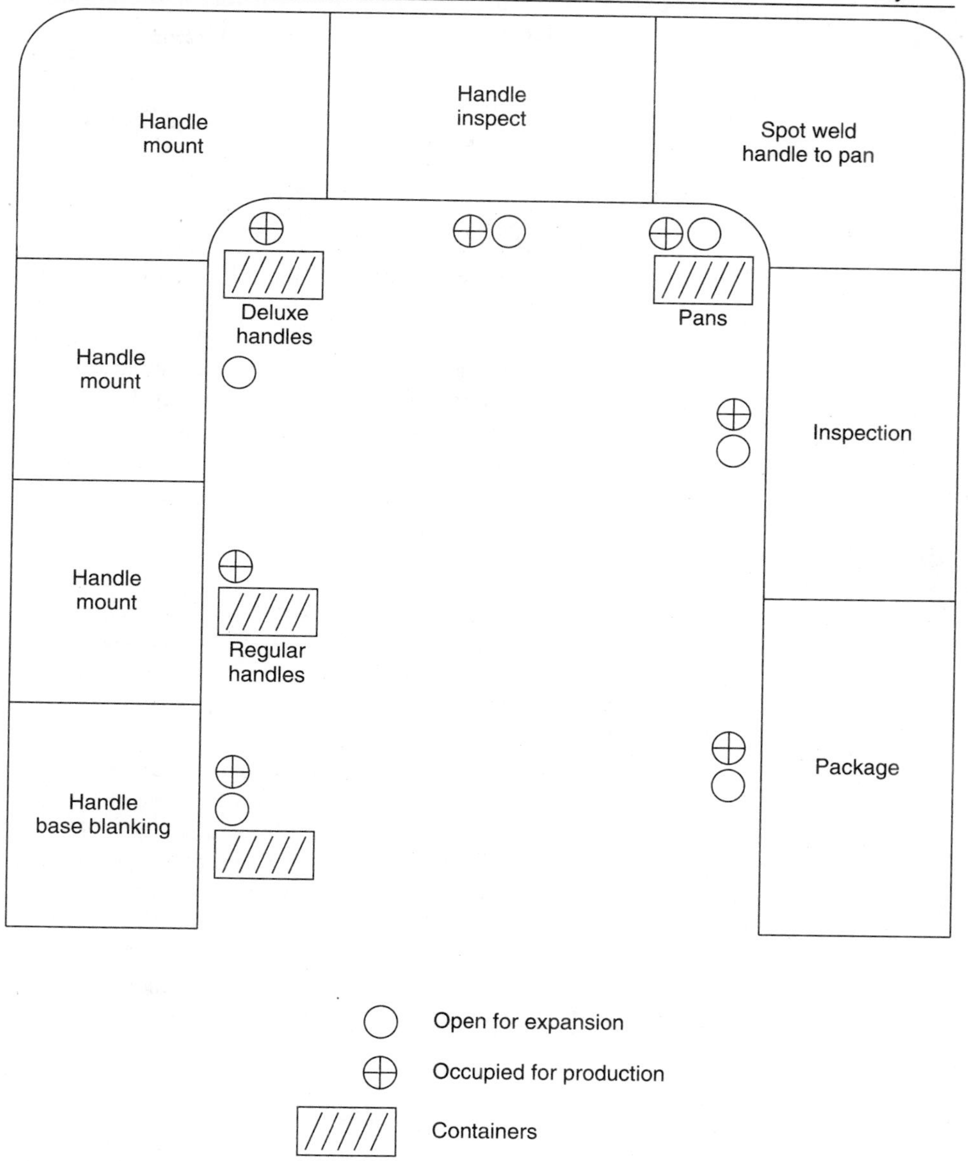

ple, no significant inventories are anywhere on the line, and both handle base material and plastic handle parts are replenished with a pull system based on containers.

Figure 3.13 also illustrates the band width concept. Several open stations along the line would permit adding personnel if volume were increased. Moreover, perhaps Muth would like to establish different production rates for certain times. For example, perhaps this pan might be manufactured in higher volumes near the Christmas season. What's needed is the capacity at the cell to move from one level of output to another. This added capacity probably means the dedicated equipment will not be highly utilized.

The cell is designed to permit variations in staffing to better respond to actual customer demands. If an unexpected surge in demand for executive handle pots comes through, the cellular approach will allow Muth to make the necessary changes faster—and to live with this kind of problem with smaller finished goods inventories. Over time, perhaps this cell can be further expanded in terms of band width and flexibility to produce handles for other Muth products.

The value of quality improvement can be seen in Figure 3.13. The inspection station takes up valuable space that could be used for production. It adds cost to the product. If bad products are being culled by inspection, buffer stocks will be required to keep the final assembly line going. All of this is waste to be eliminated.

Bill of Material Implications

The product redesign results in a streamlined bill of materials. The number of options from the customer's point of view has been maintained, but the number of parts required has gone down (e.g., components have been reduced from 14 to 10). With the cellular layout shown in Figure 3.13, the handle base and handle assembly no longer exist as inventoriable items. They are "phantoms" that won't require direct planning and control with MRP. The product structure given as Figure 3.6 now will look like Figure 3.14.

Several observations can be made about Figure 3.14. One is that handle assemblies have ceased to exist as part of the product structure. If we wanted to maintain the handle assembly for engineering and other reasons, it could be treated as a phantom. Figure 3.15 shows what the MRP record would look

FIGURE 3.14   Simplified Product Structure

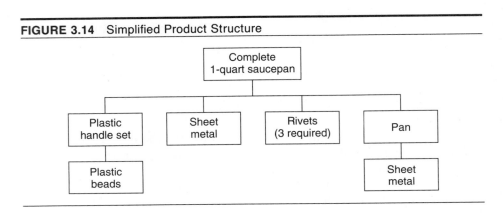

FIGURE 3.15 MRP Record for Phantom Part 137

		Week									
		1	2	3	4	5	6	7	8	9	10
Gross requirements		4050	4050	4050	4050	4050	4050	4050	4050	4050	4050
Scheduled receipts											
Projected available balance	15000	10950	6900	2850							
Planned order releases					1200	4050	4050	4050	4050	4050	4050

Q = lot for lot; LT = 0; SS = 0.

like in this case. In Figure 3.15, there's some existing inventory to use up;
phantom treatment allows this to occur, and will always use this inventory
before making more.

Another observation is that pans *do* remain as inventoriable items. Elimination of these two part numbers and their associated inventories may well
be the next goal for product and process redesign. Still another is to understand the magnitude of the reduction in transactions represented by the JIT
approach illustrated in Figure 3.13. All MRP planning for the eliminated part
numbers (or phantom treatment) is now gone. This impacts MRP planning as
well as stockrooms—and all other indirect labor associated with MRP
control.

Finally, we need to consider the impact on lead times, the resultant ability
to better respond to market conditions, the reductions in work-in-process inventories, and the greater velocity with which material moves through the
factory. If the combined lead times are computed using the product structure
in Figure 3.6 and lead time data in Figure 3.7, five weeks are required for the
flow of raw materials into pots. The JIT approach cuts that to just over two
weeks, which could be reduced even further.

JIT APPLICATIONS

Toyota is the classic JIT company in that it has gone further than any other
discrete manufacturing firm in terms of truly making the production process
into a continuous flow. Much of the basic terminology and philosophy of JIT
have their origins at Toyota. A key issue in JIT at Toyota is understanding
that automobile manufacturing is done in very large factories that are much
more complex than our simplified example. Parts will flow from one work
center to many others with intermediate storage, and flows into work centers
will also come from many work centers with intermediate storage. JIT systems have to reflect this complexity. Toyota uses a **"two-card" kanban system;**
before delving into how this system works, it's useful to first see how a **single-
card kanban system** functions in a manufacturing environment with many
work centers and intermediate storage.

Single-Card Kanban

Figure 3.16 depicts a factory with three work centers (A, B, and C) producing
component parts, three work centers (X, Y, and Z) making assemblies, and
an intermediate storage area for component parts. A single component (part
101) is fabricated in work center C and used by work centers Y and Z. To
illustrate how the system works, suppose work center Z wishes to assemble
a product requiring component 101. A box of part 101 would be moved from
the storage area to work center Z. As the box was removed from storage, the

FIGURE 3.16 Single Kanban System

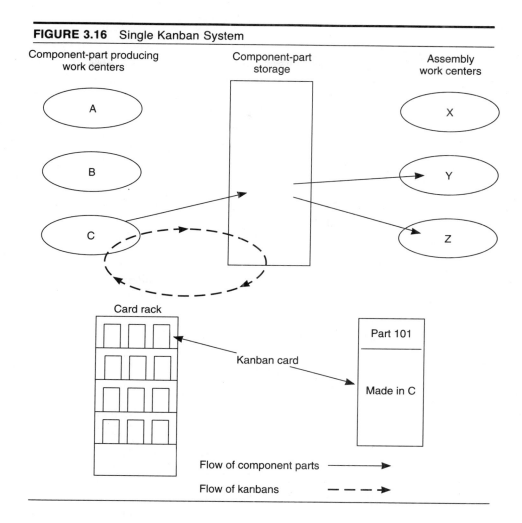

accompanying kanban card would be removed from the box; shortly thereafter, the card would be placed in the card rack at work center C. The cards in the rack at any work center represent the authorized production for that work center.

The greater the number of kanban cards in the system, the larger the inventory, but also the greater the autonomy that can be attained between the component-producing work centers and the assembly work centers. Some priority system can be implemented in the component work centers, such as working on a first-come/first-served basis or imposing some time requirements (such as all cards delivered in the morning will be returned with filled containers in the afternoon of the same day and all afternoon cards will be delivered the next morning).

Toyota

The production system at Toyota is in many ways the most advanced JIT system in the world. Its results are seen on the highways of the world. By virtually any yardstick, Toyota is truly a great manufacturing company. For example, Toyota turns its inventories 10 times as fast as U.S. and European car manufacturers and about 50 percent faster than its Japanese competitors. It's also very competitive in price, quality, and delivery performance.

Figure 3.17 shows the Toyota production system and where JIT fits within the overall approach. To some extent, the role given to JIT in Figure 3.17 may appear less encompassing than that just described. For example, the "elimination of unnecessaries" is seen as fundamental. All of the objectives and building blocks for JIT listed in Table 3.1 are in basic agreement with those in Figure 3.17. The box for production methods is basically the same as process design in Figure 3.3. Included under this heading is the multifunctional worker, which matches with several aspects of the human/organizational element building block. Also included is "job finishing within cycle

FIGURE 3.17   Toyota's Production System

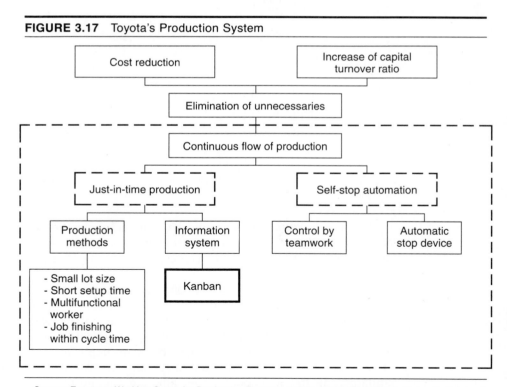

Source: European Working Group for Production Planning and Inventory Control, Lausanne, Switzerland, July 1982.

time"; this is consistent with the dominance of material flow velocity and the subservient role of direct labor utilization.

Toyota's Kanban System. The Toyota view of just-in-time production shown in Figure 3.17 includes "Information system," with "Kanban" below it. The information system encompasses the MPC activities necessary to support JIT execution. Kanban is the Toyota technique for controlling material flows. The situation at Toyota is much more complex than that illustrated in the single-card kanban example. Toyota has intermediate storage after production of components and additional intermediate storage in front of assembly work centers. This means the work flows from a producing work center into an inventory, then to another inventory, and then to the next work center.

Toyota uses a two-card kanban system. The first is a transport, or conveyance, card, which moves containers of parts from one stock location to another. The second is a production card, which authorizes production. An example of the two cards, both for the same part number, is shown in Figure 3.18. Figure 3.19 shows the flow of the kanban cards and the resultant "pull" approach to authorizing production. The two-card system is more complicated than the one-card kanban system in Figure 3.16 because there are many more parts and work centers where parts are produced and consumed.

Starting at the right side of Figure 3.19, work center K123 decides to make some product requiring one container (50 pieces) of part number 33311-3501. (It did so because a production kanban has just been received in the K123 box.) Someone at work center K123 then removed a container of parts awaiting production from stock location A-12. When this container is removed from the location, a conveyance kanban is taken from the container (the top half of Figure 3.18) and placed in the A-12 box. It authorizes someone to go to stock location A-07 and get a replacement container to put in inventory at stock location A-12 (with the conveyance kanban card). This container, while at stock location A-07, would have a production kanban attached (the lower half of Figure 3.18). This production kanban card is removed before the container is moved to stock location A-12, and is placed in the A-07 box. It then flows to the Y321 box where it becomes the authorization for work center Y321 to remove two containers of components from their input stock locations. These locations aren't shown in Figure 3.19, but the production kanban tells us that, to make 50 units of part number 33311-3501, work center Y321 needs material 33311-3504 (location A-05) and part number 33825-2474 (location B-03). In each of these locations, we find containers with exactly 50 pieces.

The kanban cards replace all work orders and move tickets. To the extent that work-in-process is significantly reduced, the problem of sequencing jobs at work centers is also diminished. The system is visual and manual in execution. The chain of dual kanban cards can extend all the way back to the suppliers. Several of Toyota's suppliers receive their authorizations to produce via kanban cards.

FIGURE 3.18   Kanban Cards

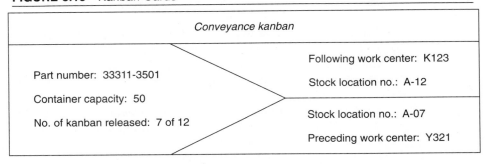

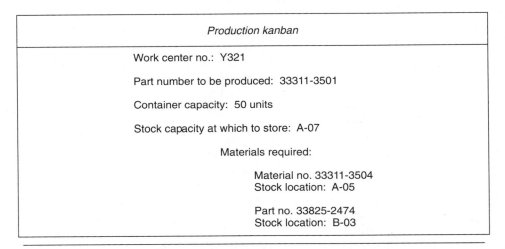

Source: R. W. Hall, *Driving the Productivity Machine: Production Planning and Control in Japan* (Falls Church, Va.: American Production and Inventory Control Society, 1981), p. 37. Reprinted with permission, American Production and Inventory Control Society, Inc.

The kanban system is a pull system, because the work centers are only authorized to produce when they have a production kanban. They only get one when a downstream work center pulls a completed container of work from the producing work center's output storage area. No work center is allowed to process input to output merely to keep workers busy. Nor is a work center allowed to transport (push) work to a downstream work center. All movements are pulled, and workers are paced by the flow of kanban cards.

The number of kanban card sets in the system directly determines the level of work-in-process inventory. The more kanban cards, the more containers filled and waiting to be used at a work center. Note that the conveyance kanban in Figure 3.18 is numbered as the seventh of 12 conveyance kanbans for that particular part and conveyance sequence. Figure 3.20 gives the for-

FIGURE 3.19 Flow of Kanban Cards

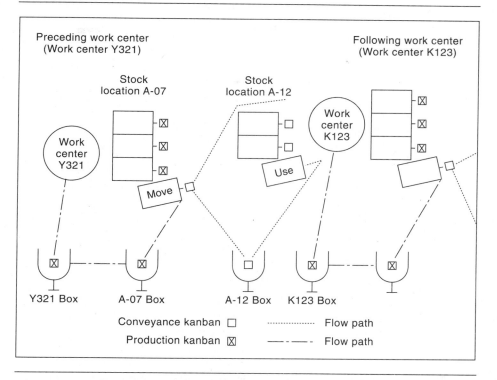

Source: R. W. Hall, *Driving the Productivity Machine: Production Planning and Control in Japan* (Falls Church, Va.: American Production and Inventory Control Society, 1981), p. 38.

FIGURE 3.20 Calculating the Number of Kanbans

$$y = \frac{DL(1 + \alpha)}{a}$$

where:

y = Number of kanban card sets.
D = Demand per unit of time.
L = Lead time.
a = Container capacity.
α = Policy variable (safety stock).

mula used to calculate the number of kanban cards needed. In this formula, there's a factor for including safety stock, which Toyota says should be less than 10 percent. Using the formula, no safety stock, and a container size of 1, we can see the philosophy of the system. If a work center required eight units per day (one per hour) and it took one hour to make one unit, only one set of two kanban cards would be theoretically necessary; that is, just as a unit was finished, it would be needed at the subsequent operation.

The container sizes are kept small and standard. Toyota feels that no container should have more than 10 percent of a day's requirements. Since everything revolves around these containers and the flow of cards, a great deal of discipline is necessary. The following rules keep the system operating:

- Each container of parts must have a kanban card.
- The parts are always pulled. The using department must come to the providing department and not vice versa.
- No parts may be obtained without a conveyance kanban card.
- All containers contain their standard quantities and only the standard container for the part can be used.
- No extra production is permitted. Production can only be started on receipt of a production kanban card.

These rules keep the shop floor under control. The execution effort is directed toward flawless following of the rules. Execution is also directed toward continual improvement. In kanban terms, this means reducing the number of kanban cards and, thereby, reducing the level of work-in-process inventory. Reducing the number of cards is consistent with an overall view of inventory as undesirable. It's said at Toyota that inventory is like water that covers up problems that are like rocks. Figure 3.21 depicts this viewpoint. If the inventory is systematically reduced, problems are exposed—and attention can be directed to their solution. Problems obscured by inventory still remain.

FIGURE 3.21   Toyota's View of Inventory

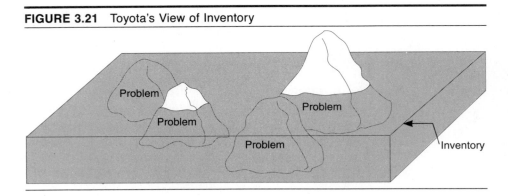

Figure 3.17 shows "continuous flow of production" as hierarchically above just-in-time production. At Toyota, a great deal of care is put into stabilizing production rates. The production plan covers a one-year horizon and is updated monthly. The master production schedule *is* the final assembly schedule. It's frozen for one month. The following two months are specified by model family. All three months are stated in daily buckets. The daily schedule is identical for each day in the month. This degree of stability is critical to using the kanban system with minimal inventory levels.

The stabilized schedule is also very useful for capacity planning purposes, both within Toyota and for its vendors. The plant is run as much as possible as a continuous flow. Cycle times are easily determined, and problem areas from a capacity point of view can be isolated.

Not everything at Toyota is controlled via kanban cards. Engines, for example, are planned somewhat differently, because there are too many varieties. If done with the standard kanban card approach, it would be necessary to have at least one engine of every variety at both the final assembly station and at the engine assembly output storage area. The number of engine possibilities makes this unwieldy. Instead, the final assembly schedule is transmitted to the engine assembly area for its own scheduling. The schedule is only frozen for several hours, and it can reflect necessary changes due to a body being pulled off the line for rework, and so on. These changes fall within the parameters of the frozen monthly schedule; they aren't unchecked reactions to changes in the marketplace.

The same approach, often called "broadcasting" of the final assembly schedule, is used to schedule all items having many options. In some cases, such as for seats, this schedule is electronically transmitted back to vendors, who also can respond to changes within hours. At the Nissan plant in Tennessee, vendor trucks are loaded in the right sequence for final assembly. With about two hours notice, schedules are made, based on approximately four deliveries per hour. The net result is that neither Nissan nor the vendor retains inventories of seats. Raw materials are converted into the exact sequence of required seats for the final assembly line.

Hewlett-Packard

Hewlett-Packard (HP) is one of the more successful users of JIT in the United States. An interesting approach to JIT has been implemented at its Medical Electronics Division in Waltham, Massachusetts. JIT is being used for assembling two major patient monitoring products called Pogo and Clover. Figure 3.22 shows the assembly area layout for these products. Clover is the older, more expensive product, with a larger number of customer-specified options. Pogo was designed as a lower-cost alternative with JIT manufacture in mind. The Clover assembly process is made up of four feeder subassemblies (A–D) and a final assembly and test area (E). Pogo is designed

FIGURE 3.22 Hewlett-Packard Waltham Division Pogo/Clover Production "U"s

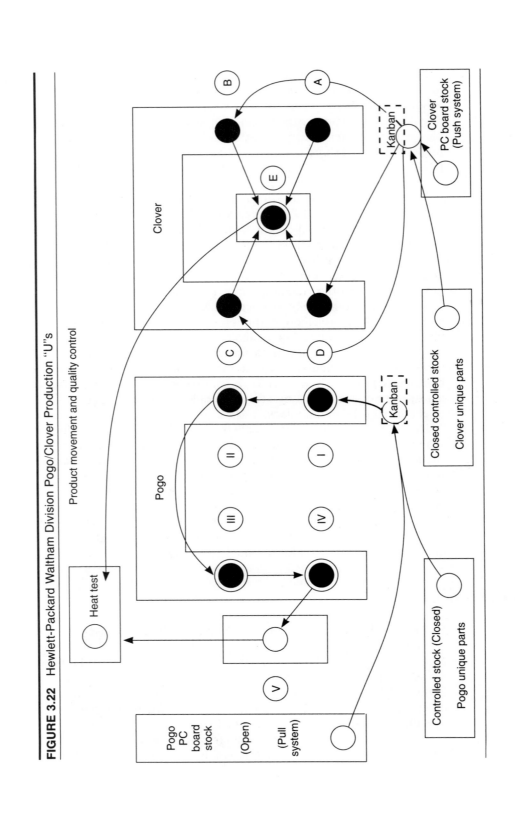

to be built in four successive assembly stations in a U shape where a test is performed in each. A final test is performed at station V. Both Clover and Pogo go into a heat test area, shown at the top of Figure 3.22. The series of tests performed on Pogo at each station (I–IV) has allowed HP to reduce the failure rate in heat testing more quickly than it did for Clover.

Both Clover and Pogo are supported by dedicated component stock areas. Each is also supported with a printed circuit board stock. In Pogo's case, 12 types of circuit boards are maintained, with a single-card kanban approach. They are supplied in lot sizes of four, with coded clothespins acting as the single kanban. On the other hand, the printed circuit boards for Clover are maintained with traditional MRP lot sizes. Both Pogo and Clover use a single-card kanban approach to pull kits of parts from the controlled stock areas.

The HP Waltham JIT system is supported by several MRP-based computer systems. These companywide systems were developed for Hewlett-Packard. At Waltham, they're mainly used for component-part planning. As time goes by, however, the semimonthly MRP explosion, weekly allocation quantities, and daily release against these allocation quantities are becoming more cumbersome for JIT manufacturing. JIT operates in a very different time frame.

A more profound issue concerns HP Waltham's overall philosophy in adopting JIT. The primary emphasis has been on stability. In Pogo, for example, the goal has been to make 10 units per day, each and every day. This goal has been achieved, and it's possible to get 10 good units between 8 A.M. and 1:30 P.M. most days. To concentrate on stability, the Pogo assembly area has been buffered on both ends. Extra supplies of component materials have been held, as have extra finished goods inventories.

At present, stability has been achieved. Relatively flawless output can be achieved with regularity. The continuous improvement attention is now shifting to reducing buffers and increasing responsiveness. If the assembly area can produce 15 units on a particular day when necessary, the finished goods can be reduced. With flawless production, lower component inventories can also be attained.

As productivity rises, new alternatives are presented. For example, perhaps a rebalance of the Pogo line tasks can free up the first worker, so stock picking can be done by that worker instead of another person. Still another opportunity is for cross training workers, so an absence or transfer of employees can be better tolerated. JIT results at Waltham are impressive. Total plant inventory fell from $56 million to $40 million in 15 months. Work-in-process inventory for the Pogo line was reduced from 50 units to 4, and the assembly floor space required decreased 65 percent. Quality increased substantially by the JIT approach, as well. But all has not been done. There are still new avenues for improvement. Included is further reduction in the printed circuit board inventories and tackling what appear to be bottlenecks in circuit board production.

NONREPETITIVE JIT

Many JIT principles that we've described for high-volume repetitive manufacturing apply in low-volume production environments as well. However, most low-volume manufacturers have balked at two basic problems presented by the "classical" approaches to JIT: (1) the requirement of setting up high-volume flow lines dedicated to a few products and (2) level loading. A merging may be taking place since, even for the classical repetitive manufacturer, it's increasingly important to respond to customer pressures for greater flexibility in volume, product mix, and other service features. High-volume repetitive manufacturers are necessarily learning to cope with greater product variety; the lower-volume job shop manufacturers are in turn learning to adapt JIT to their environments.

A Service-Enhanced View of Manufacturing

An examination of service operations can provide insights particularly germane to producing products with greater service enhancements. Service operations have limited ability to buffer customer demand with inventories. In service operations, the customer is typically more closely involved in the actual service creation process. Both phenomena are present in the environment increasingly faced by manufacturing companies. Rapid response is critical, the number of possible product/service combinations continues to grow so end item forecasting is more difficult, and large buffer inventories aren't acceptable. Customers are more actively involved in product/service definition, particularly when we take into account logistical coupling of firms as vendor and customer.

All of this argues for a JIT mode of manufacture—one whose objective is to be able to accept any customer order and turn it out right behind any other, and one that can have the flexibility to handle surges in volume or mix changes. All of this has to be done on a routine basis. Once again, service industries provide an example. McDonald's can handle two busloads of Boy Scouts or an unexpected shift from Big Macs to fish sandwiches without resorting to some "panic mode" of operation. Manufacturing firms tend to call in the indirect labor "shock troops" and overpower the formal system whenever a significant unexpected event comes along.

Fast-food operations provide still another example. Most successful fast-food chains have seen an evolution toward a broader product line (greater band width). McDonald's now serves chicken salad and other food items, for example. The objective is to increase market appeal while maintaining maximum responsiveness, minimal inventories, small lot sizes, and short lead times.

The traditional JIT view of level capacity also changes in adapting to nonrepetitive situations. Responsiveness to fickle demand requires a large band width in terms of surge capacity. No one wants the fire department to be operated at high capacity utilization; immediate response is essential. Surge capacity must be in place in both equipment and labor. When carrying these ideas into nonrepetitive manufacturing, a different view of asset management and labor utilization is required. Fixed assets (both capital and people) will be less intensively utilized to increase material velocity and overall system responsiveness.

Labor capacity has to be available to handle surges. To the extent that the whole person concept is achieved, cross training and other investments in personnel development can be focused on better ability to handle surges and more useful application of "excess" time to other enterprise objectives. More and more work now handled by staff personnel can and should be handled by people traditionally considered direct labor.

Capital utilization also has to be reexamined in light of overall objectives. An example is found in a large electronics firm with two factories making similar products. In one factory, automatic insertion equipment was purchased that operates close to its capacity. The other plant purchased considerably more insertion capacity relative to expected needs. The first plant initially thought it had done a better job—it was using capital assets more intensively. After several months, plant managers changed their minds. They now see the equipment as relatively inexpensive, compared with an ability to respond to surges brought on by changes in requirements. By having "excess" capacity in equipment and people, changes in schedules and design are much easier to handle. The result is routine execution without leveling or use of complex systems in a rapidly changing manufacturing environment.

Another example of problems brought on by level loading is seen in a manufacturer of small engines sold to OEM manufacturers. The company installed a JIT system based on standard level loading principles. Unfortunately, demand for products is seasonal and the number of exact end items sold is high (although there's considerable part commonality). Results of running this JIT system were good efficiencies in manufacturing, but they were achieved at the expense of large seasonal inventories and product mix problems; the wrong end items were frequently in finished goods. Clearly needed here is a series of level loaded schedules, where capacity isn't utilized at the same rate over the entire year, and where the basics of JIT are maintained.

Flexible Systems

Leading-edge firms are coming to understand requirements for volume and product flexibility. Some have had experience in repetitive manufacturing applications of JIT and are now moving into nonrepetitive applications. An example is a telecommunications equipment manufacturer, which began JIT in its high-volume telephone handset operations. The firm only made six models;

in two years its inventory turns were tripled, work-in-process was reduced by 75 percent, failure rates in manufacturing were cut in half, and setup times fell 50 percent. Thereafter, the firm turned to its low-volume CBX plant, where more than 150 basic circuit boards are manufactured, and every end item is somewhat of a custom order. The company has learned it must go back to the basics of JIT—product engineering, process engineering, and the whole person concept—to successfully implement JIT for its nonrepetitive products.

The firm developed cellular designs, began cellular manufacturing with great flexibility, and cross trained people with an emphasis on being able to handle volume surges in the CBX plant. MRP is still used for overall planning, but far fewer transactions are processed by the hidden factory of indirect labor. In the first six months, first pass yields on circuit boards improved 27 percent, work-in-process fell 31 percent, manufacturing cells under JIT hit 100 percent of schedule, and then the people helped out other parts of the company that were behind schedule.

Simplified Systems and Routine Execution

A major issue in any JIT firm, repetitive or nonrepetitive, is flow times. Work must flow through the factory so quickly that detailed tracking is not required. A related idea is the systems' responsiveness. In several JIT systems, we've seen in nonrepetitive environments, the firm installed what might be called a "weekly wash." In its simplest form, **weekly wash** means week 1's sales orders become week 2's production schedule.

As an example, Stanley Hardware in New Britain, Connecticut, is a make-to-stock firm for most of its items, but some are unique to a particular customer. It has applied JIT with the weekly wash concept to three different production areas. In each case, weekly sales for a particular week were determined on Friday, and the resultant quantities were manufactured the next week within some change parameters. In one case, the week-to-week variation in production could be plus or minus 20 percent. For a second product group, the swing was plus or minus 35 percent, and for a third group of products any adjustment could be handled. Because response times have been shortened, customer service has been enhanced.

Products must be manufacturable in this time frame, and manufacturing processes must have the necessary band width to take on necessary volume and mix changes for this approach to work. At Stanley, this was easier for some product lines than for others. We noted differences among the three production areas in their ability to take week-to-week output variations. This is the band width concept applied to volume. Band width as applied to product diversity was also different across the three lines. In one case, there were fewer than a dozen end items. For another, end item possibilities were about two dozen. However, the third product group encompassed several hundred end items. Moreover, the mix among end items varied significantly week to week, making this the most difficult line to design for mix band width.

Sometimes required flexibility in volume and mix was provided by product design; other times it was supported more by the processing unit's inherent flexibility. Creative use of people also supported the weekly wash at Stanley Hardware. For example, when they aren't required to make products, personnel of one department are utilized in packaging other items. A large amount of hardware packaging is done at Stanley, and required equipment is not expensive. Having extra equipment available for these workers enhances flexibility.

The weekly wash approach to JIT for nonrepetitive manufacturing shifts the emphasis from scheduling material to scheduling time blocks. The focus is on what's scheduled in the next time frame, rather than on when we'll make product X. This focus is driven by the actual requirements, rather than a forecast of needs. It's as though we were scheduling a set of trains or busses. We don't hold the train until it's full, and we can always cram a few more people onto a car, within reason. By scheduling trains on a relatively frequent basis, attempting to keep capacity as flexible as possible, and only assigning "passengers" to a time frame, responsiveness to actual demand can be increased, and detailed scheduling can be made more simple.

JIT IN PURCHASING

JIT has been applied and misapplied in purchasing. Some firms simply ask their suppliers to buffer poor schedules. On the other hand, when done well, a joint JIT approach can lead to greater bottom-line results for both firms and increased competitiveness in the marketplace.

The Basics

The first prerequisite to JIT in purchasing is a scheduling system producing requirements that are reasonably certain. Without predictability, JIT for vendors is merely a case of their customers exporting the problems. Although this may work in the short run, in the long run it can't. We've seen a factory where JIT benefits were extolled, only to find a new paving project—for vendors' trucks. Inventory had been moved from the warehouse to trailer trucks! Similar war stories abound about warehousing firms in Detroit that are there to buffer suppliers from demands made on them by auto companies as they implement JIT.

The features of an integrated MPC system are essential for JIT. As we've noted, JIT's primary emphasis is on execution activities. Good planning precedes excellence in execution. At a more detailed level, we believe effective internal use of JIT is also a prerequisite for most JIT efforts in purchasing. As a firm becomes used to JIT operations itself, it will produce the right kinds of signals for implementation with vendors. The firm will also pick up the language and understanding of how to solve problems. When people in the firm can truly view inventory as covering up problems, understand the whole person concept, adopt a true zero defects mentality and know what that implies, and adopt a fanatical drive toward continually improved performance, they can be much more effective in extending JIT into vendor firms.

With this understanding, communications with vendor firms will be fundamentally different than without it. Vendors can learn by observing JIT first-hand in their customer's factory. Talking to counterparts at a detailed level is critically important. Instead of exporting problems, with an internally functional JIT program the customer now accepts joint responsibility for solving problems. Vendor and vendee learn together; both are willing to change ways they do business to achieve the overall improvements that JIT offers.

Another requirement of JIT for purchasing is to achieve, to whatever extent possible, a stable schedule. This is consistent with level schedules for the repetitive manufacturer. To the extent that a firm makes the same products in the same quantities every day without defects and without missing the schedule, a supplier firm's schedule is extremely simple. For the nonrepetitive manufacturer, the issue isn't leveling as much as it is no surprises. The level schedule may be violated in nonrepetitive environments, but the cost is a greater need for coordinated information flows and, perhaps, for larger buffer inventories. However, there's a major difference between a *stable* (albeit nonlevel) schedule and one that's simply uncertain. The only cure for the latter case is buffer inventories.

Certainty is a relative commodity. A vendor might be able to live fairly well with a schedule that's unpredictable on a daily basis but very predictable on a weekly basis. A weekly MRP-based total, with some kind of daily call off of exact quantities, could be reasonably effective. In fact, some firms have developed "electronic kanbans" for this purpose. The notion of weekly wash as practiced by Stanley Hardware could also be used; that is, an inventory equal to some maximum expected weekly usage could be maintained and replenished on a weekly basis. For high-value products, it might be worth it to go to some kind of twice-weekly wash, or to obtain better advance information from the customer via MRP and some electronic data interchange.

Other "basics" for JIT in purchasing include all of the objectives and building blocks discussed earlier in the chapter. These are necessary both inside the customer plant and for the vendors! Setup time reduction, error-free production, statistical process control, no work-in-process tracking, worker involvement, cellular manufacturing, cross training of workers, and all the issues associated with product design, process design, human/organizational elements, and MPC must be actively pursued. Pursuing them on a joint basis and sharing experiences will only improve the speed at which true JIT gets implemented.

A JIT "basic" uniquely associated with purchasing relates to pruning the number of vendors. It follows that, if the customer firm is to work closely with supplier firms, it's important to limit the number of suppliers. Many companies have reduced their vendor base by more than 90 percent to achieve an environment where it's possible to work on a truly cooperative basis with the remaining vendors. Hidden factory issues have to be considered in vendor relations as well. Some people feel the secret to JIT is to run MRP in daily or hourly buckets. This isn't the best idea for most firms. The resultant transaction and indirect labor costs associated with such a system could be enormous unless some key changes were made. A better approach might be to use blanket orders, MRP to provide weekly quantities, agreed upon safety stocks, or amounts by which the sum of daily quantities can exceed weekly totals, and a daily phone call or an electronic kanban to the supplier for the next day's in-shipment. All this could be done without intervention of indirect labor personnel.

A computer disk drive manufacturer has such a system where, each day at about 4 P.M., an assembly line worker calls a key vendor and tells it how many of a particular expensive item to deliver the day after tomorrow. The units delivered never go into any stockroom or into any inventory records. They're delivered directly to the line without inspection and are assembled that day. The vendor is paid on the basis of deliveries of finished disk drives into finished goods inventory. Stability is handled by the customer providing the vendor with weekly MRP projections, using time fences that define stability guarantees.

Another issue relating to frequent deliveries is the cost of transportation. Some firms have approached this by asking vendors to build factories in near proximity. Although this is possible for very large companies, where a plant's output is largely dedicated to a single customer, this clearly won't be feasible for most vendors.

A different solution is for the customer to pick up goods from vendors on some prearranged schedule. This is increasingly done for several reasons. The most obvious is the savings in transportation costs over having each vendor deliver independently. A second reason relates to stability and predictability. If the customer picks up the material, some of the uncertainty inherent in vendor deliveries can be eliminated. Finally, pickup offers more chances to directly attack hidden factory costs. The customer can, for example, provide containers that hold the desired amounts and that will flow as kanbans through the plant. Savings in packaging materials as well as costs of unpacking are helpful to both parties. Items can also be placed on special racks inside the truck to minimize damage. Defective items can be returned easily for replacement without the usual costly return-to-vendor procedures and paperwork. Other paperwork can similarly be simplified when third parties aren't involved and when the loop is closed between problem and action in a short time frame.

Pickup can also be done in geographic areas beyond the factory. A Hewlett-Packard factory in Boise, Idaho, has outputs from its suppliers in Silicone

Valley, California (about 600 miles away), pooled by a trucking firm. Thereafter, the entire shipment is moved to Boise on a daily basis. New United Motor Manufacturing Inc. (NUMMI) in Freemont, California, does the same thing with its midwest suppliers: a Chicago-based trucking company collects trailer loads for daily piggyback shipments to California about 2,000 miles away. NUMMI started off holding a three-day safety stock of these parts, with plans to reduce it to one day after experience had been gained.

Lessons

The primary lesson that had to be learned in JIT purchasing is to not shift the burden for holding inventory from the company back to its suppliers. At an early point in its JIT efforts, Xerox made this error. When the consequences became known, emphasis shifted to joint problem identification and solution, a need for stabilized schedules, a true partnership, and help for Xerox suppliers to implement JIT in their suppliers. The results for Xerox have been most impressive: overcoming a 40 percent cost disadvantage, reducing its vendor base from 5,000 to 300, and winning several important awards for excellence in manufacturing.

Another lesson is to keep the relationship with the vendors simple—and to strive for ever greater simplicity and transparency. Harley-Davidson made the mistake of attempting to tie up its vendors with complex legal documents in an environment where schedules were constantly being expedited and deexpedited. Again, joint problem solution was implemented, simple agreements were put in place, and a spirit of mutual trust was developed. Harley-Davidson has also achieved extraordinary results. From a company that was being forced out of business by Japanese competition, it has come back to be a strong competitor known for high-quality products and widely recognized as a great turnaround example.

JIT SOFTWARE

Several software packages are available to support JIT. A well-known one is the Hewlett-Packard system. As we'll see, this package has much in common with the MPC system shown as Figure 3.1. To support JIT's material velocity goal, the MPS is generated as a rate, and postdeduct logic (backflush) is used for updating component inventory balances.

HP JIT Software

Figure 3.23 depicts key blocks of Hewlett-Packard's just-in-time manufacturing software. Rate-based master production scheduling is used for the front end of HP's JIT system. A monthly production plan is converted into the

FIGURE 3.23 Block Diagram of HP's JIT Software

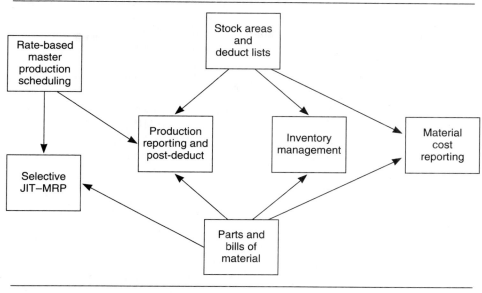

MPS (a daily build schedule) with a spreadsheet program and a personal computer (PC). The resultant schedule is matched against backlogs of customer orders and forecasted orders. Problems highlighted by the program can lead a master scheduler to make changes in the build schedule. The software is designed so data from the PC can be directly transferred to the HP JIT programs.

Stock areas and deduct lists define the manufacturing process: that is, a **stock area** is where any part is to be found. A **deduct list** is the set of parts (and quantities) to be decremented from their respective stock areas when a particular stage of product assembly (a deduct point) has been achieved. "Achievement" is transmitted to the system by production reporting. Deduct points in the process are those steps where parts are now to be associated with a higher-level assembly and no longer as individual parts. When production at a deduct point has been realized, backflushing of the component inventories (according to the deduct list) is done. In the HP software, this can occur at more than one stage in the assembly process. A deduct list is associated with each assembly for which a deduct point backflush is specified. The list also specifies from which stock areas deductions are to be made. The sum of all deduct lists has to equal the overall bill of material for the end item.

Production reporting and postdeduct (backflushing) procedures are the subsystem shown in Figure 3.23 to accomplish the deductions from each stock area's inventory balance. Inventory management maintains the current status of inventories in each designated stock location. The numbers, however, aren't precisely correct, because of the postdeduct procedures' timing. Any material that's already removed from a stock area but hasn't yet reached a deduct point will incorrectly be treated as on hand in the location by the inventory management subsystem. Material cost reporting summarizes materials consumption during defined accounting periods, subject to the same limitations of postdeduct. These errors are small (and constant) when material velocity is high and the schedules are fairly level.

Selective JIT–MRP means that the HP JIT software can be integrated with an HP MRP package. Shared data bases can be used in both systems, and bills of material can be linked. Parts common to both systems are integrated in the HP MRP package, with the JIT-driven requirements treated as independent demand. Some products can be planned with traditional MRP, while others can utilize JIT approaches. This means that adoption of JIT can be a steady evolution from MRP.

We can see how this works by returning to our earlier HP example, shown in Figure 3.22. Concentrating on the Pogo example, let's suppose that originally this product was built as five separate subassembly and test steps (I–V), going into inventory after each step, and was planned and controlled with MRP at each stage of assembly or test. For each discrete step, the prior assembly would be picked from inventory, along with the unique components associated with the particular subassembly stage. Withdrawals of subassemblies and component parts would be deducted from on-hand inventory balances, as they occurred, in classic MRP tradition.

Under HP JIT, the entire flow might be considered one step from I to V; that is, the deduct point would be when step V is completed. This is the only step that needs to be reported through production reporting. When this occurs, a deduct list would have all of the components required for steps I–V; all would be deducted at that point in time from the on-hand balance data associated with the Pogo printed circuit (PC) board stock and the controlled stock area for unique Pogo parts. We also noted that the PC boards for Pogo were controlled with a JIT system. This means receipt of boards into this area would also be a deduct point. When production reporting reports that particular circuit boards have reached the Pogo PC stock area, the deduct lists for the PC board components will decrement the respective parts and locations.

Parts and bills of material in HP JIT are the equivalent of the usual product structure data files in MRP, but they are defined according to deduct points. The JIT-based product structures have only a few levels, as was the case in the redesigned one-quart saucepan of Figure 3.14. Each level corresponds to a deduct point and the components are the deduct list. Standard cost data are associated with the product structures in this JIT format. As products are converted to JIT from MRP, it is necessary to reformat the data according to the deduct points.

The HP JIT system also supports customization by users to match certain individual needs. Some of the customization can be accomplished without additional programming. Included are definitions of system transaction screens and output reports. Advanced customization can also be supported by using predetermined exit points for transactions. This approach to customization allows JIT software to be connected with other operating systems in the company, but the firm can still use updated versions of HP JIT software.

JIT Using MRP Software

Roger Brooks describes a JIT approach used by a firm that had successfully implemented MRP. This company wanted to use JIT for some of its products, while maintaining the MRP systems. The firm schedules these products in daily buckets using firm-planned orders (FPOs). The FPOs will, of course, generate dependent demand for the component parts. The firm doesn't use a stockroom for the components, the FPOs are never converted into scheduled receipts, and components aren't issued to any shop orders. FPO data are communicated to the floor for build purposes and are used to sequence the assembly line. As assembled products are completed, they're received into inventory using normal procedures for MRP inventory transactions. These receipts are also netted against the FPO quantities. Consequently, requirements for corresponding component parts are also reduced.

These steps result in a backflush for inventory balances. The firm has high levels of data integrity in the bills of material, it keeps meticulous records of stock balances, and it cycle counts at least once per week. This attention to detail is needed for the backflushing to work properly. The JIT approach of backflushing runs completely counter to standard notions in MPC systems. Backflushing *will* work, but speed in conversion is critical, and so are other methods to ensure data integrity.

MANAGERIAL IMPLICATIONS

The vision of JIT presented in this chapter is much broader than one constrained to only manufacturing planning and control. JIT is best seen as an integrated approach to achieving continued manufacturing excellence. A holistic view of JIT encompasses a set of programs, as well as a process where human resources are continually redeployed in better ways to serve company objectives in the marketplace. In the balance of this chapter, we feel compelled to speculate a bit on what this implies for manufacturing planning and control and related areas.

Information System Implications

Since JIT requires changes in the ways manufacturing is managed and executed, so are changes required in how computer-based systems are designed to support manufacturing. The changes run counter to many existing approaches to the use of computers in manufacturing.

JIT calls for reducing the number of transactions and the size of the hidden factory. This means large centralized computer systems will tend to be less used than MRP-based execution systems. To the extent that JIT is used for nonrepetitive manufacturing, personal computers may be used by people on the floor to provide whatever detailed scheduling information is required.

The whole person concept also supports this trend. Shop people can and will learn how to use spreadsheet programs and other user-friendly computer systems. They will use this knowledge to solve execution problems of whatever variety are important. Moreover, if we accept the notion of constant evolution, systems will change as needed. For example, at some point a detailed statistical process control program might be needed to solve a quality problem. But, when the problem is clearly solved, the detailed computer system might well be abandoned. Workers no longer need it; moreover, the computer and knowledge resources might be better utilized on a new project.

Current trends in computer technology indicate that powerful computers with ever-increasing user friendliness will become quite inexpensive. Workers will apply this power to problems as perceived at a point in time. To some extent, the result will be an evolutionary approach to CIM (computer-integrated manufacturing). Computer power will be applied to problems of an integrative nature, including those crossing company boundaries. The service-enhanced view of manufacturing means firms will increasingly use their computer power in ways of value to their customers and to their customers' customers.

Manufacturing Planning and Control

JIT has profound implications for production control and other material planning and control activities. JIT offers the potential for eliminating or sharply reducing stockrooms, incoming quality control, receiving, kitting, paper processing associated with deliveries and shipments, detailed scheduling done by central staff, and all the detailed tracking associated with a shop-floor control system. It is vital to estimate these costs as the MPC system is enhanced to embody JIT. Many of them are "well hidden." The large benefits clearly justify costs associated with a JIT program in manufacturing.

It won't be easy to change systems in some companies. Organizations have grown around them, cost accounting and other areas seem to require data generated by these systems, and many jobs are involved. However, the potential is there, and some people presently working in manufacturing planning and control will rise to the challenge and make the company more competitive by implementing JIT.

Scorekeeping

The firm adopting JIT in its fullest context will need to think very carefully about reward systems and managerial scorekeeping. Traditional measurement systems tend to focus attention on costs associated with producing the products, using cost accounting systems that have changed little since the Industrial Revolution. These systems are based on an era when direct labor was the products' major cost source. Now, in many companies, material costs dominate, with direct labor cost (using traditional definitions) continually decreasing in relative importance.

Firms are more and more providing a set of "service-enhanced" products to their customers. Services include logistical support, fast response, rapid design changes, and hidden factory cost reductions. In most cases, these services are provided with knowledge work, rather than by the usual definitions of direct labor. A chemical firm is now providing a chemical in 3.5-pound bags to a particular customer who uses that amount in each batch of a particular process. The old measurements based on tons produced aren't adequate for helping this firm determine the value of manufacturing investments to provide these service enhancements.

JIT focuses on material velocity. This focus is consistent with the objectives of inventory reduction and lead time compression. These programs tend to be cost related. However, under JIT, we must be very careful about the way "costs" are measured and the resultant implications for decision making. The values of band width, flexibility, responsiveness, and worker skill enhancement need to be recognized. None of these is incorporated in traditional accounting systems.

The entire approach to capacity utilization will need to be rethought by the JIT firm. Utilization of capital assets may not be as important as responsiveness and material velocity. Being able to take any customer order, even when vastly different from forecast, and doing so with short lead times and no use of "shock troops" is the goal. However, improved responsiveness to marketplace needs will separate successful firms from the also-rans.

What all of this means for cost accounting is that many traditional views will need to be scrapped. For example, some Hewlett-Packard factories have given up the cost category of direct labor. They simply have labor. The distinction between direct and indirect isn't useful, and basing product costs on multiples of direct labor cost leads to more erroneous implications than some other scheme.

If we follow the idea of lifetime employment or commitment, then labor is a fixed cost. The whole person concept can lead to the conclusion that this labor pool is an asset that can be enhanced. It also points to the use of direct labor for activities not normally associated with direct labor. Trying to apportion the labor into various categories may just not be worth the trouble. Even if it were, the continual goal of learning/improvement means that apportionment would be constantly changing. A final scorekeeping issue is the challenge to top management to create the organizational climate where the JIT journey can best take place. We believe JIT is, in the last analysis, a key means for survival in the years ahead. Leadership will be required to guide manufacturing firms through the necessary changes.

Pros and Cons

Just-in-time may look so appealing that every company should be rushing to implement it. In fact, there are significant benefits for some companies—but there are also risks. If we quickly reduce inventories to "expose the rocks" as shown in Figure 3.21, the result may well be to hit the rocks! Careful planning is required, and the building blocks in Table 3.1 are prerequisites for JIT operations.

There are situations where JIT will work well and ones where it won't. Many authorities believe JIT is what all Japanese companies strive for. In fact, many Japanese firms with complex product structures are now actively working to implement MRP-based systems. However, JIT's realm seems to be expanding. At one time JIT was thought to apply only to repetitive manufacturing with simple product structures and level schedules. Increasingly, companies are applying JIT to nonrepetitive schedules; product complexity is being partially overcome with decentralized computing on the shop floor; and make-to-order schedules are being adapted to JIT.

Some companies ask if they need to install MRP before adopting JIT, since JIT implementation often means they must dismantle parts of the MRP-based system. Our response is that although it's conceptually possible to implement JIT without first implementing MRP, for firms that can benefit significantly from MRP, it's usually not done. Unless we can find some other way to develop the discipline of MRP, JIT operations are at great risk. In the discipline's absence, when JIT takes away the buffers, there will usually be costly disruptions of the manufacturing process, poor customer service, and panic responses to the symptoms rather than to the underlying problems.

CONCLUDING PRINCIPLES

This chapter is devoted to providing an understanding of JIT and how it fits into MPC systems. Our view of JIT encompasses more than MPC-related activities, but there's a significant overlap between JIT and our approach to MPC systems. In summarizing this chapter, we emphasize the following principles:

- Stabilizing and in some cases leveling the production schedules are prerequisites to effective JIT systems.
- Achieving very short lead times supports better customer service and responsiveness.
- Reducing hidden factory costs can be at least as important as reducing costs more usually attributed to factory operations.
- Implementing the whole person concept reduces distinctions between white- and blue-collar workers and taps all persons' skills for improving performance.
- Cost accounting and performance measurements need to reflect the shift in emphasis away from direct labor as the primary source of value added.
- To achieve JIT's benefits in nonrepetitive applications, some basic features of repetitive-based JIT must be modified.
- JIT is not incompatible with MRP-based systems. Firms can evolve toward JIT from MRP-based systems, adopting JIT as much or as little as they want, with an incremental approach.

REFERENCES

Brooks, Roger. "Backflush with Caution." *Oliver Wight Newsletter*. Essex Junction, Vt.: Oliver Wight Companies, 1986.

Crosby, Philip B. *Quality Is Free: The Art of Making Quality Certain*. New York: McGraw-Hill, 1979.

Gelb, Tom. "Harley-Davidson: A Company That's Taking a Different Route." *Target* 9 (October 1985).

Goddard, Walter. "Just-in-Time Needs Shop Floor Control." *Modern Materials Handling,* May 7, 1984.

———. *Just-in-Time: Surviving by Breaking Tradition*. Essex Junction. Oliver Wight, Ltd., 1986.

Hall, R. W. *Driving the Productivity Machine*. Falls Church, Va.: American Production and Inventory Control Society, 1981.

———. *Zero Inventories*. Homewood, Ill.: Dow Jones-Irwin, 1983.

———. H. T. Johnson; and P. B. B. Turney. *Measuring Up: Charting Pathways to Manufacturing Excellence*. Homewood, Ill.: Business One Irwin, 1990.

Karmarkar, Uday S. "Getting Control of Just-in-Time." *Harvard Business Review,* September–October 1989.

Krajewski, L. J.; B. E. King; L. P. Ritzman; and D. S. Wong. "Kanban, MRP and Shaping the Manufacturing Environment." *Management Science* 33, no. 1 (January 1987).

Lubhen, R. T. *Just-in-Time Manufacturing*. New York: McGraw-Hill, 1988.

Miller, J. G., and T. E. Vollmann. "The Hidden Factory." *Harvard Business Review*. September/October 1985, pp. 141–50.

Monden, Yasuhiro. "Adaptable Kanban System Helps Toyota Maintain Just-in-Time Production." *Journal of Industrial Engineering,* May 1981, pp. 29–46.

Nakane, J., and R. W. Hall. "Management Specs for Stockless Production." *Harvard Business Review,* May/June 1983, pp. 84–91.

Ohno, Taiichi. *Toyota Production System: Beyond Large-Scale Production*. Cambridge, Mass.: Productivity Press, 1988.

————., and S. Mito. *Just-in-time for Today and Tomorrow*. Cambridge, Mass.: Productivity Press, 1986.

Plossl, George W. (ed.). *Just-in-Time: A Special Roundtable*. Atlanta, Ga.: George Plossl Educational Services, 1985.

Schonberger, Richard J. *Japanese Manufacturing Techniques: Nine Hidden Lessons in Simplicity*. New York: Free Press, 1982.

————. "Some Observations on the Advantages and Implementation Issues of Just-in-Time Production Systems." *Journal of Operations Management* 3, no. 1 (1982).

Shingo, Shigeo. *A Revolution in Manufacturing: The SMED System*. Cambridge, Mass.: Productivity Press, 1985.

Suzaki, Kiroshi: *The New Manufacturing Challenge: Techniques for Continuous Improvement*. New York: Free Press, 1987.

Voss, C. A., and L. Okazaki-Ward. "The Transfer and Adaptation of JIT-Manufacturing Practices by Japanese Companies in the U.K." *Operations Management Review* 7, nos. 3 and 4 (1990).

Wilson, Glen T. "Kanban Scheduling—Boon or Bane?" *Production and Inventory Management* 26, no. 3. (1985).

DISCUSSION QUESTIONS

1. Some people have argued that just-in-time is simply one more inventory control system. How do you think they could arrive at this conclusion?
2. Take a common system at a university—for example, registration. Describe what kinds of "waste" can be found. How might you reduce this waste?
3. "Surge" capacity is the ability to take on an extra number of customers or product requests. Tell how the following three facilities handle surges: a football stadium, a clothing store, and an accounting firm.
4. List several ways to signal the need for more material in a "pull"-type system. Distinguish between situations where the feeder departments are in close proximity and those where they're at some distance.
5. Some companies contend that they can never adopt JIT because their suppliers

are located all over the country and distances are too great. What might you suggest to these firms?

6. "We just can't get anywhere on our JIT program. It's the suppliers' fault. We tell them every week what we want, but they still can't get it right. I've been over to their shops, and they have mountains of the wrong materials." What do you think is going on here?

7. One manager at a JIT seminar complained that he couldn't see what he would do with his workers if they finished their work before quitting time. How would you respond to this comment?

8. How do you go about creating the organizational climate for successful JIT? Where do you start?

PROBLEMS

1. The McDougall Manufacturing Company has just finished a product line analysis. Forecasted annual demand for their 10 lines is given below (in units of product). Using a 250-day work year and batch sizes of 100 per product, develop a level minimum batch (100 units) schedule for the assembly line. (The line can produce 300 units of any product per hour, and works eight hours per day.)

Product line	Forecast
1	200,000
2	125,000
3	100,000
4	75,000
5	50,000
6	25,000
7	12,500
8	6,250
9	3,125
10	3,125

2. Calculate the number of kanbans required for the following:

	A	B
Usage	120 per week	100 per day
Lead time	1 week	2 weeks
Container size	2 units	50 units
Safety stock	20 percent	0

3. The Yakima Lash Company produced four models. The forecasts of annual demand for each of the four are as follows:

	Model			
	I	*II*	*III*	*IV*
Forecast of annual demand	500	1,000	2,500	5,000

a. Using a 250-day year and an eight-hour day, determine the mixed-model level master schedule for a daily batch and hourly batch with minimum batch sizes.
b. What would the schedule of production look like for an eight-hour day using mixed-model minimum batch size production?

4. The Elk Sock Company has implemented a JIT program using kanbans to signal the movement and production of product. The average inventory levels have been reduced to where they're roughly proportional to the number of kanbans in use. For one of its products, usage averages 100 units per day, the container size is 20 units, there's no safety stock, and lead time has been one week (five days). The process engineers have been hard at work improving the manufacturing process. They have a proposal to reduce lead time from five days to three days. What would the percentage change in average inventory be?

5. The Arcane Appliance Company had the following layout for part of the production of its appliances:

Inventory and receiving	Wash room	Subassembly department
Aisle		
Frame shop	Plastic trim	Final assembly

Recently Arcane implemented a JIT delivery system using a single unit lot size for the frame members for one of its popular appliances. This means two frame members are welded into a single subassembly. Two subassemblies are joined to produce either a top or bottom unit, and both top and bottom units are used to make a completed appliance. The company instructed the supplier to deliver in lots of 40 frame members every two hours. The lots are delivered to the receiving area, where they're first inspected and then counted and a receipt notice is issued. For simplicity, assume *no* defective units are delivered so there's no inspection. Next they are issued in pairs (with an issue ticket) to the subassembly shop, where they are welded into a single subassembly. The subassembly is returned to inventory, where a receipt ticket is issued. Later, an issue transaction will be processed to release two subassemblies to the frame shop, where top and bottom frame units are made. The frame unit is returned to inventory, generating a receipt ticket.

To finish a product, the final assembly shop is issued a top and a bottom unit (now comprising four frame members each) for assembly into a completed appliance. The final step is to receive the appliance back into inventory, which requires another transaction.

a. If each receipt and each issue by the inventory clerk requires a transaction, how many transactions are there for a lot of 40 frame members (i.e., for completing five appliances in total)?

b. If the supplier sticks to the delivery schedule of one lot every two hours, how much time can be spent on an average transaction if one clerk is responsible for all transactions?

c. How might the transactions be significantly reduced?

6. Fredrick Machine Company prided itself on the ability to keep up with the state of the art in its line of equipment. Although a small firm with only 300 factory workers, it had a fairly extensive product line. The engineering staff generated an average of 10 engineering change orders a week. Over several years, the firm had developed a procedure for processing these changes each week. A group of six people met each Friday morning to make the approvals. About 80 percent of the changes were approved and went on to a scheduling meeting in the afternoon (three people). The company was considering batching the changes for monthly processing (i.e., every four weeks). The company wasn't sure of the savings for doing this, but it gathered the following information:

It took each person who attended a meeting about one half hour to prepare for the meeting, regardless of the number of changes.

It averaged about 10 minutes per engineering change to make either the approval or scheduling decision during the meetings.

Each Monday, there was a general loss of productivity due to the scheduling changes. It affected only about one-third of the factory workers and amounted to about a half hour for each person. Again, this was independent of the number of changes.

In addition to the general loss, workers affected by the changes (about 10 workers per change) lost about an hour of productive time over three days.

a. What are the weekly batch system's costs over a 12-week period in terms of labor hours? What would be the change by going to a monthly system? (A spreadsheet might help with this and subsequent analyses.)

b. Suppose that experience with monthly batching indicated the number of changes dropped. (Some changes that would have been made in the weekly system were superseded in the monthly system.) Specifically, the average number of changes per week that were batched for approval dropped to eight and the average number scheduled dropped to seven. What would the quarterly costs be now?

c. The company was considering extending the batching to a quarterly basis. What would the costs be if the number of changes per week to be considered for approval dropped to seven and the number scheduled dropped to six?

7. Develop the MRP record for the faucet subassembly (part no. 356) shown in the following bill of material. The refrigerators are assembled at the rate of 480/week, lead time is one week, there's no safety stock, lot size is 500, and 200 units are in inventory at the moment. (Note: The faucet subassembly is only used on every fourth refrigerator.)

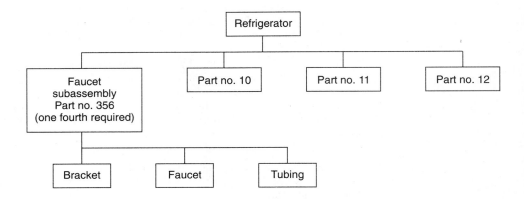

8. Suppose that the faucet subassembly in Problem 7 had been "phantomed" (i.e., the faucet would be assembled onto the refrigerator directly from the bracket, faucet, and tubing parts).
   a. What would the bill of material look like now?
   b. Construct the phantom MRP record and compare it to Problem 7's MRP record.
   c. Construct the MRP record for the bracket (lot size = 200, no safety stock, 250 on hand, lead time = 2 weeks).

9. Produce a "simulation" of the cold water faucet subassembly activities on a mixed-model refrigerator assembly line. The station where the cold water faucet is installed also installs several other options in refrigerators. Every fourth unit requires the cold water faucet installation. Cycle time is five minutes (i.e., every five minutes a new unit appears at the assembly line station).

   Faucets are supplied by a subassembly department in batches of two in a single container. When the assembly line has used both units, the container is immediately returned to the subassembly department, where two more faucet subassemblies are prepared and returned to the assembly line. It takes five minutes to perform the subassembly activities on the faucets. The subassembly group is told to start immediately on the faucet subassemblies whenever the container appears in the area. It takes a total of five minutes to travel to and from the subassembly area.

   The starting conditions as of 8 A.M. are shown below. Refrigerator number 4 is in the options assembly area. (They completed units number 1, 2, and 3 the day before.) Units number 4, 8, 12, and so on (those marked as Os) require cold water faucets. The container with two subassembly units is at the assembly line.

Refrig. no.		12	11	10	9	8	7	6	5	4	3	2	1
		0	X	X	X	0	X	X	X	0	X	X	X
Subassembly container										Yes			
Subassembly inventory										2			

Using five-minute intervals, record the number of the refrigerator being assembled in the options area, how many water faucet subassemblies are at the station, the location of the container, and whether the subassembly area is working on the faucet subassemblies. Do the "simulation" for more than an hour.

10. Graham Manufacturing has completed the following ABC analysis of the 10 products it makes:

Product	Daily sales forecast
1	800
2	500
3	400
4	300
5	200
6	100
7	50
8	25
9	12.5
10	12.5
Total	2,400.0

Graham has an assembly line that can produce 300 products per hour and works eight hours per day:

a. Prepare a daily level schedule for Graham, assuming a batch (container size) of 100 for each product.

b. Calculate the number of kanban cards required for each product, assuming a 0.5 day lead time and a 20 percent safety stock. What is the total number of kanban cards for all products?

c. Assuming storage space is proportional to the number of kanban cards, by what percentage will storage be reduced if the lead time can be reduced to one hour?

d. What are the benefits, if any, of being able to cut all batch and container sizes to 50 units instead of 100?

11. Larry Dolinsky, foreman of Graham Manufacturing's assembly department, has proposed replacing the two-card kanban system with what he calls "American kanban." Assembly is fed by two departments, with each producing parts unique to each of the 10 products. This means that the total number of kanban card sets and containers of parts required would be the double of the calculations in Problem 10. Based on a safety stock of zero and a container size of 100, Larry has calculated the total number of kanban card sets for the two supplying departments as 36. Further, Larry has observed that for products with low sales volumes (such as products 8, 9, and 10), parts sit in inventory for many days before use.

American kanban consists of red, white, and blue containers, a different size for each of the two feeder departments. There is one of each color for each department (a total of six containers). Each container holds 100 parts, so a color is one hour's worth of work in final assembly. The rule there is to rotate in the order of red, white, and blue. The feeder departments receive the empty containers as Larry empties them, and fill them according to a daily schedule. They can fill them faster than Larry can assemble.

a. Diagram the flow of parts between the feeder departments and assembly.
b. What are the strengths and weaknesses of Larry's approach?

12. Mark Davis, foreman of a feeder line in Problems 10 and 11, was discussing the American kanban with one of his workers, Howie Oden. Howie said that he didn't like the red, white, and blue system's rigidity. He would rather have the overall daily schedule and be free to build it in whatever order he saw fit. He believed economies could be gained by having some different order to the sequence of products produced in the feeder departments. What he really wanted was to be able to be a full day ahead of assembly, which should give Larry greater flexibility as well. Mark agreed with Howie and suggested that either Howie's one-day-ahead schedule be used or a fourth color (at the risk of being unpatriotic) be added.

a. Diagram the new flow of parts for a one-day-ahead schedule.
b. What are the one-day schedule's strengths and weaknesses?
c. What are the strengths and weaknesses of the fourth color suggestion?

13. Larry has been working hard to improve the assembly process described in Problem 10. He now believes that assembly rates of 350 per hour can be achieved, instead of 300.

a. What should the schedule be now?
b. What, if any, are the benefits of this improvement?
c. What are the implications?

14. Going back to the data in Problem 10, let's suppose that Graham has finished goods inventories of 500 units for products 1 through 5, and 100 units for products 6 through 10. Let's assume a schedule for products 1 through 6 that produces the exact sales forecast each day; products 7 through 10 are produced less frequently than every day.

a. On day 1, only products 1 through 6 were produced. At the end of the day, what are the inventories for products 1 through 6, assuming no forecast errors?

b. What are the inventories if 900 units of product 1 are sold, but only 300 units of product 2? What does this condition imply for the schedule for day 2?

c. What happens if less than 2,400 units in total are sold? More than 2,400?

15. Merrill's Markers has just installed an expensive machine to produce two components for the final assembly line. (It has only been 25 percent depreciated.) The machine produces either component at the rate of 100 units per hour. Setup time is one hour (to go from one product to the other or back again). Lot size is 50 units. The assembly line requires 40 units per hour of each component.

   The company has just learned of an inexpensive second-hand machine that could produce either component at a rate of 50 units per hour. The machine costs $3,500 and no additional people would be required to run it. Would it be worthwhile to purchase the machine and dedicate it to one of the components (the other machine would be dedicated to producing the other component) if each component was worth $10 at this stage?

Chapter 4

Capacity Planning

In this chapter we discuss issues of matching available capacity to the firm's manufacturing requirements. We focus primarily on techniques for determining capacity requirements implied by the production plan, master production schedule, or detailed material plans. The managerial objective in planning capacity is to ensure the match between capacity available in specific work centers and capacity needed to achieve planned production. The techniques described here enable you to estimate needed capacity. If sufficient capacity can't be made available either inside or outside the firm, the only managerial alternative is to change the material plans to conform to available capacity.

The chapter is organized around five topics:

- Capacity planning's role in manufacturing planning and control (MPC) systems: How does it fit and what role does it play?
- Capacity planning and control techniques: How can a manufacturing plan's capacity implications be estimated and its usage be controlled?
- Management and capacity planning: How can managers decide which technique(s) to use and how to use them?
- Data base requirements: How should the data base be designed for a capacity planning system?
- Examples of applications: How are capacity planning techniques applied and used for managing capacity?

Some techniques in this chapter are closely related to analogous work in advanced production planning in Chapter 15. Although finite loading appears here as a member of the capacity planning family, it's described in Chapter 5 because of its use as a scheduling model. Chapter 7's discussion of production planning contains useful managerial considerations for resource planning. The master production schedule described in Chapter 6 is a primary data source for capacity planning. Optimized production technology (OPT) and related concepts appear in Chapter 19; they focus on managing bottleneck operations' capacities.

CAPACITY PLANNING'S ROLE IN MPC SYSTEMS

A critical activity that parallels developing material plans is developing capacity plans. Without our providing adequate capacity or recognizing excess capacity's existence, we can't fully realize benefits of an otherwise effective MPC system. On the one hand, insufficient capacity quickly leads to deteriorating delivery performance, escalating work-in-process inventories, and frustrated manufacturing personnel. On the other hand, excess capacity might be a needless expense that we can reduce. Even firms with advanced material planning capability have found that their inability to provide appropriate work center capacities is a major stumbling block to achieving maximum benefits. This underscores the importance of developing the capacity planning system in concert with the material planning system.

Hierarchy of Capacity Planning Decisions

Figure 4.1 relates capacity planning decisions to an MPC system's other modules. It shows the scope of capacity planning, starting from an overall plan of resources, proceeding to a rough-cut evaluation of a particular master production schedule's capacity implications, moving to a detailed evaluation of capacity requirements based on detailed material plans, then continuing to finite loading procedures, and ending with input/output techniques to help monitor the plans.

These five levels of capacity planning activities range from large aggregations of capacity for long time periods to very detailed machine scheduling for time intervals of an hour or less. This chapter's major focus is rough-cut capacity planning procedures. These are central to establishing a correspondence between capacity plans and material plans in virtually every business, including companies using just-in-time (JIT) methods. We also discuss capacity requirements planning, a commonly used technique in companies that prepare detailed material plans using time-phased material requirements planning (MRP) records. Finally, we present input/output analysis as a method to control capacity plans in companies using MRP.

Many authorities distinguish between long-, medium-, and short-range capacity planning and control horizons as indicated in Figure 4.1. This is a useful view, but the time dimension varies substantially from company to company. This chapter emphasizes capacity planning decisions involving a planning horizon ranging from one week to a year or more in the future, depending on the firm's specific needs.

FIGURE 4.1   Capacity Planning in the MPC System

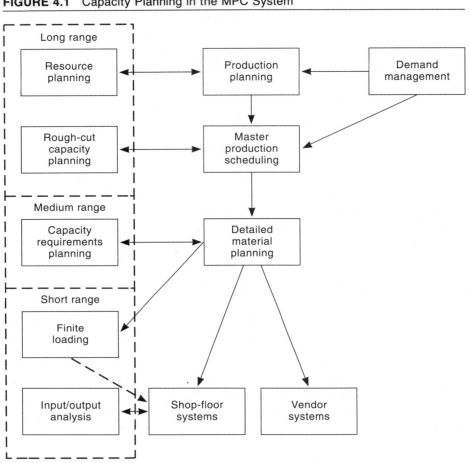

Links to Other MPC System Modules

System linkages for the capacity planning modules follow the basic hierarchy shown in Figure 4.1. **Resource planning** is directly linked to the production planning module. It's the most highly aggregated and longest-range capacity planning decision. Resource planning typically involves converting monthly, quarterly, or even annual data from the production plan into aggregate resources such as gross labor-hours, floor space, and machine-hours. This level of planning involves new capital expansion, bricks and mortar, machine tools, warehouse space, and so on, which requires a time horizon of months or years.

The master production schedule is the primary information source for **rough-cut capacity planning.** A particular master schedule's rough-cut capacity requirements can be estimated by several techniques: *capacity planning using overall planning factors (CPOF), capacity bills,* or *resource profiles.* These techniques provide information for modifying the resource levels or material plan to ensure execution of the master production schedule.

For firms using material requirements planning to prepare detailed material plans, a much more detailed capacity plan is possible using the *capacity requirements planning (CRP)* technique. To provide this detail, time-phased material plans produced by the MRP system form the basis for calculating time-phased capacity requirements. Data files used by the CRP technique include work-in-process, routing, scheduled receipts, and planned orders. Information provided by the CRP technique can be used to determine capacity needs for both key machine centers and labor skills, typically covering a planning horizon of several weeks to a year.

Resource planning, rough-cut capacity planning, and capacity requirements planning link with the production plan, master production schedule, and MRP systems, respectively. Linkages are shown as double-headed arrows for a specific reason. There must be a correspondence between capacity required to execute a given material plan and capacity made available to execute the plan. Without this correspondence, the plan will be either impossible to execute or inefficiently executed. We don't claim capacity must always be changed to meet material plans. In fact, whether this is worthwhile or whether plans should be changed to meet capacity is a managerial judgment. Capacity planning systems provide basic information to make that a *reasoned* judgment.

The *finite loading* technique also relates to firms using time-phased detailed material plans, but it can be better viewed as a shop scheduling technique. More than any other capacity planning technique, it makes clear the relationship between scheduling and capacity availability. Finite loading starts with a specified capacity level for each work center or resource grouping; this capacity is then allocated to work orders. Hence, finite loading is a method for scheduling work orders. Procedures described in this chapter don't deal with scheduling questions. They estimate capacity requirements only, with scheduling orders considered as a separate problem. The finite loading process requires linkages to the same files as the CRP technique, as well as to files specifying work center capabilities.

Input/output analysis provides a method for monitoring the actual consumption of capacity during execution of detailed material plans produced by time-phased MRP systems. It's necessarily linked to the shop-floor systems and data base for shop-floor control. Input/output analysis can indicate the need to update capacity plans as actual shop performance deviates from current plans, as well as the need to modify planning factors used in the capacity planning techniques.

This overview of capacity planning's scope sets the stage for the techniques this chapter discusses. The primary interaction among these techniques is hierarchical: long-range planning sets constraints on medium-range capacity planning, which in turn constrains detailed scheduling and execution on the shop floor.

CAPACITY PLANNING AND CONTROL TECHNIQUES

Here we describe four procedures for capacity planning. The first technique is *capacity planning using overall factors (CPOF)*. The simplest of the four techniques, CPOF is based only on accounting data. The second, *capacity bills,* requires more detailed product information. The third, *resource profiles,* adds a further dimension—specific timing of capacity requirements. The first three procedures are rough-cut approaches and are applicable to firms with or without MRP systems. The fourth, *capacity requirements planning,* is used in conjunction with time-phased MRP records and shop-floor system records to calculate capacity required to produce both open shop orders (scheduled receipts) and planned orders. We'll also discuss the input/output analysis technique for monitoring and controlling the capacity plans.

To describe the four planning techniques, we use a simple example. The example allows us to clearly see differences in approach, complexity, level of aggregation, data requirements, timing, and accuracy between the techniques. We'll then illustrate the input/output analysis procedure, which could be utilized with any of the four planning procedures. Although the appropriate unit of measure for capacity will vary, depending on a particular firm's key resources, we'll use labor hours for our examples.

Capacity Planning Using Overall Factors (CPOF)

Capacity planning using overall factors (CPOF), a relatively simple approach to rough-cut capacity planning, is typically done on a manual basis. Data inputs come from the master production schedule (MPS), rather than from detailed material plans. This procedure is usually based on planning factors derived from standards or historical data for end products. When these planning factors are applied to the MPS data, overall labor or machine-hour capacity requirements can be estimated. This overall estimate is thereafter allocated to individual work centers on the basis of historical data on shop workloads. CPOF plans are usually stated in terms of weekly or monthly time periods and are revised as the firm changes the MPS.

The top portion of Figure 4.2 shows the MPS that will serve as the basis for our example. This schedule specifies the quantities of two end products to

FIGURE 4.2 Example Problem Data

Master production schedule (in units):

End product								Period						
	1	2	3	4	5	6	7	8	9	10	11	12	13	Total
A	33	33	33	40	40	40	30	30	30	37	37	37	37	457
B	17	17	17	13	13	13	25	25	25	27	27	27	27	273

Direct labor time per end product unit:

End product	Total direct labor in standard hours/unit
A	.95 hours
B	1.85

Source: W. L. Berry, T. G. Schmitt, and T. E. Vollmann, "Capacity Planning Techniques for Manufacturing Control Systems: Information Requirements and Operating Features." Reprinted with permission, November 1982 *Journal of Operations Management*, Journal of the American Production and Inventory Control Society, Inc.

be assembled during each time period. The first step of the CPOF procedure involves calculating capacity requirements of this schedule for the overall plant. The lower portion of Figure 4.2 shows direct labor standards, indicating the total direct labor-hours required for each end product. Assuming labor productivity of 100 percent of standard, the total direct labor-hour requirement for the first period is 62.80 hours, as shown in Figure 4.3.

The procedure's second step involves using historical ratios to allocate the total capacity required each period to individual work centers. Historical percentages of the total direct labor-hours worked in each of the three work centers the prior year were used to determine allocation ratios. These data could be derived from the company's accounting records. In the example, 60.3 percent, 30.4 percent, and 9.3 percent of the total direct labor-hours were worked in work centers 100, 200, and 300, respectively. These percentages are used to estimate anticipated direct labor requirements for each work center. The resulting work center capacity requirements are shown in Figure 4.3, for each period in the MPS.

The CPOF procedure, or variants of it, is found in a number of manufacturing firms. Data requirements are minimal (primarily accounting system data) and calculations are straightforward. As a result, CPOF approximations of capacity requirements at individual work centers are only valid to the extent that product mixes or historical divisions of work between work centers remain constant. This procedure's main advantages are ease of calculation and minimal data requirements. In many firms, data are readily available and computations can be done manually.

Capacity Bills

The **capacity bill procedure** is a rough-cut method providing more direct linkage between individual end products in the MPS and the capacity required for individual work centers. It takes into account any shifts in product mix. Consequently, it requires more data than the CPOF procedure. Bill of material and routing data are required, and direct labor-hour or machine-hour data must be available for each operation.

To develop a bill of capacity for the example problem, we use the product structure data for A and B shown in Figure 4.4. We also need the routing and operation time standard data in the top portion of Figure 4.5 for assembling products A and B, as well as for manufacturing component items C, D, E, and F. The bill of capacity indicates total standard time required to produce one end product in each work center required in its manufacture. Calculations involve multiplying total time per unit values by the usages indicated in the bill of materials. Summarizing the usage-adjusted unit time data by work center produces the bill of capacity for each of the two products in the lower

FIGURE 4.3 Estimated Capacity Requirements Using Overall Factors (in Standard Direct Labor-Hours)

Work center	Historical percentage	Period													Total hours
		1	2	3	4	5	6	7	8	9	10	11	12	13	
100	60.3	37.87	37.87	37.87	37.41	37.41	37.41	45.07	45.07	45.07	51.32	51.32	51.32	51.32	566.33
200	30.4	19.09	19.09	19.09	18.86	18.86	18.86	22.72	22.72	22.72	25.87	25.87	25.87	25.87	285.49
300	9.3	5.84	5.84	5.84	5.78	5.78	5.78	6.96	6.96	6.96	7.91	7.91	7.91	7.91	87.38
Total required capacity		62.80*	62.80	62.80	62.05	62.05	62.05	74.75	74.75	74.75	85.10	85.10	85.10	85.10	939.20

*62.80 = (.95 × 33) + (1.85 × 17) using the standards from Figure 4.2.

Source: W. L. Berry, T. G. Schmitt, and T. E. Vollmann, "Capacity Planning Techniques for Manufacturing Control Systems: Information Requirements and Operating Features." Reprinted with permission, November 1982 *Journal of Operations Management*, Journal of the American Production and Inventory Control Society, Inc.

FIGURE 4.4   Product Structure Data

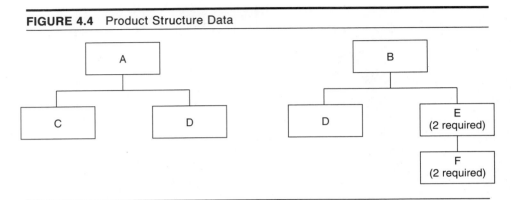

Source: W. L. Berry, T. G. Schmitt, and T. E. Vollmann, "Capacity Planning Techniques for Manufacturing Control Systems: Information Requirements and Operating Features." Reprinted with permission, November 1982 *Journal of Operations Management,* Journal of the American Production and Inventory Control Society, Inc.

portion of Figure 4.5. The bill of capacity can be constructed from engineering data, as we've done here; similar data might be available in a standard cost system. Some firms' alternative approach is to prepare the bill of capacity only for those work centers regarded as critical.

Once the bill of capacity for each end product is prepared, we can use the master production schedule to estimate capacity requirements at individual work centers. Figure 4.6 shows the determination of capacity requirements for our example. The resultant work center estimates differ substantially from the CPOF estimates in Figure 4.3. The differences reflect the period-to-period changes in product mix between the projected MPS and historical average figures. Estimates obtained from CPOF are based on an overall historical ratio of work between machine centers, whereas capacity bill estimates reflect the actual product mix planned for each period.

It's important to note that the total hours shown for the MPS (939.20) are the same in Figure 4.3 and Figure 4.6; the differences are in work center estimates for each time period. These differences are far more important in firms that experience significant period-to-period product mix variations than in those that have a relatively constant pattern of work.

Resource Profiles

Neither the CPOF nor the capacity bill procedure takes into account the specific timing of the projected workloads at individual work centers. In developing **resource profiles,** production lead time data are taken into account to

FIGURE 4.5 Routing and Standard Time Data

End products	Lot sizes	Operation	Work center	Standard setup hours	Standard setup hours per unit	Standard run time hours per unit	Total hours per unit
A	40	1 of 1	100	1.0	.025*	.025	.05†
B	20	1 of 1	100	1.0	.050	1.250	1.30
Components
C	40	1 of 2	200	1.0	.025	.575	.60
		2 of 2	300	1.0	.025	.175	.20
D	60	1 of 1	200	2.0	.033	.067	.10
E	100	1 of 1	200	2.0	.020	.080	.10
F	100	1 of 1	200	2.0	.020	.0425	.0625

Bill of capacity

	End product	
	A	B
Work center	Total time/unit	Total time/unit
100	.05	1.30
200	.70‡	.55§
300	.20	0.00
Total time/unit	.95	1.85

* .025 = Setup time ÷ Lot size = 1.0/40.
† .05 = Standard setup time per unit + Standard run time per unit = .025 + .025.
‡ .70 = .60 + .10 for one C and one D from Figure 4.4.
§ .55 = .10 + 2(.10) + 4(.0625) for one D, two Es, and four Fs.

Source: W. L. Berry, T. G. Schmitt, and T. E. Vollmann, "Capacity Planning Techniques for Manufacturing Control Systems: Information Requirements and Operating Features." Reprinted with permission, November 1982 *Journal of Operations Management*, Journal of the American Production and Inventory Control Society, Inc.

FIGURE 4.6 Capacity Requirements Using Capacity Bills

Work center	Period													Total hours	Projected work center percentage
---	1	2	3	4	5	6	7	8	9	10	11	12	13		
100	23.75*	23.75	23.75	18.90	18.90	18.90	34.00	34.00	34.00	36.95	36.95	36.95	36.95	377.75	40%
200	32.45	32.45	32.45	35.15	35.15	35.15	34.75	34.75	34.75	40.75	40.75	40.75	40.75	470.05	50
300	6.60	6.60	6.60	8.00	8.00	8.00	6.00	6.00	6.00	7.40	7.40	7.40	7.40	91.40	10
Total	62.80	62.80	62.80	62.05	62.05	62.05	74.75	74.75	74.75	85.10	85.10	85.10	85.10	939.20	100%

*23.75 = (33 × .05) + (17 × 1.30) from Figures 4.2 and 4.5.

Source: W. L. Berry, T. G. Schmitt, and T. E. Vollmann, "Capacity Planning Techniques for Manufacturing Control Systems: Information Requirements and Operating Features." Reprinted with permission, November 1982 *Journal of Operations Management*, Journal of the American Production and Inventory Control Society, Inc.

provide time-phased projections of the capacity requirements for individual production facilities. Thus, resource profiles provide a somewhat more sophisticated approach to rough-cut capacity planning.

In any capacity planning technique, time periods for the capacity plan can be varied (e.g., weeks, months, quarters). However, when time periods are long relative to lead times, much of the time-phased information's value may be lost in aggregating the data. In many firms, this means time periods longer than one week will mask important changes in capacity requirements.

To apply the resource profile procedure to our example, we use the bills of material, routing, and time standard information in Figures 4.4 and 4.5. We must also add the production lead time for each end product and component part to our data base. In this simplified example, we use a one-period lead time for assembling each end product and one period for each operation required to produce component parts. Since only one operation is required for producing components D, E, and F, lead time for producing these components is one time period each. For component C, however, lead time is two time periods: one for the operation in work center 200 and another for work center 300.

To use the resource profile procedure, we prepare a time-phased profile of the capacity requirements for each end item. Figure 4.7's operation setback charts show this time phasing for end products A and B. The chart for end product A indicates that the final assembly operation is to be completed during period 5. Production of components C and D must be completed in period 4 prior to the start of final assembly. Since component C requires two time periods (one for each operation), it must be started one time period before component D (i.e., at the start of period 3). Other conventions are used to define time phasing, but in this example we assume the master production schedule specifies the number of units of each end product that must be completed by *the end* of the time period indicated. This implies *all* components must be completed by the end of the preceding period.

For convenience, we've shown the standard hours required for each operation for each product in Figure 4.7. This information is summarized by work center and time period in Figure 4.8, which also shows the capacity requirements the MPS quantities generated in time period 5 from Figure 4.2 (40 of end product A and 13 of end product B). The capacity requirements in Figure 4.8 are only for MPS quantities in period 5. MPS quantities for other periods can increase the capacity needed in each period. For example, Figure 4.8 shows that 7.9 hours of capacity are needed in period 4 at work center 200 to support the MPS for period 5. The MPS for period 6 requires another 27.25 hours from work center 200. This results in the total of 35.15 hours shown in Figure 4.9, which provides the overall capacity plan for the current MPS using the resource profile procedure.

FIGURE 4.7 Operation Setback Charts for End Products A and B

End product A

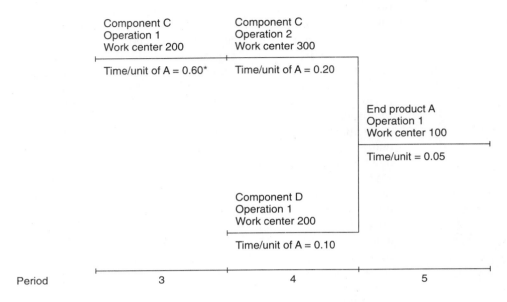

Component C
Operation 1
Work center 200

Time/unit of A = 0.60*

Component C
Operation 2
Work center 300

Time/unit of A = 0.20

End product A
Operation 1
Work center 100

Time/unit = 0.05

Component D
Operation 1
Work center 200

Time/unit of A = 0.10

Period   3   4   5

End product B

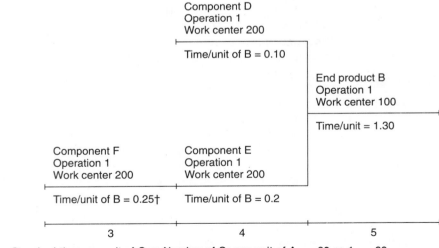

Component D
Operation 1
Work center 200

Time/unit of B = 0.10

End product B
Operation 1
Work center 100

Time/unit = 1.30

Component F
Operation 1
Work center 200

Time/unit of B = 0.25†

Component E
Operation 1
Work center 200

Time/unit of B = 0.2

Period   3   4   5

*.60 = Standard time per unit of C × Number of Cs per unit of A = .60 × 1 = .60.
†.25 = Standard time per unit of component F × Number of Fs per unit of B = .0625 × 4 = 25.

Source: W. L. Berry, T. G. Schmitt, and T. E. Vollmann, "Capacity Planning Techniques for Manufacturing Control Systems: Information Requirements and Operating Features." Reprinted with permission, November 1982 *Journal of Operations Management,* Journal of the American Production and Inventory Control Society, Inc.

FIGURE 4.8   Resource Profiles by Work Center

Time required during preceding periods for one end product assembled in period 5:

	Time period		
	3	4	5
End product A			
Work center 100	0	0	.05
Work center 200	.60	.10	0
Work center 300	0	.20	0
End product B			
Work center 100	0	0	1.30
Work center 200	.25	.30	0

Time-phased capacity requirements generated from MPS for 40 As and 13 Bs in time period 5:

	Time period		
	3	4	5
40 As			
Work center 100	0	0	2
Work center 200	24	4	0
Work center 300	0	8	0
13 Bs			
Work center 100	0	0	16.9
Work center 200	3.25	3.9	0
Work center 300	0	0	0
Total from period 5 MPS			
Work center 100	0	0	18.9
Work center 200	27.25	7.9	0
Work center 300	0	8.0	0

Source: W. L. Berry, T. G. Schmitt, and T. E. Vollmann, "Capacity Planning Techniques for Manufacturing Control Systems: Information Requirements and Operating Features." Reprinted with permission, November 1982 *Journal of Operations Management,* Journal of the American Production and Inventory Control Society. Inc.

Comparing the capacity plans produced by the capacity bills and the resource profile procedures (Figures 4.6 and 4.9), we see the impact of the time-phased capacity information. Total workload created by the master production schedule (939.2 hours) remains the same, as do the work center percentage allocations. But the period requirements for work centers 200 and 300 projected by the two techniques vary somewhat. A capacity requirement of 8 hours was projected for work center 300 in time period 6 using capacity bills versus 6 hours using resource profiles, a difference of more than 30 percent. This change reflects the difference in the timing of resources required to produce the component parts, which is taken into account by the resource bill procedure.

FIGURE 4.9 Capacity Requirements Using Resource Profiles

Work center	Past due*	1	2	3	4	5	6	7	8	9	10	11	12	13	Total hours	Work center percentage
									Period							
100	0.00	23.75	23.75	23.75	18.90	18.90	18.90	34.00	34.00	34.00	36.95	36.95	36.95	36.95	377.75	40%
200	56.50	32.45	35.65	35.15	35.15	32.15	34.75	34.75	39.45	40.75	40.75	40.75	11.80	0	470.05	50
300	6.60	6.60	6.60	8.00	8.00	8.00	6.00	6.00	6.00	7.40	7.40	7.40	7.40	0	91.40	10
Total	63.10	62.80	66.00	66.90	62.05	59.05	59.65	74.75	79.45	82.15	85.10	85.10	56.15	36.95	939.20	100%

*This work should be completed already for products to meet the master production schedule in periods 1 and 2. (If not, it's past due and will add to the capacity required in the upcoming periods.)

Source: W. L. Berry, T. G. Schmitt, and T. E. Vollmann, "Capacity Planning Techniques for Manufacturing Control Systems: Information Requirements and Operating Features." Reprinted with permission, November 1982 *Journal of Operations Management*, Journal of the American Production and Inventory Control Society, Inc.

Capacity Requirements Planning (CRP)

Capacity requirements planning (CRP) differs from the rough-cut planning procedures in four respects. First, CRP utilizes the time-phased material plan information produced by an MRP system. This includes consideration of all actual lot sizes, as well as lead times for both open shop orders (scheduled receipts) and orders planned for future release (planned orders). Second, the MRP system's gross to net feature takes into account production capacity already stored in the form of inventories of both components and assembled products. Third, the shop-floor control system accounts for the current status of all work-in-process in the shop, so only the capacity needed to *complete the remaining work* on open shop orders is considered in calculating required work center capacities. Fourth, CRP takes into account demand for service parts, other demands that may not be accounted for in the MPS, and any additional capacity that might be required by MRP planners reacting to scrap, item record errors, and so on. To accomplish this, the CRP procedure requires the same input information as the resource profile procedure (bills of material, routing, time standards, lead times) plus information on MRP planned orders and the current status of open shop orders (MRP scheduled receipts) at individual work centers.

As a medium-range capacity planning procedure, CRP exploits MRP information so as to calculate only the capacity required to complete the MPS. By calculating capacity requirements for actual open shop orders and planned orders in the MRP data base, CRP accounts for the capacity already stored in the form of finished and work-in-process inventories. Since MRP data include timing of both these open and planned orders, the potential for improved accuracy in timing capacity requirements is realized. This accuracy is most important in the most immediate time periods. Rough-cut techniques can overstate required capacity by the amount of capacity represented in inventories. In Figure 4.9, for example, the past due or already completed portion of the capacity requirements is 63.1 hours—about a full time period's capacity. This work should already have been completed if we expect to meet the MPS in periods 1 and 2. CRP's potential benefits aren't without cost. A larger data base is required, as well as a much larger computational effort.

The process of preparing a CRP projection is similar to that used for resource profiles. The major difference is detailed MRP data establish exact order quantities and timings for calculating capacity required. The resultant capacity needs are summarized by time period and work center in a format similar to Figure 4.9. The CRP results would differ from those of the other techniques, primarily in the early periods, but would be a more accurate projection of work center capacity needs. Since calculations are based on all component parts and end products from the present time period through all periods included in the MRP records (the planning horizon), we can see the enormity of the CRP calculation requirements. Some firms have mitigated this cost by collecting data as the MRP explosion process is performed.

Figure 4.10 presents one of the MRP records that drive the CRP procedure for our example. To simplify the presentation, we only show the MPS for end product A and the MRP record for one of its components, component C. We've used these data to calculate capacity requirements for work center 300. These capacity requirements incorporate the influence of lot sizes, inventories, and scheduled receipts for component C. Since item C is processed at

FIGURE 4.10 CRP Example: Detailed Calculations

Product A MPS

Period	1	2	3	4	5	6	7	8	9	10	11	12	13
	33	33	33	40	40	40	30	30	30	37	37	37	37

Component C
Lot size = 40
Lead time = 2

		1	2	3	4	5	6	7	8	9	10	11	12	13
Gross requirements		33	33	33	40	40	40	30	30	30	37	37	37	37
Scheduled receipts			40											
Projected available balance	37	4	11	18	18	18	18	28	38	8	11	14	17	20
Planned order releases		40	40	40	40	40	40		40	40	40	40		

Work Center 300 Capacity Requirements Using CRP

Period	1	2	3	4	5	6	7	8	9	10	11	12	13
Hours of capacity*	8	8	8	8	8	8	8	0	8	8	8	8	

Total = 88

*The eight hours of capacity required is derived from the scheduled receipt and planned order quantities of 40 units multiplied by the time to fabricate a unit of component C in machine center 300, 0.20 hours. (See Figure 4.7.)

Source: W. L. Berry, T. G. Schmitt, and T. E. Vollmann, "Capacity Planning Techniques for Manufacturing Control Systems: Information Requirements and Operating Features." Reprinted with permission, November 1982 *Journal of Operations Management,* Journal of the American Production and Inventory Control Society, Inc.

work center 300 during the second period of the two-period lead time, the planned order for 40 units due to be released in period 1 requires capacity in period 2 at work center 300. Required capacity is calculated using the setup and run time data from Figure 4.5 for component C.

For a lot size of 40 units, total setup and run time in work center 300 is eight hours (1.0 + [40 × .175]). Each planned order for component C in Figure 4.10 requires eight hours of capacity at work center 300, one period later. Similarly, the scheduled receipt of 40 units due in period 2 requires eight hours of capacity in week 1. Note the eight hours of capacity required for the scheduled receipt may not, in fact, be required if this job has already been processed at work center 300 before the beginning of period 1. The shop order's actual status is required to make the analysis.

In comparing CRP to the other capacity planning procedures, we shouldn't expect total capacity requirements for the 13 periods or the period-by-period requirements to be the same. Comparing capacity requirements for work center 300 developed by the resource profile procedure (Figure 4.9) and CRP (Figure 4.10) indicates estimated total capacity requirements for the 13 periods are less using CRP than resource profiles (88 versus 91.4 hours) and vary considerably on a period-by-period basis. Differences are explained by the initial inventory and use of lot sizing. Any partially completed work-in-process would reduce the capacity requirements further.

Input/Output Control

Each capacity planning technique's basic intent is to project capacity needs implied by either the production plan, the MPS, or the detailed material plan, so timely actions can be taken to balance capacity needs with capacity available. Once decisions are made concerning additions to and deletions of capacity, or adjustments to the material plan, a workable capacity plan can be created. Next we monitor this plan to see whether the actions were correct and sufficient. Monitoring also provides the basis for an ongoing correction of capacity planning data.

The basis for monitoring the capacity plan is **input/output control,** meaning planned work input and planned output at a work center will be compared to the actual work input and output. The capacity planning technique used delineates the planned input. Planned output results from managerial decision making to specify the capacity level; that is, planned output is based on staffing levels, hours of work, and so forth. In capacity-constrained work centers, planned output is based on the rate of capacity established by management. In non–capacity-constrained work centers, planned output is equal to planned input (allowing for some lead time offset).

Capacity data in input/output control are usually expressed in hours. Input data are based on jobs' expected arrivals at a work center. For example, a

CRP procedure would examine the status of all open shop orders (scheduled receipts), estimate how long they'll take (setup, run, wait, and move) at particular work centers, and thereby derive when they'll arrive at subsequent work centers. This would be repeated for all planned orders from the MRP data base. The resultant set of expected arrivals of exact quantities would be multiplied by run time per unit from the routing file. This product would be added to setup time, also from the routing file. The sum is a planned input expressed in standard hours.

Actual input would use the same routing data, but for the *actual* arrivals of jobs in each time period as reported by the shop-floor control system. Actual output would again use the shop-floor control data for exact quantities completed in each time period, converted to standard hours with routing time data.

The only time data not based on the routing file are those for planned output. In this case, management has to plan the labor-hours to be expended in the work center. For example, if two people work 9 hours per day for five days, the result is 90 labor-hours per week. This value has to be reduced or inflated by an estimate of the relation of actual hours to standard hours. In our example, if people in this work center typically worked at 80 percent efficiency, then planned output is 72 hours.

A work center's actual output will deviate from planned output. Often deviations can be attributed to conditions at the work center itself, such as lower than expected productivity, breakdowns, absences, random variations, or poor product quality. But less than expected output can occur for reasons outside the work center's control, such as insufficient output from a preceding work center or improper releasing of planned orders. Either problem can lead to insufficient input or a "starved" work center. Another reason for a variation between actual input and planned input was shown by our capacity planning model comparisons—some models don't produce realistic plans!

Input/output analysis also monitors backlog. Backlog represents the cushion between input and output. Backlog decouples input from output, allowing work center operations to be less affected by variations in requirements. Arithmetically, it equals prior backlog plus or minus the difference between input and output. The planned backlog calculation is based on planned input and planned output. Actual backlog uses actual input and output. The difference between planned backlog and actual backlog represents one measure of the total, or net, input/output deviations. Monitoring input, output, and backlog typically involves keeping track of cumulative deviations and comparing them with preset limits.

The input/output report in Figure 4.11 is based on work center 200 for our example problem, now shown in weekly time buckets measured in standard labor-hours. The report was prepared at the end of period 5, so the actual values are current week-by-week variations in planned input. These could result from actual planned orders and scheduled receipts; that is, for example,

---

FIGURE 4.11   Sample Input/Output for Work Center 200*
(as of the End of Period 5)

		Week				
		1	*2*	*3*	*4*	*5*
Planned input		15	15	0	10	10
Actual input		14	13	5	9	17
Cumulative deviation		−1	−3	+2	+1	+8
Planned output		11	11	11	11	11
Actual output		8	10	9	11	9
Cumulative deviation		−3	−4	−6	−6	−8
Actual backlog	20	26	29	25	23	31

Desired backlog: 10 hours

*In standard labor-hours.

---

if the input were planned by CRP, planned inputs would be based on timings for planned orders, the status of scheduled receipts, and routing data. The *actual* input that arrives at work center 200 can vary for any of the causes just discussed.

Work center 200's planned output has been smoothed; that is, management decided to staff this work center to achieve a constant output of 11 hours per week. The result should be to absorb input variations with changes in the backlog level. Cumulative planned output for the five weeks (55 hours) is 5 hours more than cumulative planned input. This reflects a management decision to reduce backlog from the original level of 20 hours. The process of increasing capacity to reduce backlog recognizes explicitly that flows must be controlled to change backlog; backlog can't be changed in and of itself.

Figure 4.11 summarizes the results after five weeks of actual operation. At the end of week 5, the situation requires managerial attention. The cumulative input deviation (+ 8 hours), cumulative output deviation (− 8 hours), current backlog (31 hours), or all three could have exceeded the desired limits of control. In this example, the increased backlog is a combination of more than expected input and less than expected output.

One other aspect of monitoring backlog is important. In general, there's little point in releasing orders to a work center that already has an excessive backlog, except when the order to be released is of higher priority than any in the backlog. In general, the idea is to not release work that can't be done, but to wait and release what's really needed. Oliver Wight sums this up as one of the principles of input/output control: "Never put into a manufacturing facility or to a vendor's facility more than you believe can be produced. Hold backlogs in production and inventory control."

The Capacity "Bath Tub"

Figure 4.12 depicts a work center "bath tub" showing capacity in hydraulic terms. The input pipe's diameter represents the maximum flow (of work) into the tub. The valve represents MPC systems like MPS, MRP, and JIT, which determine **planned input** (flow of work) into the tub. Actual input could vary because of problems (like a corroded valve or problem at the water department) and can be monitored with input/output analysis. We can determine **required capacity** to accomplish the planned input to the work center with any of the capacity planning techniques. The output drain pipe takes completed work from the work center. Its diameter represents the work center's planned or **rated capacity,** which limits planned output. As with actual input, actual output may vary from plan as well. It too can be monitored with input/output analysis. Sometimes planned output can't be achieved over time even when it's less than maximum capacity and there's a backlog to work on. When that occurs, realized output is called **demonstrated capacity.** The "water" in the tub is the **backlog** or **load,** which can also be monitored with input/output analysis.

FIGURE 4.12   The Capacity Bath Tub

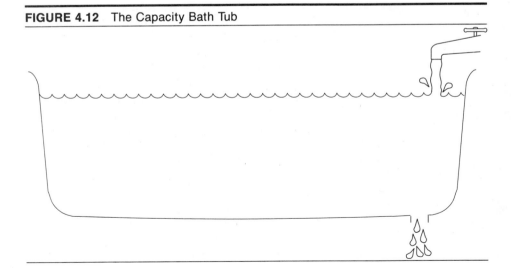

MANAGEMENT AND CAPACITY PLANNING

In designing and using the capacity planning system, we must take several management considerations into account. Here we discuss designing the systems from the perspective of the MPC framework, consider the capacity measure, look at choices in tailoring the capacity planning system to a particular firm, and consider how to use capacity planning data. Key elements of management's commitment in the design and use of the system are emphasized.

Capacity Planning in the MPC System

Figure 4.1 shows the relationship between MPC framework and various capacity planning modules. The five modules range from long-range resource planning to day-to-day control of capacity utilization. There's a vertical relationship among the capacity planning modules as well as the horizontal relationship with the material planning modules of the MPC system. These relationships can affect managerial choices for capacity planning systems' design and use in a specific firm.

To illustrate the importance of the interrelationships in designing and using the capacity planning system, let's consider the impact of production planning and resource planning decisions on shorter-term capacity planning decisions. To the extent that production planning and resource planning are done well, problems faced in capacity planning can be reduced, since appropriate resources have been provided. If, for example, the production plan specifies a very stable rate of output, then changes in the master production schedule (MPS) requiring capacity changes are minimal. If the material planning module functions effectively, the MPS will be converted into detailed component production plans with relatively few unexpected execution problems.

Toyota and several other Japanese auto manufacturers are good cases in point. These firms' production plans call for a stable rate of output (cars per day). Product mix variations are substantially less than for other auto companies because they carefully manage the number and timing of option combinations. Order backlogs and finished-goods inventories also separate factories from actual customer orders. The result is execution systems that are simple, effective, and easy to operate with minimal inventories and fast throughput times. The careful resource and production planning, and resultant stability, means that these firms have little need for either rough-cut capacity planning procedures or CRP.

A quite different but equally important linkage that can affect capacity planning system design is the linkage with the shop-floor systems. A key relationship exists in scheduling effective use of capacity. With sufficient capacity and efficient use of that capacity ensured by good shop-floor procedures, we'll see few unpleasant surprises requiring capacity analysis and change. Effective shop-floor procedures utilize available capacity to process orders according to MRP system priorities, provide insight into potential capacity problems in the short range (a few hours to days), and respond to changes in material plans. Thus, effective shop-floor systems reduce the necessary degree of detail and intensity of use of the capacity planning system.

In providing effective control of capacity plans, the shop-floor system again is key. Good shop-floor control makes plans more likely to be met. The result is a better match between actual input/output and planned input/output. Again we see attention to the material planning side of the MPC system, in this case the shop-floor control module, having an effect on the capacity planning side.

Choosing the Measure of Capacity

Choice of capacity measure is an important management issue. Alternatives run from machine-hours or labor-hours to physical or monetary units. The choice depends on the constraining resource and the firm's needs. In any manufacturing company, the "bundle of goods and services" provided to customers increasingly includes software, other knowledge work, aftersale service, and other customer services. In every case, providing these goods and services requires resources—"capacities" that must be planned, managed, and developed. Appropriate measures of capacity must be established—and changed as evolution in the bundle of goods and services occurs.

Several current trends in manufacturing have a significant bearing on the choice of capacity measures. Each can have a major impact on what's important to measure in capacity. One important trend is considerable change in the concept of direct labor. Direct labor has been shrinking as a portion of overall manufacturing employment. Distinctions between direct and indirect labor are shrinking. The ability to change labor capacity by hiring and firing (or even using overtime) has been reduced; notions of "lifetime employment" have further reduced this form of capacity adjustment.

An increasingly important direct labor issue is the whole person concept of JIT. It has labor capacity changing qualitatively as more staff work is incorporated into their jobs. Since an objective in JIT systems is continued improvement, the basis for labor capacity planning is constantly changing. This mandates control procedures for identifying and changing the planning factors as improvements take place.

Another important trend is decreased internal fabrication and increased emphasis on outside purchasing. This trend can alter the conception of what capacity requirements are important. Procurement analysis, incoming inspection, and engineering liaison may become the critical capacities to be managed, as well as planning and scheduling the capacities in vendor firms.

For many firms engaged in fabrication, machine technology is changing rapidly. Flexible automation has greatly increased the range of parts that can be processed in a machine center. Future product mixes are likely to be much more variable than in the past, with a marked effect on the equipment capacity required. Moreover, as equipment becomes more expensive, it may be necessary to plan and control the capacity of key pieces of equipment at a detailed level.

To the extent that cellular technologies are adopted as part of JIT manufacturing, the unit of capacity may need to change. Usually the entire cell is coupled and has only as much capacity as its limiting resource. Often, the cell is labor limited, so the unit of capacity is labor-hours (continually adjusted for learning). Sometimes, however, the capacity measure needs to be solely associated with a single aspect of the cell. Also, when dissimilar items are added to the cell for manufacture, it's necessary to estimate each new item's capacity requirements in terms of individual processing steps.

The first task in choosing a capacity measure is to creatively identify resources that are key and in short supply. Capacity control is too complicated to apply to all resources. The next step is to define the unit of measure. If the key resource is people, then labor-hours may be appropriate. In other instances, such measures as tons, gallons, number of molds, number of ovens, hours of machine time, square yards, linear feet, lines of code, customer calls, and cell hours have been used. In some cases, these are converted to some "equivalent" measure to accommodate a wider variety of products or resources.

After the resources and unit of measure have been determined, the next concern is to estimate available capacity. The primary issue here is theory versus practice. The engineer can provide theoretical capacity from the design specifications for a machine or from time studies of people. A subissue is whether to use "full" capacity or some fraction thereof (often 75 percent to 85 percent). A further issue is "plasticity" in capacity. For almost *any* resource, if it's *really* important, more output can be achieved. We've seen many performances that fall short of or exceed capacity calculations.

Choice of capacity measure follows directly from the objective of providing capacity to meet production plans. The appropriate measure of capacity most directly impacts meeting these plans. The measure, therefore, should be appropriate to the key limited resources and be based on what's achievable, with allowances for maintenance and other necessary activities. It must be possible to convert the bundle of products and services into capacity measurement terms. The results must be understood by the people responsible, and they should be monitored.

Choice of a Specific Technique

In this chapter's discussion, the capacity planning techniques for converting a material plan into capacity requirements include three different methods for rough-cut capacity planning (CPOF, capacity bills, and resource profiles) and capacity requirements planning. The choice of method depends heavily on characteristics of the manufacturing process.

The three rough-cut methods are most general, being applicable even in companies using just-in-time methods for shop-floor control. Rough-cut approaches can be useful in JIT operations to estimate the impact of changes in requirements called for by revisions to the master production schedule. For example, under level scheduling conditions, a change from a production rate of 480 units per day (one unit per minute) to 528 units per day (1.1 units per minute) might be needed. A rough-cut procedure could be used to examine the impact on each work center or manufacturing cell through which this volume would pass. Any indicated problems or bottleneck conditions could be addressed *before* the crisis hits. Similarly, a planned reduction in MPS could be evaluated to determine resources that might be freed to work on other tasks.

Rough-cut approaches do vary in accuracy, aggregation level, and ease of preparation. There's a general relationship between the amount of data and computational time required, and the quality and detail of the capacity requirements estimated. The issue is whether additional costs of supporting more complex procedures are justified by improved decision making and subsequent plant operations.

The capacity bills procedure has an advantage over capacity planning using overall factors (CPOF) because it explicitly recognizes product mix changes. This can be important in JIT operations, particularly where the level schedule is based on assumptions of product mix—and where different products have different capacity requirements. On the other hand, if changes in mix are easily accommodated, and there are minimal differences in capacity requirements for different products, then CPOF's simplicity can be exploited. Under JIT operations, however, there's often little need to incorporate the added sophistication of the resource profile procedure. There simply won't be any added advantage to making lead time offsets in the planning process. Work is completed at virtually the same time as it's started.

Capacity requirements planning, however, is only applicable in companies using time-phased MRP records for detailed material planning and shop-order–based shop scheduling systems. CRP is unnecessary under JIT operations anyway since minimal work-in-process levels means there's no need to estimate the impact on capacity requirements of partially processed work. All orders start from "raw materials" with virtually no amount of "capacity" stored in component inventories. Also, under JIT, there's no formal shop-floor control procedure. There are no work orders. Thus, there are no status data on work orders.

Input/output control isn't usually an issue under JIT operations because attention has been shifted from planning to execution. As a result, actual input should equal actual output. Actual input becomes actual output with an insignificant delay. The backlog is effectively a constant zero. However, planned input can indeed vary from actual input—and so can planned output vary from actual output. These variations should be achievable without violating the equality between actual input and actual output—with backlog remaining at zero. To the extent that plan-to-actual variations are possible, the result reflects the flexibility or "band width" of the JIT unit.

Using the Capacity Plan

All the techniques we've described provide data on which a manager can base a decision. The broad choices are clear—if there's a mismatch between available capacity and required capacity, either the capacity or the material plan should be changed. If capacity is to be changed, the choices include overtime/ undertime authorization, hiring/layoff, and increasing/decreasing the number of machine tools or times in use. Capacity requirements can be changed by alternate routing, make or buy decisions, subcontracting, raw material changes, inventory changes, or changing customer promise dates.

Choice of capacity planning units can lead to more effective use of the system. Capacity units need not be work centers as defined for manufacturing, engineering, or routing purposes. They can be groupings of the key resources (human or capital) important in defining the factory's output levels. Many firms plan capacity solely for key machines (work centers) and gateway operations. These key areas can be managed in detail, while other areas fall under resource planning and the shop-floor control system.

Capacity planning choices dictate the diameter of the manufacturing pipeline. Only as much material can be produced as there's capacity for its production, *regardless of the material plan*. Not understanding the critical nature of managing capacity can lead a firm into production chaos and serious customer service problems. In the same vein, the relationship between flexibility and capacity must be discussed. You can't have perfectly balanced material and capacity plans *and* be able to easily produce emergency orders! We know one general manager who depicts his capacity as a pie. He has one slice for recurring business, one for spare parts production, one for downtime and maintenance, and a final specific slice for opportunity business. He manages to pay for this excess capacity by winning lucrative contracts that require rapid responses. He *does not add* that opportunity business to a capacity plan fully committed to the other aspects of his business.

DATA BASE REQUIREMENTS

We've seen that each capacity planning technique requires different systems linkages. These linkages imply different data elements and data base considerations. Managers' use of the different capacity planning techniques and enhancements to improve utility can have data base implications as well.

Data Base Design Considerations

The CPOF, capacity bills, resource profiles, and CRP procedures all use MPS data to develop capacity requirements. The CPOF procedure calculates overall direct labor requirements for the MPS and allocates this capacity to the work centers on the basis of historically observed workload patterns. By contrast, the capacity bill procedure uses bill of material and routing information to calculate capacity at work centers and, thus, more accurately reflects the particular mix of end items shown in the MPS. The resource profile procedure time phases capacity requirements by the lead times for component parts and assemblies used to manufacture end items. The CRP procedure employs additional information from MRP and shop-floor control systems to account for the exact timing, quantities, and status of component part and end item production orders.

We see that there's an increasing requirement for data starting with relatively simple accounting system data required by CPOF. The capacity bills, resource profiles, and CRP techniques each require increased amounts of production/inventory control, industrial engineering, and shop-floor control data. These latter procedures incur successively increasing computational cost as well. The specific data linkages with the material planning modules are shown in Figure 4.1.

Although data base size for capacity bills is much smaller than for CRP, the CRP data base already largely exists if the firm has a working MRP system. For this reason, many firms use CRP systems to answer questions that could be analyzed at far lower computational cost with rough-cut techniques. They do so because there are no additional data base requirements. All that's needed is the additional computer run time, which can be expensive but is usually available. However, they usually don't do what-if analysis. This would be exceedingly expensive. The result is only *one* capacity plan—the one associated with the current MRP-based material plan.

Several other factors influence the data base's design and maintenance. The level of detail appropriate for capacity management implies a corresponding level of detail in the data base and in data base maintenance. If capacity assessments are made in terms of sales dollars or average labor-hours, data may be extracted from the financial accounting data base. This reduces the

MPC data base's complexity but requires some coordinating data base maintenance.

Input/output analysis requires a communication link with the shop-floor system to gather data for analysis. A closely related issue affecting data base complexity is labor productivity. Many firms have standard time data that differ widely from actual practice. What's more, this difference can vary between work centers. This fact can greatly complicate the data base design and maintenance problem, since it's critical to keep track of actual production rates to make an accurate conversion of material plans into capacity needs.

Extended Capabilities and Data Base Design. We have discussed the desirability of incorporating a what-if capability into the capacity planning system. This capability can create severe demands on data base design. The consideration of computer time has already been raised. The need to evaluate a number of alternative material plans means the ability to easily change the MPS must be designed in. Also, changes must be isolated from current actual MPS and MRP records, both for the sake of recovery and to not create false signals on the shop floor before appropriate analysis and approval have been accomplished.

Along with the capability of what-if testing runs a parallel set of questions on detailed implementation decisions. The choice of the level of aggregation for capacity planning, the size of the time period for analysis, the number of future periods to be analyzed, and the number and composition of machine centers all influence the data base's size, complexity, and maintenance. If capacity plans are based on one set of numbers and time periods, and another set is used to make implementation decisions, mismatches and other problems can occur. On the other hand, designing the system to support any kind of question and any kind of decision may be prohibitively expensive.

Perhaps the ultimate design objective for the data base and its use is to be able to identify and plan for the key work centers as they change over time. This would require careful attention to design of the input/output module and the tolerance limits used to trigger attention. It would also require flexibility in the data base to permit analysis of different possible groupings over time and groupings that might not correspond to current work centers or labor categories.

EXAMPLES OF APPLICATIONS

Here are examples of capacity planning techniques in practice. Specifically, we look at the Black & Decker Company's use of CRP, the Twin Disc Company's use of capacity bills, and Applicon's rough-cut capacity planning system for JIT operations.

Capacity Planning at Black & Decker

The Black & Decker Company produces a broad line of consumer workshop, garden, and household products. Production is mostly to stock. Capacity planning is largely based on CRP. Figure 4.13 shows the weekly CRP report for one key machine group (KMG 073) in department 8–01 of the Hampstead, Maryland, plant. It's called the BH "group" but contains a single critical machine. The time periods are weeks (from 741 through 775). For each week, the projection is based on the combined open shop orders (MRP scheduled receipts) and MRP planned orders. Times include setup hours ("S/U HRS.") and run-time hours ("OP HRS.").

This machine center's weekly capacity is 106 hours. Since there's only one machine, 18 hours per day, five days per week, plus 8 hours each on Saturday and Sunday account for all the capacity. The report shows the percent of capacity ("% OF CAP") required by the projected arrival of work in each week (planned input). This is shown on Figure 4.13, where 30 of the 35 weeks shown are loaded to over 80 percent. Black & Decker keeps track of the number of weeks for which projected needs exceed 80 percent of capacity. This can indicate potential serious capacity problems. Moreover, the average level for the total 35-week period is shown to be 97 percent.

The capacity problem may be significantly greater than the percentage of capacity indicates. The report shows that the current WIP (work-in-process) or backlog is 352 standard hours. This work is presently at the machine, with 109 standard hours scheduled to arrive in the upcoming week. The work center is already more than three weeks behind schedule. Given projected load and present capacity, the backlog won't decrease for the foreseeable future.

To complete the capacity planning picture, Black & Decker uses the MRP data base to analyze capacity loads on a quarterly basis and to prepare a four-week, detailed day-by-day capacity report for each work center. The four-week report is the basis for daily capacity planning decisions. Figure 4.14 shows this daily load report for the next four weeks for the KMG 073 key machine group.

Total values for the weeks show the same capacity problem as Figure 4.13. Figure 4.14 also provides last week's actual performance as part of the input/ output control information.

As useful as these capacity planning reports have been to Black & Decker over the years, they don't allow for the desirable level of what-if analysis. Subsequently, Black & Decker designed two new systems to support capacity planning. One, a capacity bill approach to rough-cut capacity planning, produces total dollar output levels by divisions and product groups, work center loadings, and critical machine group capacity requirements directly from the MPS.

The other aid to capacity planning is called **alternations planning.** This approach to what-if analysis allows use of the MRP data base to determine the

FIGURE 4.13 Black & Decker CRP Report

KMG WEEKLY LOADS, HAMPSTEAD

SEQ 1033 KMG 073 DEPT 8-01 CST/CN 063 KMG NAME BH GROUP MACH QTY 1 WEEKLY CAP 106.0

MFG. WEEK	WIP	741	742	743	744	745	746	747	748	749	750	751	752	753	754	755
S/U HRS.	38	5	4	2	5	2	3	6	4	5	2	4	5	3	4	1
OP HRS.	314	104	102	105	95	94	107	111	84	92	41	101	128	100	72	87
TOTAL	352	109	106	107	100	96	110	117	88	97	43	105	133	103	76	88
% OF CAP		103	100	100	94	90	104	110	83	91	40	100	125	97	72	83

MFG. WEEK	756	757	758	759	760	761	762	763	764	765	766	767	768	769	770	771
S/U HRS.	8	1	3	4	3	3	4	5	3	4	9	1	5	5	2	3
OP HRS.	91	92	107	151	65	96	140	86	68	97	117	62	93	98	83	132
TOTAL	99	93	110	155	68	99	144	91	71	101	126	63	98	103	85	135
% OF CAP	93	88	104	146	64	93	136	86	67	96	119	59	92	97	80	125

MFG. WEEK	772	773	774	775
S/U HRS.	3	6	8	2
OP HRS.	91	121	134	104
TOTAL	94	127	142	106
% OF CAP	89	120	134	100

AVERAGE FOR
1st 10 Wks 92%
2d 13 Wks 99%
3d 12 Wks 98%
Tot 35 Wks 97%

NUMBER WKS OVER 80%
30

Source: R. W. Hall and T. E. Vollmann, "Black & Decker: Pioneers with MRP," Case Studies in Materials Requirements Planning, ed. by E. W. Davis (Falls Church, Va.: American Production and Inventory Control Society, 1978), p. 38.

FIGURE 4.14 Black & Decker Daily Machine Load Report

```
09/15/   Week 741-1              HAMPSTEAD MACHINE LOAD REPORT                DEPT. 8-01
COST CTR 002   KMG 073     BHG ROUTING MACH     1 MACH @ 18 HRS/DAY     18 HRS/DAY AVAIL
```

741

SCHED OP	HRS	TOT SCH	TOT AVAIL	PCT LOAD	CUM AVAIL HRS
-1 -2 -3 -4 -5 -6 -7*	18 19 19 19 8 8	109	106	103	-3

742

SCHED OP	HRS	TOT SCH	TOT AVAIL	PCT LOAD	CUM AVAIL HRS
-1 -2 -3 -4 -5 -6 -7	18 18 18 16 8 10	106	106	100	-3

743

SCHED OP	HRS	TOT SCH	TOT AVAIL	PCT LOAD	CUM AVAIL HRS
-1 -2 -3 -4 -5 -6 -7	19 19 18 17 7 7 8	107	106	100	-4

744

SCHED OP	HRS	TOT SCH	TOT AVAIL	PCT LOAD	CUM AVAIL HRS
-1 -2 -3 -4 -5 -6 -7	20 16 18 19 17 7 4	100	106	94	2

HRS PRODUCED LAST WEEK

AHEAD	CURR	BHND	TOTAL	BACKLOG
	15	93	108	352

*Days of the week.

Source: R. W. Hall and T. E. Vollmann, "Black & Decker: Pioneers with MRP," Case Studies in Materials Requirements Planning, ed. by E. W. Davis (Falls Church, Va.: American Production and Inventory Control Society, 1978), p. 37.

effect of changing the timing or quantities of selected MPS values. Changes can be evaluated in terms of time-phased capacity requirements on particular machine centers without disturbing the operative data base. The report's output shows both the current plan and revised plan in a format similar to Figure 4.13. This report permits a quick assessment of the effect of MPS changes.

Capacity Planning at Twin Disc

The Twin Disc Company manufactures gears, transmissions, and other heavy components for the farm implement and heavy-equipment industries. It's primarily a make-to-order firm. As part of its capacity-planning system, Twin Disc uses capacity bills for rough-cut capacity planning. Figure 4.15 shows an example output from this system. For each machine or work center, we see the percentage of available capacity required by each of the nine product lines that Twin Disc uses for master production scheduling purposes. For example, the MPS for product line A requires 22 percent of the 1,561 hours of weekly capacity at the 2AC Chucker. Total capacity requirements for all nine product lines indicate a total load of 95 percent of the 2AC Chucker capacity.

The most important aspect of this report may be the last column. We see managerial actions taken to overcome capacity problems. For example, the present MPS loads the Maag gear grinder to 113 percent of rated capacity. One Reishauer gear grinder, however, is only loaded to 69 percent of its capacity. A decision has been made to shift some work, taking into account differences in the two grinders' capacities (roughly a 3:1 ratio in machine-hours required). Other managerial actions in the report include moving an additional machine in from another factory and shifting more capacity. The point is that this is a working document used to make effective capacity decisions.

Capacity Planning at Applicon

Applicon, a division of Schlumberger, designs and manufactures computer-aided engineering (CAE), computer-aided design (CAD), and computer-aided manufacturing (CAM) systems. Applicon has implemented numerous JIT concepts and replaced some of its MRP system modules. Already its dramatic results include a reduction in lead time (20 weeks to 4 days), inventory reduction of over 75 percent, virtual elimination of obsolescence costs, little or no inspection, and a decline in MPC personnel (86 to 14).

Figure 4.16 shows an Applicon "Capacity Status Report." Applicon has divided the factory into 17 capacity groupings (work centers) for planning purposes. It uses actual customer orders as a monthly MPS to drive capacity planning. Capacity bills are used to convert the MPS into the present "load" over the next month (20 working days) in standard hours (the second column

FIGURE 4.15 Twin Disc Capacity Bill Report

Center '03–05	Type	Qty.	No. of shifts	Cap. (hrs/wks)	A	B	C	D	E	F	G	H	I	TOTAL	Remarks
BD	2AC chucker	4	3	1561	22	22	16	3	8	11	11	2		95	
BR	3AC Warner & Swasey	8	3	2966	3	16	46	1	13	2	1	2	—	84	
CA	Reishauer gear grinder	2	3	900			38		22	—	12		—	72	
CAB	Reishauer gear grinder	2	3	950		10	43		13	2			1	69	Off load to CAB 3:1 ratio
CD	P. & W. gear grinder	1	3	544			59		4					63	
CEA	Maag gear grinder	4	3	3044		10	76		14	5			8	113	Off load to CA, CD
CG	P. & W. gear grinder	4	3	1190			120		8					128	
CI	Pfauter hobber	5	3	2374	6	22	41		9	2	4		2	86	
CJ	Barber colman hobber	1	3	620	27	39	50	1	29	25		8	1	180	Off load to CI, CW
CN	Gear shaver	3	2.5	700	14	37	15		8	9	4			87	
CQ	Gear pointer	1	1	22	13	56	—		12	3				84	
CS	Fellows shaper	1	3	549		37			8	—				45	
CW	Barber colman hobber	3	3	1530	7	12	26	2	8	9	6			70	
CX	Barber colman shaper	1	3	546	—	38	25	7	6	2	24			102	Off load to CS
CY	Fellows shaper	1	3	514	15	77	17		10	12			2	133	Off load to CS
FD	Internal grinder	1	3	285	8	8	10	5	8	4	6			49	Relieve FI and
FI	Internal grinder	1	3	275	25	59	32	11	25	10	9			171	move machine from PLI2
FJ	Surface grinder	1	1	328	6	41	20	1	11	14				93	
FM	Vertical internal grinder	1	2	368		38	57	1	16	1			3	116	Add ½ shift
FY	Gear hone	2	2	528	19	28	11	12	9	5				84	
H	Engine lathe	1	3	427	26	42	29	—	15	8	6		—	126	Off load to HES
HES	W. & S. Lathe—Special	2	2	234	5	43	18		11	12				89	Add 1 shift
JA	Horizontal broach	1	1	90	14	3	53	2	12	3				87	
NH	Gear chamfer	1	3	307	18	42	10		12	11	5			98	
PC	Magnaflux	1	1	240	3	35	48		12	2	2		—	102	Add ½ shift

Work center description

Percent utilization of capacity by product line

Source: E. S. Buffa and J. G. Miller, *Production-Inventory Systems: Planning and Control*, 3rd ed. (Homewood, Ill.; Richard D. Irwin, 1979), p. 598.

FIGURE 4.16 Applicon Capacity Status Report

```
0612
```

Work Center ******	Load (Std. Hours) ************	Capacity (Standard) **********	Capacity (Adj. Std.) **********	Capacity (Maximum) *********
ALF-A	70	480	336	528
ALF-T	5	80	56	88
HLT-A	438	800	560	880
HLT-T	85	160	112	176
MIS-A	270	800	560	880
MIS-T	14	80	56	88
MVX-A	399	1120	784	1232
MVX-T	79	80	56	88
OLD-P	81	0	0	0
PCB-A	52	160	112	176
PCB-H	44	160	112	176
PCB-I	124	320	224	352
PCB-M	441	480	336	520
PCB-P	408	960	672	1056
PCB-T	918	1680	1176	1848
PCB-V	123	160	112	176
PCB-W	56	160	112	176
TOTALS	3634	7680	5376	8440

```
     20 TOTAL WORKDAYS INCLUDED
```

in Figure 4.16). The capacities in column 3 are based on a total work force of 48 people (e.g., ALF-A has three workers who work 8 hours per day for 20 days in the month = 480 standard hours). Work center OLD-P's zero capacity indicates no worker is presently assigned to this activity.

The fourth column reduces the capacity amounts by 30 percent (the desired rate of direct production activity for Applicon workers). The remaining time is used for "whole person" activities. The last column provides a "maximum" capacity value based on 10 percent overtime.

This report was run on June 12 for the next 20 days. Differences between "load" and the three capacities represent Applicon's ability to take on additional work in the next month. Large orders can be included into a trial run of the MPS to examine the orders' impact in terms of existing capacity availabilities. Total load (3,634 hours) represents 47 percent of standard capacity, 68 percent of adjusted capacity, and 43 percent of maximum capacity. Management reviews these numbers carefully—particularly if possible large orders are under negotiation. It's relatively easy to make trial runs with those orders included to examine the impact of accepting the orders.

Of all the work centers in Figure 4.16, MVX-T appears to be in the most trouble. However, this can easily be fixed. MVX-T only has one half person allocated to it (80 hours/month). Reducing MVX-A (or some other work center) by one half person and increasing MVX-T's capacity to one person for the month solves the problem. OLD-P similarly needs to have a person allocated to it.

This rough-cut capacity planning system serves Applicon well. Problems can be anticipated. JIT operations mean results are very current, with little or no bias because of work-in-process inventories. Results of the capacity planning are given to shop personnel who make their own adjustments as they see fit, making allowances for absenteeism, particular workers' relative strengths, and other local conditions.

CONCLUDING PRINCIPLES

Clear principles for design and use of the capacity planning system emerge from this chapter:

- Capacity plans must be developed concurrently with material plans if the material plans are to be realized.
- The particular capacity planning technique(s) chosen must match the level of detail and actual company circumstances to permit making effective management decisions.
- Capacity planning can be simplified in a JIT environment.
- The better the resource and production planning process, the less difficult the capacity planning process.
- The better the shop-floor system, the less short-term capacity planning is required.
- The more detail in the capacity planning system, the more data and data base maintenance are required.
- It's not always capacity that should change when capacity availability doesn't equal need.
- Capacity not only must be planned, but use of that capacity must also be monitored and controlled.
- Capacity planning techniques can be applied to selected key resources (which need not correspond to production work centers).
- The capacity measure should reflect realizable output from the key resources.

REFERENCES

Aherns, Roger. "Basics of Capacity Planning and Control." *APICS 24th Annual Conference Proceedings,* 1981, pp. 232–35.

Belt, Bill. "Integrating Capacity Planning and Capacity Control." *Production and Inventory Management,* 1st quarter 1976.

Berry, W. L.; T. Schmitt; and T. E. Vollmann. "Capacity Planning Techniques for Manufacturing Control Systems: Information Requirements and Operational Features." *Journal of Operations Management* 3, no. 1 (November 1982).

————. "An Analysis of Capacity Planning Procedures for a Material Requirements Planning System." *Decision Sciences* 15, no. 4 (Fall 1984).

Blackstone, J. H., Jr. *Capacity Management.* Cincinnati: Southwestern, 1989.

Bolander, Steven F. "Capacity Planning through Forward Scheduling." *APICS Master Planning Seminar Proceedings,* Las Vegas, April 1981, pp. 73–80.

Burlingame, L. J. "Extended Capacity Planning." *APICS Annual Conference Proceedings,* 1974, pp 83–91.

Capacity Planning Reprints. Falls Church, Va.: American Production and Inventory Control Society, 1986.

Chakravarty, A., and H. K. Jain. "Distributed Computer System Capacity Planning and Capacity Loading," *Decision Sciences Journal* 21, no. 2, Spring 1990, pp. 253–62.

Hall, R. W., and T. E. Vollmann. "Black & Decker: Pioneers with MRP." *Case Studies in Materials Requirements Planning.* Falls Church, Va.: American Production and Inventory Control Society, 1978, p. 38.

Karmarkar, U. S. "Capacity Loading and Release Planning with Work-in-Progress (WIP) and Leadtimes." *Journal of Manufacturing and Operations Management* 2, no. 2, 1989, pp. 105–23.

Lankford, Ray. "Short-Term Planning of Manufacturing Capacity." *APICS 21st Annual Conference Proceedings,* 1978, pp. 37–68.

Solberg, James J. "Capacity Planning with a Stochastic Flow Model." *AIIE Transactions* 13, no. 2 (June 1981), pp. 116–22.

Wemmerlöv, Urban. "A Note on Capacity Planning." *Production and Inventory Management,* 3rd quarter 1980, pp. 85–89.

————. *Capacity Management Techniques for Manufacturing Companies with MRP Systems.* Falls Church, Va.: American Production and Inventory Control Society, 1984.

Wight, O. W. "Input-Output Control, A Real Handle on Lead Time." *Production and Inventory Management,* 3rd quarter 1970, pp. 9–31.

DISCUSSION QUESTIONS

1. The hierarchy of capacity planning activities in Figure 4.1 doesn't show any direct relationships between them, yet the higher-level activities impose constraints on the lower-level activities. What are those constraints?
2. A variety of resources must be provided in sufficient quantity to meet the material plans. Name some resources that may be planned in a capacity planning system.
3. Provide some examples of a university's capacity planning activities. What would happen if they weren't accomplished well?
4. How might you measure planning factors to use the CPOF procedure?

5. What does the expression, "Hold backlogs in production and inventory control," mean?
6. The CRP technique requires a substantially greater amount of computation than the other techniques. Cite examples where it might be important to have this level of detail for capacity planning, even for the longer run.
7. Contrast what-if analysis and input/output control.
8. How does the JIT concept of band width relate to capacity planning?

PROBLEMS

1. Finster Farmware has gathered data on labor-hour and machine-hour requirements for producing its Farmhelper models A and B:

	Year 1	Year 2	Year 3
Production A (units)	1,000	1,200	1,500
Production B (units)	500	580	700
Labor-hours A	330	360	420
Labor-hours B	60	65	70
Machine-hours A	100	120	150
Machine-hours B	110	116	125

a. What planning factors should it use for year 4?
b. What capacity requirements for labor- and machine-hours would you project for year 4 if 50 percent of the labor- and machine-hours each were worked in departments 101 and 102? Use the quarterly summaries in the following master schedule to do the projections:

	Year 4 quarter				
	1	2	3	4	Total
Product A	500	800	200	500	2,000
Product B	200	100	300	200	800

2. Tom Swift, master scheduler at Grove Manufacturing Company, prepared the following master production schedule for one of the firm's major end products, the 101 Spray Gun:

Week #	1	2	3	4	5	6
MPS	100	200	—	120	80	240

Tom is concerned about this schedule's impact on the Final Test Department. The Final Test Department manager has indicated that testing each 101 spray gun requires one 10th of an hour of skilled labor capacity.

   a. Prepare a rough-cut capacity analysis for the Final Test Department using the bill of capacity technique.

   b. What are the major advantages and disadvantages of the bill of capacity technique?

3. Sarah Reed, master scheduler at Walnut Hill, has developed a master production schedule for the XYZ boom box:

Week	MPS
1	250
2	400
3	575
4	980

The XYZ is fabricated in several departments, but the circuit board department and assembly areas are the potential bottlenecks. Capacity is 65 hours per week in circuit board and 80 hours in assembly. Each boom box takes 0.1 hours in circuit board and 0.2 hours in assembly. Prepare a rough-cut analysis of capacity for the MPS.

4. Bray Manufacturing makes a gizmo, which takes one half hour to assemble plus one hour of welding time for the parts. Bray also makes thingamajigs, which require one hour of assembly and two hours of welding. What are the capacity requirements in assembly and welding for the following MPS?

	Quarter			
	1	2	3	4
Gizmos	250	400	300	700
Thingamajigs	480	450	150	700

5. Management at the Green Valley Furniture Company has just approved the following master production schedule for its make-to-stock products:

End product	Week 1	Week 2	Week 3	Week 4
A	30	0	0	10
B	0	40	36	0
C	10*	0	0	30

*The remaining 10 units from a batch of 30 started last week.

The production control manager is concerned about capacity requirements for one of the automatic machines in the firm's wood shop—the #10 molder whose fixed capacity is 40 hours per week. The manager has prepared a rough-cut capacity plan for the #10 molder using a resource profile.

	Hours per unit of end product produced	
Product	#10 molder	Lead time offset
A	0.1	1 week
B	1.5	1 week
C	0.1	1 week

Weekly forecast of final product sales	
Product	Forecast (units/week)
A	10
B	25
C	15

Estimated capacity requirements per week (#10 molder)

Product	Forecast	Resources/unit	Capacity requirements
A	10	0.1 hour	1.0 hours/week
B	25	1.5 hour	37.5 hours/week
C	15	0.1 hour	1.5 hours/week
			40.0 hours/week

Evaluate the rough-cut capacity planning procedure used by the production control manager.

6. The Ticky Tacky Knickknack Company produces a knickknack shelf from two end panels, three shelves, fasteners, and hangers. The end panels and shelves have the following data and each requires a setup:

End panel			
Operation	Machine	Run time	Setup time
1	Saw	5 min.	.8 hr.
2	Planer	2 min.	.1 hr.
3	Router	3 min.	.8 hr.

Shelf			
Operation	Machine	Run time	Setup time
1	Saw	2 min.	.3 hr.
2	Molder	3 min.	1.2 hrs.
3	Router	4 min.	.7 hr.
4	Sander	1 min.	.1 hr.

The Knickknack cabinet master schedule for the next three weeks is 25, 40, and 10 units, respectively. There's a setup for each MPS quantity.

a. What's the total number of hours required on each of the five machine centers for this three-week master schedule?

b. If each of the two parts is started into production one week (five days) before needed in assembly, and it takes one day per operation, generate the week-by-week load on the routing machine.

c. If, in question b, the two parts were to arrive at the router on the same day, which would you process first? Why?

7. In jolly England, a specialized company grew pure germanium crystals for its medical and atomic research instruments. It took about 24 hours for the firm's only crystal puller to grow one usable centimeter of product for the fabrication departments. Consequently, the firm used a two-week lead time to produce a "batch" of 10 usable centimeters of crystal. The current MRP record for the crystals is as follows:

		Week					
		1	2	3	4	5	6
Gross requirements		2	8	5	8	6	5
Scheduled receipts			10				
Projected available balance	2	0	2	7	9	3	8
Planned order releases		10	10		10		

Q = 10; LT = 2; SS = 2.

a. There was real concern whether the puller's capacity had been managed correctly, especially since the firm was having difficulty meeting customer delivery date promises. The puller machine engineers had said the machine was capable of pulling two usable centimeters a day (24 hours) while operating. The company engineers had said it wasn't correct to count on the "theoretical" capacity but to use 75 percent as the expected output. On the other hand, only over the past few months had the company been able to consistently get one usable centimeter per 24-hour period. Which capacity value do you think should be used? Why?

b. Given the preceding record, what are the capacity requirements over the next five weeks? (You can assume the open order has been in progress for almost a week and is about five usable centimeters.) How do they compare to the three possible capacity measures? What advice can you give the firm's management?

8. The Single Square Company summarizes the capacity requirements for three of its key resources from each of its three product lines. A typical report (before any action is taken) is as follows:

Key resource type			Percent capacity by line				
Machine	Number	Shifts	A	B	C	Total	Remarks
Drill 1	1	3	22	13	24	59	
Drill 2	2	3	36	48	36	120	
Drill 3	1	1	52	24	21	97	
Filer 1	2	3	2	0	14	16	
Filer 2	3	3	12	20	8	40	
Dryer 1	2	2	28	36	52	116	
Dryer 2	2	2	72	51	43	166	
Dryer 3	2	1	84	27	24	135	

a. What actions would you recommend? (Assume each machine type is equivalent in terms of capacity.)

b. What other observations would you have for management, based on the preceding report?

9. The following partial input/output report was prepared at the end of week 6 for work center 101 at the Benton Plastics Company:

Week	1	2	3	4	5	6

	1	2	3	4	5	6
Planned input*	40	40	40	40	40	40
Actual input*	48	49	42	40	38	32
Cum. deviation						

	1	2	3	4	5	6
Planned output*	40	40	40	40	40	40
Actual output*	45	44	43	42	41	40
Cum. deviation						

Actual backlog						

*In standard hours.
Beginning backlog = 10 hours

The company would like to maintain the backlog between 10 and 20 hours at this work center.

a. Complete the input/output report for this work center.

b. What recommendations would you make to the manufacturing manager regarding this work center? Why?

10. Roy Harris is evaluating the palm line at the end of week 5 with input/output analysis. Complete the input/output analysis and describe any problems you see.

	1	2	3	4	5
Planned input	65	70	75	80	80
Actual input	63	63	66	67	67
Cumulative deviation					

	1	2	3	4	5
Planned output	75	75	75	75	75
Actual output	70	69	65	68	70
Cumulative deviation					

Actual backlog	20				

11. The finishing department of the Ragged Edges Company has just upgraded the finishing machine. The firm hoped the capacity would now be 20 units a day. The machine has been tested one full day. Output fell two units short of what was planned, while input was exactly as planned. To evaluate the machine's performance, Ragged Edges used a spreadsheet program to create an input/output control report five days later, starting with the deviations and backlog (5 units) from the end of the first day. The firm hoped to maintain a 4-unit backlog.

    a. What would the input/output control report look like if the firm planned inputs of 23, 14, 17, 20, and 22 units for each of the five days, and actually did input 22, 15, 21, 23, and 19 units, while output was 22, 20, 20, 18, and 21 units? What observations do you have?

    b. Suppose the actual input had been 25 units on the last day. What would the report look like (assuming no other changes)? Would your comments change? What if the department reported backlog (instead of output) and told you that the actual backlog was zero for each of the past four days?

    c. What would the input/output control report be if the input had actually been 14, 18, and 20 in the past three days? What observations can you make?

12. Suppose, in the Twin Disc example of Figure 4.15, it was decided to move all production for product lines F and I from the Maag grinder (CEA) to the Reishauer grinder (CAB). What's the resulting total percentage of capacity for each machine and the amount for product lines F and I? (Note the Reishauer is three times faster than the Maag grinder so the Maag takes three times as many hours to complete a job.) Assume setup time is negligible.

13. Applicon has the following capacity bills for its popular items 207 and 208:

207		208	
Work center	Hours/unit	Work center	Hours/unit
ALF-A	0.5	ALF-T	0.3
HLT-A	1.0	HLT-T	0.8
MIS-A	0.8	MIS-T	0.6
MVX-A	0.8	MVX-T	0.5
PCB-A	0.5	PCB-P	0.9
PCB-P	1.0	PCB-T	1.4

    a. It's now June 15. An important customer wants to know if Applicon can deliver 100 units *each* of items 207 and 208 during the next month (20 working days). Assume there are adequate materials, work center MVX-T's crew has been increased to one full person (making standard capacity = 160), the increase has come by reducing MVX-A to 6.5 persons, and no other orders have been booked since June 12. Use the standard capacity data and conditions in Figure 4.16 as the basis for your analysis.

    b. Suppose the customer in part a decided to delay the order for several months. How many units of item 207 alone could be added to the MPS in the next month? How about item 208?

14. Instant Antiques produces a series of antique knickknacks carefully mounted on a knickknack shelf. One component of the knickknack shelf is an ornate mounting bracket, which the firm manufactures. The bracket requires forming, cleaning and inspection, welding, and another cleaning and inspection. Two weeks of lead time are allowed. The first two operations (forming and the first cleaning and inspection) are normally done in the first week and the remaining operations in the second. The company is anxious to try capacity requirements planning to help them manage the shop.

The MRP record for the mounting bracket for the next few weeks is shown below, along with capacities and operations times (in hours) for each operation on the mounting bracket. In addition, loads on each of the three manufacturing areas from other items in the shop are given for the next four weeks. A report from the shop floor says the open order (to be finished this week for use next week) is a little behind. It has gone through forming but is waiting to be cleaned and inspected.

Mounting bracket		Week 1	2	3	4	5	6
Gross requirements		2	17	15	12	6	15
Scheduled receipts			20				
Proj. avail. balance	5	3	6	11	19	13	18
Planned order releases		20	20		20		

Q = 20; LT = 2; SS = 2.

Operations times and capacities (in hours) are as follows:

	Setup	Hours/unit	Capacity
Forming	1.00	1.00	35
Cleaning and inspection–1	0.50	0.25	40
Welding	0.50	0.60	35
Cleaning and inspection–2	0.20	0.50	40

Hours of work from other jobs in shop are:

		Week			
	Past due	1	2	3	4
Forming	10.00	15.00	5.50	24.00	6.00
Welding	0.00	20.00	24.00	14.50	22.00
Cleaning and inspection	10.50	31.00	27.00	14.00	16.50

a. What's the total capacity required in forming, cleaning and inspection, and welding during the next four weeks? What percentage of the capacity available does it represent?

b. Management has gotten some customers to agree to delay receipt of their goods. The efforts resulted in a shift of 7 units of gross requirements from week 2 (from 17 to 10) to week 5 (from 6 to 13) of the mounting bracket. What does this do to capacity utilization in each shop?

c. Still not satisfied, management decided also to increase the lot size from 20 to 40, since the managers understood this would save on capacity requirements. What's the effect of this change?

d. Management is interested in evaluating its capacity under different scenarios of demand. Use a spreadsheet program to evaluate capacity implications for forming, cleaning and inspection, and welding if gross requirements are 10 percent greater in each week. (Round up to the next integer value in all cases.) Repeat the analysis for 20 percent, 30 percent, and 50 percent.

e. Redo part d but add to the analysis similar increases in capacity requirements for work coming from other jobs in the shop; that is, now assume that capacity increases are general across all Instant Antiques' products.

15. The machine shop at the Northern Steel Company consists of the following equipment:

Plate shear (PS)   One shear for cutting steel plate up to ⅝″ thick.

Burnout table (BO)   One burnout table for cutting shaped parts out of steel plate using a template.

Deburring (DB)   One machine for removing sharp corners and rough edges from metal parts.

Milling (ML)   One machine for producing finished surfaces on metal parts.

Grinding (GR)   One grinding machine for producing precision surfaces on flat metal parts.

Welding (WD)   One arc welding machine for producing fabricated subassemblies from metal parts.

Orders this shop receives are processed through these work centers in different sequences, depending on the specifications of the part to be manufactured. All orders involve taking material, such as bar or plate, from the warehouse inventory and processing this material to meet the customer's specifications. Some orders involve cutting metal parts out of steel plate on the burnout table, removing sharp corners and rough edges, and grinding a precision finish on the part's top and bottom surfaces. Still other orders involve cutting steel plate into individual pieces on the plate shear, removing the sharp corners, and machining a finished edge on the part, using the milling machine. Finally, some orders involve producing metal parts at the shear, milling, and grinding machines and then welding these parts into fabricated subassemblies.

a. The sample of orders in Exhibit A can be divided into two types: orders that are first processed at the plate shear (PS) and orders that begin processing at the burnout table (BO). Using this sample of orders, prepare a capacity bill for each type of order, indicating the work centers required for processing and the processing time (in hours per order) for a typical customer order in each order type.

b. Currently, the machine shop anticipates the incoming order rate to be at a level of 50 orders per week over the next six weeks. Sixty percent of these orders are forecast to be plate shear orders and the remainder burnout orders. Using the capacity bill prepared in part a, estimate weekly capacity requirements for each work center in the shop. Enter your results in the capacity planning worksheet in Exhibit B. What are your conclusions?

c. At the start of the current week, the machine shop scheduler received a request from marketing to schedule a special rush order of 180 welded assemblies for the Ajax Company. These welded assemblies are to be delivered in three lots at 60 units each during the next three weeks and are in addition to the regular orders forecast in part b. The Ajax order would require processing at the plate shear, milling, and welding work centers. Processing time per unit for this order at each operation is: plate shear, .2 hours/unit; milling, .2 hours/unit; and welding, .6 hours/unit. Where is there insufficient capacity to schedule the production of the Ajax order and to meet the forecast requirements? What should be done to provide the additional capacity?

EXHIBIT A Sample Orders

JOB: 1	Due date: Monday	
Operation	Machine	Hours
1	PS	1

JOB: 11	Due date: Tuesday	
Operation	Machine	Hours
1	BO	2

JOB: 2	Due date: Monday	
Operation	Machine	Hours
1	BO	1

JOB: 12	Due date: Tuesday	
Operation	Machine	Hours
1	PS	2

JOB: 3	Due date: Monday	
Operation	Machine	Hours
1	PS	2

JOB: 13	Due date: Tuesday	
Operation	Machine	Hours
1	BO	2

JOB: 4	Due date: Monday	
Operation	Machine	Hours
1	BO	2

JOB: 14	Due date: Tuesday	
Operation	Machine	Hours
1	PS	3

JOB: 5	Due date: Monday	
Operation	Machine	Hours
1	PS	1

JOB: 15	Due date: Wednesday	
Operation	Machine	Hours
1	BO	2
2	DB	2
3	ML	2

JOB: 6	Due date: Tuesday	
Operation	Machine	Hours
1	PS	1
2	WD	6

JOB: 16	Due date: Wednesday	
Operation	Machine	Hours
1	PS	3
2	DB	2
3	GR	4

JOB: 7	Due date: Tuesday	
Operation	Machine	Hours
1	BO	2
2	DB	3
3	ML	7

JOB: 17	Due date: Wednesday	
Operation	Machine	Hours
1	PS	3
2	DB	2
3	GR	5
4	WD	5

JOB: 8	Due date: Tuesday	
Operation	Machine	Hours
1	PS	2
2	GR	9
3	WD	5

JOB: 18	Due date: Wednesday	
Operation	Machine	Hours
1	PS	2

JOB: 9	Due date: Tuesday	
Operation	Machine	Hours
1	BO	3
2	DB	1
3	ML	4
4	GR	5

JOB: 19	Due date: Wednesday	
Operation	Machine	Hours
1	BO	2

JOB: 10	Due date: Tuesday	
Operation	Machine	Hours
1	PS	3

JOB: 20	Due date: Wednesday	
Operation	Machine	Hours
1	PS	2

EXHIBIT A   *(concluded)*

JOB: 21	Due date: Wednesday		JOB: 24	Due date: Thursday	
Operation	Machine	Hours	Operation	Machine	Hours
1	BO	2	1	BO	2
			2	DB	1
			3	WD	3

JOB: 22	Due date: Wednesday		JOB: 25	Due date: Thursday	
Operation	Machine	Hours	Operation	Machine	Hours
1	PS	1	1	BO	3
			2	DB	1
			3	ML	3

JOB: 23	Due date: Wednesday		JOB: 26	Due date: Thursday	
Operation	Machine	Hours	Operation	Machine	Hours
1	BO	2	1	PS	3
2	DB	1	2	GR	4
3	ML	3			

JOB: 27	Due date: Wednesday	
Operation	Machine	Hours
1	BO	1

EXHIBIT B   Capacity Planning Worksheet

	Projection
Orders	
Shear orders (60%)	
Burnout orders (40%)	
Capacity	
Plate shear (80 hrs/wk)*	
Burnout table (40 hrs/wk)	
Deburring (40 hrs/wk)	
Grinding (40 hrs/wk)	
Welding (40 hrs/wk)	
Milling (40 hrs/wk)	

*Capacity available on the plate shear is 80 hrs./wk.

Chapter 5

Production Activity Control

This chapter concerns the execution of detailed material plans. It describes the planning and release of individual orders to both factory and outside vendors. Production activity control (PAC) also concerns, when necessary, detailed scheduling and control of individual jobs at work centers on the shop floor, and it concerns vendor scheduling. An effective production activity control system can ensure meeting the company's customer service goals. A PAC system can reduce work-in-process inventories and lead times as well as improve vendor performance. A key element of an effective PAC system is feedback on shop and suppliers' performance against plans. This loop-closing aspect provides signals for revising plans if necessary.

The chapter is organized around four topics:

- A framework for production activity control: How does PAC relate to other aspects of material planning and control and how do just-in-time production or individual firm decisions impact PAC system design?
- Production activity control techniques: What basic concepts and models are used for shop-floor and vendor scheduling and control?
- Production activity control examples: How have PAC systems been designed and implemented in several different kinds of companies?
- The production activity control data base: What data elements are necessary to support a PAC system?

Chapter 5 is linked closely to Chapter 2 in that many PAC techniques are designed to execute the detailed material plans produced by material requirements planning (MRP) systems. Much of the detail order tracking of PAC is not required in a just-in-time (JIT) environment, so many of the appropriate JIT shop-floor systems are described in Chapter 3. Chapter 13 deals with advanced scheduling techniques. Chapter 19 describes the optimized production technology (OPT) approach to manufacturing planning and control with its impact on shop-floor operations.

A FRAMEWORK FOR PRODUCTION ACTIVITY CONTROL

Production activity control (PAC) concerns execution of material plans. It encompasses activities within the shaded areas of Figure 5.1. The box entitled "Shop-floor scheduling and control," which we refer to as shop-floor control, falls completely within PAC. Vendor scheduling and follow-up is depicted as largely being part of production activity control, but not completely. Many firms, particularly those with JIT material control approaches,

FIGURE 5.1 Production Activity Control in the MPC System

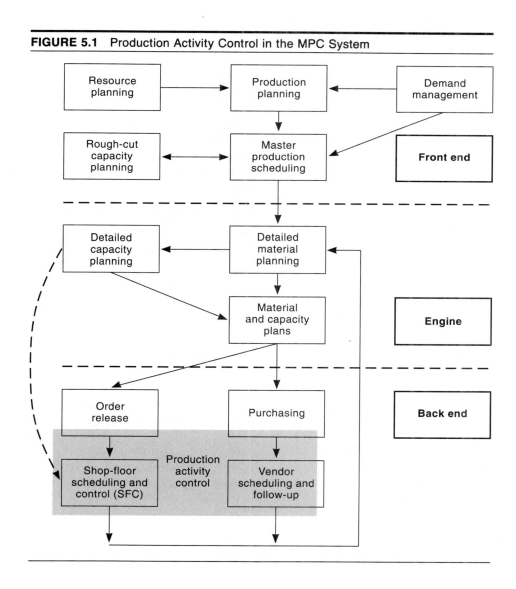

assign most vendor scheduling to PAC. Order release (which authorizes release of individual orders to the factory and provides accompanying documentation) is similarly becoming more a part of PAC. In purchasing, **procurement** is seen as a professional activity where information networks, relationships, terms, and conditions are established with vendor companies outside of PAC, while release of individual orders and follow-up activities are a part of PAC.

The extension of the definition of *production activity control* is accentuated by the growing use of computers on the shop floor and **electronic data interchange (EDI)** with vendors. As more and more traditional staff work is integrated into the basic manufacturing infrastructure, it will expand PAC as well.

MPC System Linkages

The primary connection between PAC and the rest of the MPC systems shown in Figure 5.1 comes from the box marked "Material and capacity plans." The capacity plan is especially critical to managing the detailed shop-floor flow of materials. In essence, the capacity provided represents resource availabilities for meeting material plans.

Capacity's importance for shop-floor control (SFC) is illustrated by considering two extremes. If insufficient capacity is provided, no SFC system will be able to decrease backlogs, improve delivery performance, or improve output. On the other hand, if more than enough capacity exists to meet peak loads, almost *any* SFC system will achieve material flow objectives. It's in cases with bottleneck areas and where effective utilization of capacity is important, that we see the utility of good SFC systems.

A related issue is the extent to which good capacity planning is done. If the detailed capacity planning activity in Figure 5.1 provides sufficient capacity, with relatively level loading, shop-floor control is straightforward. On the other hand, if peaks and valleys in capacity requirements are passed down to the back end, execution becomes more complex and difficult. The same general issues apply to vendor follow-up systems: vendor capacity must be carefully planned to ensure effective execution.

The material plan provides information to the SFC and vendor follow-up systems and sets performance objectives. The essential objective of both execution systems is to achieve the material plan—to provide the right part at the right time. This will result in being able to hit the master production schedule and to satisfy customer service objectives.

The Linkages between MRP and PAC

The shop-floor and vendor scheduling activities begin when an order is released. A critical information service provided by MRP is appraising the SFC systems of all changes in material plans. This means revising due dates and quantities for scheduled receipts so correct priorities can be maintained. The job thereafter might be likened to a duck hunter following a moving target. Control and follow-up systems must keep each order lined up with its due date—one that's moving—so overall MPC is supported.

Linkages between PAC and the engine aren't all one-way. There's important feedback from the shop-floor control and vendor follow-up systems to material and capacity planning. Feedback is of two types: status information and warning signals. Status information includes where things are, notification of operational completions, count verifications, order closeout and disposition, and accounting data. The warning signals help flag inadequacies in material and capacity plans: that is, will we be able to do what was planned?

Just-in-Time Impact on PAC

Shop-order–based systems are founded on the premise of job shop (now more frequently called *batch*) manufacturing where parts are routed to different parts of the factory for processing steps, with relatively long lead times, high work-in-process inventories, and high utilization of work center capacities. JIT has none of these. Manufacturing takes place in facilities, often in cells, where jobs are easily kept track of; work is completed quickly; work-in-process inventory levels are insignificant; and work centers have surge capacity or else are level loaded—in either case, capacity utilization is not a key issue.

Formal systems for shop-floor control are largely unnecessary under JIT. Release of orders is still part of PAC, but the typical "shop order" with associated paperwork isn't maintained. Therefore, the PAC functions in Figure 5.1 are greatly simplified. Order release can be accomplished with kanbans or other pull system methodologies, and work-in-process inventories in the factory are severely limited. Detailed scheduling is also unnecessary since orders flow through cells in predictable ways where workers know the sequence of conversion operations. Work is completed fast enough that "order scheduling" isn't required. Detailed scheduling of workers and equipment is similarly not an issue since design of the JIT system itself determines schedules. There's no need for data collection or monitoring since JIT basically assumes only two kinds of inventories: raw materials and finished goods. Receipt of finished goods is used to "back flush" required raw materials from inventory.

Vendor scheduling under JIT can be a bit more complex than shop-floor control, but if the relationship with the vendors is good, differences are very small. Many firms use some form of electronic kanban to authorize work at the vendor factories, and excellent vendors don't build inventories in anticipation of orders from their customers. Well-run auto companies, for example, transmit an exact build schedule to their seat vendors several times a day, as actual cars are started. By the time these cars are ready for seats to be installed, seats will be delivered by the vendor—in the exact sequence required.

The Company Environment

The primary PAC objective is managing the materials flow to meet MPC plans. In some firms, other objectives relate to efficient use of capacity, labor, machine tools, time, or materials. Under JIT and time-based competition, the objective is material *velocity*. A firm's particular set of objectives is critical to PAC design.

The choice of objectives for PAC reflects the firm's position vis-à-vis its competitors, customers, and vendors. It also reflects the company's fundamental goals and the constraints under which it operates. In many countries firms find changing capacity to be much more difficult than in the United States. This viewpoint colors the view of PAC. Similarly, some firms have much more complex products and/or process technologies than others. The result can be a much more difficult shop-floor management problem and a resultant difference in the appropriate PAC system. As a result PAC system design must be tailored to the particular firm's needs.

PRODUCTION ACTIVITY CONTROL TECHNIQUES

This section begins by describing basic concepts for production activity control for a job shop manufacturer with an MRP system. It covers priorities, loading of a particular job onto a machine center, elements of lead time, and data inputs. It then examines three approaches to shop-floor control. The first, **Gantt charts,** provides graphic understanding of the shop-floor control problem; moreover, Gantt chart models can be used in manual shop-floor control systems. The second approach is based on **priority sequencing rules** for jobs at a work center. Under the third approach to shop-floor control, **finite loading,** an exact schedule of jobs is prepared for each work center. We next look at vendor scheduling where the concepts are applied to supplier operations. To close the section, we comment on lead time management.

Basic Shop-Floor Control Concepts

Figure 5.2, an example product structure for end item A, demonstrates basic concepts underlying shop-floor control techniques. One essential input to the SFC system is the routing and lead time data for each piece. Figure 5.3 presents this for parts D and E of the example. The routing specifies each oper-

FIGURE 5.2 Example Product Structure Diagram

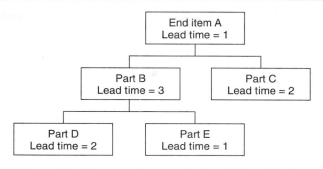

ation to be performed to make the part and which work center will perform the operation.

Production of part D, for example, requires three operations of 4, 5, and 1 days, respectively, for a total of 10 days, or two weeks. Part E requires four operations of 1, 1, 2, and 1 days, respectively, for a total of 5 days, or one week. The remaining lead times in Figure 5.2 are all derived the same way. Lead times used for MRP should match those in the routing file. If the MRP time for part E was set at two weeks instead of one week, orders would constantly be released one week early.

Lead times are typically made up of four elements:

Run time (operation or machine run time per piece × lot size).

Setup time (time to prepare the work center—independent of lot size).

Move time (from one work center to the next).

Queue time (time spent waiting to be processed at a work center, which depends on workload *and* schedule).

Queue time (the critical element) frequently accounts for 80 percent or more of total lead time; it's the element most capable of being managed. Reducing queue time means shorter lead time and, therefore, reduced work-in-process inventory. This reduction requires better scheduling.

The bottom of Figure 5.3 shows an **operation set-back chart** based on each part's lead times. Here we clearly see the implications of incorrect MRP lead time. If the MRP lead time for part E isn't the one week calculated from the routing data, the part will be released either early or late to the shop. Neither of these is a desirable outcome. Note that Figure 5.3 shows that both parts D and E go through work center 101 for their first operation. The top of Figure 5.4 shows the partial schedule for work center 101, with parts D and E scheduled according to the timing in Figure 5.3.

FIGURE 5.3 Routing Data and Operation Setback Chart

Part D routing

Operation	Work center	Run time	Setup time	Move time	Queue time	Total time	Rounded time
1	101	1.4	.4	.3	2.0	4.1	4.0
2	109	1.5	.5	.3	2.5	4.8	5.0
3	103	.1	.1	.2	.5	.9	1.0

Total lead time (days) 10.0

Part E routing

Operation	Work center	Run time	Setup time	Move time	Queue time	Total time	Rounded time
1	101	.3	.1	.2	.5	1.1	1.0
2	107	.2	.1	.3	.5	1.1	1.0
3	103	.3	.2	.1	1.5	2.1	2.0
4	109	.1	.1	.1	.5	.9	1.0

Total lead time (days) 5.0

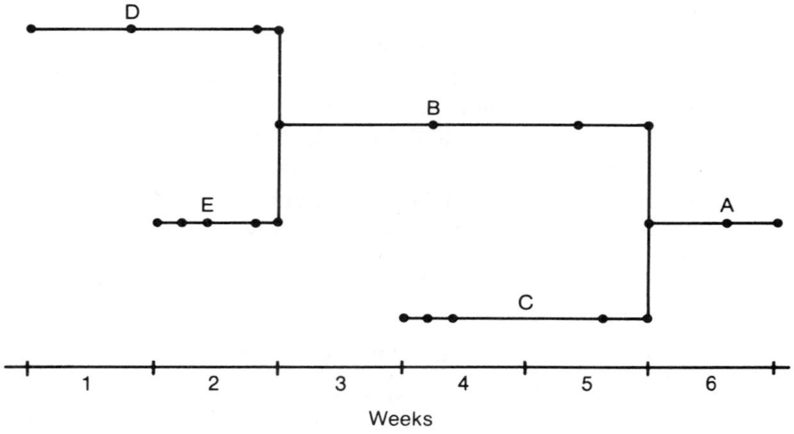

The bottom of Figure 5.4 shows two alternative detailed schedules for part D in week 1 at work center 101. The cross-hatched portion represents the 1.8 days of lead time required for setup and run time. The early schedule has part D loaded as soon as possible in the four days. The late schedule loads part D into the latest possible time at work center 101.

The key differences between the top and bottom of Figure 5.4 are in lead times. The top half includes queue time. Queue time represents slack that permits the choice of alternative schedules—a form of flexibility. This slack can be removed by good SFC practice; that is, this schedule allows 4 full days

FIGURE 5.4   Work Center 101 Schedules

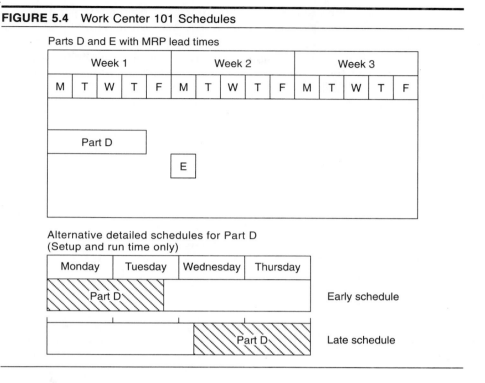

Parts D and E with MRP lead times

	Week 1					Week 2					Week 3			
	M	T	W	T	F	M	T	W	T	F	M	T	W	T	F

Part D

E

Alternative detailed schedules for Part D
(Setup and run time only)

Monday	Tuesday	Wednesday	Thursday

Part D — Early schedule

Part D — Late schedule

to complete part D, when actual time on the machine is only 1.8 days. For the remaining 2.2 days, the part waits in a queue or is moving between work centers.

The bottom of Figure 5.4's detailed schedules contain no queue time. These schedules represent loading a particular job onto a particular work center for a particular time period. The two alternatives in the bottom of Figure 5.4 are different loadings; you typically choose between alternative loadings to utilize the machine center effectively.

Gantt Charts

Gantt or **bar charts,** like those in Figure 5.4, show a schedule. The operation setback chart in Figure 5.3 is very similar. It too is a schedule for when to make each of the five parts based on lead times that include move and queue times.

One form of shop-floor control is to prepare operation setback charts similar to Figure 5.3 for each job, and use them plus the kind of data in Figure 5.3 to prepare Gantt charts, such as those in Figure 5.4; that is, the objective is to prepare a schedule for each machine center. This schedule can be based on the assumptions in either the top or bottom of Figure 5.4; that is, the schedule may or may not use lead times that include queue and move times.

The more usual practice is to prepare the detailed work center schedule *without* move and queue times. Many firms' systems do this. The typical approach is a **schedule board** with racks to hold pieces of paper. Each paper is a job and its length represents the time required.

The primary problem with this kind of system is updating. Actual data must be captured and integrated into an ongoing replanning cycle. Moreover, a means to communicate with the shop floor is usually required since schedule boards typically reside in planning offices. However, with personal computers on the shop floor, some firms have in essence created a fairly dynamic version of the "schedule board."

Priority Sequencing Rules

Priority sequencing rules concern which job to run next at a work center. To some extent, these rules can be seen as producing a loading of jobs onto individual machines, but usually only one job is committed at a time; that is, the job to run *next* is only determined near the time when the prior job has been completed. The priority (sequencing) rule is just what the name suggests: a *rule* for what job to process next.

Many different priority rules have been established. A fairly common one is to base priorities on the type of data in Figure 5.3. The lower half of that figure contains scheduled due dates for parts and operations. These due dates can be used as priorities. For example, a priority rule could be: the job to process next is the job with the earliest operation due date. An alternative is to next process the job with the earliest *part* due date. Four other commonly used sequencing rules are:

- *Order slack:* Sum the setup times and run times for all remaining operations, subtract this from the time remaining (now until the part due date), and call the remainder slack. The rule is to work on that job with the least slack. This rule addresses the problem of work remaining.
- *Slack per operation:* A variant of order slack is to divide the slack by the number of remaining operations, again taking next the job with the smallest value. The reasoning behind slack per operation is that it will be more difficult to complete jobs with more operations because they will have to be scheduled through more work centers.

- *Critical ratio:* A rule based on the following ratio:

$$\frac{\text{Time remaining}}{\text{Work remaining}}.$$

For calculation, the rule is expressed as

$$\frac{\text{Due date} - \text{Now}}{\substack{\text{Lead time remaining (including setup,} \\ \text{run, move, and queue)}}}.$$

If the ratio is 1.0, the job is on time. A ratio below 1.0 indicates a behind-schedule job, while a ratio above 1.0 indicates an ahead-of-schedule condition. The rule is to always process that job with the smallest critical ratio next.
- *Shortest operation next:* This rule ignores all due date information as well as all information about work remaining. It simply says, take as the next job the one that can be completed in the shortest time at the work center. This rule maximizes the number of shop orders that go through a work center—and minimizes the number waiting in queue.

In an MRP system, each shop order would be a scheduled receipt for the part. As such, the scheduled receipt has a due date. From this due date, operational due dates could be established by backing off expected operation times, if these data are needed to establish priority sequence. The great advantage of this computer-based system is that, whenever the due date for a scheduled receipt changes, operation due dates can be changed accordingly. These changes, in turn, lead to priority changes for shop-floor control, resulting in an execution system that works on the most-needed shop orders first. The objective is for high-priority jobs to move through the shop very quickly, while low-priority jobs are set aside. In this way, the shop-floor control system can indeed execute the dictates of the detailed material plan.

In recent times, many companies have developed a preference for sequencing rules that are easy to understand. One straightforward approach is to develop operation start and operation due dates, and use them for determining priority sequence decisions.

In a computer-based shop-floor control system, due dates wouldn't be printed on any shop paper that travels with the work-in-process inventory. The shop paper would show the routing or sequence of operations (static data), but no due dates. The changing (dynamic) due date information would be printed daily or be displayed on-line in the form of a work center schedule or dispatch list. It's the dispatch list, not the traveling paper, that shows the priority sequence. The dispatch list can be updated as rapidly as transactions are processed to the MRP data base.

Finite Loading

Finite loading systems simulate actual job order starting and stopping to produce a detailed schedule for each shop order and each work center; that is, finite loading does, in fact, *load* all jobs in all necessary work centers for the length of the planning horizon. The result is a set of start and finish dates for each operation at each work center. This schedule is based on the finite capacity limits at each work center. In essence, it's the equivalent of one of the alternative schedules in the bottom of Figure 5.4, but includes *all* the jobs scheduled for that work center.

Finite loading explicitly establishes a detailed schedule for each job through each work center based on work center capacities and the other scheduled jobs. Priority sequencing rules don't consider work center capacities or other jobs to be scheduled. For this reason, priority rules are employed with capacity planning techniques based on infinite loading, techniques like resource profiles, or capacity requirements planning.

Figure 5.1 shows the relationship between detailed capacity planning and shop-floor scheduling and control systems. The dashed line is for finite loading. That is, data similar to those used for capacity requirements planning (CRP) are used to schedule detailed shop operations—with finite loading techniques.

Figure 5.5 contrasts finite loading and infinite loading. The top shows infinite loading for work center 101 as it might be produced by CRP. The lower half of Figure 5.5 shows a capacity profile resulting from finite loading. This result (the capacity profile) is *not* the objective of finite loading; it's the result of scheduling each job through work centers, but never scheduling more work in a center than the center's capacity.

The finite loading approach will only schedule work up to a work center's capacity. Thus, the 75 hours of work shown as past due in the top of Figure 5.5 would be scheduled in week 1 under finite loading techniques. Finite loading doesn't solve the undercapacity problem in the top of Figure 5.5. If capacity isn't increased, only 80 hours of work will come out of this work center each week, regardless of the scheduling procedure. Finite loading will determine *which* jobs will come out, based on priorities.

One output of finite loading is a simulation of how each machine center operates on a minute-by-minute basis for whatever time horizon is planned. Suppose we want to finite load work center 101 beginning Monday morning of week 1. A job is already in process and 150 pieces are left with a standard time of one minute per piece. This order consumes the first 150 minutes of capacity; if work starts at 8 A.M., the machine is loaded until 10:30 A.M. The finite loading system then selects the next order to schedule on the machine, taking account of setup time and run time. This process is repeated to simulate the entire working day, then the next day, and so on.

FIGURE 5.5 Infinite versus Finite Loading (CRP Profile for Work Center 101)

CRP profile for work center 101

Finite load capacity profile

◤ Open shop orders

☐ Planned orders

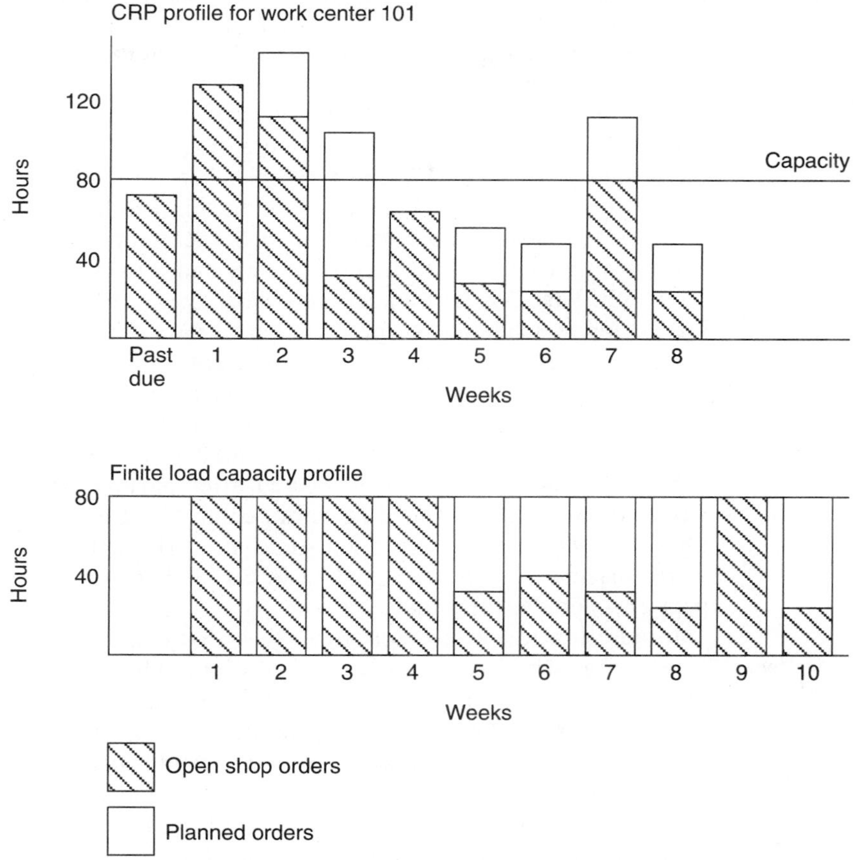

Selection of the next job to schedule is not just based on those jobs physically waiting at the work center. Most finite loading systems look at jobs coming to the work center, when they'll be completed at the prior centers, and these jobs' priorities to decide whether to leave the work center idle and immediately available for a particular job's arrival. Also, some systems overlap operations, so a job might start at a work center before completion of all work at the prior center.

The approach of filling a work center job by job is called **vertical loading.** It's consistent with how most job shop scheduling research is conducted, as well as with the priority scheduling viewpoint; that is, we look at a work center, such as our example center 101, and decide which of a set of jobs (both those at the work center and those scheduled to have had their prior operations processed by the time of the loading) to load next. Another approach is often used for finite loading. It's called **horizontal loading.** In this case, one entire shop order or job (that job with the highest priority) is loaded for all its operations, then the next highest priority job, and so on. This horizontal approach may well be at odds with a criterion of using a machine center to its maximum capacity. By creating detailed schedules with horizontal loading, the net result can be "holes" in the capacity for a work center. What this implies is that at times a machine is to sit idle, even if a job is available, because a more important job is coming.

Returning to part D in Figure 5.3, we can illustrate the horizontal loading approach. Part D is first loaded onto work center 101 whenever the previously loaded job is scheduled to be completed. Part D is scheduled for completion 1.8 days later. At that time, it's loaded onto work center 109, again as soon as this work center is available. If the prior job at work center 109 is already completed, this work center remains idle until part D is completed at work center 101. Finally, part D is similarly loaded onto work center 103.

Suppose the next job to be loaded after part D was for part Q, and the work on part Q commenced in work center 109. Let's also say that, in scheduling part D, the result was idle time at work center 109. *If* (and only if) it were possible to schedule part Q into that idle time *without* disturbing the part D schedule, this would be done.

In addition to the horizontal–vertical distinction, there's also the issue of **front scheduling** versus **back scheduling.** The back-scheduling approach would take the job backward from its due date. If the resultant schedule indicates the *start* date has already passed, we have an infeasible schedule for that job. The front-scheduling approach would load the order in as soon as capacity was available in each work center, and thereby determine the expected *completion* date. If the date is beyond the due date, infeasibility is again indicated. The net result is the need for corrective action.

In either horizontal or vertical loading, there's always the choice of which job to schedule next at a work center. The computer uses priority information somewhat like that used in priority scheduling. The same data (due dates, lead times, and the like) are available, but most finite loading systems can also use managerial weighting factors to augment the other data. These factors are typically used to reflect the particular orders' urgency, the customer's importance, the nature of the work centers through which the job will travel, and so on.

If the finite loading system were run on a Sunday night, schedules set for a particular work center for Monday should be reasonably accurate. Part of a particular machine center's capacity may already be allocated to a job in process, and perhaps the other jobs to be scheduled are already waiting in queue at the machine center. However, schedules for Tuesday must deal with Tuesday's expected starting conditions. Moreover, jobs scheduled for Tuesday are more likely to come from another work center, rather than already being at the work center. Differences between simulated operations and actual operations in those centers will be reflected in the actual job arrivals at the work center. These uncertainties on the shop floor mean there will be a decay in information validity that grows as the schedule's time horizon is lengthened.

One way to overcome this decay is to redo the planning (finite loading) more often. If it's done each night, the next day's plan will be more accurate. Even with today's computer speeds, the problem with doing finite loading every night is the cost. Finite loading can easily take 10 times as much computer run time as priority scheduling. However, new finite loading systems are being offered, many able to be run at work stations on the shop floor.

Because finite loading only schedules a work center up to its capacity (and tries to schedule it up to the limit), matching parts will often have inconsistent due dates. In our example, assembly A is made up of parts B and C. B and C's due dates would be identical under MRP. If part B is delayed two weeks at some work center, there's no reason to rush completion of part C. Some newer approaches to finite scheduling can partially take into account these product structure relationships.

Proponents of finite loading instead of CRP point to its better forecasts of the actual load on each machine center in the near term, say the next few weeks. Although there will be inaccuracies, by simulating the exact flow of orders, the result should be far superior to that based on the average queue times of MRP. Moreover, the finite loading system will be better than priority scheduling systems for looking out a week or two, because finite systems schedule orders one or more work centers away as well as those already at the work center.

The **optimized production technology (OPT)** approach enables us to combine MRP priorities and finite loading. All jobs are sorted based on whether or not they use "bottleneck" work centers. (A bottleneck work center is one loaded to a high level of its capacity.) Small subsets of jobs that cross bottlenecks are finite loaded, while most orders are scheduled using MRP due dates and priority approaches. The inconsistency between MRP due dates and those of finite loading is much less severe. Moreover, because OPT only finite loads a small subset of jobs, the computational burden is reduced.

Vendor Scheduling and Follow-Up

The **vendor scheduling** and **follow-up** aspects of PAC are the direct analog of the shop-floor scheduling and control systems. There are some important differences, however. From the vendor's perspective, each customer is usually only one of a number of demand sources. Customer demands are managed in the vendor's plant with its MPC system. The MPC relationship is largely through information exchanged between vendor and customer, often from the back-end activities of the customer directly to the vendor's MPC system.

From the customer's standpoint, the objectives of vendor scheduling are the same as those for internal work center scheduling: keep the orders lined up with the correct due dates from the material plan. This means the vendor must have a continually updated set of relative priority data. A typical approach to providing this information is a weekly updated report reflecting the current set of circumstances in the customer's plant and, sometimes, the final customer requirements that dictate them. Increasingly, computer-to-computer communication is used to transmit this information.

Since the vendor follow-up system is often concerned with changes to the schedule and keeping priorities correct, there must be limits to the amount of change the vendor will be asked to accommodate. Contractual agreements with the vendor typically define the types and degree of changes that can be made, time frames for making changes, additional elements of flexibility required, and so on. In addition, the agreement specifies procedures for transmitting needs to the vendor plus the units in which the vendor's capacity is planned and controlled. This sets the stage for vendor PAC including order release, scheduling, and follow-up.

Lead Time Management

Many people think of **lead time** as a constant, such as π. In fact, it's not a value to be measured as much as a parameter to be managed. Of the four elements of lead time (setup, processing, move, and queue), the last two can be compressed with good PAC design and practice.

Lead time and work-in-process (WIP) are directly related. Moreover, some critical feedback linkages operate. The longer the lead time is perceived to be, the longer the time between order launching date and due date. The longer this time, the more orders in the shop. The more orders in the shop, the longer the queue time (and WIP); we have a self-fulfilling prophecy.

Some WIP is needed at work centers where high utilization is important. However, a basic principle of MPC systems is to *substitute information for inventory*. The firm doesn't need to have jobs physically in front of machines. Orders can be held in a computer and converted to physical units only as needed. Plossl and Welch report that, for many plants, setup and run time only constitute 10 to 20 percent of total lead time. The rest is slack that can be substantially cut.

One interesting question is how to manage lead time. This means changing data base elements for both SFC and MRP. One alternative is to go through the data base and systematically change all lead times. Reducing them could result in a transient condition of dry gateway work centers. This might be a reasonable price to pay for the resulting WIP reduction.

Changing lead time data elements naturally leads to the question of how they're established in the first place. For most firms, lead time data are usually an input from some functional area, such as production control. An alternative is to *calculate* lead time. When we think about changing lead times as part of a management process, and when we remember that SFC lead time must be in tune with MRP lead time offset data, this approach has increasing appeal. One firm calculated lead times as follows:

Nonqueue time for each operation was set equal to setup plus run time (time per piece × lot size) plus move time.

Nonqueue time was converted to days by dividing total hours by (7 × number of shifts per day), assuming seven productive hours per shift.

Queue time was set equal to two days if the next work center in the routing was in another department, one day if it was in the same department but a different work center, and zero days if it was on the same machine.

Lead time for the total order was the sum of queue and nonqueue times. This time was calculated with an average order quantity, rounded up to a weekly lead time, and used for MRP lead time offsetting.

Selecting queue time is the critical element in this formula. Values were chosen by taking a sample of 50 parts and using different queue time estimates to yield lead times consistent with production control personnel opinions. The initial estimates were padded, but the company was not very concerned. Once the system was in operation, estimates for queue times were systematically reduced a bit at a time. The result was a managed approach to shorter lead times and reduced work-in-process.

Before leaving this discussion, let's look at one firm's results. David A. Waliszowski says a $25 million division of Hewlett-Packard reduced lead time 70 percent and increased customer service levels 80 percent. This amounted to a $1.7 million reduction in work-in-process inventory that was achieved in three months.

PRODUCTION ACTIVITY CONTROL EXAMPLES

This section applies production activity control techniques in several quite different examples.

The first example covers the Twin Disc Company. Twin Disc has a large number of component parts, complex product structures, long lead times, complex part commonality, and expensive work centers the firm wishes to heavily utilize. It has implemented a critical-ratio–based priority sequencing system and an analogous vendor scheduling system.

The second example is at Swissair. The Swissair engine maintenance shop has a finite loading system for shop-floor control. Each engine to be overhauled is brought in, disassembled, and inspected. Only at this time can the required work be determined. Swissair creates a unique routing file for each piece part, based on the inspection decisions. As a result, resulting capacity requirements vary widely at each work center. The firm employs a large, highly paid work force that isn't easily expanded or contracted. Moreover, overtime is more difficult to schedule than at most U.S. firms.

The third example is vendor scheduling at Steelcase. Here MRP-based planning is extended into vendor follow-up with excellent results. Finally, PAC procurement practice under "world class manufacturing" is examined in several company contexts.

Critical Ratio Shop-Floor Control at Twin Disc

Twin Disc, Inc., is a manufacturer of small-lot, heavy-duty transmission equipment. Products are designed to customer specifications, produced in small lots, and range in unit price from several hundred dollars to over $10,000. Annual sales were approximately $200 million at the time of this study, with more than one half being to four large customers.

Data base size for shop-floor scheduling at Twin Disc is quite large. Open shop orders usually number about 3,500. Each part typically passes through 10 to 15 operations; there are over 300 different machine centers. The average product is made up of approximately 200 parts, and the total part master data base includes about 60,000 separate part numbers.

Twin Disc installed the critical ratio priority system six years after implementing MRP. The MRP system was well understood at that time, and it provided the proper basis for updating priorities for individual shop orders. Now let's see how this shop-floor scheduling system works.

Figure 5.6 is the daily work center schedule. This report is printed during each night, so it's available to foremen at the beginning of every working day. The sample report in Figure 5.6 was printed for February 6 (Wednesday of manufacturing week 446). (The manufacturing calendar is based on a five-day week.) It's for the BH machine center in plant 3, department 5. This machine center works two shifts and has a weekly rated capacity of 110.9 hours.

The report is divided into two parts. The top portion shows six orders presently at the BH work center; the lower portion shows orders that are one work center away from BH; that is, the shop orders at the bottom of the page are not now physically at BH, but their routings indicate that, when they're completed at their present work center location, they'll be moved to BH for the next operation.

The first two columns on the report are the number and part name. The third column is the shop order number for the part. The fourth and fifth columns are the operation number (its routing sequence) and a description of the operation. The priority numbers in columns 6 and 7 are more easily explained after we discuss the other data. Column 8 shows the quantity associated with the shop order, while column 9 shows the quantity physically at the work center. This distinction facilitates operation overlapping. Column 10 shows the hours each shop order is expected to take in the BH center. Notice this column is totaled both for jobs in the machine center and for jobs one

FIGURE 5.6 Twin Disc Shop-Floor Control Report

DATE 02/06/   DAILY WORK CENTER JOB SCHEDULE   WEEK 446   DAY WEDNESDAY

PLANT 03   DEPT 05   MACH. CTR. BH   SHIFTS WORKED 2.0   CAPACITY 110.9

PART # (1)	PART NAME (2)	ORDER # (3)	OP # (4)	OPER DESC (5)	PO (6)	PI (7)	QTY OF OP (8)	QTY AT OP (9)	HOURS (10)	NEXT LOCATION (11)	WORK REM. (12)	TIME REM. (13)
209335H	IMP WHL	@438C34	020	FIN	.436		142	142	11.1	0316NB	18.3	8.0
216140A	SPINNER	445C22	010	TURN	1.236		88	88	6.4	0305BQ	18.6	23.0
209308C	IMP WHL	445C67	020	FACE		.430	212	212	16.7	0316NB	18.5	8.0
A 4639A	CARRIER	445B45	010	TURN		2.675	54	54	5.4	**SAME**	8.5	23.0
A 4639A	CARRIER	445B45	020	FACE		2.675	54	54	5.4	0305YE	6.3	23.0
B 1640A	RETAINER	441B22	010	FACE		4.106	108	108	7.3	0305EG	10.4	43.0

TOTAL HOURS IN THIS MACHINE CENTER   52.3

PARTS IN PREVIOUS WORK CENTER

										PREVIOUS		
208346	IMP WHL	443C31	010	TURN	.437		27	27	4.7	0316NBR	18.2	8.0
203587E	FW PILOT	@444C98	010	SEMI-TURN	.462		28	28	4.3	0316NBR	17.3	8.0
208346G	IMP WHL	446A09	010	TURN	.742		250	250	15.4	0316NBR	24.2	18.0
208346A	IMP WHL	446A07	010	TURN	.907		1234	1234	62.2	0316NBR	36.3	33.0
209335H	IMP WHL	@446A10	010	TURN	1.388		141	141	11.1	0316NBR	20.1	28.0
B 5164	RETAINER	445C90	020	TURN		2.006	98	98	6.1	0305BQ	11.4	23.0
A 4639B	CARRIER	446B17	010	TURN		2.215	255	255	12.6	0316NBR	10.3	23.0
208457B	IMPELLER	444A44	010	TURN		3.632	10	10	4.1	0316NBR	10.4	38.0
208346C	IMP WHL	446A08	010	TURN		4.105	50	50	5.8	0316NBR	20.2	83.0

TOTAL HOURS FOR THIS MACHINE CENTER IN PREVIOUS CENTERS   126.3

Source: E. S. Buffa and J. G. Miller, *Production-Inventory Systems: Planning and Control*, 3rd ed. (Homewood, Ill: Richard D. Irwin, 1979), p. 597.

center away. Column 11 shows where each shop order will go when it leaves BH for those jobs at BH, and where each order is for those jobs coming to BH.

Columns 12 and 13 show work remaining and time remaining for each shop order. In both cases, figures are stated in days. Time remaining is calculated by subtracting today's date from the due date shown for each shop order as a scheduled receipt in the appropriate MRP record. For example, if we go to the time-phased MRP record for part number 209335H, we find at least two scheduled receipt quantities (one for 142 pieces). Although the record might be printed in weekly time buckets, the convention would be to give it a due date of Friday in week 447. The five days of week 447 plus Wednesday, Thursday, and Friday of week 446 yield eight days remaining until this shop order is due to be closed out into inventory. This scheduled receipt for 142 pieces would be pegged to shop order 438C34. A second scheduled receipt for part 209335H would be for 141 pieces, due on Friday of week 451. This order (446A10) is shown as the fifth job in the list of orders one machine center away, with 28 days of time remaining. There could be other open shop orders for part number 209335H as well, but, if so, they aren't at BH or at one work center previous to BH.

The "Work remaining" column (12) represents lead time remaining to complete each order (18.3 days for the first order). This includes setup time, run time, move time, and queue time between operations. We can now define the critical ratio priorities shown as columns 6 and 7:

$$\text{Priority} = \frac{\text{Time remaining}}{\text{Work remaining}}.$$

We see there that for the first shop order ($8.0/18.3 = .436$). This means this shop order will have to be completed in 43.6 percent of normal lead time. If this job isn't run today, tomorrow's schedule will show a time remaining of 7 and a critical ratio of $7/18.3 = .383$. Any order with a critical ratio priority greater than 1.0 is ahead of schedule; any priority less than 1.0 indicates a behind-schedule condition (based on total lead time values).

There is a distinction between columns 6 and 7, PO versus PI. PO is the case when the shop order is pegged all the way up through product structures, an actual customer order depends on this particular shop order. It's a priority for *orders*. PI, on the other hand, is a priority for *inventory*. It means at present, no customer order promise depends upon timely completion of this shop order. The order was issued based upon a forecast of customer orders that hasn't yet materialized.

The work remaining and associated priorities for orders one work center away are based on completing prior operations; that is, the priorities are those that would exist *if* the job were to arrive at BH today. This allows both sets of jobs to be evaluated on a common base.

Jobs are arranged on the daily work center schedule in priority sequence, PO before PI. This is the sequence that jobs should be run in, all other things being equal; that is, the company believes in running jobs to support customer orders before those to go into stock to support a forecast, and, by running the smallest critical ratio job first, relative priorities are maintained.

In interpreting the shop-floor control report, a foreman knows that a critical ratio of .436 isn't a severe problem, providing that *all* the critical ratios aren't less than 1.0. What will happen is this order will be near the top of the list in each work center schedule as it passes through its routing steps. Since this means it will be run shortly after arriving at the work center, or perhaps even be started *before* all parts are finished at the prior center (i.e., operation overlapping), queue time will be small and the job should be completed on schedule.

The ability to see jobs that are coming enhances this ability to meet schedules. If the first job in the list of jobs one work center away (part 208346) had a priority of, say, .1, the BH foreman could go to the foreman in 0316NBR (the current location of the job) to see whether the job could be started in BH before all parts have been finished in 0316NBR, and perhaps try to overlap the operation following BH as well.

The report can also be used to sequence jobs to reduce setup times. If the order (446A10) for part number 209335H can be speeded up in 0316NBR, it can be combined on the same setup with order 438C34. Or perhaps running all (or most) of the impeller wheels in sequence makes sense. The report provides relative priority information to the foreman but doesn't preclude intelligent decision making on his or her part. The extent to which foremen can make decisions at variance with the shop-floor control report should be carefully defined. The key is to provide discretion—but not at the expense of missing due dates.

We've already shown how priority data change on a daily basis, as the time remaining (numerator) grows smaller, while the lead time remaining (denominator) stays constant. If a job isn't completed, its relative priority increases as it competes against other jobs for the available work center capacity.

Vendor Scheduling at Twin Disc. The vendor report in Figure 5.7 is analogous to the shop-floor control report for the Twin Disc Company; the firm calls it the "Open P.O. Buyer Fail-Safe Report." It lists all open purchase orders (scheduled receipts in MRP records), sorted by vendor.

In Figure 5.7, all purchase orders are for a single vendor. The first column of the report lists the buyer placing the purchase order. The second column is the vendor number (52487); the third column is the particular part number on order. The fourth column is the purchase order number for the particular scheduled receipt quantity. Notice, for example, that the sixth and seventh orders are for the same part number, but these are on different purchase orders. The report is printed in part-number sequence by vendor, so *particular* orders can be identified. The fifth column on the report is the due date assigned to the purchase order when it was issued. Note that the report was printed at the beginning of week 343. The sixth column is the quantity on the

---

FIGURE 5.7   Twin Disc Company's Open P.O. Buyer Fail-Safe Report

02/05		OPEN P.O. BUYER FAIL-SAFE REPORT.				WEEK-343	
BUYER	VENDOR #	PART #	ORDER #	WEEK #	QTY.	FWEEK	FQTY.
D	52487	# 9670A	791930	345	5	345	1
D3	52487	# 9670B	819371	360	50		
D1	52487	# 9682	789410	344	50	338	19
D1	52487	# 9700B	808601	347	35	347	3
D3	52487	# 9753A	819380	352	100		
D3	52487	# 9791A	789561	345	25	348	25
D3	52487	# 9791A	810201	351	65	351	1
D1	52487	# 9813	810211	354	50		
D3	52487	# 9815B	788760	343	15		
D3	52487	# 9824	819390	350	25		
D3	52487	# 9825	793490	346	50	349	15
D1	52487	# 9841	793730	345	50		
D3	52487	# 9870A	758611	347	50		
D1	52487	# 9957	810220	348	25		
D1	52487	#201522	825880	352	1000		
D3	52487	#203717A	822100	354	250		
D1	52487	#205826	819330	349	100	349	38
D3	52487	#205896	826850	358	25		
D3	52487	#205896L	825890	357	50		
D3	52487	#206207	793770	348	200	346	108
D1	52487	#206331	791841	351	50	350	13

Source: E. S. Buffa and J. G. Miller, *Production-Innventory Systems: Planning and control*, 3rd ed. (Homewood, Ill.: Richard D. Irwin, 1979), p. 589.

purchase order. Columns 7 and 8 (the most important for vendor scheduling) are the fail-safe columns. Column 7 is the fail week (the date this order is needed to meet a higher-level assembly's planned start date). Column 8 (the fail quantity) is precisely how many are needed to keep from failing to meet this need.

The first order was issued for five units of part 9670A, with a due date of week 345. As of week 343, this due date is still valid. Failure occurs if it's not met. However, it's not essential that all five pieces be delivered. The company can get by if only one of the five is delivered.

"Getting by" has a definite meaning. The fail week and fail quantity are related to actual customer orders; that is, if we were to peg the scheduled receipt for five pieces under purchase order 791930, for part number 9670A, up through product structures, we'd find a customer order depending upon one of the five parts being received in week 345. This concept is the direct analog of separating customer orders (Priority PO) from inventory orders (priority PO) for jobs in the shop.

This means that, for example, the ninth job on the list (part number 9815B) has a due date of the current week. Since there's no information in the "FWEEK" and "FQTY." columns, this order, when pegged up, won't be tied to a customer order. It will only go into Twin Disc's inventory.

Many of Twin Disc's vendors have been so well educated that this report can be sent directly to them. It provides a means for them to give priorities to all orders from Twin Disc, with continuous updating of their priorities. This information can then be integrated with data on other firms' competing needs in their own MPC systems.

Figure 5.7 indicates the third order in the list (part number 9682) is now critical. It's not due until next week, but based on Twin Disc's present conditions, it was needed five weeks ago. *Why* this is true is not important. If we believe the records, this job is now very urgent. Twin Disc will have to shrink five weeks off the combined lead times for all assemblies above this part in the product structure to meet the customer promise date. Other purchase orders can be delayed if the vendor wishes to do so. However, Twin Disc is ready to accept the parts on the due dates, even though they aren't needed until later. This will fulfill its contractual obligations.

The CAPOSS System at Swissair

Swissair operates approximately 50 aircraft on a worldwide basis. Major maintenance is centralized in Zürich, Switzerland, in the Engineering and Maintenance Department (E&M). Approximately 2,600 people are employed in E&M. Swissair also does considerable maintenance work for other airlines.

CAPOSS is an IBM finite scheduling·program. The acronym stands for "capacity planning and operations sequencing system." Swissair uses the CAPOSS system for scheduling the engine shop (which is responsible for overhauling all engines and repairing other parts that require operations on engine shop machine tools). The engine shop employs approximately 400 people.

The primary problem in scheduling the engine shop is over 60 percent of the repair work on an engine isn't definitively known until the engine has been disassembled and inspected. CAPOSS helps to respond to this inherent level of uncertainty.

Before examining the CAPOSS system itself, we should briefly overview some other systems and activities that precede it. When an engine comes in for overhaul, it's disassembled, cleaned, and inspected in the minimum time possible. Inspection of a subassembly involves physical measurement and other activities as dictated by detailed maintenance procedures. These procedures originate with the equipment manufacturers. They're translated into German and modified to match Swissair equipment standards. The entire process uses computerized text editing and is integrated with the overall maintenance control system (MCS) data base.

An inspector uses a video terminal to review the necessary maintenance procedures. Based on tests and measurements, the inspector determines the necessary repair steps for the subassembly and components. This process is, in essence, the determination of a unique routing file for each part and assembly. These data are essential for CAPOSS. Interactive computing allows routing files to be established quickly. A set of shop papers is immediately printed and attached to the work piece. MCS also keeps track of which parts belong to which engine.

A production control group working for the engine shop decides whether a particular part is to be reworked to go back into the same engine or a replacement is to be used. This decision is based upon rework requirements, time availability, capacity loads, replacement part availability, and dictates of the manufacturer.

An engine is defined as being made up of 10 modules. Based on the completion date for the overall engine, due dates are established for all parts of any module. These dates are one key input to CAPOSS.

Another key input to CAPOSS is the **external priority,** a number from 0 to 9, indicating a production control person's priority to be assigned to this order. It represents a subjective input as to how critical this shop order is relative to other orders.

The CAPOSS system first loads each order with infinite capacity loading assumptions, based on operation times plus standard interoperation times (representing move and queue times). If the resultant completion time exceeds the due date for the reworked parts, a delay factor is computed to augment the external priority. Another step is to reduce the move and queue times, based on how late the parts are. This can shorten interoperation times up to 50 percent. After this, external priorities are combined with move and queue time reductions to produce a new priority number. This priority number is used to pick the job sequence for horizontal finite loading for all the jobs through their required work centers.

The system is run daily on a large IBM computer. One output is a schedule of all jobs in each worker center. (Since outputs are in German, they aren't reproduced here.) It serves the same function as a daily dispatch report for shop-floor control. The listing goes to a production control person assigned to each work center grouping. This person's job is to actually schedule the work. He or she attempts to combine similar jobs (a shop order is issued for each part), attempts to assign jobs to workers best able to do the work, and so on. The schedule shows which jobs are in the work center as well as those not there. The work center production control person also makes sure all transactions are accurately entered into the data collection system.

The CAPOSS data base is also sorted by engine. An engine shop production control group gets these data. As a result, they can foresee part shortages. When this occurs, one typical action is to increase the external priority assigned to the particular shop order so that order will be scheduled earlier on each of the next day's work center schedules.

An engine is always scheduled for overhaul in 21 working days. This time is guaranteed to the customer. The CAPOSS schedule is prepared for 15 days into the future, which is more than overall machine shop lead time. For the work center scheduler, only the first few days in the schedule are important. For the engine scheduler, all of the time frame is important; relative priority changes are made constantly.

At the end of the 15-day schedule, any jobs that haven't been scheduled are listed in a separate report, along with their capacity requirements. These data indicate potential work center capacity problems. They are studied closely by the production control supervisor. Remedial actions include overtime, alternate routing, and increased use of replacement parts instead of reworking parts. The production control supervisor discusses these actions with the engine shop manager.

Vendor Scheduling at Steelcase

Steelcase is a large manufacturer of office furniture. The firm made a fundamental change to its purchasing organization, separating procurement from the PAC activities of order release, vendor scheduling, and other clerical functions. An early user of MRP-based systems, Steelcase has since gone on to implement JIT manufacturing approaches. Figure 5.8 shows a vendor scheduling report for one of Steelcase's vendors, Cannon Mills, which supplies fabrics for upholstered furniture. In Figure 5.8 all orders through the week of 8/26 are asterisked, which means that they're firm commitments on Steelcase's part. Also shown are orders for the next four weeks out. With many vendors, Steelcase would commit on these to the extent of the vendor's investment in raw material.

In this example, the two firms agree on production of a particular cotton cloth, with a later decision on the color it's to be dyed. The job of purchasing is to negotiate the commitment's form, when and how these "time fences" (cotton versus color) are to be crossed, prices, lot sizes, and so on.

The report in Figure 5.8 can be sent directly to the vendor. Steelcase has eliminated formal purchase orders for all but occasional purchases. This saves

FIGURE 5.8 Steelcase Requirements for Cannon Mills for the Week Ending 07/22

PAGE 1 DATE 07/22 RUN TIME 04-15

ALL TAGGED ORDERS (*) ARE FIRM
OTHER ORDERS ARE EXPECTED DATES AND QUANTITIES

PART NUMBER DIV	FINISH CODE NO.	DESCRIPTION	DATE	BUYER	REC'D LAST WEEK	CURRENT & PAST DUE	REQUIREMENTS							ON-HAND	ISSUED LAST WEEK
							7/29	8/05	8/12	8/19	8/26	NEXT 4 WEEKS	FOLLOWING 4 WEEKS		
904550000 4	5350	RED COTTON	9-0553 A 05/08/	010				400*		400*		400		442	
904550000 4	5351	RED RED ORANGE COTTON	9-0553 A 05/08/	010		800*	800*	800*	1200*	1200*	1200*	1200	1600	359	215
904550000 4	5352	RED ORANGE COTTON	9-0553 A 05/08/	010	415	785*	800*	800*	400*	800*	800*	400	1200	415	
904550000 4	5353	YELLOW ORANGE COTTON	9-0553 A 05/08/	010		800* 400*	400*	400*	1200*	800*	400*	800	1200	50	118
904550000 4	5355	YELLOW COTTON	9-0553 A 05/08/	010	402	331*			400*		400*	400	400	1804	120
904550000 4	5356	YELLOW YELLOW GREEN COTTON	9-3553 A 05/08/	010		200*	400*		400*		400*	400		237 237H	
904550000 4	5358	GREEN COTTON	9-0553 A 05/08/	010				400*	400*		400*	400	400	384	64
904550000 4	5360	BLUE COTTON	9-0553 A 05/08/	010	416	384*	400*		400*	400*		400	400	416	
904550000 4	5368	TAN VALUE 1 COTTON	9-0553 A 05/08/	010	1445	1600* 1600*	800*	400*		800*	800*	1600	1600	1502	234
904550000 4	5369	TAN VALUE 2 COTTON	9-0553 A 05/08/	010	725	75*				400*	400*	400	400	1305	197

Source: P. L. Carter and R. M. Monczka, "Steelcase, Inc.: MRP in Purchasing," in *Case Studies in Materials Requirements Planning*, edited by E. W. Davis (Falls Church, Va.: American Production and Inventory Control Society, 1978), p. 215.

on paperwork, cuts response time, and simplifies PAC. As actual orders come in for particular color fabrics, adjustments are made by clerical interactions between Steelcase and Cannon Mills.

PAC Procurement Practices under "World Class" Manufacturing

Xerox Corporation is a prime example of world class manufacturing based on purchasing. At a critical point Xerox discovered it had a 40 percent cost disadvantage compared to its major competitor (Canon). Xerox's cost structure was highly dependent upon material costs; over 70 percent of manufacturing cost was in materials. The result was a need to drastically reduce costs of purchased components. Xerox did this by reducing its worldwide vendor base from 5,000 to 300, by developing single sourcing agreements, and, most importantly, by helping its vendors become "world class" manufacturing firms—ones that could compete against anyone. Xerox also overhauled the ways it dealt with vendors. Gone are the typical adversarial relationships and narrow definitions of how Xerox would work with its suppliers. Joint design, collective efforts on new products, continual search for new methods, sharing innovations among all suppliers, open access to information, and an ongoing search for ways to reduce the "hidden factory" of transactions between Xerox and its suppliers are now Xerox's priorities. The result is cost parity with any competitor!

Production activity control is part of this search for excellence. Transactions have been minimized in Xerox and other companies following a similar path. Dealing with a smaller set of suppliers encourages commonality of procedures. Common definitions of terms and conditions, data protocols, and procedures for the release/authorization of vendor shipments are some of the main changes in PAC. Additional improvements follow. For example, JIT auto companies simply pay their suppliers for radiators on the basis of the total number of cars shipped each month. They never ship cars without radiators or with two. There aren't enough radiators in work-in-process to worry about, and any defects are returned and replaced. All this is done without paper or transactions—no purchase orders, invoices, shipping papers, debit memos, and so on. The PAC system provides the build rate to the supplier, cars are made on a level schedule, and deliveries arrive several times each day.

A computer manufacturer follows a slightly more complex PAC relationship in one of its factories. A key component is purchased from a vendor about two hours away by truck. The schedule for the computer manufacturer isn't totally level; the product is somewhat customized so individual end items are built to order. Rather than carry a large amount of inventory, the computer manufacturer has developed a simple procedure with this key component vendor. Each day at noon, someone on the assembly line calls the vendor to specify the exact number of components to deliver the following morning.

The vendor has the flexibility to make those components during the afternoon and deliver them the next morning. Again the approach is based on minimal transactions.

As the number of purchased components increases, many firms adopt **electronic data interchange (EDI)** to communicate with their suppliers on a routine basis. The objective under JIT operations is to create electronic forms of kanban or other pull system methodologies to reduce the inventories sharply—as a goal in itself as well as to eliminate the need for transactions between the companies.

Caterpillar Tractor Company illustrates using EDI in PAC. Nearly 400 of the firm's domestic and offshore supplier locations are connected using standards developed by the Automotive Industry Action Group (AIAG). Each supplier is assigned an electronic mailbox on the network. Suppliers retrieve information from their mailboxes and can forward information through the mail boxes to Caterpillar. By adopting AIAG, the "mail" is read in a language common to computers of Caterpillar and its vendors.

Caterpillar's EDI network is called "Speed" for suppliers and purchasers electronically exchanging data. Although Caterpillar has been building the Speed network for several years, it began as an outgrowth of the company's existing MPC systems. The company recognized the need to add effective electronic communications with the outside suppliers to facilitate JIT. The company viewed the Speed network's advantages to be applicable to both Caterpillar and its vendors. Direct benefits include:

- Reducing transaction time to a few hours, compared with processing through the mail.
- Quicker responses to changes for revised material requirements.
- Reduced paper-handling expenses.
- Fewer errors involving transmitting information.

THE PRODUCTION ACTIVITY CONTROL DATA BASE

Here we consider data bases and interactions of PAC with other data files and systems. We also raise some managerial issues. As we've seen, the data base for shop-floor scheduling typically includes open shop orders (scheduled receipts from MRP) and their due dates, routing files, engineering standards or other time estimates for operation times, move and queue time data, and work center information.

Each file's size and complexity varies a great deal from company to company. Some firms, such as Ethan Allen factories, deal with an open order file of less than a few hundred shop orders, whereas Twin Disc typically has 3,500 open shop orders. Similarly, the average number of operations per shop order also varies widely. The same is true for the number of work centers to be scheduled. As a result, the inherent size of the data base and the number of transactions to be processed each day vary significantly from firm to firm; the resultant systems and supporting data bases have to reflect these complexities.

Figure 5.9 depicts the relationship of the PAC data base to other parts of the MPC system. If the company uses MRP and batch (job shop) manufacturing methods, PAC is based on the shop-floor control methodologies described in this chapter, and the PAC data base will necessarily reflect the company's inherent manufacturing complexity. If, on the other hand, JIT methods are used for manufacturing, the process is greatly simplified and so are the detailed data requirements for PAC. Figure 5.9 shows the MRP and other engine systems providing input to procurement. Depending on the extent that the MPS can be leveled, and appropriate agreements with suppliers can be negotiated by procurement, detailed release of orders and vendor scheduling are greatly simplified—as are feedback and order closing.

FIGURE 5.9   The PAC Data Base

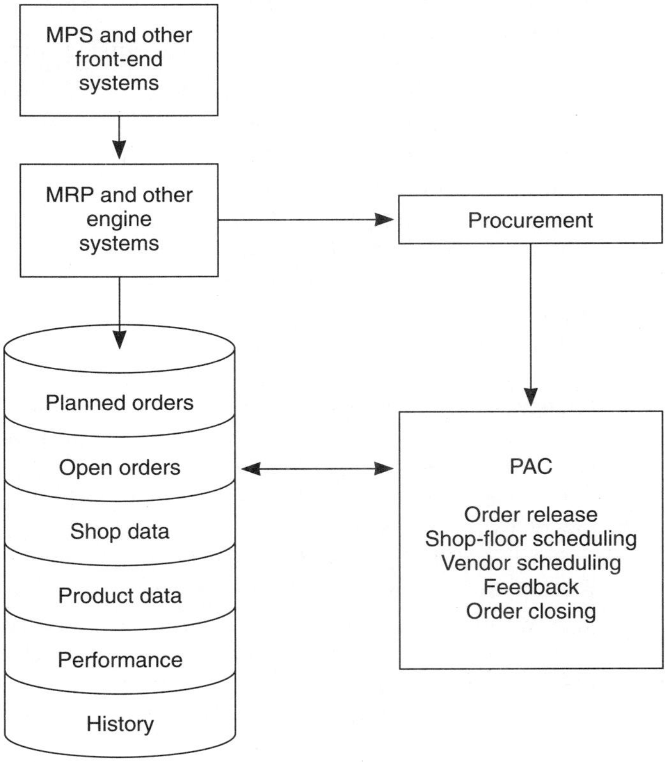

Data Acquisition and Feedback

One key requirement of a shop-floor control system is collecting status information on work-in-process. This includes processing transactions on the completion of a batch at a work center, movement of a batch to another work center, scrap losses, and processing by nonstandard methods.

Here we need data integrity and promptness of reporting. The ability to update priorities, inform customers, perform capacity requirements planning (CRP) calculations, and otherwise adjust plans all depends on effective data collection. Fortunately, technology is increasingly available. Distributed processing and automatic character recognition help in data acquisition as well as in changing the balance of effort between a centralized MPC function and one where the shop floor has greater autonomy and responsibility.

Bar coding has enabled us to automate a great deal of data acquisition. Hand-held bar code readers or wands, including those based on laser technology, can quickly and accurately read such information as part numbers and routing steps. While the UPC codes we see on products in grocery stores generally contain only a limited amount of information (such as the product number), an increasing variety of data is being put in bar code format. Not only is the bar code data base changing, the reader technology is improving, too. In some cases, the wand itself can perform local processing enabling some data editing and processing right at the point of input.

Growing computational power on the shop floor itself is a somewhat different phenomenon. An evolution from paper systems to "dumb" terminals to "smart" terminals has taken place. Small computers' growing power and shrinking cost are part of this evolution. It's now practical to have much of the mainframe computers' computational power and data storage on the shop floor. Some firms run the entire MPC systems, including MRP planning on the shop floor itself.

The combination of improved bar coding, data acquisition, and decentralized computer power means the character of shop-floor transactions has changed, and direction for shop-floor management is less in the hands of centralized production control personnel. Some firms with a complex shop environment now have dedicated minicomputers that keep track of the shop data base on a real-time basis. Paper is minimized, and only summary data are transmitted to the mainframe computer.

Shop scheduling based on actual transactions as they occur is a viable approach. In one real-time minicomputer system we've seen, first-line supervisors wand bar-coded information from a job card when the job is completed and enter the quantity completed. The computer system does the usual cross-checking of data for integrity, and then presents a prioritized list of jobs to work on next, based on up-to-the-minute conditions. The expectation is that the job at the top of the list is chosen, but this choice can be overridden. (Perhaps another job is more within the capabilities of the particular operator or machine.) If the job at the top of the list isn't chosen on three consecutive occasions, then the supervisor is blocked from choosing an alternative without higher approval. Jobs passed over, late jobs, and other problem conditions are highlighted for the plant manager. The system uses finite loading for each machine center for a one-week horizon. Beyond that time period, an approach similar to CRP is used. Alternative routing is readily accommodated, based on local decision making.

A further phase of development in PAC systems is to expand their use beyond traditional MPC system definitions. Quality reporting and equipment maintenance are two areas where data can be collected in combination with PAC. Another direction is to integrate shop-floor control systems with computer-aided design (CAD) and computer-aided manufacturing (CAM). Computer-based analysis can be used to develop the routing data base. By integrating these systems, the time frame from product conception to part production can be reduced substantially. A PAC system that routinely plans and controls day-to-day operations is critical for achieving these further levels of system integration.

Of course, the ultimate objective in PAC is often to eliminate the need for data acquisition and feedback. JIT, total quality control, fool-proof operations (often referred to by the Japanese term *poka-yoke*), and other aspects of world class manufacturing support this process.

CONCLUDING PRINCIPLES

We see the following principles emerging from this chapter:

- Production activity control system design must be in concert with the firm's needs.
- The shop-floor control system should support users and first-line supervisors, not supplant them.
- Vendor capacities should be planned and scheduled with as much diligence as are internal capacities.
- Lead times are to be managed.
- Organizational goals and incentives must be congruent with good PAC practice.
- Discretion and decision-making responsibilities in production activity control practice need to be carefully defined for both the shop and vendors.
- PAC performance should be defined and monitored.
- Feedback from PAC should provide early warning and status information to other MPC modules.
- Automated reading systems and distributed computers should facilitate data acquisition and shop-floor decision making.
- Data base design and integrity must be assessed for PAC systems to be effective.

- The ongoing evolution in PAC systems as firms increasingly adopt world class manufacturing methods is reduced detail, smaller data bases, and simpler systems.

REFERENCES

APICS. Operations Scheduling Seminar Proceedings, January 1979.

Baker, Eugene F. "Flow Management: The 'Take Charge' Shop Floor Control System." *APICS 22d Annual Conference Proceedings,* 1979, pp. 169–74.

Belt, Bill. "Input–Output Planning Illustrated." *Production and Inventory Management,* 2nd quarter 1978.

Biggs, Joseph R. "Priority Rules for Shop Floor Control in a Material Requirements Planning System under Various Levels of Capacity." *International Journal of Production Research* 23, no. 1 (1985), pp. 33–46.

Bihun, T. A. "Electronic Data Interchange: The Future," *American Production and Inventory Control Society 1990 Annual Conference Proceedings,* pp. 605–9.

Burlingame, L. J., and R. A. Warren. "Extended Capacity Planning." *1974 APICS Conference Proceedings,* pp. 83–91.

Carter, J. R., and L. D. Fredendall, "The Dollars and Sense of Electronic Data Interchange," *Production and Inventory Management Journal* 31, no. 2, 2nd quarter 1990, pp. 22–26.

Carter, P. L., and R. M. Monczka. "Steelcase, Inc.: MRP in Purchasing." In *Case Studies in Materials Requirements Planning.* Falls Church, Va.: American Production and Inventory Control Society, 1978, p. 125.

Fogarty, D. W.; J. H. Blackstone, Jr.; and T. R. Hoffmann. *Production & Inventory Management.* 2nd ed. Cincinnati: Southwestern 1989.

Goddard, W. E. "How to Reduce and Control Lead Times." *1970 APICS Conference Proceedings,* pp. 198–205.

Greenstein, Irwin. "Caterpillar Erects Paperless Network." *Management Information Systems Week,* January 20, 1986.

Hoffmann, T. R., and G. R. Scudder. "Priority Scheduling with Cost Considerations." *International Journal of Production Research* 21, no. 6 (1983), pp. 881–89.

Lankford, R. I. "Input/Output Control: Making It Work." *1980 APICS Conference Proceedings,* pp. 419–20.

————. "Scheduling the Job Shop." *1973 APICS Conference Proceedings,* pp. 46–65.

Melnyk, S. A.; P. L. Carter; D. M. Dilts; and D. M. Lyth. *Shop Floor Control.* Homewood, Ill.: Dow Jones-Irwin, 1985.

————. *Production Activity Control.* Homewood, Ill.: Dow Jones-Irwin, 1987.

Melnyk, S. A.; S. Ghosh; and G. L. Ragatz. "Tooling Constraints and Shop Floor Scheduling: A Simulation Study." *Journal of Operations Management* 8, no. 2, April 1989, pp. 69–89.

Melnyk, S. A.; S. K. Vickery; and P. L. Carter. "Scheduling, Sequencing, and Dispatching: Alternative Perspectives." *Production and Inventory Management,* 2nd quarter 1986, pp. 58–68.

Nellemann, D. O. "Shop Floor Control: Closing the Financial Loop." *1980 APICS Conference Proceedings,* pp. 308–12.

Philbrook, S. D. "Competency in Bar Code Menus and Templates." *P and IM Review and APICS News,* November 1983.

Philipoom, P.; R. E. Markland; and T. D. Fry. "Sequencing Rules, Progress Milestones and Product Structure in a Multistage Job Shop." *Journal of Operations Management* 8, no. 3, August 1989, pp. 209–29.

Plossl, G. W., and W. E. Welch. *The Role of Top Management in the Control of Inventory.* Reston, Va.: Reston Publishing, 1979, p. 78.

Production Activity Control Reprints. Falls Church, Va.: American Production and Inventory Control Society, 1986.

Raffish, Norm. "Let's Help Shop Floor Control." *Production and Inventory Management Review* 1, no. 7 (July 1981), pp. 17–19.

Schonberger, Richard J. "Clearest-Road-Ahead Priorities for Shop Floor Control: Moderating Infinite Capacity Loading Unevenness." *Production and Inventory Management,* 2nd quarter 1979, pp. 17–27.

Schorr, J. E., and T. F. Wallace. *High Performance Purchasing.* Brattleboro, Vt.: Oliver Wight Ltd., 1986.

Waliszowski, David A. "Lead Time Reduction in Multi Flow Job Shops." *1979 APICS Annual Conference Proceedings.*

Wassweiler, W. L. "Fundamentals of Shop Floor Control." *1980 APICS Conference Proceedings,* pp. 352–54.

————. "Material Requirements Planning: The Key to Critical Ratio Effectiveness." *Production and Inventory Management,* 3rd quarter 1972, pp. 89–91.

————. "Shop-Floor Control." *Annual APICS Conference Proceedings,* 1977, pp. 386–91.

Williamson, R. F., and S. S. Dolan. "Distributed Intelligence in Factory Data Collection." *P and IM Review and APICS News,* February 1984.

DISCUSSION QUESTIONS

1. Suppose an average of 20 students per year were in your major and the number of graduates varied between 15 and 30 each year. One required course, open to both juniors and seniors, is offered once a year and has a capacity of 20 students. What system would you use to assign students to that course? Would it make any difference if the course capacity was 40 students or 10 students per year?

2. What kind of warning signals would you like fed back from PAC to the MRP system?

3. Lead times have sometimes been called "rubbery." What accounts for this concept of elasticity in lead times?
4. In the list of priority rules, there is no first-come/first-served (FCFS) rule, yet most banks, cafeterias, and theater ticket booths use this rule in waiting on patrons. Why isn't it a suggested rule for the shop floor?
5. Will vertical loading produce a different schedule than horizontal loading in a finite scheduling system?
6. What's your feeling about having separate priorities for customer orders and inventory orders as Twin Disc does?
7. Why would a vendor be interested in shipping just one unit to help Twin Disc "get by," as Figure 5.7 shows for part 9670A?
8. How would you determine how much discretion to give the foreman in combining and reprioritizing jobs?
9. What benefits might be gained by having clerks in a retail store use a wand to read tags on items?
10. Discuss the merits of separating the daily transaction people (schedulers) from the buyers in the purchasing organization.

PROBLEMS

1. The Bob Row Ski Company uses three machines to manufacture specialty ski boots. Incoming jobs follow different routes through the shop. For example, a job may have to be processed on machine 1 first, then machine 3, and finally machine 2. The following table covers the next four jobs to be scheduled at the company. (It's currently April 2.)

Job	Job arrival date	Required job/machine routing	Machine processing times (days)		
			Machine 1	Machine 2	Machine 3
A	April 2	2–1–3	1	2	2
B	April 2	3–1–2	1	1	4
C	April 4	1–3–2	3	1	2
D	April 6	3–2–1	2	2	2

Assume that material for all jobs is ready for processing as soon as the jobs arrive and that a first-come/first-served sequencing rule is used. Assume also that all three machines are idle as work begins on April 2.
a. Construct an appropriately labeled Gantt chart depicting the processing and idle times for the three machines for these four jobs.
b. How many days does each job wait in the queue for processing on machine 3?

2. The production manager at the Knox Machine Company is preparing a production schedule for one of the fabrication shop's machines—the P&W grinder. He has collected the following information on jobs currently waiting to

be processed at this machine. (There are no other jobs and the machine is empty.)

Job	Machine processing time (in days)*	Date job arrived at this machine	Job due date
A	4	6–23	8–15
B	1	6–24	9–10
C	5	7–01	8–01
D	2	6–19	8–17

*Note: This is the final operation for each of these jobs.

a. The production manager has heard about three dispatching rules: the Shortest Operation Next Rule, the First-Come/First-Served Rule, and the Earliest Due Date Rule. In what sequence would these jobs be processed at the P&W grinder if each rule was applied?

b. If it's now the morning of July 10 and the Shortest Operation Next Rule is used, when would each of the four jobs start and be completed on the P&W grinder? (Express your schedule in terms of the calendar dates involved, assuming that there are 7 working days each week and 30 days per month.)

3. Jobs A, B, and C are waiting to be started on machine center X and then be completed on machine center Y. The following information pertains to the jobs and work centers:

Job	Hours allowed for machine center X	Hours allowed for machine center Y	Day when due
A	55	16	10
B	128	22	16
C	90	20	20

Machine center X has 60 hours of capacity per week (5 days) and machine center Y has 40 hours of weekly capacity. Two days are allowed to move jobs between machine centers.

a. Scheduling these jobs by earliest due date, can they be completed on time?

b. Can they be completed on time using the critical ratio technique?

c. Can the jobs be completed on time (using the earliest due date or critical ratio technique) if 30 hours of overtime are run in work center X each week?

d. Can the jobs be completed on time (using the earliest due date or critical ratio technique) if only one day is required between operations?

4. The customer for job B in Problem 3 has agreed to take half the order on day 16 and the rest on day 25. Use earliest due date to schedule the four jobs (A, B1, B2, C). Can they be completed on time assuming no extra setup time is required for splitting job B?

5. Ms. Mona Hull is in charge of a project to build a 50-foot yacht for a wealthy industrialist from Jasper, Indiana. The Yacht, the *Nauti-Lass* is scheduled to compete in the famous Lake Lemon Cup Race. Assume a 6-week lead time for constructing the yacht. Assume also that each week consists of 5 work days, for a total lead time of 30 days. The work required to complete the yacht is comprised of 10 operations, 3 days for each.
   a. On Tuesday morning of week 2, 2 of the 10 operations had been completed and the *Nauti-Lass* was waiting for the third operation. What's the critical ratio priority?
   b. What's the critical ratio priority if only 1 of the 10 operations is completed by Tuesday morning of week 2?

6. The Ace Machine Company is considering using a priority scheduling rule in its fabrication shop and must decide whether to use: (1) the Critical Ratio Rule, (2) the Order Slack Rule, (3) the Shortest Operation Next Rule, or (4) the Slack Per Operation Rule. Exhibit A (on page 203) shows the company's current inventory and shop status.
   a. State the formula for calculating the priority index for each of the four sequencing rules given previously.
   b. Compute the scheduling priority for each order in Exhibit A, using each of the four sequencing rules.

7. Figure 5.7 is the Twin Disc Open P.O. Buyer Fail-Safe Report. Suppose that vendor 52487 decides to work only on jobs with entries in the F-week column, and these in order of F week. Generate the list of part numbers in the order in which they will be produced. What are the implications of the sequence to the supplying firm and to Twin Disc?

8. Figure 5.8 shows a Steelcase requirements report for Cannon Mills. The quantities are in yards of fabric for the periods shown, for each color or style. Determine the total yards of fabric Steelcase requests over each of the next 14 weeks. (Assume the 4-week requirements are evenly distributed.) What would you think if you were Cannon?

9. The Bedford Machine Shop's production scheduler has just received the following order for a machined part:

Order No. 6243				
Operation number	1	2	3	4
Machine number	1	4	3	2
Estimated time*	6	4	2	8

*In standard hours.

   a. Bedford allows two days between machine operations for queue time, material handling time, and so on. The shop works five days per week on a single work shift, and no overtime is planned for the near future. Given the following loads at each machine center, determined by finite scheduling in the machine shop, set up a schedule for each operation on the new job and indicate when this order can be shipped. (None of the currently scheduled jobs can be changed.)

EXHIBIT A

	Inventory information					Shop information				
Order number	On-hand inventory	Reorder point	Safety stock	Average daily usage	Manufacturing lead time remaining	Shop time accumulated to date	Time remaining until due date	Current operation processing time	Number of operations remaining	Total processing time remaining
1	80	160	10	10	15	1	7	2	3	8
2	120	160	10	10	15	5	11	3	4	8
3	−10	210	10	10	20	14	−2	5	2	8
4	−40	160	10	10	15	5	−5	4	3	8
5	40	160	10	10	15	13	3	1	12	8

Note: The inventory data is measured in units, and time is measured in days. Shop time accumulated to date (and also manufacturing lead time remaining) includes both the machine processing time and the length of time orders spend moving and waiting to be processed in a machine queue. Current operation processing time (and also total processing time remaining) includes no move or queue times.

Machine #	Weekly capacity*	Number of machines	Load in week #						
			1	2	3	4	5	6	7
1	40	1	40	40	30	20	15	10	5
2	80	2	80	50	80	80	70	60	40
3	40	1	20	15	10	5	2	—	—
4	120	3	120	120	120	120	110	102	90

*In standard hours.

b. Given the information displayed in the shop load table, what capacity recommendations would you make to management?

10. The XYZ Company uses MRP to plan and schedule plant operations. The plant operates five days per week with no overtime, and all orders are due at 8 A.M. Monday of the week required. It's now 8 A.M. Monday of week 1, all machines are currently idle, and the production manager has been given the information in Exhibit B.

a. Assuming that the Shortest Operation Next Rule is used to schedule orders in the shop, how should the open orders (scheduled receipts) for items A, B, and C be sequenced at their current operations? What are the implications of this schedule?

b. Assuming that the Critical Ratio Rule is used to schedule orders in the shop, how should the open orders for items A, B, and C be sequenced at their current operations? What are the implications of this schedule?

EXHIBIT B System Data

MRP system data

Item A

Week		1	2	3	4	5	6	7	8
Gross requirements		3	16	8	11	5	18	4	2
Scheduled receipts			30						
Projected available balance	10	7	21	13	2	27	9	5	3
Planned order releases			30						

Q = 30; LT = 2; SS = 0.

Item B

Week		1	2	3	4	5	6	7	8
Gross requirements		2	5	12	4	18	2	7	10
Scheduled receipts				30					
Projected available balance	12	10	5	23	19	1	29	22	12
Planned order releases			30						

Q = 30; LT = 4; SS = 0.

EXHIBIT B *(concluded)*

Item C

Week		1	2	3	4	5	6	7	8
Gross requirements		14	4	12	7	8	3	17	2
Scheduled receipts			20						
Projected available balance	17	3	19	7	0	12	9	12	10
Planned order releases		20		20					

Q = 20; LT = 4; SS = 0.

Shop-floor control system data

Item	Routing					Current operation number
A	Operation number	1	2	3	4	3(Machine 2)
	Machine number	4	6	2	3	
	Processing time*	1	1	1	1	
B	Operation number	1	2	3	4	2(Machine 2)
	Machine number	4	2	6	1	
	Processing time*	4	2	5	3	
C	Operation number	1	2	3	4	4(Machine 2)
	Machine number	6	3	1	2	
	Processing time*	3	4	6	1.5	

*In days. Assume that there are 1.5 days of move and queue time associated with each operation for computing lead times.

c. Suppose that 32 additional units of item B have just been found in the stockroom as a result of a cycle count. What actions are required on the part of the MRP planner?

11. This morning, Pete Jones, integrated circuit buyer at Flatbush Products, Inc., received the following purchased part MRP record:

CIRCUIT #101

Week		1	2	3	4	5	6	7	8	9	10	11	12
Gross requirements		50									50	50	50
Scheduled receipts													
Projected available balance	65												
Planned order releases													

Q = 50; LT = 3; SS = 2.

a. Complete the circuit #101 MRP record.

b. Pete noted that the vendor for the circuit #101 recently indicated his plant has limited production capacity available. At most, this supplier can provide 50 units per four-week period. Pete wondered what action, if any, needs to be taken on this item as a result of this capacity limitation.

12. On Monday morning of the week before the annual shutdown, the Limited Hours Company had orders in the shop for five products (cleverly called A–E), which had arrived in alphabetical order. Management decided not to take any more orders until after the shutdown, and the five orders on hand were assigned priorities in the order of arrival (i.e., A–E). Delivery promises hadn't yet been made on any of the orders. Each order went through the same three machine centers, but not necessarily in the same sequence. Each order had to be finished at a machine center before another could be started. They couldn't be split.

    The company worked a demanding three-hour day, so it was concerned about whether there was enough total machine time (capacity) to finish the five orders in the five days remaining before the shutdown. Data on each order are as follows. Assume the time to move between work centers is negligible.

Order	Machine center routing	Hours at machine center 1	2	3
A	1–3–2	2	4	4
B	2–1–3	2	6	3
C	3–1–2	5	1	4
D	1–2–3	3	1	2
E	3–2–1	2	3	1

a. Is there enough machine time to finish the orders?
b. Using the horizontal loading technique to schedule each order through the machine centers, on what days can the deliveries be promised?
c. What changes occur if you use the vertical loading method?

13. The materials manager and vice president of manufacturing at the SCM Transmission Company are currently assessing the capabilities of the firm's suppliers to meet the company's production plan for the next year. The marketing department forecasts the total sales volume to be $70 million, of which 20 percent will be sold in the first quarter, 36 percent in the second quarter, 29 percent in the third quarter, and the remaining 15 percent in the fourth quarter. The average selling price is $1,000/transmission, and the company plans to produce an amount equal to the sales forecast each quarter because of the product's make-to-order nature and the cost of financing finished-goods inventories.

    One major purchased item is steel castings, from which machine parts are manufactured. Currently, the firm buys all its steel castings from the Kewanee Foundry. To evaluate Kewanee Foundry's capacity to handle SCM's business for the next year, the materials manager has prepared the following capacity bill for steel castings:

> *Steel casting capacity bill*
>
> .12 lbs. of steel castings per transmission sales $
> .012 machine molds per transmission sales $
> .024 cast parts per transmission sales $

The materials manager has said the Kewanee Foundry has an overall capacity to pour 80 tons of metal per day, and it works 90 production days per quarter. At most, 15 percent of the daily foundry capacity can be processed through the automated machine molding line. (Equivalently, the automated machine molding line can produce up to 2,400 molds per day, operating 3 shifts per day, 90 days per quarter.) The automated machine molding line is the only process at Kewanee that's capable of producing steel castings at a low enough cost to meet SCM's requirements. SCM has used Kewanee as a sole source for its steel casting for several years; this vendor's quality and delivery performance are excellent. SCM has achieved substantial economies in the purchase of steel castings by consolidating all its business with Kewanee.

a. Evaluate Kewanee's capability to meet SCM's total requirements for the next year.

b. SCM's master scheduler wants to develop the daily production rates for the next four quarters. Express the master schedule in terms of the number of molds required per day. (The Kewanee Foundry manager plans capacity in terms of the number of molds produced per day on the molding machine line.)

c. Can Kewanee meet the MPS requirements on a quarterly basis? What alternatives can be discussed with Kewanee?

14. On a busy day in June, the Framkrantz Factory had five jobs lined up for processing at machine center 1. Each job went to machine center 2 after finishing at machine center 1. After that they had different routings through the factory. It's now shop day 83 and due dates have been established for each job. Machine time includes setup time, but it takes one day to move between centers and two days of queue time at each center (including Finish).

	Machine center sequence for the jobs and days of machine time required						
Shop day = 83							
Job	1	2	3	4	5	Finish	Due date
A	1	4				X	102
B	3	2	8			X	123
C	3	8		2		X	104
D	6	2	1		3	X	98
E	4	1		8	2	X	110

a. Use a spreadsheet program to calculate the priorities for each job using the Critical Ratio Rule.

b. All jobs were at machine 2 by the morning of shop day 96. (Job A took 2 days instead of 1 and Job D took 1 day instead of 6 on machine 1.)

Unfortunately, there was a long job on machine 2 and none of the five jobs had started yet. What would their priorities be for machine 2?
c. What would priorities be if Job A's due date was 96?

15. Shown below is the MRP record for part number 483. The current shop day is 100 (with 5-day weeks); it's now Monday of week 1. Open orders (scheduled receipts) are due Mondays (shop days 100, 105, 110, etc.) of the week for which they're scheduled.

   The shop floor has just reported that the batch of 40 on shop order number 32 has just finished at machine center A43 and is waiting to be moved to C06. It takes one day to move between machine centers (or to I02, the inventory location) and one day of queue time at the machine centers. (The inventory location doesn't require the queue time, but one day of "machine time" is shown for clearing the paper work.) Part number 483's routing and status are also given below.

Part No. 483

Week		1	2	3	4	5	6
Gross requirements		14	4	10	20	3	10
Scheduled receipts			40*				
Projected available balance	20	6	42	32	12	9	39
Planned order releases					40		

Q = 40; LT = 2; SS = 5
*Shop Order Number 32

Part No. 483

Routing (mach. cent.):	A12	B17	A43	C06	I02
Machine time (days)	4	1	3	1	1
Status shop ord. 32:	Done	Done	Done		

---

a. Use a spreadsheet to replicate the MRP record and calculate the critical ratio for part number 483. Should the planner take any action?
b. What would the priorities be if the inventory was 23 instead of 20? What action should be taken now?
c. What if inventory was 17 instead of 20?

Chapter 6

Master Production Scheduling

In this chapter, we discuss constructing and managing a master production schedule, a critical module in the manufacturing planning and control system. An effective master production schedule (MPS) provides the basis for making customer delivery promises, utilizing plant capacity effectively, attaining the firm's strategic objectives as reflected in the production plan, and resolving trade-offs between manufacturing and marketing. The prerequisites are to define the master scheduling task in the organization and to provide the master production schedule with the supporting concepts described in this chapter.

The chapter is organized around eight topics:

- The master production scheduling activity: What is the role of master production scheduling in manufacturing planning and control and its relation to other business activities?
- Master production scheduling techniques: What are the basic MPS tasks and what techniques are available to aid this process?
- Bill of material structuring for the MPS: How can nonengineering uses of the bill of materials improve master production scheduling?
- The final assembly schedule: How is the MPS converted into an actual build schedule?
- The master production scheduler: What does a master production scheduler do and what are the key organizational relationships?
- Examples: How do some actual MPS systems work in practice?
- Master production schedule stability: How can a stable MPS be developed and maintained?
- Managing the MPS: How can MPS performance be monitored and controlled?

Chapter 6 closely relates to Chapter 14, which presents advanced concepts in master production scheduling. Chapter 7 discusses the production plan from which the MPS is derived. Chapter 4 describes managing the day-to-day demands for plant capacity. Chapter 9 describes how the MPS approach must be based on the firm's approach to the market and competition. Chapter 2's time-phased record concepts are building blocks for the material in this chapter.

THE MASTER PRODUCTION SCHEDULING ACTIVITY

We begin with a brief overview of the master production scheduling (MPS) process. What is the MPS activity and how does it relate to other manufacturing planning and control (MPC) system functions, and other company activities? What's the sequence of tasks performed by the master production scheduler?

At an operational level, the most basic decisions concern how to construct and update the MPS. This involves processing MPS transactions, maintaining MPS records and reports, having a periodic review and update cycle (we call this "rolling through time"), processing and responding to exception conditions, and measuring MPS effectiveness on a routine basis.

On a day-to-day basis, marketing and production are coordinated through the MPS in terms of **order promising.** This is the activity by which customer order requests receive shipment dates. The MPS provides the basis for making these decisions effectively, as long as manufacturing executes the MPS according to plan. When customer orders create a backlog and require promise dates that are unacceptable from a marketing viewpoint, trade-off conditions are established for making changes.

The Anticipated Build Schedule

The master production schedule is an anticipated build schedule for manufacturing end products (or product options). As such, it's a statement of production, not a statement of market demand. That is, the MPS is *not* a forecast. The sales forecast is a critical input into the planning process that's used for determining the MPS, but the MPS differs from the forecast in significant ways. The MPS takes into account capacity limitations, as well as desires to utilize capacity fully. This means some items may be built before they're needed for sale, and other items may not be built even though the marketplace could consume them.

The master production schedule forms the basic communication link with manufacturing. It's stated in product specifications—in part numbers for which bills of material exist. Since it's a build schedule, it must be stated in terms used to determine component-part needs and other requirements. The MPS can't, therefore, be stated in overall dollars or some other global unit of measure. Specific products in the MPS can be end-item production designations. Alternatively, specific products may be groups of items such as models instead of end items. For example, a General Motors assembly plant might state the MPS as so many thousand J-body cars per week, with exact product mix (e.g., Chevrolet, four-door, four-cylinder) determined with a **final assembly schedule (FAS),** which isn't ascertained until the latest possible moment. If the MPS is to be stated in terms of product groups (e.g., J-body cars), we must create special bills of material (planning bills) for these groups (e.g., an average J-body car planning bill).

Linkages to Other Company Activities

Figure 6.1 presents our schematic for an overall manufacturing planning and control system. The detailed plan or schedule produced by the MPS drives all the engine and back-end systems, as well as the rough-cut capacity planning. All feedback linkages aren't shown in the figure. As execution problems

FIGURE 6.1   Manufacturing Planning and Control System

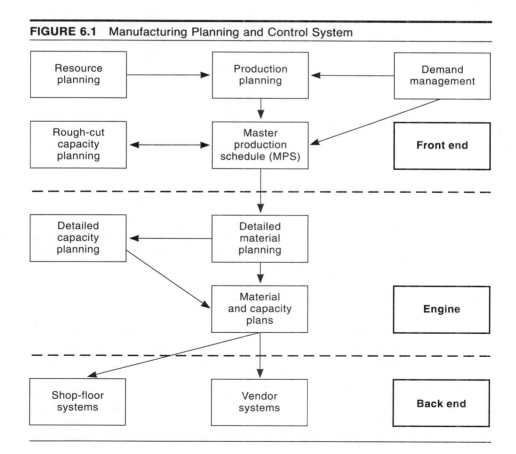

are discovered, there are many mechanisms for their resolution, with feedback both to the MPS and to other MPC modules.

The demand management block in Figure 6.1 represents a company's forecasting, order entry, order promising, and physical distribution activities. This includes all activities that place demand (requirements) on manufacturing capacities. These demands may be actual and forecast customer orders, branch warehouse requirements, interplant requirements, international requirements, and service part demands. The resultant capacity needs must be coordinated with the MPS on an ongoing basis.

The production plan represents production's role in the company's strategic business plan. It reflects the desired aggregate output from manufacturing necessary to support the company game plan. In some firms, the production plan is simply stated in terms of the monthly or quarterly sales dollar output for the company as a whole, or for individual plants or businesses. In other firms, the production plan is stated in terms of the number of units to be produced in each major product line monthly for the next year. This aggregate plan constrains the MPS, since the sum of the detailed MPS quantities must always equal the whole dictated by the production plan.

Rough-cut capacity planning involves an analysis of the master production schedule to determine the existence of manufacturing facilities that represent potential bottlenecks in production flow; that is, the linkage provides a rough evaluation of potential capacity problems from a particular MPS.

The MPS is the basis for key interfunctional trade-offs. The most profound of these is between production and marketing in terms of exact product definition in the MPS. A request to increase production for any item usually results in the need to reduce production on some other item. If production for no item can be reduced, by definition, the production plan and resultant budget for production must be changed.

Since the MPS becomes the basis for the manufacturing budget, it follows that financial budgets should be integrated with production planning/MPS activities. When the MPS is extended over a time horizon sufficient to make capital equipment purchases, a better basis is provided for capital budgets. On a day-to-day basis, both cash flow and profits can be better forecast by basing these forecasts on the planned production output specified in the MPS.

The Business Environment for the MPS

The business environment, as it relates to master production scheduling, encompasses the production approach used, the variety of products produced, and the markets served by the company. Three classic types of MPS approaches have been identified: make-to-stock, make-to-order, and assemble-to-order. The choice between these alternatives is largely one of the unit used for the MPS; that is, is the MPS to be based on end items, specific customer orders, or some group of end items and product options?

The **make-to-stock** company produces in batches, carrying finished goods inventories for most, if not all, of its end items. The MPS is the production statement of how much of and when each end item is to be produced. Firms that make to stock are often producing consumer products as opposed to industrial goods, but many industrial goods, such as supply items, are also made to stock.

The choice of MPS unit for the make-to-stock company is fairly straightforward. All use end-item catalogue numbers, but many tend to group these end items into model groupings until the latest possible time in the final assembly schedule. Thus, the Ethan Allen Furniture Company uses a **consolidated item number** for items identical except for the finish color, running a separate system to allocate a lot size in the MPS to specific finishes at the last possible moment. Similarly, the Black & Decker tool manufacturing firm groups models in a series, such as sanders, which are similar, except for horsepower, attachments, and private-brand labels. All products so grouped are run together in batches to achieve economical runs for component parts and to exploit the learning curve in the final assembly areas.

The **make-to-order** company, in general, carries no finished-goods inventory and builds each customer order as needed. This form of production is often used when there's a very large number of possible production configurations, and, thus, a small probability of anticipating a customer's exact needs. In this business environment, customers expect to wait for a large portion of the entire design and manufacturing lead time. Examples include a tugboat manufacturer or refinery builder.

In the make-to-order company, the MPS unit is typically defined as the particular end item or set of items comprising a customer order. The definition is difficult since part of the job is to define the product; that is, design takes place as construction takes place. Production often starts before a complete product definition and bill of materials have been determined.

The **assemble-to-order** firm is typified by an almost limitless number of possible end-item configurations, all made from combinations of basic components and subassemblies. Customer delivery time requirements are often shorter than total manufacturing lead times, so production must be started in anticipation of customer orders. The large number of end-item possibilities makes forecasting exact end-item configurations extremely difficult, and stocking end items very risky. As a result, the assemble-to-order firm tries to maintain flexibility, starting basic components and subassemblies into production, but, in general, not starting final assembly until a customer order is received.

Examples of assemble-to-order firms include General Motors, with its endless automobile end-product combinations; the Hyster Company, which makes forklift trucks with such options as engine type, lift height, cab design, speed, type of lift mechanism, and safety equipment; and Tennant Company, which makes industrial sweeping machines with many user-designated options.

The assemble-to-order firm typically doesn't master production schedule end items. The MPS unit is stated in **planning bills of material,** such as an average lift truck of some model series. The MPS unit (planning bill) has as its components a set of common parts and options. The option usages are based on percentage estimates, and their planning in the MPS incorporates buffering or hedging techniques to maximize response flexibility to actual customer orders.

The primary difference between make-to-stock, make-to-order, and assemble-to-order firms is in the definition of the MPS unit. However, many master production scheduling techniques are useful for any kind of MPS unit definition. Moreover, choice of MPS unit is somewhat open to definition by the firm. Thus, some firms may produce end items that are held in inventory, yet still use assemble-to-order approaches. Also, some firms use more than one of these approaches at the same time, so common systems are important.

MASTER PRODUCTION SCHEDULING TECHNIQUES

This section presents some useful basic techniques for master production scheduling. We start with the time-phased record to show relationships between production output, sales forecast, and expected inventory balance. This record provides an integration of information that is often scattered throughout the firm. We then show how plan revisions are made as you roll through time during a review cycle and take account of actual conditions. Finally, we present the process of order promising. We show how the entry of actual orders consumes the forecast in this process.

The Time-Phased Record

Using time-phased records as a basis for MPS preparation and maintenance means that they can be produced easily by computer, and they're consistent with MRP record formats. Figure 6.2 shows a highly simplified example of a master production schedule involving an item with a beginning inventory of 20 units, sales forecast of 10 units per week, and MPS of 10 units per week as well. The MPS row states the timing for *completion* of units available to meet demand. Data in this record show the expected conditions as of the current week (the first week in the master production schedule). The record covers a 12-week period (planning horizon) for which total sales forecast is 120 units. The total MPS is also 120 units.

The projected available inventory balance (available) is shown in the second row of the record in Figure 6.2. The "Available" row represents the expected inventory position at the end of each week in the 12-week schedule. It results from adding to the starting inventory of 20 units the MPS of 10 units per

FIGURE 6.2   MPS Example

						Week number						
	1	*2*	*3*	*4*	*5*	*6*	*7*	*8*	*9*	*10*	*11*	*12*
Forecast	10	10	10	10	10	10	10	10	10	10	10	10
Available	20	20	20	20	20	20	20	20	20	20	20	20
MPS	10	10	10	10	10	10	10	10	10	10	10	10
On hand	20											

week, and subtracting the sales forecast of 10 units per week. Any negative values in the projected available inventory row represent expected back orders.

There are several reasons for maintaining a positive projected inventory balance. Forecasts involve some degree of error, and the MPS is a plan for production that may not be exactly achieved. The projected inventory balance provides a tolerance for errors that buffers production from sales variations. For example, in Figure 6.2, if actual sales in week 1 were 20 units and the MPS was achieved, there would be no back order. Furthermore, if the marketing department still expected total sales for the overall 12 weeks to be 120 units (implying some week's sales would be less than the forecast of 10), production can continue at the rate of 10 units per week and still end up with the same planned inventory at the end of week 12.

The MPS row indicates the quantity and time of completion of production. Details for starting production of the various components and assembly of the product are taken care of by the MRP system. In this sense, the MPS drives the MRP system, as shown in Figure 6.1. We'll discuss several of the many alternative MPS plans. All start from the basic logic used to project the expected available inventory balance.

Figure 6.3 presents a different sales forecast from that in Figure 6.2. In Figure 6.3, the marketing department expects sales of 5 units per week for the first 6 weeks and 15 units per week for the next 6 weeks. The overall

FIGURE 6.3   A Level Production MPS Approach to Seasonal Sales

						Week number						
	1	*2*	*3*	*4*	*5*	*6*	*7*	*8*	*9*	*10*	*11*	*12*
Forecast	5	5	5	5	5	5	15	15	15	15	15	15
Available	25	30	35	40	45	50	45	40	35	30	25	20
MPS	10	10	10	10	10	10	10	10	10	10	10	10
On hand	20											

result is the same: total sales of 120 units during the 12-week period, but sales are seasonal. Figures 6.3 and 6.4 show two different master production schedules to meet this sales forecast. The MPS in Figure 6.3 represents a level 10-unit-per-week production rate over the 12-week planning horizon. The MPS in Figure 6.4, however, adjusts for the difference in sales forecasts, calling for 5 units of production per week for the first 6 weeks, and 15 units per week for the next 6 weeks.

Comparing the projected available rows in Figures 6.3 and 6.4 indicates the difference in inventory between the two MPS plans during the 12-week period. They start and end with the same inventory; but the MPS in Figure 6.3 builds up inventory during the first six weeks, which is gradually depleted during the last six weeks, while the MPS in Figure 6.4 maintains a constant inventory. These two master production schedules represent two extreme strategies. The MPS in Figure 6.3 is a **"leveling" strategy;** the MPS in Figure 6.4 is a **"chase" strategy.** The level MPS calls for no hiring, firing, or capacity adjustments. The chase MPS, on the other hand, requires production adjustments to chase the demands of the marketplace. There are, obviously, many alternative MPS plans between these two extremes. The goal is to find that plan that best balances the cost and benefits.

Figure 6.5 presents the same sales forecast as Figure 6.3, but it incorporates a lot size of 30 units. In Figure 6.5, a lot of 30 units is scheduled for

---

FIGURE 6.4   A Chase Sales MPS Approach to Seasonal Sales

						Week number						
	1	2	3	4	5	6	7	8	9	10	11	12
Forecast	5	5	5	5	5	5	15	15	15	15	15	15
Available	20	20	20	20	20	20	20	20	20	20	20	20
MPS	5	5	5	5	5	5	15	15	15	15	15	15
On hand	20											

FIGURE 6.5   Lot Sizing in the MPS

						Week number						
	1	2	3	4	5	6	7	8	9	10	11	12
Forecast	5	5	5	5	5	5	15	15	15	15	15	15
Available	15	10	5	30	25	20	5	20	5	20	5	20
MPS				30				30		30		30
On hand	20											

completion in any week when projected available balance would fall below 5 units. This trigger quantity of 5 units reflects a managerial trade-off between carrying inventory and incurring possible back orders.

The projected available balance starts at the beginning inventory position of 20 units and would drop below the 5-unit trigger in the fourth week, so a 30-unit order is scheduled for the fourth week. This order lasts until the eighth week, when the 5-unit level again would be broken. Figure 6.5 shows a total of 4 batches of 30 units being produced over the 12-week planning horizon. The first batch lasts for 6 weeks, while subsequent batches last only for 2 weeks.

Manufacturing in batches of 30 units produces inventories that last between production runs. This inventory, called **cycle stock,** is part of the projected available inventory row in Figure 6.5. The cycle stock could be cut by reducing the lot size for the whole schedule or even just during the first 6 weeks. Similarly, if the company felt overall inventory investment was too high for this MPS, the 5-unit trigger could be reduced. This would provide less **safety stock** protection against forecast errors or manufacturing problems.

Rolling through Time

Now let's turn to Murphy's law: If anything can go wrong, it will. Rolling through time requires updating the record to define how the MPS reflects actual conditions. It's necessary not only to construct the MPS but also to process actual transactions and modify the MPS.

Figure 6.6 shows the situation at the start of the second week (now for weeks 2 through 13) using the original MPS for the 12-week period as given in Figure 6.5. No material was received during the first week, since none was planned by the MPS. But actual sales were 10 units instead of 5 units, and actual inventory at the end of the first week (also the start of week 2) is 10 units (instead of 15).

In light of the higher than expected sales during the first week, it's reasonable to ask whether the sales forecast is still valid; that is, does the marketing

---

FIGURE 6.6   Using the Revised Forecast after One Week

	Week number											
	2	3	4	5	6	7	8	9	10	11	12	13
Forecast	10	10	10	10	10	15	15	15	15	15	15	15
Available	0	−10	10	0	−10	−25	−10	−25	−10	−25	−10	−25
MPS			30				30		30		30	
On hand	10											

department still believe that total sales for weeks 1 through 12 will be 120 units? What's the forecast for week 13? Let's say the marketing department has decided that the original forecast was incorrect. A new forecast at the end of the first week is for 10 units per week for the next 5 weeks (2 through 6) and 15 units per week for the following 7 weeks (7 through 13). This would total 155 units for the new 12-week planning horizon. Since the new 12-week forecast incorporates 35 more units than the original 12-week forecast, it's greater than the planned production indicated by the MPS of Figure 6.5. Figure 6.6 shows the implications of the new forecast without a revised MPS. Clearly, some adjustment to the MPS is required if anticipated customer needs are to be met. The first potential problem is seen in week 3 of the MPS, where projected available inventory goes negative. The original master production schedule called for the first batch of 30 units in week 4 and the next batch in week 8. That's not sufficient to meet the revised forecast made at the end of week 1.

The revised MPS in Figure 6.7 uses the same 5-unit trigger inventory logic used to establish the lot-sized master schedule in Figure 6.5. Figure 6.7 calls for five batches of 30 units to be produced during the MPS plan, instead of the four batches in Figures 6.5 and 6.6. This revision solves the problem of projected negative available inventory but puts in clear focus the question of feasibility. Does the company have the capacity to produce five batches during the next 12 weeks, or to immediately deliver a batch that was planned for 2 weeks hence? The capacity issue must be resolved before the new MPS is put into effect. Furthermore, high costs are typically associated with making production changes. The master production schedule should be buffered from overreaction, with changes made only when essential.

Order Promising

For many products, customers don't expect immediate delivery, but place orders for future delivery. The delivery date (promise date) is negotiated through a cycle of order promising, where the customer either asks when the

FIGURE 6.7   MPS Revisions to Accommodate Revised Forecast after One Week

	Week number											
	2	3	4	5	6	7	8	9	10	11	12	13
Forecast	10	10	10	10	10	15	15	15	15	15	15	15
Available	30	20	10	30	20	5	20	5	20	5	20	5
MPS	30			30			30		30		30	
On hand	10											

order can be shipped or specifies a desired shipment date. If the company has a backlog of orders for future shipments, the order promising task is to determine when the shipment can be made. These activities are illustrated in Figures 6.8 and 6.9.

Figure 6.8 builds upon the lot-sized MPS depicted in Figure 6.5. The original sales forecast and MPS as of the beginning of week 1 are shown. In addition, we now consider the sales forecast row to be for shipments. That is, we're forecasting when items will be shipped; and we're closing out the forecast with shipments, not with sales. The distinction separates various forms of sales (e.g., receipt of order or billing) from the manufacturing concern with actual physical movement of the goods.

The row labeled "Orders" represents the company's backlog of orders at the start of the first week. Five units were promised for shipment in the first week, three more for week 2, and an additional two units were promised for delivery in week 3. Thus, the cumulative order backlog is 10 units over the three weeks. The **available-to-promise (ATP)** value of 10 units for week 1 is calculated in the bottom portion of Figure 6.8. The on-hand inventory (20 units) has to cover all existing customer orders until the next scheduled MPS (5 + 3 + 2).

As in our previous example, we assume actual shipments in week 1 were 10 units. This means five of the units shipped in week 1 weren't on the books as sold orders at the week's start; that is, 50 percent of the orders shipped during week 1 were received during the week. This percentage varies greatly among companies.

Figure 6.9 shows the status as of the start of week 2. Figure 6.8's sales forecast and MPS have been revised in the same way as the revision from Figure 6.5 to Figure 6.7. Furthermore, additional customer orders were received during the first week for shipping in weeks 2 through 4. At the start of week 1, we had three units due to be shipped during week 2. Two additional units have been booked during week 1 for week 2 shipment, so the total backlog at the beginning of week 2 for shipment during week 2 is five units. An

---

FIGURE 6.8 Order Promising Example: Week 1

	Week number											
	1	*2*	*3*	*4*	*5*	*6*	*7*	*8*	*9*	*10*	*11*	*12*
Forecast	5	5	5	5	5	5	15	15	15	15	15	15
Orders	5	3	2									
Available	15	10	5	30	25	20	5	20	5	20	5	20
ATP	10			30				30		30		30
MPS				30				30		30		30
On hand =	20											

$$ATP = 20 - (5 + 3 + 2) = 10$$

FIGURE 6.9   Order Promising Example: Week 2

					Week number							
	2	*3*	*4*	*5*	*6*	*7*	*8*	*9*	*10*	*11*	*12*	*13*
Forecast	10	10	10	10	10	15	15	15	15	15	15	15
Orders	5	5	2									
Available	30	20	10	30	20	5	20	5	20	5	20	5
ATP	28			30			30		30		30	
MPS	30			30			30		30		30	
On hand =	10											

$$ATP = (30 + 10) - (5 + 5 + 2) = 28$$

additional three units have been booked for shipment during week 3 and an additional two units for week 4 shipment. We see then that the cumulative order backlog at the beginning of week 2 is 12 units over the 12-week planning horizon. The increase in the cumulative order backlog from 10 to 12 units over a three-week period may well have been one of the key inputs the marketing department used in revising its sales forecasts.

Orders booked for shipment in week 2 in Figure 6.9 are 5 units, while forecast total shipment in this week is 10 units. We expect to receive orders for 5 additional units during week 2 to be shipped during week 2. Carrying this analysis further, we see the cumulative backlog for weeks 2 and 3 to be 10 units, and the cumulative forecast for the same 2 weeks to be 20 units. This implies that, between the start of week 2 and the end of week 3, we expect to receive orders for 10 additional units to be shipped during that two-week period.

A more interesting relationship is seen between the order backlog and the MPS. The order backlog for week 2 in Figure 6.9 is 5 units, and the anticipated production plus beginning inventory is 40 units. This suggests we still have 35 units to use to meet additional customer requests; that is, it looks like we could make total shipments of up to 35 units in addition to what's already promised in week 2.

This isn't true; we can only accept 28 *additional* units for shipment during week 2; that is, we only have 28 units "available to promise." The reason is that the next scheduled production for this item isn't until week 5; therefore, as shown in the bottom portion of Figure 6.9, the beginning inventory of 10 units plus the 30 units in the master schedule for week 2 have to cover all existing orders for weeks 2 through 4. Since we already have orders for 12 units on the books for shipment during that period, we can only accept up to 28 additional units. That is why Figure 6.9 shows a value of 28 units in ATP for week 2. Those 28 units could be shipped any time during weeks 2, 3, or 4, or any other week in the future.

An important convention about the format of the time-phased MPS record in Figures 6.8 and 6.9 concerns the available row. The available row is the expected ending inventory. A frequently encountered convention is to use the greater of forecast or booked orders in any period for projecting the available inventory balance. In our example in Figures 6.8 and 6.9, actual customer orders never exceed forecasts for the periods. The general calculation for the available row is: Previous available + MPS − (Greater of forecast or orders).

The available-to-promise logic is a bit harder to state neatly. An ATP value is calculated for each period in which there is an MPS quantity. In the first period, it's the on-hand plus any first-period MPS minus the sum of all orders until the next MPS. For later periods, it's the MPS minus all orders in that and subsequent periods until the next MPS. Both of these rules, however, have to be modified to reflect subsequent-period ATP deficiencies. Figure 6.9 now shows 30 units as ATP in weeks 5, 8, 10, and 12. These quantities match the MPS quantities and reflect the lack of customer orders beyond week 4. Suppose an order for 35 units was booked for week 10 in Figure 6.9. The ATP for week 10 would be zero, and the ATP for week 8 would fall to 25. That is, the 35-unit order in week 10 is to be satisfied by 30 units produced in week 10 and 5 from the MPS produced in week 8.

Some companies choose to show the available-to-promise row as cumulative (58 in week 5). However, keeping the additional increments of ATP separate makes order promising easier and also has the advantage of not overstating the availability position; that is, there are not really 58 available to promise in week 5 as well as 28 in week 2. Some software packages provide ATP both as indicated in Figure 6.8 *and* in cumulative format.

In many firms, accurate order promising allows the company to operate with reduced inventory levels; that is, order promising allows the actual shipments to be closer to the MPS. Companies in effect buffer uncertainties in demand by their delivery date promises. Rather than carry safety stocks to absorb uneven customer order patterns, those firms "manage" the delivery dates.

Consuming the Forecast

One authority on master production scheduling, Richard Ling, originated the idea that actual customer orders "consume" the forecast; that is, we start out with an estimate (the forecast), and actual orders come in to consume (either partially, fully, or over) the estimate. We see this in Figure 6.9. Of the 10 units forecast for week 2, 5 have been consumed. For week 3, 5 of 10 have been consumed, as have 2 of the 10 for week 4.

Let's consider Figure 6.9 and see if, during week 2, we can accept the following hypothetical set of customer orders, assuming they were received in the sequence listed:

Order number	Amount	Desired week
1	5	2
2	15	3
3	35	6
4	10	5

The answer is yes to all but order number 4. Since the total amount requested is 65, and the cumulative amount available to promise is only 58 for weeks 2 through 6 (28 in week 2 plus 30 in week 5), only 3 units of order number 4 could be shipped in this period. Let's say the customer would not accept a partial delivery, so we negotiated for delivery in week 8. Figure 6.10 shows the time-phased record at the beginning of week 3 if no more orders were received in week 2 and the forecast for week 14 is incorporated.

Obviously, the set of orders received during week 2 represents a major deviation from the forecast. However, it does allow us to see clearly how we can use the record to make decisions as we roll through time. Let's first review the process's arithmetic. To calculate on-hand inventory at the beginning of week 3, we start with week 2's beginning inventory of 10. We add week 2's MPS of 30; then we subtract the orders for 5 units shipped in week 2 (shown in Figure 6.9) and we subtract the order for week 2 just promised (5 units). The result is $10 + 30 - 5 - 5 = 30$.

The available row provides the master production scheduler with a projection of the item availability throughout the planning horizon in a manner analogous to projecting an on-hand inventory balance in an MRP record. The convention of subtracting the "greater of forecast or orders" and adding the MPS quantities to calculate the available row has an effect here. For week 3 in Figure 6.10, for example, the available is the 30 units of inventory minus the 20 units on order, for a total of 10. In week 4, the available quantity is 0, the difference between the week 3 available of 10 and the forecast of 10. The use of the greater of forecasts or orders for calculating

---

FIGURE 6.10 Order Promising Example: Week 3

	\multicolumn{12}{c	}{Week number}

	3	4	5	6	7	8	9	10	11	12	13	14
Forecast	10	10	10	10	15	15	15	15	15	15	15	15
Orders	20	2		35		10						
Available	10	0	20	−15	−30	−15	−30	−15	−30	−15	−30	−45
ATP	3		0			20		30		30		
MPS			30			30		30		30		
On hand	30											

the available row is consistent with forecast consumption. If actual orders exceed forecast, there has been an "over" consumption that should be taken into account. As actual orders are less than forecasts, the result will appear in the on-hand balance (a gross-to-net process); this also impacts the available calculations.

The available position at the end of the planning horizon is important information for managing the master production schedule, as is the existence of negative available data during the planning horizon. During the planning horizon there is typically some length of time in which changes are to be made only if absolutely essential, to provide stability for planning and execution. At the end of this period, master production schedulers have maximum flexibility to create additional MPS quantities. If the projected available is positive in the time bucket at the end of this period, then scheduling more production of the item may not be necessary. Note, for example, at the start of week 2 in Figure 6.9 the week 13 available was 5 and no MPS quantity was entered for week 13. At the start of week 3 in Figure 6.10, the available for week 14 is −45, due to consumption of the forecast by orders booked in week 2. This indicates that in week 3 the master production scheduler should consider scheduling production for completion in week 14. The negative available numbers for weeks 6 through 14 indicate desire for more MPS during the planning horizon. Whether this can be achieved is a matter of response time, availability of materials, and competing needs (other items).

The convention of the greater of forecast or orders means large orders will be immediately reflected in the availability position. This signals the master production scheduler to consider responding to the order by increasing future item availability, which can be used for booking future orders. Separately the available-to-promise row controls the actual order promising. A related question for sales is whether a large order "consumes" forecast for only one period.

To calculate the available-to-promise position, we consider only actual orders and the scheduled production, as indicated by the MPS. We calculate only the incremental available to promise. Note that in Figure 6.10 the 30 units on hand must cover actual orders for weeks 3 and 4, since no additional production is scheduled. In week 5, 30 additional units are scheduled, but none of them is available to promise. The 3 units available to promise in week 3 come from the 30 on hand minus the 20 units ordered for week 3 and the 2 units ordered for week 4. Another 5 units of those on hand are needed for the order of 35 in week 6, since the MPS of 30 units in week 5 isn't sufficient. This leaves only 3 units available to promise for weeks 3 through 7. Of the 30 units to be produced in week 8, we see 10 will be used for the order in week 8. This leaves 20 units available to promise in week 8.

Note that the later customer orders are covered by the later MPS quantities. The 10-unit order for week 8 could have been covered by 3 units in week 3 plus 7 units in week 8 instead of all 10 in week 8. This would have left no units for promising from week 3 until week 8, greatly reducing promise flexibility. The convention is to preserve early promise flexibility by reducing the available to promise in as late a period as possible.

The use of both the "Available" row and the "ATP" row is the key to effective master production scheduling. Using the ATP to book orders means that no customer promise will be made that can't be kept. Note this may mean some orders must be booked at the end of a planning horizon concurrent with creating an additional MPS quantity. As actual orders are booked (the "Order" row), or anticipated (the "Forecast" row), or shipped (on-hand inventory), the "Available" row provides a signal for the creation of an MPS quantity. Once created, the MPS quantity provides the items available to promise for future orders.

The final item of interest in Figure 6.10 is to again focus on the negative available quantities from weeks 6 through 14. These negative quantities indicate potential problems—but only *potential* problems. However, costly MPS changes should not be made to solve "potential" problems. But a condition that created a negative ATP represents a "real" problem.

The time-phased records in Figures 6.9 and 6.10 are similar to MRP records. In fact, the same data can be integrated with standard MRP formats. The primary advantage of doing so is to obtain standard record processing. However, it's necessary to keep track of actual customer orders and the timings of MPS quantities to make ATP calculations. The result is the MRP data base will need to be expanded.

BILL OF MATERIAL STRUCTURING FOR THE MPS

The assemble-to-order firm is typified by an almost limitless number of end-item possibilities made from combinations of basic components and subassemblies. For example, the number of unique General Motors automobiles runs into billions! Moreover, each new product option for consumers tends to double the number of end-item possibilities. This means the MPS unit in the assemble-to-order environment can't feasibly be based on end items. Defining other units for master production scheduling means creating special bills of material. In this section, we present a few key definitions to clarify what a bill of material is and is not. Thereafter, we discuss modular bills of material and planning bills of material that aid MPS management. With this background, it's possible to see how master production scheduling takes place in the assemble-to-order environment.

Key Definitions

The **bill of material** is narrowly considered to be an engineering document that specifies the ingredients or subordinate components required to physically make each part number or assembly. A **single-level bill of material** comprises only those subordinate components that are immediately required, not the components *of* the components. An **indented bill of material** is a list of components, from the end item all the way down to the raw materials; it does show the components of the components.

The **bill of material files** are those computer records designed to provide desired output formats. The term **bill of material structure** relates to the architecture or overall design for the arrangement of bill of material files. The bill of material structure must be such that all desired output formats or reports can be provided. A **bill of material processor** is a computer software package that organizes and maintains linkages in the bill of material files as dictated by the overall architecture (bill of material structure). Most bill of material processors use the single-level bill of material and maintain links or chains between single-level files. It's the bill of material processor that's used in MRP to pass the planned orders for a parent part to gross requirements for its components.

The single-level bill and the indented bill are two alternative output formats of the bill of material. Alternative output formats are useful for different purposes. For example, the single-level bill supports order launching by providing the data for component availability checking, allocation, and picking. Industrial engineers often use the fully indented bill to determine how to physically put the product together; accountants use it for cost implosions. A fundamental rule is that a company should have one, and only one, set of bills of material or **product structure** records. This set should be maintained as an entity and be so designed that all legitimate company uses can be satisfied.

The rest of this section presents concepts providing another way of thinking about the bill of material. The traditional approach is from an engineer's point of view; that is, the way the product is *built*. The key change required to achieve superior master production scheduling for assemble-to-order products is to include bill of material structures based on the way the product is *sold*. In this way, the bill of material can support some critical planning and management activities.

Constructing a bill of material structure or architecture based on how the product is sold, rather than how it's built, offers important advantages. Achieving them, however, isn't without cost. The primary cost is that the resultant bills of material may no longer relate to the way the product is built. Activities based on *that* structure (e.g., industrial engineering) will have to be based on some new source of data; that is, if the description of how the parts physically go together isn't found in the bill of material, an alternative set of records must be maintained. Providing alternative means to satisfy these needs can be costly in terms of both file creation and maintenance.

The Modular Bill of Material

A key use of bill of material files is in translating the MPS into subordinate components requirements. One bill of material structure or architecture calls for maintaining all end-item buildable configurations. This bill of material structure is appropriate for the make-to-stock firm, where the MPS is stated in end items. For each end item, a single-level bill is maintained, which contains those components that physically go into the item. For General Motors, with its billions of possible end items, this bill of material structure isn't feasible.

Figure 6.11 shows the dilemma. A solution is to establish the MPS at the option or module level. The intent is to state the MPS in units associated with the "waist" of the hourglass. This necessitates that bill of material files be structured accordingly; each option or module will be defined fully in the bill of material files as a single-level bill of material. Thus, the modular bill of material structure's architecture links component parts to options, but it doesn't link either options or components to end-item configurations. If the options are simply buildable subassemblies, then all that's required for the new architecture is to treat the subassemblies as end items; that is, designate them as level zero, instead of level 1. In most cases, however, the options

FIGURE 6.11   The MPS Hourglass

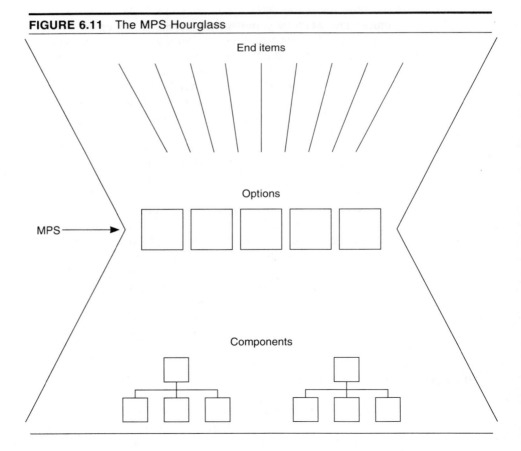

aren't stated as buildable subassemblies but as options that provide features to the customer.

Consider, for example, the air-conditioning option for a car. The single-level bill of material would show this option or module as consisting of a particular radiator, fan, hoses, compressor, and interior knobs and levers. These items are not, however, assembled together. They are assembled with still other parts as subassemblies, which eventually are assembled into the automobile.

Using the air-conditioning option as a bill of material will pass demand from the customer who wants this option down to the necessary parts. It can also be used to forecast demand for air conditioners. However, this bill of material isn't useful in the physical building process for air conditioners. For example, the air-conditioning knobs are planned by the bill for the air-conditioning option. They aren't planned by the bill of the dashboard assembly where they're installed. Thus, the industrial engineer needs other means to say how the dashboard is to be assembled and from what components.

Using the modular bill of material structure for a firm with a situation similar to that in Figure 6.11 permits the MPS to be stated in fewer different units. The MPS is stated in the terms in which the product is *sold*, not in terms in which it's built; that is, air conditioning, two doors, automatic transmission, fancy trim, and so on. The approach is compatible with marketing perceptions of models, options, and trends in options (e.g., more people buying cars with air conditioning), which tends to improve forecasting. Master scheduling may be made easier by using modular bills, but order entry tasks are more complex since each option must be evaluated.

Once the individual customer order (representing a unique collection of options) is entered, it serves the function of a one-time, unique single-level bill of material; that is, it specifies which options or modules are to be included for the particular customer order. It's controlled by a separate final assembly schedule.

The Planning Bill of Material

Restructuring the bill of material to better perform MPS activities has led many people to see that alternative bill of material approaches have additional applications. An example is the planning bill of material, which is any use of bill of material approaches for planning only, as opposed to use for building the products. The modular bill of material approach just described involves one form of a planning bill, since it's used for developing material plans and modules not all of which are buildable.

The most widely used planning bill of material is the **super bill.** The super bill describes the related options or modules that make up the *average* end item. For example, an average General Motors J-body car might have 0.6 Chevrolet unique parts, 2.6 doors, 4.3 cylinders, 0.4 air conditioners, and the like. This end item is impossible to build; but using bill of material logic, it's very useful for planning and master production scheduling. Bill of material processing dictates that the super bill be established in the product structure files as a legitimate single-level bill of material. This means that the super bill will show all the possible options as components, with their average decimal usage. The logic of bill of material processing permits decimal multiples for single-level component usages. The super bill combines the modules, or options, with the decimal usage rates to describe the average car. The bill of material logic forces arithmetic consistency in the mutually exclusive options; for example, the sum of two possible engine options needs to equal the total number of cars.

The super bill is as much a marketing tool as a manufacturing tool. With it, instead of forecasting and controlling individual modules, the forecast is now stated in terms of total average units, with attention given to percentage breakdowns—to the single-level super bill of material—and to *managing* module inventories using available-to-promise logic on a day-to-day basis as actual customer orders are booked.

Let's consider an artificially small example. The Garden Till Company makes rototillers in the following options:

Horsepower: 3HP, 4HP, 5HP.
Drive train: Chain, Gear.
Brand name: Taylor, Garden Till, OEM.

The total number of end-item buildable units is 18 (3 × 2 × 3). Management at the end-item level would mean each of these would have to be forecast. Figure 6.12 shows a super bill for four-horsepower tillers. Using this artificial end item, an average four-horsepower tiller, only one forecast is needed from marketing. More important, the MPS unit can be the super bill. The entry of 1,000 four-horsepower super bill units into the MPS would plan the appropriate quantities of each of the options to build 1,000 four-horsepower units in the average option proportions. Actual orders may not reflect the average in the short run, however.

Figure 6.12 shows the use of safety stocks for the options to absorb variations in the mix. No safety stock is shown for the common parts. This means protection is provided for product mix variances but not for variances in the overall MPS quantity of four-horsepower tillers. A commitment to an MPS quantity for the super bill means exactly that number of common parts will be needed. In Figure 6.12's example, if 1,000 four-horsepower super bills were entered, the bill of materials would call for 1,000 common parts modules, 600 gear options, 400 chain options, 400 Taylor options, 500 Garden Till options, and 100 OEM options. The safety stocks allow shipments to customers to vary from the usages specified in the bill of material percentage usage.

FIGURE 6.12   The Four-Horsepower Super Bill

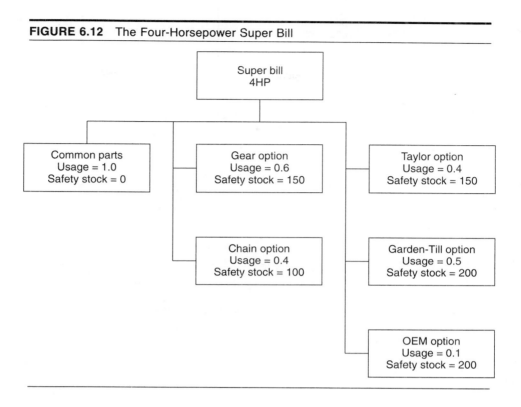

Although 600 of the 1,000 four-horsepower tillers are expected to be finished in the gear drive option, as many as 750 can be promised because of the safety stock. Similar flexibility exists for all other options. Safety stocks are maintained with MRP gross to net logic, so appropriate quantities are maintained as actual conditions become reflected in replenishment orders. Moreover, the safety stock will exist in matched sets because of the modular bill of material structure. Matched sets occur because when one unit of the module is specified for safety stock, *all* parts required for that unit will be planned. Furthermore, costs of all safety stocks are readily visible; marketing can and should have the responsibility to optimize the mix.

Order entry using planning bill of material concepts tends to be more complex than when the structure is end-item based. To accept a customer order, the available-to-promise logic must be applied to each option in the order, meaning it's necessary to check each of the affected modules. Figure 6.13 shows the flow for a particular customer order, in this case for 25 Taylor four-horsepower in the gear option (T4G). The safety stocks are available for promising and will be maintained by the gross to net logic as additional MPS quantities are planned.

FIGURE 6.13   Available-to-Promise Logic with Modular Bill Architecture (order for 25 T4Gs)

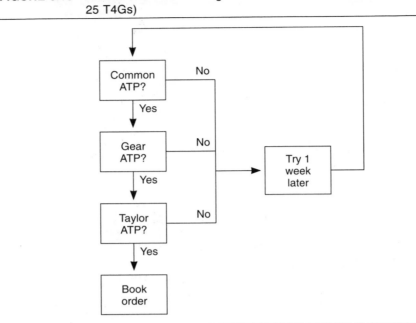

THE FINAL ASSEMBLY SCHEDULE

The final assembly schedule (FAS) states the exact set of end products to be built over some time period. It's the schedule that serves to plan and control final assembly and test operations; included are the launching of final assembly orders, picking of component parts, subassembly, painting or other finishing, scheduling the fabrication or purchase of any component items not under MPS control but needed for final assembly, and packing. In short, the FAS controls that portion of the business from fabricated components to completed products ready for shipment. It may be stated in terms of customer orders, end product items, serial numbers, or special assembly order numbers.

Relation to the MPS

The master production schedule represents an anticipated build schedule. The FAS is the actual build schedule. The MPS disaggregates the production plan into end items, options, or groups of items, whereas the FAS is the last disaggregation—into exact end-item definitions. The distinction is that the MPS generally incorporates forecasts or estimates of actual customer orders in its preparation, with actual orders thereafter imperfectly consuming these forecasts; the FAS represents the last possible adjustment that can be made to the MPS; therefore, it's advisable to make that adjustment as late as possible. Any unsold items on the FAS will become part of the firm's finished-goods inventory.

The FAS is distinct and separate from the MPS. The distinction is most clearly seen in the assemble-to-order environment. There, the MPS is typically stated in super bills and options, whereas the FAS must be stated in terms of the exact end-item configurations. However, even in make-to-stock firms (such as Ethan Allen and Black & Decker), the MPS is stated in consolidated groups of items, such as all models of a table that differ only in finish, or all models of an electric drill that differ only in speed or gearing. In both cases, flexibility is so maintained that the final commitment to end items can be made as late as possible.

It's important to note that in make-to-stock firms a single-level bill of material is typically maintained for each end item. This means that the conversion from MPS to FAS is simply the substitution of one end item part number for another. Both are valid, and both explode to components in the same way. For some make-to-stock firms, the MPS is stated in terms of the most common or most complete end item. As actual sales information is received, other end items are substituted. This process continues until a time is reached when all final substitutions are made.

For assemble-to-order and make-to-order firms, end item bills of material are not maintained. If the FAS is stated in terms of customer orders, these orders must be translated into the equivalent of a single-level bill of material; that is, these orders must lead to bill of material explosion for order release, picking, and so on. This is easily accommodated if the customer order is stated in the same modules as the planning bill. For the tillers, this means that the customer order is stated in brand name, horsepower, and drive train terms.

Avoidance of firming up the FAS until the last possible moment means the time horizon for the FAS is only as long as dictated by the final assembly lead time (including document preparation and material release). Techniques that help to delay the FAS commitment include bill structuring, close coupling of order entry/promising systems, partial assembly, stocking subassemblies, and process/product designs with this objective.

The Hill-Rom FAS

The Hill-Rom Company, a division of Hillenbrand Industries, manufactures hospital furniture and other health-care equipment. One product, an over-bed table, comes in four different models, 10 alternative color high-pressure laminate tops, and four different options of chrome "boots" (to protect the base) and casters. The result is 160 (10 × 4 × 4) end-item possibilities. Let's examine a super bill approach to master production scheduling and final assembly scheduling in this environment.

Figure 6.14 shows a super bill of material for this group of products. Manufacturing lead time for over-bed tables is 20 weeks, which means that the MPS must extend at least that far into the future. This means that an MPS time-phased record must be maintained over at least this time horizon for each of the 19 common part and option bills of material shown in Figure 6.14.

The final assembly lead time for this product is four weeks. This involves part availability checking, order launching, component part release, welding of subassemblies, snag grinding to smooth welded surfaces, degreasing, painting, subassembly, and final assembly.

Hill-Rom is basically an assemble-to-order company, but some finished-goods inventory is held. This means, for each of the 160 end-item over-bed

FIGURE 6.14   Over-Bed Table Super Bill

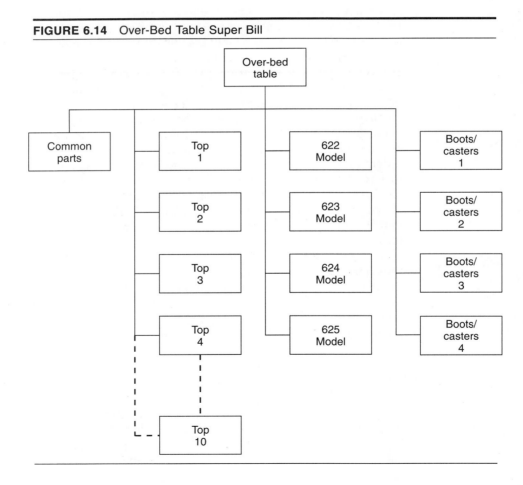

tables, an on-hand balance record must be maintained. This is incorporated in a time-phased FAS record. The overall need is for 19 time-phased MPS records (1 for each option and the common parts) maintained for week 5 through at least week 20, plus 160 time-phased FAS records, maintained for each end item for weeks 1 through 4.

The FAS job is to convert MPS records into FAS records as we roll through time. This is done by the master production scheduler interacting with marketing, since the FAS represents the final culmination of the MPS process.

A related task is order promising, since customer orders may be promised out of the FAS system (on hand or in final assembly) or out of the MPS system (option by option, as done in Figure 6.13). Figures 6.15, 6.16, and 6.17 show this process. Figure 6.15 shows an FAS record for one of the 160 end items; it's maintained only for the length of the FAS lead time, four weeks. The record is for the 622 model, with "gunstock"-colored top, and 01B boots/caster combination; the finished-good item number, which identifies this configuration, is 17123-01B GUN.

FIGURE 6.15 FAS Record for 17123-01B GUN Table

Part no.	Item	FAS lead time	On hand	
7123-01B GUN	Over-bed Table	4	120	

	Week	1	2	3	4	
	Orders	10			30	Before
	Available	160	160	210	230	booking order
	Available to promise	160		50	20	F 5264
	FAS	50		50	50	

	Week	1	2	3	4	
	Orders	10		200	30	After
	Available	160	160	10	30	booking order
	Available to promise	10			20	F 5264
	FAS	50		50	50	

MPS pegging detail				Actual order pegging detail			
Week	Shop Order	Quantities	Action	Week	Quantity	Customer Order	Code
1	011	50		1	10	F 5117	F
3	027	50		3	200	F 5264	F
4	039	50		4	30	F 5193	F

Note the record contains *only* orders (no forecasts), and they're used to compute the "Available" row. This convention recognizes that the FAS is finishing out products in a specific configuration. The 50 units that will be completed this week aren't subject to uncertainty. If no customer order is received for these, they'll go into stock; essentially, the company has written a "sales order."

Figure 6.15 also shows an order for 200 units being booked. The customer has requested shipment of the complete order as soon as possible. The available-to-promise (ATP) logic leads to putting the order into week 3, since only 160 units can be promised prior to week 3. This is shown in the "before" and "after" sections of Figure 6.15. The bottom section gives supporting pegging data for the orders. Customer orders are pegged with an F code (satisfied from the FAS system).

Next we assume another customer order is received, requesting shipment in week 6. Since this is outside the FAS, it can be satisfied from MPS. Figure 6.16 shows the MPS record for the common-parts option. Note the MPS quantities are for common and option part numbers. Moreover, inventories for these options can exist, even though the physical inventory would be only a collection of parts. (The on-hand balance for common parts is 10.) The MPS option quantities can also be committed for final assembly of specific end items. When final assembly starts, it takes four weeks to finish out the end

FIGURE 6.16 MPS Record for Common Parts

Part no.	Item	MPS lead time							
1234	Common parts	20							
Week		5	6	7	8	20	21		
Forecast		75	75	75	75	75	75		Before
Orders		10							booking
Available		−15	210	235	160	125	50		order
Available to promise		50	300	100					
MPS		50	300	100	0				

On hand = 10

Week		5	6	7	8	20	21		
Forecast		75	75	75	75	75	75		After
Orders		10	200						booking
Available		−15	85	110	35	0	−75		order
Available to promise		50	100	100					
MPS		50	300	100					

On hand = 10

FIGURE 6.17 Available-to-Promise Logic

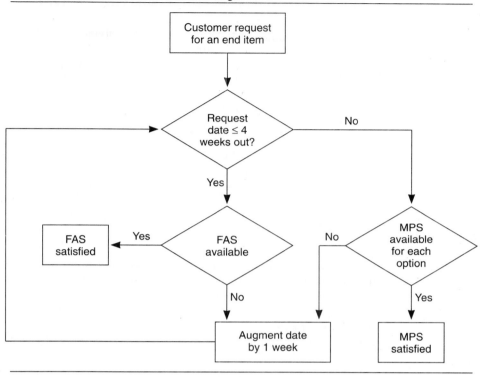

item. Figure 6.17 shows the ATP logic required to book any customer order in the Hill-Rom system.

THE MASTER PRODUCTION SCHEDULER

We turn now to the fifth of our eight topics in master production scheduling: Who's the master production scheduler, what does he or she do, and what's the appropriate job description? First, we briefly examine the use of some MRP concepts for the master scheduler.

The MPS as a Set of Firm Planned Orders

An interesting advantage of using standard MRP records to manage the master production schedule derives from the firm planned order concept. The firm planned order is similar to any planned order in that it explodes through product structures. However, it's *not* changed in either timing or amount as a result of MRP record processing. It's firm, and it can only be changed as the result of an action taken by a responsible person.

It's useful to think of the MPS as a set of firm planned orders. Thereafter, the master production scheduler's job is to convert planned orders to firm planned orders, and to *manage* the timing and amounts of the firm planned orders. The "available" row in the time-phased record provides the primary signal for performing this task. Standard MRP exception codes can provide indications of when and to what extent firm planned orders might not meet the needs.

Managing the timing and amounts of the firm planned orders means that any changes to the MPS have to be carefully evaluated in terms of their impact on material and capacity plans. The key need is to clearly understand trade-offs between customer needs and other MPC system objectives.

The Job

The master production scheduler has the primary responsibility for making any additions or changes to MPS records. He or she also has the primary responsibility for disaggregating the production plan to create the MPS and for ensuring that the sum of the detailed MPS production decisions matches the production plans. This involves analyzing trade-offs and telling top management about situations requiring decisions beyond the scheduler's authority level.

As part of the general feedback process, the master production scheduler should monitor actual performance against the MPS and production plan and distill operating results for higher management. The master production scheduler can also help in the analysis of what-if questions by analyzing the impact on the MPS of changes in plans.

The master production scheduler is often responsible for launching the final assembly schedule. This schedule represents the final commitment, taken as late as possible, to exact end items; that is, the final assembly schedule has to be based on specific finished-good items. Other master production scheduler activities include interface with order entry plus an ongoing relationship with production control to evaluate the feasibility of suggested changes.

Much of this activity involves resolving competing demands for limited capacity. Clearly, if several records for the master production schedule show a negative available at the end of the planning horizon, some trade-offs must be made. Not everything can be scheduled at once. Management of the firm planned orders must be done within capacity constraints. The available position indicates the priority for making those trade-offs. The lower the number of periods of supply or the larger the number of periods of backlog the available position shows, the more urgent the need. If too many are urgent, feedback to marketing may be necessary to change budgets.

Figure 6.18 shows the job description for the master production scheduler at Hyster-Portland. This formal job description makes it clear that the master

FIGURE 6.18 Hyster-Portland Master Production Scheduler Job Description

NAME					JOB TITLE	MASTER SCHEDULER

DEPARTMENT NAME					DATE ASSIGNED TO PRESENT POSITION
CAPACITY AND MATERIAL PLANNING					

LOCATION NAME		LOC. CODE	DATE OF LAST EVALUATION
PORTLAND PLANT		02	

PREPARED BY: Roger B. Brooks	**JOB DESCRIPTION**	8-23

RATING					
INADEQUATE	MARGINAL	SATISFACTORY	GOOD	SUPERIOR	

BASIC FUNCTION:

Responsible for planning, organizing, and controlling the activities of the Master Scheduling Section within the Material and Capacity Planning Department.

REPORTS TO: Capacity and Material Planning Manager

SUPERVISES: Order Entry Administrator, Assembly Schedulers, Master Scheduling Planners

RESPONSIBILITIES:

1. Responsible for creating and maintaining a realistic and valid Master Schedule for all Portland Products. The Master Schedule should reflect requirements for Customer Orders, Stock Orders, the Depot, Product Availability Plan and Option Forecast, Interplant and Export Orders with economic and time consideration for Plant Inventory, Manufacturing Efficiencies, Customer Service and Plant Capacity. Proper Master Scheduling will allow the plant to operate at a steady state during periods of oversold order bookings without a build up of past due job orders and inventories, while simultaneously experiencing no idle capacity on bottleneck work centers.

2. Responsible for accurate, timely and organized order entry of Sales Orders, including IMT's and Export Orders, for all Portland manufactured lift trucks, carriers, winches, sold alones and production parts. Responsible for maintaining communications with the Industrial Truck and Tractor Attachment Sales Order desks to provide accurate and timely customer sales order shipping commitments.

3. Responsible for obtaining shipping instructions for all customers orders via communications with the Industrial Truck and Tractor Attachment Sales Order desks to prevent shipping delays.

4. Responsible for scheduling industrial truck, carrier, sold alone and winch assembly to support customer commitments within the constraints of the Master Schedule, Assembly Department and Parts Bank. Responsible for scheduling major weldments and front ends to support the assembly schedule.

5. Responsible for publishing the Daily Production Report and Final Unit Shipment Report in an accurate and timely manner.

6. Select, develop, and evaluate employees so they, as a group, accomplish the foregoing tasks in a businesslike, efficient, and professional manner.

Source: W. L. Berry, T. E. Vollmann, and D. C. Whybark, *Master Production Scheduling: Principles and Practice* (Falls Church, Va.: American Production and Inventory Control Society, 1979), p. 142.

production scheduler needs to constantly balance conflicting objectives and make trade-offs. The position requires maturity, an understanding of both marketing and finance, and an ability to communicate. Computer software can greatly aid the master production scheduler, but judgments will always be required.

Managing the MPS Data Base.   For the master production scheduler to operate effectively, it's also critical that there be one single unified data base for the MPS, that it link to the production plan and to detailed material planning systems, and that clear responsibilities for all transactions be established. This involves not only the usual data integrity issues but also some organizational issues.

In the case of the MPS, many transactions occur in different functional areas. For example, receipts into finished goods may come from completed assemblies (production), shipments from order closing (marketing), or bills of lading (finance). It's critical that exact responsibilities be established for transaction processing, and that data linkages to MPS systems and files be rigorously defined and maintained.

Another critical data base requirement for the MPS is proper control over both engineering and nonengineering changes to the bill of material data base. The MPS is often stated in planning bill units that may not be buildable (e.g., an average J-body car). This requires a more complex bill of material or product structure data base. The result is a greater need to procedurally control all changes to the bill of material and to evaluate the impact of changes both from an engineering point of view and in terms of the effect on nonengineering bills of material.

To support the master production scheduler, the time-phased MPS-record–oriented software system must produce the time-phased records to maintain the data base, provide the linkages to other critical systems, provide MPS monitoring and exception messages, and provide for all MPS transactions. Included are entering of order quantities into the MPS, firm planned order treatment, removing MPS order quantities, changing the latter's timing or amount, converting MPS quantities to final assembly schedule (FAS) quantities, launching final assemblies, monitoring FAS scheduled receipts for timing or quantity changes, closing out FAS receipts into finished-goods inventory, and providing for all customer order entry pegging and promising activities.

EXAMPLES

We turn now to two actual MPS examples. First we'll show Ethan Allen's approach to master production scheduling. The approach uses a form of time-phased record. We'll see how standard MRP system approaches can be usefully applied and highlight aspects of the master production scheduler's job.

The second example illustrates how Jet Spray uses packaged software with an MPS module. The software incorporates available-to-promise logic and other features.

The Ethan Allen Master Production Schedule

The Ethan Allen Furniture Company produces case goods (wood furniture) in 14 geographically dispersed factories. Its total product line is 980 consolidated item numbers. (Different finishes for the same item make the number of end items about 50 percent larger.) Each consolidated item number is uniquely assigned to a particular assembly line or building station in one of the 14 factories. For each assembly line in each factory, a capacity is established in hours such that, if hours of capacity are fully utilized on all lines, overall company objectives as stated in its production plan will be met.

A forecast of demand is made for each consolidated item number, using statistical forecasting methods. A lot size for each item is also determined, based on economic order quantity concepts. For each assembly lot size, hours required on the assembly line are estimated. For each product, expected weekly priorities are established by dividing the expected beginning inventory by the weekly forecast. In weeks after the first, expected beginning inventory takes account of production and expected sales. The assembly line is loaded to capacity in priority sequence, smallest to largest. Figure 6.19 provides a simplified example for an assembly line with 35 hours of weekly capacity. The simplified example is based on only four products. Actual lines typically manufacture from 15 to 100 different items.

The top section of Figure 6.19 provides the basic data for each of the four products: the beginning inventory, weekly forecast of sales, lot size, and estimated hours to assemble one lot. Note that for product C, there's a beginning back order or oversold condition.

The middle portion of Figure 6.19 is the set of time-phased priority data. For product A in week 1, the beginning inventory of 20 is divided by the weekly forecast of 5, yielding a priority of 4; that is, at the beginning of week 1 there are four weeks of inventory for product A. Similar priority calculations are made for products B, C, and D in week 1.

The rule for assigning products to the assembly line is to take that product with the smallest priority—the most urgent need. Thus, product C is scheduled for production first. The assignment of product C to the assembly line in week 1 consumes all of the 35 hours of capacity in that week plus 25 hours in week 2. This is so since it takes 60 hours to assemble a batch of 150 of product C.

Moving to week 2, the expected beginning inventory for product A is 15, since forecast sales for week 1 is 5. Divide 15 by the weekly forecast (5) to get the expected priority for week 2 ($15 \div 5 = 3$). Alternatively, if four weeks

---

FIGURE 6.19   Simplified Ethan Allen MPS Example

Basic data:

Product	Beginning inventory	Weekly forecast	Lot size	Hours per lot size
A	20	5	50	20
B	50	40	250	80
C	−30	35	150	60
D	25	10	100	30

Priorities:

Product	P_1	P_2	P_3	P_4	P_5	P_6	P_7	P_8
A	4	3			0		−2	4.5
B	1.25	.25			3.5		1.5	.5
C	−.86	.64			−.57		.29	.71
D	2.5	1.5			−1.5		6.5	5.5

Schedule:

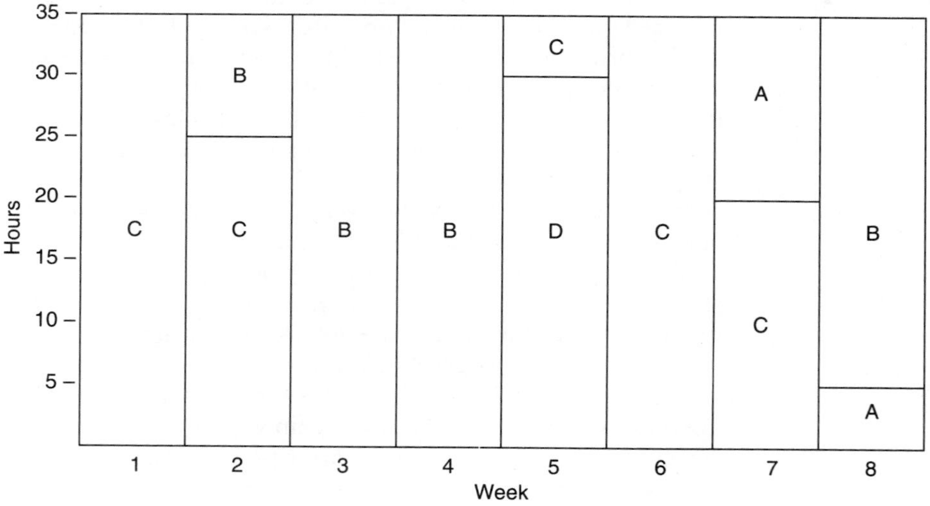

of sales are in inventory at the beginning of week 1, we'd expect to have three weeks of sales at the beginning of week 2 if no production of product A takes place. This means, for each product not produced, its priority number in the succeeding week is reduced by 1. The expected priority for product C at the start of week 2 can be computed by finding 35/60 of 150, adding this to the beginning inventory of −30, subtracting 35 units of forecast demand for week 1, and dividing the result by the forecast of 35 to give a value of 0.64.

The lowest-priority product for week 2 is B (.25). Since a lot size of the product takes 80 hours, capacity is fully utilized until the end of week 4. This is why no priority data are given for weeks 3 and 4. A similar situation is true for week 6 when product C, started in week 5, uses the full week's capacity. By loading each line to its weekly capacity, no more and no less, the match between the production plan dictated for each assembly line and detailed MPS decision making is maintained. Calculations in subsequent weeks involve adding in any production and reducing inventories by expected sales. For example, product C's priority in week 5 can be calculated as follows:

$$
\begin{aligned}
\text{Beginning inventory} \quad &= \quad -30 \\
\text{Production} \quad &= \quad \underline{150} \\
& \quad 120 \\
-4 \text{ weeks' sales at } 35 \quad &= \quad \underline{140} \\
& \quad -20 \\
-20/35 \quad &= \quad -.57
\end{aligned}
$$

Figure 6.20 shows another way to create the Ethan Allen MPS. Here, the same four products are used. For each, a **time-phased order point (TPOP)** record is developed. The same schedule shown in the bottom portion of Figure 6.19 is achieved when the line is loaded to capacity in the sequence of when planned orders occur in the TPOP records; that is, product C has the first planned order, then B, then D, and so on. Of course, the planned order for D in week 3 isn't placed in week 3 because capacity isn't available until week 5. There's also a tie shown in week 8. Both products B and C have planned orders in that week. The tie-breaking decision could produce a schedule that differs slightly from that shown in Figure 6.19.

The great advantage to using TPOP approaches to developing the MPS is that specialized MPS software development is reduced. TPOP records are produced with standard MRP logic using the forecast quantities as gross requirements. The master production scheduler's job is to convert these planned orders to firm planned orders, so capacity is properly utilized. At Ethan Allen, conversion of TPOP planned orders to firm planned orders is largely an automatic activity, so it has been computerized. Note, however, the objective is to load the assembly stations to their absolute capacity, in priority sequence. Other firms might use other criteria, such as favoring those jobs with high profitability, favoring certain customers, or allowing flexibility in the definition of capacity. If so, the master production scheduler's detailed decisions would be different.

Master Production Scheduling at Jet Spray

Jet Spray Corporation manufactures and sells dispensers for noncarbonated cold beverages and hot products (coffee and hot chocolate). Jet Spray uses an integrated on-line system encompassing MRP, capacity planning, shop-floor control, master production scheduling, inventory management, and other

FIGURE 6.20   Ethan Allen MPS Example Using Time-Phased Order Point

Week		1	2	3	4	5	6	7	8	
Gross requirements		5	5	5	5	5	5	5	5	
Scheduled receipts										A
On hand	20	15	10	5	0	45	40	35	30	
Planned orders						50				

Week		1	2	3	4	5	6	7	8	
Gross requirements		40	40	40	40	40	40	40	40	
Scheduled receipts										B
On hand	50	10	220	180	140	100	60	20	230	
Planned orders			250						250	

Week		1	2	3	4	5	6	7	8	
Gross requirements		35	35	35	35	35	35	35	35	
Scheduled receipts										C
On hand	-30	85	50	15	130	95	60	25	140	
Planned orders		150			150				150	

Week		1	2	3	4	5	6	7	8	
Gross requirements		10	10	10	10	10	10	10	10	
Scheduled receipts										D
On hand	25	15	5	95	85	75	65	55	45	
Planned orders				100						

functions. The software package (Data 3) allows the user to designate part numbers as either being MRP or MPS; that is, MRP part numbers are driven by MPS numbers (but not the opposite). Bills of material need to be designed accordingly (MPS part numbers are parents to MRP part numbers).

Any MPS part may also be designated as either a make-to-stock or make-to-order item. The distinction is that, when a shop order is created for a make-to-stock MPS item, the order is closed into finished-goods inventory. Subsequent shipment to a customer requires another transaction to remove the items from finished goods into an area awaiting shipment. The inventory in turn is reduced by actual shipment. For a make-to-order item (which includes assemble-to-order), the shop order is driven by an actual customer order and is closed directly into the area awaiting shipment.

One of Jet Spray's best known products is a two-product cold beverage dispenser, the Twin Jet 3 (TJ3), which is made to stock. Figure 6.21 is a portion of the basic MPS record for the TJ3. It shows the weeks of 11/17 through 1/12, but the system has data to support a one-year planning horizon. Notice that the available-to-promise rows are shown in the format just described as well as in cumulative format. The cumulative data are always the sum of the prior period cumulative plus any ATP in the present period.

The available-to-promise for 11/17 is determined by taking the on-hand quantity (279), subtracting the unavailable (21), and thereafter subtracting the actual demand (9) in the week starting 11/17. The ATP for 11/24 requires looking all the way out to 12/29. In that week, an actual order for 963 completely consumes all MPS quantities for the month of December (286 + 155 + 290 + 225 = 956). Thus, 7 units from the MPS for 11/24 must be promised to the order in the week of 12/29. Additionally, the other actual demand quantities for December (8 + 5 + 4 + 4 = 21) and 11/24 (26) will have to be covered. Thus, the ATP for 11/24 is the MPS (225) − 7 − 21 − 26 = 171. The other ATP figures are relatively straightforward.

The projected balance row (available) for the record is based on a different convention than we've described. The forecast values shown are the *original* forecasts, whereas the software uses the *unconsumed* forecast plus the actual demand in decrementing the projected balance. The unconsumed forecast values are not shown in Figure 6.21. We'd argue for more transparency in the system, but in actual practice, absence of the unconsumed forecast data isn't a great problem. The system is on-line, so this number can be obtained whenever it's needed.

The target inventory balance row in Figure 6.21 is worth explaining. The production plan for the TJ3 wasn't shown here. The actual production plan for the TJ3 is constantly monitored against the MPS. There are tolerances for inventory variations, and, when projected inventory data exceed limits set by the production plan, this is where those differences are recorded.

This MPS software is an on-line system. Figure 6.21 is only one of several documents available to the master scheduler. Most of the master scheduler's work is supported with a video screen, not paper, and the MPS is reviewed on a daily basis.

MASTER PRODUCTION SCHEDULE STABILITY

A stable master production schedule translates into stable component schedules, which mean improved performance in plant operations. Too many changes in the MPS are costly in terms of reduced productivity. However, too few changes can lead to poor customer service levels and increased inventory.

FIGURE 6.21 Jet Spray Corporation Master Planning Schedule

PART NUMBER	DESCRIPTION	QUANTITY ON HAND	SAFETY STOCK	QUANTITY UNAVAIL	LEAD TIME	CUM L/T	FAM GRP	LOT HOR	PLN	TIME FENCE	MASTER SCHEDULE	TYPE
S3568	TJ3 DOM. TWIN JET	279	0	21	0	92	TJ	5	001	50	MAKE TO STOCK	

BEGIN DATE	11/17/	11/24/	12/01/	12/08/	12/15/	12/22/	12/29/	1/05/	1/12/
DAYS/PERIOD	7	7	7	7	7	7	7	7	7
FORECAST	28	63	147	147	147	146	146	181	181
ACTUAL DEMAND	9	26	8	5	4	4	963	4	1
PROJECTED BAL	240	402	680	830	1,116	1,112	374	493	312
TARGET INV BAL	0	0	0	0	0	0	0	0	0
AVAIL TO PROM	249	171						295	
CUMULATIVE ATP	249	420	420	420	420	420	420	715	715
MPS	0	225	286	155	290	0	225	300	0

The objective is to strike a balance where stability is monitored and managed. The techniques most used to achieve MPS stability are firm planned order treatment for the MPS quantities, frozen time periods for the MPS, and time fencing to establish clear guidelines for the kinds of changes that can be made in various periods.

Firm Planned Order Treatment at Black & Decker

As noted already, a firm planned order is a planned order with timing and quantity that will not be changed automatically, and the entire MPS can be considered as a set of firm planned orders. That is, once the MPS is established, processing any change in it should be the sole responsibility of the master production scheduler, after careful analysis of the change's implications. The stability provided by firm planned order treatment of the MPS is particularly important to the rest of the firm's fabrication and purchasing activities. It ensures that all supporting activities aim at the same target.

The firm planned order is particularly useful to ensure that all changes in physical distribution aren't transferred directly back to manufacturing through the MPS. Black & Decker creates the firm planned orders and the MPS in a way that isolates logistics activities from the shop. First, marketing owns the finished-goods inventories, so they aren't considered in the MPS time-phased records by manufacturing. Second, the marketing/logistics group does use the finished-goods inventories and time-phased order point (TPOP) records to determine when they next need to request an MPS quantity. Third, times and quantities are negotiated with the master production scheduler. In this way, physical distribution problems are *first* reviewed by marketing/logistics before negotiating for MPS quantities or changes that require expediting in manufacturing.

Figures 6.22 and 6.23 show the Black & Decker approach. The first shows the MPS time-phased record for a model series of belt sanders. Note that it's exactly in MRP format and, in fact, is maintained as any other MRP record. Figure 6.22 does, however, show creation of the firm planned orders. Gross requirements come down directly as planned orders. This is done by specifying zero lead time, lot-for-lot order sizing, and no on-hand inventories. Any lot can be assembled in the same week it's launched; lot sizes put into the gross requirements are those negotiated with logistics. The result is a status like firm planned orders for these planned orders which are the MPS quantities. Just like firm planned orders, the only way they can change is by a manual shift in the gross requirements.

Figure 6.22 is for the model series 80070581–00. Each item in the series is a belt sander but with different accessories, and one is sold under a private brand name. In the pegging data ("STATUS") portion of Figure 6.22, the end-item model numbers can be seen as the last two digits of the reference number. This overall model series is comprised of four individual end-item

FIGURE 6.22 Black & Decker Time-Phased MPS Record

WEEK 140 **INVENTORY STATUS** CODE- A FOR 80070581-00 (PART NUMBER OR RAW MATERIAL CODE; CLASS OR FAMILY) I N D I V I D U A L - P A R T (TITLE OF INQUIRY) REQUESTED BY VANCE (NAME) 843 01 (DEPT NO)

CATEGORY	PART NUMBER OR RAW MATERIAL CODE	DESCRIPTION	UNIT OF MEASURE	ABC CODE	FAMILY CODE	CLASS CODE	BUYER CODE	STANDARDIZATION CODE	LAST EFFECTIVE ECN NUMBER	PACKAGE ASSEMBLY	UNIT FAMILY MASTER	UNIT ASSEMBLY	LINE CODE	PHANTOM	SERVICE ONLY	SPECIAL CUTTER	DATE OF LAST C/C	DATE LAST TRANS	ECN PEND	LOC
1	80070681-00	PA BELT SANDERS	50			P07			46812	X			2				000	138	P	8

ORDER CODE	LEAD TIME S/A	LEAD TIME COMP	PRODUCTION LOT SIZE	SCRAP FACTOR	SAFETY STOCK	MINIMUM PURCHASE QUANTITY	PURCHASE CONVERSION QUANTITY	NUMBER ORDERS LAST YEAR	NUMBER ORDERS YEAR TO DATE	USAGE LAST YEAR	EST	USAGE YEAR TO DATE	TOTAL STANDARD COST	SET-UP COST PER LOT	RUN COST (COST EACH LESS SET-UP)	QUANTITY BALANCES ON HAND	QUANTITY BALANCES RECEIVED	QUANTITY BALANCES STAGED	LOCATION PLANT	LOCATION AREA
P				00				0	0	38002		45291	NOT AVAIL						8	

	PAST DUE	CURRENT WEEK	WEEK - 141	WEEK - 142	WEEK - 143	WEEK - 144	WEEK - 145	WEEK - 146	WEEK - 147	WEEK - 148	WEEK - 149	WEEK - 150	WEEK - 151	WEEK - 152
REQUIREMENT	6500	6500			9000				9000				6500	
ALLOCATIONS														
ON ORDER	6500	6500												
AVAILABILITY					9000-	9000-	9000-	9000-	18000-	18000-	18000-	18000-	24500-	24500-
PLANNED ORDERS					9000				9000				6500	

	WEEK - 153	WEEK - 154	WEEK - 155	WEEK - 156	WEEK - 157	WEEK - 158	WEEK - 159	WEEK - 160	WEEK - 161	WEEK - 162	WEEK - 163	WEEK - 164	WEEK - 165	WEEK - 166
REQUIREMENT			8500				8000					9000		
ALLOCATIONS														
ON ORDER														
AVAILABILITY	24500-	24500-	33000-	33000-	33000-	33000-	41000-	41000-	41000-	41000-	41000-	50000-	50000-	50000-
PLANNED ORDERS			8500				8000					9000		

	WEEK - 167	WEEK - 168	WEEK - 169	WEEK - 170	WEEK - 171	WEEK - 172	WEEK - 173	WEEK - 174	WEEK - 175	WEEK - 176	WEEK - 177	WEEK - 178	WEEK - 179	WEEK - 180
REQUIREMENT		9000			9000				9000				9000	
ALLOCATIONS														
ON ORDER														
AVAILABILITY	50000-	59000-	59000-	59000-	68000-	68000-	68000-	68000-	77000-	77000-	77000-	77000-	86000-	86000-
PLANNED ORDERS		9000			9000				9000				9000	

	WEEK - 181	WEEK - 182	WEEK - 183	WEEK - 184	WEEK - 185	WEEK - 186	WEEK - 187	WEEK - 188	WEEK - 189	WEEK - 190	WEEK - 191			TOTALS
REQUIREMENT			9000											101500
ALLOCATIONS														
ON ORDER														6500
AVAILABILITY	86000-	86000-	95000-	95000-	95000-	95000-	95000-	95000-	95000-	95000-	95000-			95000
PLANNED ORDERS			9000											95000

ORDERS OUT OF PHASE

* A - ALLOCATION; B - SERVICE/SUBSIDIARY ALLOCATION; H - PURCHASE ORDER ON HOLD; I - TRANSFER ALLOCATION; J - MFG. ORDER - FIRST OPERATION COMPLETE; M - MANUFACTURING ORDER; P - PURCHASE ORDER; R - REQUIREMENT; S - MFG. ORDER - SHOP PAPER ISSUED; T - TRANSFER ORDER; U - PLANNED ORDER

† AUTH: P - PART, U - UNIT; CO-ORD: S - SCHEDULE SHEET, C - COORDINATED; CONTROL ECN: A - DROP IN ON USE-UP, B - PKG. ASSEM. CONTROLS ECN, C - A + B ABOVE

STATUS

* CODE	WEEK	REFERENCE NUMBER (LOC / MFG. NO. / LOT / MOD)	QUANTITY	† CODES	* CODE	WEEK	REFERENCE NUMBER (LOC / MFG. NO. / LOT / MOD)	QUANTITY	† CODES	* CODE	WEEK	REFERENCE NUMBER (LOC / MFG. NO. / LOT / MOD)	QUANTITY	† CODES	* CODE	WEEK	REFERENCE NUMBER (LOC / MFG. NO. / LOT / MOD)	QUANTITY	† CODES
		REQUIREMENTS			R	139	000706818209	5500	U	R	139	000706818210	1000	U	R	143	000706818309	9000	U
R	147	000706818409	9000	U B	R	151	000706818609	5500	U	R	151	000706818610	1000	U	R	155	000706818801	500	U B
R	155	000706818809	7000	U B	R	155	000706818810	1000	U B	R	159	000706819009	8000	U B	R	164	000706819201	500	U
R	164	000706819209	7500	U	R	164	000706819210	1000	U	R	168	000706819401	500		R	168	000706819409	7500	
R	168	000706819410	1000		R	171	000706819601	1000		R	171	000706819609	6000		R	171	000706819610	2000	
R	175	000706810209	8000		R	175	000706810210	1000		R	179	000706810409	9000		R	183	000706810601	500	
R	183	000706810609	7000		R	183	000706810610	1500											
		OPEN-ORDERS			S	139	000706818209	5500	UC	S	139	000706818210	1000	UC					
		PLND-ORDERS			U	143	000706818309	9000	U	U	147	000706818409	9000	U	U	151	000706818609	5500	U
U	151	000706818610	1000	U	U	155	000706818801	500	U	U	155	000706818809	7000	U	U	155	000706818810	1000	U
U	159	000706819009	8000	U	U	164	000706819201	500	U	U	164	000706819209	7500	U	U	164	000706819210	1000	U
U	168	000706819401	500		U	168	000706819409	7500		U	168	000706819410	1000		U	171	000706819601	1000	
U	171	000706819609	6000		U	171	000706819610	2000		U	175	000706810209	8000		U	175	000706810210	1000	
U	179	000706810409	9000		U	183	000706810601	500		U	183	000706810609	7000		U	183	000706810610	1500	
NO WHERE USED					HAS COMPONENTS					HAS OP FILE									

BLACK & DECKER MFG. CO. FORM NO 30020

Source: W. L. Berry, T. E. Vollmann, and D. C. Whybark, *Master Production Scheduling: Principles and Practice* (Falls Church, Va.: American Production and Inventory Control Society, 1979), p. 67.

models: 01, 00, 09, and 10. This section is read left to right: requirements first, then open orders, then planned orders. Thus, the past-due requirement of 6,500 (also shown as current) is made up of 5,500 model 09 and 1,000 model 10, both due last week (week number 139).

Figure 6.23 shows the same item from the marketing/logistics viewpoint. Marketing/logistics monitor the finished-goods inventory for these make-to-stock items, and they use data like those in Figure 6.23 in the process. Figure 6.23 is a portion of the TPOP record, which is printed in monthly time buckets. The exact dates, kept inside the computer, are the same as in the manufacturing system except for a lead time offset.

In Figure 6.23, the "ORDERS" columns show the model 09 order of 5,500 and the model 10 order of 1,000 sanders. The equivalent marketing record

FIGURE 6.23 Black & Decker Marketing/Logistics MPS Record

07/01/　　　MONTHLY TOOL INVENTORY PLANNING REPORT—PLANNING DIVISION 02　　　PAGE 0167

FAMILY IDENT	PROD CODE	CATALOG NUMBER	DESCRIPTION	MFG-MOD NUMBERS	E C	P-L-A-N-T CUR	LT	FUT	PROD P/GRP	UNIT GSV	UNIT A STD C	LOT SIZE	SAFETY STOCK	END-OF-MONTH BOH	B/O
020560	444	7450-01	BELT SANDER 220V	070681-01	60	00	00	040020	58.20	29.86	25		M 1176	6
	450	7450-06	BELT SNDR 240V	000000-00	60	00	00	020560	59.75	27.14	500		M	6
	451	7451	BEST BELT SANDER W/	070681-09	60	00	00	020560	50.15	31.01	5000		M 907	
	461	7461	2SPD DUST COLL SNDR	070681-10	60	00	00	020560	61.06	38.23	1000		M 898	6844

PROD CODE	NET BOH	PAST DUE	JUL 140 ORDERS	SALES	EOM-INV	AUG 143 ORDERS	SALES	EOM-INV	SEP 148 ORDERS	SALES	EOM-INV	OCT 152 ORDERS	SALES	EOM-INV
444	1170	14		95	1089		620	469		25	444		305	139
450														
451	5937-	16	5500	4333	4754-	19000	8278	13032-	19000	5759	901-	5500	5038	329-
461	898	640	1000	443	2095		715	1380		872	508	1000	605	903
TOTALS	3869-	670	6500	4871	1570-		9613	11183-	18000	6656	161	6500	5948	713

Source: W. L. Berry, T. E. Vollmann, and D. C. Whybark, *Master Production Scheduling: Principles and Practice* (Falls Church, Va.: American Production and Inventory Control Society, 1979), p. 68.

uses the product codes of 451 and 461 for 09 and 10, respectively. Each remaining order in Figure 6.23 can be matched with planned orders in Figure 6.22, except that the printout grouping into monthly time buckets combines two manufacturing planned orders for 9,000 each into one marketing order of 18,000 in the month of September.

Note in Figure 6.23 that the inventories ("NET BOH") include a −5,937 for model 451 along with positive inventories for two of the other models, and no activity at this time for the 450 model. The negative inventory is for the private-brand product. This product is made to order, not made to stock, and the negative balance is an unfilled order. The key point is that the level and composition of finished-goods inventory are marketing decisions. At Black & Decker, that's where these decisions are made; the result is a stabilized set of firm planned orders in manufacturing.

Ethan Allen Stability

Construction of the Ethan Allen MPS is based on TPOP records, with assembly lines loaded up to exact capacities, and the sequence of MPS items determined by the date sequence of planned orders. This process might seem to lead to a great deal of repositioning of MPS quantities; that is, as actual sales occur, forecast errors will tend to rearrange the MPS. In fact, this doesn't occur, because planned orders from TPOP records are "frozen," or firm planned, under certain conditions.

Ethan Allen uses three types of firm planned orders for the MPS, as Figure 6.24 shows. In essence, any firm planned order is frozen in that it won't be automatically repositioned by any computer logic. All MPS quantities for the next eight weeks are considered to be frozen or firm planned at Ethan Allen. In addition, any MPS quantity used to make a customer promise (i.e., a customer order is pegged to that MPS batch) is also a firm planned order. The third type of firm planned order used in Ethan Allen's MPS is for what they call the **manual forecast.** Included are contract sales (e.g., items to a motel chain), market specials (i.e., items to go on special promotion), and new items (MPS here being when the product is to be introduced). Finally, all blank space in Figure 6.24 is filled with the TPOP-based scheduling technique discussed earlier. Computerized MPS logic fills in the holes up to the capacity limit without disturbing any firm planned order.

Freezing and Time Fencing

Figure 6.24 shows the first eight weeks in the Ethan Allen MPS as **frozen.** This means *no* changes inside of eight weeks are possible. In reality, "no" may be a bit extreme. If the president dictates a change, it will probably happen, but such occurrences are rare at Ethan Allen.

FIGURE 6.24 Ethan Allen Firm Planned Order Approach

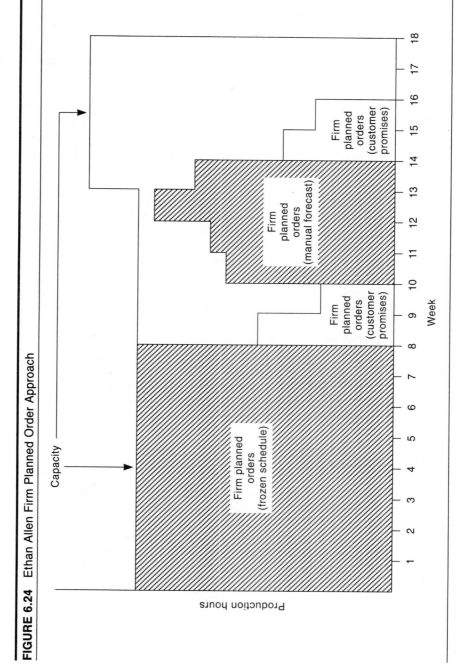

Source: W. L. Berry, T. E. Vollmann, and D. C. Whybark, *Master Production Scheduling: Principles and Practice* (Falls Church, Va.: American Production and Inventory Control Society, 1979), p. 51.

Many firms don't like to use the term *frozen,* saying that anything is negotiable—but negotiations get tougher as we approach the present time. However, a frozen period provides a stable target for manufacturing to hit. It also removes most alibis for missing the schedule!

Time fencing is an extension of the freeze concept. Many firms set time fences that specify periods in which various types of change can be handled. For example, Black & Decker has three time fences: 26, 13, and 8 weeks. The implication is that beyond 26 weeks the marketing/logistics people can make any change as long as the sum of all MPS records is synchronized with the production plan. From weeks 13 to 25, substitutions of one end item for another are permitted, provided required component parts will be available and the production plan isn't violated. From weeks 8 to 13, the MPS is quite rigid; but minor changes within a model series can be made if component parts are available. The eight-week time fence at Black & Decker is basically a freeze period similar to that at Ethan Allen, but occasional changes are made within this timing. In fact, assembly lines have been shut down to make changes—but it's so rare that everyone in the factory remembers when this happens! To achieve the productivity necessary to remain competitive, stability in short-range manufacturing plans is essential.

Two common fences are the **demand fence** and the **planning fence.** The demand fence is the shorter of the two. Inside the demand fence, the forecast is ignored in calculating the available. The theory is that customer orders—not the forecast—matter in the near term. The planning fence indicates the time at which the master production scheduler should be planning more MPS quantities. Within the demand fence it is very difficult to change the MPS. Between the demand fence and the planning fence, management trade-offs must be made to make changes; outside the planning fence, changes can be made by the master production scheduler. Some firms refer to these as the ice, slush, and water zones.

MANAGING THE MPS

We turn now to managing the MPS: How do we measure, monitor, and control detailed day-to-day performance against the MPS? The first prerequisite for control is to have a realistic MPS. Most basic management textbooks say it's critical to only hold people accountable for performance levels that are attainable. This means the MPS can't be a wish list, and it shouldn't have any significant portion that's past due. In fact, we claim that significant past due is a major indication of a sick manufacturing planning and control system.

Stability and proper buffering are also important, because the objective is to remove all alibis and excuses for not attaining the performance for which the proper budget has been provided. Successful companies hit the production plan every month, and they do the best job possible to disaggregate the plan to reflect actual product mix in the MPS.

The Overstated MPS

Most authorities have warned that the MPS must not be overstated. To do so destroys the relative priorities developed by MRP and shop-floor control; more important, the overstated MPS erodes belief in the formal system, thereby reinstituting the informal system of hot lists and black books. Walter Goddard, a well-known MPS expert, no longer tells companies not to overstate the MPS, because at some point the temptation is overwhelming; he now tells them to learn from the experience so they won't do it again!

A key to not overstating the MPS is to always force the sum of the MPS to equal the production plan. Then, when someone wants to add something, the question is, "Of what do you want less?" The company must give up what's referred to as *the standard manufacturing answer*. The standard manufacturing answer to whether more output of some product is possible is: "We don't know, but we'll try!"

The company *must* know. There should be an overall output budget for manufacturing. Capacity should be in place, and should match (not be more or less than) the budget. Manufacturing and marketing should work diligently to respond to product mix changes, but within the overall budgetary constraint. The correct response to whether more output of some product is possible is, "What else can be reduced?" If nothing, then the answer is either "No" or "The output budget and concomitant resources will have to be changed to increase capacities."

MPS Measures

There's an old Vermont story about the fellow who was asked: "How is your wife?" His answer: "Compared to what?" Likewise, measuring MPS has to be in concrete terms that reflect the firm's fundamental goals. This isn't as easy as it might seem. At one time, Ethan Allen evaluated each factory on the basis of dollar output per week. At one plant, an assembly line produced both plastic-topped tables and all-wood tables. Plastic-topped tables sold for more and could be assembled in roughly half the time, since the top was purchased as a completed subassembly. Obviously, the factory favored plastic-topped tables, even when inventories were high on those items and low on wood tables.

Ethan Allen had to change the measure for evaluating plant performance. Each line in each plant is now scheduled by the techniques we've described, and performance is based upon hitting the schedule.

Another important measure of MPS and other MPC system functions is customer service. In virtually every company, customer service is an area of concern. However, in many firms, a tight definition of precisely how the measure is to be made is lacking. Measurement is a critical step in control, and each firm will need to express how this important aspect of its operation is to be measured.

It's to be expected that whatever measure for customer service is chosen, the firm may have problems similar to those of Ethan Allen when evaluating plants using dollar output. However, the way to find the problems and thereafter eliminate them is, in fact, to start with *some* measure, no matter how crude, and evolve.

Appropriate measures vary a great deal from firm to firm, reflecting the type of market response typical in the industry and the particular company. Ethan Allen measures customer service in terms of hitting the order acknowledgment or promise dates. Jet Spray measures manufacturing performance against the MPS, as well as monthly performance in "equivalent units" of output versus the budget. The goal is a cumulative performance of at least 95 percent. Black & Decker measures its customer service in distribution, in terms of a 95 percent ability to deliver any customer order from inventory. Some assemble-to-order firms evaluate production against the production plan, which is to deliver a specific number of each model to marketing in the agreed upon time frame. They also evaluate customer service in terms of how long customers have to wait until they can get a specific end item. This indicates how well the production plan is being disaggregated.

Monitoring the MPS at Ethan Allen

Figure 6.24 shows Ethan Allen's planned order approach to its MPS. We also know that each plant is evaluated on hitting the MPS, so it should be useful to see how detailed monitoring takes place and how overall company operations have been affected.

Every Tuesday morning, Ethan Allen's vice president of manufacturing gets a report detailing each factory's performance in the prior week. Figure 6.25 shows this overall performance report. Figure 6.26 is the detail for one factory at Beecher Falls, Vermont.

Figure 6.25 shows one plant, Boonville, as having had poor performance. The last two comments about packed production and outside suppliers show total production achieved in these two categories.

The "STATIONS" in Figure 6.26 are the assembly lines; their capacities are stated in hours. The "PRIORITY" data reflect expected operating conditions 18 weeks in the future. If each priority number were 18, we'd expect the plant to exactly meet customer needs 18 weeks in the future. This particular factory will be seriously behind in terms of meeting anticipated customer demands. This is a question of capacity that's reflected in the comments about each station. For example, for the 06 station, the detail indicates 10 extra weeks of capacity are required to catch up. Note, however, lack of capacity doesn't mean the plant isn't meeting its schedule. All eight lines are reported as "no misses." This means, in the week covered by this report, the

---

FIGURE 6.25 Ethan Allen Summary MPS Performance

April 5,

To: Bill Morrissey

From: Marty Stern

Production schedule review
4/1

Summary:

Nine of the 14 factories operating hit their schedules
100 percent.

Performance against schedule was poor this week at
one of the factories: Boonville – 75 percent

Packed production was 196 million over total scheduled.
11 of the factories reporting met or exceeded their schedule.

The outside suppliers produced 171 million under their schedule
bringing total production to 25 million over scheduled goal.

cc: Marshall Ames
Barney Kvingedal
Ray Dinkel
Walter Blisky
Andy Boscoe
Steve Kammerer
Tom Ericson
Bob Schneble
Hank Walker

Source: W. L. Berry, T. E. Vollmann, and D. C. Whybark, *Master Production Scheduling: Principles and Practice* (Falls Church, Va.: American Production and Inventory Control Society, 1979), p. 54.

plant met its schedule exactly. Jobs in the schedule are being run. The schedule is loaded up to capacity—not in excess of capacity. Lack of adequate capacity means longer delivery times to customers, but not missed schedules.

By evaluating the reports in Figures 6.25 and 6.26, it's clear which plants are performing according to expectations. Life in the factories is much more calm with performance more clearly defined. No longer do salespeople, customers, marketing people, and executives call the factories. The interface between functions is reflected in the master production schedule, and each factory's job is to hit its MPS. In a sense, the entire master scheduling effort has enabled Ethan Allen to achieve centralized management of decentralized

---

FIGURE 6.26 Ethan Allen Detailed MPS Performance (for one factory)

SUMMARY OF SCHEDULE REVIEW

PLANT: __BEECHER FALLS__

FROM: MARTY STERN

SCHEDULE DATE __3/27/78__
THRU WEEK OF __8/28/78__

STATION	CAPACITY	PRIORITY	PROD.SCHED. DOLLARS	PRODUCTION GOAL
06-Cases	175.0	1	$ 240.1	232.0
07-C/Hutch	10.0	11	10.0	9.0
08-Hutch	15.0	14	17.9	19.0
20-Beds	60.0	1	85.7	84.0
21-Misc.	3.0	11	3.7	4.0
22-Bookstack	70.0	3	33.0	31.0
24-Desks	10.0	8	8.8	10.0
26-Mirrors	25.0	12	23.0	19.0
PLANT TOTAL			$ 422.2	408.0

Station 06 -- Cases: Other than misses caused by reporting date change (Canbury closed Good Friday) no misses. Items such as 10-4017, 4066, 4512P, 4522P, 11-5215, 5223 delayed 1 to 4 weeks. Nine items' service position improved. Some jobs outside frozen schedule should have been made "A" jobs and shifting would not have occurred. Priority down to 1 from 2. To schedule through priority 18 requires 10 weeks capacity, slightly higher than last months' 9 3/4 weeks.

Station 07 -- C/Hutch: No misses. Delayed 3 of 5 items on this line. Priority down to 11 from 14. 7 weeks capacity will schedule thru priority 18, up from 5 weeks.

Station 08 -- Hutch: No misses. Delayed 2 items, pulled 3 ahead. (Plant comments base the shift on purchase parts). Priority up to 14 from 3. To schedule thru priority 18 requires 3 weeks capacity, down from 6 weeks.

Station 20 -- No misses. Delayed the 11-5632-5. Pulled a number of beds ahead. Priority unchanged at 1. 8 weeks capacity, down from 10 3/4 weeks will schedule thru priority 18.

Station 21 -- Misc.: No misses. Some shifting but orders are O.K. Priority up from 10 to 11. 3 weeks capacity, same as last month, will schedule thru priority 18.

Station 22 -- Bookstack: No misses. Built 1 assembly ahead. Priority up to 3 from 2. To schedule thru priority 18 requires 4 1/2 weeks capacity down from 6 weeks.

Station 24 -- Desks: No misses. Built 1 assembly ahead. Some shifting, no delays. Priority up to 8 from 4. 3 1/2 weeks capacity, down from 5 1/2 weeks, will schedule thru priority 18.

Station 26 -- No misses. Some shifting but ahead only. Priority up to 12 from 1. 2 3/4 weeks capacity, down from 8 weeks will schedule thru priority 18.

Source: W. L. Berry, T. E. Vollmann, and D. C. Whybark, *Master Production Scheduling: Principles and Practice* (Falls Church, Va.: American Production and Inventory Control Society, 1979), p. 55.

operations. Factory operations are geographically dispersed over wide areas, but those operations are carefully evaluated in the corporate offices. Execution responsibility and criteria are unambiguously defined for each plant.

One of the master production scheduling system's most important benefits for Ethan Allen is its upward compatibility; that is, the system is transparent and will work with 5 factories or 25 factories, with new ones easily added. Centralized coordination is maintained, and performance is very clear, with the result being an important tool to support orderly growth for the company. The company has roughly tripled in size since the start of the master production scheduling effort.

CONCLUDING PRINCIPLES

The master production schedule plays a key role in manufacturing planning and control systems. In this chapter, we've addressed what the MPS is, how it's done, and who does it. The following general principles emerge from this discussion:

- The MPS unit should reflect the company's approach to the business environment in which it operates.
- The MPS is one part of an MPC system—the other parts need to be in place as well for a fully effective MPS activity.
- Time-phased MPS records should incorporate useful features of standard MRP record processing.
- Customer order promising activities must be closely coupled to the MPS.
- Available-to-promise information should be provided to both the master scheduler and the sales department.
- A final assembly schedule (FAS) should be used to convert an anticipated build schedule (MPS) into an actual build schedule.
- The master production scheduler must keep the sum of the parts (MPS) equal to the whole (production plan).
- The MPS activity must be clearly defined organizationally.
- The MPS can be usefully considered as a set of firm planned orders.
- Stability must be designed into the MPS and managed.
- The MPS should be evaluated with a formal performance measurement system.

REFERENCES

Berry, W. L.; T. E. Vollmann; and D. C. Whybark. *Master Production Scheduling: Principles and Practice*. Falls Church, Va.: American Production and Inventory Control Society, 1979.

Blevins, Preston. "MPS—What It Is and Why You Need It—Without the Jargon and Buzz Words." *APICS Master Planning Seminar Proceedings*, Las Vegas, March 1982, pp. 69–76.

Brongiel, Bob. "A Manual/Mechanical Approach to Master Scheduling and Operations Planning." *Production and Inventory Management*, 1st quarter 1979, pp. 66–75.

Dougherty, J. R., and J. F. Proud. "From Master Schedules to Finishing Schedules in the 1990s." *American Production and Inventory Control Society 1990 Annual Conference Proceedings,* pp. 368–70.

Ford, Q. "Secrets of the 5 Percent that Make MPS Work." *American Production and Inventory Control Society 1990 Annual Conference Proceedings,* pp. 368–70.

Funk, P. N. "The Master Scheduler's Job Revisited." *American Production and Inventory Control Society 1990 Annual Conference Proceedings,* pp. 374–77.

Garwood, R. D. "The Making and Remaking of a Master Schedule—Parts 1, 2, and 3." *Hot List,* January/February and March/June 1978.

Gessnez, R. *Master Production Schedule Planning.* New York: Society of Manufacturing Engineers, 1986.

Hoelscher, D. R. "Executing the Manufacturing Plans." *1975 APICS Conference Proceedings,* pp. 447–57.

Kinsey, John W. "Master Production Planning—The Key to Successful Master Scheduling." *APICS 24th Annual Conference Proceedings,* 1981, pp. 81–85.

Ling, R. C., and K. Widmer. "Master Scheduling in a Make-to-Order Plant." *1974 APICS Conference Proceedings,* pp. 304–19.

Proud, John F. "Controlling the Master Schedule." *APICS 23rd Annual Conference Proceedings,* 1980, pp. 413–16.

————. "Master Scheduling Requires Time Fences." *APICS 24th Annual Conference Proceedings,* 1981, pp. 61–65.

Sulser, Samuel S., "Master Planning Simulation: Playing the 'What-If' Game." *APICS Annual Conference Proceedings,* 1986, pp. 91–93.

Tincher, Michael G. "Master Scheduling and Final Assembly Scheduling: What's the Difference?" *APICS Annual Conference Proceedings,* 1986, pp. 94–96.

Vollmann, Thomas E. *Master Planning Reprints.* Falls Church, Va.: American Production and Inventory Control Society, 1986.

Ware, N., and D. Fogarty. "Master Schedule/Master Production Schedule: The Same or Different?" *Production and Inventory Management* 31, no. 1, 1st quarter 1990, pp. 34–38.

DISCUSSION QUESTIONS

1. Why does the text stress that the master production schedule is *not* a forecast?
2. Some companies try to increase output simply by increasing the MPS. Discuss this approach.
3. What do you feel would be the key functions of the MPS in each of the following three environments: make-to-stock, make-to-order, and assemble-to-order?
4. What are the similarities and differences in "rolling through time" for MPS and MRP?
5. Much of the order promising logic is directed at telling customers when they can honestly expect delivery. Several firms, on the other hand, simply promise their customers delivery within X weeks (often an unrealistic claim) and then deliver when they can. Contrast these two approaches.

6. One characterization of the available-to-promise record is that the "Available" row is for the master scheduler and the "Available-to-promise" row is for customers. What's meant by this contention?

7. Explain how determining the "Available" row with the "greater of forecast or orders" logic can overstate required production.

8. Many companies have come to view their master production schedulers as key people in profitably meeting the firm's strategic goals. Would you agree? What qualities would you look for in a master production scheduler?

9. Ethan Allen's approach to master production scheduling puts capacity into a primary position instead of a secondary consideration. What does this statement mean?

10. Discuss the relationship between stability and firm planned orders.

PROBLEMS

1. Excelsior Springs, Ltd., schedules production of one end product, Hi-Sulphur, in batches of 60 units whenever the projected ending inventory balance in a quarter falls below 10 units. It takes one quarter to make a batch of 60 units. Excelsior currently has 20 units on hand. The sales forecast for the next four quarters is:

		Quarter		
	1	2	3	4
Forecast	10	50	50	10

a. Prepare a time-phased MPS record showing the sales forecast and MPS for Hi-Sulphur.

b. What are the inventory balances at the end of each quarter?

c. During the first quarter, no units were sold. The revised forecast for the rest of the year is:

	Quarter		
	2	3	4
Forecast	20	40	60

How does the MPS change?

2. Neptune Manufacturing Company's production manager wants a master production schedule covering next year's business. The company produces a complete line of fishing boats for both salt and fresh water use and manufactures most of the component parts used in assembling the products. The firm uses

MRP to coordinate production schedules of the component part manufacturing and assembly operations.

The production manager has just received the following sales forecast for next year from the marketing division:

	Sales forecast (as measured in standard boats)			
Product lines	1st quarter	2nd quarter	3rd quarter	4th quarter
1000 series	8,000	9,000	6,000	6,000
2000 series	4,000	5,000	2,000	2,000
3000 series	9,000	10,000	6,000	7,000
Total	21,000	24,000	14,000	15,000

The sales forecast is stated in terms of "standard boats," reflecting total sales volume for each of the firm's three major product lines.

Another item of information supplied by the marketing department is the target ending inventory position for each product line. The marketing department would like the production manager to plan on having the following number of standard boats on hand at the end of each quarter of next year:

Product line	Quarterly target ending inventory (in standard boats)
1000 series	3,000 boats
2000 series	1,000 boats
3000 series	3,000 boats

The inventory position for each product is:

Product line	Current inventory level (in standard boats)
1000 series	15,000 boats
2000 series	3,000 boats
3000 series	5,000 boats

The master production schedule is to specify the number of boats (in standard units) to be produced for *each product line in each quarter* of next year on the firm's single assembly line. The assembly line can produce up to 15,000 standard boats per quarter (250 boats per day during the 60 days in a quarter).

Two additional factors are taken into account by the production manager in preparing the master production schedule: the assembly line changeover cost and the inventory carrying cost for the finished goods inventory. Each assembly line changeover costs $5,000, reflecting material handling costs of changing the stocking of component parts on the line, adjusting the layout, and so on. After some discussion with the company comptroller, the production manager concluded that the firm's inventory carrying cost is 10 percent of standard boat cost per year. The item value for each of the product line standard units is:

Product line	Standard boat cost
1000 series	$100
2000 series	150
3000 series	200

The master production scheduler has calculated the production lot sizes as 5,000, 3,000, and 4,000 units, respectively.

a. Develop a master production schedule for next year, by quarter, for each of Neptune's fishing boat lines. Identify any problems.

b. Verify the lot size calculations using the EOQ formula.

3. The Zoro Manufacturing Company has a plant in Murphysboro, Georgia. Product A is shipped from the firm's plant warehouse in Murphysboro to satisfy East Coast demand. Currently, the sales forecast for product A at the Murphysboro plant is 30 units per week.

The master production scheduler at Murphysboro considers product A to be a make-to-stock item for master scheduling purposes. Currently, 50 units of product A are on hand in the Murphysboro plant warehouse. Desired safety stock level is 10 units for this product. Product A is produced on a lot-for-lot basis. Currently, an order for 30 units is being produced and is due for delivery to the plant warehouse on Monday, one week from today.

The master production scheduler has heard that an MRP record which uses the forecast for gross requirements and has a lead time of zero can be used for master production scheduling. Complete the following MRP record. How can this be used for master production scheduling?

Product A Week	1	2	3	4	5	6
Gross requirements	30	30	30	30	30	30
Scheduled receipts SS = 10						
Projected available balance 50	10	10	0	0	0	0
Planned order releases		30	20	30	30	30

Q = lot for lot; LT = 0; SS = 10.

4. The Spencer Optics Company produces an inexpensive line of sunglasses. The manufacturing process consists of assembling two plastic lenses (produced by the firm's plastic molding department) into a finished frame (purchased from an outside supplier). The company is interested in using material requirements planning to schedule its operations and has asked you to prepare an example to illustrate the technique.

The firm's sales manager has prepared a 10-week sales forecast for one of the more popular sunglasses (the Classic model) for your example. The forecast is 100 orders per week. Spencer has customer orders of 110 units, 80 units, 50 units, and 20 units in weeks 1, 2, 3, and 4, respectively. The sunglasses are assembled in batches of 300. Presently, three such batches are scheduled: one in week 2, one in week 5, and one in week 8.

a. Complete the following time-phased record:

Classic model MPS record

Week		1	2	3	4	5	6	7	8	9	10
Forecast		100	100	100	100	100	100	100	100	100	100
Orders		110	80	50	20						
Available	140	30	250	200	180	380	280	180	380	280	180
Available to promise		30	180	180	180	180	180	180	180	180	180
MPS			300			300			300		

b. Prepare the MRP record for the assembly of the sunglasses using the following record. The final assembly quantity is 300, lead time is 2 weeks, and there's a scheduled receipt in week 2. Note that no inventory is shown for the assembled sunglasses in this record, since it's accounted for in the MPS record.

Week		1	2	3	4	5	6	7	8	9	10
Gross requirements											
Scheduled receipts			300		300						
Projected available balance	0										
Planned order releases											

Q = 300; LT = 2; SS = 0.

5. The MPC system at the Dansworth Manufacturing Company is run weekly to update the MPS and MRP records. At the start of week 1, the MPS for end products A and B is as follows:

	Master production schedule			
	Week 1	*Week 2*	*Week 3*	*Week 4*
Product A	10	—	25	5
Product B	5	20	—	20

To manufacture one unit of either end product A or B, one unit of component C is required. Purchasing lead time for component C is two weeks, a fixed order quantity of 40 units is used, and no (zero) safety stock is maintained for this item. Inventory balance for component C is 5 units at the start of week 1, and there's an open order (scheduled receipt) for 40 units due to be delivered at the beginning of week 1.

a. Complete the MRP record for component C as of the beginning of week 1:

Week		1	2	3	4
Gross requirements		15			
Scheduled receipts		40			
Projected available balance	5				
Planned order releases					

Q = 40; LT = 2; SS = 0.

b. During week 1 the following transactions occurred for component C:
  1. The open order for 40 units due to be received at the start of week 1 was received on Monday for a quantity of 30. (Ten units of component C were scrapped on this order.)
  2. An inventory cycle count during week 1 revealed that five units of component C were missing. An inventory adjustment of -5 was therefore processed.
  3. Ten units of component C were actually disbursed (instead of the 15 units planned for disbursement to produce end products A and B). (The MPS quantity of 5 in week 1 for product B was canceled due to a customer order cancelation.)
  4. The MPS quantities for week 5 include 15 units for product A and zero units for product B.
  5. Due to a change in customer order requirements, marketing has requested that the MPS quantity of 25 units for product A scheduled in week 3 be moved to week 2.
  6. The MRP planner released an order for 40 units, due in week 3.
    Given this information, complete the MRP record for component C as of the start of week 2:

Week		2	3	4	5
Gross requirements					
Scheduled receipts					
Projected available balance					
Planned order releases					

Q = 40; LT = 2; SS = 0.

    c. What action(s) are required by the MRP planner at the start of week 2 as a result of the transactions given in question b? What MPS policy issues are raised?

6. Peter Ward has constructed the following (partial) time-phased MPS record:

Weeks	1	2	3	4	5	6
Forecast	20	20	20	20	20	20
Orders	12	6	3			
Available						
Available to promise						
MPS	50		50			50

On hand = 5.

    a. Complete the record.
    b. Are there any problems?
    c. What's the earliest Peter can promise an order for 33 units?
    d. Assume that an order for 10 is booked for week 4. Assume the order for 33 units in part c is *not* booked, and recompute the record.

7. The Cedar River Manufacturing Company produces a line of furnishings for motels and hotels. Among the items manufactured is an Executive water pitcher with the following product structure:

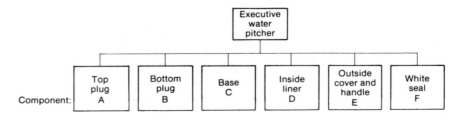

Component items A, B, C, and F are manufactured by the plastic molding shop; components D and E are purchased from a vendor. The Executive water pitcher is completed by the final assembly department.

Currently the following open shop orders for Executive water pitcher components are waiting to be processed at the #101 injection molding press in the plastic molding shop:

Plastic molding shop-floor control report

Shop order number	Component item	Order quantity	Order due date	Machine time (weeks)
10–XYZ	F	15	ASAP*	2
10–XXX	B	25	ASAP*	2
10–XZV	A	30	ASAP*	1
10–XXY	C	20	ASAP*	2

*As soon as possible.

All the orders just shown are made complete in one operation at the #101 injection molding press and are subsequently ready for the final assembly department. (Final assembly time is negligible.)

a. Given the preceding information, complete the MPS and MRP records for all of the Executive water pitcher items.

MPS record: Executive water pitcher

Week	1	2	3	4	5	6	7
Forecast	10	10	10	10	10	10	10
Orders	12	5	2				
Available							
Available to promise							
MPS							

On hand = 15.
MPS lot size = 20.

MRP records

Component A

Gross requirements	Week	1	2	3	4	5	6	7
Scheduled receipts		30						
Projected available balance	12							
Planned order releases								

Q = 30; LT = 5; SS = 0.

Component B

Gross requirements	Week	1	2	3	4	5	6	7
Scheduled receipts		25						
Projected available balance	22							
Planned order releases								

Q = 25; LT = 5; SS = 0.

Component C

Gross requirements	Week	1	2	3	4	5	6	7
Scheduled receipts		20						
Projected available balance	45							
Planned order releases								

Q = 20; LT = 5; SS = 0.

Component F

Gross requirements	Week	1	2	3	4	5	6	7
Scheduled receipts		15						
Projected available balance	70							
Planned order releases								

Q = 15; LT = 5; SS = 0.

    b. What conclusions can you make regarding the validity of the current order due dates for the open shop orders at the #101 injection molding press?

8. As a consultant in manufacturing control systems for the R&E accounting firm, you've been invited to do some work for the Green Valley Furniture Company. It would like you to install a master production scheduling system similar to

the one Ethan Allen Furniture Company uses. Please prepare an illustration of how this system works using the following example data:

End product item	Forecast sales per week	Current on-hand inventory	Final assembly production lot size
A	10	5 units	30 units
B	25	−10*	40
C	15	30	30

*10 units back-ordered.

Currently the Green Valley plant has one final assembly line. This line is run exactly 40 hours per week on a single shift with no overtime. During each week, 40 units of any end product item can be produced in 40 hours on the final assembly line (one hour per unit).

In addition to selling products directly out of finished-goods inventory, the Green Valley Furniture Company produces a few items on a make-to-order (custom order contract) basis. Currently, there are two such customer orders: one for 20 units of Item D scheduled for final assembly in week 2 and another for 15 units of Item E scheduled for final assembly in week 4. (Today is Monday of week 1.) Production of these units must be included in the master production schedule.

a. Illustrate the steps in preparing the master production schedule covering the next four weeks using the Ethan Allen procedure. Give the MPS for each end item.

b. Given the master production schedules developed in part a, recommend changes, if any, that should be made in Green Valley Furniture's final assembly line capacity.

9. Figure 6.19, the Ethan Allen example, is based on the following data:

Product	Beginning inventory	Weekly forecast	Lot size	Hours per lot size
A	20	5	50	20
B	50	40	250	80
C	−30	35	150	60
D	25	10	100	30

Priorities are calculated by dividing expected beginning inventory by forecast. In weeks after the first, expected beginning inventory takes account of production and expected sales.

a. Calculate weekly priorities and determine the master production schedule for weeks 1 through 8 for these data. Check your answers against Figure 6.19.

b. Assume the actual sales in week 1 were as follows:

Product	Sales
A	10
B	30
C	25
D	25

Given these actual sales data, calculate the weekly priorities and determine the MPS for weeks 2 through 9, assuming the forecasts remain unchanged. What impact do these changes have?

c. Given the actual sales data in part b, calculate the priorities for weeks 7, 8, and 9. Determine the MPS for weeks 2 through 9, assuming the forecasts remain unchanged and weeks 2 through 6 are frozen; that is, the schedule in part a can be revised, but only from week 7 on. What impact would the frozen schedule have on the inventory and customer service levels for products A through D?

d. Assume that the time horizon date is extended to 18 weeks resulting in the following scheduling priorities for week 18:

Product	Priority
A	2
B	1.25
C	−1
D	1

Hours	Assembly Load (week 18)
35	
30	
25	C
20	
15	
10	
5	B
0	

What capacity information can be inferred from this schedule?

10. Lou Chin is the master scheduler at Amdur Products. Assembly D is a critical assembly used in products A, B, and C. Assembly D is made in a dedicated work center with a 50-unit-per-week capacity. Lou has the following demand information and feels that the firm planned order approach to master scheduling will be helpful. Complete the time-phased record for Assembly D.

	Product demand		
Period	A(1D)	B(2Ds)	C(1D)
1		10	20
2			
3			
4			
5	50		60
6		40	
7			
8			
9	50	30	40
10			

Week		1	2	3	4	5	6	7	8	9	10
Gross requirements											
Scheduled receipts			30	30							
Projected available balance	20										
Planned orders											
Firm planned orders											

11. The master production scheduler at the XYZ Company is concerned with determining the impact of using different MPS freezing intervals on component part shortages and inventory levels in the firm's fabrication shop. Currently, the firm's end products are produced on a make-to-stock basis. A four-period MPS planning horizon is used. Lot sizing is performed at the start of every period covering all four future periods. Assembly orders are issued at the start of each period. The assembly lead time equals zero periods; the beginning finished product inventory is zero.

Table A shows the forecast, projected inventory, and MPS for three consecutive periods for one of the firm's products, using the current freeze policy of one period and assuming perfect forecasts. Table B provides similar information using a two-period MPS freeze policy. Assuming the component part 1234 is only used on this end product, with a usage rate of one unit per unit of end product, prepare MRP records for this part as of the beginning of each of the three consecutive periods under both MPS freezing policies. Assume that com-

TABLE A

Period*	1	2	3	4
Forecast[†]	177	261	207	309
Available	261	0	309	0
MPS	438	0	516	0

Period	2	3	4	5
Forecast	0	207	309	64
Available	0	373	64	0
MPS	0	580	0	0

Period	3	4	5	6
Forecast	207	309	64	182
Available	0	246	182	0
MPS	207	555	0	0

TABLE B

Period*	1	2	3	4
Forecast[†]	177	261	207	309
Available	261	0	309	0
MPS	438	0	516	0

Period	2	3	4	5
Forecast	0	207	309	64
Available	0	373	64	0
MPS	0	580	0	0

Period	3	4	5	6
Forecast	207	309	64	182
Available	373	64	0	0
MPS	580	0	0	182

*The forecast is net of beginning inventory.
[†]The available is the closing inventory balance.

ponent part 1234 has a planned lead time of two periods, that its lot size is the net requirement for the next two periods whenever a net requirement is observed, that it has zero safety stock and a beginning inventory of 450 units in period 1, and that there are no scheduled receipts.

   a. What conclusions can you draw about the two different freezing policies' effectiveness?

   b. What other freezing policies should be considered?

   c. What are the appropriate time fences?

12. The Avon Power Tool Company makes a line of snowblowers on a make-to-order basis as well as other types of products. The snowblower product line's

eight different end products (catalog numbers) vary according to horsepower, drive unit, and starting mechanism. Avon expects to sell one third of the snowblowers in two horsepower and two thirds in four horsepower. The expected breakdown for drive units is 40 percent and 60 percent, for chain and gear, respectively. Similarly, the breakdown on starters is 50–50.

Catalog number	1000	1100	1200	1300	1400	1500	1600	1700
Horsepower	2	2	2	2	4	4	4	4
Drive unit	Chain	Chain	Gear	Gear	Chain	Chain	Gear	Gear
Starter	Auto	Manual	Auto	Manual	Auto	Manual	Auto	Manual
Component parts	401	401	401	401	801	801	801	801
	200	200	200	200	200	200	200	200
	150	150	160	160	150	150	160	160
	130	135	130	135	130	135	130	135
	115	115	120	120	115	115	120	120
	101	101	101	101	101	101	101	101
	170	175	170	175	170	175	170	175
	600	600	600	600	800	800	800	800

a. Group the component parts. Which are common? Which are associated with horsepower? Drive unit? Starter?

b. Show how super bills can be used to master schedule the snowblower product line, indicating how requirements for the component items would be exploded. Please prepare sample MRP and MPS records for the 2-HP option and item #401 as illustrations. In preparing these records, assume that the safety stock is zero and the order quantity is 100 for both items. Current on-hand quantity for the 2-HP option is 65 units and the on-hand for item #401 is 140 units. Booked orders accepted for the 2-HP option total 25 in week 1, 18 in week 2, 6 in week 3, and 4 in week 4. Assume the total forecast for all types of snowblowers is 100 units per week. Also, assume item #401's lead time is two weeks.

c. What are the advantages and disadvantages of this approach?

13. The Ace Electronics Company produces printed circuit boards on a make-to-order basis.

a. Prepare an MPS record for one of its items (catalogue #2400), including the available-to-promise information, using the following data:

Final assembly production lead time = 2 weeks.
Weekly forecast = 100 units.
Current on-hand quantity = 0.
Booked customer orders (already confirmed):
   95 units in week 1.
   105 units in week 2.
   70 units in week 3.
   10 units in week 5.

Master production schedule:

200 units to be completed at the start of week 1.
200 units to be completed at the start of week 3.
200 units to be completed at the start of week 5.
200 units to be completed at the start of week 7.

b. Suppose the cumulative lead time for the item (catalogue #2400) is eight weeks. What decision must the master scheduler make this week?

14. Consider the time-phased information relating to the #R907 outboard motor in Exhibit A. The following transactions occurred during the week of 5/2:

1. The MPS for 9/19 was released (360 units).
2. Customer order #6005 requested 3 units in week of 5/9.
3. Customer order #6015 requested 5 units in week of 6/6.
4. Customer order #6028 requested 5 units in week of 6/13.
5. Customer order #6042 requested 15 units in week of 7/4.

a. Can all orders received during the week of 5/2 be accepted? Update Exhibit A as it would appear at the beginning of the week of 5/9.

b. During the week of 5/9, the following additional orders were received:

1. Customer order #6085 requested 20 units in week of 6/6.
2. Customer order #6093 requested 10 units in week of 5/30.
3. Customer order #6142 requested 250 units in week of 8/1.
4. Customer order #6150 requested 100 units in week of 7/25.

Discuss these orders' effect on the revised Exhibit A.

15. The Parker Corporation produces and sells a machine for tending golf course greens. The patented device trims the grass, aerates the turf, and injects a metered amount of nitrogen into the soil. Machines are marketed through the Taylor Golf Course Supply Company, under the Parker Company's own original equipment brand, and, recently, through a lawn and garden supply house (Brown Thumb), which serves both commercial and consumer accounts. Addition of the lawn and garden outlet and requests to add a 3-HP version have called into question the production planning and control process for the machines.

Forecasting the products to be produced is difficult. There are now two drive mechanisms (chain and gear), three different body styles (one for each outlet), and two sizes of engine (4- and 5-HP). This gives a total of 12 end items, all of which had some demand. (See Exhibit B.) The 3-HP motor would add six more end items. Forecasting demand for these new items would add to the difficulty of forecasting demand.

Lead times for some of the castings and for 5-HP motors have increased to the extent that it's not possible to wait until firm orders are received for the end items before the castings and motors have to be ordered. In addition, the firm's business is growing; it anticipates selling about 120 units next year. Consequently, the production manager has arranged to purchase enough material for 10 units per month. There's plenty of capacity for the small amount of parts fabrication required, but assembly capacity must be carefully planned. The current plan calls for assembly capacity of 10 units per month.

With regard to the specific issue of forecasting, the marketing manager summarized the data on the sales of each end item over the past year. (See Exhibit

EXHIBIT A Time-Phased Requirements

Item #	Description	Assembly lead time	On hand	Lot size	Safety stock	Time fences — Demand	Time fences — Plan	Run date
R907	Outboard motor	1	100	180	0	4	20	5/2

Week	5/2	5/9	5/16	5/23	5/30	6/6	6/13	6/20	6/27	7/4	7/11	7/18	7/25
Forecast	13	14	13	14	13	14	13	14	13	14	13	14	13
Orders	20	5	10	8	10	15	10	8	20	5	10	10	5
Available	80	75	65	57	44	29	196	182	162	148	135	121	288
Available to promise	32						117						175
Master production schedule							180						180

Week	8/1	8/8	8/15	8/22	8/29	9/5	9/12	9/19	9/26	10/3	10/10	10/17	10/24
Forecast	290	290	290	290	290	290	290	290	290	14	13	14	13
Orders		200	200	200	200	200		200					
Available	178	68	138	28	98	168	58	−232	−522	−536	−549	−563	−576
Available to promise			340		360	160	180						
Master production schedule	180	180	360	180	360	360	180						

EXHIBIT A *(concluded)*

Due date	Shop order	Quantity	Action	Required date	Quantity	Customer order	Required date	Quantity	Customer order
	MPS pegging detail			Actual demand pegging			Actual demand pegging		
6/13	M1234	180	Released	5/2	20	5678			
7/25	M1278	180	Released	5/9	5	5789			
8/1	M1347	180	Released	5/16	8	5890			
9/19	—	360	Planned						
9/26	—	180	Planned						

EXHIBIT B   Last Year's Sales by Catalog Number

Catalog number			Sales	Key
Body	Horsepower*	Drive	Total = 100	
T	4	C	1	
T	4	G	4	Body:
T	5	C	18	T = Taylor Supply
T	5	G	17	O = "OEM" Parker's own
O	4	C	13	B = Brown Thumb
O	4	G	9	
O	5	C	10	
O	5	G	8	
B	4	C	6	Drive
B	4	G	7	C = Chain
B	5	C	2	G = Gear
B	5	G	5	

*Note: The 3-HP machine would be offered in all three body styles and both drives.

B.) He felt that a 20 percent growth in total volume was about right, and that 3-HP machines will perhaps account for half that growth. Of course, once the forecasts were made, the production manager had to determine which motors and castings (gear or chain) to order. Each machine was made up of many common parts, but the motors, chain or gear drive subassemblies, and bodies were different (though interchangeable).

a. Suggest an improved method of forecasting demand for the firm's products.

b. As the production manager contemplated the difficulty of forecasting demand for the firm's products and determining exactly what to schedule into final assembly, two customers called. The first, from Taylor, wanted to know when the company could deliver a model T3G; the second wanted as early delivery as possible of one of Parker's own machines, an O4C. The Taylor representative said he felt that the three-horsepower models might "really take off."

Before making any commitment at all, it was necessary to check the material availability and get back to the two customers. It was the firm's practice not to promise immediate delivery, since units scheduled for final assembly were usually already promised. The planned assembly schedule called for assembly of three units next week, two units the following week, and alternating three and two thereafter. As a matter of practice, all parts for assembly and delivery in any week would need to be ready at the start of that week.

Exhibit C shows current inventory and on-order positions for the common part "kit," the motors, and the drives. (The production manager didn't concern himself with the body styles since all three styles can be obtained in a week.) Exhibit D lists all booked orders promised for delivery over the next few weeks. Organize this information to respond to the delivery promise requests. (Assume that no safety stocks are held.) What should the delivery promises be?

EXHIBIT C   Inventories and On-Order

Item	Inventory	On order	Due date	Lot size
Common "kit"	0	5 each	weeks 1,3,5,7,9	5
5-HP motor	10	5	week 5	5
4-HP motor	3	5	weeks 3,8	5
3-HP motor	0	5	week 1	5
Chain drive	14	—	—	10
Gear drive	6	5	week 3	10

EXHIBIT D   Booked Orders and Delivery Dates

		Delivery week				
	1	2	3	4	5	6
Models	2T4G*	1B3G	1T5C	1O5G	1T5G	1O4G
	1O5C	1T4C	1B4C			
		1O4G				
Total	3	3	2	1	1	1

*2 units of T4G.

Chapter 7

Production Planning

Production planning is probably the least understood aspect of manufacturing planning and control. However, payoffs from a well-designed and executed production-planning system are large. Here we discuss the process for determining aggregate levels of production. The managerial objective is to develop an integrated game plan whose manufacturing portion *is* the production plan. The production plan, therefore, links strategic goals to production and is coordinated with sales objectives, resource availabilities, and financial budgets. If the production plan isn't integrated, production managers can't be held responsible for meeting the plan, and informal approaches will develop to overcome inconsistencies.

Our discussion of production planning is organized around four topics:

- Production planning in the firm: What is production planning? How does it link with strategic management and other MPC system modules?
- The production-planning process: What are the fundamental activities in production planning and what techniques can be used?
- The new management obligations: What are the key responsibilities for ensuring an effective production-planning system?
- Operating production-planning systems: What is the state of the art in practice?

This chapter focuses on managerial concepts for integrated planning. Advanced concepts in production planning are emphasized in Chapter 15. Useful background for the concepts in this chapter appear in Chapter 6 on planning bills of material, Chapter 8 on demand management, and Chapter 14, which deals with advanced master production scheduling (MPS) concepts.

PRODUCTION PLANNING IN THE FIRM

The production plan provides key communication links from top management to manufacturing. It determines the basis for focusing the detailed production resources to achieve the firm's strategic objectives. By providing the framework within which the master production schedule is developed, we can plan and control subsequent MPS decisions, material resources, and plant capacities on a basis consistent with these objectives. We now describe the production plan in terms of its role in top management, necessary conditions for effective production planning, linkages to other MPC system modules, and payoffs from effective production planning.

Production Planning and Management

The production plan provides a direct and consistent dialogue between manufacturing and top management, as well as between manufacturing and the other functions. As Figure 7.1 shows, many key linkages of production planning are *outside* the manufacturing planning and control (MPC) system. As such, the plan necessarily must be in terms that are meaningful to the firm's nonmanufacturing executives. Only in this way can the top-management game plan in Figure 7.1 become consistent for each basic functional area. Moreover, the production plan has to be stated in terms MPC system modules can use, so detailed manufacturing decisions are kept in concert with the overall strategic objectives reflected in the game plan.

FIGURE 7.1   Key Linkages of Production Planning

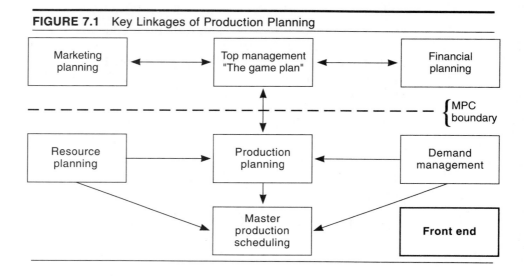

The basis for consistency of the functional plans is resolution of broad trade-offs at the top-management level. Suppose, for example, there's an opportunity to expand into a new market, and marketing requests additional production to do so. With a specified production plan, this could only be accomplished by decreasing production for some other product group. If this is seen as undesirable (i.e., the new market is to be a direct add-on), by definition a new game plan is required—with an updated and consistent set of plans in marketing, finance, and production. The feasibility of the added volume must be determined and agreed upon before detailed execution steps are taken.

The production plan states the mission manufacturing must accomplish if the firm's overall objectives are to be met. How to accomplish the production plan in terms of detailed manufacturing and procurement decisions is a problem for manufacturing management. With an agreed-upon game plan, the job in manufacturing is to "hit the production plan." Similar job definitions should exist in marketing and finance.

An interesting chicken-and-egg question sometimes arises about the production plan and detailed plans that result from the MPC system. Conceptually, production planning should precede and direct MPC decision making. Production planning provides the basis for making the more detailed set of MPC decisions. In some firms, however, it's only after the other MPC systems are in place that resultant production-planning decisions are clearly defined. In these cases, the first production plans are no more than a summation of the individual detailed plans. They're the *result* of other detailed decisions, rather than an input to those decisions. Even so, they provide the basis for management review.

The planning performed in other MPC system modules is necessarily detailed, and the language is quite different from that required for production planning. The production plan might be stated in dollars or aggregate units of output per month, while the master production schedule (MPS) could be in end product units per week. The MPS might be stated in units that use special bills of materials to manage complicated options and not correspond to the units used to communicate with top management.

To perform the necessary communication role, the production plan must be stated in commonly understood, aggregated terms. In some companies, the production plan is stated as the dollar value of total monthly or quarterly output. Other firms break this total output down by individual factories or by major product lines. Still other firms state the production plan in terms of total units for each product line. Measures that relate to capacity (such as direct labor-hours and tons of product) are also used by some firms. The key requirement is that the production plan be stated in some commonly understood homogeneous unit that thereafter can be kept in concert with other plans.

The production plan needs to be expressed in meaningful units, but it also needs to be expressed in a manageable number of units. Experience indicates that 5 to 15 family groups seems to be about right for a top-management group to handle. Each has to be considered in terms of expectations on sales, manufacturing, and resultant inventories and backlogs. The cumulative result, expressed in monetary units, also has to be examined and weighed against overarching business plans.

The overall context within which trade-offs are made and the production plan developed is increasingly called **game planning.** The game plan reflects the strategy (e.g., increased market share) and tactics (e.g., increased inventory for improved service) that are *doable* by the firm. It's not a set of uncoordinated wishes that some people would like to see realized. The manufacturing part of the game plan is the production plan.

The production plan is *not* a forecast of demand! It's the planned production, stated on an aggregate basis, for which manufacturing management is to be held responsible. The production plan is not necessarily equal to a forecast of aggregate demand. For example, it may not be profitable to satisfy all demands, in which case production would be less than forecast. Conversely, a strategic objective of improved customer service could result in aggregate production in excess of aggregate demand. These are important management trade-offs.

The production plan for manufacturing is a result of the production planning process. Inputs to the process include sales forecasts; but these need to be stated on the basis of shipments (not bookings) so the inventory projections match physical inventories and so demands on manufacturing are expressed correctly with respect to time.

Production Planning and MPC Systems

Up to this point, we've emphasized production planning's linkages to activities outside MPC system boundaries. Because of these linkages, the production plan is often called "top management's handle on the business." To provide execution support for the production plan, we need linkages to the MPC systems. The most fundamental linkage is to the master production schedule, which is a disaggregation of the production plan. The result drives the detailed scheduling through detailed material planning and other MPC modules.

The MPS must be kept in concert with the production plan. As the individual daily scheduling decisions to produce specific mixes of actual end items and/or options are made, we must maintain parity between the sum of the MPS quantities and the production plan. If the relationship is maintained, then "hitting the schedule" (MPS) means the agreed-upon production plan will be met as well.

Another critical linkage shown in Figure 7.1 is the link with demand management. Demand management encompasses order entry, order promising, and physical distribution coordination as well as forecasting. This module must capture every source of demand against manufacturing capacity, such as interplant transfers, international requirements, and service parts. In some firms, one or more of these demand sources may be of more consequence than others. For the firm with distribution warehouses, for example, replenishing those warehouses may create quite a different set of demands on manufacturing than is true for other firms. The contribution of demand management, insofar as production planning is concerned, is to ensure that the influence of all aspects of demand is included and properly coordinated.

As a tangential activity, the match between actual and forecast demand is monitored in the demand management module. As actual demand conditions depart from forecast, the necessity for revising the production plan increases. Thus, the assessment of changes' impact on the production plan and the desirability of making a change depends on this linkage. It's critical for top management to change the plans, rather than to let the forecast errors per se change the aggregate production output level.

The other direct MPC linkage to production planning shown in Figure 7.1 is with resource planning. This activity encompasses long-range planning of facilities. Involved is the translation of extended production plans into capacity requirements, usually on a gross or aggregate basis. In some firms, the unit of measure might be constant dollar output rates; in others, it might be labor-hours, head counts, machine-hours, key-facility–hours, tons of output, or some other output measure. The need is to plan capacity, at least in aggregate terms, for a horizon at least as long as it takes to make major changes.

Resource planning is directly related to production planning, since, in the short term, the resources available provide a set of constraints to production planning. In the longer run, to the extent that production plans call for more resources than available, financial appropriations are indicated. A key goal of the linkage between production planning and resource planning is to answer what-if questions. Maintaining current resource-planning factors, related to the product groupings used for planning, is the basis for performing this analysis.

Much of the very near term production plan is constrained by available material supplies. Current levels of raw material, parts, and subassemblies limit what can be produced in the short run, even if other resources are available. This is often hard to assess unless information links from the detailed material planning and shop status data bases are effective.

Links through the MPS to material planning and other MPC modules provide the basic data to perform what-if simulations of alternative plans. Being able to quickly evaluate alternatives can facilitate the game-planning process. This is *not* an argument to always change the production plan. On the contrary, having the ability to demonstrate the impact of proposed changes may reduce the number of instances in which production "loses" in these negotiations.

The value of the production-planning activity is certainly questionable if there's no monitoring of performance. This requires linkages to the data on shipments/sales, aggregated into the production-planning groupings. Measuring performance is an important input to the planning process itself. Insofar as deviations in output are occurring, they must be taken into account. If the plan can't be realized, the entire value of the production-planning process is called into question.

One final performance aspect where effort must be expended is in the reconciliation of the MPS with the production plan. As day-to-day MPS decisions are made, it's possible to move away from the production plan unless constant vigilance is applied. Like other performance monitoring, it requires a frequent evaluation of status and comparison to plan.

Payoffs

Game planning is top management's handle on the business. It provides important visibility of the critical interactions between marketing, production, and finance. If marketing wants higher inventories, but top management decides there's not sufficient capital to support the inventories, the production plan will be so designed. Once such critical trade-off decisions are made, the production plan provides the basis for monitoring and controlling manufacturing performance in a way that provides a much more clear division of responsibilities than is true under conventional budgetary controls.

Under production planning, manufacturing's job is to hit the schedule. This can eliminate the battle over "ownership" of finished-goods inventory. If actual inventory levels don't agree with planned inventory levels, it's basically not a manufacturing problem, *if* they hit the schedule. It's either a marketing problem (they didn't sell according to plan) or a problem of product mix management in the demand management activity (the wrong individual items were made).

The production plan provides the basis for day-to-day, tough-minded trade-off decisions as well. If marketing wants more of some items, it must be asked "Of what do you want less?" There's no other response, because additional production without a corresponding reduction would violate the agreed-upon production plan. In the absence of a new, expanded production plan, production and marketing must work to allocate the scarce capacity to the competing needs (via the master production schedule).

The reverse situation is also true. If the production plan calls for more than marketing currently needs, detailed decisions should be reached about which items will go into inventory. Manufacturing commits people, capacities, and materials to reach company objectives. The issue is only how best to convert these resources into particular end products.

Better integration between functional areas is one of the major payoffs from production planning. Once a consistent game plan between top levels of the functional areas is developed, it can be translated into detailed plans that are in concert with top-level agreements. This results in a set of common goals, improved communication, and transparent systems.

Without a production plan, the expectation is that somehow the job will get done—and in fact, it does get done, but at a price. That price is organizational slack: extra inventories, poor customer service, excess capacity, long lead times, panic operations, and poor response to new opportunities. Informal systems will, of necessity, come into being. Detailed decisions will be made by clerical-level personnel with no guiding policy except "get it out the door as best we can." The annual budget cycle won't be tied in with detailed plans and will probably be inconsistent and out of date before it's one month old. Marketing requests for products won't be made so as to keep the sum of the detailed end products in line with the budget. In many cases, detailed requests for the first month are double the average monthly volume. Only at the end of the year does the reconciliation between requests and budget take place; but in the meantime it has been up to manufacturing to decide what's really needed.

We've seen many companies with these symptoms. Where are these costs reflected? There's no special place in the chart of accounts for them, but they'll be paid in the bottom-line profit results. More and more firms are finding that a well-structured monthly production-planning meeting allows the various functional areas to operate in a more coordinated fashion and to better respond to vagaries of the marketplace. The result is a dynamic overall plan for the company, one that changes as needed and fosters the necessary adaptation in each function.

THE PRODUCTION–PLANNING PROCESS

This section views aids to managing the production-planning process. Specifically, we'll be concerned with routinizing the process, the game-planning output, cumulative charting, and the tabular display. We examine these techniques with an example.

Routinizing Production and Game Planning

The game- and production-planning process typically begins with an updated sales forecast covering the next year or more. Any desired increases or decreases to inventory or backlog levels are added or subtracted, and the result is the production plan. The most immediate portion of this plan won't be capable of being changed, since commitments to labor, equipment, and materials already will have been made. An effective production-planning process will typically have explicit time fences for when the aggregate plan can be increased or decreased; there also may well be tight constraints on the amounts of increase or decrease. An example might be no changes during the most immediate month, up to $+10$ percent or -20 percent in next month, and so on. Effective production planning also implies periodicity; that is, it's useful to perform the production- and game-planning process on a regular, routine cycle.

Performing the game-planning function on a regular basis has several benefits. It tends to institutionalize the process and force a consideration (which might otherwise be postponed) of changed conditions and trade-offs. The routine also keeps information channels open for forecast changes, different conditions, and new opportunities. Performing the task routinely helps ensure a separation between forecasts and the plan.

The cycle's frequency varies from firm to firm, depending on the firm's stability, cost of planning, and ability to monitor performance. The trade-off is difficult. There's a high cost to planning involving data gathering, meetings, what-if analysis, and other staff support activities. On the other hand, delaying the planning process increases the chance of significant departures of reality from plan, creating the opportunity for informal systems to take over. Ironically, the more successful firms can use a less frequent cycle because of their formal systems' ability to keep the firm on plan and warn of impending problems. A common schedule among successful firms is to review plans monthly and revise them quarterly or when necessary. Figure 7.2 shows a monthly cycle at Ethan Allen.

Ethan Allen is a make-to-stock furniture manufacturer so much of its demand forecasting is based on routine extrapolation of historical data. The first event in Figure 7.2, "Determine (manual forecast)," is to forecast the nonroutine items. These include large contract sales (such as to a motel chain), new items for which there are no historical data, and market specials.

The second event in the figure is the "End of month." Following immediately is a "6-month economic review," a process that attempts to summarize opinions as to future economic conditions in the furniture industry. At the same time, the sales screening report is prepared. It's a review of the actual sales in light of the forecast, to suggest changes before the preparation of the new routine statistical forecast.

The sixth event in Figure 7.2 ("Set output level for next 6 months") is the production plan. A group of executives reviews the economic outlook, present inventory levels, statistical forecasts, and other factors. The net result is a rate of output in total dollars for Ethan Allen production. Subsequent activities involve preparing reports based on this production plan, allocating the total to individual factories, and preparing the detailed MPS that supports the production plan.

The discipline required in routinizing the production- and game-planning process is to replan when conditions indicate it's necessary. If information from the demand management module indicates differences between the forecast and actual have exceeded reasonable error limits, replanning may be necessary. Similarly, if conditions change in manufacturing, a new market

FIGURE 7.2 Example of a Monthly Cycle for Production Planning

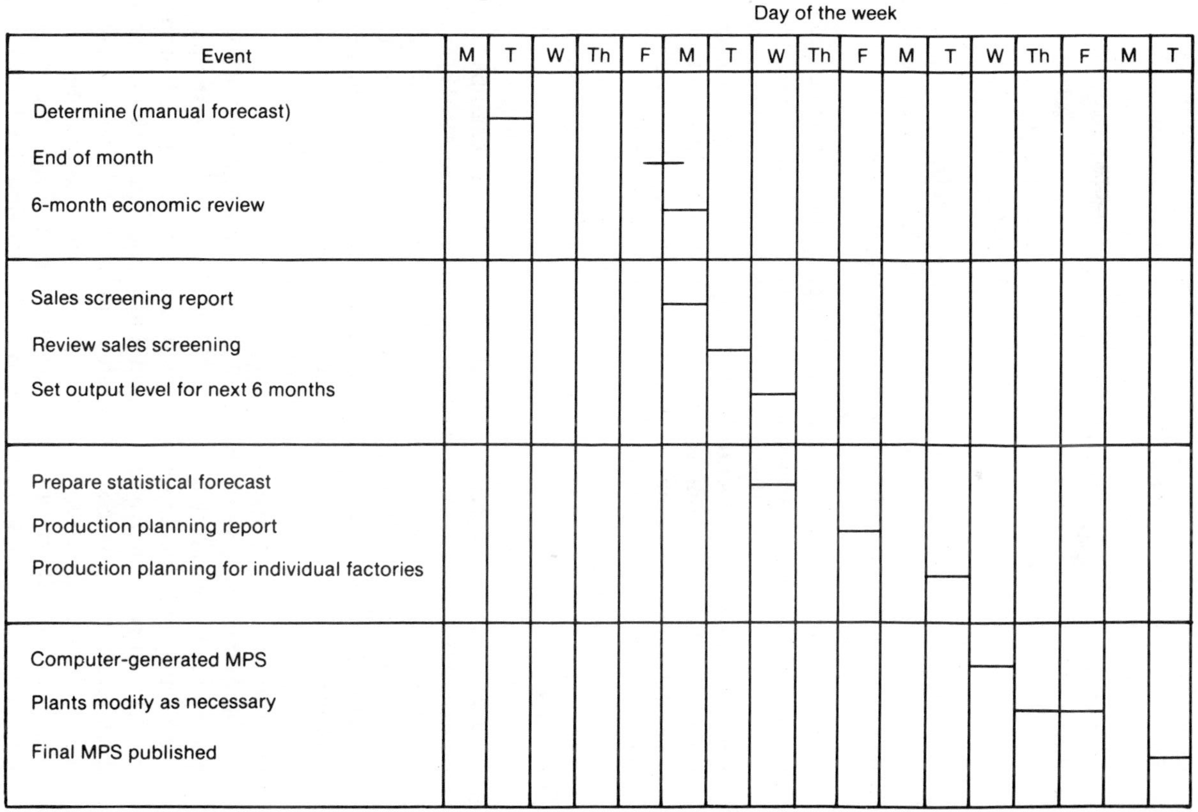

Day of the week

Event	M	T	W	Th	F	M	T	W	Th	F	M	T	W	Th	F	M	T
Determine (manual forecast)																	
End of month																	
6-month economic review																	
Sales screening report																	
Review sales screening																	
Set output level for next 6 months																	
Prepare statistical forecast																	
Production planning report																	
Production planning for individual factories																	
Computer-generated MPS																	
Plants modify as necessary																	
Final MPS published																	

Source: W. L. Berry, T. E. Vollmann, and D. C. Whybark, *Master Production Scheduling: Principles and Practice* (Falls Church, Va.: American Production and Inventory Control Society, 1979), p. 45.

opportunity arises, or the capital market shifts, replanning may be needed. So, though regularizing has its advantages, slavishly following a timetable and ignoring actual conditions isn't wise management practice.

Since the purpose of the planning process is to arrive at a coordinated set of plans for each function (a game plan), mechanisms for getting support for the plans are important. Clearly, a minimum step here is to involve the top functional officer in the process. This does more than legitimize the plan; it involves the people who can resolve issues in the trade-off stage. A second step used by some firms is to virtually write contracts between functions on what the agreements are. The contracts serve to underscore the importance of each function performing to plan, rather than returning to informal practices.

To illustrate the nature of production-planning decisions, we now turn to an example based on a firm with a seasonal sales pattern. We raise the issues in the context of a single facility. In this context, the problem is to find a low-cost combination of inventories, overtime, changes in work force levels, and other capacity variations that meet the company's production requirements. Our first example presents the **cumulative charting** approach. We then examine a tabular representation of the alternative strategies.

The Basic Trade-Offs

Figure 7.3 shows aggregate sales forecasts for our example, the XYZ Company, for the year. Monthly totals vary from a high of $15.8 million to a low of $7 million. Figure 7.4 shows these monthly sales data in the form of a cumulative chart (solid line). In addition, the dashed straight line represents the cumulative production plan at a constant rate of production. The production-planning problem is to choose a low-cost cumulative production plan depicted by a line on the cumulative chart that's always on or above the cumulative forecast line.

The cumulative chart shows clearly the implications of alternative plans. For example, the vertical distance between the dashed line and the solid line represents the expected inventory at each point in time. If no inventory is to be held, the cumulative production line would equal the cumulative sales line. This policy is a **chase strategy** where production output is changed to chase sales. The opposite extreme is a **level policy,** where production is at a constant uniform rate of output, with inventory buildups and depletions. Changing production output incurs the costs of changing the work force level, hours worked, and subcontracting. Keeping production at a constant rate incurs inventory holding and backorder costs.

To convert aggregate output into capacity for planning purposes, a planning factor is used. In our example, the XYZ Company keeps a planning statistic to convert aggregate sales dollars into total labor capacity requirements. This

FIGURE 7.3 XYZ Company Aggregate Monthly Sales Forecast

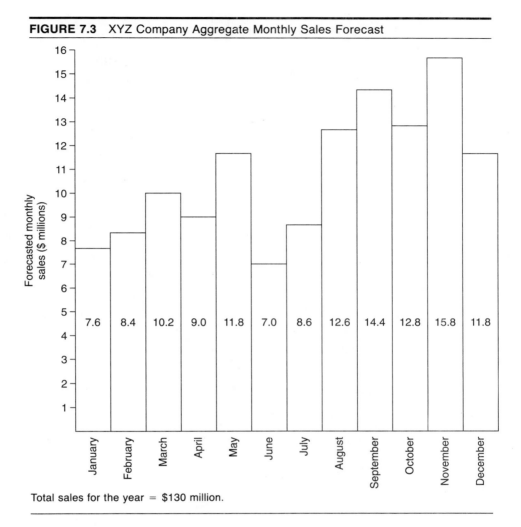

Total sales for the year = $130 million.

statistic, obtained from accounting records, indicates each direct labor-hour produces $30 in sales on the average. This factor is used to convert the sales forecast data in Figure 7.3 into a labor-hour forecast. The first column in Figure 7.5 shows the conversion.

The second column of Figure 7.5 presents the working days in each month for the year. This is an important addition, since the number varies sharply from month to month. The lowest number occurs in July, which only has 10 working days due to the XYZ Company's annual two-week summer shutdown.

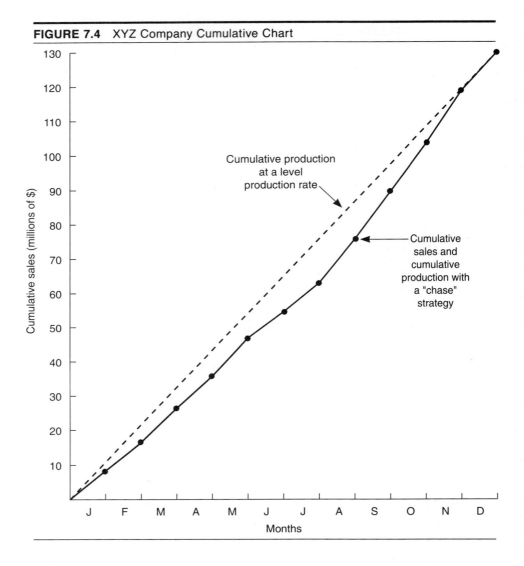

FIGURE 7.4   XYZ Company Cumulative Chart

If the total labor-hour requirements for the year's sales (4,333,333 hours) is divided by the number of working days (243), the result (17,832) is the number of labor-hours necessary to work each day if level production is to be employed. Dividing the result by eight working hours per day gives an implied constant labor force of 2,229 workers required to produce the year's forecast.

Figure 7.5's third column represents one approach to a pure chase strategy. For this example, the working week is held at 40 hours, and labor force size is varied as needed to only produce the forecast sales. For example, January forecast sales are $7.6 million. When divided by $30, the resultant labor-hour

FIGURE 7.5   XYZ Company Labor Capacity Requirements

	Chase strategy (Inventory = 0)				Level production
Month	Sales in labor-hours (000s)	Working days	Variable work force	Variable workweek	Variable inventory ($000)
January	253	20	1,583	28.4	3,099
February	280	21	1,667	29.6	5,933
March	340	23	1,848	33.1	8,037
April	300	20	1,875	33.7	9,737
May	393	22	2,235	40.2	9,706
June	233	22	1,326	23.9	14,475
July	287	10	3,583	64.4	11,224
August	420	23	2,283	41.0	10,929
September	480	20	3,000	53.8	7,228
October	427	22	2,424	43.5	6,197
November	527	20	3,292	59.1	1,096
December	393	20	2,458	44.2	0‡
Total	4,333	243	2,229*	40.0*	7,035†

*Weighted averages.
†Average inventory.
‡The month-by-month detailed calculations will result in a December ending inventory of approximately −$5,000. This is because a work force of 2,229 workers is slightly less than necessary to create $130 million of sales in 243 working days.

capacity requirement is 253,333. Dividing this result by 20 working days, we obtain a need for 12,667 labor-hours each day. If each worker puts in an eight-hour day, the implied work force for January is 1,583 workers.

Month-to-month differences between implied work force levels represent hire–fire decisions. For example, the decision for February would be to hire 84 more workers over the January total (1,667 − 1,583). The hire–fire decision's impact is especially severe during summer. It would be necessary to fire 909 workers in June, but to hire 2,257 in July.

In most circumstances, only a firm like a summer resort or farm that harvests an agricultural commodity could consider such a high level of hiring and firing. In many Western countries, it's difficult to hire and fire workers. In Western Europe and parts of South America, it's virtually impossible to fire workers—the cost is very high.

The next to last column in Figure 7.5 is another approach to a chase strategy. In this case, the labor force is kept at a constant level of 2,229 workers, but the length of the working week is varied as necessary to provide the labor-hours dictated by the first column. Again, the same kinds of variations are seen, but in the hours worked this time. Clearly the July plant shutdown causes severe overtime problems if this strategy is followed. This could be mitigated against by continuing to work at a more normal level of output in June.

The last column in Figure 7.5 shows a "pure" level strategy. This is essentially the dashed line in Figure 7.4, with allowances for the exact numbers of working days in each month. In this example, 2,229 workers would need to be employed for eight hours on each of the 243 working days in the year to meet the forecast sales levels.

In January, the 2,229 workers, working 20 days of eight hours each, create $10,699,200 of goods, using the planning factor of $30 of output per labor-hour. Since the sales forecast for January is $7.6 million, the expected inventory at the end of January is $3,099,200. Figure 7.5's value for February, $5,933,000, includes the January ending inventory plus the net addition to inventory created during February. (Figure 7.5 shows all values rounded to the nearest $1,000.) Note that inventory increases during each month that the variable workweek (next to last column) is less than 40 hours, and decreases if it exceeds 40 hours.

This example clearly shows the basic trade-offs in production planning. They involve inventory accumulations, hiring and firing, undertime and over-time, and alternative capacity forms, such as outside contracting. Evaluating these trade-offs is very firm-specific.

Evaluating Alternatives

The cumulative chart in Figure 7.4 and tabular presentation in Figure 7.5 show the implications of the pure chase and level production strategies. So far in the example, we've said nothing about evaluating the trade-offs involved. The management issue is how to choose between them or how to construct an alternative that's superior to either pure strategy. To do this rigorously, we must establish cost data that relate to the alternative production-planning methods. But in many firms, relevant cost data aren't readily available. In such cases, analysis could be done using executive opinion.

Suppose, for example, no explicit cost data exist for production planning at the XYZ Company. In that case, XYZ executives could evaluate such data as those in Figure 7.5 to point out situations they don't like. Implications of revised plans could be quickly calculated, using a personal computer, for subsequent evaluation by the executives. For example, the work force could be allowed to build up as indicated from January through May for the chase strategy and be held at 2,235 for June, instead of dropping to 1,326. The resulting inventory would be included in the analysis performed for July and beyond. If this process of revising and evaluating were continued until the

---

FIGURE 7.6 XYZ Company Production-Planning Data

Hiring cost	$200 per employee
Firing cost	$500 per employee
Regular labor cost	$5 per hour
Overtime premium cost	$2.50 per hour
Undertime premium cost	$3 per hour
Inventory carrying cost	2% per month (applied to the monthly ending inventory)
Beginning inventory	0 units
Beginning labor force	1,583 persons

---

managerial group was satisfied, it could be possible to imply, from the choices made, the relative importance or costs assigned to various conditions.

For illustrating the analysis when cost data are available, we assume Figure 7.6's cost data were provided for the XYZ Company. The cost to hire an employee is estimated to be $200, whereas the cost to fire is $500. The average labor cost at XYZ is $5 per hour, and any overtime work has a 50 percent premium; that is, for overtime work, the average person earns $7.50 for each hour. "Undertime" cost is harder to assess, but clearly there's a heavy morale cost associated with working less than a normal workweek (40 hours). In some firms, undertime may be used for cross-training or other useful whole person or continual improvement activities that actually can provide value. At the XYZ Company, people are sent home but the company estimates the cost is still about $3 per hour, even though no direct payment is made to the work force. The final cost element, inventory carrying cost, is estimated to be 2 percent per month, based upon the monthly ending inventory value.

We must determine the starting conditions before beginning the production-planning analysis. That is, are there any inventories and what's the beginning work force level? From Figure 7.6 we see there's zero beginning inventory and a work force of 1,583 (the desired level for January) at the beginning of the year.

In Figure 7.7, the "Hire–fire" column for the first alternative specifically states the necessary additions and deletions to the work force to comply with the levels in Figure 7.5. The "Cost"column is simply the hire and fire cost for these actions. No regular labor costs are shown in Figure 7.7, since they'll be the same for each alternative plan considered; that is, the same number of working hours will be used for every plan, although some will be paid an overtime premium. This results in an incremental cost of a pure hire–fire policy, with the given cost values of $2,708,300.

The second alternative in Figure 7.7 is to also maintain a zero inventory level, but with a constant work force of 2,229 persons. The workweek's length

FIGURE 7.7 Costs of Alternative Production Plans

Month	Zero inventory/hire-fire as required		Zero inventory/overtime-undertime/constant work force = 2,229		Level production/constant work force = 2,229		Mixed strategy*				
	Hire-fire	Cost	Hire-fire	Cost	Hire-fire	Cost	Hire-fire	Work force	Workweek	Inventory ($000)	Cost
January	—	—	+646	$ 439,477	+646	$ 191,180	—	1,583	40.0	$ 0	$ —
February	+84	$ 16,800	—	292,088	—	118,660	+84	1,667	40.0	0	$ 16,800
March	+181	36,200	—	212,245	—	160,740	+181	1,848	40.0	0	36,200
April	+27	5,400	—	168,512	—	194,740	+27	1,875	40.0	0	5,400
May	+360	72,000	—	4,904	—	194,120	+360	2,235	40.0	0	72,000
June	-909	454,500	—	473,707	—	289,500	—	2,235	40.0	1,081	96,020
July	+2257	451,400	—	271,938	—	224,480	—	2,235	40.0	1,565	31,300
August	-1300	650,000	—	25,633	—	218,580	—	2,235	40.0	1,302	26,040
September	+717	143,400	—	307,602	—	144,560	+494	2,729	40.0	1	98,820
October	-576	288,000	—	85,817	—	123,940	—	2,729	40.0	1,610	33,200
November	+868	173,600	—	425,739	—	21,920	—	2,729	43.3	0	90,875
December	-834	417,000	—	93,618	—	0	-271	2,458	40.0	0	135,000
Total		$2,708,300		$2,801,280		$1,882,420					$641,155

*Hire until May. Constant work force until September (build inventory). Build inventory in October. Overtime in November. Fire in December.

will be adjusted according to dictates of the next to last column of Figure 7.5, which shows the workweek length required to produce the sales with 2,229 people. As noted, we again begin with a work force of 1,583, so the first and only hire–fire action is to add 646 workers.

The cost for January ($439,477) is based on the cost of hiring 646 workers (646 × 200 = $129,200) plus the cost of having each worker idle for 11.6 hours (40 − 28.4) per week at $3 per hour. This means each worker will average undertime work in the amount of 2.32 hours per day (11.6/5). At $3 per hour, 20 working days per month, and 2,229 workers, January's undertime premium cost is $310,277. When this is added to the hiring cost of $129,200, January's total cost is $439,477.

An illustration of overtime cost can be seen for July. The required work-week is 64.4 hours, or 4.88 hours of overtime per worker each day, on the average. Since there are 10 working days in the month, each overtime hour has a premium cost of $2.50, and there are 2,229 workers, total overtime premium for the month is $271,938. Total expected cost for the year, with an overtime/undertime strategy, is $2,801,280.

The third alternative in Figure 7.7 is the level strategy. In this case, enough people are added to the work force in January to raise it from 1,583 to 2,229 workers. Each worker puts in a constant 40-hour week, and inventories are varied, as Figure 7.5's last column shows. When each of these planned inventory values is multiplied by 2 percent, the costs in Figure 7.7 are obtained (e.g., February = $5,933,000 × .02 = $118,660).

The last alternative evaluated in Figure 7.7 is a mixed strategy calling for adding employees as necessary from January through May, but thereafter keeping the resultant work force constant (2,235) until September. This means some inventories are held during June, July, and August. In September, the work force is again expanded, this time by 494 workers, to provide the necessary output levels. This is determined by subtracting the August ending inventory of $1,302,000 from the September forecast of $14.4 million to get a net requirement of $13,098,000. At eight hours per day for 20 days and $30 per hour, it takes 2,729 workers to produce this amount.

The plan calls for this work force level to be maintained during October, which results in an addition to inventory of $1,610,000, which is needed in November. To meet the sales forecast of $15 million in November, however, overtime work of 3.3 hours each week is needed. The lower level of forecast in December results in a layoff of 271 workers.

This plan may not be the best plan possible. For example, it's less costly to carry the extra inventory produced by 271 workers in December ($26,016) than to lay off all 271 workers ($135,500). The analysis thus far hasn't considered the desirable ending conditions in terms of work force levels. The valuation of any particular ending work force level must be made in light of the following year's sales forecast. The determination of mixed strategies that improve costs over those obtained by employing pure strategies can be guided by mathematical models.

THE NEW MANAGEMENT OBLIGATIONS

Implementing production planning requires major changes in management, particularly in top-management coordination of functional activities. If the production plan is to be the game plan for running a manufacturing company, it follows that top management needs to provide the necessary direction.

Top-Management Role

Top management's first obligation is to commit to the game-planning process. This means a major change in many firms. The change involves the routine aspects of establishing the framework for game planning: getting forecasts, setting meetings, preparing plans, and so on. The change may also imply modifications of performance measurement and reward structures to align them with the plan. We should expect at the outset that many existing goals and performance measures will be in conflict with the integration provided by a working game-planning system. These should be rooted out and explicitly changed. Enforcing changes implies a need to abide by and provide an example of the discipline required to manage with the planning system. This implies even top management must act within the planned flexibility range for individual actions and must evaluate possible changes that lie outside the limits.

As part of the commitment to the planning process, top management *must force* the resolution of trade-offs between functions prior to approving plans. The production plan provides a transparent basis for resolving these conflicts. It should provide basic implications of alternative choices even if it doesn't make decisions any easier. If trade-offs aren't made at this level, they'll be forced into the mix of day-to-day activities of operating people who'll have to resolve them—perhaps unfavorably. If, for example, manufacturing continues long runs of products in the face of declining demand, the mismatch between production and the market will lead to increased inventories.

Game-planning activities must encompass *all* formal plans in an integrated fashion. If budgeting is a separate activity, it won't relate to the game plan and operating managers will need to make a choice. Similarly, if the profit forecast is based solely on the sales forecast (revenue) and accounting data (standard costs) and doesn't take into account implications for production, its value is doubtful. The game-planning process's intention is to produce complete and integrated plans, budgets, objectives, and goals that are used by managers to make decisions and provide the basis for evaluating performance. If other planning activities or evaluation documents are in place, the end result will be poor execution. An unfortunate but frequent approach is to invest management time in the production-planning activity, but thereafter allow the company to be run by a separate performance measurement system or budget.

Some firms find the term *game planning* more acceptable than *production planning,* which connotes a functional focus that's not accurate. Lately, the term *sales and operations planning* is being used to describe the interfunctional nature of the process. Whatever the term, the production-planning process is interfunctional and needs to be coordinated at the top-management level.

Functional Roles

The primary obligation under game planning is to "hit the plan" for all functions involved: manufacturing, sales, engineering, finance, and so on. A secondary obligation is this need to communicate when something will prevent hitting the plan. The sooner a problem can be evaluated in terms of other functional plans, the better. The obligation for communication provides the basis for keeping *all* groups' plans consistent when changes are necessary.

The process of budgeting usually needs to change and to be integrated with game planning and subsequent departmental plans. In many firms, budgeting is done on an annual basis, using data that aren't part of the manufacturing planning and control system. Manufacturing budgets are often based on historical cost relationships and a separation of fixed and variable expenses. These data aren't as precise as data obtained by utilizing the MPC system data base. By using the data base, we can evaluate tentative master production schedules in terms of component part needs, capacities, and expected costs. We can then analyze the resultant budgets for the effect of product mix changes as well as for performance against standards.

Another important aspect of relating budgeting to the game-planning activity and underlying MPC systems and data base is that the cycle can be done more frequently. We won't need to collect data—they always exist in up-to-date form. Moreover, inconsistencies are substantially cut. The budget should always agree with the production plan, which, in turn, is in concert with the disaggregated end-item and component plans that support the production plan. As a result, an operating manager should have to choose between a budget and satisfying the production plan far less often.

With budgeting and production planning done on the same basis with the same underlying dynamic data base, it's natural to incorporate cost accounting. This enables us to perform detailed variance accounting as well as cross-check transaction accuracy.

The most obvious need for integrated planning and control is between marketing and production. Yet it's often the most difficult to accomplish. Firms must ensure product availability for special promotions, match customer orders with specific production lots, coordinate distribution activities with production, and deal with a host of other cross-functional problems.

The marketing job under integrated game planning is to sell what's in the sales plan. We must instill the feeling that overselling is just as bad as underselling. In either case, there will be a mismatch with manufacturing output, financial requirements, and inventory/backlog levels. If an opportunity arises to sell more than the plan, it needs to be formally evaluated via a change in the game plan. By going through this process, we can time this increase so it can be properly supported by both manufacturing and finance. And once the formal plan has been changed, it's again each function's job to achieve its specified objectives—no more and no less.

Similarly, it's manufacturing's job to achieve the plan—exactly. Overproduction may well mean that too much capacity and resources are being utilized. Underproduction possibly means the reverse (not enough resources) or means poor performance. In either case, performance against the plan is poor. This can be the fault of either the standard-setting process or inadequate performance. Both problems require corrective action.

When manufacturing is hitting the schedule, it's a straightforward job for marketing to provide good customer order promises and other forms of customer service. It's also a straightforward job for finance to plan cash flows and anticipate financial performance.

If the production-planning results can't be achieved, whoever can't meet their plan must be clearly responsible for reporting this condition promptly. If, for example, a major supplier can't meet its commitments, the impact on the detailed marketing and production plans must be quickly ascertained.

Integrating Strategic Planning

An important direction-setting activity, strategic planning can be done in different ways. Some companies approach it primarily as an extension of budgeting. Typically, these firms use a bottom-up process, which is largely an extrapolation of the departmental budgets based on growth assumptions and cost-volume analysis. One key aspect of these firms' strategic plans is to integrate these bottom-up extrapolations into a coherent whole. Another is to critically evaluate overall outcome from a corporate point of view.

A more recent approach to strategic planning is to base the plan more on products and less on organizational units. The company's products are typically grouped into strategic business units (SBUs), with each SBU evaluated in terms of its strengths and weaknesses vis-à-vis competitors' similar business units. The budgetary process in this case is done on an SBU basis rather than an organizational unit basis. Business units are evaluated in terms of their competitive strengths, relative advantages (sometimes based on learning curve models), life cycles, and cash flow patterns (e.g., when does an SBU need cash and when is it a cash provider?). From a strategic point of view, the objective is to carefully manage a portfolio of SBUs to the firm's overall advantage.

Game planning and departmental plans to support these strategic planning efforts can be important. In the case of the production plan, the overall data base and systems must ensure that game plans will be in concert with disaggregated decision making. In other words, the MPS and related functions ensure that strategic planning decisions are executed!

All advantages of integrating production planning with budgeting also apply when the SBU focus is taken. It makes sense to state the production plan in the same SBU units; that is, rather than using dollar outputs per time unit, the production plan should be stated in SBU terminology.

Controlling the Production Plan

A special responsibility involves control of performance against the plan. As a prerequisite to control, the game-planning process should be widely understood in the firm. The seriousness with which it's regarded should be communicated as well as the exact planned results that pertain to each of the organization's functional units. In other words, the planning process must be transparent, with clear communication of expectations, to control actual results. For the production plan, this means wide dissemination of the plan and its implications for managers.

Another dimension of control is periodic reporting. Performance against the production plan should also be widely disseminated. When actual results differ from plans, we must analyze and communicate the source of these deviations.

The Tennant Company provides an example of this communication. Some of its more important measures of performance and reporting frequency are:

Measure	Reporting
Conformity of the master production schedule to the production plan	Weekly
Capacity utilization	Weekly
Delivery performance	Daily
Actual production to master production schedule performance	Weekly
Inventory/backlog performance	Weekly

Recently in its history, Tennant hadn't missed a quarterly production plan for the previous 2.5 years. Moreover, it had met the monthly production plan in 10 out of 12 months for each of the previous years. These results are well known inside the company and widely disseminated outside too. All levels of the firm understand the production plan's importance.

Key issues in production planning are when to change the plan, how often to replan, and how stable to keep the plan from period to period. No doubt, a stable production plan results in far fewer execution problems by the detailed master production scheduling, material planning, and other execution modules. Stability also fosters achievement of some steady-state operations, where capacity can be more effectively utilized.

At Tennant, production plan changes are batched until the next review, unless they're required to prevent major problems. In other companies, stability in the plan is maintained by providing time fences for changes and permissible ranges of deviation from plan. Tennant provides flexibility within the plan by planning adequate inventories or other forms of capacity to absorb deviations within an agreed-upon range.

Increasingly, companies are using just-in-time (JIT) concepts, with many aspects of the system based on manual controls. One key to making JIT work is a stable production plan. The output rate is held constant for long time periods and is only modified after extensive analysis. This means the production rate at each step of the manufacturing process can be held to very constant levels, providing stability and predictability.

We can see the other side of this coin by reviewing one U.S. auto manufacturer's approach. In the face of diminishing sales, the company continued to produce in excess of sales. This led to a buildup of finished-goods inventory exceeding 100 days of sales. The results on the financial statements were significant; adjustments in manufacturing were even more severe. Finished-goods inventories and order backlogs can buffer manufacturing from day-to-day shocks, but long-run changes have to be reflected in the basic production plan itself.

OPERATING PRODUCTION–PLANNING SYSTEMS

In this section, we show examples of production-planning practice. In particular, we present organizational aspects of production planning at the Compugraphic Corporation, the entire process for the Mohawk Electric Company, and the Hill-Rom Company's use of SBU-related bills of material for tying the production plan to its strategic business units.

Production Planning at Compugraphic

Compugraphic Corporation makes typesetters and related equipment for the printing industry in five separate factory locations in Boston's northern suburbs. Figure 7.8 shows how production planning fits in the corporation. Figure 7.9 details its relationship with master production scheduling.

The production-planning committee is made up of the company's top-management group representing all functional areas. It develops and monitors the production plan that determines the manufacturing resources to support

FIGURE 7.8 Compugraphic Production Planning

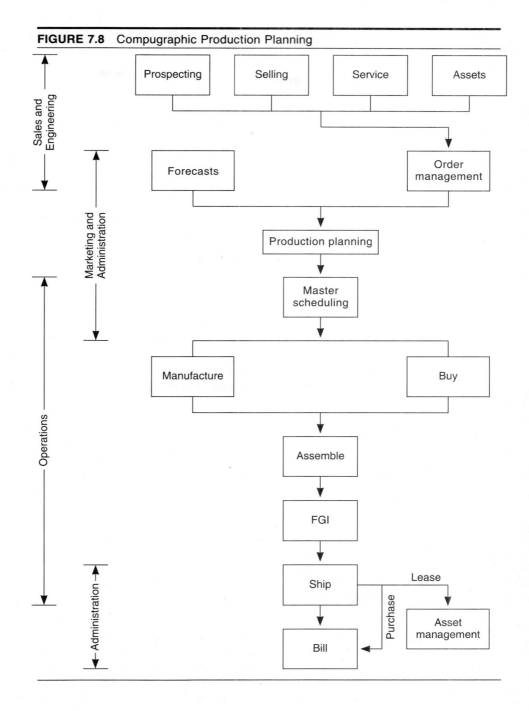

FIGURE 7.9 Compugraphic's Production Planning and the MPS

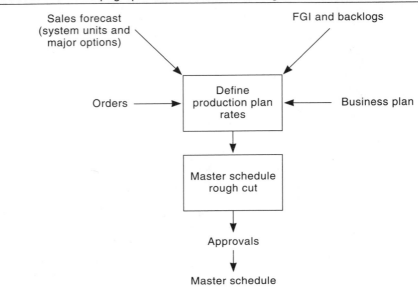

the business plan and corporate objectives. The committee's mission is further delineated as:

- To assure that sales forecasts and production plans are consistent with the annual business plan.
- To establish performance measurements for evaluating the production-planning process.
- To communicate sales forecasts for product families on a monthly basis.
- To assure that manufacturing capabilities are consistent with the production plan.
- To monitor actual results against plans and make adjustments as required.
- To manage the finished-goods inventory with the targets established in the production and business plans.
- To provide direction to the development and execution of the master production schedule.

At a monthly meeting, each of the 11 product family groups are reviewed. For each family, forecast and order performance for the current month are reviewed as well as the outlook for the coming 12 months. Next, manufacturing performance is reviewed. Was the production plan for the past month met? Are there any expected sales or manufacturing deviations in the future? If so, are they to be compensated for in other months? The other major review point is the finished-goods inventories in relation to plans, including consideration of established safety stock levels. Projected inventory levels are based on netting the sales and manufacturing data, but they're reviewed independently to see if revisions in either sales or manufacturing are needed.

Also reviewed at the meeting are several other items that could impact a particular family group's performance. This includes customer backlogs, customer service levels achieved, new marketing plans (such as price changes or sales incentives), and product requirements for demonstration or other non-sales purposes.

Figure 7.10 shows records for one product family. Data were collected during the month of May. The production-planning process is essentially made up of three parts, each corresponding to one of the records. The top third of Figure 7.10 is devoted to this product family's marketing outlook. At the end of April, actual shipments and backlog to be shipped were exactly equal to the base line plan (43 units). The original forecast for May was 128 units, but as of the date the data were collected, it appeared the actual shipments would exceed this plan by 7 units. June was expected to fall short by 54 units (47 cumulative), but marketing expected to catch up during July.

The bottom third of Figure 7.10 gives production performance. In April, manufacturing overbuilt the schedule by 14 units. However, the plan shows them underbuilding in May to get back on the base line schedule for the year.

The middle third of Figure 7.10 nets shipment plans against manufacturing plans to project inventories. This allows top management to examine the overall impact of both forecast errors and manufacturing performance deviations.

Compugraphic's production-planning process has had an important impact, particularly on its MPC systems. When the firm implemented a new on-line MRP-based system, it became clear that the production plan was critical to provide the necessary direction to master production scheduling and resultant MPC modules. Prior to developing production planning, there was a tendency to not react rapidly to market shifts (particularly downturns). The result was larger than necessary inventories, which, in turn, required more radical adjustments in production. The monthly production-planning process's goal was to make more frequent smaller adjustments on a regular basis, based on top management's ongoing assessment of where the business was headed.

Mohawk's Integrated Planning Process

The Mohawk Electric Company (a disguised name for an actual Midwest firm) manufactures electrical switches, controls, and measurement instruments for industrial applications. Three main product lines (energy management devices, tachographs, and data systems) represent annual sales of $25 to $30 million, with a price range of $15 to $50,000 per unit. About 40,000 units are sold each year, involving some 5,000 unique final-product catalogue numbers. Approximately 70 percent of sales volume is shipped directly from the firm's finished-goods inventory.

FIGURE 7.10 Compugraphic's Production Planning: Sample Family

Product family _Sunline 25_   Month _May_

Shipment performance

		Apr	May	June	July	Aug	Sept	Oct	Nov	Dec	Jan	Feb	Mar		
Baseline (plan)	–	43	128	228	222	311	356	299	288	383	232	232	308		
Actual & backlog	–	43	–	–	–	–	1	–	1	–	–	–	1		
Shipment outlook		–	135	174	269	311	356	288	288	383	232	232	309		
Variance		Ø	7	<54>	47	Ø	Ø	Ø	Ø	Ø	Ø	Ø	Ø		
Cum variance		Ø	7	<47>	Ø	Ø	Ø	Ø	Ø	Ø	Ø	Ø	Ø		

Inventory performance

		Apr	May	June	July	Aug	Sept	Oct	Nov	Dec	Jan	Feb	Mar		
Baseline (plan)	–	89	193	32	8	27	1	33	65	2	10	18	10		
Actual*	15	103	–	–	–	–	–	–	–	–	–	–	–		
Outlook		–	86	79	8	27	1	33	65	2	10	18	10		
Variance		<14>	7	<47>	Ø	Ø	Ø	Ø	Ø	Ø	Ø	Ø	Ø		

Production performance

		Apr	May	June	July	Aug	Sept	Oct	Nov	Dec	Jan	Feb	Mar		
Baseline (plan)	–	117	132	167	198	330	320	320	320	320	240	240	300		
Actual	131	–	–	–	–	–	–	–	–	–	–	–	–		
Outlook		–	118	167	198	330	330	320	320	320	240	240	300		
Variance		14	<14>	Ø	Ø	Ø	Ø	Ø	Ø	Ø	Ø	Ø	Ø		

April Invty
Perpetual   103*

The production plan and master production schedule are established quarterly as a part of the firm's regular budgetary planning activities. For most of Mohawk's business, the master production schedule is stated in terms of the number of units to be produced for each end product (catalogue number) during the next four quarters.

Overall production lead time (covering the purchasing, fabrication, and assembly operations) generally exceeds the delivery time quoted to customers. Thus, the production plan and master production schedule are primarily based on sales forecasts and financial plans—instead of on actual customer orders. Figure 7.11 depicts the budgetary planning process. Once every quarter, the company's sales, finance, and manufacturing executives prepare an overall business game plan covering the next four quarters, including: (*a*) a sales forecast for each of the firm's three product lines and (*b*) a detailed financial operating plan specifying a forecast of plant shipments (Mohawk calls this the "delivery plan"; in fact, it's the production plan), inventory level targets (the inventory plan), and a capacity plan (covering budgets for manpower and materials). Once these plans are prepared to produce an over-

FIGURE 7.11   Mohawk Company's Quarterly Budgeting Cycle Activities

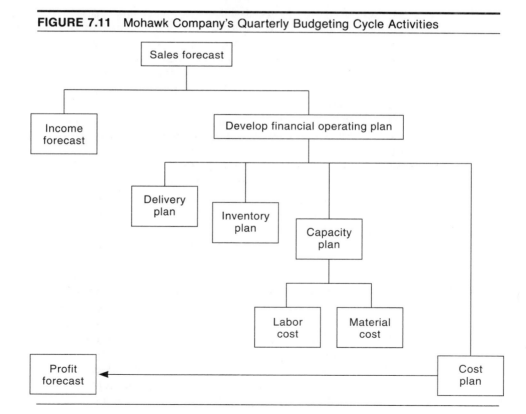

all profit forecast for the firm, work can begin on preparing (revising) the master production schedule that also covers the next four quarters.

The budgeting cycle, performed at each quarter's midpoint, begins with the preparation of a sales forecast for each of the firm's three product lines. Sales forecasting is the responsibility of a general manager, who has the profit responsibility for a particular product line. In preparing the sales forecast, the general manager, financial staff, and marketing organization work closely with a separate field sales organization. The initial sales forecast is for one year in the future. This forecast (stated in terms of both dollar sales and unit sales) corresponds to the product groupings the financial staff uses to value inventory and measure gross profit levels.

As an example, the energy management product line (with annual sales of $15 to $18 million) includes some 75 individual product categories, each representing annual sales of $25,000 to $2.5 million. While a separate forecast is prepared for each product category, some of these sales forecasts are combined to reduce the number of product categories considered in the budgeting cycle. In fact, only 21 product groupings are considered in the energy management product line budgeting cycle. They are shown in Figure 7.12, for which the sales forecast was prepared at the midpoint of the second quarter in year 1. In this figure, the sales forecast is expressed in dollars for each product grouping on a monthly basis for the next quarter, and a quarterly basis for the following three quarters. In producing the data in Figure 7.12, sales for the second quarter of year 1 are treated as actual, even though the quarter isn't yet finished. As a result, the monthly forecasts are for the third quarter of year 1, and the quarterly forecasts have been prepared through the second quarter of year 2.

Once the sales forecast is made, financial and manufacturing representatives become involved in the cycle to prepare a game plan for the product line. One of the first steps is to translate the sales forecast into a delivery plan (production plan) for manufacturing and an income forecast for finance. The company uses a tabular presentation form for preparing these plans.

The delivery plan is a statement of the total planned factory deliveries to customers, to finished-goods inventory and other company locations. Representatives from manufacturing, finance, and sales develop the delivery plan. The sales forecast, desired changes to inventory, potential capacity constraints, vendor deliveries, cash requirements, personnel available, and so on, are considered; adjustments to the sales forecasts (plan) and/or inventory plan are negotiated if necessary. If, for example, manufacturing can't produce the volume necessary to satisfy both the sales plan and an increase in inventory, the cycle stops and a new sales and/or inventory plan is agreed upon. This production plan, therefore, is an integral part of the entire process; subsequent planning does *not* proceed until there's complete agreement between the sales, finance, and manufacturing representatives.

The next step in the cycle is converting the delivery plan into a capacity plan for each product line and for the plant in total. Figure 7.13 illustrates the

FIGURE 7.12 Mohawk Company's Summary Sales Forecast, Energy Management Products ($000)*

Product Grouping	History			1st Qtr.	2d Qtr.	Year 1			4th Qtr.	Year 2 1 Qtr.	Year 2 2d Qtr.
	Year 2	Year 1	Year 0			July	Aug.	Sept.			
Singlephase	886	700	265	51	23	13	14	14	41	41	39
Polyphase	4059	1699	349	46	32	20	20	20	60	60	50
Con-Ed	402	108	—								
Lincoln Billing	1609	1354	1412	451	351	102	161	146	409	527	435
Meter Timeswitch	331	224	188	63	76	20	30	27	77	78	69
Sockets	41	57	84	13	14	2	2	2	6	6	10
	7328	4142	2298	624	496	157	227	209	593	712	603
Lincoln Nonbilling	2301	2721	1837	725	560	149	176	204	529	528	586
Line Controls	615	698	358	97	120	15	21	22	58	83	90
Timeswitch	882	1107	708	186	185	43	55	76	174	262	202
	3798	4526	2903	1008	865	207	252	302	761	873	878
Transformers 600V	2767	3139	2248	666	559	193	254	281	728	756	677
Transformers 15KV	383	528	410	232	110	45	50	60	155	201	175
D.C. Meters	143	200	102	70	59	7	18	18	43	53	56
	3293	3867	2760	968	728	245	322	359	926	1010	908
Survey Recorders	695	826	901	321	355	99	162	132	373	398	367
S.R. Systems				8	95	—	178	45	223	237	141
Digital Pulse Rec.	460	436	225	119	174	43	48	50	141	126	140
C.M.E.	663	448	407	153	168	6	31	57	94	145	145
	1818	1710	1533	601	792	148	419	284	851	906	788
Parts—Winchester & Memphis	1348	1466	1027	405	413	75	135	135	345	395	390
Misc.	407	307	276	224	82	24	32	33	89	120	129
Repairs (Replacement Parts)	37	46	53	16	10	3	3	4	10	10	12
	444	353	329	240	92	27	35	37	99	130	141
Resale:											
Demand Control	—	2	151	64	56	42	52	53	147	172	110
Sigma-form	482	443	—								
	482	445	151	64	56	42	52	53	147	172	110
Total	18511	16509	11001	3910	3442	901	1442	1379	3722	4198	3818

*Prepared at the midpoint of the second quarter of year 1.

Source: W. L. Berry, R. A. Mohrman, and T. R. Callarman, "Master Scheduling and Capacity Planning: A Case Study" (Bloomington: Indiana University Graduate School of Business, Discussion Paper No. 73, 1977).

FIGURE 7.13 Mohawk Company's Aggregate Production and Inventory Plan ($000)*

Labor forecast	Year 0		Year 1						Year 2	Year 3
	8/31 Actual	12/31 4th Qtr.	1/31 Jan.	2/29 Feb.	3/31 Mar.	6/30 2d Qtr.	9/31 3d Qtr.	12/31 4th Qtr.	12/31 1977	12/31 1978
Energy management:										
Finished-goods inventory	$ 244	$ 240	$ 250	$ 270	$ 275	$ 310	$ 300	$ 300	$ 300	$ 300
Work-in-process inventory	862	825	825	805	800	800	800	800	800	800
Subtotal	$1,106	$1,065	$1,075	$1,075	$1,075	$1,110	$1,100	$1,100	$1,100	$1,100
Net change		(41)	10	—	—	35	(10)	—	—	—
Labor in sales forecast		599	176	168	210	540	531	542	2,196	2,146
Transfer from Memphis		(54)	(12)	(12)	(12)	(36)	(36)	(36)	(144)	(144)
Transfer to Memphis		72	25	25	15	13	13	13	52	52
Total labor input		$ 576	$ 199	$ 181	$ 213	$ 552	$ 498	$ 519	$2,104	$2,054
Plant total:										
Labor input	$ 210	$ 829	$ 290	$ 273	$ 342	$ 847	$ 745	$ 830	$3,243	$3,706
Days per period	20	61	21	20	25	62	52	57	236	236
Average labor per day	$ 10.5	$ 13.6	$ 13.8	$ 13.6	$ 13.7	$ 13.7	$ 14.3	$ 14.6	$ 13.7	$ 15.7

*Measured in direct labor-dollars.

Source: W. L. Berry, R. A. Mohrman, and T. R. Callarman, "Master Scheduling and Capacity Planning: A Case Study" (Bloomington: Indiana University Graduate School of Business, Discussion Paper No. 73, 1977).

development of the capacity plan for the energy management product line considering the sales forecast's labor content, the forecast labor content of the two inventories (finished goods and work in process), and the labor content of the forecast interplant transfers both to and from this plant. The bottom line in the energy management section of Figure 7.13 ("Total labor input") indicates plant capacity requirements for this product line for the periods indicated where capacity is stated in terms of direct labor dollars. For example, $199,000 of direct labor input is planned for January, year 1. This represents about two thirds of the total direct labor input (capacity) for the Mohawk plant in January ($290,000)—which is shown on the next line of Figure 7.13. Since January has 21 working days, this means an average of $13,800 of direct labor input per day. This translates into a total manpower level for the plant, using planning factors for the number of dollars of direct labor per person/day.

Mohawk considers three factors in arriving at the capacity plan. First, it determines the sales forecasts' direct labor content using standard cost system data. Figure 7.13 shows the sales forecasts' direct labor content for the energy management product line (labeled "Labor in sales forecast"). Note that $176,000 of direct labor is required to support the sales forecast for January of year 1. Next, labor-dollars in the sales forecasts are modified to account for any inventory buildup or depletion planned for the coming year. Figure 7.13 indicates desired levels over the next year, including both finished-goods and work-in-process inventories. The levels (also measured in terms of direct labor-dollars) indicate cash requirements to finance the inventory during the next year as well. Note that a $41,000 inventory reduction is planned for the fourth quarter of year 0, while an increase of $10,000 is planned for January, year 1. These changes must be considered in planning the direct labor input for the energy management product line. Thus, the $10,000 increase in January means $186,000 of direct labor is needed in this month, instead of $176,000. The third factor in determining plant capacity is interplant sales of equipment. Figure 7.13 shows in January the Memphis plant will expend $12,000 of labor for products sold by this plant. An additional $25,000 in labor will be expended here for products sold by Memphis. Thus, there's a net addition of $13,000 in direct labor required above and beyond sales by this plant. When added to the $10,000 inventory increase and the $176,000 of labor in the sales forecast, total labor input for January, year 1, is $199,000.

After all negotiations are complete, the budgeting cycle produces an overall game plan for the business that includes an approved sales plan, a delivery plan, an inventory plan, and a capacity plan. Additional steps are performed in the budgeting cycle. They involve preparing a direct material budget of the purchasing dollars required to support the delivery plan, and a cost plan specifying a budget of indirect manufacturing and administration expenses. Engineering and marketing are also included in the budgeting cycle. All plans are then combined by the finance staff to produce a profit forecast for each product line (profit center).

Hill-Rom's Use of Planning Bills of Material

The use of planning bill of material concepts can be very helpful in the production-planning process. An example is the application developed at Hill-Rom, a manufacturer of hospital beds, related equipment, and accessories for hospitals and nursing homes.

Hill-Rom has expanded the planning bill concept to what it calls the "super-duper" bill. Figure 7.14 shows an abbreviated example. Using this approach, only one item is forecast, total bed sales. All other forecasts are treated as bill of material relationships. For example, the forecast for the super bill group, over-bed tables, is a percentage of overall bed sales.

One marketing person at Hill-Rom found the super-duper bill concept ideal for implementing an idea he'd been thinking about for some time. He believed the company makes trigger products and trailer products. Beds are trigger products, whereas over-bed tables, chairs, and add-ons (such as trapezes or intravenous fluid rods) are trailer products. Purchase of trailer products is dependent upon purchase of trigger products in somewhat the same relationship as component sales depend on end-item sales. This relationship means that, rather than forecasting demand for over-bed tables, Hill-Rom tracks and maintains the percentage relationship between sales of beds and over-bed tables.

This bill of material relationship will probably be a better estimate than a direct forecast of over-bed tables. If we expect bed sales to go up or down, by treating over-bed tables as a trailer product with a bill of material linkage, we have an automatic adjustment in over-bed table forecasts, as in all trailer products.

Using bill of material approaches to forecasting also forces a logical consistency. At one time the forecast for 84-inch mattresses at Hill-Rom exceeded combined forecasts for beds using 84-inch mattresses. Treating these relationships with bill of material approaches reduces these inconsistencies, which always result from independent estimating.

The production-planning unit for these products at Hill-Rom is total beds. Furthermore, the percentage split into hospital beds and nursing home beds is not only estimated, it's managed. Sales personnel are held to specified tolerance limits on this split because the capacity and net profit implications of the percentage split are important.

Below each of these two super-duper bills are "super" bills for the various model series. Finally, there's another trigger–trailer relationship between total hospital bed sales and sales of hospital furniture such as cabinets and flower tables. The same kind of bill of material relationship is used to forecast nursing home furniture sales. These various bill of material relationships pass the planning information down through the MPC system in a logically consistent way.

Finally, this entire approach is consistent with the way the firm does its strategic planning, which is in terms of strategic business units (SBUs). SBUs

FIGURE 7.14 Hill-Rom's "Super-Duper" Bill

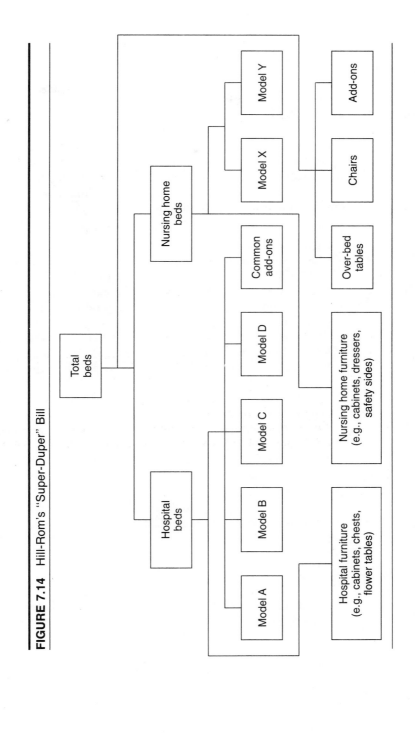

are established as super bills. The result is a very close integration of MPC and strategic planning.

CONCLUDING PRINCIPLES

Production and game planning are key inputs to MPC systems. They represent management's handle on the business. This chapter emphasizes the key relationships of top management and functional management in developing and maintaining an effective production plan. The following important principles summarize our discussion.

- The production plan is not a forecast; it must be a managerial statement of desired production output.
- The production plan should be a part of the game-planning process so it will be in complete agreement with the other functional plans (sales plan, budget, and so on) that make up the game plan.
- The trade-offs required to frame the production plan must be made *prior* to final approval of the plan.
- There must be top-management involvement in the game-planning process, which should be directly related to strategic planning.
- The MPC system should be used to perform routine activities and provide routine data, so management time can be devoted to important tasks.
- The MPC system should be used to facilitate what-if analyses at the production-planning level.
- Reviews of performance against manufacturing plans and sales forecasts are needed to prompt replanning when necessary.
- The production plan should provide the MPS parameters, and flexibility should be specifically defined. The sum of the detailed MPS must always equal the production plan.
- The production plan should tie the company's strategic activities directly through the MPS to the MPC's execution modules.

REFERENCES

Berry, W. L.; R. A. Mohrman; and T. Callarman. "Master Scheduling and Capacity Planning: A Case Study." Bloomington: Indiana University Graduate School of Business, Discussion Paper No. 73, 1977.

Bitran, Gabriel R.; Elizabeth A. Haas; and Arnoldo C. Hax. "Hierarchical Production Planning: A Single Stage System." *Operations Research* 29, no. 4 (July–August 1981), pp. 717–43.

Dougherty, John R. "Getting Started with Production Planning." *Readings in Production and Inventory Control and Planning,* APICS 27th Annual Conference, 1984, pp. 176–79.

Goldratt, Eliyanu. "The Unbalanced Plant." *APICS 24th Annual Conference Proceedings,* 1981, pp. 195–99.

Hall, Robert. "Driving the Productivity Machine: Production Planning and Control in Japan." Falls Church, Va.: APICS, 1981.

Hayes, R. H., and S. C. Wheelright. *Restoring Our Competitive Edge.* New York: Wiley, 1984.

Hill, Terry. *Manufacturing Strategy.* London: MacMillan Education Ltd., 1985.

Hodgson, T. J.; R. E. King; and C. U. King. "Development of a Production Planning System: A Case History." *Production and Inventory Management* 31, no. 4, 4th quarter 1990, pp. 18–24.

Holt, C. C.; F. Modigliani; J. F. Muth; and H. A. Simon. *Planning Production, Inventories, and Workforce.* New York: Prentice-Hall, 1960.

Ling, R. C., and W. E. Goddard. *Orchestrating Success.* Essex Junction, Vt.: Oliver Wight Ltd., 1988.

Leong, G. K.; M. D. Oliff; and R. E. Markland. "Improved Hierarchical Production Planning." *Journal of Operations Management* 8, no. 2, April 1989, pp. 90–114.

Peterson, Rein, and Edward A. Silver. *Decision Systems for Inventory Management and Production Planning.* 2nd ed. New York: Wiley, 1985.

Sari, John F. "Why Don't We Call It Sales and Operations Planning, Not Production Planning?" *APICS 29th Annual Conference Proceedings,* 1986, pp. 22–24.

Shirley, G. V., and R. Jaikumar. "Production Planning in Flexible Transfer Lines." *Journal of Manufacturing and Operations Management* 2, no. 4, 1989, pp. 249–67.

Singhal, K., and V. Adlakha. "Cost and Shortage Trade-Offs in Aggregate Production Planning." *Decision Sciences Journal* 20, no. 1, Winter 1989. pp. 158–65.

Skinner, C. W. *Manufacturing in the Corporate Strategy.* New York: Wiley, 1978.

Vollmann, T. E. "Capacity Planning: The Missing Link." *Production and Inventory Management,* 1st quarter 1973.

DISCUSSION QUESTIONS

1. The production plan is sometimes called "management's handle on manufacturing." Discuss.
2. What would you guess would be the aggregate terms used in managing a university as a whole? The computer center? Buildings and grounds? An individual major?
3. The production plan is stated in aggregate terms; therefore, there will be some error in resource planning. Wouldn't it be better to use the MPS where the specific product mix is either anticipated or incorporated directly?
4. What are the differences between game planning and budgeting?
5. Some experts argue the production/game-planning process ought to be done only on exception; that is, only when conditions have changed enough to warrant replanning. Others argue it should be done on a periodic basis *and* when required by exception. What's your view?

6. Discuss the relative merits of cumulative charts (e.g., Figure 7.4) and tabular plans (e.g., Figure 7.5) for production planning.
7. What are some implications of not making the trade-offs explicit in the game-planning process?
8. In incorporating the firm's strategic objectives into the game-planning process, what differences exist between such firms as coal mines or cardboard manufacturers versus fashion or cosmetic manufacturers versus manufacturers of machine tools?
9. In the university setting, what is an analogue to the Hill-Rom super-duper bill and its trigger-and-trailer notions?

PROBLEMS

1. A production planner for Lavella Chase is developing a production plan that involves backorders. The company's demand and production rates for the next four periods are as follows:

Period	Demand	Production
1	4,000	3,800
2	7,500	6,000
3	2,200	4,100
4	6,400	6,000

Given beginning inventory at the start of period 1 of 200 units, calculate beginning inventory, ending inventory, average inventory, and the backorder amount, if any, for each of the next four periods.

2. A company with a policy of level production has zero beginning inventory and must meet the following demand quantities for its major product:

Period	Demand
1	10
2	2
3	11
4	9
5	8

a. What production rate per period will result in zero ending inventory for period 5?
b. When, and in what quantities, will backorders occur?

c. What level production rate per period will avoid backorders?

d. Using the production rate computed in part c, what will be the ending inventory for period 3?   2

3. The VP of manufacturing for the N. L. Hyer Company is ready to decide on her aggregate output plan for next year. The firm's single product (known throughout the adult puzzle trade as "Hyer's Hexagon") is "red hot," according to the VP of marketing, so she insists that backorders aren't acceptable next year. Since the United Puzzlemakers Union is trying to organize the plant, the manufacturing VP decides the number of employees must remain constant (i.e., level) throughout all four quarters.

Quarter	Production days	Demand forecast
1	60	10,000
2	58	12,000
3	62	15,000
4	70	9,000

Beginning inventory = 1,000 units.
Regular time output rate = 10 units/day employee.

a. What daily production rate and number of employees will be required?

b. How many units will be in inventory at the end of quarter 4?

4. Below is the plotted cumulative demand for Joan's Joyous Nature Food (in pounds) for the next four months. Beginning inventory is 10 pounds.

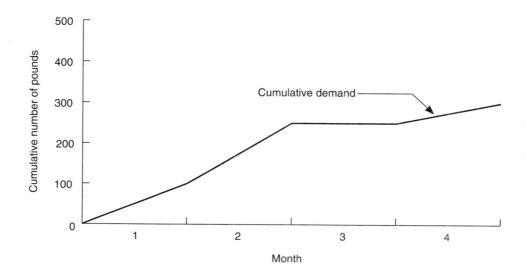

Month	Demand
1	100
2	150
3	0
4	40

a. How much should Joan produce each month if she wishes to have a level production plan with no backorders or stockouts? Plot your cumulative production on the preceding graph.

b. What is ending inventory for month 4 under this plan?

c. Joan decides to have a level production, level employment production plan with no ending inventory at the end of the planning horizon. How much should she make each month? What are the monthly backorders?

d. Given inventory carrying costs of $2/pound/month (on the average inventory) and backorders of $4/pound/month (based on month-end backorders), calculate the cost of backorders and inventory for the plan in part c.

5. The Oro del Mar Company's forecast for the first three months of next year is: January—100, February—0, March—300 (in 1,000 pounds). Beginning inventory is 100,000 pounds.

a. Plot cumulative demand and a level aggregate production plan that meets demand with no backorders and no ending inventory in March.

b. What production each month is needed to meet part a's conditions?

6. A portion of the Warmdot Game Company's aggregate production plan appears below. Assuming the graph shows cumulative production and demand, is the company expecting an inventory or backorder at the end of August? How many units is this inventory or backorder expected to be?

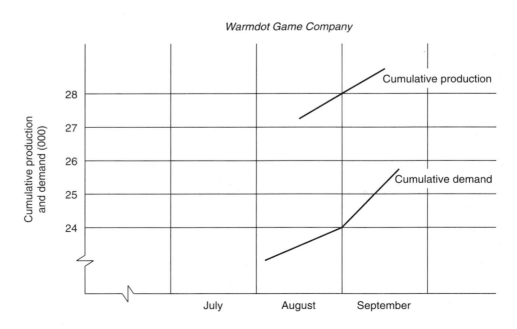

Warmdot Game Company

7. The Crazy Tubb Thumpers uses a level plan to produce its snowshoe covers. Beginning inventory for January is zero. Demand for covers in January, February, and March is as follows:

Month	Demand
January	5,000
February	0
March	20,000

a. Graph cumulative production for a level production plan of 10,000 units per month. Show cumulative demand for the months January through March and indicate the ending inventory for March.

b. Consider the following additional data for Crazy Tubb Thumpers:

Employment = 9 people.
Production = 1,000 units/month/employee (regular time).
Standard hours/month = 166 hours (regular time).
Regular time labor rate = $1,200/month.
Overtime rate = 1½ times regular time rate.

Assuming a production rate of 10,000 units per month, what's the labor cost of one month's production at Crazy Tubb Thumpers?

8. The Bi-Product Company produces two products (A and B) that are similar in terms of labor content and skills required. Company management wishes to "level" the number of employees needed each day so no hiring or layoffs will be required during the year. A complication to this problem is that the number of working days in each quarter varies.

	Demand		
Quarter	Product A	Product B	Working days
1	9,800	14,500	68
2	12,000	30,000	56
3	14,000	19,500	62
4	31,000	25,000	58

Beginning inventory: 2,400 units of Product A.
900 units of Product B.
Inventory holding cost: $10 per unit per quarter (either product).
No backorders allowed.
No variations in size of work force allowed.
Output rate = 25 units of either product per day per employee.

a. What daily production rate will be required to meet the demand forecast and yield zero inventory at the end of quarter 4?

b. How many employees will be required each day?

9. Two students who share an apartment decide to use MRP for control over supplies. They shop every day and have the following "master production schedule" for tomorrow: breakfast—6 Bloody Marys, 4 Pop-Tarts; lunch—5 peanut butter and jelly sandwiches, 4 cans beer; dinner—3 vodka martinis, 4 hamburgers, 8 oz. potato chips, 4 Hostess Twinkies; snack—18 cans beer, 16 oz. potato chips.

a. Given the following recipes, on-hand inventories, and lot sizes, prepare tomorrow's shopping list:

Recipes

Bloody Mary
    2.5 oz. vodka
    4 oz. tomato juice
    dash Worcestershire (⅛ oz.)
    dash Tabasco (⅛ oz.)
Vodka martini
    4 oz. vodka
    .001 oz. dry vermouth
    ice

Peanut butter and jelly sandwich
    2 slices bread
    1 oz. peanut butter
    .5 oz. jelly
    .5 oz. butter
Hamburger
    4 oz. ground beef
    2 slices bread
    1 oz. catsup

Pantry inventory	*On-hand*	*Lot size*
Vodka	12 oz.	Quart
Tomato juice	3 oz.	46-oz. can
Worcestershire	6 oz.	8-oz. bottle
Tabasco	0	4-oz. bottle
Pop-Tarts	3	8-tart package
Peanut butter	6 oz.	12-oz. jar
Jelly	2 oz.	12-oz. jar
Bread	8 slices	20-slice loaf
Butter	4 oz.	1 lb.
Vermouth	5 oz.	16 oz.
Ice	Plenty	—
Ground beef	.8 lb.	1 lb.
Catsup	6 oz.	8-oz. bottle
Potato chips	2 oz.	16-oz. package
Hostess Twinkies	24	1-dozen package
Beer	9 cans	6 pack

---

b. Student A has just learned that the Bloody Marys must be made in batches of 24. Assuming no change in ingredients or amounts, how does this quantity affect the shopping list?

10. Assume the two students in Problem 9a eat the same set of "meals" daily.
    a. What are their daily "financial" capacity requirements, given the following prices?

Item	Price
Vodka	$.20 per ounce
Tomato juice	.02 per ounce
Worcestershire	.01 per dash
Tabasco	.01 per dash
Pop-Tarts	.20 each
Peanut butter	.04 per ounce
Jelly	.03 per ounce
Bread	.03 per slice
Butter	.05 per ounce
Vermouth	.12 per ounce
Ice	No charge
Ground beef	.05 per ounce
Catsup	.02 per ounce
Potato chips	.03 per ounce
Hostess Twinkies	.15 each
Beer	.25 each

    b. What is the out-of-pocket cost of tomorrow's shopping list?
    c. What is the out-of-pocket cost if the Bloody Marys are made in batches of 24 instead of one at a time?

11. Mike Blanford, master scheduler at General Avionics, has the following demand forecast for one line in his factory:

Quarter	Unit sales
1	5,000
2	10,000
3	8,000
4	2,000

At the beginning of January, there are 1,000 units in inventory. The firm has prepared the following data:

Hiring cost per employee = $200
Firing cost per employee = $400
Beginning work force = 60 employees
Inventory carrying cost = $2 per unit per quarter of ending inventory
Stockout cost = $5 per unit
Regular payroll = $1,200 per employee per quarter
Overtime cost = $2 per unit

Each employee can produce 100 units per quarter. Demand not satisfied in any quarter is lost and incurs a stockout penalty. If Mike produces exactly enough to meet demand each quarter, with no inventories at the end of quarters and no overtime, how much will he produce each quarter, and what is the overall cost? (Use a spreadsheet model for the calculations.)

12. Use the data in Problem 11 to calculate production amounts and costs for a level rate of output with no ending inventory. (Stockouts are allowed.)

13. The marketing people at General Avionics (Problem 11) feel somewhat unsure of their predictions of demand. They want to devise plans that allow for a 10 percent overage in sales in each quarter (i.e., total sales potential of 27,500 units for the year). Prepare two plans (one a chase strategy and one with level production) and determine the costs. (No stockouts allowed.)

14. At the end of the first quarter, actual sales were 4,000 units. Revise both plans for Problem 13 to provide for the 10 percent overage (i.e., the total sales for the remaining three quarters = 22,000).

15. For the data in Problem 11, at the end of the first quarter, inventory is 2,000 units. Marketing has revised the forecasts for quarters 2, 3, and 4 with a 20 percent reduction (i.e., remaining sales = 16,000). Develop a level and chase plan for the revised forecast that provides for an inventory of at least 1,000 units at the end of each quarter. (Assume 50 people were in the work force in the first quarter.)

Chapter 8

Demand Management

This chapter covers the highly integrative activity we call demand management. Through demand management all potential demands on manufacturing capacity are collected and coordinated. This activity manages day-to-day interactions between customers and the company. A well-developed demand management module within the manufacturing planning and control (MPC) system brings significant benefits. Proper planning of all externally and internally generated demands means capacity can be better planned and controlled. Timely and *honest* customer order promises are possible. Physical distribution activities can be improved significantly. This chapter shows how to achieve these benefits. The focus is less on techniques than on management underpinnings and concepts necessary to perform this integrative activity.

This chapter is organized around four topics:

- Demand management in manufacturing planning and control systems: What role does demand management play in the MPC system?
- Demand management techniques: What techniques have proven useful for demand management in different market environments?
- Managing demand: How do we live with demand management on a day-to-day basis.
- Company examples: Effective demand management in practice.

Several chapters in this book closely relate to demand management. Chapters 6 and 14 deal with master production scheduling, which is intimately tied to demand management. The production-planning process discussed in Chapter 7 concerns overall levels of demand. Chapter 18 details distribution requirements planning. The book also addresses technical material as it relates to forecasting (Chapter 16) and to inventory management (Chapter 17).

DEMAND MANAGEMENT IN MANUFACTURING PLANNING AND CONTROL SYSTEMS

Demand management encompasses forecasting, order entry, order-delivery-date promising, customer order service, physical distribution, and other customer-contact–related activities. Demand management also concerns other sources of demand for manufacturing capacity, including service-part demands, intracompany requirements, and pipeline inventory stocking. All quantities and timing for demands must be planned and controlled.

For many firms, planning and control of demand quantities and timings are a day-to-day interactive dialogue with customers. For other firms, particularly in the process industries, the critical coordination is in scheduling large inter- and intracompany requirements. For still others, physical distribution is critical, since the factory must support a warehouse replenishment program, which can differ significantly from the pattern of final customer demand.

Demand management is a gateway module in manufacturing planning and control providing the link to the marketplace. Activities performed here provide coordination between manufacturing and the marketplace, sister plants, and warehouses. Through demand management, we maintain a channel of communication between MPC systems and their "customers." Specific demands initiate actions throughout MPC, which ultimately result in product delivery and consumption of materials and capacities.

Figure 8.1 depicts external aspects of the demand management module as the double-ended arrow connected to the marketplace outside the MPC system. One implication of this connection is the need to forecast demand as a prerequisite to the other MPC activities. An important aspect is providing forecasts at the appropriate level of detail. It also may imply constraining forecasts to meet certain overall requirements of the company. Both considerations are taken into account in the demand management process. Techniques for aggregating and disaggregating demand, such as **pyramid forecasting,** can facilitate this process.

FIGURE 8.1   Demand Management in the MPC System

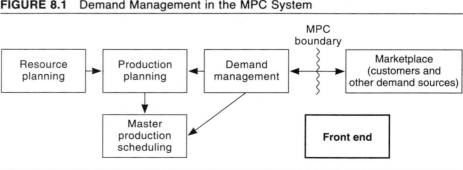

The importance of identifying all sources of demand is obvious, but sometimes overlooked. If material and capacity resources are to be planned effectively, we must identify *all* sources of demand: spare parts, distribution, inventory changes, new items, promotions, and so on. Only when we've accounted for all demand sources can we develop realistic MPC plans.

Demand Management and Production Planning

The exact linkage of demand management and production planning depends to some extent on how the firm handles production planning. If the production plan is a quarterly statement of output in dollars or some other financial measure, then the key requirement for demand planning is for synchronization with this target. If delivery timings for significant customer orders affect the production plan, this information must be communicated to production planning. Similarly, a major change in distribution inventory policy might influence the production plan.

In addition to the role of synchronization and communication between market activities and the production plan, a key activity in the demand management module is to assure the completeness of demand information. All sources of demand for manufacturing resources must be identified and incorporated in the production and resource planning processes. Sometimes this is more difficult than it appears. The difficulty seems to be greatest for companies with a significant number of interplant transfers. We've often heard plant managers complain that their worst customer is a sister plant or division.

To get a complete picture of the requirements for manufacturing capacity and material, we collect such sources of demand as spare parts demand, intercompany transfers, promotion requirements, pipeline buildups, quality assurance needs, exhibition or pilot project requirements, and even charitable donations. The principle is clear, though specifics differ from firm to firm: We must take all sources of demand into account. All must be included in the production plan to provide synchronization with other MPC activities.

Demand Management and Master Production Scheduling

Interactions of demand management and master production scheduling (MPS) are frequent and detailed. Details vary significantly between make-to-stock, assemble-to-order, and make-to-order environments. In all instances, however, the underlying concept is that forecasts are consumed over time by actual customer orders as Figure 8.2 shows. In each case, forecast future orders lie to the right and above the line, while actual customer orders are to the left and below the line. (These areas are labeled for the make-to-order example.)

Observe in Figure 8.2 that the three lines' positions are quite different. For make-to-stock environments, there are very few actual customer orders, since

FIGURE 8.2   MPS Time Fences—Forecasts Consumed by Orders

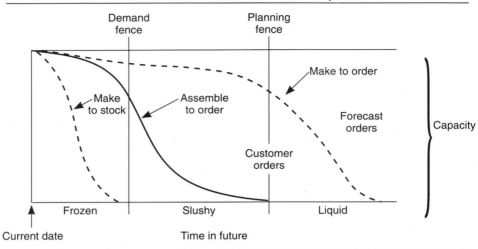

demand is generally satisfied from inventory. Thus, the master production scheduling task is one of providing inventory to meet forecasted future customer orders.

In the assemble-to-order environment, a key scheduling task is to provide viable customer promise dates. Usually, there are customer orders already booked for several periods into the future. The master production scheduler uses the **available-to-promise** concept for each module or customer option to manage the conversion from forecasts to booked orders.

Still different demand management/MPS problems confront the firm with a make-to-order environment, even though there's a relatively larger backlog of customer orders. Some orders can be in progress, even though they aren't completely specified and engineered. This means the master production scheduler is concerned with controlling these custom orders as they progress through all steps in the process. This involves engineering activities. Those activities impact on resource requirements as well as on manufacturing. All this has to be coordinated with customers as the orders become completely specified.

In each company environment, the demand management module's objective is to bridge the firm and the customer. This is facilitated by the time fences (the demand and planning fences) in Figure 8.2. The two fences result in three areas sometimes termed "frozen," "slushy," and "liquid." The authority for making changes, as well as the way the MPS is stated, differs among these areas. Fences provide guidelines to help the master production scheduler as forecasts become actual orders.

Types of uncertainty also differ from company to company. One aspect of the relationship between master production scheduling and demand management is facilitating the buffering against this uncertainty. In the make-to-stock case, uncertainty is largely in the demand variations around the forecast at each of the inventory locations. In this case, levels of safety stock (and/or safety lead time) must be set in order to provide the service levels required.

In the assemble-to-order case, the uncertainty involves not only the quantity and timing of customer orders but product mix as well. Safety stocks can be used, and hedging is a valuable technique. For make-to-order environments, the uncertainty is often not the timing or quantity of the customer order but, rather, how much of the company's resources will be required as the engineering is finally completed and exact requirements are determined.

The demand management task of specifying all sources of demand on manufacturing impacts master production scheduling in several ways. Some sources are handled directly in the master production schedule, such as pipeline inventory buildup, special exhibition requirements, and interplant transfers. Others, such as spare parts, may not be.

The spare-part issue can be complex in many firms. Demand for spare parts is typically forecast on an item-by-item basis and added to the gross requirement data in the corresponding material requirements planning (MRP) records. There may not be an explicit treatment of each spare part by a master production schedule. The actual spare-part demand, of course, varies around the forecast. Safety stocks can buffer variability. This simplified approach works reasonably well in firms where it would take too long to manage each spare part by an MPS record.

A key problem might arise if we don't incorporate spare-part demand at the MPS level; the resulting capacity requirements won't be reflected in rough-cut capacity planning models. The requirements would, however, be seen by the detailed capacity requirements planning models driven from MRP records.

To adequately perform rough-cut capacity planning, we must include the service-part demand. This could be done on a rough basis by estimating service-part demand in monetary units and using some base to convert it to the capacity measure. Alternatively, we could prepare a bill of capacity for each service part and multiply by the forecast service-part demand levels. The principle is clear: Capacity must be provided for *all* sources of demand.

Outbound Product Flow

Distribution activities are planned on the basis of the information developed in the demand management function. Customer delivery promise dates, inventory resupply shipments, interplant shipments, and so on are all used to develop short-term transportation schedules. Information used for master production schedules can be integrated with distribution planning as well. We can use the information to plan and control warehouse resupply. Moreover, transportation capacity, warehouse capacity, and the other resources within which the day-to-day distribution function operates can also be better planned and controlled with this information.

Integration of distribution with master production scheduling can have high payoffs for some firms. In essence, resupply shipping decisions are demand inputs, which the MPS must satisfy. Conversely, the MPS provides a set of product availabilities that we can use in the distribution planning system.

It's in demand management that we explicitly define service levels and resultant safety stocks. The requisite degree of flexibility for responding to mix or engineering design changes is set here as well. This is done through the determination of buffer stocks and timings. The master scheduler is then responsible for maintaining the required level of buffer stocks and timings.

Through conversion of day-to-day customer orders into product shipments we realize the company's service levels. Careful management of actual demands can provide the stability needed for efficient production: that stability provides the basis for realistic customer promises and service. Booking actual orders also serves to monitor activity against forecasts. As changes occur in the marketplace, demand management can and should routinely pick them up, indicating when managerial attention is required.

Data Capture

Data capture and monitoring activities of demand management fall into two broad categories: the overall market and the detailed product mix. The activity most appropriate for production planning is overall market trends and patterns. Data should correspond to the units used in production planning. The intent is to determine on an ongoing basis the general levels of actual business for input to the production-planning process.

The second activity concerns managing the product mix for master production scheduling and customer order promising. Since final demand is in catalog numbers or stockkeeping units, day-to-day conversion of specific demands to MPS actions requires managing the mix of individual products.

For both the overall market and the detailed product mix, it's important to capture *demand* data where possible. Many companies use sales instead of demand for purposes of making "demand" projections. Unless all demands have been satisfied, sales can understate actual demand. In other instances, we know of firms that use shipments as the basis for projecting demands. In one such instance, the company concluded its demand was increasing since its shipments were increasing. Not until they had committed to increased raw-material purchases did they realize the increased shipments were replacement orders for two successive overseas shipments lost at sea.

Dealing with Day-to-Day Customer Orders

A primary function of the demand management module is converting specific day-to-day customer orders into detailed MPC actions. Through the demand management function, actual demands consume the planned materials and capacities. Actual customer demands must be converted into production actions regardless of whether the firm manufactures make-to-stock, make-to-order, or assemble-to-order products. Details may vary depending on the nature of the company's manufacturing/marketing conditions.

In make-to-order environments, the primary activity is controlling customer orders to meet customer delivery dates. This must be related to the master production schedule to determine the impact of any engineering changes on the final customer requirement. While firms often perform this function the same way for assemble-to-order products, communication with the final assembly schedule may also be needed to set promise dates. In both these environments, there's communication from the customer (a request) and to the customer (a delivery date) through the demand management module. These aspects of demand management have such names as order entry, order booking, and customer order service.

In a make-to-stock environment, demand management doesn't ordinarily provide customer promise dates. Since material is in stock, the customer is most often served from inventory. If there's insufficient inventory for a specific request, the customer must be told when material will be available or, if there's allocation, told what portion of the request can be satisfied. Conversion of customer orders to MPC actions in the make-to-stock environment triggers resupply of the inventory from which sales are made. This conversion is largely through forecasting, since the resupply decision is in anticipation of customer orders.

In all these environments, extraordinary demands often must be accommodated. Examples include advance orders in the make-to-stock environment, unexpected interplant needs, large spare-part orders, provision of demonstration units, and increased channel inventories. These all represent "real" demands on the material system.

DEMAND MANAGEMENT TECHNIQUES

This section begins with a general discussion of aggregating and disaggregating forecasts on a consistent basis for demand management at all company levels. Thereafter, three subsections deal with demand management techniques most useful in the MPS environments of make-to-stock, assemble-to-order, and make-to-order.

Aggregating and Disaggregating Forecasts

Often forecasts are prepared for a variety of business decisions by many different people in a company. In addition to forecasts for *individual products* (units) used in scheduling operations, top managers utilize forecasts of *overall sales activity* (dollars) to plan and prepare budgets, and product managers frequently prepare forecasts for *groups of products*—for example, by horsepower (units), package size (pounds), or product brand (dollars)—in developing marketing plans for sales promotions, advertising, and distribution. The fact that sales forecasts have so many different decision-making purposes presents a significant coordination problem in many firms. Often these *independent* forecasts are never brought together within the organization to ensure that they're consistent and that the whole equals the sum of the parts.

One means of providing a consistency of forecasts for various purposes is through specialized bills of material (BOM). The **super bill** is an example. A super bill forces a consistency to the forecasts by structuring the BOM in the way the product is sold (e.g., in basic products and options). The result is a consistency between end items (e.g., cars) and options (e.g., air conditioners).

Another method of constraining the forecasts is **pyramid forecasting.** It provides a means of coordinating, integrating, and forcing consistency between forecasts prepared in different parts of the organization. We can then use these consistent forecasts to meet the needs in marketing, manufacturing, distribution, and so on. We can implement techniques to constrain forecasting in a number of ways, but we'll focus on a specific example of pyramid forecasting.

The procedure used in implementing the pyramid forecasting approach begins with individual item forecasts at level 3, which are rolled up into forecasts for product groups shown as level 2 in Figure 8.3. We then aggregate forecasts for product groupings into a total business forecast (in dollars) at level 1 in the product structure. Once the individual item and product grouping forecasts have been rolled up and considered in finalizing the top-management forecast (plan), the next step is to force down (constrain) the product grouping and individual item forecasts, so they're consistent with the plan.

In the example, the 11 individual items are divided into two product groupings. Two of these items, X_1 and X_2, form product group X (which we'll study in detail), while the remaining products, Z_1 through Z_9, are included in product group Z. These two product groups, X and Z, represent the firm's entire line of products. Figure 8.4 shows unit prices and initial forecasts for each level.

The roll-up process starts by summing the individual item forecasts (level 3) to provide a total for each group (level 2). For the X group, the roll-up forecast is 13,045 units (8,200 + 4,845). The sum of the individual Z group

FIGURE 8.3 Pyramid Forecasting Example

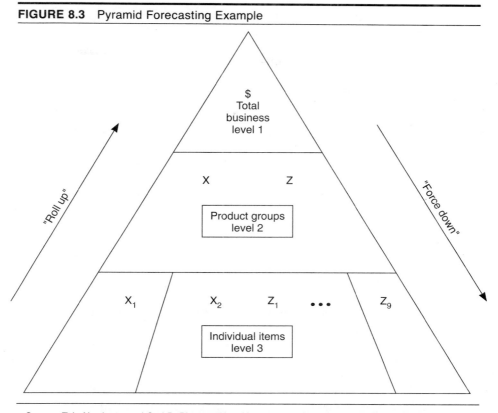

Source: T. L. Newberry and Carl D. Bhame, "How Management Should Use and Interact with Sales Forecasts," *Inventories & Production Magazine,* July–August 1981.

items gives a forecast of 28,050 units. Note that the X group roll-up doesn't correspond to the independent group forecast of 15,000 units. If there's substantial disagreement at this stage, a reconciliation could occur or an error might be discovered. If there's to be no reconciliation at this level, we needn't prepare independent forecasts for the groups. If dollar forecasts are required at level 2, prices at level 3 can be used to calculate an average price.

To roll up to the level 1 dollar forecasts, the average prices at the group level are combined with the group roll-up forecasts. The total of $778,460 [(13,045 × 16.67) + (28,050 × 20.00)] is less than the independent business forecast of $950,000. For illustrative purposes, we'll assume management has evaluated the business forecast *and* the roll-up forecast and has decided to use $900,000 as the forecast at level 1. The next task is to make the group and individual forecast consistent with this amount.

FIGURE 8.4   Initial and Roll-Up Forecasts

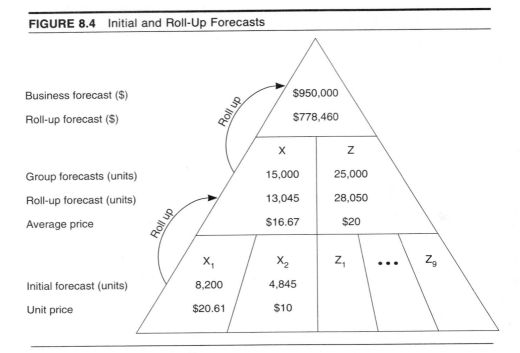

	X	Z
Business forecast ($)	$950,000	
Roll-up forecast ($)	$778,460	
Group forecasts (units)	15,000	25,000
Roll-up forecast (units)	13,045	28,050
Average price	$16.67	$20
	X₁ — 8,200 — $20.61	X₂ — 4,845 — $10

To bring about the consistencies, we use the forcing-down process. The ratio between the roll-up forecast at level 1 ($778,460) and the management total ($900,000) is used to make the adjustment.

The forecasts at all levels appear in Figure 8.5. The results are consistent forecasts throughout the organization, and the sum of the parts is forced to equal the whole. Note, however, the process of forcing the consistency needs to be approached with caution. In the example, forecasts at the lower levels are now higher than they were originally. Even though the sum of the parts equals the whole, it's possible the people responsible for the forecast won't "own" the number. They mustn't be made to feel they're simply being given an allocation of someone else's wish list.

There are several reasons for aggregating product items in both time and level of detail for forecasting purposes. We must do it with caution, however. Aggregating individual products into families, geographical areas, or product types, for example, must be done in ways that are compatible with the planning systems. Product groupings must also be developed, so that the forecast unit is sensible to forecasters. Provided we follow these guidelines, we can use product groupings to facilitate the forecasting task.

It's a well-known phenomenon that *long-term or product-line forecasts are more accurate than detailed forecasts.* This merely verbalizes a statistical verity. Consider the example in Figure 8.6. Monthly sales average 20 units but

FIGURE 8.5 Forcing Down the Management Forecast of Total Sales

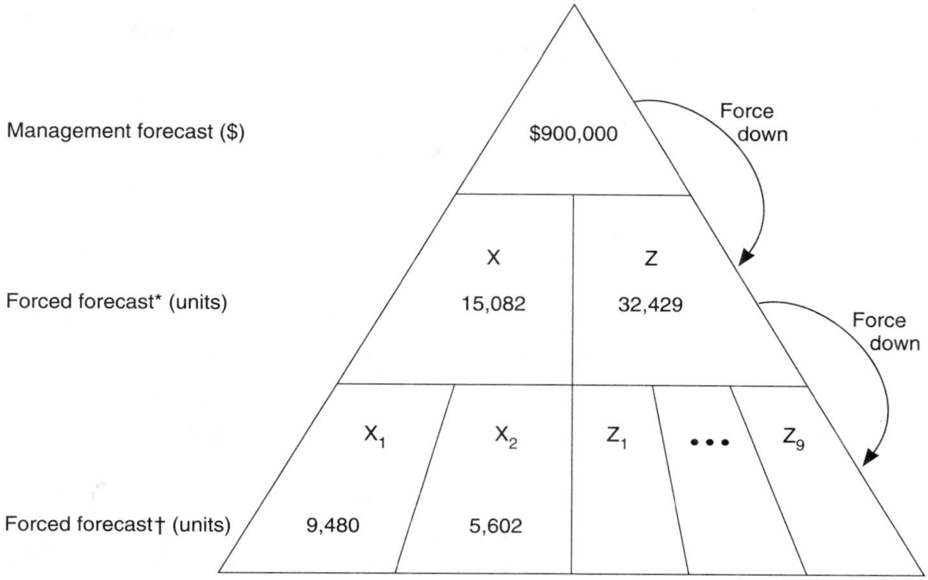

Management forecast ($) $900,000   Force down

Forced forecast* (units)

| X | Z |
| 15,082 | 32,429 |

Force down

Forced forecast† (units)

| X_1 | X_2 | Z_1 | ... | Z_9 |
| 9,480 | 5,602 | | | |

$$\text{*Forced forecast } (X) = \frac{\$900,000}{\$778,460} \times (13,045) = 15,082 \text{ units}$$

$$\text{*Forced forecast } (Z) = \frac{\$900,000}{\$778,460} \times (28,050) = 32,429 \text{ units}$$

$$\text{†Forced forecast } (X_1) = \frac{15,082}{13,045} \times (8,200) = 9,480 \text{ units}$$

$$\text{†Forced forecast } (X_2) = \frac{15,082}{13,045} \times (4,845) = 5,602 \text{ units}$$

vary randomly with a standard deviation of 2 units. This means 95 percent of the monthly demands lie between 16 and 24 units (assuming a normal distribution). This corresponds to a forecast error of plus or minus 20 percent around the forecast of 20 units per month.

Now suppose, instead of forecasting demand on a monthly basis, we prepare an annual forecast of demand—in this case, 240 units for the year. The resulting standard deviation is 6.9 units (assuming monthly sales are independent). This corresponds to a 95 percent range of 226 to 254 units or a plus or minus 5.8 percent deviation. The reduction from plus or minus 20 percent to plus or minus 5.8 percent is due to using a much longer time period. The same

---

FIGURE 8.6 Effect of Aggregating on Forecast Accuracy

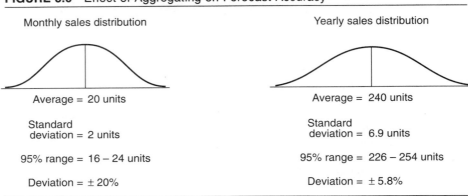

Monthly sales distribution	Yearly sales distribution
Average = 20 units	Average = 240 units
Standard deviation = 2 units	Standard deviation = 6.9 units
95% range = 16 – 24 units	95% range = 226 – 254 units
Deviation = ± 20%	Deviation = ± 5.8%

---

effect can be seen in forecasting demand for product families instead of for individual items.

Use of aggregate product forecast data can be illustrated with a survey concerning forecast accuracy, taken by Polysar International of plant, distribution, scheduling, sales, and transportation managers. Figure 8.7 shows the survey's results. Even though the persons surveyed used detailed information to make labor, inventory, scheduling, traffic, and warehousing decisions, they recognized the magnitude of the forecast error in the detailed product-item forecasts. Their conclusion was to develop methods to live with the forecast errors arising from the short-term, detailed forecasts.

There's one final point about forecast errors: Many efforts to reduce them are wasted—since they won't go away. A better alternative to improved forecasting is improved systems for *responding* to forecast errors. The gross to net logic of MRP prevents errors from compounding, since net requirements

---

FIGURE 8.7 Realistic Levels of Deviation of Actual versus Forecast (Percent)*

		1 month	1 quarter	1 year
	Total volume	± 12	8	8
	Family	15	10	8
	Type in family	15		12
	Grade in type	30		
	SKU†	50		

Time ⟶

Item detail ↓

*Average of a Polysar International Survey.
†Stockkeeping unit.

are simply changed from period to period as actual inventory deviates from projected inventory.

Make-to-Stock Demand Management

Providing adequate inventory to meet customers' needs throughout the distribution system and to maintain desired customer service levels requires detailed forecasts. Despite our arguments for the increased accuracy of aggregated forecasts, the need in make-to-stock situations is for item-level forecasts by location and time period. This means we still face the problem of providing this level of detail.

One means to deal with the inherent difficulty of forecasting individual items is to forecast ratios or percentages of aggregated forecasts, rather than stockkeeping units (SKUs) directly. Figure 8.8 shows the reduction in detail made possible by the technique. Jag's Coffee Company has four grinds, four package sizes, and three brands. This is a maximum total of $4 \times 4 \times 3 = 48$ possible combinations. As indicated in Figure 8.8, only 20 had sales in the period covered by the example. If percentages of total coffee sales by each

FIGURE 8.8 Sales in Tons and Percentages by Brand, Size, and Grind of Jag's Coffee

Brand and size	*I*	*R*	*D*	*S*	Total		Total	Percent
		Grind*						
Jingle						Coffee	440	100%
Packet	10				10	Brand		
Regular	20	40	30		90	Jingle	130	30
Giant		20	10		30	Jangle	80	18
						Jungle	230	52
Jangle						Size		
Packet	10				10	Packet	20	4
Regular	15	10	20		45	Regular	225	52
Promotion	5	5			10	Promotion	10	2
Giant		5	10		15	Giant	185	42
Jungle						Grind		
Regular		40	40	10	90	Grind *I*	60	14
Giant		50	50	40	140	Grind *R*	170	39
						Grind *D*	160	36
						Grind *S*	50	11
Total	60	170	160	50	440			

*I = instant; *R* = regular; *D* = drip; *S* = special.

brand, size, and grind are computed, there are only 12 separate items (11 mix ratios and total coffee sales) to deal with; that is, we could forecast total coffee sales and break them down into brands, sizes, or grinds on a percentage basis.

For many firms, product-mix ratios remain fairly constant over time. In such situations, managerial attention can focus directly on forecasting overall sales. Percentage forecasting can be used to routinize the individual item forecasting problem. Moreover, if the master scheduler keeps MPS records for each product option, it may not be necessary to forecast individual items. The important job is not *forecasting* individual end items; it's *building* them and providing adequate customer service. Demand management is designed to facilitate the process.

Figure 8.8 also shows the need for applying judgment in forecasting. If the promotion package size was for a one-time promotion, it should be dropped from the forecast. Some manufacturing step is subject to the greatest forecast error. It can help if that step can be postponed as long as possible. For example, if we can postpone labeling or filling the package, we might reduce forecast error because some specific customer orders will be on hand before we make commitment to any particular item.

Once we've developed forecasts, we can use distribution requirements planning (DRP) to coordinate replenishment of distribution inventories. The basic DRP record has the same format and logic as an MRP record but uses the detailed forecasts as the requirements. Figure 8.9 shows an example DRP record—for Jag's Jingle brand, regular size, drip grind coffee at the Omaha warehouse. In the DRP record, in transit is used to indicate material on its way to the warehouse, and planned shipments indicate plans to send material from central stock to resupply the warehouse. Planned shipments provide a key input to the master production schedule. When summed for all warehouses, they give the demands on the central stock.

FIGURE 8.9   DRP Record for Jag's Jingle Brand, Regular Drip Coffee

Omaha Warehouse

		Period				
		1	*2*	*3*	*4*	*5*
Forecast requirements		10	10	10	10	10
In transit		20				
Projected available balance	6	16	6	16	6	16
Planned shipments			20		20	

LT = 1, Q = 20, SS = 3

In Figure 8.9's example, we see the use of safety stock in the distribution warehouse as a form of buffer against larger than forecast demands. Use of safety stock in inventories is quite common. Less common is the use of safety lead time. In instances where finished product is transported to a distant inventory location (such as a warehouse or distribution center), there's often a range of time when delivery can take place. The forecast error in this case is in terms of time of arrival, not the quantity to arrive. In these circumstances, dispatching shipments earlier than would be necessary on the average provides a safety lead time buffer against late-arriving shipments. In this case, we advance both shipment date and scheduled delivery date by the amount of safety lead time so the system is driven by the correct data.

Assemble-to-Order Demand Management

In the assemble-to-order environment, we're concerned about making accurate promise dates to customers. This requires MPS stability and predictability. The master scheduler can help to achieve this by using the time fences shown in Figure 8.2. The "frozen" zone holds schedules firm; the time fences specify the management level required to formally approve a change request. For assemble-to-order products, stability is also related to the discipline applied in booking customer orders. Key in accomplishing this is the available-to-promise concept. Figure 8.10 provides an example from the Jag's Coffee Company problem, using the 60 tons per period of instant coffee from Figure 8.8 as the basis. We treat Jag's instant coffee as an assemble-to-order product for this example.

The example shows all instant coffee available in the first two periods (50 + 120 = 170 tons) is committed to customer orders already booked. No additional promises for delivery in the first two periods are possible without

---

FIGURE 8.10 Application of Available-to-Promise Logic to Jag's Instant Coffee Production

	Period				
	1	2	3	4	5
Forecast	60	60	60	60	60
Booked orders	90	80	50	20	0
Available	80	0	60	0	60
Available to promise	0		50		120
Master schedule production	120		120		120

On hand = 50, Q = 120

changing the delivery date on an order already booked. There are 120 tons of production planned for period 3, of which 70 tons are already promised. The remaining 50 tons are available for promise as early as period 3. If, for example, Rosita's Deli wanted over 50 tons delivered in one shipment, she'd have to wait until at least period 5, unless 50 tons of orders presently promised for delivery in periods 1, 2, or 3 could be rescheduled to period 5 or later. We might change the MPS itself, but the production of 120 tons of instant coffee currently scheduled in period 1 must be nearly completed, and 120 tons for period 3 might be too far along in the process.

If customers for the instant coffee want a unique packaging to meet their specifications, a separate schedule may be needed. Suppose, for example, that lead time for packaging is one period. A separate packaging schedule (equivalent to a final assembly schedule) can be established for managing this additional level of detail. The packaging schedule need extend only far enough into the future, say two to three periods, to allow for planning the detailed packaging sequences. The booked order entries in the MPS record would have to be offset by one period to allow time for the packaging. The principle of holding the integrity of the customer order promises remains however, and this needs the predictability and discipline of available-to-promise logic.

The focus on stability in the information system and MPS shouldn't be taken as an argument for inflexibility. Indeed, as we argued earlier, mechanisms for dealing with forecast errors are often more important than attempts to reduce forecast errors. This means we need a form of flexibility, particularly to accommodate short-term variations in the mix of SKUs being sold.

For assemble-to-order products, hedging provides an effective, easily managed buffering technique. Figure 8.11 gives an example, Jag's regular grind coffee. It shows records for three steps of the production process. Green coffee beans are removed from storage and prepared for roasting. This takes one period. Next, prepared beans are roasted over three periods. Finally, roasted beans are ground. Lead time for the final step is one period.

The records trace the planned orders through the process from the master schedule for regular grind coffee to green bean preparation. Since we calculate demand for regular grind coffee as a percentage of the total coffee forecast, and variations can occur in the mix between instant, regular, drip, and special grinds, the company wants flexibility to respond to product mix changes. However, the firm doesn't want to hold excess regular grind coffee in inventory.

The approach is to establish a hedge of 20 tons five periods from now. We do this by creating a hedge time fence at period 5 and introducing a 20-ton master schedule entry. In this example we assume this has been done and the system has been operating for some time. Currently, we are at the start of period 121 and the hedge of 20 tons has just come over the time fence into period 125. Note previously placed orders (scheduled receipts) and planned orders balance out the requirements and no inventory is shown except for prepared green coffee beans.

FIGURE 8.11   An Example of Hedging for Regular Grind Jag's Coffee

Regular grind coffee   Hedge time fence = 5 periods

Period	121	122	123	124	125	126
Master schedule	340	0	340	0	340	0
Hedge quantity					20 ⟶ 20	
Gross Requirements	340		340		340 ⟶ 20	
Scheduled receipts	340					
Projected available balance	0	0	0	0	0	0	0
Planned order releases		340		340 ⟶ 20		

   Q = lot-for-lot, LT = 1.

Roasted beans

Period	121	122	123	124	125	126
Gross requirements	0	340	0	340 ⟶ 20		0
Scheduled receipts		340				
Projected available balance	0	0	0	0	0	0	0
Planned order releases	340 ⟶ 20					

   Q = lot-for-lot, LT = 3.

Prepared green beans

Period	121	122	123	124	125	126
Gross requirements	340 ⟶ 20	0	0	0	0	
Scheduled receipts	340	0				
Projected available balance	20	20	0	0	0	0	0
Planned order releases						

   Q = lot-for-lot, LT = 1.

If there's no indication that the mix has changed, the hedge quantity isn't required and the 20 hedge tons are pushed back over the fence to period 126 (top arrow in Figure 8.11). This reduces the gross requirement for regular grind coffee from 360 to 340 tons in period 125 and changes the planned order in period 124. The effect is passed through the roasted bean record to the prepared green beans. At this point, the scheduled receipt of 340 and the inventory of 20 would have been enough to satisfy the gross requirement for 360, but it's not necessary, so the 20 tons are held in prepared green bean inventory. Pushing out the hedge unit in subsequent periods leaves the 20 tons of prepared green beans in inventory.

If the hedge quantity had been allowed to remain within the time fence, we would have issued a planned order for 360 tons of roasted coffee beans and started the 20 tons of prepared green beans on the way to becoming roasted coffee beans. If we'd required the hedge quantity, it would be ground during period 124 and would be available in period 125. The hedge fence's placement indicates how quickly we can respond to product volume changes.

Setting the time fence and managing the hedge units must take into account both economic trade-offs and current conditions. Setting the time fence too early means inventory will be carried at higher levels in the product structure, which often decreases alternative uses of basic materials. Also, too short a time fence may not provide enough time to evaluate whether the mix ratio is changing. Similarly, it makes no sense to provide flexibility where it's not needed. Over the period when all planned available coffee is committed to specific customer orders, there's no need for flexibility since we know the exact product mix (e.g., in the frozen part of Figure 8.2). In this circumstance, we should push out the hedge until there's still some forecast usage that hasn't been consumed by actual orders.

Make-to-Order Demand Management

In the make-to-order environment, we're very much concerned with control of customer orders after they're entered into the system. In some cases, a plant will have weeks or months of order backlog. In many of these orders, a great deal of detailed engineering will take place before the order is completed and sent to the customer. This occurs because orders aren't completely specified when they're booked. Indeed, for products that take many months to manufacture, technology could even change while the product is in process. Thus, even though we may know the number and timing of customer orders for some time into the future, there can still be much uncertainty concerning these orders.

In this situation, demand management has to track these orders through all phases of plant activity, including the engineering-related functions. There are three primary reasons for this tracking. First, the master scheduler needs to know the impact of final specifications on parts design, component lead times, and customer promise date for the product as a whole. The second reason is to manage overall lead time so satisfactory customer service results. The third reason concerns the impact that engineering resources' use will have on *other* customer orders in the factory.

Hedging will buffer some of this uncertainty, just as it will for the assemble-to-order product case. This may not be sufficient to cover all changes, however. What-if analysis can help the master production scheduler determine the impact of changes and the resolution of design uncertainties. We can assess the overall impact in terms of shop load (and load on engineering) and the resultant completion dates for other products. But other tools are helpful in managing the make-to-order product situation. Some of the newer technologies for manufacturing—especially co-engineering and computer-aided design/computer-aided manufacture (CAD/CAM)—can help us reduce engineering lead times and improve the pass-off from engineering to manufacturing. Also valuable are specialized planning bills of material—the topic that we'll now address.

Since general parameters for design of a make-to-order product are known at the time of order entry, we can create a planning bill of materials at that time. We can pattern it after the product made in the past that's believed to be most like the present product, or it could be a generic BOM. The intention is to get something close to what the final product will be into the MPC system as soon as possible, so management of both the design and manufacturing processes can all be done with one integrated system.

Initial planning must be based on a combination of what's known and what's estimated. Parts of the product that aren't completely specified are carried with part number codes indicating that they're "temporary." The planning bill that represents the customer order is carried in the MPS. We enter modifications to the BOM as the engineering process takes place and temporary codes are removed from individual part numbers. Overall knowledge of the status of engineering is provided, and the "critical path" for engineering and manufacturing can be determined and managed.

Figure 8.12 illustrates this approach. The initial product structure was developed from a similar product that had been built before. Three of the com-

Figure 8.12   Planning Bill of Material before and after Specification of the Parts

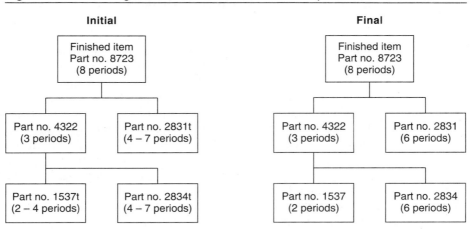

Note: All parts require one each of their components.

ponents aren't completely specified, as shown by the "t" designation in the part numbers. In each case, there's also a range of time given—an estimate of the lead time for design and fabrication. Fully specified parts have the standard MPC lead time connotation.

After we've completed engineering and indicated changes in the final product structure, we no longer need to hedge against initial design uncertainty. Any subsequent engineering changes, however, need to be approved by the customer and managed by the demand management module.

MANAGING DEMAND

In this section we look at managerial issues related to the performance of day-to-day demand management tasks.

Organizing for Demand Management

Most companies already perform many, if not all, the activities we've associated with demand management. In many instances, organizational responsibility for these activities is widely scattered throughout the firm. The finance or credit department performs credit checks and order screening associated with customer orders. Sales or customer service departments handle order entry or booking. Outbound product activities are associated with the distribution, traffic, or logistics departments of firms.

Some companies establish a materials management function to coordinate demand management activities. Organizational responsibility for demand management tends to be a function of the organization's history and nature. It's much less important, however, to have a unified organizational home for all activities, than to appropriately define and coordinate them with one integrated data base.

In marketing-oriented firms (where success requires close contact with demand trends and good customer relations), demand management might well be performed by the marketing or sales organization. In firms where product development requires close interaction between engineering and customers, a technical services department might manage demand. The materials management organization has grown up in firms that feel it important to manage the flow of materials from purchasing raw materials through the production process to the customer. In such firms, which span both industrial and consumer products, the demand management function can be part of materials management. In all instances, we must clearly assign responsibilities to make sure nothing is left to chance.

If flexibility is a key objective, then management must carefully design and enforce rules for interacting with the system and customers so the system can provide this flexibility. By this we mean customer order processing must be established and enforced through the master production scheduling system. It involves carefully establishing rules for serving particular special customers. For example, if an extraordinarily large order is received at a field warehouse, procedures need to be established for determining whether that order will be allowed to consume a large portion of the local inventory or be passed back to the factory. We must define and enforce limits within which changes can be made. If any of these procedures are violated by a manager who says, "I don't care how you do it, but customer X must get his order by time Y," demand management is seriously undercut.

A useful technique for defining and managing these areas of responsibility is to tie them to time fences. Abbott Laboratories, Ltd., of Canada has developed a highly formalized set of time fences. Figure 8.13 shows the firm's four levels of change responsibility. As a change request affects the MPS nearer to the current date, responsibility for authorizing the change moves up in the organization. This procedure doesn't preclude a change but does force a higher level of review for schedule changes to be made in the near term.

The underlying concept for approval procedures is to take the informal bargaining out of the system. By establishing and enforcing such procedures for order entry, customer delivery date promising, changes to the material system, and responses to mix changes in the product line, everyone plays by the same rules. In the Abbott example, flexibility is part of the change procedure, but the difficulty of making a change increases as the cost of making that change increases. Clearly this is more a matter of management discipline than technique. The ability to respond: "What don't you want?" to the "I have to have it right away" for a particular customer request helps establish this discipline.

Managing Service Levels

One way to help the organization live with a formal system for placing demands on the manufacturing organization is to explicitly set levels of service and to publicize them throughout the organization. Substantial theoretical work has been performed concerning setting service levels for finished-goods inventory. This work indicates inventory investment increases exponentially as service-level objectives are increased. More important is the need for discipline in the management of service levels. Simply stated, this means understanding that less than a 100 percent service-level target implies occasionally there *will* be a stock-out. Truly understanding and living with that can be difficult. Often a stock-out or late delivery focuses so much attention on a given transaction that people respond to prevent its recurrence. This is frequently the origin of the impossible order given to many inventory clerks: "Keep the inventory low but don't stock-out."

Determining appropriate service levels requires careful, considerate trade-offs. With increasingly high costs of carrying inventory, levels of service pro-

FIGURE 8.13   Approval Fences for Master Scheduling Change at Abbott
Laboratories, Ltd.

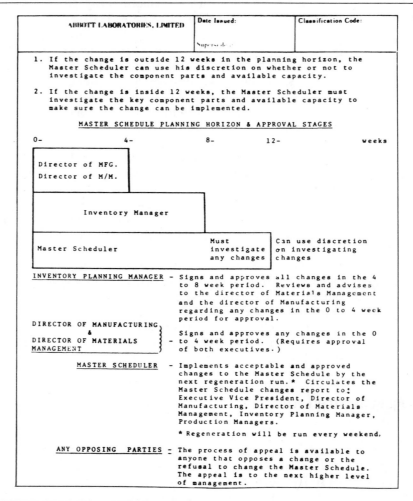

Source: W. L. Berry, T. E. Vollmann, and D. C. Whybark, *Master Production Scheduling: Principles and Practice* (Falls Church, Va.: American Production and Inventory Control Society, 1979), p. 83.

vided to customers from a finished-goods inventory must be reevaluated very
honestly. This means assessing the value of maintaining service levels versus
savings from reduced inventory. Statistical methods developed to solve the
technical aspects of this trade-off don't solve the difficult managerial problem.
Most firms recognize that 100 percent service (i.e., meeting every customer
demand from inventory or at the time the customer requests it) is simply

beyond the realm of financial possibility. For make-to-stock firms, 100 percent service implies huge inventories. For make-to-order firms, immediate delivery implies substantial idle capacity.

Lest we interpret these remarks as a plea for poor customer service, let's state emphatically that such is not the case. We firmly believe major improvements are possible, but emotional responses aren't the answer. MPC systems are designed to trade information for inventories and other kinds of slack— including poor delivery performances. By using systems well, we can be close order coupled with customers; that is, demand management can often lead to substantial improvements in customer service *without* massive inventories or idle capacity.

Increasing use of just-in-time provides significant help as well. A CAD/ CAM terminal manufacturer reduced lead time from over 15 weeks to four days. At the same time, the company went from make-to-stock to make-to-order. All products are now built to exact customer order.

Using the System

An effective demand management module will gather marketing information, generate forecast information, screen and monitor performance information, and provide detailed action instructions to the material planning and control system. Once implemented, we can use the system for routine tasks. A specific example is forecasting. The system can break out item sales within a product family, and management's attention can be focused on demand for the family itself. Focusing on the broader category both brings attention to bear where it's most needed and prevents squandering human resources on trying to reduce forecast error, which is unlikely. Only through support of the system can we redirect human resources.

The management control function also runs through formal MPC activities. Gathering intelligence on actual conditions in the marketplace provides the basis for deciding whether to change the organization's game plan (production plan, sales plan, budget, and so on) and for determining the appropriate level of flexibility. Again, the approach is to use the system to gather this information and then apply management talent where it's needed.

Perhaps the most important change that improved management in this area can effect is the ability to be honest with customers. In our experience, customers prefer honest answers (even if they're unpleasant) to inaccurate information. An effective manufacturing planning and control system with discipline in order promising and service-level maintenance provides the basis for honest communication with customers. They can be told when to expect delivery or when inventory will be replenished—and they can count on it. Providing the basis for honest communication with customers can pay handsome dividends in terms of customer loyalty.

COMPANY EXAMPLES

In this section, we illustrate actual demand management practice as well as records that demonstrate key concepts discussed in this chapter. These illustrations use material gathered from Abbott Laboratories, the Tennant Company, and the Elliott Company.

Abbott Laboratories, Inc.: Forecasting

Abbott Laboratories, Inc., in North Chicago, Illinois, is a multinational health-care firm. Its product line includes pharmaceutical products for professional and personal use, medical electronics, cosmetics, and related chemical/pharmaceutical products. Products in this example are produced on a make-to-stock basis; the company has a reputation for maintaining a high level of customer service. Figure 8.14 shows the forecasting process for these products.

Abbott Laboratories' forecasting procedure uses many techniques suggested earlier in this chapter. In this example, the first step in the process is to develop monthly forecasts by product. The initial input is a computer-developed forecast that uses data on customer demand and provides a basis for marketing review and approval. Figure 8.14 shows marketing did not modify the forecast for April and May but changed it for June. Management judgment is used in reviewing the forecast for these monthly totals. Next, the forecast is broken down by distribution center using a computer program and the demand history data base.

The process of dividing up the forecast by distribution center begins with determining the historical percentage of each product sold by each distribution center. Step III in Figure 8.14 shows this. The computer then applies these percentage breakdowns to the approved total forecast to develop the monthly forecast by distribution center.

At Step IV in Figure 8.14, we break down the forecast further by week within the month. Again, the computer does this by taking into account split weeks, vacations, and so on. The result is a weekly forecast for each product at each distribution warehouse.

Abbott then utilizes these forecasts as gross requirements data for the time-phased order point records for each warehouse. The result of thereafter applying MRP logic is the planned shipments for replenishing inventory from the factory. Data on actual demand are captured at the warehouses. The gross to net logic and safety stock are used to absorb fluctuations between sales and forecasts.

An interesting reaction occurred when warehouse supervisors first received these weekly projections of product sales at each warehouse. In many in-

FIGURE 8.14 Developing Detailed SKU Forecasts at Abbott Laboratories

Illustration of Weekly Forecast
Development for a Product
(by distribution center)

	Month		
	April	May	June
Week* Step I: Computer developed forecast by product (preliminary forecast)	1–4 520	5–8 648	9–12 712
Step II: Marketing revision and/or approval (final forecast)	520 (OK)	648 (OK)	620 (Revised by Mkt.)

Step III: Computer proration of
monthly forecast by distribution center (DC)

DC	Forecast[†]	Percent of total FC
#218	155	[‡]31%
#233	310	62%
#244	35	7%
Total	500	100%

April	May	June
[§]160	200	192
320	400	394
40	48	44
520	648	620

Step IV: Development of weekly forecast

DC		Week:								
		1	2	3	4	5	6	7	8	9
	#218	[‖]40	40	40	40	50	50	50	50	48
	#233	80	80	80	80	100	100	100	100	96
	#244	10	10	10	10	12	12	12	12	11

*Four weeks per month used to simplify example.
†Each DC forecast is done independently using an exponential smoothing technique based
on past DC sales history.
  Sample of calculations (within computer):

$$‡155 \div 550 = 31\%.$$
$$§31\% \times 520 = 160 \text{ (rounded)}.$$
$$‖ 160 \div 20 \text{ days/mo.} \times 5 \text{ days/wk.} = 40/\text{wk.}$$

Source: D. C. Whybark, "Abbott Laboratories, Inc.," in *Studies in Material Requirement Planning*, ed.
E. W. Davis (Falls Church, Va.: America Production and Inventory Control Society, 1977), p. 17.

stances they could use their knowledge of the local purchasing patterns to
adjust the planned distribution of the forecast within each month to better
plan inventory resupply. Even though the information was quite detailed, they
could review weekly patterns and adjust them within the monthly totals provided by the computer breakdown of the marketing approved forecast.

Tennant Company: Order Promising and Flexibility Management

The Tennant Company manufactures industrial floor-maintenance equipment and associated cleaning products. Equipment varies from small walk-behind cleaners to larger operator-driven units. Machines are used for both indoor and outdoor applications. They are divided into families for planning and scheduling purposes. Within each family are a large number of customer-specified options and accessories for each machine.

Customer order promising is done explicitly from the machine availability plan prepared by the material planning and control system. Specifically, the order entry function does it using a report of machine status. Figure 8.15 provides an example. The first line of the document shows the status of machines in the E2 family. This report includes machines that have already been assembled, machines in various stages of preparation for customer delivery, as well as machines scheduled for future production. It shows that all but one of the machines scheduled for production in the next seven weeks are promised to customers. The earliest possible customer delivery date promise is for a machine in week 7 (manufacturing day 211, June 5). Although this report has a different format than the available-to-promise logic we saw earlier, it contains the same basic information and is used the same way.

The next two lines indicate two of the options available for the E2 family: the gas and LP motor options. This provides a second level of order entry testing. The only machine that will be available in the week of June 5 is an LP machine. We see this because there's no commitment to the LP motor available in the week of June 5. This logic of testing product availability illustrates the process of order entry that matches customer requests to the availability of scheduled machines and options at Tennant.

Figure 8.16 illustrates the production version of the same document in somewhat greater detail. This MPS control report has basic parts, unique gas parts, and unique LP parts, as did the order entry document. It's important to notice that exactly the same information is available to manufacturing as was available to order entry. Additional information in this report is of interest in illustrating flexibility management. First, we see specifically identified time fences for the basic parts at manufacturing day 276 and 436 (note the vertical lines of slashes), and different fences identified on the records for the gas and LP parts. These time fences are defined to indicate time frames in which management wants flexibility. For the basic parts, the fence set at manufacturing day 276 is a **volume hedge** fence. Its purpose is to provide flexibility to accommodate an increase in overall demand for the E2 family. The fence for the gas and LP engines provides flexibility for adapting to *mix* changes, which occur between these two options.

The hedge concept is applied as illustrated in Figure 8.11. For example, in the basic parts, the hedge unit indicated in period 276 has just crossed the time fence and the master scheduler will move it out unless a management

FIGURE 8.15 Tennant's Report of Machine Status: The Customer Order Promising Document

DAILY MACHINE RESERVATION REPORT

MFG DATE 182

	FINAL	1 182	2 186	3 191	4 196	5 201	6 206	7 211	8 216	9 221	10 226	11 231	12 236	13 241	14 246	15 251	LATER
	LOGS	AP26	MY01	MY08	MY15	MY22	MY29	JN05	JN12	JN19	JN26	JL03	JL10	JL17	JL24	JL31	CMTMNT

16100 PARTS BASIC E2

E2LINE #16100
PARTS-BASIC
CRTS-UNCH TEST RWRK ASSY PULD
SCHEDULED   1
STOCK CMT
CUST CMT   1 OF

17005 FINAL ASSY E2

E2LINE #16102
PARTS-UNIQUE-GAS
CRTD UNCR TEST RWRK ASSY PULD
SCHEDULED
STOCK CMT
CUST CMT   1 OF

17006 FINAL ASSY E2 LP

E2LINE #16103
PARTS-UNIQUE- LP
CRTD UNCR TEST RWRK ASSY PULD
SCHEDULED
STOCK CMT
CUST CMT

Source: W. L. Berry, T. E. Vollmann, and D. C. Whybark, *Master Production Scheduling: Principles and Practice* (Falls Church, Va.: American Production and Inventory Control Society, 1979), p. 163.

FIGURE 8.16 Tennant Flexibility Management Document

MASTER SCHEDULE CONTROL REPORT

TIME 02.45
PAGE 21
LAST REVIEW 417

| 16103 | PARTS-UNIQUE- LP | E2LINE | D.RATIO 35% | SHORT RANGE HEDGE 45% 50% 55% | MFG DATE 182 TIME FENCES 00 03 16 00 | BACKLOG 2 | LONG RANGE HEDGE % % % |

SEQ 0580

ADJ.
PLANNED: -APR-----/--------MAY-----/--------JUN-----/--------JUL-----/--------AUG-----/--------SEP----OCT
181 186 191 196 201 206 211 216 221 226 231 236 241 246 251 256 261 266 271 276 281 286 291 296 301 306

HEDGE
M/S
COMMITTED

PLANNED: --------/--------NOV-----/--------DEC-----/--------JAN-----/--------FEB-----/--------MAR----APR
311 316 321 326 331 336 341 346 351 356 361 366 371 376 381 386 391 396 401 406 411 416 421 426 431 436

HEDGE
M/S
COMMITTED

PLANNED: --------/--------MAY-----/--------JUN-----/--------JUL-----/--------AUG-----/--------SEP----OCT
441 446 451 456 461 466 471 476 481 486 491 496 501 506 511 516 521 526 531 536 541 546 551 556 561 566

HEDGE
M/S

PLANNED: --------/--------NOV-----/--------DEC-----/--------JAN-----/--------FEB-----/--------MAR----
571 576 581 586 591 596 601 606 611 616 621 626 631 636 641 646 651 656 661 666 671 676 681 686 691 696

HEDGE
M/S

Source: W. L. Berry, T. E. Vollmann, and D. C. Whybark, *Master Production Scheduling: Principles and Practice* (Falls Church, Va.: American Production and Inventory Control Society, 1979), p. 166–67.

decision is made to increase the volume for the E2 family to include the hedge quantity. Fences for the options indicate the flexibility range for product mix changes. The number of gas and LP options within the time fences exceeds the basic parts scheduled. This provides flexibility to adjust to day-to-day swings in demand for product options without having too large an inventory.

Tennant manages the hedge units explicitly, which means the amount of flexibility is highly visible and reviewed continually. A great deal of computer information on time fences, hedge percentages, and current status is available to the master production scheduler to assist in managing the hedges. An important concept is to provide flexibility in volume and product mix only where necessary. This is noted by the fact that no hedge units are provided prior to manufacturing day 211. There's no need for flexibility before that time, since all available machines and options are covered by customer orders.

With all this system's formality and the support it provides planners, there's still a major element of management discipline that makes this system work. When asked what would happen if a sales person tried to promise delivery of an E2 prior to week 211, Doug Hoelscher (director of manufacturing at the time these reports were gathered) replied, "We'd fire the person."

The discipline at Tennant runs through to manufacturing as well. Under the new responsibilities created by formal systems, the manufacturing mandate is to produce the scheduled products. If no customer order is available for an item approaching final assembly, management will release a stock commitment for that item. If no customer order is received by the time the item goes to final assembly, it's produced in an easily retrofitted model and goes into inventory. *The top-management committee* of Tennant owns this inventory. Top management feels that the commitment to meeting the plans, be they in manufacturing or marketing, is important enough that *they* will own any unsold finished goods. They recognize that, if marketing is to meet the sales plan, any currently unsold machines will be sold in the future.

Make-to-Order Products at Elliott Company, Division of Carrier Corporation

The Jeanette, Pennsylvania, plant of the Elliott Company manufactures large air and gas compressors and steam turbine devices. Products are highly engineered using state-of-the-art manufacturing techniques and materials. Engineered apparatus products are designed and built to customer specifications to accomplish a specific function. Products typically weigh 50 tons and take a year or more to produce. Over half this lead time consists of order processing, design engineering, and purchasing.

Scheduling each customer order is based upon the assignment of an imaginary (planning) bill of material to each major piece of equipment. This bill is established by using elements of previously built products that are similar to the product on the customer order. This imaginary bill of material is then processed by standard MRP logic. Lead times to produce the components on the imaginary bill of material include estimated times to perform the engineering design and do the necessary drafting, in addition to manufacturing lead times. The result is proper ordering of when each component should be designed, priorities for all customer orders relative to due dates, and a capacity requirement profile for each work center. Capacity profiles are produced for engineering and drafting work centers on a routine basis. Figure 8.17 is one of the imaginary bills of material.

Figure 8.18 is part of an exception report showing behind-schedule project activities. For example, the first item on the list is nine weeks behind schedule, has project engineer A in charge, and is presently in engineering department H3P. The report is printed in order of those jobs with the worst delays. Elliott uses shop-floor control and other MPC system modules to plan and control each customer order during the several months each is in progress. In two years of using these systems, performance against customer promise dates improved 50 percent; inventory levels fell 23 percent; and meanwhile, sales volume rose 32 percent. The advantages stemmed from better planning and control of *all* aspects of the business, from order entry through engineering, to the shop floor. Both hard and soft activities are planned/controlled, with the MPC-based systems for project management.

CONCLUDING PRINCIPLES

This chapter has focused on the integrative nature of demand management. It's necessary to capture all sources of demand, to maintain a proper demand management data base, and to carefully integrate demand management both with production planning and with the detailed MPS decision making. We see the following key principles as important to accomplishing these objectives:

- MPC systems must take into account *all* sources of demand, properly identified as to time, quantity, location, and source.
- Order promising must be done using available-to-promise concepts.
- Customer-service standards must be developed and maintained.
- Management of outbound product flows must be coordinated through the master production schedule.
- Attaining more accurate forecasts may be an impossible dream. Management attention should focus on appropriate responses to actual conditions and forecast errors.
- The wide variety of demand forecasts prepared by an organization should be coordinated so detailed forecasts reflect top-management business plans and objectives.

FIGURE 8.17 Imaginary Bill of Material for Elliott Company

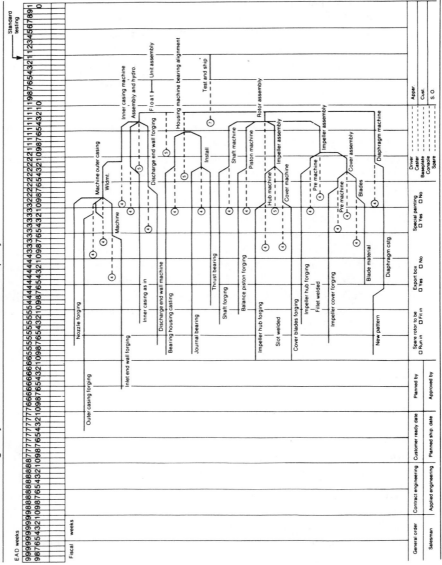

Source: W. L. Berry, T. E. Vollmann, and D. C. Whybark, *Master Production Scheduling: Principles and Practice* (Falls Church, Va.: American Production and Inventory Control Society, 1979), p. 124.

FIGURE 8.18   Late Project Status Report for Elliott Company

Report no. ELCH08391

Engineered apparatus project engineering records scheduled for action

Shop order	Description	Quantity	Rel. no.	Project engineer	EAC Project engineering schedule	Status (weeks)
A528156000	Piping Agreement	1.0	H3P	A	356	−9
A529157000	Piping Agreement	1.0	H3P	A	376	−7
A628502000	Purchase Response	1.0	P3		326	−6
A628503000	Purchase Response		P3		326	−6
A628505000	Coupling	1.0	S5	V	326	−6
A528196000	Major Components	1.0	S5	A	466	−5
A628505000	Release of S2	1.0	S3	V	336	−5
A528164000	Coupling	1.0	S2T		356	−3
	Coupling	1.0	S2T		356	−3
A528187000	Oil Schematic	1.0	M1P	A	356	−3
	Piping Agreement	1.0	H3P	A	416	−3
V025094000	Pipe Agreement	1.0	M3P	O	376	−3
A528175000	Piping Agreement	1.0	H3P	A	366	−2
A528027000	Place Pr	1.0	S2		376	−1
A528043000	Purch Response	1.0	P2P		376	−1
	Major Components	1.0	S5	N	376	−1
A528131000	Lube Information	1.0	SAP		376	−1
	Firm Incomplete	1.0	S1A		376	−1
A528142000	Purchase Response	1.0	P3		376	−1

Source: W. L. Berry, T. E. Vollmann, and D. C. Whybark, *Master Production Scheduling: Principles and Practice* (Falls Church, Va.: American Production and Inventory Control Society, 1979), p. 125.

- Flexibility must be explicitly planned into the system.
- To provide helpful stability in the factory, demand management and MPS activities need to be closely coordinated.
- Reliable customer promises should be the rule of the day, rather than wishful thinking.
- Clear definitions of authority and responsibility for demand management activities must be made to obtain the attendant benefits.

REFERENCES

American Production and Inventory Control Society. *Interfaces Seminar Proceedings,* March 1980.

Berry, W. L.; T. E. Vollmann; and D. C. Whybark. *Master Production Scheduling: Principles and Practice.* Falls Church, Va.: American Production and Inventory Control Society, 1979.

Borgendale, Mac. "Spare Parts: Deciding What to Stock at Each Location." *APICS Service Parts Seminar Proceedings,* Las Vegas, April 1981, pp. 1–4.

Christopher, Martin. "Creating Effective Policies for Customer Service." *International Journal of Physical Distribution and Materials Management* 13, no. 2 (a special edition of the journal devoted to customer service), 1983.

Haskins, Robert E. "Demand Management by Exception (Not by Exception Report)." *Inventories and Production Magazine,* July–August 1981.

Kern, Gary M., and Hector H. Guerrero. "A Conceptual Model for Demand Management in the Assemble-to-Order Environment." *Journal of Operations Management* 9, no. 1 (January 1990), pp. 65–84.

Kuehne, W. A., and P. Leach. "A Sales Forecasting Pyramid for Dow Corning's Planning Endeavors." *Production and Inventory Management Review.* August 1984, pp. 6–11.

Ling, Richard C. "Demand Management: Let's Put More Emphasis on This Term." *26th Annual Conference Proceedings,* American Production and Inventory Control Society, 1983, pp. 11–12.

————."Sales and Operations Planning." *American Production and Inventory Control Society 1990 Annual Conference Proceedings,* pp. 161–64.

Martin, Andre. "DRP: Another Resource Planning System." *Production and Inventory Management Review,* December 1982.

————.*DRP: Distribution Resource Planning.* Essex Junction, Vt.: Oliver Wight Ltd., 1983.

Perry, W. "The Principles of Distribution Resource Planning (DRP)." *Production and Inventory Management,* December 1982.

Pyke, D. F., and M. A. Cohen. "Push and Pull in Manufacturing and Distribution Systems." *Journal of Operations Management* 9, no. 1, January 1990, pp. 65–84.

Shycon, H. N., and C. R. Sprague. "Put a Price Tag on Your Customer Service." *Harvard Business Review,* July–August 1975.

Tucker, Frances G. "Creative Customer Service Management." *International Journal of Physical Distribution and Materials Management* 13, no. 3, 1983.

Vaughn, O.; T. Perez; and B. Stemwedel. "Short Cycle Replenishment at 3M." *American Production and Inventory Control Society 1990 Annual Conference Proceedings,* pp. 515–18.

DISCUSSION QUESTIONS

1. Discuss the statement, "Demand management is the customer's handle on our business."
2. What is implied by the frozen, slushy, and liquid areas of Figure 8.2?
3. In lower levels of the pyramid forecasting system, how would you prevent abdication of responsibility for forecasting?
4. Can a grocery store capture "demand" data? How would a warehouse capture demand data?
5. Which do you have more confidence in: a forecast of the number of credit hours you'll have accumulated by the time you graduate or the specific courses that you'll have taken? In general, what principle does this illustrate?
6. Describe the relationship between percentage forecasting as illustrated in Figure 8.8 and the BOM.

7. How can an MPS incorporate both stability and flexibility?
8. What part of the organization might have responsibility for demand management in a steel mill, industrial products firm, or children's toy manufacturing firm?
9. How can the demand management forecasting activity incorporate judgment?

PROBLEMS

1. The Tarmack and Pothole Company has capacity to produce 50 potholes per week. The firm currently has booked orders as follows:

Week	Orders
1	43
2	51
3	32
4	21
5	24
6	10
7	7
8	5
9	0
10	2
11+	0

   a. Plot booked orders against capacity.
   b. Assume the following transactions. In week 1, 45 potholes were shipped. Orders for two potholes were canceled in week 2, and two more were canceled in week 5. Additional orders were booked for 5 in week 2, 20 in week 3, 10 in week 4, 5 in week 6, 4 in week 7, 2 in week 9, and 1 in week 11. What does the plot look like as of week 2?
   c. What problems do you foresee?

2. Five individual products in a product family have identical sales patterns. Each averages 100 units per month, with a standard deviation of 10 units. Assuming normal distributions and independent demands:
   a. What is the yearly sales distribution of each product?
   b. What is the monthly sales distribution for all products together?
   c. What is the yearly sales distribution for all products together?
   d. Using plus or minus three standard deviations for the values obtained in parts a, b, and c, compare your results to those of Polysar International in Figure 8.7.

3. Hortense Frobisher has attempted to improve customer delivery performance for the Deluxe Duplicator Company. In her latest effort, she has looked at applying the available-to-promise logic to the most profitable product—the Destructo Deluxe. Sales forecast is 40 units per week for the next 5 weeks.

Hortense will use this time period for her analysis. No units are on hand; 87 units are in the master schedule for week 1; and 80 more are in week 3.

a. If the actual customer orders booked for delivery are 62 for week 1 (past due and week 1 combined), 33 for week 2, and 28 for week 3, can she book an order for 10 units in week 2? Week 3?

b. What actions should she take?

4. A master production scheduler manages product 1 that has a current inventory of 120 units, a forecast of 50 units per period, a lead time of two periods, no safety stock, and a lot-sizing rule that orders two periods of requirements at a time. Management would like flexibility to respond to increased demand of one unit in six weeks (i.e., in period 7). A second item (part 2) is used to manufacture product 1 at the rate of two units of part 2 per single unit of product 1. For part 2, there's no safety stock, no inventory, the lot sizing is lot-for-lot, there's a scheduled receipt of 200 due in the current period, and lead time is two periods.

a. Prepare the MRP records for product 1 and part 2 covering the next seven periods. Be sure to indicate how you'd respond to management's request for flexibility.

b. What would the record look like at the start of the second period if demand is exactly 50 in the first period, any planned shipments or orders for period 1 are released, and management wants to increase output by one unit in week 7?

c. What would the records look like in the second period if management said there was no need to increase output by one unit in week 7 (but wanted to maintain the flexibility to increase by one unit in 6 weeks), demand equaled 50, and planned shipments or orders for period 1 are released?

A.

Product 1		1	2	3	4	5	6	7
Forecast		50	50	50	50	50	50	51
Scheduled receipts								
Projected available balance	120	70	20	70	20	70	20	69
Planned shipments				100		100		100

Q = 2 periods supply; LT = 2; SS = 0.

Part 2		1	2	3	4	5	6	7
Gross requirements		100	100	100	100	100	100	102
Scheduled receipts		200		200		200		102
Projected available balance	0	100	0	100	0	100	0	0
Planned order releases								

Q = lot-for-lot; LT = 2; SS = 0.

B.

Product 1	2	3	4	5	6	7	8
Forecast							
Scheduled receipts							
Projected available balance							
Planned shipments							

Q = 2 periods supply; LT = 2; SS = 0.

Part 2	2	3	4	5	6	7	8
Gross requirements							
Scheduled receipts							
Projected available balance							
Planned order releases							

Q = lot-for-lot; LT = 2; SS = 0.

C.

Product 1	2	3	4	5	6	7	8
Forecast							
Scheduled receipts							
Projected available balance							
Planned shipments							

Q = 2 periods supply; LT = 2; SS = 0.

Part 2	2	3	4	5	6	7	8
Gross requirements							
Scheduled receipts							
Projected available balance							
Planned order releases							

Q = lot-for-lot; LT = 2; SS = 0.

5. What are the weekly forecasts by distribution center (DC) if the monthly total forecast is 750 units; and the DC percentages of sales are 30, 20, and 50, for DCs A, B, and C, respectively? Assume a four-week month and then recalculate with a five-week month.

6. Suppose the forecast for a particular DC is 20 units per week in one month and 40 per week the next. What recommendations would you make about displaying the forecast in the last week of the first month and the first week of the second month?

7. Frank Stewart operates a series of pharmaceutical warehouses in Arkansas that are served from his central distribution center in Little Rock. Three warehouses are in Pine Bluff, Texarkana, and Fort Smith. Over the last few years, the concentration on service and the increasing number of products carried in the line have swelled inventories at the warehouses and distribution center to the point where efficiency and profitability dropped substantially. It's important for Frank to develop good forecasts of the weekly demand at each of his locations to provide better plans for managing the inventories and staffing the warehouses.

   For the past few weeks, Frank has kept detailed records of demand for two of his products. One, a burn ointment, had nearly constant demand each week across the state, although there was some variation from warehouse to warehouse, as the nature and number of customers served by the warehouses shifted. The second product—a vitamin tablet sold on promotion to many drug stores—has very high seasonal demand peaks during the two times a year when the product is heavily promoted. Exhibit A shows the information Frank collected for each of the products at each of the three warehouses.

   Frank discussed the forecasting problem with the sales force and requested that they develop forecasts by week for each warehouse for each product. But the sales crew found this difficult, and the resulting forecast errors were too large for Frank. Even though Frank needed weekly forecasts by item at each warehouse, this involved too great a level of detail for the sales force. On the other hand, the sales staff was doing a reasonably good job of forecasting total demand for each product from all warehouses. They were able to include such influences as the special promotion for the vitamin tablet. These forecasts

---

EXHIBIT A   Past Demand for Two Products at All Three Warehouses for Frank's Pharmaceuticals

| | Burn ointment | | | | Vitamin tablet | | | |
| | Warehouse | | | | Warehouse | | | |
Week	FS*	T*	PB*	Total	FS*	T*	PB*	Total
1	20	15	12	47	8	4	2	14
2	18	17	11	46	12	7	0	19
3	22	14	10	46	10	8	5	23
4	23	14	8	45	9	4	1	14
5	19	16	11	46	11	2	3	16

*FS = Fort Smith; T = Texarkana; PB = Pine Bluff

covered a four-week accounting period (the company's "month") and served as the basis for developing the marketing effort, setting sales quotas, and submitting the sales budget. Frank agreed the monthly aggregated forecasts were much more accurate than the forecasting attempts at the item level.

Frank studied the data on the past five weeks' demand for each of the two products for each of the warehouses (shown in Exhibit A) and considered how he could reconcile his need for detailed forecasts of each product's demand by week at each warehouse, with forecasts of monthly demand for all of Arkansas provided by the sales staff. The disparity between their projections and his needs is dramatized by the sales staff's forecasts for the next four weeks (weeks 6, 7, 8, and 9). Forecast demand for the burn ointment is 180 units during the next four weeks; for the vitamin tablet, the demand projection is for 200 units during the next four weeks.

a. Determine how Frank can use the marketing information to make his weekly forecasts at each warehouse, and how he should modify the forecasts as information on actual demand becomes available.

b. The information on the actual demand for week 6 has just been received for the two products at each of the warehouses. Exhibit B summarizes this information. In addition to this data, marketing has just told Frank they felt their projections for the four-week period were still valid for each of the two products. How should Frank incorporate this information in his forecasts?

EXHIBIT B Actual Demand for Week 6 for Frank's Pharmaceuticals

	Burn ointment			Vitamin tablet		
	FS	*T*	*PB*	*FS*	*T*	*PB*
Demand	20	17	9	27	12	4

---

8. The general sales manager at Knox Products Corporation has just received next year's sales forecast (in units) for two of the firm's major products (Bad and Worse) from sales managers of the Eastern and Western sales regions:

Eastern region forecast		Western region forecast	
Bad	*Worse*	*Bad*	*Worse*
100	200	200	300

Bad sells for $1 per unit and Worse sells for $2 per unit.

a. The corporate economist has forecast a total corporatewide sales volume for these two products of $2,000 for next year. What's the disparity between the two forecasts at the *item* level?

b. If top management agrees to a total corporatewide sales forecast volume of $1,500, what's the sales forecast at the *item* level?

9. Imogene Imaginary has created the following planning bill based on similar products from the past. Parts A, C, and E have had all engineering completed, and parts B and D have just been estimated.

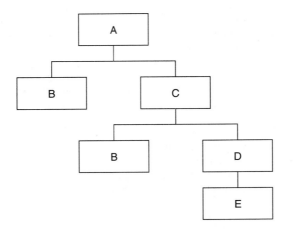

Part	Engineering	Time required
A	Done	4 weeks
B	Estimated	2–5 weeks
C	Done	5 weeks
D	Estimated	1–3 weeks
E	Done	3 weeks

a. What would be the delivery time if B could be done in two weeks and D in one week?
b. What would it be if B would take five weeks and D could be done in one week?
c. Which part would you want to watch most closely? Why?

10. Use a spreadsheet to create a DRP warehouse record for a product that has forecast demand of 20 units a period for the first three periods, 25 for period four, and 30 for the next three periods. The shipping quantity is 50 units; lead time is one period. There are 46 units in inventory. What's the average ending inventory balance per period if six units of safety stock are used?

11. Using the data in Problem 10, what would average ending inventory be if one period of safety lead time were used instead of safety stock?

12. The Gonzales Electric Company's 12 products are further grouped into four product families. Products A, B, and C comprise family 1; D, E, and F comprise family 2; G, H, and I comprise family 3; and products J, K, and L make up family 4. Dick Gonzales has the following exponential smoothing forecasts of monthly demand for each product.

Family	Product	Forecast	$/unit	Family	Product	Forecast	$/unit
1	A	10	$ 1,000	3	G	100	$ 250
	B	15	1,200		H	180	100
	C	20	900		I	220	100
2	D	5	5,000	4	J	2	10,000
	E	3	7,000		K	4	9,000
	F	2	9,000		L	3	8,000

Gonzales's sales force has also come up with the monthly forecasts of sales for each product family.

Family	$ sales
1	$50,000
2	50,000
3	75,000
4	75,000

   a. Top management has independently set a $300,000 overall monthly sales goal for the company. Roll up the individual product forecasts and compare them with the family data. Use a spreadsheet and the family forecasts to revise the individual item forecasts (in both dollars and units).

   b. Roll up the family forecasts to the top level, compare these to the overall forecasts, and roll the forecasts back down to families and to individual unit forecasts (dollars and units).

13. Suppose a top-management meeting produces a decision that the family forecast for product family 3 (products G, H, and I) in Problem 12 can't be increased. At the same meeting, the company's overall forecast ($300,000) is maintained. Roll the forecasts up and down to determine the dollar and unit forecasts.

14. Suppose the family forecast data in Problem 12 are assumed to be correct, except in any case where the sum of the exponential smoothing forecast data for items exceeds the family forecast. In these cases, the sum of the item forecasts will be used instead of the family forecast. Roll up the resultant overall forecast and roll down the resultant item forecasts.

15. Using Problem 12's data, suppose a major customer order has just been received for 10 units of product J. This order wasn't expected and is in addition to any other forecasts for product J. The company still wants to plan a total monthly sales volume of $300,000. Use the family forecasts (revised) to roll the forecasts up and down to get revised individual product forecasts.

Chapter 9

Integrated MPC Systems

This chapter concerns two integration issues in designing manufacturing planning and control (MPC) systems. The first is linking the design of a firm's MPC system with its corporate strategy for competing in the marketplace. As the investment in an MPC system is large and remains fixed over considerable time, getting it correct is critical to short- and long-term prosperity. Many companies make costly mistakes when their MPC system doesn't support their basic mission in the marketplace. The second issue concerns integrating manufacturing requirements planning (MRP) and just-in-time (JIT) in existing or new MPC systems. The chapter centers around four topics:

- MPC design options: What are critical alternatives in designing an MPC system to meet a firm's evolving needs?
- Choosing the options: How should the options be selected to best support the corporate strategy and to fit with production process design?
- The choices in practice: How have manufacturing firms with different competitive missions gone about designing their MPC systems?
- Integrating MRP and JIT: How can these different approaches be linked in a company's MPC system?

Integrated MPC design issues are connected to nearly every chapter in this book, especially Chapters 2 and 3 (the basic chapters on MRP and JIT) and Chapters 11 and 12 (on MRP and JIT's advanced features). Chapter 10's coverage of operational issues concerning MPC system implementation in organizations complements the strategic view of Chapter 9.

MPC DESIGN OPTIONS

A wide range of alternatives are available in designing MPC systems. These include such basic approaches as MRP, MRPII, JIT, OPT (optimized production technology), periodic control systems, and finite scheduling systems.

Moreover, there are a wide variety of options for designing the individual modules of the MPC system shown in Figure 9.1. The next three sections illustrate the variety of options for master production scheduling, detailed material planning, and back-end activities.

Master Production Scheduling Options

Several different approaches can be taken to designing the master production schedule: *make-to-order (MTO), assemble-to-order (ATO),* and *make-to-stock (MTS).* Figure 9.2 shows the major differences between these alternatives. A **make-to-order** approach to master production scheduling is typical when the product is custom-built to individual customer specifications. In this case the MPC system needs to encompass preproduction engineering design activities as well as manufacturing and supplier operations. For MTO, the customer

FIGURE 9.1   Basic MPC System

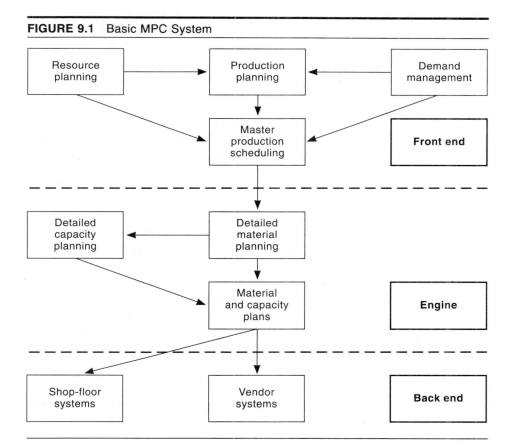

FIGURE 9.2 Features of Master Production Scheduling Approaches

	Master scheduling approach		
Basis for planning and control	*MTO*	*ATO*	*MTS*
Control point	Order backlog	FAS	Forecast
MPS unit	Customer orders	Options	End items
Product level	End product	End to intermediate product	End product
MPS features			
Customer order promising	High requirement	⟶	Low requirement
Need to monitor forecast accuracy	Low requirement	⟶	High requirement
Use of planning bills	Yes	Yes	No
Need to cope with design and process uncertainty	High	⟶	Low
Basis of delivery to customer	Make to customer order on time	Make to customer order on time	Make to stock replenishment order or to customer call-off schedule

---

order represents the unit of control in the MPS; the backlog of customer orders forms part of the overall lead time for the product. Overall, the order backlog is a critical measure for estimating material and capacity requirements. Customer order promising is based on the backlog plus estimates for each design, procurement, and manufacturing step for a particular job. Planning bills of material are extensively utilized to estimate times and to prioritize design efforts on the "critical path." There's an inherently large degree of uncertainty associated with the time requirements, since each order requires a unique approach.

An **assemble-to-order** approach is typically used when overall manufacturing lead time exceeds that desired by the customer, where the variety and cost of end products preclude investment in finished-goods inventory, and where engineering design has created modules or options that can be combined in many ways to satisfy unique customer requirements. Here component (or product option) inventory is held to reduce overall manufacturing lead time, and end products are assembled to meet the scheduled delivery dates for individual customer orders. As Figure 9.2 shows, a key control point is the final assembly schedule (FAS), which converts "average" products into unique products in response to actual customer orders. Planning bills of material are based on average products and on optional features. The planning bills reflect how the product is sold, rather than how it's manufactured. They are often used to simplify data requirements in preparing and maintaining the master production schedule. The uncertainty underlying an ATO business is fundamentally one of product mix, rather than one of product volume. The MPS and FAS are designed to hold off commitment to unique product configurations until the last possible moment and yet to offer wide configuration choices to customers.

Under **make-to-stock (MTS),** the MPS is stated in end items, and these end products are produced to forecast demand; customer orders are filled directly from stock in order to provide short delivery lead times for standardized products. While customer order promising records are not normally required, we must provide procedures for monitoring demand forecasts' accuracy since manufacturing plans are mostly based on forecast information. This means the type of uncertainty inherent in the MTS environment is one of forecasting errors; the manufacturing function needs to recognize errors on a timely basis and to make corrective responses.

Detailed Material Planning Options

We can accomplish detailed material planning in several ways. Two popular alternatives are *time-phased* and *rate-based material planning*. Use of these approaches depends importantly on the production process's design characteristics. Figure 9.3 shows key differences between these approaches.

Time-phased planning for individual product components is typically carried out with material requirements planning approaches. The production process design is usually based on **batch (job shop) manufacturing** and materials also purchased in batch orders. Preparation of time-phased plans requires a manufacturing data base that includes information on: MPS quantities stated in bill of material terminology to determine gross requirements; on-hand inventory balances and open shop (or purchase) orders to determine net requirements; production lead times, supplier lead times, and safety stocks to determine order release dates; and lot size formulas to determine order quantities. Under MRP, plans are typically updated on a periodic (daily or weekly) basis to develop priorities for scheduling manufacturing and supplier operations.

As Figure 9.3 indicates, time-phased material planning is based on explosion of requirements, where shop and purchase orders are created for batches of components. The schedule for any work center varies depending on the batches that arrive at that work center; work-in-process is kept at high levels to effectively utilize work center capacities. Planning is carried out on a level-by-level basis corresponding to the levels in the bill of materials (BOM), with material going into and out of inventory at each level. Detailed planning is required for each level in the BOM, and lead time offsetting is utilized at each level.

FIGURE 9.3   Features of Detailed Material Planning Approaches

Basis for planning and control	Material planning approach	
	Time phased	Rate based
Control point	Shop/purchase orders	Planning bills
Control unit	Batches	Kanbans
Product level	Material explosion of time-phased net requirements for product components	Material explosion of rate-based requirements for product components
Material planning features		
Fixed schedules	No	Yes
Use of WIP to aid planning	High	Low
Updating	Daily/weekly	Weekly/monthly
Inventory netting	Performed	None
Lead time offsetting	Performed	None
Lot sizing	Performed	None
Safety stock/safety lead time	Considered	Not considered
Container size	Not considered	Considered
Bill of material	Many levels	Single level

A different approach is taken to detailed material planning under **rate-based planning.** Examples of firms using rate-based planning include repetitive manufacturing, assembly lines, just-in-time, and other flow systems. The primary intent in rate-based scheduling is to establish rates of production for each part in the factory. Realizing these rates allows the company to move material through the manufacturing system without stopping, in the shortest time possible. Typically, single-level planning bill of material information is used to convert rate-based master production schedules into material plans that specify the appropriate daily or hourly flow rates for individual component items. Planning of intermediate items in the bill of materials is not usually required, because the number of intermediate-level items is too small to be of concern. Because of MPS stability, high rates of material flow, negligible work-in-process inventory levels, short manufacturing lead times, and a relatively small variety of final products in the MPS, we don't need detailed status information on work-in-process items. This reduces the manufacturing data base's size, the number of transactions, and the number of material planning personnel in comparison with time-phased detailed material planning.

Shop-Floor System Options

A wide variety of manual and computer-based shop-floor scheduling systems exist. The two basic approaches (material planning driven by MRP and material planning driven by JIT) depend greatly on the manufacturing process's characteristics. Figure 9.4 distinguishes between these approaches.

The MRP-based approach supports batch manufacturing operations where shop orders are released against a schedule developed by the material planning function, based on lead times for component and subassembly items largely comprised of queue or waiting time. The shop-floor scheduling system's objective is to coordinate the sequencing of orders at individual work

---

FIGURE 9.4   Features of Shop-Floor System Approaches

Basis for planning and control	*Shop-floor system approach*	
	MRP	*JIT*
Control basis	Work center capacity utilization	Overall product flow times
Unit of control	Shop orders	Kanban cards or containers
Product level	Individual operations scheduled at each work center	Production on an as-required basis to replenish downstream stocks that support end item requirements
Shop-floor system features		
Control of material flow	Work center dispatching rules	Initiated by downstream kanban cards
Sequencing procedure	Due-date oriented dispatching rule	Not an issue
Order tracking	Shop-floor transactions by operation and stocking point	None (paperless system)
Monitoring and feedback	Input/output and shop load reports	Focus on overall result
Order completion	Shop order close-out in stockroom	None
Achieving delivery reliability	Batch order status reports	Through flow of material
Lot size	Large	Small
Work-in-process and safety stock	Large	Negligible

---

centers with customer delivery requirements. A large manufacturing data base requiring a substantial volume of shop transactions is needed to provide control reports for order tracking, dispatching, and work center monitoring.

In MRP-based shop-floor systems, one objective is to utilize each work center's capacity effectively. This form of manufacturing is based on relatively large batches of each component and significant work-in-process inventories to support independence among the work centers. This shop-floor approach is based on scheduling shop orders that dictate the set of detailed steps or operations necessary to make each component part. The flow of materials is controlled with dispatching rules establishing the order in which all jobs in a particular work center are to be processed. The primary criterion in establishing this order are the due dates for the parts, which are continually reestablished through MRP planning. Shop orders are tracked as they progress through the factory by processing detailed transactions of work at every work center. Shop orders are opened as part of MRP planning, and they're closed out as components are received into a stockroom. Problems are highlighted through input/output analysis and shop load reports.

In JIT-based shop-floor scheduling systems, the approach is based on minimal flow times for the entire product. That is, the emphasis is on end items, with the scheduling of individual operations, and even component parts, in a subservient position. Cellular manufacturing techniques are typically employed, where detailed scheduling is accomplished as part of the basic manufacturing task. Kanban cards, containers, and other signals of downstream need for components serve as the authorization to produce, typically in small lot sizes. The sequencing procedure isn't an issue because work is only started on an as-needed basis, with little or no competition for work center capacity. Similarly, order tracking is nonexistent since work-in-process is minimal, and material moves through the factory quickly enough to negate the need for tracking. The only close-out is of finished items. Often the close-out transaction generates a computer-based "back flush" of the requisite component parts. The very short queue times, small lot sizes, and relatively narrow product range in JIT can result in a paperless shop-floor scheduling system. The manufacturing data base requirements, volume of shop transactions, and number of shop scheduling personnel are minimal.

Many authorities have attempted to use the terms *push* and *pull* to distinguish between MRP-based and JIT-based shop-floor systems. The argument is, under JIT, when a customer "pulls" some product out of inventory it pulls some replacement inventory from the factory, which pulls some parts from the shops, which pulls some materials from the store rooms, and so on. On the other hand, MRP-based systems "push" components into the factory, then into inventory, then back into manufacturing, and so on. We find this terminology to be not very helpful. It has spawned debates over whether MRP is a push or pull system, whether kanbans are a part of a pull system when the company is make-to-stock with inventory, or if a JIT system is push-based when the need for an end item is exploded into raw materials that are then sent through the factory without any kanban type of replenishment. These debates simply aren't very helpful. The distinction we believe is useful pertains to whether individual work centers are allowed to utilize capacity ("to keep busy") without being driven by a specific end item schedule. Increasingly, JIT is being utilized in nonrepetitive environments, where specific product configurations are moved through manufacturing in short lead times without tracking or other transactions and capacity utilization is a result, not an objective.

The key distinction we're trying to make is these two approaches' characteristics must match the manufacturing process and infrastructure in which they operate. Activities in the MRP-based systems are triggered by paperwork authorizing production quantities, routings, due dates, and so forth. JIT-based systems produce in response to downstream use of the item, which may be work center by work center or may be in response to demand for the overall end item. For systems installed to date, relatively constant demands are required for the JIT-based approach to function.

CHOOSING THE OPTIONS

There's a temptation to view some MPC design options as a continuum where movement toward JIT is "good." This isn't the correct conclusion. We must match MPC system design with the ongoing needs of a company's market, the task in manufacturing, and the manufacturing process. An MPC system represents a major investment in a business, and as such it must be designed to support the firm's competitive strategy. The framework for accomplishing this was developed by Berry and Hill. Let's turn to how this matching takes place.

Market Requirements

Figure 9.5 shows how MPC system design is influenced by a company's market requirements and the resultant manufacturing task. Figure 9.5 labels these last two factors "business specifications." The point is these determine, from a business point of view, what has to be done in manufacturing to serve the chosen markets. Then technical requirements are defined. This involves the interaction of the manufacturing task, MPC system, and manufacturing process. Each of these three areas needs to be carefully considered before the choices can be made in the approaches in master production scheduling, detailed material planning, and shop-floor scheduling. Moreover, the three areas must be seen as constantly changing: new customer requirements, new process technology, and new strategic goals in manufacturing. Any of these can mandate a change in the MPC system design.

FIGURE 9.5   MPC System Design Choices

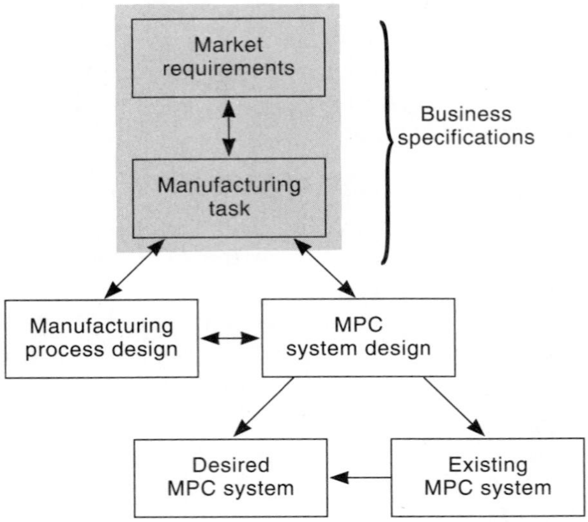

Figure 9.5 also shows the MPC system design as influenced by the desired MPC system and existing MPC system. In some cases improvements can be made by investing in the evolution of the existing system design. In other cases, we need to start afresh.

The first step in the development of market requirements is to review the customers and market segments targeted by the business, their present needs with regard to the company's products and services, competitors' products and services, and existing sales growth opportunities. Many companies face dynamic markets where customer requirements and global competition are changing dramatically. We must continuously review market requirements and adapt marketing strategies to exploit opportunities. For example, many companies increasingly see the need to enhance their products with services to help their customers solve problems. *Market focus, customer prosperity,* and *delighting the customer* are common phrases. But if these phrases are to be more than hype, we must redefine the manufacturing task to create the desired results. Thereafter, we well may have to redesign the MPC system as well as the manufacturing process. To illustrate, the manufacturing organization in a packaging materials firm supplying the food industry suddenly had to deliver products in small quantities on a twice weekly basis to support its major customer's new JIT program. Neither the production process nor the MPC system was designed to support the changed business requirements. More fundamentally, the firm's manufacturing strategy had to be revised to support this kind of customer requirement.

The Manufacturing Task

The next step in choosing MPC system design options is to develop a statement of the manufacturing task that's consistent with (and that supports) the marketing strategy. If the company decides to satisfy customers on a just-in-time basis, this has to be reflected in the manufacturing task. Similarly, if quality is now the way to win orders, it too must be reflected in changed manufacturing values, process investments, improvements in the quality support function, and revised manufacturing performance measurements. If the targeted customers are moving toward more highly customized products, again, this needs to be captured in the manufacturing task.

Hill points out that stating the manufacturing task for the business is critical to ensuring manufacturing capabilities are developed to support the different targeted market segments. Developing the manufacturing task involves characterizing the markets targeted by the company in terms of the requirements they place on manufacturing as described by Hill's manufacturing strategy framework. Such requirements may, for example, include volume and delivery flexibility, low-cost production, critical product quality specifications, and other manufacturing-related capabilities—whatever is required to win orders in different market segments.

A clear statement of the manufacturing task enables management to recognize major changes may be required in the design of both production processes and the MPC system. Figure 9.5 shows this by the two-headed arrows linking the manufacturing task to the design of both manufacturing processes and MPC systems.

Manufacturing Process Design

Most firms have large investments in production processes, employee capabilities, and other elements of infrastructure in manufacturing. As a consequence they tend to remain fixed over long periods of time. This establishes the manufacturing capabilities of a company according to Hill's process choice framework.

The arrow linking manufacturing process design and MPC design indicates the interdependency between MPC option choices and manufacturing process features. For example, installing a JIT process with cellular manufacturing and short production lead times means rate-based detailed material planning approaches may be much more appropriate than time-phased approaches.

A more subtle example of manufacturing process design impacting MPC system design occurs in the case of quality improvement programs. Many companies use complex scheduling procedures because the firms suffer from poor quality and the resultant unpleasant surprises. Quality is usually improved through investments in better manufacturing processes. Where quality is enhanced significantly, there are fewer surprises, the company is better able to execute routine plans, and MPC systems can be more straightforward.

Finally, in some cases there are simultaneous changes in marketplace requirements, manufacturing processes, and manufacturing task definitions. For example, Digital Equipment (DEC) and other computer manufacturers at one time faced a very long lead time to make a computer; they achieved customization by individual wiring and other hardware features. New computers were "announced" in the marketplace long before they were available for shipment, customers would place orders just to get their place in the queue of orders, and the MPC system had to manage a fictitious backlog of orders. Moreover, each order's configuration would constantly change and delivery dates would be extended or canceled. The net result was a very complex set of requirements for the MPC system. Now computers are relatively easy to make, most customization is done with software, and orders are rapidly shipped. Moreover, computers per se are becoming a commodity; DEC and other companies increasingly view their manufacturing strategy as solving problems for their customers. The resultant changes in end "products"—and the processes that produce them—dictate a completely different set of design requirements for the MPC system.

MPC System Design

Because of the magnitude of the investment in MPC systems and the time required to implement MPC system changes, we must recognize differences between desired and existing MPC system options and features. Figure 9.5 shows this by the lines connecting MPC system design with desired and existing MPC systems. A company currently using time-phased MRP records while installing a JIT process with cellular manufacturing might continue to use MRP records with some modifications until necessary investment funds and management time were available to make the MPC system changes required to implement rate-based material planning. Although the marketing strategy, manufacturing task, manufacturing process, and MPC system design specifications might have been agreed upon within the business, the opportunity to move to implementation might not yet have occurred.

This example illustrates another integration issue—consistent MPC option choices. We need to have the right choice (and consistency) in the MPS approach, the detailed material planning approach, and the shop-floor system approach. This issue frequently arises during JIT implementation in a company using MRP for detailed material planning in which batch and line production processes are appropriate for different parts of the business. Therefore, issues of how to link JIT and MRP options in MPC system design and how to maintain *one* MPC system are often difficult. Our experience indicates attention paid to marketplace requirements and to how these requirements may be changing helps you determine the dominant choices among the MPC options.

Master Production Scheduling Options. In Figure 9.6 the three MPS approaches are related to key aspects of marketplace requirements and to aspects of the manufacturing task and manufacturing process. A make-to-order (MTO) master scheduling approach supports products of wide variety and custom design, frequently involving the development of engineering specifications. They're typically produced in low unit volumes, where delivery speed is achieved through overlapping schedules for design and manufacture of the various elements comprising the customer order. Delivery reliability is somewhat difficult to guarantee, since products are customized to meet individual customer needs. This approach is frequently used to support markets characterized by high levels of product change and new product introductions, and where the firm's competitive advantage is in providing product technology requirements in line with the customer's delivery and quality requirements. Since the manufacturing task often involves providing a broad range of production capabilities, the process choice supports low-volume batch (job shop) manufacturing. One key aspect of the manufacturing task is how to respond

FIGURE 9.6 Linking Market Requirements and Manufacturing Strategy to Design of the MPS Approach

Strategic variables			Master scheduling approach		
			MTO	*ATO*	*MTS*
Market requirements	Product	Design	Custom	————————————▶	Standard
		Variety	Wide	—————————▶	Predetermined and narrow
	Individual product volume per period		Low	—————————▶	High
	Delivery	Speed	Through overlapping schedules	Through reducing process lead time	Through eliminating process lead time
		Reliability	Difficult	—————————▶	Straight-forward
Manufacturing	Process choice		Low-volume batch	—————————▶	High-volume batch/line
	Managing fluctuations in sales volume		Through order backlog	Through WIP or finished goods inventory	Through finished goods inventory

to fluctuations in sales volumes. These are typically managed through adjustments in the level of the customer order backlog.

An assemble-to-order (ATO) master scheduling approach represents an intermediate position. Products of both standard and special design are produced, and variety is accommodated by customer selection from a wide series of standardized product options. The unit production volumes are relatively high at the option level, and customer responsiveness in regard to delivery speed is enhanced by lead time reductions and short time frames for frozen final assembly schedules. Delivery reliability is well accommodated as long as overall volumes are kept within planning parameters. That is, the ATO environment is designed to be relatively accommodative of changes in product mix.

Typically, ATO manufacturing is done in batches, with more and more firms using cellular approaches for popular options and families of similar parts. Stocking components, intermediate subassemblies, or product option items can shorten customer lead time to that of the final assembly process, thereby improving delivery speed and reliability in markets where fluctuations in sales volumes are hard to anticipate.

The make-to-stock (MTS) master scheduling approach supports products of standard design produced in high unit volumes in narrow product variety for which short customer delivery lead times are critical. Delivery speed is enhanced by reducing process lead times, frequently by adopting flow-based manufacturing methods. Reliability of production schedules is relatively straightforward.

The process choice is usually line manufacturing or high-volume batch manufacturing. While an investment in finished-goods inventory can provide short, reliable delivery lead times to customers, and can buffer fluctuations in sales, it can also enable us to stabilize production levels, thereby permitting important cost improvements in manufacturing. Since products are often produced on high-volume batch or line processes, schedule stability is often critical, especially in price-sensitive markets.

Material Planning Options.   Figure 9.7 relates the two detailed material planning approaches to key aspects of marketplace requirements and to aspects of manufacturing task and manufacturing process. Time-phased detailed material planning is appropriate for custom products produced in wide variety and low volumes. It also facilitates schedule changes and revisions in customer delivery dates as well as changes in product mix. Delivery speed is enhanced through better scheduling, based on relative priorities. This approach can be applied in markets characterized by a high rate of new product introductions, rapid shifts in product technology, and custom-engineered products by using planning bill of material techniques.

Time-phased planning is often associated with batch (job shop) manufacturing and is supported by relatively high overhead and work-in-process inventory costs due to the necessary planning staff and extensive transaction

FIGURE 9.7   Linking Market Requirements and Manufacturing Strategy to the Design of the Detailed Material Planning Approach

Strategic variables			Detailed material planning approach	
			Time phased	*Rate based*
Market requirements	Product	Design	Custom	Standard
		Variety	Wide	Narrow
	Individual product volume per period		Low	High
	Ability to cope with changes in product mix		High potential	Limited
	Delivery	Speed	Through scheduling/ excess capacity	Through inventory
		Schedule changes	Difficult	Straightforward
Manufacturing	Process choice		Batch	Line
	Source of cost reduction	Overhead	No	Yes
		Inventory	No	Yes
		Capacity utilization	Yes	No

processing. This planning approach can result in higher capacity utilization and is often favored in manufacturing facilities employing expensive equipment.

Rate-based material planning is appropriate for a relatively narrow range of standard products, with stable product designs produced in high volume. Rate-based detailed material planning is much more limited in its ability to cope with changes in product mix. The limited product line permits straightforward changes in the schedule as long as they're within the product design specifications. Enhancements in customer delivery speed are typically accommodated with finished-goods inventories.

These marketplace requirements are normally best supported in manufacturing by production line processes. Use of rate-based material planning and line production processes yields an opportunity to cut work-in-process inventory and overhead costs, providing important support for price-sensitive mar-

FIGURE 9.8   Linking Market Requirements and Manufacturing Strategy to the Design of the Shop-Floor System Approach

Strategic variables			Shop-floor system approach	
			MRP based	*JIT based*
Market requirements	Product	Design	Custom	Standard
		Variety	Wide	Narrow
	Individual product volume per period		Low	High
	Accom-modating demand changes	Total volume	Easy/incremental	Difficult/stepped
		Product mix	High	Low
	Delivery	Speed	Achieved by schedule change	Achieved through finished goods inventory
		Schedule changes	More difficult	Less difficult
Manufacturing	Process choice		Low-volume batch	High-volume batch/line
	Changeover cost		High	Low
	Organizational control		Centralized	Decentralized (shop-floor based)
	Work in process		High	Low
	Source of cost reduc-tion	Overheads	Low	High
		Inventory	Low	High

kets. On the other hand, rate-based material planning doesn't support intensive utilization of capacities in the same way as time-phased approaches.

Shop-Floor System Options. In Figure 9.8 the two shop-floor system approaches are related to key aspects of marketplace requirements and to aspects of manufacturing task and manufacturing process. The MRP-based approach to shop-floor scheduling is appropriate when a wide variety of custom products is produced in low unit volumes. Changes in demand are accommodated relatively easily; volume changes are supported by overtime operations in critical work centers, and product mix change is an inherent characteristic. This approach supports markets characterized by rapid changes in product technology, high rates of new product introduction, and substantial changes in product design.

Low-volume batch or jobbing processes involve use of the MRP-based shop-floor scheduling system approach. These processes have significant changeover costs and numerous manufacturing steps, requiring a complex shop-floor scheduling system that's centrally driven, thereby limiting the reduction of overhead and inventory-related costs.

JIT-based approaches for shop-floor scheduling provide important support for standard products produced in limited variety and high volume. Such products are best supported by high-volume batch or line production processes that are able to provide short customer lead times. Accommodation of changes in product volume is limited because of the cost of production schedule and capacity changes; this increases the need for schedule stability. Delivery speed is enhanced by short manufacturing throughput times and often by finished-goods inventories.

The emphasis on inventory reduction and the simplicity of shop-floor control procedures under the JIT approach provide the potential for significant cuts in overhead and inventory-related costs, providing important support for price-sensitive markets.

THE CHOICES IN PRACTICE

Achieving a close fit between marketplace requirements, the manufacturing task and process, and the MPC system design gives a firm important competitive advantages. In this section we briefly describe marketing and manufacturing strategies of three companies (Moog, Inc., Space Products Division; Kawasaki, U.S.A.; and Applicon, Division of Shlumberger) and how they've designed their MPC systems. Figure 9.9 shows the three MPC systems' overall design. Moog uses MTO and ATO approaches to master production scheduling, a time-phased approach to detailed material planning, and an MRP-based shop-floor system. Kawasaki uses MTS master production scheduling, rate-based material planning, and JIT shop-floor scheduling. Applicon uses ATO master production scheduling, both MRP and rate-based scheduling for material planning, and a JIT-based shop-floor system.

Moog and Kawasaki represent examples of stable MPC system designs to support the requirements of a single market. Applicon, however, provides an example of an MPC system that changed in response to shifting market requirements and process design changes. Let's now see the overall pattern of decisions in each firm concerning the influence of marketing and manufacturing strategy on MPC system design, and see how the resultant systems support their businesses.

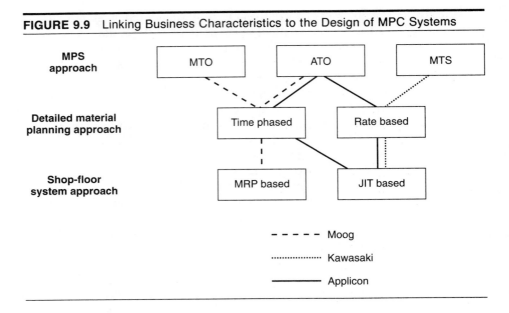

FIGURE 9.9 Linking Business Characteristics to the Design of MPC Systems

Moog, Inc., Space Products Division

This firm produces high-quality hydraulic systems of advanced design for the aerospace industry. These products cover a wide range of design types and represent a critical element in the overall production lead times for its aerospace customers. The company designs and produces the initial order for new products as well as follow-on orders. Thus, engineering design and advanced product features are key factors in obtaining sales. Other important factors that qualify the firm to compete in this market include delivery reliability, reputation for quality, and price. Figure 9.10 summarizes characteristics of the market served by Moog along with key elements of its manufacturing strategy.

The manufacturing task involves providing a broad range of equipment and employee capabilities to make high-precision, custom-designed products in low unit volumes. Substantial uncertainty exists with regard to production process yields and time estimates to produce initial orders. In addition, design changes contribute to process uncertainty. Labor cost is a significant portion of product cost since highly skilled employees and a wide variety of precision equipment are keys to the production process. Major investments have been made in numerical control (NC) and computerized numerical control (CNC) equipment as well as machining centers in a batch manufacturing process.

All manufacturing planning and control system functions in Figure 9.1 are performed at Moog. Both make-to-order and assemble-to-order master production scheduling approaches are used. The MPS is stated in terms of ac-

FIGURE 9.10 Moog, Inc., Space Products Division

Market characteristics	Manufacturing		Manufacturing strategy		
			Manufacturing planning and control system		
	Task	Features	Master production scheduling	Detailed material planning	Shop-floor systems
Customized products	Reducing process lead time	Batch manufacturing	Make-to-order/ assemble-to-order from:	Time-phased material planning	MRP-based systems
Wide product range		Long process routings			Priority scheduling of shop orders
Low volume per product	Manufacturing to engineering specifications and quality standards	High-precision work	Customer orders Anticipated orders Forecast orders	Material is particular to customer orders	System supported by dispatching and production controller personnel
Make-to-customer specifications		Accommodate delivery and design changes with reliable deliveries		High obsolescence risk	
Initial pilot orders	Delivery reliability critical	Labor cost equals 60%	Used for rough-cut capacity planning due to long lead time impact on delivery	Extra materials needed for scrapped items	Capacity requirements planning by work center
Future repeat (blanket) orders		Control of actual costs against budget		Trade-off: shorter lead time versus raw material inventory	Order tracking and status information
Key customer requirements: Design capability Delivery speed		Scrap and rework: First orders Repeat orders	Customer order promising		
Market qualifiers: Delivery reliability Quality Price		First order processing uncertainties (process unknown, time estimates) Process and product uncertainties			

tual, anticipated, and forecast customer orders with substantial emphasis on customer order promising and capacity planning activities. The master production schedule uses this information to determine requirements for component material. Time-phased material requirements planning records are used to coordinate scheduling of manufactured and purchased components, and these records are used to prepare shop load forecasts for individual departments and work centers.

At Moog the MRP-based approach is used for shop-floor scheduling and vendor scheduling. An advanced computer-based MRPII system provides priority scheduling information for sequencing and dispatching shop orders at individual work centers. The shop-floor system supports the batch manufacturing of products under high levels of process uncertainty. A variety of production reports assist supervisors in the detailed tracking of open shop orders, reporting order status, and evaluating work center performance.

Kawasaki, U.S.A.

Kawasaki produces six different types of motorcycles as well as motorized water skis at its U.S. plant. About 100 different end product items are manufactured for shipment to the firm's distribution centers. Although demand for products is highly seasonal, workload at the plant is stabilized by permitting fluctuations in the finished-goods inventory carried at the distribution centers. The company frequently introduces new product designs which represent styling changes in the product. The key elements in gaining sales are price, product styling, and product performance. Factors qualifying the firm to compete in the market are quality and delivery speed. Figure 9.11 summarizes characteristics of the market served by Kawasaki along with key elements of its manufacturing strategy.

Manufacturing's task is to produce standardized products in high volume at low cost. Since material costs are significant, major emphasis is placed on reducing plant inventories using just-in-time manufacturing methods. The production process is characterized by short setup times and small production batches using production line and high-volume batch processes. Standardized assembly operations and repetitive employee tasks characterize the production process.

All the manufacturing planning and control functions in Figure 9.1 are performed at Kawasaki; a make-to-stock master production scheduling approach is used. Customer orders for end products are filled from the finished-goods inventory held by the company's distribution division. The MPS is based on forecast information, and mixed model assembly is used in performing final assembly operations. Substantial emphasis is placed on leveling the master production schedule and freezing it over a three-month planning horizon.

A rate-based material planning approach utilizes a simple planning bill of materials to schedule the rates of flow for manufactured and purchased com-

FIGURE 9.11 Kawasaki, U.S.A.

	Manufacturing strategy					
	Manufacturing		**Manufacturing planning and control system**			
Market characteristics	**Task**	**Features**	**Master production scheduling**	**Detailed material planning**	**Shop-floor systems**	
Narrow product range	Provide a low-cost manufacturing support capability	High-volume batch and line production process	Make-to-stock	Rate-based material planning	JIT-based systems	
Standard products			Manufacture to forecast		Kanban containers	
High volume per product	Support the marketing activity with high delivery speed through finished-goods inventory	Short setup times	Level production		JIT flow of material	
Seasonal demand		Small batch size	Three-month frozen planning horizon		Low raw material, component, and WIP inventory	
Sales from finished-goods inventory at distributors		Low-cost manufacturing	Manufacture to replenish distribution inventories			
Introduction of new products		Low labor cost				
Changing product mix		High material cost				
Key customer requirements: Price		Low inventories (raw material, components, and WIP)				
Delivery speed (through finished-goods inventory in distribution divisions)		Low overheads (low MPC costs)				
Market qualifiers: Quality						
Delivery reliability						
Basic design and peripheral design changes						

ponents. A JIT shop scheduling system using kanban containers controls the flow of material between work centers. The JIT system supports low-cost manufacturing with small plant inventory levels and high-volume material flows. Very few personnel and minimal transactions are required in planning and controlling production activities.

Applicon

This firm designs and manufactures computer-aided engineering (CAE), design (CAD), and manufacturing (CAM) systems for the electronics and mechanical design markets. High-end products include systems for highly sophisticated customers in a variety of analytical engineering applications. Low-end systems use Applicon software, Sun and Tektronics work stations, and DEC VAX processors for applications in robotics and numerical control machines.

The mechanical design market represents the firm's major growth area. In this market, unlike the electronics market, the price-to-performance ratio is a critical issue to price-sensitive CAD/CAM customers. In addition, the ability to respond rapidly to changes in technology and frequent design changes is also critical. Figure 9.12 summarizes characteristics of the market served by Applicon along with key elements of the old manufacturing strategy (i.e., the one employed by the company before the process change).

The manufacturing task for the mechanical design market involves producing high-quality products having a wide range of optional features in small volumes at low cost in short customer lead times while accommodating rapid engineering changes. As Figure 9.12 shows, the previous manufacturing approach was to produce products using a batch manufacturing process where the plant was organized into functional groupings of machines, and production was planned and controlled using an MRPII system to fill customer orders directly from finished-goods inventory. Long production lead times under this strategy led to poor competitive performance. The inability to make changes in product designs didn't allow the firm to keep up with major changes in product technology; large work-in-process and finished-goods inventories created substantial write-offs of obsolete inventory; poor customer service in product delivery resulted with high manufacturing costs.

As a consequence the company changed its manufacturing strategy, investing in a JIT production process having straight line flows of material with closely coupled manufacturing cells dedicated to individual product families and short changeover times. Four cells are dedicated to the final assembly of four different product model families, while the fifth cell produces printed circuit boards (PCBs) for the final assembly cells. Thanks to this process, overall manufacturing lead time fell from 75 to 5 days, work-in-process and finished-goods inventories declined significantly, and product quality improved greatly. Likewise, the MPC system design was changed to include an assemble-to-order MPS because of the short manufacturing lead time, a new

FIGURE 9.12 Applicon's Old Manufacturing Strategy

	Manufacturing		Manufacturing planning and control system		
				Manufacturing strategy	
*Market characteristics**	*Task**	*Features*	*Master production scheduling*	*Detailed material planning*	*Shop-floor systems*
Customized products	Low-cost manufacturing	Batch manufacturing	Make-to-stock MPS	Conventional MRP system	Large shop-floor order quantities typically representing one month's usage.
Wide product range	Short production lead times	General-purpose equipment	High levels of finished-goods inventory	Stockroom kitting of assemblies prior to release of work orders at each stage in the process	
Major design changes occurring monthly	High product quality	Functional plant layout	Monthly MPS is created for each end product, using an annual build plan		Work is scheduled on the shop floor using a priority control system for work orders.
Quick response required to changes in product technology	Rapid engineering change capability	4–5-month manufacturing lead time		MPS is exploded into time-phased work orders for components and subassemblies using bill of materials, inventory data, and monthly time periods.	
Need to reflect both price sensitivity and the price/performance ratio for a sophisticated customer base		Long design change cycle			Large numbers of shop-floor and inventory transactions are processed to maintain data integrity in the MRP system.
		High rate of inventory obsolescence			
		85% plug and play rate in final inspection			
High product quality requirements		Excessive rework costs			Large overhead costs are incurred to support the MPC system, as illustrated by 83 people employed in the materials management area.
		160 actual production operators in the manufacturing areas			
Delivery responsiveness is critical		20 weeks of work-in-process inventory			
		Product family mix change flexibility			

*The market characteristics and manufacturing task are common to the old and new strategies.

MRP material planning approach that takes into account JIT plant operations, and a JIT-based shop scheduling approach. Figure 9.13 describes the new manufacturing approach.

The Driver Is the Marketplace

It might be tempting to believe the evolution in MPC system design is always toward JIT-based systems. In fact, that's not the case. Several major Japanese companies that have used JIT approaches now are trying to integrate MRP into their MPC designs. Why? Because they're entering markets where MRP approaches make sense. More and more Japanese firms are moving out of standardized product markets and into more customized areas, where MRP planning supports wider product variety. The resultant systems will almost surely be hybrids where some JIT methodologies prevail, but the benefits of MRP planning are needed as well. The bottom line is, marketplace dictates define the manufacturing task, the process, and MPC design.

INTEGRATING MRP AND JIT

As was clear with the Applicon example, there are many ways that MRP and JIT are combined and substantial need to do so. Here we discuss needs to integrate these approaches, physical changes that support the integration, and techniques for integration.

The Need to Integrate

In the majority of the cases, the need for integration arises in companies that have an installed MRP system and are in the process of implementing some aspect of JIT. The pressure of meeting world class standards, the use of global bench marking, and intimidating competition have all brought home the necessity of major changes in how manufacturing is done. The response to these concerns in the best of companies has been to implement aspects of JIT.

Often these JIT programs seem in conflict with the MRP system the firm may have in place. As lead times shrink and material velocity increases, the limiting activity can turn out to be transaction processing. Increased demand, either from customers or through increased part commonality, can compound the problem.

As an example, a European consumer electronics company significantly cut production time required to make a major high-volume component in response to increased demand. Product design changes and process capability improvements were both used to reduce setup and run times. Lot sizes were

FIGURE 9.13 Applicon's New Manufacturing Strategy

	Manufacturing		Manufacturing strategy		
				Manufacturing planning and control system	
Market characteristics*	Task*	Features	Master production scheduling	Detailed material planning	Shop-floor systems
Customized products Wide product range Major design changes occurring monthly Quick response required to changes in product technology Need to reflect both price sensitivity and the price/performance ratio for a sophisticated customer base High product quality requirements Delivery responsiveness is critical	Low-cost manufacturing Short production lead times High product quality Rapid engineering change capability	Straight-line flows of material Manufacturing cells dedicated to particular product families Short setup times Short manufacturing lead times (1 week) Short design change cycles Low work-in-process and finished-goods inventories Low flexibility to product family mix changes	An assemble-to-order MPS is stated in top-level item terms and is coded by major model number. The company plans using forecast information in the MPS, but builds product only to customer orders using a final assembly schedule. Customer order promising is a key activity. Available-to-promise records are used. Customer orders are used to convert the weekly production plan into specific daily requirements.	MPS uses monthly time periods covering 5 future months to plan and order purchased materials using family bills of material, MRP records, and bill of material explosion techniques. No stockrooms since material is located in the manufacturing cells MPC system is run weekly providing planning information to planners and buyers, and capacity planning information to plant work cells. Only two inventory transactions are recorded—from suppliers into the stock bins, and out of stock bins as finished products are shipped from the plant.	Work orders are not scheduled for internally manufactured items. Material is pulled through the production process using JIT methods. Delivery of 70% of supplier items directly onto the shop floor Customer orders, referred to as build cards, provide the basis for scheduling work cells and for pulling material through the plant.

*The market characteristics and manufacturing task are common to the old and new strategies.

reduced, but lead times were not significantly reduced. The combination of smaller lot sizes and increasing volume simply meant there were substantially more open orders on the floor being tracked by the MRP system, moving into and out of inventory, and being accounted for during the process. These "hidden factory" activities were limiting the improvements possible from the other activities.

When changes take place on the factory floor, MPC system change may be a required response. These changes can come from internal actions like implementing a JIT program or from external requirements that change the manufacturing task. In either case the need for a change in production activity control systems may be clear; the direction is most often from shop-order–based systems to kanban or other simple signals. A typical response is backflushing component usages at all levels triggered by receipt of completed items into finished-goods inventory.

Physical Changes That Support Integration

One of the first requirements to support the JIT approaches in the factory is to reduce the inventory transaction volume. Cutting the number of times a lot has to be logged into and out of an inventory location not only reduces transactions but enables material to move to the next operation more quickly. This clearly helps increase the velocity and reduce lead times. Physically, this may mean making some changes in how lots get moved from department to department and how the need for the move gets signaled, but the major improvements are in making physical changes to the production process, such as the introduction of cellular manufacturing.

Cellular manufacturing supports integrating MRP and JIT approaches. The cell allows us to accomplish several routing steps as if they were a single step and allows the shop floor to be scheduled at the level of part numbers instead of the level of routing steps. More encompassing cellular manufacturing approaches permit the cell to be planned and controlled at the level of assemblies instead of at the part number level. One key objective is to reduce the need for inventory accounting and the other hidden factory transactions. Control of the cell is straightforward and doesn't need the detailed tracking necessary when parts move all over the factory.

The choice of where to implement cellular manufacturing is important since we can create islands of velocity, like the islands of automation prevalent in the early installations of some computer-integrated manufacturing schemes. These islands might be quite successful on their own, but not be well integrated into the system as a whole. Increasingly, we've found firms in this position needing to make more than cosmetic changes to their overall MPC system.

Some Techniques for Integrating MRP and JIT

Whenever there's a combination of MRP and JIT in the shop, we need to move back and forth between the systems. A JIT cell in the middle of a process under MRP control must communicate with the MRP system. There must be a handoff from MRP to JIT at the start of the JIT process and a transfer back to MRP at the end.

One way of supporting this need is to create phantom bills for activities under JIT control. Material requirements planning records can be used to plan raw material requirements, with movement through the factory done with JIT approaches (no shop orders or tracking). The phantom bill would ignore the creation of the detailed parts and assemblies performed under JIT scheduling, while the MRP system would pick up the completed part or assembly as a part number on the bill of materials at completion.

Strategy for Combining MRP and JIT

We're often asked if you need to go through the agony of implementing MRP if thereafter the goal is to dismantle parts of it to use JIT. A typical question is, "Why can't we just go to JIT in the first place?"

For a long time our response was that conceptually JIT could be implemented in a company with no formal MPC system, but it was difficult. First, you need the discipline of a system where execution according to a schedule was part of the basic factory culture. The usual problems of month-end surges in output, inadequate data integrity, pulling dollars instead of products, and panic conditions must be eliminated before you could implement a system with virtually no buffers. Even MRP has small buffering. JIT, by design, "exposes the rocks" as a basic philosophy. Without the underlying discipline, the steady-state condition would be rocks showing; the factory would be shut down much of the time until JIT discipline was achieved. Maintaining commitment to a JIT philosophy under these circumstances would be difficult.

Now, however, we believe a more balanced view toward MRP/JIT integration is required. We need to realize JIT encompasses much more than MPC. You can start to work on cellular manufacturing approaches at a fairly early point, either concomitant with JIT implementation or not. The result can be progress toward manufacturing excellence on more than one front.

Similarly, improved quality is a fundamental underpinning to JIT. One does not need to wait to get started on the quality improvement process. As quality is improved, the "surprises" in MPC are reduced, and more simple systems suffice.

Finally, it's important not to leave this discussion with the impression that every company is evolving from MRP toward JIT. In Japan, many firms are now going in the opposite direction, because the markets in which they wish to compete dictate a manufacturing task, process design, and MPC system that are consistent. In general, these companies are moving away from high-volume standardized goods (such as consumer electronics) to customized higher-value-added products (such as machine tools). We see the key point in all this in Figure 9.5.: Market choices dictate many manufacturing choices, including MPC system design.

CONCLUDING PRINCIPLES

This chapter focused on two major integration issues in designing MPC systems: how to link the design of MPC systems to a firm's corporate strategy and the requirements of its market, and how to integrate MRP and JIT approaches in designing MPC systems. The following principles summarize the major points:

- Since investment in MPC systems is large and fixed over a long period of time, its design must support the firm's competitive strategy.
- A wide range of options are available in designing MPC systems, and the choices must be governed by the company's competitive needs.
- Business as well as technical specifications need to be considered in designing an MPC system.
- MPS system design should begin with an analysis of the market requirements to support the firm's competitive strategy.
- Understanding the manufacturing task is critical in developing the production process design, the MPC system design, and other elements of the manufacturing infrastructure.
- The manufacturing process's particular features need to be considered in choosing among the options in MPC system design.
- MRP and JIT approaches can be effectively integrated in designing MPC systems.
- Improved company performance can result from matching MPC system design to the firm's competitive strategy.

REFERENCES

Belt, B. "MRP and KANBAN—A Possible Synergy?" *Production and Inventory Management Journal* 28, no. 1 (1987), pp. 71–80.

Berry, W. L., and T. J. Hill. "Linking Systems to Strategy," Working Paper. Chapel Hill: University of North Carolina, Kenan-Flagler School of Business, 1989.

Dixon, J. R.; A. J. Nanni; and T. E. Vollmann. *The New Performance Challenge.* Homewood, Ill.: Richard D. Irwin, Inc., 1990.

Fakhoury, E. A. F., and T. E. Vollmann. "Applicon Case Study." Boston: Boston University, School of Management, 1987.

Giffi, Co.; A. V. Roth; and G. M. Seal. *Competing in World Class Manufacturing.* Homewood, Ill.: Richard D. Irwin, Inc., 1990.

Goddard, W. E. *Just-in-Time.* Essex Junction, Vt.: Oliver Wight Ltd., 1986, chapters 7, 9, and 12.

Hall, Robert W. "Kawasaki, U.S.A., Transferring Japanese Productivity Methods to the U.S.A." Case Study 08002. Falls Church, Va.: American Production and Inventory Control Society, 1982.

Hill, T. J. *Manufacturing Strategy: Text and Cases.* Homewood, Ill.: Richard D. Irwin, 1989, chapters 2 and 3.

Karmarkar, U. S. "Alternatives for Batch Manufacturing Control." Working Paper no. QM8613. Rochester, N.Y.: Graduate School of Management, University of Rochester, 1986.
————. "Integrating MRP with Kanban/Pull Systems." Working Paper no. QM8615. Rochester, N.Y.: Graduate School of Management, University of Rochester, 1986.
————. "Push, Pull, and Hybrid Control Schemes." Working Paper no. QM8614. Rochester, N.Y.: University of Rochester, 1986.
————. "Beyond MRPII: Evolution to a New Standard." Working Paper Series no. CMOM 89–1. Rochester, N.Y.: University of Rochester, 1989.
————. "Getting Control of Just-in-Time." *Harvard Business Review,* September–October 1989.

Louis, R. S. "MRPIII: Material Acquisition System." *Production and Inventory Management,* July 1991, pp. 26–27.

Melnick, S. A., and P. L. Carter. "Moog Inc., Space Products Division," in *Shop Floor Control Principles, Practices, and Case Studies.* Falls Church, Va.: American Production and Inventory Control Society, 1987.

Rao, A. "A Survey of MRPII Software Suppliers' Trends in Support of JIT." *Production and Inventory Management Journal* 30, no. 3 (Third Quarter 1989), pp. 14–17.

Stalk, G., Jr., and T. M. Hout. *Competing against Time.* New York: Free Press, 1990.

Van Dierdonck, R. J. M., and J. G. Miller. "Designing Production Planning and Control Systems." *Journal of Operations Management* 4, no. 1, 1980, pp. 37–46.

DISCUSSION QUESTIONS

1. It has been suggested that in a make-to-order company, order backlog is inventory carried by the customer. What does this statement mean?
2. Why do you suppose the terms *push* and *pull* came to be used for MRP and JIT systems?
3. What key features of the manufacturing task impact the design of the MPC system?

4. Some manufacturing executives have raised the question "Which is better, MRP or JIT?" How would you respond to this question?

5. What approaches other than finished goods inventory provide short lead times to the customers?

6. Henry Ford was said to have specified "any color as long as it's black." Today's automobiles come in literally billions of end item combinations. What are the implications of this change in market requirements on MPC system design?

7. What changes in market requirements at Applicon dictated a new production process and MPC approach?

8. What kind of changes in the products produced and markets served would enable Moog to adopt make-to-stock master production scheduling?

9. Why is discipline required for both JIT and MRP?

PROBLEMS

1. Worldwide Batteries, Inc., sells industrial batteries to both OEM and replacement market customers. About 80 end products are classified as high-volume products, 200 as medium-volume products, and 170 as low-volume products. The vice president of marketing and sales has provided the following description for two market segments that reflect the general characteristics of all market segments currently targeted by the company:

   • Replacement batteries for indoor material handling equipment for which 60 percent of the sales volume is for deliveries desired by the customers in 7 days or less (high-volume items), 30 percent is for delivery within 14 days (medium-volume items), and 10 percent is for delivery within 30 days (low-volume items). Key factors in gaining sales orders in this segment are competitive prices and delivery speed.

   • Replacement batteries for outdoor transportation equipment applications which involve low-volume product items. Customers expect delivery within two weeks. The key factors in winning orders in this segment are competitive prices and delivery reliability.

   Recently, the company has implemented a kanban system for controlling the material for manufacturing component parts for batteries. The lead time for final assembly of battery components into end products, battery charging, and transport to the customers is seven days. High- and medium-volume purchased components (used in the final assembly operation) are provided by suppliers on a JIT delivery basis with a guaranteed five-day delivery lead time. Low-volume purchased components (also used in the final assembly operation) require a four-week delivery lead time for the suppliers.

   Specify the MPC system design requirements for the company's markets using the framework shown in Exhibits 9.10, 9.11, and 9.12. Refer to Exhibits 9.2, 9.3, 9.4, 9.6, 9.7, and 9.8 for MPC design details.

2. Currently, the Cambridge Plastics Company sells industrial packaging materials to two different market segments:

- Injection-molded plastic bottles sold to soft-drink bottlers in very high volume with significant seasonal variations in sales. The customers provide call-off schedules indicating their product requirements over the next six months, specifying the delivery quantities each week. Key factors in winning this business are the ability to meet large changes in the weekly quantities and the ability to provide a major capacity increase for the increased demand during the peak summer selling season.
- Custom-molded plastic bottles sold in wide variety and low volume to manufacturers of cosmetic, pharmaceutical, and agricultural chemical products. These are produced to specific customer orders, with the key factors in winning orders in this business being competitive prices and delivery reliability.

The soft-drink bottles are produced using high-volume processes dedicated to the production of specific products, while the custom products are produced using low-volume batch processes. In both cases the tracking of actual versus planned production is very important in ensuring that the current quantities of products are delivered against the customer orders.

Specify the MPC system design requirements for the company's markets using the framework shown in Exhibits 9.10, 9.11, and 9.12. Refer to Exhibits 9.2, 9.3, 9.4, 9.6, 9.7, and 9.8 for MPC design details.

3. The Arnold Company is a make-to-stock firm. A typical part for one of the firm's products has the following data:

   Manufacturing lead time (LT) = 6 weeks
   Forecast = 25 per week
   Mean absolute deviation (MAD) of forecast errors = 5
   Safety stock = 95 percent service level (2.056 MAD $\sqrt{\text{LT}}$)

   A cellular manufacturing approach will reduce a typical part's manufacturing lead time to one week.
   a. What's the reduction in safety stock?
   b. The cell can manufacture 35 percent of the parts (the present total safety stock investment is $100,000). Assuming a lead time reduction from 6 weeks to 1 week for these parts, what's the overall reduction in safety stock investment?
   c. An engineer believes redesign of the cell can increase the percentage of parts produced therein from 35 percent to 50 percent. How much is the redesign worth?

4. The cell in Problem 3 is designed with a capacity of 500 units per day. A new product design uses many common parts; the result is a cell scheduled with a one-day lead time and an overall shift to assemble-to-order manufacturing. Customers are happy to receive one-day deliveries. Overall demand for parts made in the cell is 300 units per day with a standard deviation of 50 units. The 300 units are typically completed in approximately five working hours. Thereafter, workers in the cell perform equipment maintenance, schedule production, and manage quality.
   a. What happens to the service level provided to customers?
   b. What happens to safety stock inventory levels?

5. Dixon Plastics has 100 customers. It produces make-to-order products using time-phased material planning and an MRP-based shop-floor system. Amdur Electronics, one of its larger customers, accounts for 20 percent of total sales. It promises to double its volume with Dixon within two years and to communicate its detailed planning to Dixon via electronic data exchange (EDI). The plans would be frozen for the next week, subject to + or − 15 percent in weeks 2 through 6, and open to revision after six weeks. Any raw material purchased by Dixon to meet requirements in the next 6 weeks would be guaranteed against obsolescence by Amdur Electronics.

   a. What changes in MPC design options would Dixon require if total manufacturing lead time was 1 day?

   b. What if the total manufacturing lead time is 6 weeks?

6. Consider the following three MRP records. Product A is comprised of two part Bs. One part B requires one part C.

Product A

Week		1	2	3	4	5	6	7	8
Gross requirements		10	20	0	40	20	10	0	20
Scheduled receipts									
Proj. avail. balance	15	5	15	15	5	15	5	5	15
Planned order release		30		30	30			30	

Q = 30, SS = 5, LT = 1.

Part B

Week		1	2	3	4	5	6	7	8
Gross requirements		60		60	60			60	
Scheduled receipts			100						
Proj. avail. balance	75	15	115	55	95	95	95	35	35
Planned order release			100						

Q = 100, SS = 0, LT = 2.

Part C

Week		1	2	3	4	5	6	7	8
Gross requirements			100						
Scheduled receipts									
Proj. avail. balance	0	0	0	0	0	0	0	0	0
Planned order release		100							

Q = LFL, SS = Q, LT = 1.

a. What transactions are required to keep the shop on schedule?
b. Assume it takes one week for MRP processing for each level in the bill of materials. What's product A's total lead time?
c. If everything went according to plan in week 1, what would the records look like in week 2? (Assume a new requirement for 20 units of product A in week 9.) What transactions took place in week 1 and/or must take place in week 2 to keep the shop on schedule?
d. Suppose product A is to be built in a cell (lead time is one week), and parts B and C are to be made into phantom items. What happens to transaction counts and total lead time?

7. Product X is made from one unit of part Y, which in turn is made from one unit of Z, which is made from raw material R. Consider the following set of records:

Product X

Week		1	2	3	4	5
Gross requirements		50	50	50	50	50
Scheduled receipts		50				
Proj. avail. balance	10	10	10	10	10	10
Planned order release		50	50	50	50	

Q = LFL, SS = 10, LT = 1.

Part Y

Week		1	2	3	4	5
Gross requirements		50	50	50	50	
Scheduled receipts						
Proj. avail. balance	52	2	52	2	52	
Planned order release		100		100		

Q = 100, SS = 0, LT = 1.

Part Z

Week		1	2	3	4	5
Gross requirements		100		100		
Scheduled receipts						
Proj. avail. balance	100	0	0	0		
Planned order release			100			

Q = LFL, SS = 0, LT = 1.

Raw Material R

Week	1	2	3	4	5
Gross requirements		100			
Scheduled receipts					
Proj. avail. balance	0	0	0		
Planned order release	100				

Q=LFL, SS=0, LT=1.

---

a. How many transactions (order launches, inventory receipts, and disbursements) are indicated in the records as shown above for product X and parts Y, Z, and R?
b. How would the records look if parts Y and Z are made into phantom items? Assume the raw material (LT = 1, Q = LFL) is still delivered to inventory before being released directly to the line that produces product X. Assume the lead time is still one week for product X. How many transactions are implied now?

8. Melnick Mines uses a special type of hardened crusher ball for preparing the ore. Purchasing the balls is managed with time-phased material planning. The following records are for two successive weeks:

Part: Crusher ball

Week	1	2	3	4	5	6	7	8
Gross requirements	25	25	25	25	25	25	25	25
Scheduled receipts		80						
Proj. avail. balance	20	−5	50	25	80	55	30	85	60
Planned order release	80			80				

Q=80, SS=20, LT=3.

Part: Crusher ball

Week	2	3	4	5	6	7	8	9
Gross requirements	25	25	25	25	25	25	25	25
Scheduled receipts			80					
Proj. avail. balance	70	45	20	75	50	25	80	55	30
Planned order release			80					

Q=80, SS=20, LT=3.

a. What transactions made during week 1 would result in the record as of week 2?

b. How would this set of transactions be changed if rate-based scheduling and JIT-based shop-floor control were implemented at the vendor with lead time one day?

9. The Ronsi Rist Watch Company, producers of famous brand watches, expanded its product line from 20 to 50 models. Each model requires approximately 100 units of safety stock to provide the customer service levels the company wants from make-to-stock operations. A small engineering change in watch design and new layout in the factory made it possible to shift to an assemble-to-order approach. Each model is assembled using one item from each of five options (12 face plates, 2 movements, 4 bands, 4 cases, and 2 kits of mounting hardware). The company was disappointed to learn it took about 100 units of safety stock for each option to provide the desired customer service levels, even though it could assemble virtually any order in a matter of hours. The marketing manager said, "Since we still need about 100 units of safety stock for each item, we should go back to our familiar make-to-stock system, which is much easier for my order entry clerks to handle." Do you agree?

10. Desai Processors, Ltd., was considering a shift from time-phased material planning to quarterly rate-based planning. One component for a popular product is currently made in batches of 1,000 units. If quarterly rate-based planning were used, the average weekly requirements for the quarter would be produced each week. The requirements for a typical quarter are:

Week	1	2	3	4	5	6	7	8	9	10	11	12
	100	500	800	300	500	200	700	800	300	500	600	700

Assuming zero beginning inventory and zero lead time, what would be the average inventory levels under the two approaches?

11. In an attempt to react to a shifting market, the Isandar Wedge Company reengineered its product line for rate-based manufacturing. Previously, the company competed on product variety, custom tailoring products to meet customer needs on a make-to-order basis. The redesign allows customers to make their own adjustments and enables the company to produce a smaller number of end products to stock, thereby providing immediate delivery.

a. Current inventory for an example product is zero, but the firm wants to build up 100 units of safety stock over the next four weeks. Average demand for the product is expected to be 25 units per week. From week 5 on, the company wants to produce at the demand rate. What should the master production schedule be for this product for the next six weeks?

b. Isandar now believes the 100-unit safety stock is excessive. Standard deviation of demand per week is 10 units, so a safety stock of 30 provides a service level above 99 percent. If the company is starting with 100 units of safety stock, what MS results in a safety stock of 30 units in 5 weeks?

12. The Leone Company used to make designer underwear in red, white, and blue to stock. In response to many customer requests, it now offers them with or without three colors of exotic decals (black, green, and orange) with or without four colors of sequins (white, brown, black, and gold). If 10 units of safety stock are kept for each end item, measure the increase in safety stock from the new product offerings. Customers require next-day delivery. Manufacturing lead time is one week.

13. The Leone Company in Problem 12 is having trouble with its inventory control. When only three products were made, safety stock of 10 units per item seemed sufficient. But now the greater product variety has led to a more volatile product mix. After analyzing the most popular options, Bob Leone decided to make the following restrictions to the product range:

    • Red won't be available with green or orange decals.
    • White won't be available with black decals.
    • Blue won't be available with orange decals.
    • Red won't be available with brown or gold sequins.
    • White won't be available with white or black sequins.
    • Blue won't be available with brown or black sequins.

    Assuming 480 units of safety stock are available, what's the average safety stock that can be held for each new item?

14. If the Leone Company in Problems 12 and 13 switches to an assemble-to-order manufacturing option and holds 20 units of safety stock in each option, what's the overall safety stock (in total number of parts)? Assembly lead time is one hour.

15. As part of the final assembly process, Leone (Problems 12 through 14) decides to dye underwear to particular colors and replace decals with silk screening. Assembly lead time increases to two hours. What's the impact on safety stock requirements and the ability to offer a full product line?

Chapter 10

Implementation of MPC Systems

This chapter concerns implementing an effective manufacturing planning and control (MPC) system. The focus is almost exclusively managerial, since that's where the ultimate responsibility lies. The chapter raises fundamental issues involving preparing the organization, forming an implementation team, managing the project, and assessing the results. Addressing these issues is critical to implementing a truly effective MPC system on a timely basis with realized results at each step in the process.

The chapter focuses on seven topics:

- Initiating the project: How does a firm create the MPC vision, justify the project, and determine the effort required to get there?
- Essential prerequisites: What must be in place to ensure successful implementation?
- Organizational issues: What are the key players' roles, and how is this different from the current situation?
- The implementation project team: How should the project team be formed, and what are the requirements for making it successful?
- Education: What are the requisites for educating the users now and in the future?
- Project management: How are implementation projects successfully planned and managed?
- Auditing: Why is periodic auditing of the project essential at any stage, and what does it encompass?

Implementation issues are connected to virtually every chapter in the book. Each of the manufacturing planning and control system modules and the advanced techniques associated with them require implementation, either singly or in concert with other modules. Of particular relevance to the implementation question is Chapter 1, which discusses the differing environments for MPC systems, and Chapter 9, which describes the link between company strategy and MPC. The new concepts in Chapter 19 will also need to be implemented.

INITIATING THE PROJECT

This section presents a framework for MPC implementation. The first step is developing a vision of life at the end of the project: a yardstick against which progress can be measured and targets established. Second, it's important to clearly assess the current company environment and to maintain an evolutionary point of view on how to impact that environment during implementation. The vision provides a broad view of the benefits, while assessing the current environment helps define the costs involved.

The Yardstick of Performance

Developing a vision of life at the end of the project is helpful for a number of reasons. It provides a sales tool for enlisting support for the project within the company. It helps us identify all the benefits that can be gained. The vision clarifies the goals during implementation and helps bring new people on board the project. Perhaps most important, however, it's the yardstick against which implementation progress is measured.

Key to creating a successful vision is a clear view of the changes that will impact everyone and the advantages the firm will derive from the new MPC system. Regardless of whether the new system is based on material requirements planning (MRP), just-in-time (JIT), a combination of the two, or some other approach, a common vision of how the system will operate and its benefits is critical to motivating users. Clearly, the details of the vision will differ in different firms, and those differences will be reflected in the MPC system design.

Common to *any* MPC implementation is replacing the informal system with the formal system. Under MRP, the formal system will involve computer-generated information for making the decisions and reporting status. With JIT, information for managing the flow of materials will come from kanbans and other simplified visual signals; material status records may be minimal. In either case, all personnel must obey the formal system's dictates for the benefits to be achieved. In every implementation this means jobs will change—some dramatically.

Creating a broad vision helps identify all the MPC system's benefits. We've found Figure 10.1's general MPC schematic helpful in developing this broad view of life after the implementation. It helps show how people's roles in the system need to be related. It also helps in testing for completeness. What

FIGURE 10.1 Manufacturing Planning and Control System

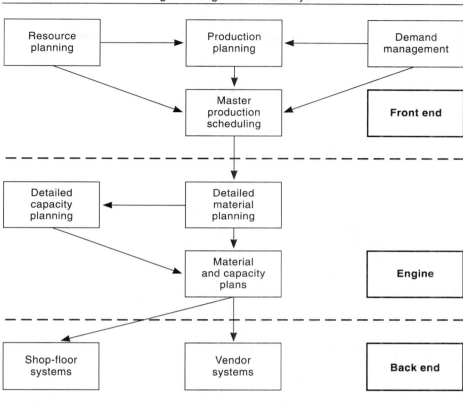

gaps in the current system need to be filled? Where are the greatest sources for improvement over current practice? How will each phase be accomplished in the new system?

In the JIT environment, the MPC vision needs to include images of kanbans, production cells and other new layouts, elimination of paper and inventory, and product flowing rapidly through the shop. The JIT view also needs to encompass changes required in processes, product design, and organizational behavior. The vision is supported by procedures to accomplish each module in the MPC framework. Finally, the vision should include an evolutionary dimension. This means mechanisms for continual production improvement, product and process design modification, and measuring performance achievements.

For MRP-based systems, a method for assessing MPC system performance was developed by Oliver Wight (one of the best-known MPC educators). He defined four categories of performance for MRP-based companies. We present his categories, not because of the heavy emphasis on MRP, but because the classifications are well known and useful in setting targets for improvement. We can use them to position the MPC system performance overall or by individual module as a basis for setting goals. The categories are:

Class A MRP users employ MRP in a closed-loop mode. They have material requirements planning, capacity planning and control, shop-floor dispatching, and vendor scheduling systems in place and being used; and management uses the system to run the business. They participate in production planning. They sign off on the production plans. They constantly monitor performance on inventory record accuracy, bill of material accuracy, routing accuracy, attainment of the master schedule, attainment of capacity plans, and so on.

In a Class A company, the MRP system provides the game plan sales, finance, manufacturing, purchasing, and engineering people all work to. They *use* the formal system. Foremen and purchasing people work to the schedules. There's no shortage list to override the schedules and answer the question. The answer to "What material is really needed when?" comes from the formal MRP system.

Companies using MRP II have gone even a step beyond Class A. They have tied in the financial system and developed simulation capabilities so that "what-if" questions can be answered using the system. In such a company, management can work with one set of numbers to run the business because the operating system and the financial system use the same numbers.

Technically then an MRP II system has the financial and operating systems married together and has a simulation capability. But the important point is the system is used as a company game plan. This is what really makes a company Class A.

A **Class B** company has material requirements planning and usually capacity requirements planning and shop-floor control systems in place. The Class B user typically hasn't done much with purchasing yet and differs from the Class A user primarily because top management doesn't really use the system to run the business directly. Instead, Class B users see MRP as a production and inventory control system. Because of this, a Class B user can become a Class C user quickly. Another characteristic of the Class B company is that it does *some* scheduling in the shop using MRP, but its shortage list is what really tells it what to make. Class B users typically see most of their benefits from MRP in inventory reduction and improved customer service because they have more of the right things going through production. Because they haven't succeeded in getting the expediting "monkey" off the backs of the purchasing people and foremen, they haven't seen substantial benefits in reduced purchase costs or improved productivity—and they still have more inventory than they really need.

A **Class C** company uses MRP primarily as an inventory ordering technique rather than as a scheduling technique. Shop scheduling is still done from the shortage list, and the master schedule in a Class C company is typically overstated. It hasn't really closed the loop. It probably will get some benefits in inventory reduction as a result of MRP.

A **Class D** company only has MRP really working in the data processing department. Typically, its inventory records are poor. If it has a defined master schedule, it's usually grossly overstated and mismanaged, and few or no results have come from installing the MRP system. Ironically, except for education costs, a Class D company will have spent almost as much (about 80 percent) as a Class A company, but not achieved the results.

Wight's description of MRP II is totally compatible with the overall MPC system in Figure 10.1. The intent is to clearly delineate the integral relationship between MPC systems, marketing systems, and financial systems. Because this view of MRP is so much broader than planning material requirements in the MPC system's engine, it's been called **manufacturing resource planning** instead of material requirements planning. The classification scheme has been useful as an implementation guide and has been applied to some research efforts.

The Company Environment

In initiating the implementation of MPC systems, one must establish priorities reflecting the company environment. It is not desirable (and sometimes not possible) to do all things at once. To establish priorities, we consider such issues as company objectives, ability to get a quick payback, and need to redress current overall operating problems. For example, in companies whose priority is the production of highly engineered products to customer order against tight delivery dates, the key issue may be master production scheduling. For an environment with lengthy engineering design requirements, the priority may be to tie engineering design efforts to the MPS. This may be more critical to promising realistic customer delivery dates than efforts to improve other aspects of the MPC system.

Likewise, shop-floor systems may be the first priority for implementation in firms having difficulties keeping track of orders in the shop. Many firms have symptoms of increasing work-in-process, missed customer delivery dates, and incomplete material for final assembly. In these environments, finding out what is on the shop floor is an absolute prerequisite to other MPC implementation activities. Indeed, in some of these situations, the payback is large and rapid.

A different sort of environmental consideration is the organization climate and receptiveness to implementation of various systems. We've seen firms where achieving accurate shop-floor reporting isn't difficult, and others where such conditions as incentive wage systems virtually preclude accurately reporting actual production or shop-floor conditions on a timely basis.

At a more fundamental level, a key environmental issue is how much management identifies with improved systems for materials flow. We list management commitment as a key implementation success variable. Has management recognized the need? Does it have the appropriate *total* level of commitment?

All these issues, from implementation priorities to management attitudes, are basic elements of the assessment process. They must be honestly addressed if a firm is to develop appropriate implementation priorities and resources.

Justification

There are several reasons for justifying MPC implementation beyond the obvious one of making sure that it's financially warranted. Justification is an integral part of the MPC project planning process. It's necessary to identify everything required to make the project succeed—both for costing purposes and for planning for the necessary resources. Performing both tasks concurrently improves implementation quality—even in cases where the project obviously is economically justified. Ineffective implementation for an obviously attractive MPC application is an all too common occurrence. As a consequence, all affected people—from top management to the factory work force—will view additional projects with skepticism.

The Magnitude of the Payoffs Can Be Huge. In many cases, anticipated MPC system benefits exceed those of any other project competing for company funds. The resultant attention can be a blessing and a curse. It puts pressure on the MPC professional to rapidly implement the project and achieve the attendant benefits. It can also create healthy skepticism that forces a complete review of the justification, the MPC system itself, and the required operational changes. In any event, thorough justification is essential. To achieve the required thoroughness, it will be necessary to actively involve people in accounting, sales, marketing, and other affected departments. They are needed initially for gathering input data and subsequently as allies when the implementation process reaches their departments.

Justification and delineation of benefits from implementing an MPC project depend on which module is involved. Any of the modules included in the front end of Figure 10.1 will have direct impacts on customer service and overall company planning processes. All front-end modules are concerned, in a broad sense, with matching future supply to demand. As these are improved, better customer service can result and overall operations can be better managed. Capacity can be more closely matched to need. This reduces overtime, subcontracting, expediting, premium transportation, and other reactionary behavior.

Implementing better systems in the engine part of MPC can greatly reduce the chaos so often characterizing the shop. Justification here must be based on savings derived from more accurate planning of material and capacity. Expected benefits include greatly reduced overtime, inventories, expediting, and other costs of chaos. They also should explicitly include provision of time for planned maintenance, workers' meetings for continuous improvement, training, and other long-term–oriented activities.

Justification where the focus is on major improvements in back-end modules can also focus on significant improvements in customer service through appropriate priority setting and follow-up. In addition, better short-term assignment of people, machines, and material to specific orders can greatly improve the shop's overall efficiency. Effective utilization of outside suppliers can result in improved relations, lower costs, better delivery performance, and reduced transaction costs.

Clearly there are direct and indirect benefits from improvements in each area of the MPC system. One of the most visible savings from an improved MPC system is inventory reduction. Reductions can be in finished goods because of the closer match between supply and demand. They can also be manifested in work-in-process because of better scheduling and work completed closer to the customer need—or even "just in time." Raw-material inventory can be reduced with better timing for vendor deliveries and intercompany coordination.

Inventory reduction is an important benefit in its own right, but often secondary benefits can be even more significant—albeit less visible. Reduced inventory results in less damage to parts in storage or on the shop floor, makes things easier to find, lowers space needs, requires fewer transactions, generates fewer records, and needs fewer people to move it around. Reduced inventory can also greatly decrease obsolescence costs.

Other benefits of reduced inventories are even more indirect, harder to quantify—and more significant. Reduced inventories can allow postponing plant or warehouse expansion. They can also make it much less expensive and easier to introduce an engineering change that can improve the product relative to the competition.

In the customer service area, there are again fewer direct benefits to include in new MPC system justification. Improved matching of supply to demand can greatly increase available capacity for additional products, product enhancements, or new customers. Improved relations with marketing and sales provide major indirect benefits: they're better able to do their jobs! Opportunity for increased market share or new markets shouldn't be underestimated. In some companies, MPC improvements have been instrumental in holding on to market share in the light of stiff international competition.

Improvement in effective use of equipment and human resources also has important indirect benefits. Saved labor-hours can be used for training and improvement work. Effective use of equipment can mean lower total maintenance costs and higher quality. More effective use of labor may help a firm attract and retain a better grade of worker. Better use of equipment can result in higher effective capacity (higher-quality parts and improved performance) and longer life.

Benefits can be assessed in a multitude of ways. We feel the commitment to some concrete statement of benefit measured on an agreed-upon basis, rather than precisely how the benefits are measured, is the key. One possibility is pro forma accounting statements. These can be generally based on certain operating ratios and improvements in those ratios, and thereafter matched with actual results. The problem with the pro forma approach, however, is that bottom-line results are a function of many causes. It may be hard to isolate the MPC system's exact contribution.

It may be necessary to assess the benefits through improvements in other surrogate measures. Possibilities include:

- Shipping budget performance.
- Labor utilization rates.
- Productivity measures.
- Expediting budget.
- Obsolete inventory write-off.
- Cycle count accuracy.
- Overtime hours.
- Purchased component costs.
- Vendor delivery performance.
- Premium shipment costs.
- Customer delivery promise performance.
- Spare-part service levels.
- Obsolete inventory reduction.
- Raw-material inventory as a percentage of sales.
- Work-in-process inventory as a percentage of sales.
- Finished-goods inventory as a percentage of sales.
- Safety-stock inventory as a percentage of sales.
- Inventory turnover.
- Annual accounting inventory adjustment.
- Hidden factory and other overhead costs.
- Lead time reduction.

Many indirect benefits can also be stated, but problems of concrete measurement are more complex. The subjective benefits that firms may wish to achieve include improvements in:

- Customer relations.
- Competitive position.
- Professionalism.
- Morale and *esprit de corps.*
- Coordination between finance, marketing, and production.
- Accounting control.
- Product quality.

Figure 10.2 indicates kinds of indirect benefits a variety of firms have realized.

FIGURE 10.2 Example Benefits from MPC System Implementation

- A North Carolina IBM plant reduced the raw-material and work-in-process inventory so much it turned planned warehouse space into additional employee parking, avoiding the need to buy more land for a parking lot.
- Hill-Rom, after implementing its MPC system, was able to evaluate detailed implications of a large unexpected, profitable order for hospital beds from the Mideast, and thereafter to deliver the order with minimal disruption to its operations.
- After implementing a JIT program, Square D converted a warehouse into an employee gym and exercise area. It expects the health program to improve workers' productivity.
- Hewlett-Packard has freed time for its labor force to make substantial inputs to both product and process design in many of its plants. In some instances this time savings has been achieved while sales were growing.
- Sunwind AB (a small supplier of parts to Volvo and Saab) became a JIT supplier to its customers, thereby expanding its business, after implementing JIT concepts inside its facility.
- By carefully matching demand to supply from a number of plants, Sara Lee Knitwear met a significant increase in demand with no increase in plant space.

The Costs of Implementing MPC Systems Vary Greatly.   They can run from a few thousand dollars to millions. Again there are direct or visible costs plus those much harder to see. Too many managers ask "How much is it?" referring simply to software a vendor is selling. Unfortunately, software (or even software plus hardware) is rarely the implementation's most costly element and most certainly isn't the most important. In fact, these costs are decreasing dramatically and will probably continue to do so.

The parts of the business requiring the greatest investment of time and money are those concerned with preparing people for the new era. This means training people who'll use the system, those who'll provide data to the system, and those who'll change their jobs because of the system. Other key elements of cost involve getting the data base correct and keeping it that way, rearranging facilities, and providing new tools and services to support the system and people.

As with the benefits, it's important to identify as many of the costs as possible, both for justification and control. Figure 10.3 shows several cost categories. An additional opportunity cost is the managerial time and effort associated with giving MPC system implementation a high priority in the company. But the biggest opportunity cost of all can be the cost of *not* investing in improved MPC systems. This could even mean going out of business! Competitors can take a firm's markets by offering better service, lower costs, and more satisfactory products. And these competitors now can come from any corner of the globe.

FIGURE 10.3   Categories of Costs in MPC System Implementation

---

- Training:
  - Training trainers and system users.
  - Exposure of top and middle management.
  - Connecting activities (order entry, purchasing, etc.).
  - Customers and vendors (and other functions).
- Cleaning up the data base:
  - Auditing the current status.
  - Developing the correct data.
  - Designing systems to keep it right.
  - Creating new forms, procedures, and incentives.
- Personnel expenses:
  - Full-time project team members.
  - Part-time project team members.
  - Systems and staff people who work on the project.
  - Outside consultants, trainers, programmers, etc.
- Support for the people:
  - New tools (terminals, kanbans, problem-solving aids).
  - Redefined jobs.
  - Revised pay schemes.
  - Changed incentives.
- Relayout of facilities:
  - Factory floor changes.
  - New inventory locations and containers.
  - Changes or elimination of office space.
  - Installation of support tools for people.
- Software:
  - Package purchase cost plus installation support.
  - Customization of output documents and screens.
  - Enhancement and implementation of vendor updates.
  - Maintenance and evaluation of new alternatives.
- Hardware:
  - Basic expansion or acquisition of computers.
  - Networking equipment.
  - Special devices (EDI, scanners, etc.).
  - Provision for growth and staged implementation.

Justification Should Include Cost–Benefit Analysis.   Far too many large system projects proceed without sufficient analysis of costs and benefits, or have their benefits computed on an after-the-fact basis. An MPC system effort represents a substantial investment for the firm; the fact that data are hard to estimate precisely shouldn't preclude rigorous analysis.

For every project, costs and benefits will be a bit unique depending on the project's scope and conditions prevailing at the time. However, to the extent the effort's critical costs and benefits can be included as a set of time-phased expectations, the steering committee can review project performance more effectively. Discounted cash flows or payback periods can be calculated, and monitoring expenditures and benefits can be made part of a periodic review cycle. Another issue is a tendency in many cases to underestimate actual expenses. To the extent major expenditures are specified up front, there will be fewer unpleasant surprises. It well may be easier to sell an entire package at one time than to continually "return to the well."

As actual work progresses, any evolution in anticipated costs and benefits should be directly discussed in progress meetings. A finalized time-phased set of expected costs and benefits should be put forth at the time of major conversion, and continuing audits should be carried out thereafter to measure actual results versus plans.

A major management responsibility is to truly control the costs. To put a team together with a year to complete the first requirement and then let it drift along will ensure cost overruns. It also will allow enthusiasm to wane, which reduces chances of achieving long-run benefits. The greatest early leverage is to hold persons responsible to hit the design schedule and in that way control timings and costs.

The timing issue may be even more critical than the cost issue. MPC systems' benefits are usually so substantial that cost overruns are easily covered. *But* if timings are allowed to slip, people resources will be used longer than originally planned. This leads to the need to reassign priorities to project work in the firm, skepticism on the part of management, and cost increases. The JIT approach would suggest getting on with the implementation, hitting the schedule and moving on to the next higher-priority improvement item.

An Evolutionary Point of View

In many instances, assessing the company environment will lead to pessimistic conclusions. What can a firm do in this case? Is the only resort to roll over and play dead? The issue always comes up at the educational seminars we hold. One participant summed up his feelings: "This is all well and good, and I really see how these systems could do marvelous things for my company. But next week I'll return to my job and within hours I will have out my gun and knife, shooting and stabbing as usual."

Our response is that the only way to eat an elephant is one bite at a time, and it doesn't make a great deal of difference if you start at the tail or the trunk. The key is to critically assess where your company is and where it wants to be! Who are you, and what authority and resources do you control? What confederations can you make that will yield results? Where can you take a bite of the elephant that will yield clear-cut results that can be used to widen the commitment? What parts of the elephant should be avoided now because success will be too long in coming?

What all of this means is that, although our general model for MPC systems applies to every firm, and the ultimate goal is effective systems in all modules, the emphasis at any point in time must be assessed on an individual company basis. George Bevis's remark, "MRP is a journey, not a destination," is very apt. There are many ways to drive from New York to California, but the key is always to safely reach some intermediate destination. The following description of an MRP journey illustrates these points.

The Ethan Allen Implementation. At a very early stage, the Ethan Allen Furniture Company's newly appointed assistant vice president of manufacturing saw the need to better plan and control production in a widely decentralized set of manufacturing plants. Assessing where the firm was at the time versus where it wanted to be revealed a vast need for change. The product structure data base yields an example of the gap. Not only did the company lack a computerized bill of materials, it didn't even have part drawings! Products were made by launching an end item in terms of rough sizes for wooden pieces, finding a sample piece hanging on a nail somewhere, and making more like it. Tolerances were expressed as "Machine to fit."

There was no uniformity of building techniques among the factories, and each plant had completely different names for parts as well as different ways to make them. Clearly the company needed to construct a product structure data base, but when it was considered at the time, management was convinced this effort couldn't succeed. The effort was too massive for the very limited resources, it would be far too long before results would be seen on the bottom line, and the effort would be the first to be set aside for fire fighting or any mandated cost reduction.

The area selected for implementation first was master production scheduling, and the original effort was confined to only one factory. When this factory could be scheduled in a way that was clearly superior to past practices, word spread throughout the other factories and plant managers were clamoring to use the new system. Once the master production scheduling system was in use, the need for shop-floor systems to move parts in sequence with assembly dictates was clear—these systems were put in with minimal effort. Thereafter, need for data accuracy was keenly felt, so locked storeroom concepts began to be applied. And so it went. Interestingly, this natural evolution eventually came to the product structure data base, the need for standardization of methods and parts, and a consistent cost-accounting system. The journey has taken 15 years and still isn't over.

ESSENTIAL PREREQUISITES

The MPC implementation process comprises a great deal more than installing MPC system modules and training people in the new approaches. Three prerequisites must be met: having a clear match between manufacturing task and MPC system design, developing and maintaining high levels of data base integrity, and creating the discipline to use the new system.

All three prerequisites are required, irrespective of the MPC system. However, concern for data base integrity is higher for MRP or other heavily computer-based approaches than for a JIT-based system. But the *need* still exists; JIT solves the problem by having such high material velocity that data integrity is ensured without the processes necessary in an MRP-based system. Indeed, one JIT objective is to greatly decrease data requirements. Moreover, until the JIT system is well implemented, there's a need for accurate data.

Matching the Manufacturing Task to System Design

It may sound obvious to suggest the system should be appropriate to the task, but too often this isn't the case. Causes for the mismatch fall into three categories: imprecise understanding of the manufacturing task, excessive focus on costs, and dynamics of the field and/or company. Usually the implementation team must address each of these, even if they were well considered in the original system design. Some concerns can only be understood at the detailed level after other issues have been resolved.

The Manufacturing Task Must Be Clear to All. An unambiguous manufacturing mission is essential to make the correct MPC system choices. Determination of the mission is a strategic issue for the business. The market segments targeted by the firm, the customer requirements, the competitive strategy, the characteristics of the manufacturing process, and available manufacturing resources all enter into this decision. The output should be a clear mission statement for manufacturing, indicating the manufacturing capabilities required to support the firm's marketing strategy. The mission statement indicates the manufacturing priorities needed to meet the firm's business objectives. There should be no ambiguity about the manufacturing task—debate, certainly, but not uncertainty about the debate's results.

An articulate statement of the manufacturing task is necessary to answer critical MPC design and implementation questions. What master scheduling approaches are needed, and for what products? Is full-blown MRP necessary for all parts? Do we need capacity detail for promising delivery dates? The manufacturing task will also guide the choice of where to begin the implementation process as well as how to monitor implementation priorities.

Cost Minimization Should Not Be the Objective. Rather, the objective should be to maximize the company's net benefits. This may well require larger rather than smaller expenditures. A common trap is to attempt to minimize software and hardware costs. Buying the lowest-cost software can produce several negative results. The most common is a company discovering too late the software isn't fully debugged and the firm is, in fact, the pilot implementation. Our advice on software selection has always been to select a package with a solid user base and competent local support. A related concept involves customizing software. A short-term view of "costs" often leads the company to conclude a packaged software product doesn't fit the way it does its business, so the package needs to be customized. Avoid this option if at all possible. Change company operations to match the software, not the reverse! This isn't an argument against customized output reports or screens, but avoid modifying the basic code for several reasons. The most important is, as the software vendor brings out improved versions, it will be very difficult to adopt them; the result is increasingly obsolete software. Moreover, knowledge of how it works is often in the head of only one person, who might leave the company. Finally, in today's computer world it's increasingly possible to operate a manufacturing company with a minimal staff of computer experts using all packaged software. A single phrase sums it up: "Idle hands write code."

The MPC World and Competitive Environment Are Not Static.   We must consider an evolutionary point of view as a basic prerequisite to MPC implementation and subsequent evolution. A firm changing from one manufacturing mission to another has to consider both the implications for MPC system design and the extent to which the existing MPC supports or inhibits the change. In a similar vein, we need to continually consider the latest changes in technology and their potential payoffs if incorporated into their MPC systems. Expert systems, local use of personal computers, EDI (electronic data interchange), and so on should be considered where appropriate. Just as we wouldn't buy a new machine just because someone else bought it, changes to a particular MPC system configuration need to be carefully evaluated in terms of the firm's individual requirements.

MPC Data Base Integrity

The bottom-line goal of MPC implementation is to change system users' day-to-day jobs. An essential aspect is for users to take the management actions indicated by the formal system and to ensure all actions are taken to update the data base; management *by* the data base, and management *of* the data base. This mandate is essential if the MPC system is to be "transparent" to all users. That is, everyone has to believe others will do their jobs as dictated by the system design, without costly cross checking and "control" procedures. The mandate is independent of the MPC system's basis—MRP, JIT, or whatever. The principle for user response is clear. When the MPC system indicates some action is required, it's the sole signal for taking action—there are no informal management dictates overriding these signals. The implication for data integrity is equally clear: information from the formal system must be correct, regardless of how it's provided—computer or kanban. Details of ensuring data integrity are greater for MRP-based systems, but no less important in JIT approaches. Several principles underpin the process for providing data integrity. Although we present them couched largely in terminology and examples of MRP, they clearly apply for each system module in any MPC system implementation.

The Data Base Must Reflect Reality. The objective here is to always have an exact match between what's physically true and the data base representation of the physical reality, namely the data entity. This means if the inventory balance in a part record indicates 27 pieces of the part are on hand, then a physical count of that part must yield 27. It also means if the location file indicates the 27 pieces are in location N–7, then that's where the 27 pieces must be found, and nowhere else. If a bill of material says three pieces are required for some assembly, there must be three pieces required. If a scheduled receipt indicates 35 pieces are being worked on in the shop, we can similarly verify that count through physical audit. If someone scraps one or more of these pieces, the system (data base) needs to be informed so the real number of pieces is always reflected. Similarly, location of the scheduled receipt, work completed, work remaining, due date, and so on all have to be reflected accurately in the system.

A strong principle that comes from experience is any action, no matter how worthy it may seem, coming at the expense of data integrity must be avoided. Costs of ignoring this principle are high indeed, including complete collapse of the system. It's more important to strive for an exact match between data entity and physical reality than to try to determine what's "good enough." Bankers spend dollars to count pennies in order to realize this concept.

Data Base Transactions Must Be Processed Rapidly. For the system to constantly mirror reality, we must process changes rapidly as well as accurately. If a scheduled receipt is received and put into stock, the open shop order must be closed out and the on-hand balance increased quickly. We need similar speed for processing customer orders, shipments, orders placed with vendors, receipts from vendors, and any adjustments necessary because of scrap or inventory losses. State-of-the-art systems now process transactions on-line as they occur. Most batch processing systems reflect transactions within 24 hours. The longer the time lag between actual physical change and concomitant change in the data base, the less the data base reflects reality—thereby creating a need for informal systems to find the truth. For example, in a large service center for distribution of structural steel, this principle has been violated. The computer batches and processes inventory transactions on a time availability basis. As a consequence, inventory clerks maintain a separate card file to know what's really going on. Their need to know on a daily basis leads them to maintain an informal system—which in turn makes the formal system less useful! Meanwhile, the company incurs the costs of both systems.

Data Base Maintenance Must Be Tightly Controlled. We must carefully maintain the critical distinction between *access* to data (which can be available to many users) and authority to *change* a given data element. If the data base is truly to be managed, then specific individuals must have sole authority to make changes in specific data elements and, similarly, be held responsible for those data's accuracy. This implies data accuracy must be a specific part of many job descriptions, and organizational changes must be made to ensure proper data maintenance. A good case in point is the bill of materials file, which the engineering department traditionally managed. Principles of data base accuracy and rapid response to change clearly apply to changes to bills of material. This means bill of material changes can't be constrained by antiquated procedures, and nonengineering uses of the bill of materials must be accommodated. In some companies, these dictates have resulted in maintenance activity for the bill of materials data being separated from the engineering department and assigned to a materials management group.

All data files required to support an MPC system must be tightly controlled. This means organizational assignment of authority to make data base changes. Other areas where reorganization is often dictated by data base maintenance considerations include functions involving customer order entry, delivery date promising, receipt of purchased parts, and quality assurance reporting.

User Actions Must Be Integrated with Data Base Transactions.   To achieve data base integrity and to maintain this integrity at high levels, it helps to achieve a high degree of congruence between the job actually performed by users and the data those users provide to the system. If data collection is simply an added burden to your job, chances for mistakes are much greater than if job performance itself relies upon accurate data. For example, if you can demonstrate to a first-line supervisor his or her job is easier if he or she enters data correctly, higher quality data input should result. To the extent we integrate transaction reporting on individual orders (scheduled receipts) with payroll, variance accounting, movements from work-in-process to finished goods, and the like, we increase pressures for data accuracy.

The System Must Tell the Truth to the Users.   This principle relates to the data base's need to reflect reality—but there's an added dimension. We must design the system, its parameters, and its data base so unnecessary cushions, personal hedges, and inconsistencies are sorted out. If the formal system is to be used to make day-to-day decisions, output data must be believable. If it is, the system can substitute information for the physical hedges so often used for buffering in informal systems. A case in point relates to due dates assigned to shop orders launched into the factory. One company used three stamps for its orders: RUSH, CRITICAL, and EMERGENCY! Clearly, here's a case where RUSH stands for *R*outine *U*sual *S*low *H*andling, since it's the lowest-order priority. The point to all this is simply, if everything is marked rush, then nothing is rush. We must establish *relative* priorities for shop orders and purchase orders. This is best done by telling the truth about due dates, lead times, and other system parameters. If everyone has to second-guess other people's hedges, the resultant mismatch between the system and reality will be great.

Discipline to Use the Formal System

The changed nature of jobs brought about by implementing an MPC system requires a new discipline in many companies. One aspect of this is eliminating the informal system and old habits. A second is developing the attitudes and understanding necessary to keep them out.

The Informal System Must Die.   Many companies have a formal system for production planning and inventory control and an informal system that gets products shipped *in spite of* the formal system. These informal systems are usually made up of "hot lists" indicating what's *really* needed to meet shipments, physical staging of the materials to make *sure* they're available, black books for how to *really* make the products, telephone calls, visits over lunch, and so on. As long as there's any vestige of an informal system, the incentive to use the formal system is reduced. The only way to kill off the informal system is to design the formal system so well that users always get better information from it than from the informal system. It's also necessary to continually educate users so they understand why the formal system produces better results and why their efforts are better devoted to solving problems *with* the system, than to inventing ways of going around it.

New Attitudes Must Be Created.   Again, this task falls to both implementation team and managers who must subsequently live with the system. The discipline to do what's required by the system and to keep the system honest is a prerequisite to any successful MPC system implementation. The JIT experience has taught us that knowing how to use the system isn't enough. We must also know it well enough to resist old behavioral patterns such as keeping workers busy by building inventories. This goes beyond training in the "how" of the system to the "why." In the long run, the "why" associated attitudes make the difference.

ORGANIZATIONAL ISSUES

In designing the implementation plan, we must consider several organizational issues: top management's role in the project, the project's focus on waste or "slack" reduction, and the job design changes needed to reduce them. Again, the basic issues are the same regardless of the type of MPC system being implemented, though specific actions will differ.

Top-Management Commitment

Key to successful implementation is top management's deep and lasting commitment. A survey of MRP companies by Anderson and Schroeder and others found a strong correlation between the Wight classification scheme and the level of top-management support. All authorities say this commitment is critical to success, and empirical evidence backs up this claim.

Another interesting finding of this survey was a strong correlation between success and level of support provided by the marketing group. Companies whose marketing departments actively participate during MPC system design and implementation seem more likely to succeed than those whose marketing personnel play a weak or nonexistent role.

Top-management commitment means a great deal more than a chief executive giving his or her blessing to the MPC systems. The key to commitment isn't even in providing the necessary funding for the effort. It's first and foremost in recognizing the MPC effort will require the sole use of some of the best people in the organization for a significant period of time. These people must be identified, they must be freed from present responsibilities (hire replacements if necessary), they must be molded into an effective team, and they must have the authority and responsibility to do the job.

Top-management commitment also relates to understanding how implementation will affect the entire company. Top management should provide leadership for the change, rather than playing a passive role. This kind of leadership means that one companywide system is the goal; that this system will be used for budgeting and for strategic planning; that marketing decisions such as customer order promising will be made within the system; that key trade-offs are made in the MPS; that the accounting and finance systems will at some point integrate with the companywide data base; that engineering will support the effort to whatever degree necessary; that manufacturing's job will be to hit the schedule; and that an ongoing companywide education program, beginning with top management, will be put in place. George Bevis has said the key to making the Tennant Company a Class A system user was when he clearly understood he personally had to make an active commitment to the MPC systems. At the time, he was executive vice president for the company. He had to change how he thought about manufacturing and carry this change through the organization.

Waste or "Slack" Reduction

The firm with poor systems has extra inventories, extra lead times, extra capacity, extra personnel, excessive overtime, and other kinds of waste built into its operations. These elements of waste or slack allow various organizational units to operate somewhat independently of each other. JIT has waste reduction as a primary goal, with both the initial thrust and the future focus, through continual improvement. With effective MPC operations in place, better scheduling and other kinds of improved communication among organizational subunits compensate for the reduced slack.

Waste or slack reduction requires new forms of dialogue supported either by formal organization changes or by other mechanisms, such as regularly scheduled meetings for processing engineering changes. We take such meetings very seriously, since they're usually part of a processing cycle created to eliminate slack. If one member doesn't show up, some "default" rule must be applied, and the no-show will be just as responsible as if he or she had attended.

One communication level often changed significantly is between first-line management and the computer department. Things go wrong, and foremen occasionally get reports with "mysterious" data. Timely resolution of these problems is critical to MPC system health. In other cases, particularly under JIT, the connection with computer systems is greatly reduced; the processing cycle for materials is much too fast to be coordinated with detailed data processing. In still other cases, we achieve slack reduction through users doing computing themselves.

Job Design

Successful implementation means some jobs will be significantly changed and some will be removed. Under both JIT and MRP approaches, many jobs will have added dimensions of "knowledge" work, work that demands an understanding of not just how, but also why. It follows, then, the means of evaluating job effectiveness must change.

We must design and evaluate jobs on a basis consistent with system transactions and system performance measures. Accountability needs to be established on a basis congruent with the new job designs. Foremen should be held accountable for meeting schedules. Stockroom personnel should be evaluated in terms of stockroom accuracy measures. Planners should be evaluated for inventory levels and for shortages of manufactured or purchased materials. Purchasing buyers should be evaluated in terms of material cost reductions, service improvements, and vendor performance. Employees in a JIT environment should be evaluated in terms of cross-training capabilities and their abilities to achieve continuous improvement.

An interesting implication of job design changes was observed at a large heavy-equipment manufacturer. The continuing addition of systems to be executed on the shop floor resulted in a fundamental change in the foreman's job. Instead of someone who moved iron and yelled a lot, the foreman became more of an information processor. One result was rethinking the career path leading to a foreman's job. The MPC-system–driven company is fundamentally different from its predecessor.

The organization will have to adapt to the system. Many successful implementers now see that the key is to change the organization to match the system, rather than vice versa. That isn't to say the system isn't tailored to meet a particular environment's needs, but only to say activities at the operational level after sucessful implementation are different. They are as different as a barnstormer's job is from a professional pilot's.

We can achieve a successful MPC system under any form of organization. Many people have asked whether it's necessary to install a materials management organization before implementing MPC systems. A materials management organization typically integrates production planning, purchasing, traffic, distribution, and all physical inventory control activities within one organizational unit. Also included are formal feedbacks associated with transaction error resolution, shop-floor control, and vendor follow-up systems.

The materials management type of formal organization is no guarantee of success and it's not a prerequisite to success. However, due to interactions among organizational subunits and the reduction of slack that formerly allowed them to operate more independently, many companies have adopted a materials management form of organization to support their systems to control materials flow.

Organizational change can be quite significant under JIT. Typically flatter organizations (fewer levels) are implemented. More profound are the changes in management style under JIT. All employees are expected to participate and to commit to continuous improvement. No one is allowed to park his or her brain at the door on the way into the factory, and everyone must ask not only did the job go according to plan—but how can it be done better tomorrow!

THE IMPLEMENTATION PROJECT TEAM

MPC system design is only one part of the initial implementation process. Organizing properly for implementation is also critical to long-run MPC system success. Carefully considering the roles each implementation project team member is to play and the hand-off to operating personnel are illustrations. Before providing an example, we'll discuss project team structure, the primary project team, and team-building activities.

Project Team Structure

We see the need for three kinds or layers of project management teams, but there are overlaps between them during the MPC system's evolution. The most important level of project management is what we call the primary project team. This is a group of about five to eight people coming from different functional areas. Their task is to define the detailed set of subsystem projects, order the priorities, make sure each subsystem project remains congruent with the overall MPC system efforts, and act as a conduit to and from their respective functional areas and user groups.

The primary project team needs to report to a special top-management steering committee. This second group serves as a decision-making group when MPC efforts require coordination that exceeds the primary project team's authority. An example might be coordinating some aspects of marketing and design engineering. The steering committee is also the source of resources for the project teams. We must earmark critical resource persons and allocate funding to the MPC system efforts. The steering committee will be needed to reconcile these needs with competing needs in the company.

The third level of project management is the team assigned to a detailed MPC subsystem, such as designing kanbans, bill of material structuring, engineering change control, manufacturing process changes, or master production scheduling. In the early stage of the MPC system effort, primary project team members often serve as members of detailed project teams. This broadens their understanding of the MPC system and helps them develop realistic expectations of how long subprojects should take, what organizational changes are required, how to define educational needs, and the like.

The Primary Project Team

The five to eight persons on the primary project team should definitely include one representative each from marketing, engineering, production planning and control, line manufacturing, and the computer department. Beyond this organizational representation, the only need is for bright, hard workers, preferably people with some history of getting projects accomplished. Many teams also include a member from finance. This is particularly important when the project's short-term scope is to incorporate cost-accounting systems or critical interfaces with such systems as inventory reporting.

The project team should be a core of people who are largely freed from other responsibilities during the project's course. They should be physically located in some separate area, with minimal contact with their previous jobs and associates. The team should consist of people from within the company. There's a tendency to believe this kind of talent can better be hired, but in reality this isn't usually the case for two reasons. First, the supply of qualified professionals in this field is severely limited; most of the best ones are happily employed. Second, even if talent were available, the organization must make the necessary adaptations to use the new system on its own. It's critical to know what the changes will be, what the major roadblocks are, who'll be supportive, and who'll need to receive special handling. Only insiders have this kind of knowledge.

A further dimension of using insider project teams is the opportunity for individual growth participation in this kind of project offers. The primary project team will evolve into a highly professional group, and materials management based on MPC systems will similarly become more professional. The result should be some very nice jobs for productive members of the project teams. Few chances for rapid advancement in status and compensation in industrial organizations outstrip those from achieving demonstrated results on an MPC project team.

Use of insider project teams doesn't mean the team shouldn't receive outside counsel. On the contrary, a qualified professional consultant can be extremely valuable. This person can help define the project scope, provide education for the project teams, define the education program for the company, and audit results. Another role an outsider can perform well is conscience. It's useful to have an outsider come around and ask whether the project is on target, whether anyone has been removed from the team or has been diverted, and whether directions are being maintained.

One person on the primary project team has to be project leader. It's tempting to believe this person needs to be able to perform miracles while serving coffee to the team. People with those qualifications are hard to find. The project leader's key attribute is a demonstrated track record for management in the company. The person should be decisive, know how to assign task authority and responsibility, provide leadership to the team, and know how to deal with top management.

The primary project team leader must be a user, definitely not someone from the computer department. Ideally, the project leader will view the post as a temporary assignment and intend to return to a position where he or she uses the system.

A source of leadership in the primary project team is from one key member of the top-management steering committee. This member serves as top management's representative and should be more knowledgeable about the MPC project than other steering committee members. A good choice might be the vice president of manufacturing. In some firms, this person goes so far as to assume personal responsibility for the MPC project's success or failure. This level of commitment places MPC system implementation in the same category as any major corporate goal. Building a new factory based on new technology is a useful analog.

Team Building

In addition to changing communication channels and people's roles on the teams, it's sometimes useful to engage in formal team-building activities. Project team members will come from different departments, and often view each other as the enemy; they may have had little experience working together on a *common* project. Indeed, they may even speak different functional languages. Communication may be difficult. The objective is to achieve a level of professionalism where team members have confidence in each other. This occurs when each feels his or her decision on a particular matter is what the group wants as a whole.

You can team build with an outside facilitator or with internal training people with this expertise. The focus is on improved communication skills, including listening skills, as well as developing tools to facilitate team approaches to problem solving. Team trust and self-confidence come with successful problem solving, but some companies use Outward Bound kinds of programs to more quickly establish the team as a cohesive entity.

The Swissair Project Team.   Swissair's engineering and maintenance department is a good example of how detailed MPC subsystem projects can be defined and managed. Maintenance on Swissair aircraft is performed in Zurich. This facility also provides maintenance service for other airlines. Total employment in the maintenance group is about 2,600.

The Swissair integrated maintenance and control system (MCS) is divided into about 50 subsystems or segments, of which about 35 have been implemented. The overall developmental effort for the remaining 15 implemented segments is 260 man-years; costs are estimated at approximately $12.5 million. These costs are more than offset by direct savings, primarily in labor costs, through increased productivity. From a strategic point of view, the Swissair MCS system represents a long-term competitive weapon.

The Swissair MCS system implementation treats each segment as a separate project, with its own project team. Segments are constrained to be no longer than approximately 20 labor-years of effort and two to three years in elapsed time. Figure 10.4 shows the overall organization and management of the MCS project. One key aspect of this organization is the general man-

FIGURE 10.4   Swissair's Project Organization

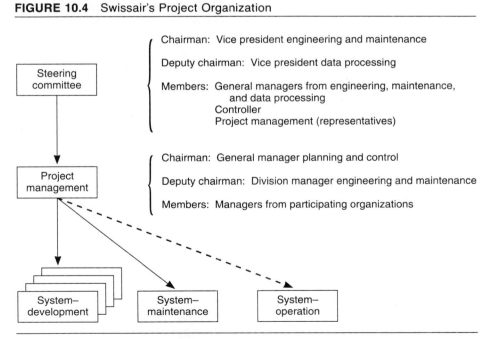

Source: Konrad Wittorf, Swissair, Zurich, Switzerland.

ager—planning and control. This person chairs the project management committee as well as serving on the steering committee as a project management representative. In essence, this manager is the top-management representative for the detailed MCS project planning and control. This person spends at least one day per week reviewing each segment project's exact status.

Figure 10.4 shows the management organization for an individual segment project. Each project encompasses the same set of six phases, which are clearly defined as the rows in the matrix of Figure 10.5. The matrix's columns show four key project team members. Entries in the matrix show which members participate in each phase and who's responsible for project leadership during that phase.

The first important observation is it's a user who'll have ultimate responsibility for implementing the system. Since he or she knows this at the outset, careful attention is given to all other phases, so a usable system will result. The second observation is the EDP (electronic data processing) person is responsible only for computer program development—the only activity involving programmers. According to the project phase's definition, functional specification isn't complete until an unambiguous programming job is defined.

The key responsibilities in Figure 10.5 are the system planner's. This is a person who's basically a user, not a computer expert. It's a career assignment for fast-track career persons. Their major job is defining a segment's exact

---

FIGURE 10.5 Swissair's Segment Teams: Participation in Development Phases

Skill / Phase	User	System specialist — System planner (user)	System specialist — System analyst (EDP)	Programmer
Outline	+	⊕	+	
Functional specialist	+	⊕	+	
Program development			⊕	+
User-procedures	+	⊕		
User-tests	+	⊕	+	
Introduction	⊕	+	+	

+ Participation

◯ Responsible

Source: Konrad Wittorf, Swissair, Zurich, Switzerland.

scope, including expected costs, timings, and benefits. They also have the major responsibility of designing new procedures, testing the system, and making it ready for introduction.

The planners have a long-run career interest in becoming managers in a user department. They don't look to careers in data processing. Neither do they have a long-run interest in being project leaders. As one developmental step in their career, they take on one or two of these projects. If results are favorable, they're eligible for a significant line-management job.

Figure 10.5's final aspect of interest is, in reality, the system planner and EDP system analyst act as a duo. This is to ensure, if either leaves the company during the effort's duration, Swissair won't suffer irreparable damage to the project.

EDUCATION

It's almost impossible to overemphasize education's role in achieving a successful system implementation. One question in the survey by Anderson and Schroeder was: "What is the major problem that your firm has faced in implementing MRP?" Topping the list of replies was "Education of personnel."

Education Levels and Requirements

We see four distinct focuses or levels of education needed for implementation. Figure 10.6 shows these four levels as a pyramid. At the top is the top-management group; this group may be identical to the steering committee. Members need to attend a short course on MPC systems, go to follow-up seminars on a regular basis, read top-management materials dealing with MPC systems, and stay abreast of current company efforts.

The next level of education is for the primary project team and the detailed project teams. The primary project team and leaders of detailed projects should become true professionals, knowledgeable about the state of the art in this field. This isn't an easy job, and it's ongoing. But the size of the MPC investment mandates having this level of knowledge within the firm.

This knowledge can be partially purchased through consultants and obtained from hardware/software vendors, but a firm should be especially wary of letting its knowledge base depend too heavily on outsiders, particularly when those outsiders are trying to sell expensive products and services.

The primary project team needs to attend seminars on a regular basis, be active in professional societies, attend conferences and meetings of the societies, read journals, and interact with other professionals in this area of interest. Leaders of detailed projects need to attend educational seminars especially devoted to their particular projects. The American Production and

---

FIGURE 10.6   Education Requirements

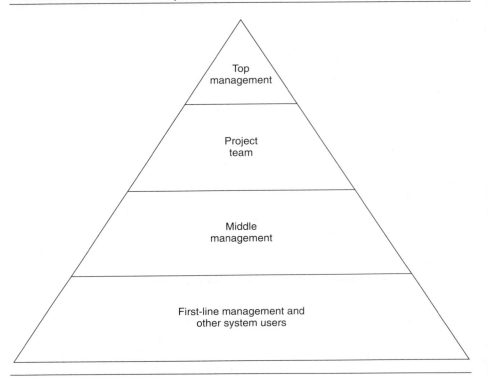

Inventory Control Society (APICS) has a professional certification program based on a series of examinations; achievement of this certification status is a reasonable goal for persons who intend to manage through MPC systems.

The third layer of education in Figure 10.6 is for the company's middle managers, all of whom should receive education. This could take the form of attending two or three seminars of three to four days' duration. This group needs to get outside the company, learn the general approach of these systems, and then see how they've been applied in several other firms. By seeing the before/after conditions through case studies and other means, they can determine the effort's general direction as well as the degree of change necessary to make them effective. They simply need to take off the blinders imposed by existing company practices to see what these systems offer.

Outside education is followed up by inside education in the systems designed for the firm. With an outside perspective, middle managers can now more actively participate in designing the systems that will affect their working lives. This participation goes a long way toward ensuring the system is *our* system instead of *the* system.

The final layer of education in Figure 10.6 is for first-line supervisors and other system users. In the last analysis, implementation will succeed or fail on the factory floor. Foremen and others must execute the system, process transactions accurately, be rewarded for working within the system, and keep the informal system dead and buried.

There has been a tendency in some companies to consider educating foremen, stockroom personnel, production planners, buyers, and other users as a narrow job of training. This focus is only on the system's use as it affects the person's individual area. A much better point of view includes broad-based education about why the system has been adopted, what it does for the firm, and, most important, how decisions made in each area (by specific people) link together. When the foreman sees exactly how unreported scrap causes an incorrect scheduled receipt that can result in a rush order back to his or her own department, he or she will be much more interested in data accuracy than if someone simply says, "We want you to always report scrap." The objective is not only to train people in procedures but to provide them with a new framework and understanding. They need to be able to see why certain procedures are necessary to help them treat some action that hasn't been previously explained.

Some sort of in-house program for this group is usually appropriate. Some firms have success with videotapes; but the best programs we've seen mix overview concepts and specific company examples. This can often best be achieved by the project team designing an education program, perhaps aided by a consultant. The program can also usefully employ middle managers, who learn greatly from the process as the teachers.

An Education Program

The successful firm may find it wise to design the entire education program as part of the project. This can be the best way to address the entire span of educational needs in Figure 10.6. Moreover, it's necessary to include redundancy. We have personal experience with users in a seminar asking questions that were clearly covered in an earlier seminar they had attended. We simply must take a fairly pessimistic view of how much can be absorbed in one session, and realize that people forget quickly. It's necessary to give the message several times, preferably in somewhat different ways.

In addition to the need for redundancy, there's a natural turnover in personnel. People leave or take other assignments, and new personnel must be trained and educated. Since many aspects of executing the new system run counter to people's work experiences, it's important not to expect very much from prior on-the-job training. The new foreman might know how to put some number into a computer terminal because a prior boss showed him or her how, but chances are slim there was a clear explanation of *why*. Often the worst possible training comes from expecting a new person to learn the ropes by watching someone else. The "why" rarely gets conveyed this way, and professionalism suffers. One further goal of a broad-based education program should be to locate personnel who can get enthusiastic about material control systems and can grow into more responsible positions. There's always need for these people, particularly those with first-hand experience.

The final dimension of an educational program we want to mention is the degree of organizational change required to implement major MPC system changes. This process can be the largest single change in people's working lives. Education can represent a much less direct threat to people than other change approaches. If educational programs and exercises clearly show how one way of doing something is better than another, acceptance can be a much easier task and changes can be much more easily made.

Figure 10.7 details an overall MPC education effort: the *initial* education program at Jet Spray Corporation. This program includes *every* company employee, even dishwashers and guards at the doors. It also includes vendors. The intensity of senior managers' education as well as production and inventory control people's education is noteworthy. This firm clearly believes strongly in education.

Problem Solving

An important aspect of the education program is empowering users with problem-solving skills. These skills are in both how to solve problems with the new MPC system and how to continuously improve basic operations. Early development of these capabilities will greatly facilitate the implementation process. Users can help tailor the system to reflect key conditions, prioritize the implementation process, identify valuable changes, modify processes in ways that complement the MPC system, and more rapidly accept ownership of the MPC system as a result.

Use the Tools that Exist for Problem Solving. JIT programs have empowered workers in many places of the world—a clear demonstration of JIT problem-solving tools' value and universality. **Shop-floor engineering** is the term people frequently apply to the tools used. This includes process analysis, industrial engineering techniques, quality circles, Pareto analysis, fishbone diagrams, and statistical analyses. Incorporating use of these tools with implementation of advanced MPC systems permits changes both in the processes themselves and in planning and control systems for those processes.

Attitudes May Need to Be Changed. Too often we hear the phrases "That's not my problem" or "What's that person doing around here?" These attitudes can seriously degrade MPC implementation efforts. The very nature of all

---

FIGURE 10.7   Jet Spray's Educational Program

Initial education—Summarized by hours

Course	Number of people	Hours	Total hours
Teacher/design	14	70	980
Top management	9	41	369
Production & inventory control	3	67	201
Purchasing	3	38	114
Manufacturing managers & supervisors	12	43	516
Manufacturing & industrial engineers	2	40	80
Direct labor employees	135	2	270
Stockroom	20	13	260
Indirect labor employees	12	10	120
Marketing/inside sales	8	32	256
Field sales	13	3	26
Design engineering	6	38	228
Clerical employees	6	13	78
Quality control	3	32	96
Management information systems	4	59	236
Finance & accounting	7	30	210
Vendors	12	7	84
Total			4,124

Initial education — Broken down by number of hours:
Over 60 hours— 17 people
40–59 hours— 27 people
20–39 hours— 27 people
0–19 hours—204 people

Note: In addition to this in-house education, which was largely based on use of video tapes, Jet Spray had 47 attendees at outside courses during the first year of implementation. These (or, if you like, executive overview) courses varied from three to five days covering a range of subjects, including top-management overview, master production scheduling, intensive MPC, purchasing, shop-floor systems, bill of material, and computer systems.

MPC projects is integrative and cross-functional. Not only are new MPC tools necessary, but the will to use them must be created as well. The project team must realize attitudes can be at least as important (if not more) than technical knowledge; the team must not assume good ideas will automatically sell themselves.

Coping with Change

A critical topic in the education program is coping with change and, better yet, leading it. We must see change, not as a one-time process, but as ongoing. Earlier views of MPC (and other) system implementation tended to be as a "conversion" from one monolithic system to another—as opposed to the current adoption of continuous improvement as a basic axiom. The entire organization needs to think of change as continuous; a particular MPC effort might represent a large change, but it shouldn't be viewed as definitive.

Not Only MPC System Elements Are Changing. The approach to most MPC system modules is changing, and the changes will continue. But these aren't the only forces of change MPC professionals must face. Computer and software technologies are on a steep learning/cost curve; increasing computational capability makes feasible more advanced techniques. The applicability of artificial intelligence, expert systems, and other system-enhancing approaches is greatly increased by reduced computational costs. But these technological enhancements may not be the most important ones from the point of view of overall change.

The business itself can be in constant change. Firms are adapting constantly to changing customer requirements. Both customer needs and customers themselves are changing rapidly. Moreover, as manufacturing competence improves, its role in achieving new business opportunities also increases. The ability to accurately specify delivery dates from the MPC system, JIT deliveries, customer partnerships for product design, and guaranteed quality are all examples of manufacturing improvements that can allow the firm to increase its competitive posture.

A greatly neglected change dimension is personnel. New people are hired; others change jobs, retire, quit, or get promoted. New and unfamiliar people become system users. The first need is to train them in their new jobs. The second is to incorporate their new ideas into the continuous learning process. Far too often training new people is left up to the person leaving, with predictable emphasis solely on the job basics. Little provision is made for integrating the new person into the problem-solving process.

Training Needs Extend Far Beyond Implementation. This concern involves more than bringing new users on board. The implementation team must understand the need for redundancy in training plus the need to plan for future education/training as new organizational imperatives are delineated. Moreover, training must extend to the broader set of skills associated with problem-solving tools and attitudes that foster continuous improvement. Without these skills, users can't adapt to the changing world. The abilities to adapt to change and to see the changes coming are vital.

Anticipating change requires educational efforts that are much broader based than those focusing on skill training for one particular job. It requires programs that develop key users' professionalism, both inside the company and outside. These users should be conversant in new approaches to MPC modules, technological improvements, new approaches to manufacturing excellence, and updates in business strategy, as well as the defined manufacturing task. Users also need to recognize symptoms of poor MPC system performance. Increasing queues of work, declining customer service, increased inventories, overtime, and the reappearance of an informal system are all such symptoms.

Reacting to Change Is Not Enough.   World-class performance demands that the firm lead the changes—making *competitors* play "catch-up ball." This expands educational requirements' scope even further. Key users need to communicate with customers, vendors, other professionals in MPC, and major players in the company—those who shape the firm's competitive posture. These activities often aren't even listed in MPC professionals' job descriptions. They require much more than the technical skills needed for system use.

These skills require professional education, usually a combination of formal education and significant in-house education too. Users need job rotations to understand key functional areas as well as business practices and systems in each area. We also need training in how best to communicate with people outside the firm, raise the right questions, sort information, and provide concrete action proposals that fit the company's particular culture.

PROJECT MANAGEMENT

We've already discussed how the MPC system must reflect problem perception and vision at a point in time. The gap between the vision and the current company situation often dictates a project management approach. This requires defining subsystems, resource requirements, sequential constraints, planning mechanisms, and control/feedback of results versus plans.

Defining the Scope

Project scope is defined by the gap between the vision and the assessment of the current set of conditions. The gap's dimensions need to express both system modules and organizational difficulties. In most instances, the gap is too large to consider as one problem to be solved by one working group; breaking it into segments or subprojects is critical to success. Every segment needs to be doable in a few months to two years, and each must have defined benefits and a clear relationship to the overall project.

Where to Start

The question of where to start is a bit more complex than the analogy of eating an elephant one bite at a time. Some idea of the task and priorities for improvement help define the starting point. For example, may companies only turn to the MPS after detailed material planning systems are in place. Without the MPS, the result often is sophisticated systems to execute insanity. Moreover, with some degree of stability at the system's front end, resulting execution problems significantly shrink. The other side of the coin is a sophisticated MPS that can't be executed won't be widely acclaimed by the organization. But any improvements in the MPS should help management, even with crude execution systems. Resolving this dilemma hinges on carefully assessing organizational readiness for change and where change is needed. Many times initial payoffs from process simplification and quality improvements can create a much more tractable shop-scheduling situation.

Since an MPC system's first impact is felt in the factory, it follows the first group of users should probably be somewhere in the manufacturing organization (including materials management). At least one key user should belong to the design team. In that way, he or she will be continually assessing the system design in terms of what's possible and what key problems are to be solved.

Project Planning

The overall plan for the project will be company specific, depending upon its circumstances. However, each plan should have visible timings and milestones, and should perform tasks in the correct sequence. Some form of project scheduling, such as PERT or CPM, is appropriate, but a bar chart approach can be used. Figure 10.8 is a partial example. It incorporates most elements necessary for a firm starting from informal systems to achieve the basic engine portion of the MPC system in Figure 10.1.

One critical aspect is to be sure all subprojects have well-defined milestones. If the project takes too long before concrete benefits are achieved, users lose interest and chances for success fall. We personally feel a reasonable approach is to restrict each project to no more than one year in length. If timing slips, and 15 months elapse, chances for success are probably still fairly good; but if two years elapse before some operating unit sees demonstrable results, the whole project can be in trouble.

A final comment on project planning is, once a detailed project plan has been agreed to, try very hard to adhere to the plan, resisting temptation to go off in other directions. Unless the steering committee has been notified and agrees with a revision of the plan, don't permit deviations. Implementation may be a journey instead of a destination, but the journey mustn't become random.

Project Control

Control of a project involves three distinct phases. The first is to ensure that the project team for a particular subproject is ready to initiate certain key actions. The second phase is to recognize and handle specific issues that con-

FIGURE 10.8 Sample Implementation Plan

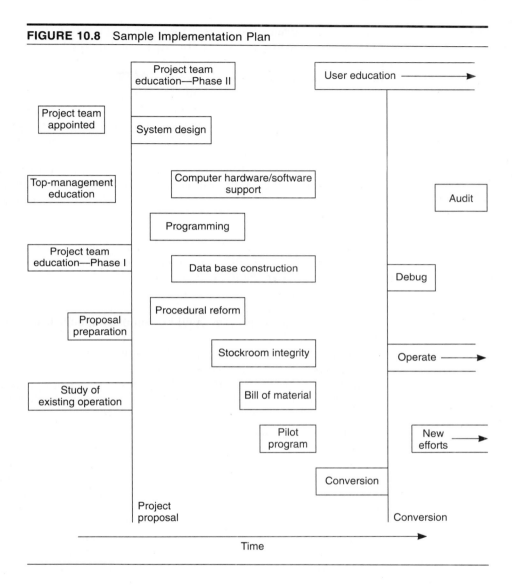

strain the key actions, and to provide mechanisms for resolving detailed operating problems. The final control activity is the overall task of ensuring each subproject is on time and within cost. This last activity can be handled by project management techniques such as PERT and CPM.

A comprehensive example illustrates one approach to accomplishing the first two phases. Compugraphic conducts a "readiness review" before implementing any major system module. The readiness review's stated purpose is to determine if the particular implementation should proceed or if it should be delayed until it's improved in certain areas. The readiness review is a formal meeting where the project team for that part of the system makes a presentation to a project review board (including an outside consultant) of whatever members think is important to implement the particular project, and to identify whatever problems they see. The review board is free to ask any question, and they are not limited to the presentation. The overall objective is to see if the project team feels comfortable with going on to implementation, or if a delay is required because cutting over to the new system will cause some major confusion for the workers or because some part of the system doesn't work properly.

The readiness review typically is scheduled for an entire afternoon. The review team takes the next morning to prepare a report on its findings, which it presents to senior management, with the project team in attendance, the afternoon of the second day. The review team prepares a report based on individual problems it sees.

Figure 10.9 is an example for Compugraphic. The problem involves the new system's ability to correctly pass requirements down from the Haverhill plant to plant 90 (a feeder plant making printed circuit boards). As we see, the review board felt this problem had an unacceptable risk level for implementation. The unacceptable level means the board felt if this problem wasn't solved prior to implementation, it was serious enough to postpone the cutover to the new system.

The readiness review is a great deal of work for the project team, and teams usually resent the effort required. However, after the fact, teams typically feel the effort was worth it. There's a new level of commitment, problems and potential pitfalls are well known, and management has an appreciation of the work done and work left to do.

Another useful implementation activity developed at Compugraphic is a "control center," one central place where any user can call to discuss any problem at any time. The intents are to create a place where anyone in the organization can turn for help and to eliminate all excuses for not reconciling a problem as soon as it occurs. A related reason for the control center is to demonstrate commitment to the MPC project. It's necessary to provide whatever resources are needed to solve problems. Among the resources are specialists in communications, particular techniques and applications, and people who, speaking the language of the shop floor, are able to identify with user frustrations. To the extent that readiness reviews are done well, there should be few surprises in terms of the demands placed on the control center.

When to Quit

The question of when to quit is partly related to the notion of a journey, rather than a trip. The journey has to be periodically redefined and refunded. Most firms have ongoing need for system improvements, reformatting of data, inclusion of different facts on output reports, evolution from paper outputs to

---

FIGURE 10.9 Compugraphic Implementation Review

Topic _____ Haverhill turn-on _____
Date _____ 4/10 _____
Location _____

| | Risk number |
Critical area: _____ Haverhill/90 (4.0 - 8.0) Integration _____   1

Risk level _____ Modest _____ High _____ Unacceptable _____ ✓ _____

Findings – consequences:

   Integration process is well defined. However, the initial test yielded many
   demand mismatches.

   Risk is in the validity and timing of subassembly demands to support
   master schedule, as well as component demands.

Supporting evidence:

   Eight issues identified for corrective action.

Corrective action under way:

   Task list prepared and responsibilities assigned for corrective action.
   Test run and match-off scheduled.

Recommendations:

   If test MRP run is successful (minimal mismatches or none), risk will be
   eliminated. If mismatches are numerous and cannot be easily reconciled
   and corrected, turn-on should not occur.

---

cathode ray tubes, new cross-checks for data integrity, and so on. These continuous improvement efforts are important and need to be encouraged so that
users develop a genuine identity with the systems and feel their needs are
being met. On the other hand, too much attention to such effort can come at
the expense of development in other subsystems. Moreover, problem symptoms will be eliminated later after other system efforts. We need a degree of
balance and an eye to the long term, while not ignoring legitimate short-term
user needs. Users' long-term commitment (obtained by solving their short-
term needs) is important in this balance.

AUDITING

An ongoing view of MPC system implementation requires some sort of periodic audit—both for systems currently being implemented and for those already installed. Top management should make sure work is progressing according to plans and existing systems are doing their intended jobs. In existing JIT-based systems, auditing can directly focus on results and on a visual inspection of the operating areas. There should be so little inventory that exceptions will be very visible. Moreover, detailed auditing of transactions will be largely unnecessary. For MRP systems, it's necessary to perform more detailed checks of the underlying systems and procedures.

Cross-Checking the Data Base

An aid to routine auditing of data base transactions is for the systems to provide the maximum cross-checking, editing, and screening possible. This includes procedures for using check digits and the like and, more importantly, any kind of computerized validation checking possible. For example, suppose employee 123 wants to report completion of 100 pieces of part 456, on shop order 789, operation 3. The cross-checking would include:

- Is employee 123 a legal employee number?
- Is he or she at work today (checked in)?
- Has this employee already checked in on shop order 789?
- Does this employee work in the department associated with operation 3 for part 456?
- Is this employee assigned to the labor grade consistent with this operation?
- Is part 456 a valid part number?
- Is there an open order (scheduled receipt) for part 456 on shop order 789?
- Is this shop order for 100 pieces?
- Has operation 2 been completed?
- Did operation 2 report 100 pieces?

Any of these cross-checks could include tolerances for errors. The objective is to use the computer as much as possible to screen out transaction errors and to focus attention on required remedial actions, such as new counts, scrap not reported, and so on.

Reconciling Data Base Errors

Any cross-check that fails results in an exception condition. In a system with detailed transaction processing, you need a group of people whose entire job is to audit and reconcile exception conditions. For example, is the "in" transaction correct or is the purchase order correct? How can we be out of stock in location N–7 when records indicate 100 pieces are there? An operator reported work on shop order 1234 that isn't valid for the department—was it really shop order 1234? Operation 3 reports 50 pieces and operation 4 reports 48 pieces—is a scrap ticket for 2 pieces needed? Operation 3 reports 50 pieces and operation 4 reports 55 pieces—who's wrong? The assembly department reports none of part 678, yet we show they have enough to last for three weeks—why?

In an on-line system, many of these discrepancies can be monitored as the transaction is being made. For systems based on batch manufacturing, the goal should be to reconcile all exceptions as soon as possible. It's simply not possible to reconcile errors when the trail gets cold. Going to a machine operator and saying "Do you remember job 1234 you ran three weeks ago?" will usually yield "No." Moreover, if someone can enter his or her telephone number instead of the part number and get away with it, the system is a joke. Someone will try this, and the objective is to catch the person quickly.

Figure 10.10 is part of an actual daily exception report for the Hill-Rom Company. It shows 17 parts that generated exception conditions, so they were included on the H/R IN TICKET EXCEPTION REPORT FOR H45. H45 is one of the company's main stockrooms. The first item, part F16774, shows, in the last two columns, 8,375 were reported as finished at the last production operation but only 7,506 have been located or put into stock.

The 4th through 10th items show invalid part numbers. As the handwriting indicates, this problem's resolution is that all seven parts were incorrectly given an F prefix (fabricated part), when in fact they should have an A prefix (assembly). Note also this exception report shows these seven parts' location (Department H45, Bay H41, Row WA, Tier LL) so they can be checked if necessary.

The Just-in-Time (JIT) Approach to Data Accuracy

We have now discussed the need for accuracy in the data base, the importance of transactions, the need for auditing, and a new level of discipline in the system. Remember we need data base accuracy; transactions, auditing, and discipline are *one* means to that end.

Many firms are beginning to use the JIT approach in their factories. JIT's relationship to information integrity is most interesting. There's little inventory in these systems so detailed monitoring of transactions is unnecessary. At the Toyota auto factory, for example, parts deliveries from vendors go out of the door as finished autos within hours. Parts have to arrive or the line stops! Moreover, every worker is also a quality controller, so parts must be of the correct quality. In essence the data integrity check for material receipt is product completion.

Cross-checks as listed previously exist in JIT, but are less complex. A vendor invoice won't be paid unless the products on which the items are used have been received into finished goods. In fact, the system might be designed to pay the vendor as products are received into finished goods—eliminating the need for invoices completely.

FIGURE 10.10 Daily Exception Report (H/R In Ticket Exception Report for H45)

CDE	TRANS. DATE	PART NUMBER NUMBER	OPTN	SSW	LAM	PL/ PNT	UPHL	ORDER NUMBER	FROM DEPT	QTY PROD	PROD. BY	U/M	QTY LOCTD LOCTD DEPT	LOCATION BAY RW TR	SPECIAL INFORMATION	LOCTD BY	TOTAL PROD	TOTAL LOCTD
	04/18	F 16774				PL		JN31318		008375		PCS	0 PARTS PRODUCED ARE NOT LOCATED				008375	007506
	04/19	F 16774				PL		JN35882		006923		PCS	0 PARTS PRODUCED ARE NOT LOCATED				006923	002893
	04/05	F 16773						JN33270		004785		PCS	0 PARTS PRODUCED ARE NOT LOCATED				004785	004122
IT	04/27	F 17127						JN31782	H41	000322	6149	PCS	000322 H45 H41 WA LL INVALID PART NUMBER. NOT ON PART MASTER			941	0	0
IT	04/27	F 17127						JN31782	H41	000228	6845	PCS	000228 H45 H41 WA LL INVALID PART NUMBER. NOT ON PART MASTER			941	0	0
IT	04/27	F 17127						JN31782	H41	000225	6912	PCS	000225 H45 H41 WA LL INVALID PART NUMBER. NOT ON PART MASTER			941	0	0
IT	04/27	F 17127						JN31782	H41	000221	6912	PCS	000221 H45 H41 WA LL INVALID PART NUMBER. NOT ON PART MASTER			941	0	0
IT	04/27	F 17127						JN31782	H41	000282		PCS	000282 H45 H41 WA LL INVALID PART NUMBER. NOT ON PART MASTER			941	0	0
IT	04/27	F 17127						JN31782	H41	000284	6149	PCS	000284 H45 H41 WA LL INVALID PART NUMBER. NOT ON PART MASTER			941	0	0
IT	04/27	F 17132						JN31074	H41	000269	3254	PCS	000269 H45 H41 WA LL INVALID PART NUMBER. NOT ON PART MASTER			941	0	0
	04/27	F 17133						JN29631		008066		PCS	0 PARTS PRODUCED ARE NOT LOCATED				008066	006450
	04/11	F 18124						M981		000518		PCS	0 PARTS PRODUCED ARE NOT LOCATED				000518	000000
	03/12	F 18855				PL		JN32242		004581		PCS	0 PARTS PRODUCED ARE NOT LOCATED				004581	000000
	04/06	F 18855				PL		JN35424		005000		PCS	0 PARTS PRODUCED ARE NOT LOCATED				005000	004581
	03/23	F 18871						HR79272		008005		PCS	0 PARTS PRODUCED ARE NOT LOCATED				008065	007855
	04/18	F 19219				PL		JN35873		004000		PCS	0 PARTS PRODUCED ARE NOT LOCATED				004000	003600
	03/29	F 19237						JN30112		002430		PCS	0 PARTS PRODUCED ARE NOT LOCATED				002430	000000

A Detailed Checklist for the MRP-Based System

Many firms have used the Wight ABCD classification scheme as a standard for judging MPC system performance. Additionally, any MPC system should be compared to the initial design specifications, factors used in justification, and actual costs incurred. Besides these general performance measures, the following questions are useful in evaluating MRP-based systems.

- Are the time buckets used for planning no longer than one week?
- Is the planning (regeneration) at least weekly?
- Is performance against schedule evaluated on a weekly or more frequent basis using concrete measurements?
- Is there a defined measure of customer service?
- Is production held responsible for hitting the schedule and not for finished-goods inventory levels?
- Is any past-due portion of the master production schedule held to less than one week's capacity?
- Are the capacity implications of the production plan and MPS evaluated on a long-term resource need and a rough-cut basis?
- Are the detailed aspects of capacity planning using capacity requirements planning a key part of manufacturing planning?
- Is there an input/output or other system in place to compare machine center results with expectations?
- Does the shop-floor control system provide daily dispatch lists for departmental foremen?
- Are daily dispatch lists the sole source of priority information for the foremen?
- Is an inventory cycle counting procedure in place with periodic measurement and evaluation?
- Are all records in the computer (no handwritten documents)?
- Are accuracy levels maintained for bills of material?
- Is there a well-functioning approach to the control of engineering changes?
- Are shop orders closed out religiously (no tag ends, counts reconciled, and so on)?
- Are other critical procedures to control transactions (such as rejected materials, rework, and the like) adequate?
- Does the system incorporate pegging?
- Does the system have firm planned order capabilities?
- Do planners use bottom-up replanning to solve material availability problems?
- Are state-of-the-art purchasing systems in use, and are vendor capacities planned with the same vigor as internal capacities?
- Is vendor performance routinely measured?
- Is the system integrated with finished-goods planning and control?
- How long has it been since the MPC system was improved?

- How does JIT fit in this firm and is it being pursued?
- Do planners *and* shop-floor people use on-line systems?
- What approaches are being used to reduce lead times?

This formidable list of questions is, however, not all-inclusive. Moreover, some negative responses may be more tolerable than others for a given company. Let's now turn to the approach we've found useful in auditing a company's MPC systems.

The Audit Process

One part of a routine audit is to examine the general level of effectiveness of each of the firm's particular systems; that is, whatever the stated system design, do the systems in fact work? Are they in use? What else is needed for users to do their jobs?

Our approach to this issue is a routine audit involving several steps. First, we take whatever amount of time is necessary to understand the systems in place, how documents are created, what data files are used to process records, and so on. This is not to say, however, that we're auditing the computer program or the computerized data base. Effective use of the computer is a separate technical issue. The MPC audit's intent is to clearly understand the set of output documents, how they are linked, who uses them, and what the users do with them.

It's important to adopt a viewpoint of substantial ignorance at the outset. We ask to be shown actual current output data and examine these documents in considerable detail. We take the position you need to understand what all of the data items are on each report or screen, where they come from, and what arithmetic operations are performed to get them. We examine several actual records to see whether we can understand the numbers, and whether the stated logic can be applied to produce the sample results.

You often find logical inconsistencies in this process, such as negative inventory values, records with all zeros in some field, and the like. These are clues that all is not well. It's also of interest to reconcile documents that presumably should have been produced from a common set of data. An example is shipments for some common time period from records in marketing, finance, and manufacturing.

As a side issue of this analysis, we sometimes find the in-house experts can't explain how documents are created, what they truly contain, who'll use them, and how they're used. Inconsistencies in their beliefs must similarly be explained in sufficient detail.

After gaining an understanding of the system, the audit should turn to those who use the output documents. If a foreman who's asked for the daily dispatch list replies, "My daily what?", you begin to wonder whether systems are really being used. On the other hand, if it's in his or her pocket, with pencil lines and fingerprints, you begin to feel system outputs are used and the system matches reality. You need similar checks in other areas. Is it possible for one of us as an outsider to look important and gain entry into the stockroom? Is the door locked? Who has the keys? What is done on the night shift? Ask a stockroom person to produce the current inventory listing and pick parts at random. Can each of these parts be found in the exact amounts in the exact location? How do stockroom people feel about data accuracy? Ask them how many times parts aren't there. Ask whether engineers get parts for R&D without proper paperwork. Check out the receiving area. Are there boxes of goods there? Ask how long they've been there. Find out how the system is notified of arrivals. How are counts verified? Look at the documents that accompany work-in-process. Do they make sense? Are all fields properly filled out?

Visit the assembly department. Ask whether they're working on the jobs the system has indicated they should be working on. Ask whether there's an end-of-the-month bulge in shipments. Ask how many times they run out of parts. Do they use their terminal as indicated?

Talk to the master scheduler. Ask him or her to show you how the MPS is prepared. Ask about the support obtained from marketing. Ask if they get silly forecast data they have to override.

Audit the level of education used in the company. Who designed the education program? Who gets it? How often? What's the program? Is an overall level of understanding considered important? Does the foreman know how a shop order is created? Is top management knowledgeable about the MPS system, or do they believe education is important—only for their subordinates! Are members of the design team active professionally? Do they attend seminars, workshops, and conferences?

In sum, the routine audit starts with the idea you believe nothing unless proven. The objective is to find out what's supposed to be the case and whether actuality matches the system. In the last analysis, a system is only effective when used.

The MPC system audit is not a witch hunt. Whatever systems are presently in use is simply a matter of fact. It's not the users' fault it isn't any better. Making it better requires action plans, project teams, and resources. The MPC audit attempts to help the company assess where it truly is, where it might be, and how to get there.

Continual Improvement

MPC system implementation includes putting continuous improvement teams in place. These teams implement methods to reduce waste, improve performance, and enhance quality. Often ideas come from observation of the workplace and process knowledge—ideas that are essentially internally oriented. Teams respond to the question "How can we get better?" but not to the question "How good should we be?" The approach to the latter question is externally focused, not internally. There are two components to the second question: "What do the customers want?" and "How can we be better than our competitors?"

Customer-driven manufacturing is a popular phrase. It implies manufacturing will respond to customer needs, which in turn implies systems are in place to ascertain those needs. Linkages between manufacturing and marketing and with customers themselves are needed, although Whybark has shown that these linkages are in short supply. These requirements can stimulate changes in the systems for order entry, MPS, priority setting on the factory floor, and other linkages. New customer services often are the first requirement for defining MPC system improvements.

The second input into defining the MPC improvement process is **benchmarking.** This involves finding out how well other organizations are doing a particular task. The objective should be to find the best practice *in the world.* Benchmark organizations can be competitors, other firms in the industry, or totally different companies. For example, some manufacturing firms look to mail order catalog companies for the best practice in customer order entry. Companies that win internationally recognized trophies, like the Demming or Malcolm Baldrige awards, set benchmark standards for other firms to attain.

The objective here is to find external measures of performance against which to audit current performance levels. Things might be getting better but at too slow a pace—or not in the appropriate direction. Continued internal focus can miss a major shift in competitive capability. Both these concerns involve looking outside for standards of performance. External benchmarks also remove the excuse the goal is impossible—after all, someone is doing it.

CONCLUDING PRINCIPLES

In this chapter we have focused attention of the major problem in MPC systems—how to get implemented results. The following principles summarize the major points:

- MPC system implementation efforts should start with an assessment of where the firm is and what can be done.
- An objective assessment of costs and benefits should be part of any MPC system implementation program.
- The total MPC system implementation effort should be divided into manageable subtasks that each yields concrete benefits.
- Implementation should start with a subsystem that can be finished quickly and that will be of recognized use to some readily identified user group.
- An early goal in an MPC system implementation should be data integrity because of its fundamental importance.
- Management should be prepared to install new organizational forms to facilitate MPC implementation and operation.

- Key users must be on project teams.
- Education programs should be designed for all levels of the organization.
- The design scope for each particular project should be carefully defined and maintained unless broad agreement on redefinition is reached.
- Project definition and project management techniques should be used to manage the implementation.
- Management should audit the implementation process and systems already installed.
- External benchmarks should be used to provide goals and demonstrate that results can be achieved.

REFERENCES

Anderson, J. C., and R. G. Schroeder. "Getting Results from Your MRP System." *Business Horizons,* May/June 1984.

Bevis, G. E. "Closed Loop MRP at the Tennant Company." Report. Minneapolis, Minn.: Tennant Company.

Camp, R. C. *The Search for Industry Best Practices that Lead to Superior Performance.* Quarterly Research, 1989.

Cerveny, Robert P., and Lawrence W. Scott. "A Survey of MRP Implementation." *Production and Inventory Management* 30, no. 3 (3rd quarter 1989), pp. 31–34.

Fisher, Kenneth. "How to Implement MRP Successfully." *Production and Inventory Management 22,* no. 4 (4th quarter 1981), pp. 36–54.

Gaither, N.; G. V. Frazier; and J. C. Wei. "From Job Shops to Manufacturing Cells." *Production and Inventory Management Journal* 31, no. 4 (4th quarter 1990), pp. 33–37.

Goddard, Walter. "Fast Track to Success." *APICS International Conference Proceedings,* 1990, pp. 439–42.

Golhar, Damodar Y.; Carol L. Stamm; and Wayland P. Smith. "JIT Implementation in Small Manufacturing Firms." *Production and Inventory Management* 31, no. 2 (2nd quarter 1990), pp. 44–48.

Hall, R. W., and T. E. Vollmann. "Planning Your Material Requirements." *Harvard Business Review,* September/October 1978.

Inman, R. Anthony, and Satish Mehra. "Potential Union Conflict in JIT Implementation?" *Production and Inventory Management* 30, no. 4 (4th quarter 1989), pp. 19–21.

Lee, S. A., and M. Ebrahimpour. "Just-in-Time Production System: Some Requirements for Implementation." *International Journal of Production Research* 24, no. 4 (1984), pp. 3–15.

Malley, John C., and Ruthann Ray. "Informational and Organizational Impacts of Implementing a JIT System." *Production and Inventory Management* 29, no. 2 (2nd quarter 1988), pp. 66–70.

McManus, Joseph P. "Developing a Detailed MRP Implementation Plan." *Production and Inventory Management* 30, no. 2 (2nd quarter 1989), pp. 75–78.

Meredith, J. R. "The Implementation of Computer-Based Systems." *Journal of Operations Management* 2, no. 1 (1981).

Neeley, Parley S. "Taking the Pulse of MRP: Using Systems Performance Software to Monitor MRP System Usage." *Production and Inventory Management* 30, no. 3 (3rd quarter 1989), pp. 61–65.

Nicholas, John M. "Developing Effective Teams for Systems Design and Implementation." *Production and Inventory Management,* 3rd quarter 1980, pp. 37–47.

Ormsby, Joseph G.; Susan Y. Ormsby; and Carl R. Ruthstrom. "MRPII Implementation: A Case Study." *Production and Inventory Management* 31, no. 4 (4th quarter 1990), pp. 77–82.

Ross, David F. "The Role of Information in Implementing MRPII Systems." *Production and Inventory Management* 30, no. 3 (3rd quarter 1989), pp. 49–52.

Schroeder, R. G. *Materials Requirements Planning: A Study of Implementation and Practice.* Falls Church, Va.: American Production and Inventory Control Society, 1981.

Schroeder, R. G.; J. C. Anderson; S. E. Tupy; and E. M. White. "A Study of MRP Benefits and Costs." *Journal of Operations Management* 2, no. 1 (October 1981).

Sower, V. E., and P. R. Foster. "Implementing and Evaluating Advanced Technologies: A Case Study." *Production and Inventory Management Journal* 31, no. 4 (4th quarter 1990), pp. 44–48.

Voza, J. A.; J. V. Saraph; and D. C. Peterson. "JIT Implementation Practices." *Production and Inventory Management Journal* 31, no. 3 (3rd quarter 1990), pp. 57–59.

Wallace, Thomas F. *MRPII: Making It Happen.* Essex Junction, Vt.: Oliver Wight Ltd., 1985.

Walleigh, R. C. "What's Your Excuse for Not Using MRP?" *Harvard Business Review* (March 1986), pp. 38–50.

White, E. M.; J. C. Anderson; R. G. Schroeder; and S. E. Tupy. "A Study of the MRP Implementation Process." *Journal of Operations Management* 2, no. 3 (May 1982).

Whybark, D. C. "Are Markets Linked to Manufacturing?" Chapel Hill: University of North Carolina Center for Manufacturing Excellence, November 1990.

Wight, Oliver W. "MRPII—Manufacturing Resource Planning." *Modern Materials Handling,* September 1979, pp. 78–94.

————. *MRPII: Unlocking America's Productivity Potential.* Essex Junction, Vt.: Oliver Wight Ltd., 1981.

DISCUSSION QUESTIONS

1. How might the implementation vision differ and the yardsticks change for each of the three parts of the system in Figure 10.1?
2. How would you recognize a "Class D" user or someone in need of improved MPC systems?

3. Where would you put the implementation emphasis for an integrated paper mill, a company that buys cloth and sews fashion garments, or a company manufacturing a new kind of candy bar?
4. How can you liken the implementation process to getting a college degree?
5. Why is there emphasis on data accuracy in implementation?
6. What makes coordinated planning in a company (or any enterprise) so difficult? Give some examples from your experience.
7. What are some of the difficulties in securing top-management support and co-operation for an MPC installation?
8. What is a "user"? Why is it important to have users on the project teams?
9. Who do you think should be the project leader of an implementation? How would you cope with turnover on the project team during a very long implementation period?

PROBLEMS

1. Critique the following implementation strategies:
    a. A coal mining company that ships unit trains of coal to utility companies started its MPC system implementation with engineering change control.
    b. A toy manufacturer started its effort in finished goods and distribution inventory management.
    c. A custom yacht builder started in the finished-goods area.

2. After reading the following ad, would you take the job?

   We need you. The boss says our inventory is too high, our service is too low, and the shop is a mess. He wants to see some changes in six weeks. He bought a new computer and has hired two computer scientists. They are on the implementation committee, chaired by the chief programmer from our MIS department. Will you be the manufacturing representative on the committee?

3. Ivor Morgan was team leader for implementing an MPC system at Rosenthal Inc. He had to decide between two alternatives for the company. The first would require an investment in a small computer ($50,000) and software ($20,000). The software company said it could do the training and provide manuals for a total of $10,000. The system would require maintenance and update training costing about $20,000 per year. Steve Rosenthal said only direct benefits could be used in justification. The direct benefits that could be identified were a one-time inventory reduction of $10,000, payroll savings of $20,000 per year, shop and material savings of $10,000 per year, and purchase savings of $8,000 per year.

   The other alternative was substantially more expensive. The computer would cost $100,000 and software $50,000. Even training would be more, $100,000. Maintenance and training costs would be $40,000 per year to keep the system current. Under this alternative, inventories would be reduced $40,000. Other direct benefits include payroll savings of $40,000 per year, shop and materials savings of $40,000 per year, purchase savings of $25,000 per year, and distribution savings of $15,000 per year.

   Which alternative should he choose?

4. Comment on the following implementation schedule and status:

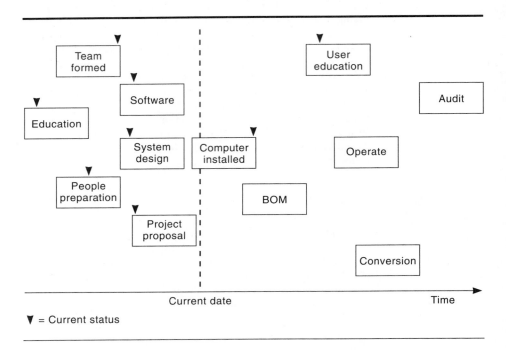

5. Aleda Roth was concerned about a project in her MPC system implementation. Three subprojects were involved: hardware acquisition, software development, and user training. The hardware specification was planned for February, the hardware ordering for mid-March with delivery and testing during July. The software specification was to be worked out during March and April, with programming running May through July. Software testing was to start when the hardware was fully tested and would last one month. Training was to take place three times, for a month each time (with different groups training each period). The first training was to be in January, the second in March, and the third in July.

It's now June 1. Implementation is to be complete by September. Aleda learns the hardware is nearly completely tested, the second month of training is about one half complete, and programming is just beginning.

Develop a bar chart of the implementation plan and indicate the current status. What should Aleda do?

6. Should the following project be approved if the cost of capital is 25 percent?

Benefits		Costs	
Reduced inventory	$1,000,000	Computer	$500,000
Reduced personnel	100,000/year	System development	750,000
Scrap rework	200,000/year	Ongoing support	400,000/year
Purchase savings	400,000/year	Training	200,000
Improved flow time	—	Forms redesign	50,000
Improved quality	—	New people	200,000/year

7. The Sumpump Company evaluated benefits of proposed improvements in its manufacturing planning and control system. The materials manager estimated the direct cost of manufacturing the firm's products ($50/unit) can be reduced by $1/unit through a reduction in organizational slack made possible by the new system. Assume:
   a. The firm's profit before taxes is 2 percent of sales.
   b. A sales volume of 100,000 units is budgeted for next year.
   c. The product selling price is $100/unit.
   What increase in sales volume would provide the same profit increase as the savings resulting from the proposed system?

8. A plant manager at Black Manufacturing estimates that fabrication shop lead times can be reduced from six to three weeks by proposed improvements to the firm's manufacturing planning and control system. He also estimates that, at the present sales volume, a one-week reduction in the manufacturing lead time provides a corresponding $1 million reduction in the work-in-process inventory investment. If the firm's out-of-pocket inventory carrying cost is 15 percent of item value and the average employee earns $25,000, what equivalent reduction in employment does the new system provide?

9. The Ace Widebelt Company is contemplating investing $500,000 in an improved manufacturing planning and control system. The firm estimates the new system would improve final assembly labor efficiency from the current level of 45 percent to a new level of 85 percent. If the firm's annual sales volume is $500 million and final assembly labor cost is currently only 1 percent of total dollar sales volume, would such an investment be worthwhile?

10. The steering committee for an MPC implementation project has identified several measurable costs and benefits of its proposal. The company has asked the committee to perform the financial and what-if analysis using estimates for the next five years. The committee felt this was useful because not all costs or benefits were constant for all five years. For example, inventory savings would be greatest in the second year—building up from $10,000 to $25,000 and then

going down to $15,000, $5,000, and 0. On the other hand, scrap savings would be $15,000 per year. Some costs also varied from year to year. Software costs were high in the first two years ($20,000 and $30,000) and dropped to $5,000, $1,000, and $1,000 in the remaining three years. But wage costs were $10,000 for each of the five years.

The remaining costs were for education ($100,000, $20,000, $10,000, $10,000, and $10,000) and hardware ($5,000, $80,000, $5,000, $2,000, and $2,000). Other benefits were in labor savings ($15,000, $30,000, $75,000, $80,000, and $80,000) and purchase savings, which would end in three years ($35,000, $30,000, and $10,000) because of changes already planned for the product line.

a. In which year would the project break even? (Use a spreadsheet if you're assigned parts b and c of the problem.)

b. Would this change if the purchase savings had been left at $35,000 per year?

c. What if education costs had been underestimated by 20 percent? (Leave purchasing at $35,000 per year.)

11. Jose Zott is trying to cost justify introducing electronic data interchange (EDI) at his company. Inventory of purchased items is now $4.5 million, with an expected reduction to $4.3 million after implementation. There are currently four people in incoming inspection (at wages of $40,000 per year per person), and average lead time is six weeks. Jose feels that with good quality information from the vendor, the inspection staff could be cut in half. In addition, he expects to reduce the clerical staff (annual wages are $30,000 per person) by two people. A lead time reduction of two weeks should provide savings of $50,000 in in-transit inventory. The EDI investment would be $400,000, and annual operating costs would be $40,000. If the company requires a two-year payback on investments, should Jose take the deal?

12. Suppose Jet Spray executives estimate the value of the time of their average employee involved in MPC education at $25 per hour. If the average outside course lasts four days and costs $2,000 including travel expenses, what's the cost of the initial education program described in Figure 10.7?

13. Andrews Manufacturing Company estimates it costs $5,000 per year to administer each of its 1,000 vendors. As Andrews implements just-in-time manufacturing, the company expects to spend $1 million per year for vendor education and training.

a. By how much does Andrews have to reduce the vendor base to pay for the education and training?

b. Suppose the company reduces the vendor base to 300 companies, but continues to spend $1 million per year in developing better vendor relationships, improved vendor quality, and so on. How much do these efforts have to return per vendor to break even?

14. The following simplified income statement covers a Swiss manufacturing firm's numerical control division:

	SFr. (000)*	Percent
Revenue	18,934	100%
Cost of goods sold	7,384	39
Gross profit	11,550	61
Sales expense	2,896	15
Marketing expenses	1,441	8
Inventory expense†	1,136	6
Corporate overhead	1,057	5
Product development	5,963	31
Profit (loss)	(943)	(5)

*Thousands of Swiss francs.
†Carrying and obsolescence costs combined.

Use a spreadsheet program to determine if you'd accept the following project when the firm has a policy of not accepting any project without at least a 20 percent return on investment. The project requires an investment of SFr. 24 million. It would increase overhead expenses by SFr. 200,000 and would increase revenues by SFr. 2 million. The project will reduce cost of goods sold by about 5 percent and will improve inventory turnover from $3\frac{1}{3}$ times per year to 5 times per year.

15. Anne Shepley is in charge of implementing a new MPC module. There's considerable grumbling about the new system module. She wonders if it's just the usual process of adapting to a new system or if something is going wrong. At the end of the first two weeks, the control center has the following log of calls to resolve problems:

Day	Calls
1	95
2	125
3	140
4	130
5	140
6	175
7	160
8	180
9	190
10	180

Weekly data for four other MPC modules that have been implemented are:

	Module (number of calls)			
Week	A	B	C	D
1	400	800	25	150
2	300	900	15	95
3	150	400	5	60
4	50	100	0	20
5	5	50	0	10

What should Anne do?

Chapter 11

Advanced Concepts in Material Requirements Planning

This chapter concerns some advanced issues in material requirements planning (MRP). Some concepts and conventions discussed here can improve well-functioning basic systems. Most concepts are of a "fine-tuning" nature and can provide additional benefits to the company.

We feel the first, most important phase in MRP is to install the system, make it part of an ongoing managerial process, get users trained in the use of MRP, understand the critical linkages with other areas, achieve high levels of data integrity, and link MRP with other modules of the front end, engine, and back end of manufacturing planning and control (MPC) systems. Having achieved this first phase, many firms then turn to the advanced issues discussed in this chapter.

Chapter 11 is organized around five topics:

- Determining manufacturing order quantities: What are the basic trade-offs in lot sizing in the MRP environment, and what techniques are useful?
- Determining purchase order quantities: How should lot sizes be chosen when purchasing discounts are in effect?
- Buffering concepts: What are the types of uncertainties in MRP, and how can we buffer against these uncertainties?
- Nervousness: Why are MRP systems subject to nervousness, and how do firms deal with system nervousness?
- Other advanced MRP concepts: What conventions have been successfully applied? How are they useful?

Chapter 11 is linked with Chapter 2 in that this chapter presupposes understanding of MRP systems, record processing, the MRP data base, and so on. Chapter 11 is also linked to Chapter 17, which treats several inventory concepts (lot sizing, buffering, and service levels). The focus here is on the dependent demand (MRP) environment, whereas Chapter 17, largely deals with systems for independent demand. Additional buffering concepts appear in Chapters 6 and 14 (master production scheduling) and in Chapter 8 (demand management).

DETERMINING MANUFACTURING ORDER QUANTITIES

The MRP system converts the master production schedule into a time-phased schedule for all intermediate assemblies and component parts. Detailed schedules consist of two parts: scheduled receipts (open orders) and planned orders. Each scheduled receipt's quantity and timing (due date) have been determined prior to release to the shop. We determine quantities and timings for planned orders via MRP logic using the inventory position, the gross requirements data, and a specific procedure for making the decisions.

A number of procedures have been developed for MRP systems, ranging from **ordering as required (lot-for-lot),** to simple decision rules, and finally to extensive optimizing procedures. This section describes five such lot-sizing procedures.

The primary consideration in the development of lot-sizing procedures for MRP is the nature of the net requirements data. The demand dependency relationship from the product structures and the time-phased gross requirements mean the net requirements for an item might appear as illustrated in Figure 11.1. First, it's important to note that the requirements do *not* reflect the key independent demand assumption of a constant uniform demand. Second, the requirements are *discrete,* since they're stated on a period-by-period basis (time-phased), rather than as a rate (e.g., an average of so much per

---

FIGURE 11.1   Example Problem: Weekly Net Requirements Schedule

Week number	1	2	3	4	5	6	7	8	9	10	11	12
Requirements	10	10	15	20	70	180	250	270	230	40	0	10

Ordering cost $= C_P =$ \$300 per order.
Inventory carrying cost $= C_H =$ \$2 per unit per week.
Average requirements $= \overline{D} = 92.1$

Source: W. L. Berry, "Lot-Sizing Procedures for Requirements Planning Systems: A Framework for Analysis," *Production and Inventory Management,* 2nd quarter 1972, p. 21.

month or year). Finally, the requirements can be *lumpy;* that is, they can vary substantially from period to period and even have periods with no requirements.

MRP lot-sizing procedures are designed specifically for the discrete demand case. One problem in selecting a procedure is reductions in inventory-related costs can generally be achieved only by using increasingly complex procedures. Such procedures require more computations in making lot-sizing determinations. A second problem concerns local optimization. The lot-sizing procedure used for one part in an MRP system has a direct impact on the gross requirements data passed to its component parts. The use of procedures other than lot-for-lot tends to increase the requirements data's lumpiness farther down in the product structure.

The manufacturing lot-size problem is basically one of converting requirements into a series of replenishment orders. If we consider this problem on a local level—that is, only in terms of the one part and not its components—the problem involves determining how to group time-phased requirements data into a schedule of replenishment orders that minimizes the combined costs of placing manufacturing orders and carrying inventory.

Since MRP systems normally replan on a daily or weekly basis, timing affects the assumptions commonly made in using MRP lot-sizing procedures. These assumptions are as follows. First, since we aggregate component requirements by time period for planning purposes, we assume all requirements for each period must be available at the beginning of the period. Second, we assume all requirements for future periods must be met and can't be back ordered. Third, since the system is operated on a periodic basis, we assume ordering decisions occur at regular time intervals (e.g., daily or weekly). Fourth, we assume the requirements are properly offset for manufacturing lead time. Finally, we assume component requirements are satisfied at a uniform rate during each period. Therefore, we use average inventory level in computing inventory carrying costs.

In the following sections, we'll illustrate the results from applying five different ordering procedures to the example data in Figure 11.1. This example will illustrate how these procedures vary in their assumptions and how much they utilize available data in making lot-sizing decisions.

Economic Order Quantities (EOQ)

Because of its simplicity, people often use the **economic order quantity (EOQ)** formula as a decision rule for placing orders in a requirements-planning system. As the following example shows, however, the EOQ model frequently must be modified in requirements planning system applications. Since we base the EOQ on the assumption of constant uniform demand, the resulting total cost expression won't necessarily be valid for requirements planning applications.

Figure 11.2 shows the results of ordering material in economic lot sizes for the example data. In this example the EOQ formula used average weekly demand of 92.1 units for the entire requirements schedule to compute the economic lot size. Note, too, order quantities are shown when received, and average inventory for each period was used in computing the inventory carrying cost.

This example illustrates several problems with using economic lot sizes. When the requirements aren't equal from period to period, as is often the case in MRP, fixed EOQ lot sizes result in a mismatch between order quantities and requirements values. This can mean excess inventory must be carried forward from week to week. As an example, 41 units are carried over into week 6 when a new order is received.

In addition, we must increase the order quantity in those periods where the requirements exceed the economic lot size plus the amount of inventory carried over into the period. An example occurs in week 7. This modification is clearly preferable to the alternative of placing orders earlier to meet demand in such periods, since this would only increase inventory carrying costs. Likewise, the alternative of placing multiple orders in a given period would needlessly increase the ordering cost.

Finally, use of the average weekly requirements figure in computing economic lot size ignores much of the other information in the requirements schedule. This information concerns magnitude of demand. For instance, there appear to be two levels of component demand in this example. The first covers weeks 1 to 4 and 10 to 12; the second covers weeks 5 to 9. We could compute an economic lot size for each of these time intervals and place orders accordingly. This proposal, however, would be difficult to implement because determining different demand levels requires a very complex decision rule.

FIGURE 11.2 Economic Order Quantity Example

Week number	1	2	3	4	5	6	7	8	9	10	11	12
Requirements	10	10	15	20	70	180	250	270	230	40	0	10
Order quantity	166					166	223	270	230	166		
Beginning inventory	166	156	146	131	111	207	250	270	230	166	126	126
Ending inventory	156	146	131	111	41	27	0	0	0	126	126	116

Ordering cost	$1,800
Inventory carrying cost	3,065
Total cost	$4,865

$$(\text{Economic lot size} = \sqrt{2C_P\overline{D}/C_H} = \sqrt{2(300)(92.1)/2} = 166)$$

Source: W. L. Berry, "Lot-Sizing Procedures for Requirements Planning Systems: A Framework for Analysis," *Production and Inventory Management*, 2nd quarter 1972, p. 22.

Periodic Order Quantities (POQ)

One way to reduce the high inventory carrying cost associated with fixed lot sizes is to use the EOQ formula to compute an economic **time between orders (TBO).** We do this by dividing the EOQ by the mean demand rate. In the preceding example, the economic time interval is approximately two weeks (166/92.1 = 1.8). The procedure then calls for ordering *exactly* the requirements for a two-week interval. This is termed the **periodic order quantity (POQ).** Applying this procedure to the data in our example (Figure 11.1) produces Figure 11.3. The result is the same number of orders as the EOQ produces, but with lot sizes ranging from 20 to 520 units. Consequently, inventory carrying cost has been reduced by 30 percent, thereby improving the total cost of the 12-week requirements schedule by 19 percent in comparison with the preceding EOQ result.

Although the POQ procedure improves inventory cost performance by allowing lot sizes to vary, like the EOQ procedure it too ignores much of the information in the requirements schedule. Replenishment orders are constrained to occur at fixed time intervals, thereby ruling out the possibility of combining orders during periods of light product demand (e.g., during weeks 1 through 4 in the example). If, for example, orders placed in weeks 1 and 3 were combined and a single order were placed in week 1 for 55 units, combined costs can be further reduced by $160, or 4 percent.

Part Period Balancing (PPB)

The **part period balancing (PPB)** procedure uses all the information provided by the requirements schedule. In determining an order's lot size, this procedure tries to equate the total costs of placing orders and carrying inventory. We illustrate this point by considering the alternative lot-size choices avail-

---

FIGURE 11.3   Periodic Order Quantity Example

Week number	1	2	3	4	5	6	7	8	9	10	11	12
Requirements	10	10	15	20	70	180	250	270	230	40	0	10
Order quantity	20		35		250		520		270			10
Beginning inventory	20	10	35	20	250	180	520	270	270	40	0	10
Ending inventory	10	0	20	0	180	0	270	0	40	0	0	0

Ordering cost	$1,800
Inventory carrying cost	2,145
Total cost	$3,945

Source: W. L. Berry, "Lot-Sizing Procedures for Requirements Planning Systems: A Framework for Analysis," *Production and Inventory Management,* 2nd quarter 1972, p. 23.

able at the beginning of week 1. These include placing an order covering the requirements for:

1. Week 1 only.
2. Weeks 1 and 2.
3. Weeks 1, 2, and 3.
4. Weeks 1, 2, 3, and 4.
5. Weeks 1, 2, 3, 4, and 5, etc.

Inventory carrying costs for these five alternatives are shown below. We base these calculations on average inventory per period, hence the 1/2 (average for one week), 3/2 (one week plus the average for the second week), and so on.

1. ($2) · [(1/2) · 10] = $10.
2. ($2) · [(1/2) · 10] + [(3/2) · 10] = $40
3. ($2) · [(1/2) · 10] + [(3/2) · 10] + [(5/2) · 15] = $115.
4. ($2) · [(1/2) · 10] + [(3/2) · 10] + [(5/2) · 15] + [(7/2)· 20] = $255.
5. ($2) · [(1/2) · 10] + [(3/2) · 10] + [(5/2) · 15] + [(7/2) · 20] + [(9/2)· 70] = $885.

In this case, the inventory carrying cost for alternative 4 (ordering 55 units to cover demand for the first four weeks) most nearly approximates the $300 ordering cost; that is, alternative 4 "balances" the cost of carrying inventory with the ordering cost. Therefore, we should place an order at the beginning of the first week and the next ordering decision need not be made until the beginning of week 5.

When we apply this procedure to all the example data, we get the result in Figure 11.4. As seen, total inventory cost falls almost $500—it's 13 percent lower than the cost obtained with the periodic order quantity procedure. The PPB procedure permits both lot size and time between orders to vary. Thus,

---

FIGURE 11.4 Part Period Balancing Example

Week number	1	2	3	4	5	6	7	8	9	10	11	12
Requirements	10	10	15	20	70	180	250	270	230	40	0	10
Order quantity	55				70	180	250	270	270			10
Beginning inventory	55	45	35	20	70	180	250	270	270	40	0	10
Ending inventory	45	35	20	0	0	0	0	0	40	0	0	0

Ordering cost	$2,100
Inventory carrying cost	1,385
Total cost	$3,485

Source: W. L. Berry, "Lot-Sizing Procedures for Requirements Planning Systems: A Framework for Analysis," *Production and Inventory Management,* 2nd quarter 1972, p. 25.

for example, in periods of low requirements, it yields smaller lot sizes and longer time intervals between orders than occur in high demand periods. This results in lower inventory-related costs.

Despite the fact that PPB utilizes all available information, it won't always yield the minimum-cost ordering plan. Although this procedure can produce low-cost plans, it may miss the minimum cost, since it doesn't evaluate all possibilities for ordering material to satisfy demand in each week of the requirements schedule.

McLaren's Order Moment (MOM)

The **McLaren order moment (MOM)** procedure is quite similar to the part period balancing procedure. It evaluates the cost of placing orders for an integral number of future periods (e.g., for period 1 only; periods 1 and 2; periods 1, 2, and 3). However, instead of equating total costs of placing orders and carrying inventory directly in determining order quantities (as the part period balancing procedure does), the order moment procedure uses the part period accumulation principle directly.

A part period is one unit of inventory carried for one period. The total number of part periods accumulated is proportional to total inventory carrying cost. The MOM procedure determines the lot size for individual orders by matching the number of accumulated part periods to the number (target) that would be incurred if an order for an EOQ were placed under conditions of constant demand. We accomplish this by first calculating the target number of part periods and then accumulating the actual period-by-period part periods until the target is reached. We calculate the target value as follows:

$$OMT = \bar{D}\left[\sum_{t=1}^{T^*-1} t + (TBO - T^*)T^*\right] \qquad (11.1)$$

where

$$OMT = \text{Order moment target.}$$
$$\bar{D} = \text{Average requirements per period.}$$
$$TBO = \text{EOQ}/\bar{D}.$$
$$T^* = \text{The largest integer less than (or equal to) the } TBO.$$

The order moment procedure accumulates requirements from consecutive periods into a tentative order until the accumulated part periods reach or exceed OMT in period k, using the following equation:

$$\sum_{t=1}^{k} (k - 1)D_k \geq OMT \qquad (11.2)$$

In the period when the accumulated part periods first reach or exceed the OMT value, we make a second test before determining the lot size for the current order quantity.

The second test shows whether it's worthwhile to include one more period's requirement in the order. The test involves comparing the carrying cost incurred by including the requirement for period k in the current order with the cost of placing a new order for that period's requirements in period k. This comparison is made using the following equation:

$$C_H(k - 1)D_k \leq C_P \tag{11.3}$$

where

C_H = The inventory carrying cost per period.

k = The period currently under consideration.

D_k = The requirements for period k.

C_P = The ordering cost.

When the accumulated part periods exceed OMT and $C_H(k - 1)D_k \leq C_p$, the order quantity covers requirements for periods 1 through k. However, when the accumulated part periods exceed OMT and when $C_H (k - 1)D_k > C_p$, the order quantity covers requirements for periods 1 through $k - 1$.

The example in Figure 11.5 applies the MOM procedure to our example data. Note the accumulated part periods first exceed the OMT target in week 4 when $[(1 \cdot 10) + (2 \cdot 15) + (3 \cdot 20)] > 73.7$. Since the cost of carrying the

---

FIGURE 11.5   McLaren's Order Moment Example

Week number	1	2	3	4	5	6	7	8	9	10	11	12
Requirements	10	10	15	20	70	180	250	270	230	40	0	10
Order quantity	55				70	180	250	270	280			
Beginning inventory	55	45	35	20	70	180	250	270	280	50	10	10
Ending inventory	45	35	20	0	0	0	0	0	50	10	10	0
Part periods	0	10	40	100	180†	250†	270†	230†	0	40	40	70
Exceed target?	No	No	No	Yes	Yes	Yes	Yes	Yes	—	—	—	No, but end of record
$C_H(K - 1)D_k$	—	—	—	120	360	500	540	460				60

$\overline{D} = 92.1$

EOQ = 166.2

TBO = 166.2/92.1 = 1.8

$T^* = 1.0$

OMT = 92.1[0 + (1.8 − 1)(1)] = 73.7

Ordering cost	$1,800
Inventory carrying cost	1,445
Total cost	$3,245

†Measured in the following week.

Source: B. J. McLaren, "A Study of Multiple Level Lot-Sizing Techniques for Material Requirements Planning Systems," Ph.D. dissertation, Purdue University, 1977.

20 units required in week 4 is less than the cost of placing an order in week 4 (120 < 300) in this case, we place the first order in week 1 for 55 units. Then, since the accumulated part periods exceed OMT in weeks 5 through 8, and the cost of carrying the next week's requirement exceeds the ordering cost, we place weekly orders during that time interval.

The total cost of using the order moment procedure in this example is $3,245, 7 percent less than total cost for the part period balancing procedure. In fact, the MOM procedure found the optimal solution in this particular example.

Wagner-Whitin Algorithm

One optimizing procedure for determining the minimum-cost ordering plan for a time-phased requirements schedule is the **Wagner-Whitin (WW)** algorithm. Basically, this procedure evaluates all possible ways of ordering material to meet demand in each week of the requirements schedule, using dynamic programming. We won't attempt to describe the computational aspects of the Wagner-Whitin algorithm in the space available here. Rather, we'll note the difference in performance between this procedure and the part period balancing procedure.

Figure 11.6 shows the results of applying the Wagner-Whitin algorithm to the example. (Note the order quantities are identical to those in Figure 11.5). Total inventory cost is reduced by $240, or 7 percent, compared with the ordering plan produced by the part period balancing procedure in Figure 11.4. The difference between these two plans occurs in the lot size ordered in week 9. The part period balancing procedure didn't consider the combined cost of placing orders in both weeks 9 and 12. By spending an additional $60 to carry 10 units of inventory forward from week 9 to 12, we avoid the $300 ordering cost in week 12. In this case, we can save $240 in total cost. The increased

---

FIGURE 11.6   Wagner-Whitin Example

Week number	1	2	3	4	5	6	7	8	9	10	11	12
Requirements	10	10	15	20	70	180	250	270	230	40	0	10
Order quantity	55				70	180	250	270	280			
Beginning inventory	55	45	35	20	70	180	250	270	280	50	10	10
Ending inventory	45	35	20	0	0	0	0	0	50	10	10	0

Ordering cost	$1,800
Inventory carrying cost	1,445
Total cost	$3,245

Source: W. L. Berry, "Lot-Sizing Procedures for Requirements Planning Systems: A Framework for Analysis," *Production and Inventory Management,* 2nd quarter 1972, p. 26.

number of ordering alternatives considered, however, clearly increases the computations needed in making ordering decisions.

Simulation Experiments

The example problem we've used to illustrate these procedures is for only one product item, without regard for *its* components, with no rolling through time, and with only a fixed number of weeks of requirements. To better understand lot-sizing procedures' performance, we should compare them in circumstances more closely related to company dynamics. Many simulation experiments do exactly that.

Figure 11.7 presents summary experimental results. The first experiment in this figure is for a single level (i.e., one MRP record) with no uncertainty. MOM, PPB, POQ, and EOQ are compared to Wagner-Whitin. MOM produces results about 5 percent more costly, PPB about 6 percent, POQ about 11 percent, and EOQ over 30 percent greater than Wagner-Whitin. These differences may be more important than the magnitudes indicate. Total cost savings of 5 percent may not be trivial.

Moving down to the third experiment, we see results for a multilevel situation, again with no uncertainty. In this case, the comparison isn't against Wagner-Whitin, but against a dynamic programming procedure that produces

---

FIGURE 11.7   Summary Experimental Results

	Procedure				
	Wagner-Whitin	MOM	PPB	POQ	EOQ
Experiment 1: Percent over Wagner-Whitin cost; Single level, no uncertainty*	0	4.93	5.74	10.72	33.87
Experiment 2: Percent over Wagner-Whitin cost; Single level, uncertainty*	0	−0.25	−.67	2.58	.19
Experiment 3; Percent over nearly optimal procedure; Multilevel, no uncertainty†	.77	3.07	6.92	16.91	—
Computing time†	.30	.11	.10	.08	—

*These results are from U. Wemmerlöv and D. C. Whybark, "Lot-Sizing under Uncertainty in a Rolling Schedule Environment," *International Journal of Production Research* 22, no. 3 (1984).
†These results are from B. J. McLaren, "A Study of Multiple Level Lot-Sizing Techniques for Material Requirements Planning Systems," Ph.D. dissertation, Purdue University, 1977. The multilevel procedure was designed specifically to take into account the relationships of a single part to its components and parents. Computing time is the average CPU time for one sample problem.

close to optimal results in a multilevel environment. The key finding in this experiment is that the results are roughly the same as in the first comparisons, although POQ does a little worse and MOM a little better than in the first experiment.

Perhaps the most interesting result in Figure 11.7 comes from comparing the first and third experiments to the *second* experiment. The second experiment is for a single-level procedure, but *with* uncertainty expressed in the gross requirements data. The results here are quite mixed. Note PPB does *better* than Wagner-Whitin, and both MOM and POQ are within 3 percent of Wagner-Whitin.

The conditions modeled in the second experiment replicate conditions likely to be found in actual industrial situations. Moreover, other studies show as uncertainty grows increasingly larger, it becomes very hard to distinguish between lot-sizing procedures' performance. What's more, while there were statistically significant differences among procedures in the first experiment, there were none in the second.

The message is clear. Lot-sizing enhancements to an MRP system should only be done *after* major uncertainties have been removed from the system; that is, *after* data integrity is in place, other MPC system modules are working, stability is present at the MPS level, and so on. If the MPC isn't performing effectively, that's the place to start, *not* with lot-sizing procedures.

DETERMINING PURCHASE ORDER QUANTITIES

So far, we've discussed procedures for determining lot sizes for manufactured items in an MRP environment. In many firms, a high percentage of component items are purchased from external sources. The purchase quantity decision can be complex when price discounts are available for placing orders in large quantities and/or when transportation savings are available for shipping full carload quantities instead of less than carload lots. To start this section, we briefly describe the purchasing discount problem and then turn to three procedures that take into account such discounts: least unit cost, least period cost, and McLaren order moment.

The Purchasing Discount Problem

To illustrate each of these procedures, we'll use an example problem based on the first four periods' requirements in Figure 11.8. Note, in addition to the ordering and the inventory carrying costs used in the previous examples, base and discount prices and discount quantity have been added for this item.

A convention has developed in the purchasing research literature that affects the procedures we'll describe. The purchasing procedures use *period-end inventory balances* to calculate inventory carrying cost. This isn't the

---

FIGURE 11.8   Example Purchase Discount Problem

Period	1	2	3	4	5	6	7	8	9	10	11	12
Requirements	80	100	124	100	50	50	100	125	125	100	50	100

Ordering cost	$100.
Inventory carrying cost	$2/period/unit.
Base price	$500/unit.
Discount price	$450/unit.
Discount quantity	350 units.

---

same convention as that used previously for the manufacturing lot-size calculations. To be consistent with the previous purchasing lot-sizing research, we'll use the period-end inventory convention.

The increase of the quantity discount information adds complexity to this ordering problem's solution. Moreover, specification of an "all-units" discount (such as a $500/unit price for units 1 through 349 and a $450 price for a *total* order quantity exceeding 349 units) adds further computational difficulties. Alternatively, an "additional-units" discount is sometimes specified, such as $500/unit price for units 1 through 349 and a $450/unit price for any *additional* units ordered in excess of 349. Unlike the additional units discount schedule, the all-units discount applies to all units purchased, when at least the discount quantity is purchased. The all-units discount is considerably more common in industry than the additional-units discount and creates a much more difficult decision problem

The ordering procedures we now consider assume the use of an all-units discount, as well as the earlier assumptions listed for the manufacturing order quantities under MRP (except average inventory). In addition, we follow several conventions for all procedures. One is to consider orders sequentially, which would cover an increasing number of periods. For the example in Figure 11.8, this would be orders of 80 (1 period), 180 (2 periods), and so on. In addition, we consider an order for the exact discount quantity (350 units). Another convention is we make calculations for at least the number of periods needed to reach the discount quantity (period 4 in the example). The following paragraphs explain each procedure's details.

Least Unit Cost.   The **least unit cost (LUC)** procedure evaluates different order quantities by accumulating requirements, at least through the period in which the discount can be obtained, until cost/unit starts to increase. We place the order for the quantity that provides the least unit cost. This procedure involves three steps. First, requirements are accumulated through an integral number of periods until the quantity to be ordered is sufficient to qualify for the discount price. The next step is to determine whether the discount should be accepted on the basis of the least unit cost criterion. The final step is to evaluate ordering a quantity exactly equal to the discount quantity. If the least unit cost criterion indicates neither the integral number of periods nor exactly the discount quantity is the most economical order, we place the order for a quantity without the discount.

We illustrate these steps using Figure 11.9 and the example problem in Figure 11.8. The setup cost of $100 is incurred once it's determined an order will be placed. The inventory carrying cost accumulates at $2 per unit times the number of periods it will be carried until the period in which it's used. Base price per unit is $500 per unit until a point during period 4, hereafter called "period 3*," when cumulative requirements exactly equal discount quantity and when unit price drops to $450. The additional inventory carrying cost of getting to period 3* will be $(350 - 304) \times \$2 \times 3$ periods ($276) as the remaining 46 units wouldn't be used until period 4. Total inventory carrying cost for period 3* is $276 + $696 = $972. Cost per unit is the total cost divided by the cumulative requirements. When cost per unit increases (as in period 4) and the discount quantity has been surpassed, the LUC heuristic chooses as the lot size the quantity providing minimum cost per unit (i.e., $453.06, lot size = 350).

Least Period Cost.   The **least period cost (LPC)** works in the same manner as the least unit cost procedure, except the criterion for lot sizing changes. Cost calculation in LPC is the same as in the LUC procedure (Figure 11.9). The difference between LPC and LUC is LPC uses the lowest cost per *period* to determine lot size, instead of lowest cost per *unit*. The number of periods of demand considered is the divisor for the cost/period calculation. At period 3* where the exact discount quantity is considered, we determine the total time by adding the fraction of the period proportional to the quantity required to qualify for the discount $[3 + (350 - 304)/100] = 3.46$, where 100 is the requirement in the split period (4). The LPC procedure uses the cost per period as the criterion for lot sizing. For this example, the lot size would be 80, as indicated in Figure 11.10.

FIGURE 11.9   Least Unit Cost Example

Period	Require-ments	Cumulative require-ments	Setup cost	Inventory carrying cost	Unit purchase price	Cumulative total cost	Cost/ unit
1	80	80	$100	$ 0	$500	$ 40,100	$501.25
2	100	180	100	200	500	90,300[†]	501.67
3	124	304	100	697	500	152,796	502.62
3*	46	350	100	972	450	158,572	453.06
4	100	404	100	1,296	450	183,196	453.46

[†]$(180 \times 500) + \$100 + \$200 = \$90,300$.

FIGURE 11.10   Least Period Cost Example

Period	Cumulative requirements	Cumulative total cost	Cost/period
1	80	$ 40,100	$40,100
2	180	90,300	45,150
3	304	152,796	50,932
3*	350	158,572	45,930
4	404	193,196	45,799

McLaren's Order Moment.   The McLaren's order moment (MOM) procedure works somewhat differently. First, the discount's attractiveness is measured by calculating the number of part periods of inventory that must be carried to offset potential savings from the discount. Part periods are accumulated by summing the number of units to be carried times the number of periods they're carried. Next, the actual number of part periods necessary to qualify for the discount is determined from the requirements, and we decide whether to order exactly the discount quantity even if it means splitting requirements to qualify exactly for the discount. If the discount is favorable, the order placed will be for the discount quantity. If it's unfavorable, McLaren's procedure for determining a lot size without discounts is used.

Figure 11.11 presents an application of the McLaren procedure to the example purchase discount problem. The first step is to calculate the number of part periods that would exactly offset savings available from the discount. This is termed the **target level** and is the dollar savings gained by taking the discount divided by the incremental cost of carrying an additional unit for one period. Equation (11.4) expresses calculation of the target level. To perform lot sizing, the actual cumulative part periods until the discount quantity is exceeded are computed and tested against the target level, as Figure 11.11 shows.

$$\text{Target level} = \frac{(\text{Base price} - \text{Discount price}) \times \text{Discount quantity}}{\text{Inventory carrying cost per period}}$$
$$= [(\$500 - \$450) \times 350]/\$2 \qquad\qquad (11.4)$$
$$= 8,750 \text{ part periods.}$$

At period 4, cumulative requirements exceed discount quantity, and cumulative part periods are less than the target level. This indicates the discount will more than offset extra carrying costs associated with the larger quantity. We set lot size to 350. Then we reset requirements in period 4 to 404 − 350 = 54. Note if the number of part periods exceeds the target at four periods,

FIGURE 11.11 McLaren's Order Moment Example

Period	Requirements	Cumulative requirements	Part periods	Cumulative part periods
1	80	80	$80 \times 0 = 0$	0
2	100	180	$100 \times 1 = 100$	100
3	124	304	$124 \times 2 = 248$	348
4	100	404	$100 \times 3 = 300$	648

we make the comparison for exactly 350 units to determine whether splitting the period's requirements is worthwhile.

The essence of the MOM procedure is to first assess whether it's worthwhile to carry the extra inventory required to qualify for the quantity discount. The target level states how much extra inventory *can* be carried (in part period terms) and still make the discount worthwhile. If the amount we *must* carry, given the actual requirements, is less, we order the discount quantity. To decide whether to purchase even more than the discount quantity, we use the **look-ahead feature,** which we describe next.

Look-Ahead Feature

Each of the three purchase discount procedures can be used with a *look-ahead* enhancement. After the procedure has determined the initial lot size, the look-ahead feature performs a check to see whether the cost of carrying an additional period's requirements (or the remainder of a period whose requirements are split) is less than the cost of the setup required to supply that period's requirements in a separate order. If cost of carrying additional inventory is less, the requirements are added to the original lot. We repeat the look-ahead procedure for the following period's requirements and again for the next period until it no longer pays to carry the additional inventory.

Let's illustrate the look-ahead feature for the example purchase discount problem. For the least unit cost example in Figure 11.9, a lot size of 350 units has been determined. The remaining requirements in period 4 are 54. Using the look-ahead test, calculate the carrying cost of 54 units to period 4 and compare it to the setup cost:

$$54 \text{ units} \times 3 \text{ periods} \times \$2/\text{unit} = \$324.$$
$$\text{Setup cost} = \$100.$$

The look-ahead test fails if carrying costs exceed the setup cost; the lot size remains at 350. If carrying cost were less than setup cost, the additional units would be included in the lot size and the look-ahead feature would be applied to the requirements in period 5.

Performance Comparisons

Simulation experiments have been conducted to evaluate these lot-sizing procedures' performance. Figure 11.12 shows the results of one set of experiments. For this set of results, a mixed integer programming (MIP) procedure was used to provide optimal solutions to the problems as reference points for the comparisons. The Wagner-Whitin procedure was modified to accommodate the discounts and was used as another nonoptimal procedure.

The other procedures in Figure 11.12 are those described here, both with and without look-ahead. Surprisingly, performance is worse with the look-ahead option for two of the three procedures. The LPC procedure suffers in comparison to the others on virtually every count. On the other hand, the MOM procedure excels in every category but one: percentage of optimal solutions found. The trade-off between LUC and the Wagner-Whitin modified (WWM) procedure is the dispersion versus the average. WWM is much closer to the optimal solution in the worst case, but has worse overall performance.

Perhaps the biggest practical difference between the MOM and LUC procedures is the ease of understanding. Here LUC wins. Both are superior, on the average, to the other nonoptimal procedures and have very reasonable computer times. Also cost differences appear small in an absolute sense. Three points are relevant here. First, the comparisons are among reasonable

---

FIGURE 11.12   Summary of Discount Procedure Comparisons

Procedure	Average percent above optimal cost	Maximum percent above optimal cost	Percent of optimal solutions	Average computer time[4]	Maximum computer time[4]
MIP[1]	0	0	100	85.76	778.360
WWM[2]	.326	1.35	19	.0121	.014
LUC(LA)[3]	.491	5.406	53	.0017	.004
LPC(LA)[3]	2.136	11.118	6	.0016	.002
MOM(LA)[3]	.486	5.406	50	.0010	.002
LUC	.164	3.355	58	.0015	.003
LPC	2.158	11.118	6	.0017	.003
MOM	.031	.230	53	.0010	.002

[1]Mixed integer program (optimal).
[2]Wagner-Whitin modified.
[3]With look-ahead (LA).
[4]CPU seconds on a CDC 6600.

Source: T. E. Callarman and D. C. Whybark, "A Comparison of Procedures for Determining Purchase Quantities for Time-Phased MRP Requirements," *Journal of Purchasing and Materials Management*, Fall 1981.

FIGURE 11.13   Effect of Uncertainty on Performance Comparisons

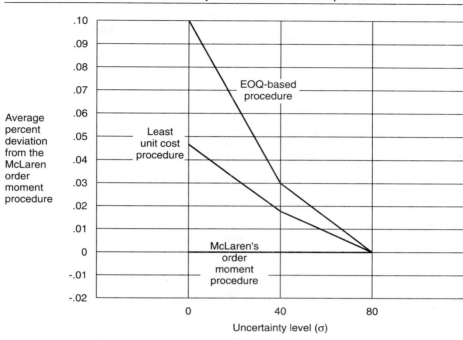

Source: W. C. Benton and D. C. Whybark, "Material Requirements Planning (MRP) and Purchase Discounts," *Journal of Operations Management* 2, no. 2 (February 1982), p. 143.

procedures and may understate the savings over practice. Second, even a small percentage saving for a company whose purchases represent a large proportion of the cost of goods sold can be a large total that passes right through to the bottom line. Finally, the savings shown are based on *total* cost. If we use just controllable costs (setup, inventory, and potential discount), percentage differences are highly magnified.

These experiments were performed without any uncertainty, using fixed scheduling horizons. In additional experiments performed under conditions of purchase discounts, rolling schedules, and uncertainty, the differences between procedures shrank and absolute costs increased. Figure 11.13 depicts the lesson. This is the same result we pointed out in Figure 11.7 for the single-price procedures. There is little difference between the procedures compared—the results are the same. The overall conclusion is: reducing surprises (uncertainty) to a minimum is worthwhile in itself and a necessary prelude to reaping the benefits of the differences in the procedures described here.

BUFFERING CONCEPTS

In this section we deal with another advanced concept in MRP, the use of buffering mechanisms to protect against uncertainties. We, however, make the same proviso as for lot sizing: Buffering is not the way to make up for a poorly operating MRP system. First things must come first.

Categories of Uncertainty

Two basic sources of uncertainty affect an MRP system: demand and supply uncertainty. These are further separated into two types: quantity uncertainty and timing uncertainty. The combination of sources and types provides the four categories of uncertainty Figure 11.14 summarizes and Figure 11.15 illustrates.

Demand timing uncertainty is illustrated in Figure 11.15 by timing changes in the requirements from period to period. For example, the projected requirements for 372 units in period 7 actually occurred in period 4. This shift might result from a change in the promise date to a customer or from a change in a planned order for a higher-level item on which this item is used.

Supply timing uncertainty can arise from variations in vendor lead times or shop flow times. Thus, once an order is released, the exact timing of its arrival is uncertain. In Figure 11.15, for example, a receipt scheduled for period 3 actually arrived in period 1. Note in this case the uncertainty isn't over the order's amount but over its timing. The entire order may be late or early.

Demand quantity uncertainty is manifest when the amount of a requirement varies, perhaps randomly, about some mean value. This might occur when the master production schedule is increased or decreased to reflect changes in customer orders or the demand forecast. It can also occur when there are changes on higher-level items on which this item is used, or when there are

---

FIGURE 11.14 Categories of Uncertainty in MRP Systems

	Sources	
Types	*Demand*	*Supply*
Timing	Requirements shift from one period to another	Orders not received when due
Quantity	Requirements for more or less than planned	Orders received for more or less than planned

Source: D. C. Whybark and J. G. Williams, "Material Requirements Planning under Uncertainty," *Decision Sciences*, October 1976, p. 598.

FIGURE 11.15 Examples of the Four Categories of Uncertainty

	Periods									
	1	*2*	*3*	*4*	*5*	*6*	*7*	*8*	*9*	*10*
Demand timing:										
Projected requirements	0	0	0	0	0	0	372	130	0	255
Actual requirements	0	0	0	372	130	0	146	255	143	0
Supply timing:										
Planned receipts	0	0	502	0	0	403	0	0	144	0
Actual receipts	502	0	0	0	0	403	0	0	144	0
Demand quantity:										
Projected requirements	85	122	42	190	83	48	41	46	108	207
Actual requirements	103	77	0	101	124	15	0	100	80	226
Supply quantity:										
Planned receipts	0	161	0	271	51	0	81	109	0	327
Actual receipts	0	158	0	277	50	0	77	113	0	321

Source: D. C. Whybark and J. G. Williams, "Material Requirements Planning under Uncertainty," *Decision Sciences,* October 1976, p. 599.

variations in inventory levels. In Figure 11.15, period 1's projected requirements of 85 actually turned out to be 103 units of usage.

Supply quantity uncertainty typically arises when there are shortages of lower-level material, when production lots incur scrap losses, or when production overruns occur. Figure 11.15 illustrates this category of uncertainty, where actual quantity received varied around planned receipts.

Safety Stock and Safety Lead Time

There are two basic ways to buffer uncertainty in an MRP system. One is to specify a quantity of safety stock in much the same manner as with statistical inventory control techniques. The second method, safety lead time, plans order releases earlier than indicated by the requirements plan and schedules their receipt earlier than the required due date. Both approaches produce an increase in inventory levels to provide a buffer against uncertainty, but the techniques operate quite differently, as Figure 11.16 shows.

The first case in Figure 11.16 uses no buffering. A net requirement occurs in period 5, and a planned order is created in period 3 to cover it. The second case specifies a safety stock of 20 units. This means the safety stock level will be broken in period 3 unless an order arrives. The MRP logic thus creates a planned order in period 1 to prevent this condition. The final case in Figure 11.16 illustrates use of safety lead time. This example includes a safety

FIGURE 11.16   Safety Stock and Safety Lead Time Buffering

Order quantity = 50 units
Lead time = 2 periods   *Period*

No buffering used	1	2	3	4	5
Gross requirements		20	40	20	0	30
Scheduled receipts			50			
Projected available balance	40	20	30	10	10	30
Planned order releases			50		

Safety stock = 20 units	1	2	3	4	5
Gross requirements		20	40	20	0	30
Scheduled receipts			50			
Projected available balance	40	20	30	60	60	30
Planned order releases		50			

Safety lead time = 1 period	1	2	3	4	5
Gross requirements		20	40	20	0	30
Scheduled receipts			50			
Projected available balance	40	20	30	10	60	30
Planned order releases		50			

Source: D. C. Whybark and J. G. Williams, "Material Requirements Planning under Uncertainty," *Decision Sciences*, October 1976, p. 601.

lead time of one period. The net result is the planned order being created in period 2 with a due date of period 4.

Most MRP software packages can easily accommodate safety stock, since we can determine planned orders simply by subtracting the safety stock from the initial inventory balance when determining the projected available balance. Safety lead time is a bit more difficult. We can't achieve it by simply inflating the lead time by the amount of the safety lead time. In our example, this approach wouldn't produce the result shown as the last case in Figure 11.16. The due date for the order would be period 5, instead of period 4. Thus, we must change the planned due date as well as the planned release date.

Both safety stock and safety lead time illustrate the fundamental problem with all MRP buffering techniques: They lie to the system. The *real* need date for the planned order shown in Figure 11.16 is period 5. If the *real* lead time is two periods, the *real* launch date should be period 3. Putting in buffers can lead to behavioral problems in the shop, since the resulting schedules don't tell the truth. An informal system may be created to tell people what's really needed. This, in turn, might lead to larger buffers. There's a critical need to communicate the reasoning behind the use of safety stock and safety lead times, and to create a working MPC system that minimizes the need for buffers.

Safety Stock and Safety Lead Time Performance Comparisons

Some simulation experiments reveal a preference for using either safety stock or safety lead time, depending on the category of uncertainty to be buffered. These results show a distinct preference for using safety lead time in all cases where demand or supply *timing* uncertainty exists. Likewise, the experiments show a strong preference for using safety stock in all cases where there's uncertainty in either the demand or supply *quantity*.

Figures 11.17 and 11.18 show typical results from these experiments. Figure 11.17 compares safety stock and safety lead time for simulated situations

FIGURE 11.17　Experimental Results: Average Inventory versus Service Level with Timing Uncertainty

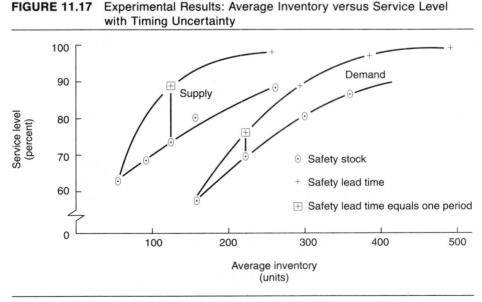

Source: D. C. Whybark and J. G. Williams, "Material Requirements Planning under Uncertainty," *Decision Sciences*, October 1976, p. 602.

FIGURE 11.18   Experimental Results: Average Inventory versus Service Level with Quantity Uncertainty

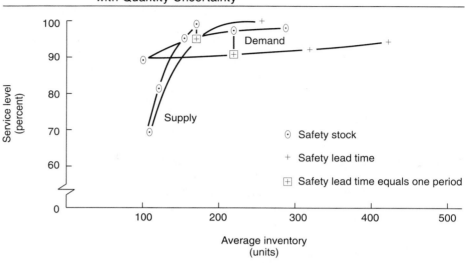

Source: D. C. Whybark and J. G. Williams, "Material Requirements Planning under Uncertainty," *Decision Sciences*, October 1976, p. 603.

similar to Figure 11.15's top two examples. The horizontal axis shows the average inventory held, and the vertical axis depicts the service level in percentage terms; that is, the horizontal axis is based on the period-by-period actual inventory values in the simulation, and the vertical axis is based on the frequency with which actual requirements were met from inventory.

For both the supply and the demand timing uncertainty cases, Figure 11.17 shows a strong preference for safety lead time buffering. For any given level of inventory, a higher service level can be achieved with safety lead time than with safety stock. For any given level of service, safety lead time can provide the level with a smaller inventory investment.

Figure 11.18 shows the comparison for uncertainty in quantities. This simulated situation is similar to Figure 11.15's bottom two examples. The results are a bit more difficult to see, since the graphs for supply and demand uncertainty overlap. Nevertheless, the results are again clear. For any given level of inventory investment, higher service levels are achieved by use of safety stocks than by use of safety lead times. This result is true for situations involving quantity uncertainty in both demand and supply.

The results of the experiments provide general guidelines for choosing between the two buffering techniques. Under conditions of uncertainty in timing, safety lead time is the preferred technique, while safety stock is preferred under conditions of quantity uncertainty. The experimental conclusions didn't change with the source of the uncertainty (demand or supply), lot-sizing technique, lead time, average demand level, uncertainty level, or lumpiness in the gross requirements data. The experiments also indicate that, as lumpiness and uncertainty levels increase, so does the importance of making the correct choice between safety stock and safety lead time.

These guidelines have important practical implications. Supply timing uncertainty and demand quantity uncertainty are the two categories with the largest differences in service levels. An obvious instance of supply timing uncertainty is in vendor lead times. Orders from vendors are subject to timing uncertainty due to variability in both production and transportation times. These experiments strongly support the use of safety lead time for purchased parts experiencing this type of uncertainty. Demand quantity uncertainty often appears in an MRP system for parts subject to service part demand. Another cause of demand quantity uncertainty is when an end product can be made from very different options or features. The experimental results support using safety stock for buffering against these uncertainties.

Other Buffering Mechanisms

Before we end our discussion of uncertainty, it's useful to consider some additional alternatives for dealing with uncertainty. First, rather than living with uncertainty, an alternative is to reduce it to an absolute minimum. In fact, that's one of the major objectives of MPC systems.

For example, increasing demand forecasts' accuracy and developing effective procedures for translating demand for products into master schedules reduces the uncertainty transmitted to the MRP system. Freezing the master schedule for some time period achieves the same result. Developing an effective priority system for moving parts and components through the shop reduces the uncertainty in lead times. Responsive shop-floor control systems can achieve better due date performance, thereby reducing uncertainty. Procedures that improve the accuracy of the data in the MRP system reduce uncertainty regarding on-hand inventory levels. Aspects of JIT manufacturing reduce lead time, improve quality, and decrease uncertainty, providing the same benefits. Other activities could be mentioned, but all focus on the reduction of the amount of uncertainty that needs to be accommodated in an MRP system.

Another way to deal with uncertainty in an MRP system is to provide for slack in the production system in one way or another. Production slack is created by having additional time, labor, machine capacity, and so on over what's specifically needed to produce the planned amount of product. This extra production capacity could be used to produce an oversized lot to allow for that lot's shrinkages through the process. We also could use slack to allow for production of unplanned lots or for additional activities to speed production through the shop. Thus, providing additional capacity in the shop allows us to accommodate greater quantities than planned in a given time period or to expedite jobs through the shop. We must understand, however, that slack costs money, but if the people can be put to good use when production is not needed, the "costs" can become investments.

NERVOUSNESS

This chapter so far has described several enhancements to MRP systems. However, we should recognize some lot-sizing procedures can contribute to the problem of "nervousness" (i.e., instability) in the MRP plans. In this section, we discuss the problem of nervousness in MRP systems and guidelines for reducing its magnitude.

Sources of MRP System Nervousness

MRP system nervousness is commonly defined as significant changes in MRP plans, which occur even with only minor changes in higher-level MRP records or the master production schedule. Changes can involve the quantity or timing of planned orders or scheduled receipts. Figure 11.19 illustrates just such a case. Here, a reduction of one unit in the master schedule in week 2 produced a significant change in the planned orders for item A. This change had an even more profound impact on component part B. It's hard to imagine a *reduction* at the MPS level could create a past-due condition, but that's precisely what Figure 11.19 shows—how the change caused by a relatively minor shift in the master schedule is amplified by use of the periodic order quantity (POQ) lot-sizing procedure.

There are a number of ways relatively minor changes in the MRP system can create nervousness and instability in the MRP plans. These include planned orders released prematurely or in an unplanned quantity, unplanned demand (as for spare parts or engineering requirements), and shifts in MRP parameter values, such as safety stock, safety lead time, or planned lead-time values. Nervousness created by such changes is most damaging in MRP systems with many levels in the product structure. Furthermore, use of some lot-sizing techniques, such as POQ, can amplify system nervousness at lower levels in the product structure, as Figure 11.19 shows.

Reducing MRP System Nervousness

There are several ways to reduce nervousness in MRP systems. First, it's important to reduce causes of changes to the MRP plan. It's important to introduce stability into the master schedule through such devices as freezing and time fences. Similarly, it's important to reduce the incidence of un-

FIGURE 11.19 MRP System Nervousness Example

Before reducing second-week requirements by one unit:
Item A
POQ = 5 weeks
Lead time = 2 weeks

Week		1	2	3	4	5	6	7	8
Gross requirements		2	24	3	5	1	3	4	50
Scheduled receipts									
Projected available balance	28	26	2	13	8	7	4	0	0
Planned order releases		14					50		

Component B
POQ = 5 weeks
Lead time = 4 weeks

Week		1	2	3	4	5	6	7	8
Gross requirements		14					50		
Scheduled receipts		14							
Projected available balance	2	2	2	2	2	2	0	0	0
Planned order releases			48						

After second-week requirement change:
Item A
POQ = 5 weeks
Lead time = 2 weeks

Week		1	2	3	4	5	6	7	8
Gross requirements		2	(23)	3	5	1	3	4	50
Scheduled receipts									
Projected available balance	28	26	3	0	58	57	54	50	0
Planned order releases			63						

Component B
POQ = 5 weeks
Lead time = 4 weeks

Week		1	2	3	4	5	6	7	8
Gross requirements			63						
Scheduled receipts		14							
Projected available balance	2	16	−47						
Planned order releases		(47)							

Past due

planned demands by incorporating spare parts forecasts into MRP record gross requirements. Furthermore, it's necessary to follow the MRP plan with regard to the timing and quantity of planned order releases. Finally, it's important to control the introduction of parameter changes, such as changes in safety stock levels or planned lead times. All of these actions help dampen the small adjustments that can trigger MRP system nervousness.

A second guideline for reducing MRP system nervousness involves selective use of lot-sizing procedures; that is, if nervousness still exists after reducing the preceding causes, we might use different lot-sizing procedures at different product structure levels. One approach is to use fixed order quantities at the top level, using either fixed order quantities or lot-for-lot at intermediate levels, and using period order quantities at the bottom level. Since the fixed order quantity procedure passes along only order timing changes (and not changes in order quantity), this procedure tends to dampen lot-size–induced nervousness. Clearly, fixed order quantity values need to be monitored, since changes in the level of requirements may tend to make such quantities uneconomical over time.

A third guideline for reducing nervousness involves using firm planned orders in MRP (or in MPS) records. Firm planned orders tend to stabilize requirements for lower-level items. The offsetting cost, however, is the necessary maintenance of firm planned orders by MRP planners.

These guidelines provide methods for reducing nervousness in MRP plans. There's a distinction, however, between nervousness in the MRP *plans* and nervousness in the *execution* of MRP system plans. Nervousness in the execution of the plans can also influence behavior. If system users see the plans changing, they may make arbitrary or defensive decisions. This can further aggravate changes in plans.

One way to deal with the execution issue is simply to pass updated information to system users less often. This suggestion argues against the use of net change MRP systems, or at least against publishing every change. An alternative is simply to have more intelligent users. A well-trained user responding to the problem in Figure 11.19 might, through bottom-up replanning, change the lot sizes to eliminate the problem. However, Figure 11.19 does indicate this isn't an easy problem to detect. Many aspects are counterintuitive. The fact still is, more intelligent users will make more intelligent execution decisions. User education may still be the best investment!

OTHER ADVANCED MRP CONCEPTS

Several additional MRP concepts or conventions are used in practice to facilitate use of the system. Some of these ideas' applicability depends on the company environment. In this section, we briefly deal with timing conventions, bucketless MRP systems, phantom assemblies, scrap allowances, and and automatic updating of MRP data base elements.

Timing Conventions

Throughout this book the timing convention we've used in the MRP records has been that a gross requirement was due at the *beginning* of the period. Other conventions are sometimes used—to some extent, any convention will work if it's consistently applied and universally understood.

If shop paper for an order specifies a due date in a particular week, we interpret this as meaning the *beginning* of that week; that is, if an order is due in week 3, it should be available on Monday to be on time to meet its gross requirement. This may not work if it takes time to close out the shop order, recognize the new on-hand balance, create shop orders for the parent item, pick the parts, move them, and begin work.

For this reason, some firms set the due date for a shop order to the week prior to the one that has the gross requirement. This timing convention has the same result as using one week of safety lead time. In fact, the shop may use all the preceding week except the weekend by not completing the order until Friday.

Bucketless Systems

To some extent, problems of timing conventions are tied to use of time buckets. When buckets are small enough, problems are reduced significantly. However, smaller buckets mean more buckets, which increases review, storage, and computation costs. A bucketless MRP system specifies the exact release and due date for each requirement, scheduled receipt, and planned order. Managerial reports are printed out on whatever basis is required, including by exact dates.

Bucketless MRP systems are a better way to use the computer. Above and beyond that, the approach can allow for better maintenance of lead time offsets and provide more precise time-phased information. The approach is consistent with state-of-the-art software. Many firms are successfully using bucketless systems.

A related concept is real-time systems. Dedicated real-time systems are now the state of the art in MPC. These systems are run on a daily basis (regeneration or net change) utilizing bucketless concepts. The major addition is that the planning cycle itself is bucketless. That is, plans are revised as necessary, not on a periodic schedule, and the entire execution cycle is also shortened.

Phantom Assemblies

Another issue related to timing has to do with **phantom assemblies.** In essence, a phantom assembly is any assembly that doesn't go into and out of inventory; that is, it's an assembly that exists physically but is normally not stored, so inventory transactions aren't posted against it. An example would be assembly of a drawer in a chest while the chest is being assembled. Still another example is the chest itself, which is assembled but thereafter finished (stain and the rest). Often there can be phantoms within phantoms.

We can potentially eliminate the need for phantom assemblies by redesigning the bill of materials and engineering drawings. Note, in the drawer/chest example, this would mean listing the drawer parts on the single-level bill of materials for the chest, with the drawer itself no longer existing. This won't always work. At times the phantom does exist; that is, the drawer doesn't get assembled directly into the chest and must be inventoried. This can occur when more subassemblies (drawers) than needed are produced, or when subassemblies are sold as service parts.

Phantom assembly treatment is widely used as MRP planning evolves into just-in-time (JIT) execution. That is, the number of "real" bill of material levels is reduced so products go into and out of inventory (requiring MRP planning) fewer times. This by itself can reduce inventory and lead times.

If subassemblies aren't treated as phantoms, bill of material processing logic will necessarily create orders for *both* chests *and* drawers. Thus, one reason firms use phantom assemblies is to avoid the work and time associated with closing out shop orders and opening new shop orders. The objective of the phantom assembly treatment of the drawers is to pull drawer parts automatically by only launching orders for chest assemblies and not drawer assemblies.

The phantom condition can also exist in areas that aren't truly assemblies. For example, many firms don't typically inventory parts coming off a paint line (even though painted parts have different part numbers than unpainted ones); they simply flow into final assembly and not through inventory, *unless* they're to be used for service parts or extras are produced.

Phantom bill treatment in MRP requires special processing. Net requirements for the phantom bills must be determined (in case there are any inventories), so they can be passed down to the components, but without creating explicit records. The trick to gross to netting the phantom requirements is to explicitly code a phantom bill (to avoid producing the record) and use a zero lead time offset. Whenever a specific need for phantom items occurs, say for service parts, the phantom itself can be treated as any other MRP record item, with usual order-launching and closeout treatment.

Scrap Allowances

A concept closely tied to buffering is use of scrap allowances in calculating the lot size to start into production to reach some desired lot size going into the stockroom. It's a fairly straightforward procedure to use any lot-sizing procedure to determine the lot size and then adjust the result to take into account the scrap allowance. One issue that arises is whether the quantity shown on the shop paper (and as a scheduled receipt) should be the *starting* quantity or the *expected finished* quantity. Practice suggests using the former. This requires, however, that each actual occurrence of scrap be transacted and reflected in updated plans.

The overall issue of the scrap allowance is clearly related to the use of safety stocks for quantity uncertainty buffering. One or both of these techniques could be used in a particular situation. The point is, if scrap losses occur, they must be planned for and buffered. It also means this may be an area where tight control can lead to performance improvements.

An advanced MRP concept related to scrap allowances is how they should be applied in a volatile learning environment, as is the case in manufacturing integrated circuits. In such cases, there are typically low initial yields, followed by learning or improvement in yields. Dolinsky et al. show substantial reductions in mean inventory levels can be realized in low-yield environments if learning is properly included in the order release logic. This finding proves to be robust with respect to modest errors in the estimation of the learning rate.

Automatic Updating

The final enhancement issue we consider is use of the computer to automatically perform a transaction normally done by MRP planners. A good case in point is updating scheduled receipt due dates as indicated by exception coding; that is, computer software tells an MRP planner a scheduled receipt's due date needs to be moved from week 7 to week 6. Why not let the computer do it automatically?

We raise this issue here, under advanced concepts, because the operational issue is similar to that of more sophisticated lot sizing: It is OK—*after* we have the basic house in order. When an MRP system is in place, data integrity is good, the other MPC systems are in place, and users are well trained, *then* automatic updating might make sense.

There are two good reasons for updating scheduled receipt data automatically. First, new data will be passed more quickly to shop-floor systems, resulting in a more responsive shop-floor control system. The second reason is to save time and effort for the MRP planners. Typically, a large percentage of the exception messages involve minor repositioning of scheduled receipt data. If this is done by computer, time can be freed for more useful pursuits.

A related issue is "back flushing" as done in JIT. There, component inventories are only reduced in quantity after the final products in which they're imbedded are received into finished goods. The benefit is reduced transaction processing; the cost is a slight mismatch between records and reality. With fast throughput and educated users, costs can be well worth it.

The key question of whether to do automatic updating is behavioral and partly related to system nervousness. Are users ready for it? Do *they* want it? Will *they* still be accountable for the results? Will *they* have a hand in specifying the logic used to automatically update? Is the system stable enough to not produce too many surprises? Once again, we see this enhancement intimately tied to developing intelligent system users.

CONCLUDING PRINCIPLES

Chapter 11 describes several advanced concepts and conventions in MRP systems. Many ideas are of research interest, but all have practical implications too. Certain kinds of enhancements can be made in a well-operating MRP system, if made by knowledgeable professionals and if implemented with knowledgeable users. The following principles are critical:

- MRP enhancements should be done *after* a basic MPC system is in place.
- Discrete lot-sizing procedures for manufacturing can reduce inventory-associated costs. The complexity should not outweigh the savings, however.
- Selecting the appropriate lot-sizing procedure for purchasing should consider quantity discounts.
- Safety stocks should be used when the uncertainty is of the quantity category.
- Safety lead times should be used when uncertainty is of the timing category.
- MRP system nervousness can result from lot-sizing rules, parameter changes, and other causes. The MPC professional should take appropriate precautions to dampen the amplitude and impact.
- MRP system enhancements should follow the development of ever more intelligent users.

REFERENCES

Axsater, Sven. "Evaluation of Lot Sizing Techniques." *International Journal of Production Research* 24, no. 1 (1986), pp. 51–57.

Baker, K. R. "Lot Sizing Procedures and a Standard Data Set: A Reconciliation." *Journal of Manufacturing and Operations Management* 2, no. 3, 1989, pp. 199-221.

Benton, W. C. "Multiple Price Breaks and Alternative Purchase Lot Size Procedures in Material Requirements Planning Systems." *International Journal of Production Research* 23, no. 5 (1985), pp. 1025–47.

————, and D. C. Whybark. "Material Requirements Planning (MRP) and Purchase Discounts." *Journal of Operations Management*, February 1982, pp. 137–43.

Berry, W. L. "Lot Sizing Procedures for Requirements Planning Systems: A Framework for Analysis." *Production and Inventory Management* 13, no. 2 (2nd quarter 1972), pp. 19–33.

Blackburn, J. D.; D. H. Kropp; and R. A. Millen. "A Comparison of Strategies to Dampen Nervousness in MRP Systems." *Management Science* 32, no. 4 (April 1986).

Bobko, P. R., and D. C. Whybark. "The Coefficient of Variation as a Factor in MRP Research." *Decision Sciences* 16, no. 4 (Fall 1985).

Callarman, T. F., and D. C. Whybark. "Determining Purchase Quantities for MRP Requirements." *Journal of Purchasing and Materials Management* 17, no. 3 (Fall 1981), pp. 25–30.

————, and R. S. Hamrin. "A Comparison of Dynamic Lot Sizing Rules for Use in a Single Stage MRP System with Demand Uncertainty." *International Journal of Operations and Production Management* 4, no. 2, 1984, pp. 39–48.

Carlson, R. C., and C. A. Yano. "Safety Stocks in MRP-Systems with Emergency Setups for Components." *Management Science* 32, no. 4 (April 1986).

Chalmet, L. C.; M. Debodt; and L. Van Wassenhove. "The Effect of Engineering Changes and Demand Uncertainty on MRP Lot Sizing: A Case Study." *International Journal of Production Research* 23, no. 2 (1985), pp. 233–51.

Christoph, Orinda Byrd. "McLaren's Order Moment Lot-Sizing Technique in Multiple Discounts." *Production and Inventory Management Journal* 30, no. 2 (2nd quarter 1989), pp. 44–47.

————, and R. Lawrence LaForge. "The Performance of MRP Purchase Lot-Size Procedures under Actual Multiple Purchase Discount Conditions." *Decision Sciences Journal* 20, no. 2 (Spring 1989), pp. 348–58.

Coleman, B. J., and M. A. McKnew. "An Improved Heuristic for Multilevel Lot Sizing in Material Requirements Planning." *Decision Sciences Journal* 22, no. 1, Winter 1991, pp. 136–56.

Collier, D. A. "A Comparison of MRP Lot-Sizing Methods Considering Capacity Change Costs." *Journal of Operations Management* 1, no. 1 (1980).

Dixon, P. S., and E. A. Silver. "A Heuristic Solution Procedure for the Multi-Item, Single Level, Limited Capacity, Lot-Sizing Problem." *Journal of Operations Management* 2, no. 1 (1981).

De Bodt, M. A., and L. N. Van Wassenhove. "Cost Increases Due to Demand Uncertainty in MRP Lot Sizing." *Decision Sciences* 14, no. 3 (July 1983).

Dolinsky, L. R.; T. E. Vollmann; and M. J. Maggard. "Adjusting Replenishment Orders to Reflect Learning in a Material Requirements Planning Environment." *Management Science* 36, no. 12 (December 1990).

Gaimon, C. "Optimal Inventory, Backlogging and Machine Loading in a Serial

Multistage, Multi-Period Production Environment." *International Journal of Production Research* 24, no. 2 (May–June 1986).

Jesse, Jr., R. R., and J. H. Blackstone, Jr. "A Note on Using the Lot-Sizing Index for Comparing Discrete Lot-Sizing Techniques." *Journal of Operations Management* 5, no. 4 (1986).

Karmarkar, U.S. "Lot Sizes, Lead Times and In-Process Inventories." *Management Science* 33, no. 3 (1987).

Kerni, R., and Y. Roll. "A Heuristic Algorithm for the Multi-Item, Lot-Sizing Problem with Capacity Constraints." *AIIE Transactions* 14, no. 4 (December 1982).

Kropp, D. H.; R. C. Carlson; and J. V. Jucker. "Heuristic Lot-Sizing Approaches for Dealing with MRP System Nervousness." *Decision Sciences* 14, no. 2 (April 1983).

Lambrecht, M. C.; J. A. Muckstadt; and R. Luyten. "Protective Stocks in Multi-Stage Production Systems." *International Journal of Production Research* 22, no. 6 (1984), pp. 1001–25.

McClelland, Marilyn K., and Harvey M. Wagner. "Location of Inventories in an MRP Environment." *Decision Sciences Journal* 19, no. 3 (Summer 1988), pp. 535–53.

McKnew, M. A.; C. Seydam; and B. J. Coleman. "An Efficient Zero-One Formulation of the Multilevel Lot-Sizing Problem." *Decision Sciences Journal* 22, no. 2, Spring 1991, pp. 280–95.

McLaren, B. J. "A Study of Multiple Level Lot Sizing Techniques for Material Requirements Planning Systems." Ph.D. dissertation, Purdue University, 1977.

Melnyk, S. A., and C. J. Piper. "Leadtime Errors in MRP: The Lot-Sizing Effect." *International Journal of Production Research* 23, no. 2 (1985), pp. 253–64.

Narasimhan, R., and S. A. Melnyk. "Assessing the Transient Impact of Lot Sizing Rules Following MRP Implementation." *International Journal of Production Research* 22, no. 5 (1984), pp. 759–72.

New, C., and J. Mapes. "MRP with High Uncertain Yield Losses." *Journal of Operations Management* 4, no. 4 (1984).

Prentis, Eric L., and Basheer M. Khumawala. "Efficient Heuristics for MRP Lot Sizing with Variable Production/Purchasing Costs." *Decision Sciences Journal* 20, no. 3 (Summer 1989), pp. 439–50.

Raturi, Amitabh S., and Arthur V. Hill. "An Experimental Analysis of Capacity-Sensitive Setup Parameters for MRP Lot-Sizing." *Decision Sciences Journal* 19, no. 4 (Fall 1988), pp. 782–800.

St. John, R. "The Cost of Inflated Planned Lead Times in MRP Systems." *Journal of Operations Management* 5, no. 2 (1985).

Schmitt, T. G. "Resolving Uncertainty in Manufacturing Systems." *Journal of Operations Management* 4, no. 4 (1984).

Smith-Daniels, D. E., and Aquilano, N. J. "Constrained Resource Project Scheduling Subject to Material Constraints." *Journal of Operations Management* 4, no. 4 (1984).

Steele, D. C. "The Nervous MRP System: How to Do Battle." *Production and Inventory Management* 16, no. 4 (April 1973).

Steinberg, E., and A. Napier. "Optimal Multilevel Lot Sizing for Requirements Planning Systems." *Management Science* 26, no. 12 (December 1980), pp. 1258–72.

Veral, E. A., and R. L. LaForge. "The Performance of a Simple Incremental Lot-Sizing Rule in a Multilevel Inventory Environment." *Decision Sciences* 16, no. 1 (Winter 1985).

Veral, Emre A., and R. Lawrence LaForge. "The Integration of Cost and Capacity Considerations in Material Requirements Systems." *Decision Sciences Journal,* 21, no. 3 (Summer 1990), pp. 507–20.

Vickery, S. K., and R. E. Markland. "Multi-Stage Lot Sizing in a Serial Production System." *International Journal of Production Research* 24, no. 3 (1986), pp. 517–34.

Wacker, John G. "A Theory of Material Requirements Planning (MRP): An Empirical Methodology to Reduce Uncertainty in MRP Systems." *International Journal of Production Research* 23, no. 4 (1985), pp. 807–24.

Wagner, H. M., and T. M. Whitin. "Dynamic Version of the Economic Lot Size Model." *Management Science,* October 1958, pp. 89–96.

Wassweiler, W., "Tool Requirements Planning." *American Production and Inventory Control Society 1990 Annual Conference Proceedings,* pp. 451–53.

Wemmerlov, Urban. "The Behavior of Lot-Sizing Procedures in the Presence of Forecast Errors." *Journal of Operations Management* 8, no. 1 (January 1989), pp. 37–47.

————, and D. C. Whybark. "Lot Sizing under Uncertainty in a Rolling Schedule Environment." *International Journal of Production Research* 22, no. 3 (1984), pp. 467–84.

Whybark, D. C., and J. G. Williams. "Material Requirements Planning under Uncertainty." *Decision Sciences* 7, no. 4 (October 1976).

DISCUSSION QUESTIONS

1. Some practitioners complain discrete lot-sizing procedures (e.g., POQ, PPB, MOM) aggravate system nervousness because of changing lot quantities. What do they mean?
2. Reviewing Figure 11.7 and using your own powers of intuition, what do you think would happen to the difference in costs between lot-sizing procedures as uncertainty gets larger and larger? What about the absolute cost values?
3. How do the quantity discount lot-sizing procedures illustrate the contention that advanced concepts are enhancements to sound, basic MPC systems?
4. Why is it necessary to change both the release date and the due date for safety lead time?
5. What are some of the difficulties with introducing "organizational slack" as a method of buffering against uncertainty?

6. One suggestion for reducing execution nervousness is to give the status information only at the time of need. Thus, the only time a foreman would need to determine job priorities would be when he or she was choosing the next job to put on a machine. Give pros and cons of this suggestion.

PROBLEMS

1. Consider the following information about an end product item:

Ordering cost = $32/order.
Average usage = 8 units/week.
Inventory carrying cost = $2/unit/week.

a. How many orders should we place per year to replenish inventory of the item based on average weekly demand?
b. Given the following time-phased net requirements from an MRP record for this item, determine the sequence of planned orders using economic order quantity and periodic order quantity procedures. Assume lead time equals zero and current on-hand inventory equals zero. Calculate the inventory carrying cost on the basis of weekly ending inventory values. Which procedure produces the lowest total cost for the eight-week period?

	Week							
	1	2	3	4	5	6	7	8
Requirements	15	2	10	12	6	0	14	5

2. A final assembly (A) requires one week to assemble and has a component part (B) requiring two weeks to fabricate. Three units of final assembly A and four units of part B are currently on hand. The gross requirements for assembly A for the next 10 weeks are as follows (one part B is used on each A):

	Week									
	1	2	3	4	5	6	7	8	9	10
Requirements	1	4	2	8	1	0	6	2	1	3

a. What are the planned order releases for part B using lot-for-lot lot sizing for both parts A and B?
b. What are the planned order releases for part B using POQ = 2 for both parts A and B?

3. A company has estimated net requirements for a particular part as follows:

	Month											
	1	*2*	*3*	*4*	*5*	*6*	*7*	*8*	*9*	*10*	*11*	*12*
Requirements	100	10	15	20	70	250	250	250	250	40	0	100

Ordering cost associated with this part is $300. Estimated inventory carrying cost is $2 per unit per month calculated on average inventory. Currently no parts are available in inventory. The company wishes to know when and how much to order over the next 12 months.

a. Apply the economic order quantity and the part period balancing procedures to solve this problem.

b. What important assumptions are involved in each of the approaches used in part a?

4. The Chan Hahn Clothing Company is trying to decide which of several lot-sizing procedures to use for its MRP system. The following information pertains to one of the "typical" component parts:

Setup cost = $80/order.
Inventory cost = $1/unit/week.
Current inventory balance = 0 units.

	Week							
	1	*2*	*3*	*4*	*5*	*6*	*7*	*8*
Demand forecast	60	40	25	5	100	20	60	10

a. Apply the EOQ, POQ, and PPB lot-sizing procedures and show the total cost resulting from each procedure. Calculate inventory carrying costs on the basis of *average* inventory values. Assume orders are received into the beginning inventory.

b. Indicate advantages and disadvantages of using each procedure suggested in part a.

5. Apply the McLaren order moment (MOM) lot-sizing procedure to the following 12 periods of requirements data, indicating order receipts' size and period. Assume order costs are $100/order placed and inventory carrying cost is $1/unit/period. Calculate the inventory carrying costs on the basis of *ending* inventory values.

	Period											
	1	*2*	*3*	*4*	*5*	*6*	*7*	*8*	*9*	*10*	*11*	*12*
Requirements	70	30	35	60	60	25	35	70	45	70	80	55

6. Use the requirements data from Problem 5 and order costs of $100/order, inventory carrying cost of $1/unit/period, and a unit cost of $50 in lots of less than 100 and $45 for lots of 100 or more, and inventory carrying costs calculated on the basis of *ending* inventory values to:
   a. Apply McLaren's order moment (MOM) lot-sizing procedure (with look-ahead).
   b. Apply the least unit cost (LUC) procedure without look-ahead.

7. Ellen Farr is responsible for purchasing forgings and castings for Farr Machine Corporation's raw material stockroom. Ellen's job is to purchase a sufficient number of castings and forgings to meet weekly demand for these items by the firm's fabrication shop in producing machined parts. She's interested in low cost and whether to take the discount offered. Forecast weekly requirements for one item (the input shaft forging) and cost information for this item are as follows:

			Week						
	1	2	3	4	5	6	7	8	9
Forecast	44	2	10	42	46	2	30	10	4

   Order cost = $50
   Inventory carrying cost = $5/unit/week.
   Item price = $100/unit ($95/unit if orders are issued for 80 units or more).

   Calculate the orders Ellen would place using the least unit cost, the least period cost, and the MOM (with look-ahead) procedures. Find total cost of each solution procedure, assuming inventory carrying cost is based on the average inventory.

8. The Fisher Products Company produces a line of children's parlor games. The production process includes two departments: fabrication and assembly. The fabrication shop produces game parts, such as plastic pieces, game markers, and special indicators. The company maintains inventories both of the raw material needed to produce the game parts and of finished game parts themselves. The assembly department consists of a single assembly line that collates and packages all parlor games in Fisher's product line to meet incoming customer orders. The company maintains no inventory of finished products (games) but produces only to customer order.

   Each week the assembly foreman schedules the assembly line and supervises withdrawal of game parts from the stockroom that are needed on the assembly line to meet the production schedule. Sometimes several games are assembled during a single week. Since the company uses MRP to plan and control production of games and game parts, the assembly foreman prepares a master production schedule for a period covering eight weeks into the future. Exhibit A shows his master schedule with two end products (games A and B).
   a. A component (game part) called the toy cup is used in producing the two

parlor games (A and B). Two units of the toy cup are needed to produce one unit of game A, and one unit of the toy cup is required to produce one unit of game B. Assuming toy cup planned lead time is one week, current on-hand inventory is 44 units, and there are no scheduled receipts, complete the MRP record for the toy cup in Exhibit A. Use the periodic order quantity ordering policy for lot sizing. Ordering cost for the toy cup is $9/order; inventory carrying cost is $.10/unit/week.

b. After completing the toy cup's MRP record in part a, complete the MRP record in Exhibit A for the plastic molding material needed to produce the toy cup. Two ounces of the plastic molding material are needed to produce one toy cup, planned lead time for the plastic material is one week, 20 ounces of plastic material are currently on hand, scheduled receipt for 90 ounces of plastic material is due in week 1 from the supplier, and the periodic ordering policy is used for this item. Ordering cost is $5/order; inventory carrying cost is $.04 per ounce per week.

c. Fisher's assembly foreman has just handed you the MRP records for the toy cup and plastic molding material in Exhibit B. These MRP records contain

---

EXHIBIT A Material Requirements Planning Worksheet

End products		Week number 1	2	3	4	5	6	7	8
Game A master schedule		21	0	0	21	20	0	15	0
Game B master schedule		2	2	9	0	6	3	0	9

Toy cup		1	2	3	4	5	6	7	8
Gross requirements									
Scheduled receipts*									
Projected available balance	44								
Planned order releases									

Plastic molding material		1	2	3	4	5	6	7	8
Gross requirements									
Scheduled receipts*		90							
Projected available balance	20								
Planned order releases									

*Received at the beginning of each week.

a different master schedule, changed on-hand inventory values, and a new scheduled receipt value. Also, planned lead time for both items is now two weeks. Complete the new MRP records for both items, assuming the part period balancing ordering policy is used for the toy cup and plastic material.

d. What problem(s), if any, are apparent after you've completed the MRP records in Exhibit B? What are the MRP planner's alternative courses of action in resolving the problem(s)? What specific course of action should be taken? Why?

EXHIBIT B   Material Requirements Planning Worksheet

					Week number				
End products		1	2	3	4	5	6	7	8
Game A master schedule			22		21	15			
Game B master schedule		9	1	7	7	3	10	1	2

Toy cup		1	2	3	4	5	6	7	8
Gross requirements (2 units/game A) (1 unit/game B)									
Scheduled receipts*									
Projected available balance	55								
Planned order releases (Lead time = 2 weeks)									

Plastic molding material		1	2	3	4	5	6	7	8
Gross requirements (2 oz./toy cup)									
Scheduled receipts*			166						
Projected available balance	10								
Planned order releases (Lead time = 2 weeks)									

*Received at the beginning of each week.

9. Using Figure 11.2's format and a spreadsheet program, compare EOQ and POQ lot-sizing approaches for ordering to satisfy the following requirements sequence. Ordering cost is $17 per order, and carrying cost is $.20 per unit per period. (Use average inventory.)

					Week					
	1	*2*	*3*	*4*	*5*	*6*	*7*	*8*	*9*	*10*
Requirements	48	12	53	91	33	8	60	55	38	22

10. The Silver brothers (Ed, Quick, and Hiho) had their very own factory. Ed was the general manager, and Quick produced the finished product from a part Hiho made. Here are selected data for the finished product and part. Requirements for the finished product for the next few periods appear as well.

	Quick	Hiho
Order quantity(EOQ)	40	100
Safety stock	5	0
Lead time	1	1
Current inventory	7	12
Scheduled receipt in period 1	40	0

				Period				
	1	*2*	*3*	*4*	*5*	*6*	*7*	*8*
Requirements	23	13	36	12	21	8	34	23

   a. Using a spreadsheet, develop MRP records for Quick and Hiho. What are the planned order releases for Hiho's part?
   b. Devise a different ordering plan for Quick and Hiho that has the same number of orders as the solution in part a but reduces inventory levels *and* has the same amount of closing inventory (period 8) as the original plan.

11. Develop a spreadsheet to calculate LUC. Purchased part usage averages 20 units per period, ordering cost is $5 per order, carrying cost is $.20 per unit per period, and purchase price is $1. Use these requirements:

			Period			
	1	*2*	*3*	*4*	*5*	*6*
Requirements	10	18	30	35	10	16

   a. What is the ordering pattern?
   b. If a quantity discount of $.05 per unit is given for orders of more than 50 units, what happens? (For simplicity, assume you can't split a period's requirements.)

12. Consider the following time-phased net requirements from an MRP record.

	Week		
	1	2	3
Net requirements	2	6	12

a. Determine *all* the possible ways orders can be placed to satisfy net requirements without incurring a stockout. Assume lead time equals zero.
b. If ordering cost equals $100 per order and inventory carrying cost equals $10 per unit per week, what's the cost for each ordering alternative?
c. What is the least cost sequence of orders?

13. The XYZ Company's production manager is investigating causes of nervousness in the firm's MRP system. His study has produced MRP schedules for item A as of the start of weeks 1, 2, 3, and 4.

LT = 0.
SS = 0.
Q calcuated using the Wagner-Whitin algorithm.
Ordering cost = $400/order.
Inventory carrying cost = $1/unit/week.

Item A

Week		1	2	3	4
Gross requirements		177	261	207	309
Scheduled receipts					
Projected available balance	0	261	0	309	0
Planned order releases		438		516	

Item A

Week		2	3	4	5
Gross requirements		261	207	309	64
Scheduled receipts					
Projected available balance	261	0	373	64	0
Planned order releases			580		

Item A

Week		3	4	5	6
Gross requirements		207	309	64	182
Scheduled receipts					
Projected available balance	0	0	246	182	0
Planned order releases		207	555		

Item A

Week		4	5	6	7
Gross requirements		309	64	182	0
Scheduled receipts					
Projected available balance	0	246	182	0	0
Planned order releases		555			

a. Assume we need one unit of raw material item B to produce one unit of product item A. Calculate gross requirements for item B covering a four-week planning horizon as of the start of weeks 1, 2, 3, and 4 (assuming we only use item B to produce item A). Compare gross requirements for item B for each week as they're calculated at the start of weeks 1, 2, 3, and 4. What differences do you observe in item B's gross requirements from week to week?

b. Given there are no changes to either item A's gross requirements or beginning inventory over the four-week interval, how do you explain changes in the gross requirements for item B observed in part a? How do you explain this situation to the production manager? What impact would such changes in gross requirements for item B have on supplier relations if B is a purchased part?

14. The following MRP record is for an item purchased from a supplier requiring a two-week lead time. The percent defective for purchased lots of this item averages 10 percent and has been as large as 20 percent. Defective items are removed and returned to the vendor when found. How would you construct the MRP record to protect against inventory shortages with the degree of defects described? An economic order quantity of 20 units is used for this item.

Week		1	2	3	4	5	6	7	8
Gross requirements		10	0	18	5	23	2	15	20
Scheduled receipts									
Projected available balance	14								
Planned order releases									

15. The following MRP record is for a purchased item having an order quantity of 30 units. Its supplier's delivery reliability history indicates average lead time is two weeks. However, delivery lead time has occasionally been four weeks. Construct the following MRP record so inventory shortages aren't incurred.

Week		1	2	3	4	5	6	7	8	9	10	11	12
Gross requirements		22	8	15	25	17	2	5	21	12	14	0	5
Scheduled receipts		30											
Projected available balance	45												
Planned order releases													

Chapter 12

Advanced Concepts in Just-in-Time

This chapter addresses advanced issues in just-in-time (JIT) manufacturing that affect the design of manufacturing planning and control systems. Substantial emphasis is being placed on implementing just-in-time manufacturing by companies, which in turn stimulates a growing body of research literature focused on fundamental JIT concepts and JIT's application in manufacturing. While research on JIT encompasses a broad range of topics, this chapter primarily concerns research affecting how manufacturing planning and control decisions are made.

Chapter 12 centers around five topics:

- A JIT research framework: What are the key areas of research in JIT and which of these relate to manufacturing planning and control?
- Scheduling: How does the need for mixed model schedules and schedule stability affect the master production schedule?
- Supplier management: How should decisions concerning the scheduling of incoming shipments be determined?
- Production floor management: How should the number of kanbans be determined, and to what extent do factors such as demand uncertainty and planning horizon length affect this determination?
- JIT performance and operating conditions: How do operating factors such as demand uncertainty, process time variance, and the number of kanbans affect JIT operating performance?

Chapter 12 is closely linked to Chapter 3, which describes basic JIT methods, and Chapter 5, which discusses design of scheduling systems under JIT. There are also links to Chapter 6 on issues concerning managing the master production schedule, and to Chapter 9, which covers MRP and JIT integration. Chapter 17 deals with lot sizing under independent demand conditions.

A JIT RESEARCH FRAMEWORK

There are many aspects to a JIT manufacturing system. Before considering those that affect design of manufacturing planning and control systems, we'll describe a framework that shows how they relate to key manufacturing functions and to other functions in a company. The framework has been developed by Sakakibara, Flynn, and Schroeder in conjunction with their measurement research on JIT. As background they conducted a comprehensive review of the published literature on JIT and visited 12 plants in the United States (7 Japanese-owned and 5 U.S.-owned) that had implemented JIT.

Based on their literature review and field research, Sakakibara, Flynn, and Schroeder found substantial agreement between academics and practitioners on 16 core JIT components (shown in the center of Figure 12.1): setup time reduction, small lot size, JIT delivery from suppliers, supplier quality level, multifunction workers, small-group problem solving, training, daily schedule adherence, repetitive master schedule, preventive maintenance, equipment layout, product design simplicity, kanban (if applicable), pull system (if applicable), material requirements planning (MRP) adaptation to JIT, and accounting adaptation to JIT.

Figure 12.1 indicates the linkage between these core JIT components and other functions and activities in companies. As an example, there's a two-way link between JIT and human resource management. Here the core JIT component of multifunction workers is directly linked to the human resource management concerns with recruiting and selection, training, and other variables. Sakakibara, Flynn, and Schroeder note that in one plant over 600 applicants were interviewed for 20 positions. The company screened the applicants to find employees willing to work as team members, and to be trained to handle many different tasks.

The framework in Figure 12.2 groups the 16 core JIT components into 6 key manufacturing activities: production-floor management, scheduling, process and product design, work force management, supplier management, and information system. This grouping provides a clear indication of JIT's impact on the design of manufacturing planning and control (MPC) systems. Eight of Figure 12.2's 16 core components affect MPC system design, as seen in the figure's shaded areas. We believe Figure 12.2's framework is useful in indicating those areas where JIT research is needed in the design of MPC systems. Accordingly, we've organized the recent JIT research findings described in this chapter to match many elements in this framework.

SCHEDULING

This section deals with advanced research relating to two aspects of the master production scheduling activity in JIT: the determination of mixed model schedules for end product items and the problem of stabilizing the final assembly schedule.

FIGURE 12.1 Components of a Just-in-Time Manufacturing System

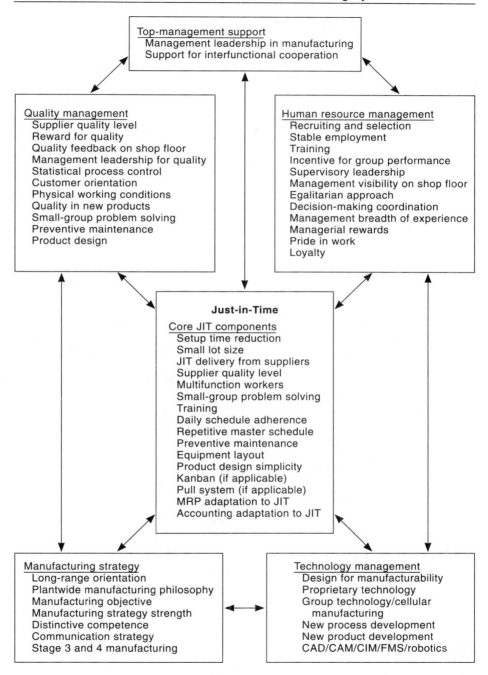

Source: S. Sakakibara, B. B. Flynn, and R. G. Schroeder, "A Just-in-Time Manufacturing Framework and Measurement Instrument," Working Paper 90–10, Curtis L. Carlson School of Management, University of Minnesota, August 1990.

FIGURE 12.2 Core Just-in-Time Manufacturing Framework

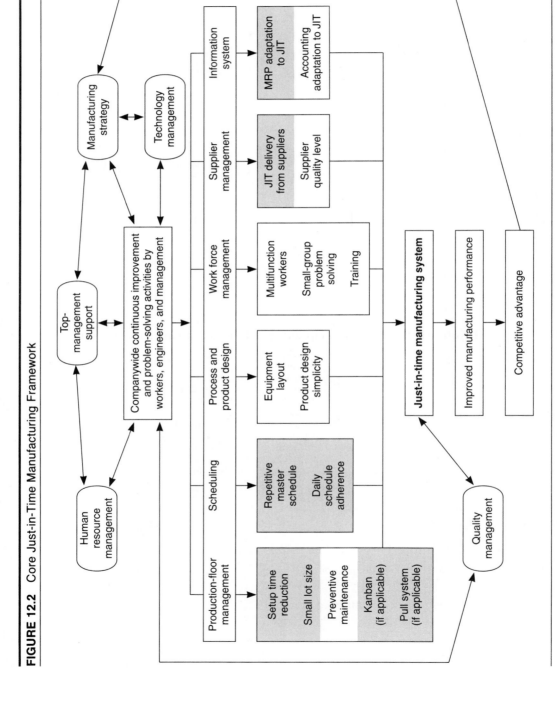

Source: S. Sakakibara, B. B. Flynn, and R. G. Schroeder, "A Just-in-Time Manufacturing Framework and Measurement Instrument," Working Paper 90–10, Curtis L. Carlson School of Management, University of Minnesota, August 1990.

Scheduling Mixed Model Assembly Lines under JIT

Mixed model assembly lines provide the flexibility to produce a wide range of end products in small lots, enabling the company to hold small inventories of finished products but still provide short customer delivery times. Use of mixed model assembly involves the traditional problems of assembly line design (i.e., determining the line cycle time, the number and sequence of stations on the line, and balancing the line). The mixed model approach also involves determining the sequence in which products will be scheduled for assembly so a level work load is maintained at each station on the line and line cycle time isn't exceeded.

An objective, under JIT, is to have a fairly constant usage rate for each component going into final assembly to facilitate the use of kanban or other JIT shop-floor systems. This means sequencing the assembly of finished products in the mixed model line to provide a fairly constant rate of use for each component. While producing end products in small lots with level model scheduling will, itself, reduce variation in component part demand, variations in component part requirements can still occur because some components are only needed for some end products or because of variations in the number required for others.

Monden reports methods for the mixed model scheduling of final assembly lines that minimize the variation in usage for individual component parts (i.e., leveling or balancing the component part schedule). These methods, used by Toyota, enable the "consumption speed" for each part used in a mixed model line to be kept as constant as possible.

Figure 12.3 illustrates the nature of the variation in component part usage that can occur in a final assembly schedule. The graph in the upper left corner is for component 1. The horizontal axis represents the sequence number (K) for the end product items as they're scheduled on the assembly line, and the vertical axis represents the actual cumulative usage of a component (X_{jK}); that is, for component 1 ($j = 1$) in this case, beginning with the first end product to be assembled ($K = 1$) and continuing through to the last end product in the assembly sequence ($K = 10$ in this case).

The objective is to minimize variations in the actual use of the component (X_{jK}) around the dashed line which represents the average use of a component (e.g., component 1). That is, the objective is to sequence the products to minimize the distance between the dashed lines and the solid lines for the graphs in Figure 12.3. We calculate average component usage by dividing the total quantity of a component required by the total quantity of all end products in the assembly schedule (i.e., N_1/Q or 5/10 in the case of component 1). Average component usage at any point on the dashed line can be computed as KN_1/Q (e.g., when $K = 5$, average component usage is 2.5).

In order to minimize variation in actual component usage around average usage, Monden suggests using a heuristic to sequence the final assembly of end products on a production line. This heuristic works by building the sched-

FIGURE 12.3   Assembly Sequence Numbers versus Component Usage

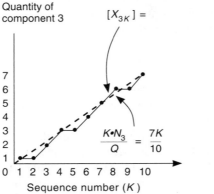
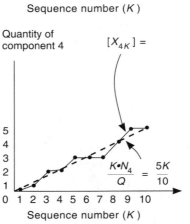

Source: Y. Monden, Toyota Production System (Norcross, Ga.: Institute of Industrial Engineers Press, 1983).

uled sequence of end products, starting with the selection of the first end product to be assembled and working toward the final end product. At each step in the scheduling sequence, we use Equation (12.1) to determine the next end product to be scheduled:

$$D_{Ki} = \sqrt{\sum_{j=1}^{\beta} \left(\frac{KN_j}{Q} - X_{j,K-1} - b_{ij} \right)^2} \qquad (12.1)$$

where:

D_{Ki} = The distance to be minimized for sequence number K and for end product i.

β = The number of different components required.

K = The schedule sequence number of the current end product.

N_j = The total number of component j required in the entire final assembly sequence.

Q = The total number of end products to be assembled in the final assembly sequence.

$X_{j,K-1}$ = The cumulative number of component j actually used through assembly sequence $K-1$.

b_{ij} = The number of component j required to make one unit of end product i.

The example in Figure 12.4 illustrates the use of the Monden heuristic. In this case three end products ($i = 1, 2,$ and 3) require four component parts. A total of 10 end products (Q) are to be produced which require the following total quantities for components 1 through 4: $N_1=5$, $N_2=8$, $N_3=7$, and $N_4=5$. Therefore, the N_j/Q ratios for components 1 through 4 are: 5/10, 8/10, 7/10, and 5/10.

The final assembly sequence for this example is built by applying the heuristic (Equation 12.1) to select each end product in the sequence. The first end product in sequence number one ($K=1$) is selected by computing D_{Ki} for each end product and selecting the minimum D_{Ki} as follows:

$$D_{1,1} = \sqrt{\left(\frac{1 \times 5}{10} - 0 - 1\right)^2 + \left(\frac{1 \times 8}{10} - 0 - 0\right)^2 + \left(\frac{1 \times 7}{10} - 0 - 1\right)^2 + \left(\frac{1 \times 5}{10} - 0 - 1\right)^2} = 1.11. \qquad (12.2)$$

$$D_{1,2} = \sqrt{\left(\frac{1 \times 5}{10} - 0 - 1\right)^2 + \left(\frac{1 \times 8}{10} - 0 - 1\right)^2 + \left(\frac{1 \times 7}{10} - 0 - 0\right)^2 + \left(\frac{1 \times 5}{10} - 0 - 1\right)^2} = 1.01. \qquad (12.3)$$

$$D_{1,3} = \sqrt{\left(\frac{1 \times 5}{10} - 0 - 0\right)^2 + \left(\frac{1 \times 8}{10} - 0 - 1\right)^2 + \left(\frac{1 \times 7}{10} - 0 - 1\right)^2 + \left(\frac{1 \times 5}{10} - 0 - 0\right)^2} = 0.79. \qquad (12.4)$$

FIGURE 12.4   Example Products and Component Requirements

End product (i)	1	2	3
Planned production quantity Q_i	2	3	5

End Product (i) \ Component i	1	2	3	4
1	1	0	1	1
2	1	1	0	1
3	0	1	1	0

Source: Y. Monden, *Toyota Production System* (Norcross, Ga.: Institute of Industrial Engineers Press, 1983).

In this case, end product 3 is selected since it has the minimum D_{Ki} ratio of .79. Next, the actual cumulative usage for each component item (X_{jK}) is updated using the $X_{jK} = X_{j,K-1} + b_{ij}$. These computations as well as the D_{Ki} values are shown on the right-hand side of Figure 12.5.

The second end product to be scheduled in the sequence ($K=2$) is selected after new values of D_{Ki} are computed:

$$D_{2,1} = \sqrt{\left(\frac{2 \times 5}{10} - 0 - 1\right)^2 + \left(\frac{2 \times 8}{10} - 1 - 0\right)^2 + \left(\frac{2 \times 7}{10} - 1 - 1\right)^2 + \left(\frac{2 \times 5}{10} - 0 - 1\right)^2} = 0.85. \quad (12.5)$$

$$D_{2,2} = \sqrt{\left(\frac{2 \times 5}{10} - 0 - 1\right)^2 + \left(\frac{2 \times 8}{10} - 1 - 1\right)^2 + \left(\frac{2 \times 7}{10} - 1 - 0\right)^2 + \left(\frac{2 \times 5}{10} - 0 - 1\right)^2} = 0.57. \quad (12.6)$$

$$D_{2,3} = \sqrt{\left(\frac{2 \times 5}{10} - 0 - 0\right)^2 + \left(\frac{2 \times 8}{10} - 1 - 1\right)^2 + \left(\frac{2 \times 7}{10} - 1 - 1\right)^2 + \left(\frac{2 \times 5}{10} - 0 - 0\right)^2} = 1.59. \quad (12.7)$$

Since end product 2 has the smallest D_{Ki} value, it's placed second in the assembly sequence. Again, the actual cumulative component usage values (X_{jK}) are updated as is shown in the right-hand side of Figure 12.5. The overall solution for the final assembly line sequence is shown as the last line in Figure 12.5. It's this solution that produced the actual usage values for components shown in Figure 12.3.

In situations where the number of end products and components is large, this approach could be applied by scheduling short periods such as an hour, instead of a day. Additional research on mixed model assembly scheduling is required. Such research can provide important benefits in stabilizing factory schedules for component operations under JIT.

FIGURE 12.5   Final Assembly Sequence

K	D_{K1}	D_{K2}	D_{K3}	Sequence of end products (i)								X_{1K}	X_{2K}	X_{3K}	X_{4K}
1	1.11	1.01	.79†	3								0	1	1	0
2	0.85	0.57†	1.59	3	2							1	2	1	1
3	0.82†	1.44	0.93	3	2	1						2	2	2	2
4	1.87	1.64	0.28†	3	2	1	3					2	3	3	2
5	1.32	0.87†	0.87	3	2	1	3	2				3	4	3	3
6	1.64	1.87	0.28†	3	2	1	3	2	3			3	5	4	3
7	0.93	1.21	0.82†	3	2	1	3	2	3	3		3	6	5	3
8	0.57†	0.85	1.59	3	2	1	3	2	3	3	1	4	6	6	4
9	1.56	0.77†	1.01	3	2	1	3	2	3	3	1	2	5	7	6	5
10	—	—	0†	3	2	1	3	2	3	3	1	2	3	5	8	7	5

†Indicates smallest distance D_{Ki}.

Source: Y. Monden, Toyota Production System (Norcross, Ga.: Institute of Industrial Engineers Press, 1983).

Schedule Stability in Implementing JIT

Stabilizing the manufacturing schedule is an important objective under JIT, especially when customer demand is erratic and unpredictable. Chapman reports a general model for maintaining final assembly schedule stability while implementing JIT. This approach is particularly applicable in make-to-stock environments where finished-goods inventories can be used to buffer final customer demand uncertainty.

Many companies cite a stable production environment as necessary to concentrate resources toward simplifying shop-floor systems, setup time reduction, product quality programs, process improvements, and machine maintenance. All these process modifications support shorter production lead times, reduced lead time variation, and increased product flexibility. However, without a stable production environment, manufacturing resources tend to be devoted to reacting to constant schedule changes instead of achieving process improvements.

Chapman indicates one solution is to create an inventory barrier between the company and its customers to absorb demand fluctuations while the internal JIT activity takes place. This "wall" of inventory would enable the company to establish a stable schedule covering the manufacturing lead time to allow allocation of resources toward JIT improvement activities. This inventory is temporary and should be reduced as JIT is implemented. There are two key questions concerning this implementation strategy:

- How does the organization determine how much inventory is enough?
- How does the organization eventually eliminate the temporary buffer?

Chapman recommends using statistics on demand and manufacturing lead time to determine the approximate size of the downstream inventories necessary to stabilize manufacturing schedules. This isolates manufacturing from demand fluctuations when finished-goods inventory is used as the barrier. Lot size is calculated using the economic order quantity (EOQ) formula. The average inventory required is calculated as:

$$\text{Average inventory} = \text{EOQ}/2 + S \tag{12.8}$$

where:

S = Safety stock = $z\sqrt{\bar{L}\sigma_D^2 + \bar{D}^2\,\sigma_L^2}$.

$\bar{D}$ = Average demand per day.

$\bar{L}$ = Average manufacturing lead time in days.

σ_L = Standard deviation of manufacturing lead time.

σ_D = Standard deviation of the demand per day.

z = Desired customer service level expressed as a multiple of the standard deviation of demand during lead time, usually between 0 and 3.

Proper implementation of JIT should reduce both $\overline{L}$ and S (by reducing σ_L), thereby reducing the average inventory calculated in Equation 12.8. The methods for achieving these improvements include reduced setup times, improved yields, and smaller queues, which provide shorter manufacturing lead times, more predictable manufacturing lead times, and increased facility flexibility. In addition, a shorter manufacturing lead time implies that the period over which the schedule must be stabilized is reduced. Since requirements for the near future can generally be forecast with much better accuracy than those for the more distant future, these changes enhance our ability to meet customer demand. All these improvements should combine to reduce lot size and safety stock, thereby reducing the "wall" of inventory.

The example in Figures 12.6 and 12.7 illustrates the impact of improving setup cost and manufacturing lead time on the average finished-goods inventory. In the example, reducing setup cost from \$350 to \$50 and manufacturing lead time from 35 days to 5 days results in a 63 percent reduction in the average finished-goods inventory $[(461 - 172)/461 = 63\%]$. This approach to stabilizing schedules while implementing JIT enables manufacturing to focus on improvement, and the stabilizing vehicle (the "wall" of inventory) is to be removed as process improvements are achieved.

SUPPLIER MANAGEMENT

A second area where MPC system design strongly impacts JIT is in developing suppliers to support JIT delivery. One aspect of inadequate JIT delivery concerns the transportation schedule for purchased items. Hill and Vollmann argue deliveries are critical in the JIT environment, and customer pickup from local vendors can be a better alternative than vendor delivery. When a manufacturer manages its own inbound local deliveries, it benefits from more timely information on upsets, reduced transportation costs, reduced transaction costs, and, most importantly, reduced uncertainty in deliveries.

In a JIT company, by definition, very small buffers exist between vendors and factory. In the best of circumstances, vendors deliver directly to points of use several times per day. In such companies we'd expect vendor deliveries to be scheduled only hours or days before they're needed on the factory floor. The exact timing depends on the manufacturer's confidence in its vendor's delivery performance. It's important to note in any true JIT firm, a stockout has very high costs, typically shutting down an entire line or factory. Hill and Vollmann report the experience of some JIT companies that still have substantial safety lead times used to protect against inventory shortages and other firms with hidden buffers. They therefore argue that the JIT firm should give serious consideration to assuming responsibility for inbound transportation because of reduced delivery uncertainty and reduced transportation costs.

FIGURE 12.6 Example of Reductions in Inventory Levels with JIT Improvements

Ordering cost	EOQ[1]	Average Manufacturing lead time, $\bar{L}$ (days)	Safety stock[2]	Average inventory[3]
350	836	35	199	617
340	824	34	193	605
330	812	33	188	594
320	800	32	182	582
310	787	31	176	570
300	774	30	171	558
290	761	29	165	545
280	748	28	159	534
270	734	27	154	521
260	721	26	148	509
250	707	25	143	496
240	692	24	137	483
230	678	23	131	470
220	663	22	126	457
210	648	21	120	444
200	632	20	115	431
190	616	19	109	417
180	600	18	103	403
170	583	17	98	389
160	565	16	92	375
150	547	15	86	360
140	529	14	81	345
130	509	13	75	330
120	489	12	70	315
110	469	11	64	298
100	447	10	58	282
90	424	9	53	265
80	400	8	47	247
70	374	7	41	229
60	346	6	36	209
50	316	5	30	188
40	282	4	25	166
30	244	3	19	141
20	200	2	13	113
10	141	1	8	78

1. EOQ = Economic order quantity.
2. Safety stock based on .95 stock out probability ($Z = 1.64$).
3. Average finished-goods inventory = ½ EOQ plus safety stock.
Notes:
  Annual demand = 5,000 units.
  Inventory carrying cost = 25% of item unit cost per year.
  Item unit cost = $20.
  Average daily demand = 13.7 units.
  Standard deviation of daily demand = 3 units.
  Standard deviation of manufacturing lead time = 0.25 $\bar{L}$ days (assumes simultaneous reductions in lead time and lead time variance).

Source: S. N. Chapman, "Schedule Stability and the Implementation of Just-in-Time," *Production and Inventory Management* 31, no. 3 (1990).

FIGURE 12.7   Average Finished-Goods Inventory versus Manufacturing Lead Time

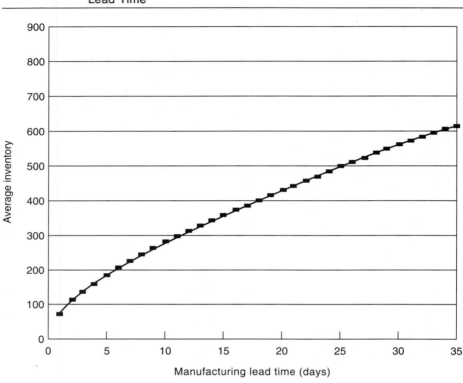

Source: S. N. Chapman, "Schedule Stability and the Implementation of Just-in-Time," *Production and Inventory Management* 31, no. 3 (1990).

Hill and Vollmann present a formulation of the "JIT Vendor Pickup Problem" and propose a scheduling heuristic. This problem is stated as follows:

> For a given number of vehicles with limited capacity find a one-week vendor pickup schedule that will minimize the total inventory carrying, travel, and setup costs subject to the constraint that the manufacturer does not run out of any item from any vendor.

A multistep heuristic scheduling algorithm is proposed to provide solutions to this problem. These steps include:

1. Determining the number of pickups per week for each vendor.
2. Assigning these pickups to each day of the week.
3. Developing an initial vehicle schedule for each vehicle for each day.
4. Applying three improvement heuristics to improve the schedules as much as possible.

Since the heuristic does not construct the vehicle schedule until after the number and day of the vendor pickups are decided, a key feature of the method involves determining an upper and a lower bound on the travel time to and from each vendor (Vendor = 1, 2 . . . k).

The upper and lower bounds on the travel time to each vendor are determined in the following manner. The upper bound, $dtmax(k)$, is assumed to be simply the time to visit the vendor and return to the manufacturer's plant, and not visit any other vendor. If $t(0,k)$ represents the travel time from the manufacturer's plant (0) to vendor k, then the complete trip takes $[t(0,k) + t(k,0)]$ hours. The example shown in Figures 12.8 and 12.9 illustrates this calculation. Using the data shown in Figure 12.9, the round-trip travel time from the plant to vendor 3 is four hours.

The lower bound on the travel time, $dtmin(k)$, to and from a vendor is assumed to be the travel time from the vendor closest to vendor k plus the time to travel from vendor k to the next-closest vendor. If j is the closest vendor and l is the next-closest vendor to vendor k, then the lower bound on the travel time is $[t(j,k) + t(k,l)]$. In the data shown in Figure 12.9, the nearest vendor to vendor 3 is vendor 6, and the next-closest is vendor 2. Therefore, the minimum travel time to vendor 3 is determined to be three hours.

Having calculated upper and lower bounds on the travel time to each vendor, the heuristic is then able to calculate upper and lower bounds on the number of pickups to be scheduled each week for each vendor. This information is needed to construct the actual vehicle schedule. Bounds on the

---

FIGURE 12.8 Example JIT Pickup Problem Map

•
Vendor 1
$\overline{D}(1) = 1,000$ units/week
c (1) = $10

•
Vendor 2
$\overline{D}(2) = 200$ units/week
c (2) = $60

•
Vendor 3
$\overline{D}(3) = 300$ units/week
c (3) = $30

•
*Production
site (0)*

•
Vendor 4
$\overline{D}(4) = 400$ units/week
c (4) = $400

•
Vendor 5
$\overline{D}(5) = 50$ units/week
c (5) = $4,000

•
Vendor 6
$\overline{D}(6) = 600$ units/week
c (6) = $200
$\overline{D}(7) = 700$ units/week
c (7) = $40

Note: Each dot (·) represents a location on the map.

Source: A. V. Hill and T. E. Vollmann, "Reducing Vendor Delivery Uncertainties in a JIT Environment," *Journal of Operations Management* 6 (1986).

FIGURE 12.9 JIT Pickup Problem: Travel Time Matrix (Times in Decimal Hours)

		To vendor					
	0	*1*	*2*	*3*	*4*	*5*	*6*
From vendor 0	0.0	1.6	0.6	2.0	1.7	1.6	1.5
1	1.6	0.0	1.9	3.5	0.7	1.3	3.1
2	0.6	1.9	0.0	1.6	2.2	2.2	1.6
3	2.0	3.5	1.6	0.0	3.7	3.6	1.4
4	1.7	0.7	2.2	3.7	0.0	0.7	3.1
5	1.6	1.3	2.2	3.6	0.7	0.0	2.7
6	1.5	3.1	1.6	1.4	3.1	2.7	0.0

Source: A. V. Hill and T. E. Vollmann, "Reducing Vendor Delivery Uncertainties in a JIT Environment," *Journal of Operations Management* 6 (1986).

number of pickups are calculated using the expression for the minimum-cost number of pickups, $n(k)$, shown in Equation 12.9:*

$$n(k) = [[.5C_r \Sigma_{i \epsilon P(k)} D(i)c(i)]/[\Sigma_{i \epsilon P(k)} C_P(i) + \text{VVC } dt(k)]]^{.5} \qquad (12.9)$$

where:

$n(k)$ = Number of stops to vendor k per week.

C_r = Inventory carrying cost (percent of unit cost/week).

$D(i)$ = Weekly demand (usage) rate at the customer for item i.

$c(i)$ = Unit cost for item i.

$C_P(i)$ = Setup cost required to pick up item i.

VVC = Variable vehicle cost ($/hour of travel time).

$dt(k)$ = Additional travel time to make a pickup at vendor k.

*Notes:

  1. With $n(k)$ pickups at vendor k at even intervals, the average quantity picked up is $\overline{D}(i)/n(k)$, and the average lot size inventory is $.5\overline{D}(i)/n(k)$ units.

  2. The demand $\overline{D}(i)$ rates are assumed to be constant.

  3. All items sourced from each vendor are picked up upon each stop to the vendor.

  4. The preceding formulation is similar to the periodic review inventory model where setup cost is $\Sigma C_P(i) + \text{VVC } dt(k)$ and weekly demand is $\Sigma \overline{D}(i)$ where we sum all over $i \epsilon P(k)$.

An upper bound for the number of pickups each week for a vendor, $n\text{max}(k)$, is calculated by substituting the lower bound on the travel time to the vendor, $dt\text{min}(k)$, for the $dt(k)$ term in Equation 12.9. Likewise, a lower bound for the number of pickups each week for a vendor, $n\text{min}(k)$, is calculated by substituting the upper bound on the travel time to the vendor, $dt\text{max}(k)$, for the $dt(k)$ term in Equation 12.9. Once these bounds have been calculated, the heuristic assigns pickups for a vendor to a particular day(s) of the week. Next the actual vehicle schedule for making the pickups is constructed.

The authors make two key assumptions in using their heuristic algorithm. First, they assume that each time a vendor pickup is scheduled to be picked up, all of the items purchased from that vendor are included in the pickup. Second, they calculate total weekly costs using an equation similar to that used in an inventory model. This cost equation is shown in Equation 12.10:

$$\text{TWC} = \text{Inventory carrying cost} + \text{setup cost} + \text{travel cost} \qquad (12.10)$$
$$= \Sigma_k \left[.5C_r \, \Sigma_{i\epsilon P(k)} \, D(i) \, c(i)/n(k) + [\Sigma_{i\epsilon P(k)} \, C_P \, n(k) + (\text{VVC})(\text{TT})] \right.$$

where:

$P(k) = $ Set of all items to be picked up from vendor k. If item i is sourced at vendor k, we denote this as $i\epsilon P(k)$.

$\text{TT} = $ Total travel time per week.

The flowchart shown in Figure 12.10 lists the eight steps in the Hill and Vollmann heuristic. Step 1 begins by initializing the variables in the algorithm. Step 2 calculates the upper and lower bounds on the travel time to each vendor, $dt\text{max}(k)$ and $dt\text{min}(k)$, and then calculates the upper and lower bounds on the number of pickups per week for each vendor, $n\text{max}(k)$ and $n\text{min}(k)$. In Step 3, a parametric range for the minimum-cost number of pickups to be made weekly at each vendor, $n(k)$, is used to generate a range of schedules by defining $n(k)$ as equaling $n\text{min}(k) + \text{alpha}[n\text{max}(k) - n\text{min}(k)]$. When alpha is zero, the number of pickups at each vendor is at the lower bound $n\text{min}(k)$. Conversely, when alpha is one, the number of pickups at each vendor is at the upper bound $n\text{max}(k)$. Next, Step 4 assigns the $n(k)$ pickups determined in Step 3 to the five days of the week. It's important to assign pickups evenly over the week to minimize inventory carrying costs. This step is also shown in Figure 12.10.

Step 5 takes a "first cut" at scheduling the vehicles on each day to efficiently make the pickups, using the Clarke-Wright "travel time saved" heuristic. This procedure allows for constraints on the vehicle capacities (weight and volume) as well as on the maximum route length and travel time per day. If the Clarke-Wright heuristic is unable to find a feasible schedule for a day because of the preceding constraints, the pickup with the lowest marginal economic value is reassigned to the next day, and the heuristic is retried. $V(k)$, the marginal economic value for a pickup at vendor k, may be estimated using Equation 12.11.

$$V(k) = .5C_r \Sigma_{i\epsilon P(k)} \left[\overline{D}(i)c(i)/n(k)^2 + C_P(i) \right] + dt(k) \, \text{VVC} \qquad (12.11)$$

where $dt(k)$ equals $n\text{min}(k) + \text{alpha} \, [n\text{max}(k) - n\text{min}(k)]$.

Step 6 attempts to improve the schedule constructed in Step 5 by applying three iterative improvement heuristics: the exchange heuristic, the insertion heuristic, and the deletion heuristic. The exchange heuristic tests all pairwise

---

FIGURE 12.10   Hill and Vollmann's Algorithm Flow Chart

Step 1:   Read input data.
Number of vehicles, $\overline{D}(i)$, $c(i)$, $C_{,}$, VVC, $P(k)$, $t(k, l)$.

Step 2:   Initialize.
Set alpha to zero.
Calculate:
   $d t$max (k)
   $d t$min (k)
   $n$max (k)
   $n$min (k)

Step 3:   Calculate $n(k)$ for all k.
$n(k) = n\text{min}(k) + \text{alpha} [n\text{max}(k) - n\text{max}(k)]$

Step 4:   Assign the $n(k)$ pickups to days of the week.
If $n(k) \geqslant 5$, then assign pickups to each day of the week and set $n(k)$ to $n(k) - 5$.
If $n(k) = 4$, then assign pickups to Monday, Tuesday, Thursday, and Friday.
If $n(k) = 3$, then assign pickups to Monday, Wednesday, and Friday.
If $n(k) = 2$, then assign pickups to Tuesday and Thursday.
If $n(k) = 1$, then assign a pickup to Wednesday.

Step 5:   Apply the Clarke-Wright heuristic to develop initial vehicle schedules for each day.

Step 6:   Apply improvement heuristics.

Step 7:   If this is the lowest-cost schedule so far, save this schedule as the best schedule.

Step 8:   Alpha = alpha + .1. If alpha ≤ 1, go to Step 3. Otherwise stop and report the best schedule.

Source: A. V. Hill and T. E. Vollmann, "Reducing Vendor Delivery Uncertainties in a JIT Environment," *Journal of Operations Management* 6 (1986).

---

exchanges of all vendors in the schedule (within a route, between routes on the same day, or between routes on different days) to find the exchange that will most reduce total incremental cost. At every iteration of the heuristic the best exchange is found and made. The exchange heuristic is continued until no further exchanges will result in a lower total incremental cost schedule. The insertion and deletion heuristics are applied in a similar manner to find the best opportunity to improve total incremental cost. If any improvement is found, the best improvement is made in the schedule and then the heuristic is run again. The three heuristics are applied repetitively in sequence until no further improvement can be made.

If the current schedule has the lowest total incremental cost so far, this schedule is saved in Step 7. Otherwise, the alpha value is increased by 0.1 in

Step 8 and the algorithm returns to Step 3. When the algorithm reaches an alpha value of 1.0, the algorithm stops and reports the best schedule found.

Figures 12.8, 12.9, and 12.11 show an example JIT pickup problem illustrating the Hill and Vollmann procedure. In this example the objective is to find a schedule for the manufacturer's one vehicle given setup cost is zero, the cost of carrying inventory per week is 1 percent of the inventory investment, and variable vehicle cost is $12 per hour per vehicle. Figures 12.12 and 12.13 detail a solution resulting from an alpha setting of 0.

In using this model, the set of vendors considered for JIT pickups must be predefined. Vendors more than one day away cannot be incorporated into daily routes for JIT pickup. Likewise, vendors that already provide inexpensive and reliable service may also be excluded. The question of whether a

---

FIGURE 12.11 Example JIT Pickup Problem: Minimum and Maximum Number of Stops per Week

Vendor k	Item i	Demand $\overline{D}(i)$	Cost c(i)	Travel times		Number of stops	
				dtmin(k)	dtmax(k)	nmin(k)	nmax(k)
1	1	1,000	$ 10	2.0	3.2	1.1	1.4
2	2	200	$ 60	1.3	1.3	2.0	2.0
3	3	300	$ 300	3.0	4.1	1.0	1.1
4	4	400	$ 400	1.4	3.4	4.4	6.9
5	5	50	$4,000	2.0	3.2	5.1	6.5
6	6	600	$ 200	2.9	3.1	4.5	4.6
6	7	700	$ 40	—	—	—	—

Source: A. V. Hill and T. E. Vollmann, "Reducing Vendor Delivery Uncertainties in a JIT Environment," *Journal of Operations Management* 6 (1986).

FIGURE 12.12 Example JIT Pickup Problem: Day Assignments (Alpha = 0)

Vendor schedule		Truck schedule	
Vendor	Visited on days	Day	Stops at vendors
1	W	M	4, 5, 6
2	T, TH	T	2, 4, 5, 6
3	W	W	1, 3, 5, 6
4	M, T, TH, F	TH	2, 4, 5, 6
5	M, T, W, TH, F	F	4, 5, 6
6	M, T, W, TH, F		

Source: A. V. Hill and T. E. Vollmann, "Reducing Vendor Delivery Uncertainties in a JIT Environment," *Journal of Operations Management* 6 (1986).

FIGURE 12.13 Example JIT Pickup Problem: Final Schedule (Alpha = 0)

Day	Sequence	Travel time
Monday	0–4–5–6–0	6.90 hours
Tuesday	0–2–6–5–4–0	7.30 hours
Wednesday	0–1–5–6–3–0	9.00 hours
Thursday	0–2–6–5–4–0	7.30 hours
Friday	0–4–5–6–0	6.90 hours

Total travel time per week = 37.40 hours.
Total travel cost per week = $448.80.

Item	Average inventory (units)	Average inventory investment	Average carrying cost
1	570.24	$ 5,702.28	$ 57.02
2	66.86	$ 4,011.43	$ 40.11
3	171.07	$ 5,132.14	$ 51.32
4	78.43	$31,373.69	$313.74
5	7.94	$31,765.14	$317.65
6	95.26	$19,052.68	$190.53
7	111.11	$ 4,444.30	$ 44.44

Total carrying cost per week = $1,014.81.
Total increment cost per week = $1,463.61.

Comments:
1. Each of the above schedules starts at 8 A.M.
2. This schedule is for alpha equal to zero. If the algorithm were completed, it would generate similar schedules for alpha equal to .1, .2, . . . , 1.0. The algorithm would then report the "best" schedule (lowest total incremental cost schedule).
3. In this example we allowed overtime for the driver on Wednesday.
4. Carrying cost is assessed on the average inventory assuming a five-day work week with a single eight-hour shift per day. In other words, carrying cost is affected by inventory levels during the nights and weekends as well as during the workday.

Source: A. V. Hill and T. E. Vollmann, "Reducing Vendor Delivery Uncertainties in a JIT Environment," *Journal of Operations Management* 6 (1986).

vendor can provide less expensive delivery is largely a matter of which firm can run the most effective distribution system given the set of vendors for the manufacturer and the set of customers for the vendor.

PRODUCTION FLOOR MANAGEMENT

Another group of core JIT components from Figure 12.2 that has a major impact on MPC systems is production floor management. In this section we discuss the results of research concerned with setup time reduction, small lot size production, and kanbans. We also report work that assesses the impact of different plant factors, such as the variability of demand rates and processing times, on the production floor management components and, subsequently, on production performance.

Setup Time Reduction

Setup time reduction is a key aspect of JIT since it supports reductions in manufacturing lead times and inventories. It also enables small lot sizes and kanban systems for material flow—achieving major improvements in production floor management. To date, however, little research has been reported concerning the impact of setup reduction programs on plant performance. Recently, Hahn, Bragg, and Shin have posed three questions for managers regarding implementing setup reduction programs:

- What are the direct costs and benefits associated with setup time, and how should they be measured?
- Under what conditions is setup time reduction most effective?
- How does setup reduction compare to other alternatives such as overtime and capacity expansion?

In pursuing these questions, Hahn, Bragg, and Shin considered the situation when demand exceeds the production capacity at an operation. They noted that under these conditions, much of the previous inventory model research focused on increasing order quantities by reducing setup frequencies to meet capacity limitations, as opposed to investigating the possible increase of available capacity to allow increased setup frequency or investigating reducing the time required per setup. Based on this observation, they developed a decision model to evaluate the setup time reduction and increased capacity alternatives in a way that allows for reductions in both order quantities and inventory levels in capacity-constrained situations.

Figures 12.14 and 12.15 illustrate Hahn, Bragg, and Shin's decision model to evaluate the effectiveness of setup time reduction activities. Three different joint lot-sizing models' solutions are compared. The IS (independent solution) model described by Buffa and Miller does not consider capacity as a limiting factor, and lot sizing is simply based on setup and inventory holding cost trade-offs. The model for the IS approach is

$$N^* = \sqrt{\frac{\Sigma_j C_{Hj} A_j \left(1 - \dfrac{d_j}{p_j} \right)}{2\Sigma_j C_{pj}}} = \frac{A_j}{Q_j^*} \qquad (12.12)$$

$$\text{TIC}^* = \sqrt{2\Sigma_j C_{Pj}\Sigma_j C_{Hj} A_j \left(1 - \frac{d_j}{p} \right)} \qquad (12.13)$$

FIGURE 12.14 Example Problem for Setup Time Reduction

Product	Annual demand A_j	Setup time S_j	Processing time/unit P_j	Inventory holding costs/unit/year
A_1	18,000	10 hours	.025 hours	$ 75.00
A_2	34,000	10 hours	.025 hours	87.50
A_3	36,000	10 hours	.025 hours	100.00
A_4	13,500	10 hours	.025 hours	62.50
A_5	25,000	10 hours	.025 hours	62.50
Total	126,500	50 hours		

Annual inventory holding costs = 25% of unit costs.
Production capacity = 16 hours/day × 250 working days = 4,000 hours.
Setup cost = 10 hours × 5 workers × $10/hour = $500/setup. (This
   assumes setup cost is solely a function of setup time.)
Additional capacity cost = $75.00/hour.
Setup time reduction = $1,000/hour of setup time reduced.
Regular capacity cost = $50/hour (for 4,000 hours per year).

Source: C. K. Hahn, D. J. Bragg, and D. Shin, "Impact of Setup Variable on Capacity and Inventory Decisions," *Academy of Management Review* 13, no. 1 (1988).

$$\text{TIC} = \Sigma_j \left[\left(C_{Pj} \times \frac{A_j}{Q_j} \right) + \left(C_{Hj} \left(1 - \frac{d_j}{p_j} \right) \right) \times \frac{Q_j}{2} \right] \qquad (12.14)$$

where:

N^* = Optimal number of order cycles.
Q_j = Order quantity in units for item j.
TIC = Total incremental costs per year.
C_{Hj} = Inventory carrying cost per unit per year for item j.
C_{Pj} = Setup costs per order for item j.
d_j = Daily demand rate in units for item j.
p_j = Daily production rate in units for item j.
A_j = Demand per year in units for item j.

The CC (common cycle) model described by Sugimori et al., as well as Pinto and Mabert, takes into account capacity limitations, but does so by increasing order quantities and reducing setup frequencies to meet the capacity limitations. The CC model is

$$T^* = \frac{\Sigma_j S_j}{W - \Sigma_j A_j P_j} \quad \text{and} \qquad (12.15)$$

$$Q_j^* = TA_j \qquad (12.16)$$

where:

W = Available capacity in hours.
S_j = Setup time per order for item j (in hours).
P_j = Processing time per unit for item j (in hours).
T^* = Length of the processing cycle time (a fraction of a period) or $1/N^*$.

In contrast, the Hahn, Bragg, and Shin optimum solution model determines the order quantities by considering whether it's better to increase available capacity to allow increased setup frequency, or to reduce the time required per setup. This approach represents a new direction in inventory model research.

In this example the capacity limitation is 4,000 hours, and the IS model exceeds this capacity by 1,214.5 hours. Hahn, Bragg, and Shin have added the assumptions that capacity can be expanded at a cost of $75 per hour (150 percent of the current cost rate of $50 per hour) and that setup time can be decreased at a cost of $1,000 per setup hour eliminated. Using their model, three steps are required to determine whether the setup time reduction or the capacity expansion alternative is best.

Step 1: Determine the optimum level of capacity given the capacity increase cost (H = $75) and current setup times:

$$W^* = \Sigma_j A_j P_j + \sqrt{\frac{\Sigma_j S_j \Sigma_j C_{Hj} A_j \left(1 - \frac{d_j}{p_j}\right)}{2H}} = 3,162.5 + 1,675.5 \quad (12.17)$$

$$= 4,838 \text{ hours.}$$

Step 2: Compute the savings rate per hour of setup time reduction (SR) made possible by shorter setup times:

$$SR = \frac{\Sigma_j C_{Hj} A_j \left(1 - \frac{d_j}{p_j}\right)}{2(W - \Sigma_j A_j P_j)} = \$5,026. \quad (12.18)$$

Step 3a: If cost of the setup time reduction ($1,000 per setup hour in this example) is greater than the hourly savings rate (SR), setup time shouldn't be reduced.

Step 3b: If cost of the setup time reduction is less than the savings rate— as is the case in the example ($1,000 < $5,026)—then the ideal condition is to eliminate the setup time entirely. (However, in reality it may be difficult to eliminate the setup time completely.)

Figure 12.15 shows the results of applying the IS, CC, and Hahn, Bragg, and Shin models. The total relevant cost of the IS solution ($193,673) includes

FIGURE 12.15 Comparison of Different Solution Methods for Capacity Constrained Situation

	Buffa and Miller (IS)	Sugimori et al., Pinto and Mabert (CC)	Optimum solution
Time between cycles	6.09 days	14.92 days	7.46 days
Number of orders/year	41.04 orders	16.75 orders	33.51 orders
Setup hours	2,052.00 hours	837.50 hours	1,675.50 hours
Processing time	3,162.50 hours	3,162.50 hours	3,162.50 hours
Total capacity requirements	5,214.50 hours	4,000.00 hours	4,838.00 hours
Inventory holding costs	$102,600	$251,360	$125,680
Capacity increase costs†	91,087	-0-	62,840
Total relevant costs	$193,687	$251,360	$188,520
Cost of existing capacity	$200,000	$200,000	$200,000
Total costs	$393,687	$451,360	$388,520

†The IS solution assumes there are no capacity limits. If we assess the cost of $75 per hour for additional capacity, the cost will increase to $91,087.

Source: C. K. Hahn, D. J. Bragg, and D. Shin, "Impact of Setup Variable on Capacity and Inventory Decisions," *Academy of Management Review* 13, no. 1 (1988), p. 26.

the cost of the additional 1,214.5 hours required to implement the smaller lot sizes at the higher setup frequency and the inventory holding costs. The total relevant cost of the CC model solution ($251,360) reflects the larger inventory carrying cost associated with the reduced setup frequency and larger order quantities because of the capacity constraint.

The total relevant cost of the Hahn, Bragg, and Shin model ($188,520) reflects the fact that when the option of increasing the capacity at $75 per hour is added to the CC model, a different lot-sizing solution results and some additional capacity is desirable (i.e., 838 hours) given the change in economic trade-offs in the model. This new capacity level is determined in Step 1 of the Hahn, Bragg, and Shin model. Steps 2 and 3 of their model indicate in this example savings can be obtained by reducing setup times at the rate of $1,000 per reduced setup hour. For example, if setup time were reduced from 10 to 5 hours per setup (i.e., from 50 to 25 hours for all five products each cycle), then total setup hours per year could be reduced from 1,675.5 to 838. This would eliminate the additional capacity cost of $62,840 since the 4,000 hours of regular capacity would be sufficient. This would require setup reduction costs of $25,000.

As a result of an analysis of the implications of their decision model, Hahn, Bragg, and Shin develop several propositions concerning setup time reduction activities. These include:

- The general effectiveness of setup time reduction as an alternative to increasing capacity improves as the ratio of total setup requirements to total capacity requirements increases. Therefore, setup time reduction is most effective when a large portion of available capacity is dedicated to setup operations.
- The incremental effectiveness of setup time reduction as an alternative to increasing capacity accelerates as a further reduction is made. Therefore, in order to assess the true impact of setup time reduction on the capacity alternative, the maximum extent of setup time reduction must be considered.

The importance of this research is to provide a more fundamental understanding of the impact of setup time reduction activities on plant performance. Further research in this area should provide managers with guidelines indicating those products and processes where implementing setup time reduction programs under JIT can provide maximum benefits.

Determining the Optimal Number of Kanbans

In many firms the focus on setup time reduction and small lot production leads directly to implementing simple shop-floor scheduling systems. Rapid transit through the shop means low work-in-process inventory where each operation in the process can be triggered by usage at the next operation, and a kanban card (or other signal) can be used to authorize additional production. The number of circulating kanban cards is important in the system's operation. Too many kanban cards produce excess work-in-process inventory, while too few lead to production-floor disturbances.

When the demand rate is constant for each period over the planning horizon, the number of kanban cards can be determined using the Toyota formula:

Number of kanban cards =

$$\frac{\text{Demand rate} \times \text{Lead time} \times [1 + \text{Policy variable (i.e., safety stock)}]}{\text{Container size}} . \quad (12.19)$$

However, when the demand rate is permitted to vary between periods over the planning horizon, further analysis is required. Bitran and Chang have proposed several mathematical models to determine the optimal number of kanban cards for each period when the demand rate varies. Their approach is deterministic and involves multiple time periods and multiple work centers. For some special cases the linear integer model can be transformed into a linear program (LP), requiring $2NT$ constraints (N is the number of work centers and T is the number of time periods) with $NT + T$ variables.

Because of the size and complexity of the optimization approaches, Moeeni and Chang propose two heuristics for determining the number of kanban cards. These heuristics can be applied manually, producing solutions that appear to be close to the LP solutions for those problems that can be solved by LP, and close to the LP approximations to the integer programming problem when LP won't work. These heuristics are based on the following assumptions:

- Each work center in the production process produces only one type of product.
- Demand must be met with no backorders.
- Production lead time is zero for all operations.
- Kanban lead time is 1 (i.e., a kanban card returned to a work center at time t can be used to initiate production at time $t+1$).
- There are no limitations on capacity.
- There is a single inventory point between any two consecutive work centers so that each kanban card is circulated for an individual item.

In addition, a feasible solution to the problem will satisfy the following constraints:

$$W_{t-1}^1 + X_t^1 \geq D_t \text{ for all } t, \tag{12.20}$$

$$W_{t-1}^j + X_t^j \geq X_t^n \text{ for all } n, t, \text{ and } j \in P(n), \tag{12.21}$$

$$U_t^n \geq X_t^n \text{ for all } n, \text{ and} \tag{12.22}$$

$$U_t^n + W_{t-1}^n = U_o^n + W_o^n \text{ for all } n \text{ and } t \tag{12.23}$$

where:

n = Work center index; $n \in \{1,...N\}$. Note that $n = 1$ corresponds to the last work center in the production sequence.

t = Time index; $t \in \{0,1,...T\}$. A relatively short time period (e.g., one hour or one-half shift) is typically used.

$P(n)$ = Set of immediately preceding work center(s) to work center n.

U_t^n = Number of kanban cards waiting to initiate production at the beginning of time t at work center n. Note that U_o^n is the number of kanban cards injected into work center n at the start of the planning horizon.

W_t^n = Number of kanban cards for completed work at work center n at the end of time t. The number of kanban cards for completed work in work center n at the beginning of the planning horizon W_o^n is the starting inventory.

X_t^n = Number of kanban cards that actually initiate production in work center n at time t.

D_t = Demand for final product in terms of an integral number of kanban card orders.

The first constraint ensures that demands are met each period while the second constraint requires production at any work center not to exceed the amount of product supplied by its immediately preceding work centers. The third constraint limits production at any work center to the number of kanbans available at this point. The fourth constraint states the total number of kanban cards for each work center must be constant over the planning horizon, that is, $U_o^n + W_o^n$.

Heuristic 1. Although not formally stated, this heuristic's objective is to minimize costs associated with work-in-process inventory. The solution process begins with values for demand (D_t) and the number of kanban cards for completed work (W_o^n), and determines values for the number of kanban cards ready to initiate production at each work center at the start of period 1 (U_o^n). (Note that $S(n)$ is the set of immediately succeeding work center(s) of work center n.) There are two steps in this heuristic:

Step 1: Set $U_o^1 = \text{Max}_t \{D_t\} - W_o^1$. \hfill (12.24)

Step 2: Set $U_o^n = U_o^{s(n)} + W_o^{s(n)} - W_o^n, n = 2,3,...N$. \hfill (12.25)

The example in Figure 12.16 is a production process with four work centers where production starts at work center 4 and moves through work centers 3 and 2 to the final operation at work center 1. There are five periods in the planning horizon with the following demand rates and starting inventory levels:

FIGURE 12.16 Solution to Example Problem Using Heuristic 1

	U_o^n	Period 1	Period 2	Period 3	Period 4	Period 5
Demand		4	15	20	5	12
Work center 1	20					
Beg. inv.		0	16	5	0	15
Prod.		20	4	15	20	5
End. inv.		16	5	0	15	8
Work center 2	15					
Beg. inv.		5	0	16	5	0
Prod.		15	20	4	15	20
End. inv.		0	16	5	0	15
Work center 3	20					
Beg. inv.		0	5	0	16	5
Prod.		20	15	20	4	15
End. inv.		5	0	16	5	0
Work center 4	18					
Beg. inv.		2	0	5	0	16
Prod.		18	20	15	20	4
End. inv.		0	5	0	16	5

Demand rates	Starting inventory
$D_1 = 4$	$W_0^1 = 0$
$D_2 = 15$	$W_0^2 = 5$
$D_3 = 20$	$W_0^3 = 0$
$D_4 = 5$	$W_0^4 = 2$
$D_5 = 12$	

Given this information, the heuristic determines the number of kanbans to be injected into the system at the beginning of the planning horizon (U_o^n):

Step 1: $U_0^1 = \text{Max}\{4,14,20,5,12\} - 0 = 20.$ (12.26)

Step 2: $U_0^2 = 20 + 0 - 5 = 15.$ (12.27)

$\qquad U_0^3 = 15 + 5 - 0 = 20.$ (12.28)

$\qquad U_0^4 = 20 + 0 - 2 = 18.$ (12.29)

Column 2 of Figure 12.16 shows this solution along with the starting values for the kanban cards waiting to initiate production (W_o^n) and the demand rates for each period (D_t). The figure also simulates the production process over the five periods, indicating the beginning and ending inventories as well as the production at each work center each period.

Heuristic 2.   While heuristic 1 produces a feasible solution, improved solutions can be obtained by using a different heuristic. The second heuristic works on reducing the number of kanban cards at the work centers upstream from the final work center (1) by applying the following steps:

Step 1: Set $U_0^1 = \text{Max}\{D_t\} - W_0^1$. For the preceding example,
$\qquad U_0^1 = \text{Max}_t\{D_t\} - W_0^1 = 20 - 0 = 20.$

Step 2: Divide the planning horizon into F smaller subplanning horizons such that every subproblem contains a time-ordered, nondecreasing demand. For example, if $F = 2$ in the preceding example, the planning horizon can be divided into $\{4,15,20\}$ and $\{5,12\}$.

Step 3: For every subproblem find the average demand $(\overline{D}^n)$ and the second largest demand $(\hat{D}^n)$. Set $\hat{D}^n = 0$ for subproblems with only one element, and round any noninteger average demand up to the next integer. For the preceding example, the $\{\overline{D}^n, \hat{D}^n\}$ values are $\{13,15\}$ and $\{9,5\}$.

Step 4: Compute $U^* = \text{Max}_n\{\overline{D}^n, \hat{D}^n\}$. For the example,
$\qquad U^* = \text{Max}\{(13,15),(9,5)\} = 15.$

Step 5: Compute $U_o^k = U^* - W_o^k$, $k \epsilon P(1)$. For the example,
$\qquad U_o^2 = 15 - 5 = 10.$

Step 6: Compute $U_o^n = U_o^{s(n)} + W_o^{s(n)} - W_o^n$, for all n without a designated number of kanbans. For the example, $U_o^3 = 10 + 5 - 0 = 15$ and $U_0^4 = 15 + 0 - 2 = 13$.

This solution is shown in Figure 12.17, whose information is similar to Figure 12.16's. Note that this solution requires fewer kanbans and there are no shortages (backorders) in any period, so we have a feasible solution again.

Experimental Results.   Moeeni and Chang report experimental results that compare the performance of heuristic 2 against the linear programming approximation Bitran and Chang proposed. These procedures were compared on the basis of the cost of work-in-process inventory for all work centers, using the percentage increase of the objective function value: (heuristic − LP)/LP. Figure 12.18's results indicate the heuristic's performance relative to the LP method improves rapidly as variability in demand is decreased. Moeeni and Chang also investigated the impact of planning horizon length on the heuristic's performance. Figure 12.18 shows that the changes in planning horizon length don't seem to affect this heuristic's performance. These results indicate that the development of heuristic methods for determining the number of kanbans when demand varies from period to period looks very promising, but their performance may well be sensitive to differences in operating conditions. The next section discusses these differences further.

FIGURE 12.17   Solution to Example Problem Using Heuristic 2

		Period 1	Period 2	Period 3	Period 4	Period 5
Demand	U_o^n	4	15	20	5	12
Work center 1	20					
Beg. inv.		0	11	5	0	10
Prod.		15	9	15	15	10
End. inv.		11	5	0	10	8
Work center 2	10					
Beg. inv.		5	0	6	0	0
Prod.		10	15	9	15	15
End. inv.		0	6	0	0	5
Work center 3	15					
Beg. inv.		0	5	0	6	0
Prod.		15	10	15	9	15
End. inv.		5	0	6	0	0
Work center 4	13					
Beg. inv.		2	0	5	0	6
Prod.		13	15	10	15	9
End. inv.		0	5	0	6	0

FIGURE 12.18   Comparison of Heuristic and Linear Programming Solutions

Planning horizon (number of periods)	Demand variability			
	$U(0,300^\dagger)$	$U(150,300)$	$U(225,300)$	Mean
5	$.125^\ddagger$	.028	.003	.052
10	.039	.004	.000	.014
15	.164	.034	.005	.068
20	.144	.048	.019	.070
Mean	.118	.029	.007	.051

†The uniform distribution was used for demand $U(x,y)$ where x,y represents the demand range.
‡Average of the percent increase in work-in-process inventory cost for the heuristic over the programming procedure for five randomly generated problems.

Source: F. Moeeni and Y. Chang, "An Approximate Solution to Deterministic Kanban Systems," *Decision Sciences* 21, no. 3 (1990), pp. 608–25.

JIT PERFORMANCE AND OPERATING CONDITIONS

Research on various aspects of JIT indicates production performance may be affected by the operating conditions under which JIT is implemented. Substantial differences in such conditions exist between plants. For example, firms that have made progress in areas such as stabilizing the MPS, developing highly skilled multifunction employees, standardizing machine processing times, reducing setup times, and eliminating variation may achieve better results overall. Therefore, other managers may be concerned with determining to what extent JIT performance is sensitive to such factors, and what guidelines exist for determining what factors should receive major attention in implementing JIT. Several recent research efforts have examined these issues using simulation analysis.

Variability in Operating Conditions

Huang, Rees, and Taylor report the results of a Q-Gert model that simulates a kanban-based MPC system with three production lines, each having between one and three work centers. They examined three factors' effects on JIT performance: variable processing times, variable master production schedules, and imbalances between different production work centers.

Figure 12.19 shows the effect of variation in the machine processing times on the mean (u) and standard deviation (σ) of four measures of operating performance: overtime, preproduction activity inventory at the final assembly work center, postproduction activity inventory at the same work center, and

FIGURE 12.19 Impact of Variability in Processing Times on JIT Performance

Processing time distribution	Overtime (minutes)		Preproduction activity inventory at operation 1		Post-production activity inventory at operation 1		End-of-day production (before overtime) at operation 1	
	μ	σ	μ	σ	μ	σ	μ	σ
Constant								
1 kanban	.00	.00	300.00	.00	.00	.00	1000.00	.00
2 kanbans	.00	.00	600.00	.00	10.00	30.00	1000.00	.00
Exponential								
1 kanban	32.80	19.39	233.90	95.10	9.82	29.76	982.74	37.84
2 kanbans	12.07	18.22	549.91	90.66	20.67	53.32	985.75	35.78
Normal (σ = 4.8)								
1 kanban	29.93	9.49	258.54	85.86	5.73	23.23	950.41	50.07
2 kanbans	19.58	8.91	572.49	69.45	15.82	44.48	952.05	50.03
Normal (σ = 24)								
1 kanban	160.04	43.87	219.68	92.66	12.25	32.79	817.53	80.99
2 kanbans	110.07	42.22	508.04	133.52	36.45	31.59	904.11	96.99

Source: P. Y. Huang, P. L. Rees, and B. W. Taylor, "A Simulation Analysis of the Japanese JIT Technique for a Multiple, Multi-stage Production System," *Decision Sciences* 14 (1983).

end-of-day production at the final assembly work center. Clearly, increases in processing time variability have a major effect on overtime, postproduction inventory, and end-of-day production performance, but not on preproduction inventory. The impact of increased processing time variance on mean over-time is nearly linear as Figure 12.20 shows. Figure 12.21 shows the effect of additional kanbans on overtime requirements. Adding a second kanban re-duces the impact of processing time variation on overtime requirements. But additional kanbans after the second have little effect on overtime requirements.

In a second set of experiments, Huang, Rees, and Taylor evaluate the effect of uncertainty in the demand rate on the plant. While very little data is re-ported on these experiments, they do indicate overtime and end-of-day pro-duction increase substantially. Changes in the other measures weren't nearly as dramatic. In a final set of experiments, Huang, Rees, and Taylor tested the impact of variance in both machine processing times and demand rate. These results appear in Figure 12.22. Again end-of-day production variance and overtime are up sharply, while all other values are similar to those we see in Figure 12.19.

These results reveal the favorable performance impact that standardizing machine processing times has under JIT. They also indicate the impact of schedule instability on production performance under JIT through the de-

FIGURE 12.20 The Effect of Increasing Uncertainty on Overtime

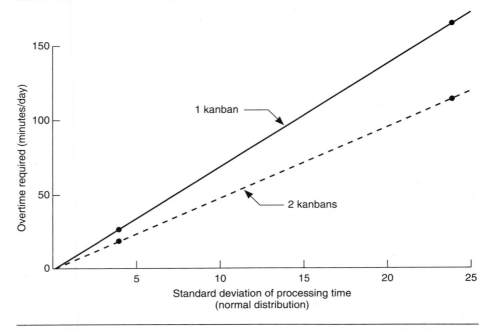

Source: P. Y. Huang, P. L. Rees, and B. W. Taylor, "A Simulation Analysis of the Japanese JIT Technique for a Multiple, Multistage Production System," *Decision Sciences* 14 (1983).

mand uncertainty experiments. The results also point to the need for further research in this area.

Lot Size

Shen's simulation research addressed another element of operating conditions that can impact JIT performance: lot size. In this case a Slam model is used to analyze a two–work-center kanban production process. The objective of Shen's research is to determine the effect of changes in production lot size on manufacturing performance in terms of unfilled demand and finished-goods inventory. One experiment simply varied lot size; the results appear in Figure 12.23. Here smaller lot sizes required a higher portion of the production capacity when the setup time doesn't change. The results indicate an unfavorable trade-off between reduced lot size and unfilled demand, although inventories (and cycle stock) drop. Further analysis of Shen's results in Figure 12.24 shows the interaction between lot size, setup time, and unfilled de-

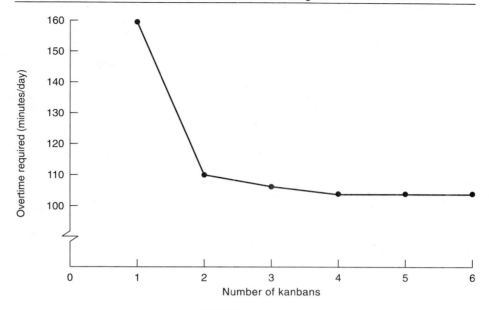

FIGURE 12.21 The Effect on Overtime of Increasing the Number of Kanbans

Processing time standard deviation = 24.

Source: P. Y. Huang, P. L. Rees, and B. W. Taylor, "A Simulation Analysis of the Japanese JIT Technique for a Multiple, Multistage Production System," *Decision Sciences* 14 (1983).

FIGURE 12.22 Impact of Uncertainty in Demand and Processing Times on JIT Performance

Distribution	Overtime (minutes)		Preproduction activity inventory at operation 1		Post-production activity inventory at operation 1		End-of-day production (before overtime) at operation 1	
	μ	σ	μ	σ	μ	σ	μ	σ
Constant								
1 kanban	.00	.00	300.00	.00	.00	.00	1000.00	.00
2 kanbans	.00	.00	600.00	.00	10.00	30.00	1000.00	.00
Exponential								
1 kanban	3.06	91.75	233.04	95.68	10.29	30.38	873.97	98.10
2 kanbans	−14.96	87.88	551.18	89.55	21.67	54.26	877.26	104.58
Normal (σ = 4.8)								
1 kanban	29.50	53.55	259.39	85.31	5.62	23.03	924.66	59.26
2 kanbans	18.99	52.93	570.64	72.67	16.18	45.23	929.59	61.13
Normal (σ = 24)								
1 kanban	227.32	302.21	221.37	93.86	16.32	39.96	731.55	185.39
2 kanbans	178.73	288.59	510.20	131.25	51.59	76.73	787.61	221.87

Source: P. Y. Huang, P. L. Rees, and B. W. Taylor, "A Simulation Analysis of the Japanese JIT Technique for a Multiple, Multistage Production System," *Decision Sciences* 14 (1983).

FIGURE 12.23 Impact of Lot Size on JIT Performance

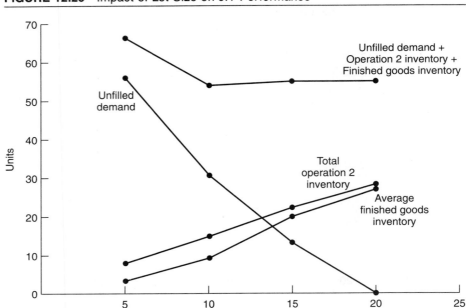

Source: H. N. Shen, "Simulation of a Two Stage Kanban System Using Slam," Working Paper CMOM-87 (Rochester, N.Y.: William E. Simon Graduate School of Business Administration, University of Rochester, 1987).

mand. When setup times were reduced in Shen's experiments by 20 percent, 40 percent, and 60 percent, unfilled demand also fell, indicating the importance of a setup reduction time program in releasing capacity for more timely production.

Comparing MPC System Approaches

Simulation research by Krajewski, King, Ritzman, and Wong indicates while kanban systems do indeed perform well under certain operating conditions, so do other MPC systems approaches like material requirements planning (MRP) and reorder point (ROP). As a result of their research they argue that the operating conditions themselves are the key to major improvements in manufacturing performance. Thus, they argue the key to improved performance is "shaping the production environment" through factors such as re-

FIGURE 12.24 Effect of Setup Time Reduction on Manufacturing Performance

Setup reduction	Lot size	Unfilled demand	Total operation 2 inventory	Average finished-goods inventory
0%	5	55	8	3
	10	30	15	8
	15	13	22	20
	20	0	28	27
20%	5	49	8	4
	10	24	15	11
	15	10	21	20
	20	0	28	31
40%	5	40	7	4
	10	17	14	12
	15	1	21	24
	20	0	28	34
60%	5	29	7	4
	10	7	14	13
	15	0	21	27
	20	0	29	35

Source: H. N. Shen, "Simulation of a Two Stage Kanban System Using Slam," Working Paper CMOM-87 (Rochester, N.Y.: William E. Simon Graduate School of Business Administration, University of Rochester, 1987).

duced setup times and lot sizes, improved product yield rates, and increased employee flexibility.

Krajewski et al. developed a discrete-event simulator capable of modeling nearly any batch manufacturing plant. Their simulator was developed over an eight-year period with the help of plant managers representing diverse manufacturing environments in a major U.S. corporation. This model's results, incorporating up to 250 inventory items and 250 work centers, were validated using the panel of plant managers.

The robustness of the JIT production system using kanbans was tested under a wide variety of plant-floor conditions. Figure 12.25 shows these experiments' results. Three clusters of operating condition factors indicate statistically significant differences in operating performance: inventory, process, and product structure. Analysis of the inventory cluster shows kanban performance (measured in terms of inventory investment) is highly sensitive to lot sizes and setup times. Like Shen's work, this underscores the importance of reduced setup times and lot sizes. Furthermore, customer service (past due demand) performance is very sensitive to changes in process

FIGURE 12.25   Kanban System Performance

Factor operating conditions	Setting	Inventory (weeks of supply)	Past due demand (% orders)
Customer influence (CI)	High	1.95	48.3
	Low	1.89	48.2
	Relative difference[†]	3%	0%
Vendor influence (VI)	High	2.06	46.5
	Low	1.78	51.9
	Relative difference	15%	−11%
Buffer mechanisms (BM)	High	2.06	46.8
	Low	1.78	49.7
	Relative difference	15%	−6%
Product structure (PS)	High	2.14	38.8
	Low	1.71	57.7
	Relative difference	23%[‡]	−39%[‡]
Facility design (FD)	High	2.14	43.6
	Low	1.70	52.9
	Relative difference	23%	−19%
Process (P)	High	1.91	56.2
	Low	1.93	40.3
	Relative difference	− 1%	33%[‡]
Inventory (I)	High	2.95	50.8
	Low	0.89	45.7
	Relative difference	107%[§]	11%
Averages	High	2.17	47.3
	Low	1.67	49.5
	Relative difference	26%	−5%

[†]Relative differences are the differences between the results from the high and low settings divided by the average performance for all kanban experiments. Effects on weekly labor requirements aren't shown. The PS and I clusters were significant at the 0.01 level, while all other clusters weren't statistically significant. Changing the PS factor changes work content by definition. The I cluster had a 34 percent impact on labor because of the added labor requirements for long setup times at the high setting.
[‡]Statistically significant at the 0.05 level. The Biomedical Package P Series (1977) was used for the fractional factorial analysis of variance. The eight interaction terms were pooled (none were statistically significant) to create the error term.
[§]Statistically significant at the 0.01 level.

Source: L. J. Krajewski, B. E. King, L. P. Ritzman, and D. S. Wong, "Kanban, MRP and Shaping the Production Environment," *Management Science* 33, no. 1 (January 1987).

factors such as scrap rates, low worker flexibility, and high equipment failure; but these factors have little effect on inventory performance.

Changes in product structure affect kanban performance in reverse directions for inventory and customer service. Inventory performance improves with pyramid product structures, shallow bills of material, and low commonality of parts, while customer service improves when an inverted pyramid structure is used with deep bills of material and high part commonality. When the pyramid structure is used, kanban tends to reduce inventories at the end-item level, exposing the plant to problems in master production scheduling due to forecast errors. Likewise, the inverted pyramid structure leads to larger component lot sizes and inventories making the plant less sensitive to market uncertainties experienced with the pyramid structure.

These results suggest the importance of the inventory, process, and product structure factors in implementing JIT systems. Indeed, a second set of simulation experiments reported by Krajewski et al. further underscores the importance of shaping the production environment by improving operating conditions. The second set of experiments compares a kanban system's performance with that of the reorder point system (ROP). The objective is to determine how much of the performance improvement is attributable to the kanban system as opposed to the operating conditions under which it was applied.

The kanban system in these experiments differs from the ROP system in that kanban has more frequent reviews of inventory positions, stages material on the shop floor, doesn't adjust lot sizes for yield losses, and uses the first-in-system/first-served dispatching rule. The simulation results, shown in Figure 12.26, for two shop configurations, indicate the kanban system and ROP systems are similar in performance. Differences in inventory and customer service exist because two kanbans were used in the kanban system experiments to assure reasonable customer service performance. This provides additional safety stock not afforded the ROP systems. Allowing this added stock would have made the performance results even closer.

FIGURE 12.26 Comparison of the Kanban and Reorder Point Approaches

Experimental setting	Labor requirements (hours per week)	Inventory (weeks of supply)	Past due demand (weeks of supply)
Shop 1:			
Kanban	21,888	4.12	0.17
Daily ROP	21,925	3.80	1.02
Shop 2:			
Kanban	6,563	14.68	0.67
Daily ROP	6,583	11.65	1.18
Weekly ROP	4,562	16.18	0.05

Source: L. J. Krajewski, B. E. King, L. P. Ritzman, and D. S. Wong, "Kanban, MRP and Shaping the Production Environment," *Management Science* 33, no. 1 (January 1987).

The authors conclude the reason the kanban system appears attractive is not the system itself. ROP systems perform just as well. The reason is the kanban system is simply a convenient way to implement a small lot production strategy and to expose operating conditions that need improvement. Thus, the key to improved performance is to shape the production environment.

CONCLUDING PRINCIPLES

Chapter 12 describes recent research on a broad range of topics where MPC system design is affected by implementing JIT manufacturing methods. These topics have been grouped into the areas of production floor management, scheduling, and supplier management. All these areas have practical JIT core components that have implications for MPC system design and performance. The following principles are critical:

- The sequence of models in a mixed model final assembly schedule must be carefully planned to achieve stability in component production.
- The release and control of the number of kanban cards must be closely managed to achieve inventory reductions without factory disruptions.
- Raw-material stocks and transportation costs can be significantly cut by shifting the supplier management strategy to include the scheduling of inbound transportation.
- The proportion of setup time to total capacity should be considered in determining the attractiveness of setup time reduction programs.
- Setup time reduction programs should be implemented to enable small-lot production.
- Small-lot production must be undertaken to achieve advantages associated with kanban MPC methods.
- MPC professionals must realize that improving operating conditions can be more important than changing the features of the MPC system to achieve improved manufacturing performance.

REFERENCES

Bitran, G. R., and L. Chang. "Mathematical Programming Approach to Deterministic Kanban Systems." *Management Science* 33, no. 4 (April 1987), pp. 427–41.

Buffa, E. S., and J. G. Miller. *Production Inventory Systems: Planning and Control.* Homewood, Ill.: Richard D. Irwin, 1979.

Chapman, S. N. "Schedule Stability and the Implementation of Just-in-Time." *Production and Inventory Management* 31, no. 3 (3rd quarter 1990), pp. 66–70.

Clarke, G., and J. W. Wright. "Scheduling of Vehicles from a Central Depot to a Number of Delivery Points." *Operations Research* 11, no. 568 (1963).

Deleersnyder, J.; T. J. Hodgson; H. Muller; and P. J. O'Grady. "Kanban Controlled Pull Systems: An Analytic Approach." *Management Science* 35, no. 9 (September 1989), pp. 1079–91.

Hahn, C. K.; D. J. Bragg; and D. Shin. "Impact of the Setup Variable on Capacity and Inventory Decisions." *Academy of Management Review* 13, no. 1 (1988), pp. 91–103.

Hill, A. V., and T. E. Vollmann. "Reducing Vendor Delivery Uncertainties in a JIT Environment." *Journal of Operations Management* 6 (1986), pp. 381–92.

Huang, P. Y.; P. L. Rees; and B. W. Taylor. "A Simulation Analysis of the Japanese JIT Technique for a Multiple, Multistage Production System." *Decision Sciences* 14 (1983), pp. 326–44.

Karmarkar, U. S., and S. Kekre. "Batching Policy in Kanban Systems." *Journal of Manufacturing Systems* 8, no. 4 (1989).

Krajewski, L. J.; B. E. King; L. P. Ritzman; and D. S. Wong. "Kanban, MRP and Shaping the Production Environment." *Management Science* 33, no. 1 (January 1987), pp. 39–57.

Kubiak, W., and S. Sethi. "A Note on 'Level Schedules for Mixed Model Assembly Lines in Just-in-Time Production Systems'." *Management Science* 37, no. 1 (January 1991), pp. 121–22.

Leib, R., and R. A. Miller. "JIT and Corporate Transportation Requirements." *Transportation Journal* 27 (1988), pp. 5–10.

Miltenburg, G. J. "Level Schedules for Mixed-Model Assembly Lines in Just-in-Time Production Systems." *Management Science* 35, no. 2 (February 1989), pp. 192–207.

———, and G. Sinnamon. "Scheduling Mixed-Model Multi-Level Just-in-Time Production Systems." *International Journal of Production Research* 27, no. 9 (1989), pp. 1487–509.

Mitra, D., and I. Mitrani. "Analysis of a Kanban Discipline for Cell Coordination in Production Lines." *Management Science* 36, no. 12 (December 1990), pp. 1548–66.

Moeeni, F., and Y. Chang. "An Approximate Solution to Deterministic Kanban Systems." *Decision Sciences* 21, no. 3 (summer 1990), pp. 608–25.

Monden, Y. *Toyota Production System.* Norcross, Ga.: Institute of Industrial Engineers Press, 1983.

Narasimhan, R., and S. A. Melnyk, "Setup Time Reduction and Capacity Management: A Marginal Cost Approach." *Production and Inventory Management* 31, no. 4 (4th quarter 1990), pp. 55–59.

Pinto, P., and V. Mabert. "A Joint Lot-Sizing Rule for a Fixed Labor Cost Situation." *Decision Sciences* 17 (1986), pp. 139–50.

Sakakibara, S.; B. B. Flynn; and R. G. Schroeder. "A Just-in-Time Manufacturing Framework and Measurement Instrument." Working Paper 90-10. Curtis L. Carlson School of Management, University of Minnesota, August 1990.

Shen, H. N. "Simulation of a Two Stage Kanban System Using Slam." Working Paper CMOM 87-03. Rochester, N.Y.: Graduate School of Business Administration, University of Rochester, 1987.

Sugimori, Y.; K. Kusunoki; F. Cho; and S. Uchikawa. "Toyota Production System and Kanban System—Materialization of Just-in-Time and Respect-for-Human System." *International Journal of Production Research* 151 (1977), pp. 553–64.

Yano, C. A., and Y. Gerchak. "Transportation Contracts and Safety Stocks for Just-in-Time Deliveries." *Journal of Manufacturing and Operations Management* 2, no. 4 (winter 1989), pp. 314–30.

DISCUSSION QUESTIONS

1. Figure 12.1 implies manufacturing strategy is part of the JIT program and is linked to quality, technology, and JIT core components. Do you agree?
2. The Monden (Toyota) heuristic for level schedule determination uses squared penalties for deviations from average requirements for component manufacturing. What other approaches might be suggested?
3. Chapman suggests using a "wall of inventory" to provide level schedules and insulation for manufacturing while implementing JIT improvements. What will it take to get this to work and assure the improvements get translated into reducing the wall?
4. "Today's kanban research is the modern equivalent of yesterday's economic order quantity research." Do you agree?
5. There's considerable evidence that getting correct operating conditions is more important than the choice between MRP, kanban, or reorder point methods in the material planning and control system. How general do you believe this situation to be?

PROBLEMS

1. TMI, Inc., produces electronics equipment for the telephone industry. At one plant a single assembly line produces memory circuits for cellular telephones using a mixed model scheduling approach. Component operations at this plant have recently been converted to a JIT material flow operation. The company wants to reduce demand variation at the component manufacturing stations using the Toyota approach. TMI has provided the following data for two end products, A and B (which use differing quantities of two components):

End product number	1	2
Planned quantity/shift	2	3

End product number	Component item usage/end item unit	
	A	B
1	4	1
2	1	2

   a. In what sequence should the five units (two of end product 1 and three of end product 2) be run on the final assembly line to minimize demand variation at the A and B component manufacturing station?

   b. Graph cumulative versus average usage for components A and B.

2. How would the sequence determined in the solution to problem 1 change if only end product 1 were produced?

3. The Brighton Electronics Company is planning to install JIT manufacturing in its hearing aid electronics plant. To stabilize operations while JIT is being implemented, it plans to invest in finished-goods inventory. The following data have been collected for the firm's Deluxe Model (#1234):

> Average daily customer demand = 2 units/day.
> Daily demand standard deviation = 0.5 units.
> Manufacturing lead time standard deviation = 2.0 days.
> Finished-goods inventory carrying cost = 25% of item value/year.
> Unit cost = $1,000.

Currently, the Deluxe Model's average manufacturing lead time is 75 days. Setup cost is $274 (or 27.4 hours at $10 per hour). The company plan is to reduce average manufacturing lead time to 5 days, the lead time standard deviation to 0.5 days, and the setup cost to $0.68 (or 4 minutes at $10 per hour).

   a. Calculate the Deluxe Model's economic order quantity for current and planned operations.

   b. Assuming that the company plans to hold sufficient safety stock to permit a .05 stockout probability for the current and planned operations, calculate the safety stock for both current and proposed operating conditions.

   c. What will the average finished-goods investment be for both current and planned operating conditions? What finished-goods inventory savings will occur?

4. During the past several months the Hanson Manufacturing Company has experienced an increased number of line shutdowns because of unreliable deliveries of incoming purchased parts from its JIT suppliers. As a result the company is considering a decision to manage the transportation of incoming purchased parts from three of its suppliers. The company has developed the following data concerning purchased items A, B, and C:

Item	Vendor	Average weekly usage	Item cost†
A	1	8,000 units	$50/unit
B	2	3,000	80/unit
C	3	40,000	20/unit

†This value reflects a transportation rebate given to Hanson if it picks up the item at the vendor's plant.

Notes: 1. Inventory carrying cost / week = 0.4% of item cost
2. Truck setup cost = $60 / item picked up
3. Variable vehicle cost = $100 / hour

Transportation time (in hours)

From vendor	To vendor			To plant
	1	2	3	
1	—	.1	.1	.85
2	.1	—	.1	1
3	.1	.1	—	.4
From plant	.85	1	.4	—

a. Calculate the minimum and maximum travel times using the Hill and Vollmann method for each item and vendor.

b. Calculate the upper and lower bounds on the number of stops per week for each item and vendor.

5. Vendors 1, 2, and 3 at the Hanson Manufacturing Company have agreed to give the firm a 10 percent rebate on the price of items A, B, and C if Hanson takes over transporting the items from the vendor's plant to Hanson's plant. The following data have been provided:

Item	Vendor	Current price/unit	New price/unit
A	1	$55.55	$50
B	2	88.88	80
C	3	22.22	20

a. Using the preceding data and the data in Problem 4, compute the combined weekly cost of purchasing and carrying inventory for items A, B, and C when vendors deliver the items to Hanson once a week.

b. Assuming Hanson picks up items A, B, and C from the vendor once a week, compute the weekly total incremental cost (TWC) of this alternative using the TWC equation in the Hill and Vollmann method. In this case one truck picks up the three items in one run, starting at the Hanson plant and visiting vendors 3, 2, and 1 (in that order) prior to returning to the plant.

c. What recommendations would you make to Hanson?

6. Given the following information:

Item	Annual demand (in units)	Setup time (in hours)	Processing time/unit	Inventory carrying cost/unit/year	Unit cost	Daily demand rate (in units)	Daily production rate (in units)
A	270,000	8	0.2 min./unit[†]	20% × Unit cost	$2.00	1,080	2,400
B	540,000	6	0.1 min./unit[‡]	20% × Unit cost	1.80	2,160	4,800

[†] or .00333 hours/unit.
[‡] or .00167 hours/unit.
Notes: 1. Setup cost = [(Item setup time) ($60) + $600 per item in setup waste].
      2. Assume 250 days per year and an eight-hour work shift per day.

a. Compute the optimal number (N^*) of times per year to produce items A and B using the IS model and assuming both items are run in each processing cycle.

b. Compute optimal order quantities for items A and B using the IS model.

c. Compute total incremental cost per year (TIC*) using the IS model.

d. How many hours per year of capacity are required to use this solution?

7. Using the data in Problem 6 and the following information:

Production capacity/year = 1,850 hours (since 150 hours per year are required for maintenance).

a. Compute the optimal processing cycle (T^*) to produce items A and B using the CC model and assuming both items are produced in each cycle.

b. Compute the optimal order quantities for items A and B using the CC model.

c. Compute TIC per year using the CC model.

d. How many hours per year of capacity are required to use this solution?

8. Using the data in Problems 6 and 7 and the following information:

Overtime capacity hours (exceeding 1,850) cost $90/hour.
Setup time can be reduced at a cost of $800 per setup hour reduced.

a. Compute the optimum level of capacity (W^*) using the Hahn, Bragg, and Shin model.

b. Compute the savings rate per hour of setup time reduction (SR) and determine whether a setup time reduction program should be implemented, using the Hahn, Bragg, and Shin model.

9. Using the Hahn, Bragg, and Shin model and the data from Problems 6 through 8:
   a. Compute the optimal order quantity for items A and B.
   b. Compute the total incremental cost per year for items A and B using the Hahn, Bragg, and Shin model.
   c. How many capacity hours per year are required to implement the Hahn, Bragg, and Shin solution?
   d. By what percentage would the setup times for items A and B need to be reduced in order to implement the Hahn, Bragg, and Shin solution with 1,850 hours per year of capacity? What would this reduction cost?

10. Complete the following table for the IC, CC, and Hahn, Bragg, and Shin model solutions computed in Problems 6 through 9.

	IS	CC	Hahn, Bragg, and Shin
Inventory carrying cost/year	___	___	___
Setup cost/year	___	___	___
Extra capacity cost/year	___	___	___
Total cost/year	___	___	___

11. Given the following information, please determine the number of kanban cards required:

> Demand rate   = 100 units/hour (constant).
> Lead time   = 2 hours.
> Safety factor = .05.
> Container size = 50 units.

12. The XYZ Electronics Company recently implemented JIT in its manufacturing plant and its major vendors. However, requirements for the subassemblies used in the final assembly operation vary considerably throughout each day. The following example shows this variation for one subassembly item.

	Hour #							
	1	2	3	4	5	6	7	8
Hourly subassembly A requirements (in terms of kanbans)	2	18	10	25	5	15	4	6

The subassembly is produced in two steps at work centers 102 and 103 as follows:

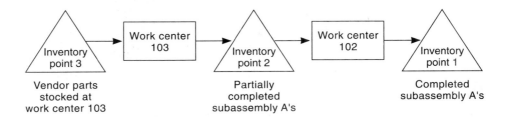

Assume there are now six kanbans of completed subassembly A's at inventory point 1 and two kanbans for partially completed subassembly A's at inventory point 2. Apply Moeeni and Chang heuristic 1 to determine the total number of kanbans at work centers 102 and 103.

13. Complete an eight-hour simulation in the following table using the solution obtained in Problem 12. Use a spreadsheet program for performing the simulation.

		Hour							
		1	2	3	4	5	6	7	8
Work center 102	Demand (in kanbans)								
	Beginning inventory								
	Production								
	Ending inventory								
Work center 103	Beginning inventory								
	Production								
	Ending inventory								

14. Using the data in Problem 12, please apply Moeeni and Chang heuristic 2 to determine the number of kanbans at work centers 102 and 103.

15. Complete the following eight-hour simulation using the solution determined with heuristic 2 in Problem 14. Use a spreadsheet program for the simulation. What differences do you observe between the number of kanbans in the solutions to Problems 12 and 14? Explain these differences.

		Hour							
		1	2	3	4	5	6	7	8
Work center 102	Demand (in kanbans)								
	Beginning inventory								
	Production								
	Ending inventory								
Work center 103	Beginning inventory								
	Production								
	Ending inventory								

Chapter 13

Advanced Concepts in Scheduling

This chapter addresses advanced issues in scheduling, with primary emphasis on detailed scheduling of individual jobs through work centers in a shop. The intent is to provide direction for the firm that has a working MPC system in place and wishes to enhance the shop-floor systems. This chapter also addresses scheduling issues associated with cellular manufacturing and flexible manufacturing systems (FMS).

The approaches in this chapter presume effective front-end, engine, and back-end systems are in place. Chapter 13 provides an application perspective to research in scheduling. It's completely beyond our scope to even summarize the vast amount of research on this topic. Rather, our interest here is to focus on some basic concepts and results, relate them to some of the newer manufacturing approaches, and show how you might apply results in certain operating situations.

Chapter 13 centers around four topics:

- A scheduling framework: What are the key definitions, performance criteria, and kinds of scheduling problems studied?
- Basic scheduling research: What are the fundamental scheduling problem structures and what are the practical implications of scheduling results that have been consistently verified in the research?
- Advanced research findings: What findings from advanced research seem to be particularly helpful in assigning jobs or labor to machines?
- Emerging issues in scheduling: What are the critical scheduling issues in cellular manufacturing and flexible manufacturing systems (FMS) and how do findings from other scheduling research apply?

Chapter 13 is most closely linked to Chapters 3 and 5, which describe just-in-time and basic shop-floor systems and their place within an overall MPC system framework. There are also indirect links to Chapters 6 and 14 for master production scheduling. Chapter 19 deals with optimized production technology (OPT), which raises additional scheduling issues.

A SCHEDULING FRAMEWORK

There are many ways to think about scheduling, as well as different kinds of scheduling problems and decisions. Before we delve into scheduling research, it's useful to first develop a brief framework for scheduling. This includes key definitions, criteria for judging scheduling performance, and important dimensions of scheduling problems.

We can define a schedule as a plan with reference to the sequence of and time allocated for each item or operation necessary to complete the item. This definition lets us think of a schedule that has a series of sequential steps, or a routing. The entire sequence of operations, the necessary sequential constraints, the time estimates for each operation, and the required resource capacities for each operation are inputs to developing the detailed plan or schedule.

This definition also allows us to think of component part scheduling based on product structures, with scheduling of components for subassemblies and, in turn, subassemblies to support end-item assembly. The detailed material planning module breaks down end-item schedules to subassembly and component schedules. Material plans also establish the associated resource capacity requirements.

We can also view scheduling as a process; that is, someone prepares a schedule either for when an end item will be completed or for what jobs are to be completed during a specified time by the work center of interest. This definition implies repetition of the scheduling task. We prepare the schedule, observe actual performance, and reschedule as uncertain events become resolved (e.g., forecasts of customer orders become actual customer orders, planned results become actual results).

Performance Criteria

Three primary objectives or goals apply to scheduling problems. The first goal concerns *due dates:* we typically want to avoid late job completion. The second goal concerns *flow times:* this objective is to minimize time a job spends in the system, from creation or opening of a shop order until it's closed. The final goal concerns *work center utilization:* we want to fully utilize the capacity of expensive equipment and personnel.

These three objectives often conflict. We can better meet due dates if more capacity is provided and if the work center capacity is less intensively utilized. Similarly, more capacity typically reduces flow time, but at reduced capacity utilization. If extra jobs are released to the shop, they tend to have longer flow times; but capacity can be better utilized and *perhaps* due date performance can be improved.

For each of the three primary scheduling objectives, we must establish exact performance measures. Moreover, for each objective there are competing measures. We might simply specify meeting due dates on a yes/no basis. More typically, however, due date performance is based on "lateness." *Average lateness* for jobs is one measure, but this raises the issue of "earliness." In calculating the average, can early jobs offset late jobs? Is earliness also undesirable? Is maximum lateness an issue? We follow the convention that lateness measures both positive and negative deviations from the due date.

Another alternative is to measure the variability of actual completion dates against due dates. The objective is to minimize the variance of lateness, thereby improving the reliability of delivery to customers. But this measure raises the question of which due date should be used to measure performance in an MRP environment when due dates are routinely revised by the MRP system.

Similar problems arise with flow times. Is the measure average flow time, variance in flow time, or maximum flow time? Should we weight flow time by some monetary value to favor shorter flow times for expensive work?

Work center utilization measures have equal problems. Are we interested in utilizing all work centers or selected centers? How do we select? How do we measure utilization?

Measurement issues are important to the MPC professional, both for understanding what particular research results mean and for establishing appropriate manufacturing performance criteria in applying the concepts; that is, an operating scheduling system must have unambiguous definitions of performance and these measures must be congruent with the firm's business objectives. We must understand performance criteria utilized in particular research studies and carefully assess the match between those criteria and what's truly important in an actual company.

Shop Structure

Another facet of a scheduling framework relates to the shop structure being studied. One important structure is a flow shop found in repetitive manufacturing and cellular production systems; that is, all jobs tend to go through a fixed sequence of the same routing steps. Other structures are more of a job shop nature involving custom-made products. Each particular job tends to have a unique routing, jobs go from one work center to another in a somewhat random pattern, and time required at a particular work center is also highly variable. Scheduling complexity and constraints in a flow shop can be quite different from those in a job shop. Particular performance criteria and scheduling systems' appropriateness should reflect those differences.

One dimension of shop structure has been clarified in scheduling research. It's tempting to think what works well in a shop with 10 work centers won't work well in a shop with 100 work centers. This, in fact, is not so. If their other attributes are the same (e.g., shop structure or percentage capacity utilization), conclusions drawn from a relatively small shop will be similar to those for a large shop.

Product Structure

Product structure is another facet of the scheduling environment that must be defined for a particular company. One question is the existence of either single part or assembly routings. The issue is whether the scheduling problem is dealing only with individual jobs or if we necessarily must worry about matched sets of parts (jobs); that is, must we schedule each component of an assembly to ensure all are done at the same time? In some companies, a large percentage of jobs are essentially one piece part per customer order. Other firms must produce parts and assemblies.

MRP partially addresses assembly scheduling in that the MRP planning process coordinates component due dates. In this context MRP is best seen as a scheduling technique. However, scheduling of MRP plans in a shop is sometimes a separate issue. To the extent that a shop scheduling procedure tries to deal with maintaining matched set due dates, the scheduling task becomes complex. Note in finite loading approaches, due dates are typically adjusted to reflect finite capacity constraints (which implies the performance criterion is work center utilization).

The type of processing time distribution is another product structure issue. Most research studies have been based on processing times which are represented by the negative exponential distribution. Some simulation studies have used empirical distributions. The practical issue is again whether research results are robust enough to apply to the wide variety of applied scheduling situations.

Another issue in both research and practice is using alternative routings. Design and maintenance of alternative routing files can be an enormous job. Moreover, decision rules for when to use alternative routings can be quite complex. On the other hand, alternative routings can aid operating performance. If one work center is overloaded and another work center is underloaded, alternative routings can improve due date performance, flow times, and work center utilization.

A related concern is use of operation overlapping. If we can start a job at a work center *before* it's completely finished on the previous work center, we can improve scheduling performance. Overlapping is a form of scheduling that increases flexibility, but it doesn't come without cost. It's necessary to have good information on time requirements and to start successive operations only when the work center won't run out of work (i.e., when operation 2 for a job requires less time per unit than operation 1, the job can only be started at the second operation when a sufficient queue exists).

Still another issue concerning product structure is the extent to which setup times remain the same, regardless of the sequence for processing jobs. Thus, in some firms, a potential setup time saving can be made by better sequencing jobs through a work center. However, only after investing in developing a data base and an appropriate scheduling system can we realize these savings.

A final issue concerning product structure is whether lot sizes are fixed or variable. For most scheduling research, lot sizes are fixed and don't vary as the sequence of operations are performed. In some recent research and practice, lot sizes have been larger for processing through some work centers than for others. Better schedule performance has been achieved.

Work Center Capacities

A final facet of a framework for scheduling relates to work center capacities. One consideration is the extent to which capacities are fixed or variable. This is analogous to the alternate routing issue. The extent to which capacity for a particular work center can be increased or decreased and the time delay to change the capacity both affect scheduling performance.

A related issue is how much machines' or labor's capacity limits a particular work center's capacity. The benefit of a labor-limited system versus a machine-limited system is the possibility of increased flexibility, since the same labor capacity can be assigned to several different machines. The extent to which a work force has multiple skills, and the degree of flexibility union contracts specify on these matters clearly influence this consideration. Most cellular manufacturing and just-in-time (JIT) approaches are based on increasing the scope of jobs and worker cross-training.

An additional issue in work center capacity is to focus attention on a subset of work centers: the bottlenecks. If we can more intensely utilize the bottleneck work center's capacities, we can improve overall schedule performance in several ways. Conversely, utilization of nonbottlenecks is *not* a high-priority action. Attempts to increase nonbottleneck utilization usually increase work-in-process inventories and average flow times.

BASIC SCHEDULING RESEARCH

Classic scheduling research has addressed two fundamental kinds of scheduling problems: static scheduling problems and dynamic scheduling problems. Some, but not all, results apply to both of these situations. The static problem consists of a fixed set of jobs to be scheduled until they're all completed. The dynamic scheduling problem deals with an ongoing situation. New jobs are continually added to the system, with emphasis on scheduling approaches' long-term performance.

This section describes static and dynamic scheduling problem formulations. We then present basic research findings detailing appropriate consequences for scheduling practice.

Static Scheduling Approaches

The **static scheduling** problem consists of a fixed set of jobs to be run. Typical assumptions are the entire set of jobs arrive simultaneously and all work centers are available at that time. Most static scheduling research has used a criterion called minimum "make-span" (minimum total time to process all jobs). This is a flow time criterion, not a due date or work center utilization criterion. Furthermore, the minimum make-span criterion isn't the same as the average flow time criterion, as we will show in discussing the one-machine scheduling case later.

Static scheduling research has been performed using deterministic processing times (known and nonvarying) and stochastic processing times (subject to random variations). Methods for dealing with deterministic times can be divided into those producing optimum results and those utilizing heuristic scheduling procedures. In general, optimization methods are only applicable to relatively small problems. The computational difficulty tends to increase exponentially with problem size.

Large-scale problems are usually treated with heuristic procedures called dispatching, or sequencing, rules. These are logical rules for choosing which available job to select for processing at a particular machine center. In using dispatching rules, scheduling decisions are made sequentially rather than all at once.

Dynamic Scheduling Approaches

Dynamic scheduling problems are those in which new jobs are continually being added over time. Processing times for these jobs can be either deterministic or stochastic, but most research has focused on the latter case. Analytic approaches have been based on queuing models that provide expected steady state conditions for certain kinds of situations and time distributions. The criteria applied in the queuing studies typically involve average flow time, average work-in-process or number of jobs in the system, and machine center utilization.

One approach in the dynamic scheduling studies is to use different scheduling (dispatching) rules at the work centers. The use of some rules gives better results for certain criteria. The original work in queuing studies concerned single-machine systems; later work extended the results to multiple machines. However, as the system size increases, simulation is the most frequently used research methodology. Moreover, simulation allows you to forgo the time-limiting assumptions inherent in most analytical queuing approaches.

Simulation studies of large-scale scheduling problems are again mainly based on dispatching rules. There are a substantial number of these studies, and the MPC professional must understand certain basic conclusions. Some results are quite counterintuitive. Furthermore, understanding some basic simulation issues (such as sample size requirements and run lengths as a function of problem complexity) is useful if you're contemplating a simulation study for a particular firm.

The One-Machine Case

Research on single-machine scheduling has been largely based on the static problem of how to best schedule a fixed set of jobs through a single machine, when all jobs are available at the start of the scheduling period. It's further assumed setup times are independent of the sequence.

If the objective is to *minimize total time* to run the entire set of jobs (i.e., the minimum make-span), it doesn't matter in which order jobs are run. In this case, the make-span equals the sum of all setup and run times for any sequence of jobs. However, if the objective is to *minimize the average time* each job spends at the machine (setup plus run plus waiting times), then we can show this will be accomplished by sequencing jobs in ascending order according to their total processing time (setup plus run time). As an example, if three jobs with individual processing times of one, five, and eight hours, respectively, are scheduled, *total time* required to run the *entire* batch under any sequence is 14 hours. If we process jobs in ascending order, the average time that each job spends in the system is $(1 + 6 + 14) \div 3 = 7$ hours. However, if we process jobs in the reverse order, average time in the system is $(8 + 13 + 14) \div 3 = 11.67$ hours.

This result has an important consequence. Average time in the system will always be minimized by selecting the next job for processing that has the shortest processing time at the current operation. This rule for sequencing jobs at a work center (called **shortest processing time,** or **SPT**) provides excellent results when we use the average time in system criterion.

SPT also performs well on the criterion of *minimizing the average number of jobs in the system.* As we've noted previously, work-in-process inventory levels and average flow time are directly related measures. If we increase or reduce one, the other changes in the same direction. Analytical work shows that the SPT rule again provides superior performance when the work-in-process criterion is applied in the single-machine case.

When the criterion is to *minimize the average job lateness,* again SPT is the best rule for sequencing jobs for the single-machine case. To introduce the lateness criterion, we first must establish due dates for the jobs. However, an interesting aspect of scheduling research is, no matter what procedure we use to establish due dates, SPT will minimize *average job lateness.*

The one-machine scheduling research is very useful in gaining insights into scheduling rules' behavior under particular criteria. The most important conclusion we can draw from the single-machine research is the SPT rule represents the best way to pick the next job to run, if the objective is to minimize average time per job, to minimize average number of jobs in the system, or to minimize average job lateness. However, if the objective is to minimize either the maximum lateness of any job or the lateness variance, then jobs should run in due date sequence.

The Two-Machine Case

Developing scheduling procedures for the two-machine case is somewhat more complex than for single-machine systems. In the two-machine case, we must schedule both machines to best satisfy whatever criterion is selected. Moreover, we have to consider job routings. We assume each job always goes from a particular machine to another machine. For analytically based research, we make additional assumptions, such as those for the one-machine case. For example, all jobs are available at the start of the schedule, and setup times are independent.

A set of rules has been developed to minimize the make-span in the two-machine case. Note while the minimum make-span doesn't depend on job sequencing in the one-machine case, this isn't true in the two-machine case. Additionally, if total time to run the entire batch of jobs is to be minimized, this doesn't ensure either the average time each job spends in the system or the average number of jobs in the system will also be minimized.

The following scheduling rules to minimize make-span in a flow shop were developed by Johnson:

Select the job with the minimum processing time on either machine 1 or machine 2. If this time is associated with machine 1, schedule this job first. If it's for machine 2, schedule this job last in the series of jobs to be run. Remove this job from further consideration.

Select the job with the next smallest processing time and proceed as above (if for machine 1, schedule it next; if for machine 2, as near to last as possible). Any ties can be broken randomly.

Continue this process until all of the jobs have been scheduled.

The intuitive logic behind this rule is the minimum time to complete the set of jobs has to be the larger of the sum of all run times at the first machine plus the smallest run time at the second machine, or the sum of all run times at the second machine plus the smallest run time at the first machine.

We can also apply these rules to larger flow shop scheduling problems. For example, Campbell, Dudek, and Smith (CDS) have developed an efficient heuristic. This procedure uses the Johnson algorithm to solve a series of two-machine approximations to the actual problem having *M* machines using the following rules:

Solve the first problem considering only machine 1 and M, ignoring the intervening *M-2* machines.

Solve the second problem by pooling the first two machines (1 and 2) and the last two machines (M-1 and M) to form two dummy machines. Processing time at the first dummy machine is the sum of the processing time on machines 1 and 2 for each order. Processing time at the second dummy machine is the sum of the processing time at machines M-1 and M for each order.

Continue in this manner until M-1 problems are solved. In the final problem, the first dummy machine contains machines 1 through M-1, and the second dummy machine contains machines 2 through M.

Compute the make-span for each problem solved and select the best sequence.

Additional procedures using branch and bound algorithms and integer-programming methods have been developed to solve static flow shop three-machine scheduling problems using the minimum make-span criterion. However, the solutions are generally feasible only for very small problems. Currently, heuristic methods such as the Campbell, Dudek, and Smith algorithm are the only means of solving larger-scale flow shop scheduling problems.

We can make several important observations from these research efforts. First, the size of problems we can treat with analytical methods is small and of limited applicability for the "real world." Second, computer time required to solve scheduling problems with analytical methods grows exponentially with the number of jobs and/or machines to be scheduled. Third, the performance measure, minimizing the make span, isn't the same as minimizing average time in the system or average number of jobs in the system. Moreover, any of these criteria aren't necessarily related to the job lateness criterion. Fourth, static scheduling assumptions (beginning with all machines idle and all jobs available, and ending with all jobs processed and all machines idle) clearly influence the results. Fifth, the machine processing times reflect no randomness, which could reduce the techniques' applicability. Finally, on the positive side, it's important to note the two-machine scheduling rules utilize the shortest processing time logic. The SPT application in the two-machine case isn't exactly the same as it was in the single-machine case, but it's clearly an essential element in producing the desired scheduling performance in both problem situations.

Queuing Model Approaches

Applying queuing models to scheduling problems allows us to relax some of the limiting constraints just mentioned. In particular, queuing approaches deal with the dynamic problem, rather than the static problem. Randomness in interarrival and service times are considered, and steady state results are provided for average flow time, average work-in-process, expected work center utilization, and average waiting time.

The queuing research first examined the single-machine case and then expanded to the multiple-machine case. The single-machine research has shown, again, the SPT rule for sequencing jobs yields the best average completion time, average work-in-process level, and average waiting time criteria. Applying queuing theory to the multiple-machine case requires such limiting assumptions that results are only interesting from a research point of view. To examine realistic, multiple-machine, dynamic scheduling situations, we often use simulation models. With simulation, we can examine various rules' performance against several criteria. We can expand the size of problems studied (work centers and jobs), consider effects of startup and ending conditions, and accommodate any kind of product structure, interarrival time patterns, or shop capacity. Simulation studies address such primary research questions as: Which dispatching rules for sequencing jobs at work centers perform best? For which criteria? Are some classes of rules better than others for some classes of criteria or classes of problems?

Sequencing Rules

Figure 13.1 illustrates a typical scheduling environment for a complex job shop. At any time, if a set of n jobs is to be scheduled on m machines, there are $(n!)^m$ possible ways to schedule the jobs, and the schedule could change with the addition of new jobs. For any problem involving more than a few machines or a few jobs, the computational complexity of finding the best schedule is beyond even a modern computer's capacity.

Figure 13.1 shows complex routings. For example, after processing at work center A, we may send jobs for further processing to work centers B, D, or F. Similarly, some jobs are completed after being processed at work center A and go directly to finished component inventories. Also note a job might flow from work center A to work center D and then back to A.

Figure 13.1 depicts a sequencing or dispatch rule between each queue and its associated work center. This indicates a dispatching rule exists for choosing the next job in the queue for processing. The question of interest is which sequencing rule will achieve good performance against some scheduling criterion.

FIGURE 13.1   The Scheduling Environment

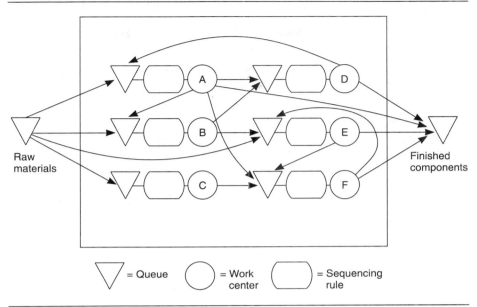

A large number of sequencing rules have appeared in research and in practice. Each could be used in scheduling jobs. Here are some well-known rules with their desirable properties:

R (random). Pick any job in the queue with equal probability. This rule is often used as a benchmark for other rules.

FCFS (first come/first serve). This rule is sometimes deemed to be "fair" in that jobs are processed in the order they arrive at the work center.

SPT (shortest processing time). As noted, this rule tends to reduce work-in-process inventory, average job completion (flow) time, and average job lateness.

EDD (earliest due date). This rule seems to work well for criteria associated with job lateness.

CR (critical ratio). This rule is widely used in practice. Calculate the priority index using (due date–now)/(lead time remaining).

LWR (least work remaining). This rule is an extension of SPT in that it considers *all* processing time remaining until the job is completed.

FOR (fewest operations remaining). Another SPT variant that considers the number of successive operations.

ST (slack time). A variant of EDD that subtracts the sum of setup and processing times from time remaining until the due date. The resulting value is called "slack." Jobs are run in order of the smallest amount of slack.

ST/O (slack time per operation). A variant of ST that divides the slack time by the number of remaining operations, again sequencing jobs in order of the smallest value first.

NQ (next queue). A different kind of rule, NQ is based on machine utilization. The idea is to consider queues at each of the succeeding work centers to which the jobs will go and to select the job for processing that's going to the smallest queue (measured either in hours or perhaps in jobs).

LSU (least setup). Still another rule is to pick the job that minimizes changeover time on the machine. In this way, capacity utilization is maximized. Note this rule explicitly recognizes dependencies between setup times and job sequence.

This list isn't exhaustive. Many other rules, variants of these rules, and combinations of these rules have been studied. In some cases, use of one rule under certain conditions and use of another under other conditions has been studied.

One issue Figure 13.1 highlights is whether we should use the same rule at each work center. We might, for example, build a case for using SPT at the "gateway" work centers and using some due date–oriented rules for downstream centers. Or perhaps selection of a rule should depend on queue size or how much work is ahead of or behind schedule.

Another issue in selecting sequencing rules is their usage cost. Some rules (such as random, first come/first serve, shortest processing time, earliest due date, and fewest operations remaining) are easy to implement, since they don't require other information than that related to the job itself. Other rules (such as the critical ratio, least work remaining, slack time, and slack time per operation rules) require more complex information plus time-dependent calculations. The next queue and least setup rules require even more information, involving the congestion at other work centers, or a changeover cost matrix for all jobs currently at a work center.

Sequencing Research Results

Extensive research has addressed different sequencing rules' performance. We'll highlight some of these efforts and draw conclusions for practice. Conway, Maxwell, and Miller tested 39 different sequencing rules, using the same set of 10,000 jobs. Results from the first 400 and last 900 weren't included in the results to eliminate startup and ending conditions. Figure 13.2 reports their results for two criteria: average time in the system and variance of the time in system. As we've noted, the average time in the system is directly related to work-in-process inventory and average number of jobs in the system. This measure also directly relates to average job lateness. The results in Figure 13.2 clearly show the SPT rule performs quite well for this set of criteria.

There is, however, a concern in using SPT. It can allow some jobs with long processing times to wait in queue for a substantial time, causing severe due-date problems for a few jobs. However, since the SPT rules can complete the average job in a relatively short time compared with other rules, it produces a much smaller average job lateness. Therefore, overall lateness performance might be much less severe than we might think.

Figure 13.3 shows simulation results using various lateness criteria. It reveals the slack time per operation (ST/O) rule is best in terms of the percentage of jobs late. However, the SPT rule is a close second, and SPT is considerably better than the earliest due date (EDD) rule. The earliest due date rule has a much lower variance of job lateness than SPT, but the *average* lateness measure of SPT might be more than enough to compensate for the high job lateness variance associated with SPT.

FIGURE 13.2   Simulation Results for Various Sequencing Rules

Sequencing rule	Average time in system	Variance of time in system
SPT	34.0	2,318
EDD	63.7	6,780
ST/O	66.1	5,460
FCFS	74.4	5,739
R	74.7	10,822

Source: R. W. Conway, W. L. Maxwell, and L. W. Miller, *Theory of Scheduling* © 1967, Addison-Wesley Publishing Co., Inc., Reading, Ma. Page 287, Table C–3.

FIGURE 13.3   Simulation Results for Other Criteria

Sequencing rule	Average job lateness*	Variance of job lateness	Percentage of jobs late
ST/O	−12.8	226	3.7
SPT	−44.9	2,878	5.0
EDD	−15.5	432	17.8
FCFS	−4.5	1,686	44.8

*Minus sign means jobs are early on average.

Source: R. W. Conway, W. L. Maxwell, and L. W. Miller, *Theory of Scheduling* © 1967, Addison-Wesley Publishing Co., Inc., Reading, Ma. Page 287, Table C–3.

In additional studies, job due dates were established using a variety of procedures. SPT's performance has been shown to be better than other rules, even under conditions where due dates are assigned a bit less "rationally"; that is, where due dates are set without regard to work content involved in processing the job. In studies where the shop utilization varied, SPT was less sensitive to changes in work centers' capacity utilization than other sequencing rules.

A simulation study by Nanot tested several rules using very large sample sizes (number of orders scheduled) and several combinations of shop load and routing (job shop versus flow shop) conditions. The general results support SPT as a superior rule in terms of average time in system and related measures. Nanot found time in system variance is higher than for some of the other rules. Nanot also concluded the results didn't seem sensitive to changes in shop load and routing conditions.

Other research efforts have tried to combine SPT with other scheduling rules to obtain most of SPT's benefits without the large time in system variance. One approach has been to alternate SPT with FCFS (first come/first serve) to "clean out the work centers" at periodic intervals. Other combinations of SPT with ST/O or with critical ratio produce similar results.

Simulation experiments have consistently demonstrated SPT is a very good dispatching rule for many criteria. This conclusion is robust and is supported by a variety of simulation studies. This result runs counter to what many practitioners believe: To ignore due dates for sequencing jobs seems irresponsible. But since SPT use reduces lead times and work-in-process inventories, the benefits might argue for its implementation. Conway et al. sum it up well:

> The priority rule under which the job with the shortest processing time is selected (SPT) clearly dominates all the other rules tested. Its performance under every measure was very good, it was an important factor in each of the rules that exhibit a "best" performance under some measure, and is simpler and easier to implement than the rules that surpass it in performance. It surely should be considered the "standard" in scheduling research, against which candidate procedures must demonstrate their virtue.

ADVANCED RESEARCH FINDINGS

This section covers several additional research studies we feel are particularly relevant to MPC practice. These studies focus on determination of lead times (management of due dates) for manufactured items and determination of labor assignments in manufacturing operations. In each case, we think important practical issues are raised and the practicing professional can use the available, though perhaps tentative, conclusions.

Due Date Setting Procedures

The scheduling procedures presented so far in this chapter have assumed order release and due dates for individual jobs are *givens*. Many firms assign setting such dates to manufacturing; it's frequently the subject of intense negotiations between manufacturing and marketing personnel. Often, due dates must be set at the time of order receipt or when bidding for an order. An effective MPC system can help by providing appropriate information regarding material availability, capacity, and resource requirements for individual jobs. As an example, we normally assign due dates for make-to-order products based on raw material and equipment capacity availabilities. Likewise, we set order release and due dates for manufactured components in MRP systems by determining length of the planned lead time for such items. Therefore, establishing lead time offsets and due dates is a vital and ongoing function in a manufacturing system. A well-functioning shop-floor control system based on good dispatching rules will help us achieve these due dates.

Although very little research has been reported on establishing manufacturing lead times and due dates, Baker and Bertrand have provided some useful insights. They analyzed three different procedures' effectiveness for estimating lead times and setting due dates for a job shop. Specifically, they set due dates for orders by adding an estimate of the manufacturing time to the date the order is received. Their three methods for establishing the estimate of manufacturing time are:

CON: A *constant* time allowance for manufacturing all jobs; that is, the same lead time is added to all jobs at receipt date to calculate the due date.

SLK: A time allowance that provides an equal (constant) waiting time or *slack* for all jobs; that is, the due date is set equal to the receipt date plus the sum of all processing times, plus a fixed additional slack time.

TWK: A time for waiting that has slack proportional to a job's *total work content;* that is, lead time to be added to the receipt date is a multiple of the sum of all processing times.

Each procedure has a single parameter (the constant time, the slack time, or the multiple) to be determined. Other informational needs are similar to those of shop-floor scheduling problems. The first procedure is easy to implement in many firms since shop-floor system data base requirements are minimal. The other two procedures, however, require an estimate of a job's processing time to set the due date.

Research in due date setting concerns the degree of improvement in manufacturing performance to be obtained by implementing the more complex due date procedures, SLK and TWK.

The evaluative criterion was "due date tightness." Here we presume tight due dates (or short lead times) are strategically more desirable than loose due dates. Tight due dates provide a competitive advantage by permitting the firm to offer an improved level of customer service, as well as achieve lower costs through reductions in work-in-process inventory. The experiments' approach was to so set each of the three parameters that *no* late deliveries occurred; that is, the parameters were chosen so the longest lead time is just sufficient. Thereafter, actual lead times are observed in the simulation. The preferred procedure is the one that achieves the smallest mean lead time.

The experiments involved a single-machine system using the shortest processing time dispatching rule for all three due date setting rules. However, they were conducted under a wide variety of operating conditions: 80 percent to 99 percent machine utilization, a variety of jobs, 20 replications, and use of both exponentially and normally distributed processing times. The exponentially distributed processing times gave a much greater degree of variability in achieved lead times (coefficient of variation, $c_v = 1.0$) than the normally distributed processing time ($c_v = .25$). Two releasing rules were used as well. The random release rule meant orders were issued to the shop as soon as received. The "controlled" release rule meant jobs were released when work-in-process inventory levels fell below a "trigger point." The trigger point was chosen to provide a specified average number of jobs in the shop.

The results indicate the SLK and TWK procedures set tighter due dates than the CON procedure. As Figure 13.4 shows, these two procedures provided as much as a 50 percent reduction in lead time required for manufacturing (in comparison with the CON procedure) under exponentially distributed processing times. Much smaller differences were noted when normally distributed processing times were used. Furthermore, there was a clear preference for the TWK procedure (as opposed to the SLK procedure) when random work releasing was used. In using controlled work releasing, preference shifts to the TWK procedure at higher levels of machine utilization.

While considerably more research is required (especially for multiple machines) and significant modifications of the procedures may be discovered, we believe this research contains some important messages. In particular, these results indicate the important potential for reductions in lead time and work-in-process inventory when dues dates are set in relation to job processing times. An important step in implementing these procedures is determining how much variability exists in a shop's processing times. An indication of high variability (e.g., coefficient of variation of 1.0 or more) suggests potential for major improvements in due date setting.

We see these results as particularly important as firms systematically reduce work-in-process inventories. As long as lead times are about 90 percent waiting time, the influence of processing times and processing time variability is masked. When lead times fall, processing time becomes a larger element of the total lead time and, thus, more important to take into account.

FIGURE 13.4   Simulation Results for Manufacturing Lead Time Estimating Procedures

Treatment	Mean number of jobs	Utilization	Mean manufacturing lead time			Frequency best*		
			TWK	SLK	CON	TWK	SLK	CON
Exponential times, random release	4.00	0.80	4.43	9.04	10.14	20	0	0
	5.67	0.85	5.63	10.37	11.39	20	0	0
	9.00	0.90	6.20	11.79	12.76	20	0	0
Exponential times, controlled release	4.00		5.26	4.53	8.79	3	17	0
	5.67		6.51	6.23	10.09	7	13	0
	9.00		8.28	9.51	13.49	17	3	0
Normal times, random release	4.28	0.90	7.20	7.70	7.72	16	2	2
	9.59	0.95	10.06	10.70	10.75	16	2	2
	52.09	0.99	10.44	10.99	11.07	20	0	0
Normal times, controlled release	4.28		6.65	5.31	5.90	0	20	0
	9.59		12.35	10.61	11.18	0	20	0
	52.09		48.53	53.10	53.64	20	0	0

*Number of times in the 20 replications that each procedure performed the best (i.e., produced the lowest mean manufacturing lead time).

Source: K. R. Baker and J. W. M. Bertrand, "A Comparison of Due Date Selection Rules." *AIIE Transactions* 13, no. 2 (June 1981), pp. 128–29.

Dynamic Due Dates

A key limitation of the research discussed so far is due dates are assigned just once and they aren't revised over time. This differs from shop-floor systems operating as part of an effective MPC system: due dates are continually revised as one portion of the MRP planner's job. Determining due dates for orders when they are released to the shop is therefore only one aspect of managing due dates in scheduling. A second aspect concerns maintaining *valid* due dates as orders progress through the manufacturing process and as new orders are added. The need for due date maintenance arises from the manufacturing environment's dynamic nature. Management actions (such as master production schedule changes, planned lead time adjustments, and bill of material modifications) can create the need to reschedule manufacturing orders and to revise priorities given to the shop. Likewise, variations in shop conditions (such as unexpected scrap and unplanned transactions) can also create the need to revise job due dates.

Many firms' systems and procedures result in changes in open order due dates. This practice is called *dynamic due date maintenance*. The primary argument for this practice is the shop should use accurate and timely information in dispatching jobs to machines to provide a high level of customer service. In spite of its widespread use, controversy surrounds the advisability of implementing dynamic due date maintenance systems. Some suggest dynamic due dates can have an adverse impact on scheduling performance because of system "nervousness." Steele, for example, argues a job shop can function effectively only if open order priorities are stable enough to generate some coherent action on the shop floor. He defines a scheduling system with *unstable* open order priorities as a nervous scheduling system, which can lead to shop floor distrust and overriding formal priorities. A second behavioral argument against dynamic due date maintenance is the volume of rescheduling messages might so inundate the production planner that he or she can't process the necessary changes in a timely fashion.

In such cases, the production planner may simply stop trying to perform an impossible task; the shop could lose faith in the priority system and revert to using an "informal" system; or ill-chosen or misleading rescheduling messages may be communicated to the shop. Any or all of these responses may cause the shop and inventory system performance to deteriorate.

While much scheduling research has focused on determining the "best" heuristics for scheduling a shop to meet fixed or open order due dates, little experimental research has addressed whether and how to respond to these issues when the actual need date for an open order changes while the order is being processed by the shop.

In one study conducted to determine the impact of using *dynamically* updated due dates on manufacturing performance, Berry and Rao evaluated use of dynamic due dates with the critical ratio scheduling rule in a make-to-stock environment. They studied scheduling products produced by a job shop for a finished-goods inventory controlled using an economic lot size/reorder point system. The *dynamic* due dates used in critical ratio scheduling were produced using the following formula:

$$\text{Due date} = \text{Current date} + \frac{\text{On-hand inventory balance} - \text{Buffer stock}}{\text{Average daily usage}}.$$

The result is that larger than expected demand for a product will move the due date for an open order for that item closer to the current date, thereby increasing the critical ratio scheduling priority.

This study's results were counterintuitive. They showed use of dynamic due dates significantly increased combined costs of setup, carrying work-in-process and finished product inventory, and inventory shortages. Nervousness in the dispatching rule priorities increased both the mean and variance of the job flow times, thereby increasing the work-in-process inventory level and reducing the customer service level.

In a related study, Hausman and Scudder evaluated dynamic inventory information's use in priority scheduling rules for a jet engine repair facility. This facility performed the disassembly, manufacture of component parts, and reassembly of jet engines. Quick turnaround times were important to the customer, so the study used time-weighted final product (engine) back orders as the measure of performance. The inventory management policy was a continuous review, one-for-one (S–1, S) inventory policy, and the product had a multilevel product structure.

The repair facility had 10 machines. The study compared back order performances of a wide range of priority sequencing rules including static rules (earliest due date, FCFS, etc.), dynamic rules (minimum slack time per operation, critical-ratio, etc.), and rules that consider additional information. The most effective priority scheduling rules tested proved to be:

Inventory-based: Select the job with the smallest value of net inventory. (Net inventory equals current on-hand inventory minus back orders for the component type.)

Multiple use: Select the job required by the largest number of modules (subassemblies) awaiting parts for assembly.

Both of these rules incorporate dynamic information regarding the expected component work-in-process inventory. In effect, developing the priority index information used in these rules is similar to gross-to-netting procedures used in an MRP system.

Use of dynamic work-in-process, finished component, and subassembly requirement information provided significant improvements in operating performance. For example, the inventory-based rule resulted in 3.67 mean delay days as opposed to 6.03 for critical ratio and 7.28 for the shortest processing time rule. For the repair facility, a two-day improvement in mean delay time represents savings of one engine from the spares inventory, $2 million.

Penlesky partly explains the seemingly contradictory results observed in the previous studies. He evaluates the use of several dynamic due date procedures as well as the use of simple procedures for selectively implementing a few of the many due date changes that would normally be implemented (filtering procedures). In particular, the study concerns determining what types of job-related information are important to consider in formulating open order rescheduling procedures and evaluating rescheduling's impact on manufacturing performance in MRP systems.

He considers three different filters for making rescheduling decisions: ability, magnitude, and horizon filters. The ability filter's purpose is to assure only attainable due date adjustments are passed along to the shop. In using this procedure:

1. All *rescheduling out* actions (when the new due date is later than the previous due date) are implemented.

2. Implementing *reschedule-in* actions depends on one of three conditions:
   a. If the machine setup and processing time remaining is less than the time until the new due date, the new due date is implemented.
   b. If the machine setup and processing time remaining is less than the time until the old due date but greater than the time until the new date, the due date is set to the present time plus the machine setup and processing time to complete the order.
   c. If the machine setup and processing time remaining exceeds the time allowed until the old due date, no change is made to the old due date.

The magnitude and horizon filters consider different information. These procedures are designed to filter out trivial due date adjustments by means of a *threshold* value. In the magnitude procedure, if the absolute value of the difference between the new and the old due dates exceeds a threshold value (T_m), the due date is changed. Similarly, the horizon procedure is designed to filter out due date changes too far out in the planning horizon to be of immediate concern to the production planner. Only if the old due date falls within the period of interest (T_H) is the new due date implemented. By setting parameter values for T_m and T_H, the number of rescheduling changes to be implemented can be adjusted. The procedures will implement all changes when $T_m = 0$ and $T_H = \infty$, providing full dynamic procedures. Static dates result when $T_m = \infty$ and $T_H = 0$.

Simulation experiments were used to investigate the effect of incorporating dynamic due date information in the sequencing rules and use of filtering procedures. These experiments were conducted using a make-to-stock job shop simulator, with both component manufacturing and assembly operations, controlled by an MRP system. Procedures were tested under differing values of machine utilization, uncertainty in the master production schedule, length of planned lead times, and size of production order quantities. The three measures of effectiveness used were end product customer service level, combined work-in-process and finished item inventory level, and number of rescheduling changes implemented.

Figure 13.5 indicates performance gains from dynamic due dates depend on shop operating conditions. These results indicate that under certain operating conditions, dynamic due date information improves customer service and total inventory level. The results help explain the apparently contradictory results reported by Berry/Rao and Hausman/Scudder. While both studies were conducted under high machine utilization conditions, Hausman and Scudder's experiments used small lot sizes (a single unit). (These correspond to experiments 6 and 8 in Figure 13.5, where performance gains were obtained.) Berry and Rao's study used much larger order quantities (economic order quantities involving several periods of demand). (These correspond to experiments 14 and 16 in Figure 13.5, where no improvements in performance were observed.)

We can draw another important conclusion from Figure 13.5's results. Dynamic due dates can reduce total inventory level while *simultaneously* improving customer service (e.g., in experiments 3, 6, 8, and 11).

Figure 13.6 compares the filtering procedures' performance for experiment number 8 of Figure 13.5. We can make two observations regarding these results. First, there's no significant difference in performance between the filtering procedures and the dynamic due date procedure without filtering. All rescheduling procedures significantly improve performance over the static procedures. Second, magnitude and horizon filters provide comparable performance to the dynamic rescheduling procedure—but with far fewer re-

FIGURE 13.5 Percentage Improvements in Service and Inventory Levels Using Dynamic Due Dates*

Periodic order quantity	Planned lead time	Performance measure	Low master schedule uncertainty		High master schedule uncertainty	
			Low machine utilization	High machine utilization	Low machine utilization	High machine utilization
Small	Low	Experiment number Customer service level Total inventory level	1 3.4 —	2 — —	3 15.2 10.5	4 — —
	High	Experiment number Customer service level Total inventory level	5 .5 —	6 9.3 5.1	7 4.8 —	8 31.8 8.3
Large	Low	Experiment number Customer service level Total inventory level	9 4.3 —	10 — —	11 14.3 8.0	12 — —
	High	Experiment number Customer service level Total inventory level	13 2.5 —	14 — —	15 6.2 —	16 — —

*[(Static − Dynamic) ÷ Static] × 100; calculated only in those cases where there was a statistically significant difference in the performance measure between the two procedures.

Source: R. J. Penlesky, "Open Order Rescheduling Heuristics for MRP Systems in Manufacturing Firms," doctoral dissertation, Indiana University, 1982.

FIGURE 13.6 Results of Applying the Filtering Procedures

Procedure	Filter level*	Customer service level		Total inventory level	
		Mean	Standard deviation	Mean	Standard deviation
Static due dates	0	.651	.084	14,357	895
Ability filter	100	.871	.048	12,990	412
Magnitude filter	53	.873	.041	12,873	919
Horizon filter	45	.831	.055	13,190	958
Dynamic due dates without filtering	100	.858	.049	13,161	999

Note: Data from experiment 8 of Figure 13.5.
*Percent of indicated reschedules that were implemented.

Source: R. J. Penlesky, "Open Order Rescheduling Heuristics for MRP Systems in Manufacturing Firms," doctoral dissertation, Indiana University, 1982, p. 148.

scheduling actions implemented. Therefore, it would seem dynamic rescheduling's benefits can be achieved by *selectively* implementing rescheduling actions. By filtering rescheduling messages, we can reduce information processing costs and adverse behavioral effects of system nervousness without an adverse effect on operating performance.

Labor-Limited Systems

The scheduling research results presented so far are useful when dispatching (sequencing) rules represent the principal means of controlling work flow in a plant. In many firms, besides assigning jobs to work centers, there's a need to make labor assignment decisions. Labor assignment decisions are important in controlling work flow when labor capacity is a critical resource in completing work. This can occur even when only one particular labor skill is the bottleneck resource. In such instances, the system is said to be labor-limited.

Labor limitations provide an additional dimension to shop-floor scheduling that's important for many JIT and cellular manufacturing situations. The controllable cost is labor, and the primary scheduling job is assigning labor to machine centers. Good labor scheduling practice enables us to vary labor capacity at work centers to better match day-to-day fluctuations in work loads. To the extent there's flexibility in assigning people to work centers, we can improve manufacturing performance (e.g., reduced flow times, better customer service, and decreased work-in-process inventory). However, the degree of flexibility in making labor assignments depends on such factors as amount of cross-training in the work force, favorable employee work rules, and costs of shifting people between work centers.

Nelson has provided a comprehensive framework for control of work flow in labor-limited systems. It lists three major elements for controlling work flow in scheduling:

1. Determining which job to do next at a work center (dispatching).
2. Determining when a person is available for transfer to another work center (degree of central control).
3. Determining the work center to which an available person is to be assigned (work center selection).

Various decision rules, using information similar to that used in making dispatching decisions, have been suggested for making the latter two decisions. The decision rules Nelson suggested for determining a person's availability for transfer utilize a central control parameter, d, that varies between 0 and 1. When $d = 1$, the person is always available for reassignment to another work center. When $d = 0$, the person can't be reassigned as long as jobs are waiting in the queue at the person's current work center assignment.

We can control the proportion of scheduling decisions in which a person is available for transfer by adjusting d's value between 0 and 1.

Fryer suggests two different approaches to transfer availability. One considers time; the other considers the queue. The time approach suggests the person must be idle for t or more minutes before a transfer can be made. The queue approach suggests making a transfer only when the person's work center queue has fewer than q jobs waiting for processing. Labor flexibility is increased by decreasing the value of t or increasing the value of q.

The third decision in the framework, deciding to which work center a person should be assigned, can be made using decision rules that resemble dispatching rules. We can determine priorities for assigning labor to unattended work centers on the basis of which work center has as its next job to process:

1. The job that was first at the current work center, first-come/first-served (FCFS).
2. The job that was in the shop, first-in-shop/first-served (FISFS).
3. The shortest job (SPT).
4. The most jobs in the queue.

We combine these decision rules with decision rules for making dispatching and labor availability decisions to control the work flow. Random assignment was used as a base line for comparison.

Simulation experiments have been conducted to evaluate the performance of the different work flow control rules suggested for labor-limited systems. These studies generally measure improvement in the job flow time performance. An interesting general finding is, while changes in dispatching rules involve a trade-off between the mean and variance in job flow times, changes in labor assignment rules often reduce both measures simultaneously. Figure 13.7 shows these results.

Experiments involving the labor flexibility factor, d, also show the importance of labor flexibility in a shop. A change between no labor flexibility ($d = 1$) and complete labor flexibility ($d = 0$) resulted in 12 percent and 39 percent reductions in the mean and variance of job flow times, respectively.

Research on labor assignment rules demonstrates the importance of cross-training and labor assignment flexibility. Moreover, it indicates both labor and job dispatching can have a major impact in controlling work flow through a shop. With an operating shop-floor control system in place, further performance improvements might come from better design of labor assignments and from operational changes that permit greater flexibility in labor assignments.

Lessons for Practice

In this section we have overviewed basic scheduling research. This research offers important insights to the professional who has an operating MPC system and is interested in further enhancements. One important practical result

---

FIGURE 13.7   Time and Number of Jobs in System

Size of labor force	Statistic: Queue discipline	Mean time and mean number in system*			Variance of time in system			Variance of number in system		
		FCFS	FISFS	SPT	FCFS	FISFS	SPT	FCFS	FISFS	SPT
4	Machine limited	17.7	17.7	.9.4	488	295	612	201	205	24
3	Labor assignment rule 0	11.0	11.0	7.0	200	125	295	76	80	17
	1	10.2			173			54		
	2		10.5			102			63	
	3			6.6			343			15
	4	10.1	10.1	6.4	169	97	281	50	53	11
2	Labor assignment rule 0	8.7	8.7	6.2	158	147	186	65	67	23
	1	8.7			153			49		
	2		8.7			147			67	
	3			5.0			285			10
	4	8.7	8.8	5.1	154	89	293	46	48	9
1	Labor assignment rule 0	8.3	8.3	5.5	157	174	176	74	69	24
	1	8.3			149			48		
	2		8.3			174			69	
	3			4.2			296			9
	4	8.3	8.3	4.4	150	174	298	45	69	8

Note: Labor assignment rules:
  0 = Random labor assignment to a work center.
  1 = FCFS labor assignment to a work center.
  2 = FISFS labor assignment to a work center.
  3 = SPT labor assignment to a work center.
  4 = Most jobs in queue labor assignment to a work center.
*Parameters so chosen that the mean time and the mean number in the system were equal.

Source: R. T. Nelson, "Labor and Machine Limited Production Systems," *Management Science* 13, no. 9 (May 1967), p. 660.

of the research on sequencing methods has been to clearly understand the shortest processing time (SPT) rule's effectiveness. This suggests the value of a combined rule, such as a critical ratio/SPT, or some other hybrid based on a combined due date–oriented rule and SPT. For example, Twin Disc (which uses critical ratio) segregates jobs into PO jobs (pegged to actual customers orders) and PI jobs (pegged to forecasted usage). A hybrid rule is to first run all PO jobs with a critical ratio less than 1.0 in critical ratio sequences and then sequence all other jobs by SPT.

Black & Decker employs a similar hybrid approach in its plant operations using dispatching. Most jobs are processed on the basis of SPT. But lateness importance grows exponentially until it becomes important enough to override SPT. Thus, no jobs can get "lost" due to long processing times at some operations. This and other practical experiments lead us to believe that other firms with complex job shop processes will develop and implement hybrid rules.

The research on due date setting suggests we can gain important reductions in manufacturing flow times and work-in-process inventory by adopting lead time setting procedures based on job-processing times. Using this information to set the lead time offset data for MRP and shop-floor control will be particularly important as we systematically reduce lead times (and work-in-process inventory levels).

Research on managing open order due dates suggests we can improve performance by implementing dynamic due date procedures under certain operating conditions. The improvements, however, are influenced by shop structure considerations such as order quantity sizes and lead times. Research on filtering procedures suggests we can improve manufacturing performance by implementing a relatively small proportion of the suggested due date changes.

Research on labor-limited scheduling shows the potential an increase in labor flexibility can have on manufacturing performance. In many firms today, work centers' combined machine capacity far exceeds labor capacity. Labor-limited scheduling is also important in cellular manufacturing, which is a major aspect of JIT.

EMERGING ISSUES IN SCHEDULING

A major shift in direction has occurred in recent research on scheduling methods. It closely relates to major current changes in production process design at many firms. These changes are motivated by availability of new process technology such as computer integrated manufacturing (CIM), the introduction of just-in-time (JIT) methods, and the intensity of worldwide competition in manufacturing. While scheduling methods described so far in this chapter have been developed for job shop production processes, many of the new processes being installed in firms today are designed to capture benefits of repetitive manufacturing and continuous flows of material. As a result, much of the recent new scheduling research concerns development of new concepts and techniques for repetitive manufacturing operations. The rest of this chapter involves two scheduling approaches. These concern two different process technologies for repetitive manufacturing: cellular manufacturing systems with manned cells and systems with limited manning, like flexible manufacturing systems (FMS).

Scheduling Cellular Manufacturing Systems

Cellular manufacturing systems are designed to process part families in dedicated production areas, referred to as manufacturing cells. Benefits associated with the use of manufacturing cells include reduced order flow time, less work-in-process inventory, smaller setup times, lower material handling costs, improved quality and productivity, improved job satisfaction and status, and simplified planning and control procedures. Wemmerlöv describes three types of material that can be processed in cells: piece parts, subassemblies, and assemblies. Cells can also be of a hybrid type, where both machining and assembly can take place in the small cell. As a result, we can refer to cells as machining, fabrication, assembly, and hybrid cells.

Another important distinction for scheduling purposes is whether a cell is manned or has limited manning. A manned cell is staffed by one or more operators who are responsible for both processing and material handling activities. Often there are fewer operators than machines, and operators move around the cell to process the highest-priority orders. In many cases, scheduling these cells is complicated by labor's limited availability. This means concepts of labor-limited scheduling are applicable.

Other cells involve applying computer-based technology to processing and material handling activities. Examples of such cells include flexible manufacturing systems, surface mount technology for assembling integrated circuits in the electronics industry, and other applications of computer integrated manufacturing. In these cells, operators perform tool loading, monitoring, and inspection tasks.

The flow pattern within a cell also impacts selecting scheduling methods in cellular manufacturing. Manufacturing cells range from pure flow shops to job shops. In a pure flow shop cell, job routing is the same for all jobs. Here, the job sequence established for the first machine in the cell is maintained for all subsequent machines. Therefore, we make a sequencing decision only once—at the point of entry into the manufacturing cell. For all other flow patterns (e.g., in job shop cells), orders may skip machines, backtrack, or enter the cell at multiple points. In such cases, we must make a sequencing decision at each machine. The next section concerns development of scheduling methods for manned cells, while the following section concerns scheduling flexible manufacturing systems.

Scheduling Manned Cellular Manufacturing Systems

Both static and dynamic scheduling approaches have been applied to scheduling flow shop cells. Some of this work uses analytical methods for solving static scheduling problems developed by Johnson and Campbell, Dudek, and Smith, which we described earlier in this chapter. These methods assume all jobs to be scheduled are available at the first machine in the cell at the beginning of the scheduling period. Recently, work has been reported on development of dynamic scheduling approaches for flow shop cells. Here, orders arrive at random time intervals for processing in a manufacturing cell. Wemmerlöv provides a survey of this research. An important part of this effort has been directed toward developing and testing sequencing heuristics similar to those this chapter earlier described for application to dynamic flow shop scheduling problems. In this section we describe two studies that report development and testing of dynamic scheduling heuristics for dynamic manned manufacturing flow shop cells dedicated to producing certain part families.

Wemmerlöv and Vakharia report dynamic scheduling heuristics for a five-stage flow shop cell, with a queue of orders in front of each stage. Each stage in this cell has one machine, and all orders have the same routing through the cell. This cell processes as many as six part families, with individual part orders arriving at random time intervals according to a Poisson process. Upon arrival at the cell, each part order is assigned a due date based on the total work content (TWK) to be performed in the cell on the order. Total work content includes combined values of the part family setup time and processing time at each stage in the cell.

Simulation studies were conducted to evaluate different scheduling rules, using a computer model of the manufacturing cell. A scheduling rule is used to sequence orders at the first stage in the cell: this same sequence is maintained at all of the remaining stages in the cell. The four scheduling rules evaluated were:

FCFS: First come/first served.

SLACK: Slack time.

CDS: Campbell, Dudek, and Smith's procedure.

NEH: Nawaz, Enscore, and Ham's procedure.

The first two rules are job shop sequencing rules and are used to maintain a priority sequence of orders in the queue at the first stage in the line. The second two rules are static scheduling rules, which were applied periodically in these experiments to develop a priority sequence of orders in the queue at the first stage of the line.

While the CDS rule was used as described earlier in this chapter, the NEH rule uses a different sequencing procedure. This heuristic starts with a partial sequence (it could be just one job) of jobs in queue at the first stage. We compute the make-span for a new job inserted in all positions without disturbing the order of the previous, partial sequence. We keep it in the position that gives the lowest make-span, and we evaluate another job until all available jobs have been considered.

In an effort to minimize setup time at each stage in the cell, a variation of each of the four rules was developed. The variation simply partitions the queue of orders at the first stage in the line according to a part family. The sequence in which we process orders within each part family grouping and the sequence in which we process part families are both established. For example, the FCFS rule processes that family having the oldest order first. After all orders in that family are processed, the part family having the next oldest order is processed next.

The family versions for the SLACK, CDS, and NEH rules are more complex. For the SLACK rule, we partition orders in the queue at the first stage into those orders having negative and positive slack. For those orders having negative slack, we first process the family having the order with the most negative slack. After we've processed all orders in this family in the order of their slack time priority, we then apply the rule to determine the next part family to process. For orders having a positive slack priority, the next part family we process is that family having the largest sum of the combined setup and processing time for all stages in the cell. Once we've selected the next part family to be processed, orders within that family are processed in order according to the smallest sum of the combined setup and processing time for all stages in the line.

The family versions of the CDS and the NEH rules also develop a sequence for the families represented in the order set, and then a sequence for the jobs in each family. These procedures proceed by first collapsing the five-stage scheduling problem into a series of two-stage scheduling problems, and then use a procedure similar to the Johnson algorithm to solve each two-stage problem. The solution with the minimum make-span is used both to sequence the part family to be processed next and to establish order sequence within each part family.

Wemmerlöv and Vakharia report simulation experiments that evaluate performance of these rules considering the following measures of performance: average order flow time and lateness, total number of early orders, total number of late orders, total number of family setups, and total number of operations processed. Several factors were varied in these experiments, including: number of part families, ratio of family setup time to order processing time, and cell utilization level.

Simulation results indicate family-oriented versions of the sequencing rules consistently outperform other rules, and the difference in performance increases as the ratio of the setup to processing time increases. However, these results are quite sensitive to number of part families processed by a cell. When a small number of part families was processed (e.g., 3), family versions of the FCFS and the SLACK rules outperformed other rules. However, when the cell processed a larger number of part families (e.g., 6), the FCFS(Family) rule produced the best due date performance while the CDS(Family) rule produced a smaller average flow time. In analyzing simulation results, differences between the FCFS(Family) and the CDS(Family) rules are apparently quite small. Therefore when administrative costs of using the CDS(Family) rule are considered, it may be advantageous to use the simpler FCFS(Family) rule.

In selecting a scheduling rule for a manufacturing cell, these results indicate the value of using a sequencing rule that works toward reducing setup time in the cell. The resulting increase in effective capacity in the cell provides an important improvement in scheduling performance. Also the FCFS(Family) is very effective, especially when we consider costs of administering the shop-floor control system. However, if a cell is designed to produce a larger number of part families than the number considered in these experiments, we should consider the CDS(Family) rule.

Mosier, Elvers, and Kelly report a second study of scheduling heuristics for manufacturing cells. They examine performance of sequencing heuristics for the cell in Figure 13.8. It contains four machines, each having three queues of orders, with a separate queue for each part family. An order may require processing on one or more of the first three machines, and all orders are processed by the fourth machine. Therefore, part routings aren't the same for all orders. Orders arrive at the cell at random time intervals, according to the Poisson process, and the due dates for the cell are assigned on the basis of total work content, using the TWK procedure.

Since this cell is organized as a job shop and orders aren't routed to all machines in the same sequence, sequencing decisions must be made at all four machines in the cell. Three decisions are made at each machine:

When to select orders from a different queue (part family).

Which of the two remaining queues (part families) to select orders from.

What order to select from the chosen queue.

Three different rules were used to make the first two decisions:

AVE: Select the part family queue having the highest average order priority and process all orders in the queue at the time of queue selection.

WORK: Select the part family queue having the largest sum of the processing times for this machine (that is, the queue having the largest work content) and process all orders in the queue.

FIGURE 13.8 Schematic of the Manufacturing Cell Used by Mosier, Elvers, and Kelly

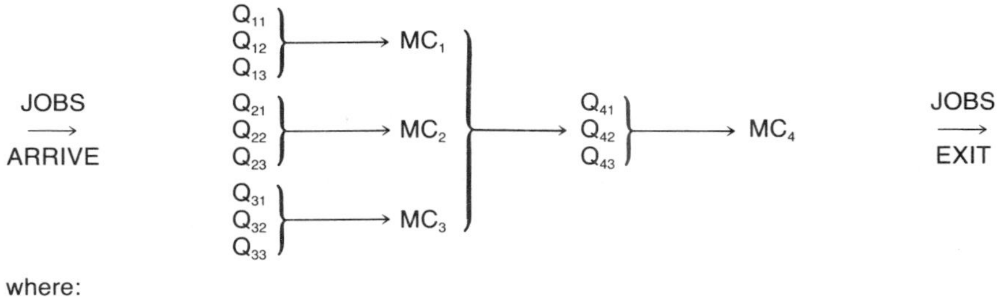

where:
Q_{ij} = Queue at machine i for part family j
MC_i = Machine type i

Source: C. T. Mosier, D. A. Elvers, and D. Kelly, "Analysis of Group Technology Scheduling Heuristics," *International Journal of Production Research* 22, no. 5, 1984.

ECON: After each order is processed at a machine, calculate the combined expected setup time for all orders in each queue. Switch to the queue having the largest total expected setup time if that value exceeds the total expected setup time for the queue currently being serviced.

Five different order sequencing rules were used to establish the dispatching priority for the orders in each queue:

1. Slack Time.
2. Modified Critical Ratio 1 (time remaining until due date divided by total processing time remaining).
3. Shortest Processing Time.
4. Modified Critical Ratio 2 (time remaining until due date divided by remaining number of operations).
5. First in the Shop, First served.

Simulation experiments were conducted to evaluate these rules using the following criteria: mean and variance of the order lateness and tardiness (only late jobs), mean and variance of the order flow times, proportion of orders failing to meet the due date, and total machine setup and idle times. These experiments were conducted using two different levels for the machine setup time and machine utilization.

The simulation experiments indicate little difference between the WORK and ECON rules with regard to mean order flow time, mean lateness, and order tardiness. The AVE rule is inferior to both other rules on these performance measures. However, the AVE rule provides better performance than the WORK and ECON rules when we consider percentage of late orders. With regard to setup time savings, the WORK and ECON rules provided much greater savings in setup time than the AVE rule.

These studies indicate important differences between different sequencing rules applied to manufacturing cells. In particular, note the flow pattern in a cell impacts the scheduling decisions that control the work flow in the cell. One clear principle emerges: We can gain important setup savings by taking family groupings into consideration in scheduling work in manufacturing cells, as well as in basic design of such cells.

In the next section, dealing with flexible manufacturing systems, the nature of the scheduling decision varies from the two cells described in this section. As research continues on scheduling manufacturing cells, a better understanding will no doubt emerge of the impact of a cell's characteristics on the scheduling decisions.

Scheduling FMS Systems

The process technology for the low- to medium-volume production of metal parts has changed dramatically in recent years with development of flexible manufacturing systems. Such systems typically have 2 to 16 machine tools and use a computer to control the various steps in the machining process, material handling activities of moving parts between machines, and scheduling of the flow of orders through the system. In effect, FMS systems permit many of the efficiencies and utilization levels of repetitive manufacturing systems while retaining the flexibility of general-purpose equipment. FMS systems provide important benefits in reducing machine setup times, shortening the overall manufacturing cycle (typically from weeks to days) with corresponding reductions in work-in-process inventory and tooling costs.

Although developing shop-floor systems for scheduling day-to-day operations in a flexible manufacturing system is still an emerging area of research, we can approach scheduling such systems using many of the scheduling concepts discussed in this chapter. For example, one view could be a static scheduling problem, where a fixed set of orders are to be scheduled, either using optimization or priority scheduling heuristics. Alternatively, we could view the problem as a dynamic scheduling problem, where orders arrive periodically for scheduling (e.g., as daily order releases from an MRP system or as individual customer orders). As a static scheduling problem, such performance criteria as minimizing the make-span or the mean order flow time may be of interest, while additional criteria, such as the mean and variance of the order lateness and tardiness, may be of concern in the dynamic version of the problem.

Under either view of the scheduling problem, three different scheduling and control decisions occur in managing day-to-day operations at the shop-floor level of an FMS:

1. Part loading timing: When to load a part into the FMS?
2. Part loading: Which part type to load into the FMS?
3. Dispatching: To which machine should a part be dispatched after it has completed its current operation?

The part loading timing decision concerns when to enter a new part into the system at the loading station of an FMS, considering such factors as system congestion and part fixture availability. The part loading decision concerns choosing the part type to enter the FMS at the loading station when it's time for a new part to enter the system, considering such factors as part characteristics and machine workload conditions. The dispatching decision concerns routing parts through the FMS at the time of actual production, such as sequencing parts at the individual machines in an FMS.

These procedures assume production planning decisions at an aggregate level concerning the types and mix of parts to be run on the FMS have been previously resolved in developing part routings and in preparing overall material and capacity plans for the company. Stecke has identified five such interrelated production planning problems that need to be solved prior to actual production: FMS part type selection, machine grouping, production ratio determination, resource allocation, and cell loading.

In this chapter, we're concerned with day-to-day scheduling of FMS operations and with how FMS performance can be influenced by the choice of the scheduling method used to make these decisions at the shop-floor level. We assume, therefore, these higher-level decisions have been made. Furthermore, other parameters that control the level of work-in-process (such as number of material handling fixtures for parts, and use of special or general purpose part holding fixtures) can impact FMS performance: we assume they have been addressed and determined. On this basis, we review some results of FMS scheduling studies.

Stecke and Solberg report a study involving scheduling a 10-machine FMS built by Sunstrand in Caterpillar Tractor Company's Peoria, Illinois, plant. Figure 13.9 diagrams this FMS. It includes four large five-axis machining centers called Omnimills (OM); three four-axis machining centers, Omnidrills (OD); two vertical turret lathes (VTL); and an inspection station. Each machine has a limited-capacity tool magazine to hold the tools needed by each operation assigned to the machine. Two transporters run on a straight track and carry parts from machine to machine. Sixteen load/unload stations provide a centralized queuing area for work-in-process inventory. This FMS is dedicated to manufacturing four parts, representing two sizes of transmission housings, with each housing including two matched parts: a transmission case and a cover. The system includes 18 holding fixtures (one fixture holds two parts) to convey parts through the system. Fixtures are dedicated to individual parts in proportion to the parts demanded.

Simulation experiments were run to evaluate different scheduling procedures using a simulation model of the Caterpillar FMS. These procedures included the following part loading timing, part loading, and dispatching rules:

1. Whenever two parts are completed (one fixture emptied), new parts are loaded into the empty holding fixture at the loading station.
2. The new parts to be loaded have to be of the same type as the completed parts.
3. Sixteen different dispatching rules were tested for deciding which parts to route to an empty machine. Figure 13.9 lists these rules.

Each simulation experiment consisted of solving a static scheduling problem, in which approximately 172 parts (including an equal number of the four part types) were scheduled during six eight-hour days of simulated time. In all, 80 simulation experiments were run, using 16 different "dispatching" rules, with five different levels of alternate routing flexibility; that is, moving from use of fixed routings to an increased number of routing alternatives for each operation. FMS performance was measured in terms of number of parts completed during the six-day period.

The SPT/TOT dispatching rule performed the best over all levels of the different alternate routing flexibility, providing an 8 percent to 24 percent improvement over the method originally used to assign orders to machines. Since the objective was to maximize machine output during a given time interval, the SPT/TOT dispatching rule's performance is consistent with pre-

---

FIGURE 13.9   Caterpillar Flexible Manufacturing System (FMS)

FMS Scheduling Procedures

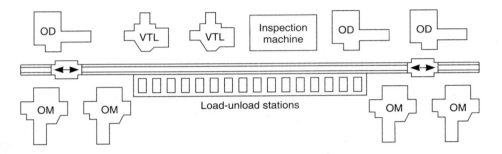

Part loading timing rule: Load next empty fixture.
Part loading rule: Load same part type as completed part type.
Dispatching rules:

1. Original—for each machine, the assigned operations (obtained at the production planning stage for all loads) are ordered according to the largest workload of the next operation's machine. This was the scheduling strategy originally applied by the FMS and the most complicated, computationally, to apply among all those tested.
2. SPT—operations are ordered according to the shortest operation time first.
3. LPT—longest operation time first.
4. FOPR—fewest operations remaining for each part.
5. MOPR—most operations remaining.
6. SRPT—shortest remaining processing time of each part.
7. LRPT—longest remaining processing time.
8. SPT·TOT—smallest operation time multiplied by total processing time for the part.
9. SPT/TOT—shortest processing time for the operation divided by total processing time for the part.
10. LPT/TOT—longest processing time for the operation divided by total processing time for the part.
11. LPT·TOT—longest processing time for the operation times total processing time for the part.
12. Part type priority—case, assembly, cover (MOPR).
13. Part type priority—cover, assembly, case (MOPR).
14. Part type priority—assembly, case, cover (MOPR).
15. Part type priority—case, assembly, cover (FOPR).
16. Part type priority—assembly, case, cover (FOPR).

Source: K. E. Stecke and J. J. Solberg, "Loading and Control Procedures for a Flexible Manufacturing System," *International Journal of Production Research* 19, no. 5, 1981.

vious research findings. In addition to giving a high priority to short operations, orders requiring a large amount of FMS time and operations that are early in a routing also receive high scheduling priority.

Denzler and Boe reported a different FMS scheduling approach in a study of a 16-machine Kearney and Trecker FMS at John Deere's plant in Waterloo, Iowa. Instead of focusing on dispatching parts to machines, this study analyzed different part loading procedures, determined the best number of in-process parts (fixtures), and took into account congestion on the material handling system.

Figure 13.10 diagrams the Deere FMS. This system is dedicated to manufacturing eight prismatic automatic transmission housings for agricultural tractors. It includes 5 head indexers, 11 moduline machining stations, 2 load stations, and 2 unload stations. A computer-controlled material handling system, with 37 carts and 51 dedicated part fixtures, is used to move the eight parts through the system. Because a part's orientation can differ between operations, routing such parts is broken into two or three segments, with each segment requiring a pass through the FMS—resulting in a total of 14 part-items to be processed by the FMS. Instead of a centralized buffer storage queue, this FMS has room for two incoming and two outgoing part fixtures at each machine, and room on the material handling system track for additional part fixtures awaiting assignment to a machine for the next operation on the routing.

A simulation model of John Deere's FMS was used to evaluate the part loading methods shown in Figure 13.10, using minimum make-span and machine utilization as the performance criteria. Each simulation experiment solved a 10-week (five eight-hour days per week) static scheduling problem in which 7,590 units of the eight parts were scheduled. Several experimental factors were varied, including the part loading rule, number of part fixtures in the system at any given time, and alternate routing flexibility for the parts. The part loading timing rule used in these experiments was the Next Empty Fixture rule, indicating a new part was loaded whenever a completed part left the FMS. Also, parts were dispatched to machines using the Least Work in Next Queue rule, since small work queues normally exist at each machine in Deere's FMS.

The simulation results indicate the smallest proportion of jobs launched (SPJL) and the next empty fixture (NEF) part loading rules performed significantly better than the other procedures, producing a shorter schedule length (minimum make-span) with a higher machine utilization. The results also indicate an interesting interaction between two experimental factors: alternate routing flexibility and number of part fixtures in the system. A 12 percent gain in machine utilization occurred when a high alternate routing flexibility and a large number of in-process part fixtures were used in contrast with the opposite case. Yet, the results also indicate decreasing marginal improvement in the machine utilization, as the number of in-process part fixtures in-

---

FIGURE 13.10   Deere & Company Flexible Manufacturing System (FMS)

FMS Scheduling Procedures

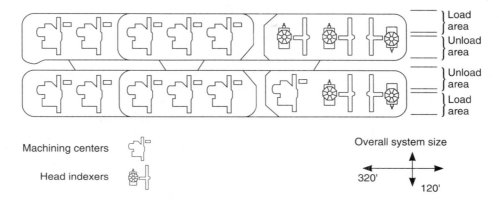

Part loading timing rule: Next empty fixture (NEF).
Part loading rules:

1. SPJL—Load that part that has the smallest proportion of job launched; that is, the smallest proportion of its batch requirement started.
2. SPT—Load that part with the shortest total processing time.
3. S&S—Load that part with the smallest ratio of loading station time divided by total processing time (Stecke and Solberg's rule).
4. NEF—As each fixture is unloaded, reload it with a like product, if possible.
5. FEM—Find the first empty machine that can do that part on it.
6. HPEM—Find the highest-priority empty machine and load that part that can be processed by that machine.

Dispatching rule: Least work in next queue (LWNQ).

Source: D. R. Denzler and W. J. Boe, "Experimental Investigation of Flexible Manufacturing System Scheduling Rules," *International Journal of Production Research* 25, no. 7, 1987.

---

creased. Since there's an upper limit on the number of part fixtures available to the system, increasing the number released into the FMS also decreases the choice of part types to be considered by the part loading rule at the loading station.

These studies illustrate use of shop-floor scheduling methods for managing day-to-day operations for an FMS dedicated to producing a relatively small family of similar parts. Taken together, the two studies provide an overall view of the performance impact of using different scheduling procedures and criteria for the three different decisions in such FMS systems. Very little work has been reported on scheduling "random" FMSs—FMS systems designed to process a much larger variety of parts that arrive continuously at random time intervals. Installing random FMSs is an important new trend and provides additional customer service possibilities.

A study by Shanker and Tzen reports early work on methods for planning workloads on a random FMS, and rules for dispatching orders to individual machines at the time of actual production. Their work considers planning balanced workloads at different machines in an FMS when setting due dates for jobs processed on an FMS. They evaluate five methods for planning FMS workloads in conjunction with four different rules for dispatching jobs to machines: first in/first served, shortest processing time, longest processing time, and most operations remaining first. The simulation experiment results Shanker and Tzen report indicate two dispatching rules (shortest processing time and most operations remaining first) perform effectively using machine utilization as the performance criterion. Furthermore, FMS performance improves substantially with a workload planning procedure that achieves a balanced workload between machines.

Shanker and Tzen's work confirms the effectiveness of the shortest processing time rule Stecke and Solberg observed in dispatching orders to FMS machines, and indicates this rule's robust nature under a wide variety of production process configurations. It may well be that, as scheduling random FMSs is further researched, other parallels will be drawn between earlier research and FMS production system applications.

A different issue in day-to-day operations of FMS systems appears in Jaikumar's work. Figure 13.11 compares FMS use by U.S. and Japanese manufacturers. The key differences are in the number of different parts processed by the FMSs, and the utilization of equipment. Figure 13.11 offers many lessons; but the key, from a scheduling point of view, is it seems increasingly

---

FIGURE 13.11 Jaikumar's Comparison of U.S. and Japanese FMS Use

	United States	*Japan*
System development time in years	2.5 to 3	1.25 to 1.75
Number of machines per system	7	6
Types of parts produced per system	10	93
Annual volume per part	1,727	258
Number of parts produced per day	88	120
Number of new parts introduced per year	1	22
Number of systems with untended operations	0	18
Utilization rate, two shifts	52%*	84%*
Average metal cutting time per day-hours	8.3	20.2

*Ratio of actual metal-cutting time to time available for metal cutting.

Source: R. Jaikumar, "Post Industrial Manufacturing," *Harvard Business Review,* November–December 1986, p. 71.

necessary to utilize FMSs at higher rates, to schedule smaller, more frequent lots, and to continually produce new parts on them.

CONCLUDING PRINCIPLES

The advanced scheduling concepts described in this chapter lead to the following concluding principles:

- We must determine the objective(s) to be achieved in scheduling before selecting a scheduling approach since different approaches provide different results.
- The shortest processing time sequencing rule, contrary to our intuition, can produce effective performance and should be considered as a standard in designing shop-floor systems.
- Introducing flexibility in scheduling (e.g., through alternate routings, adjustments in labor assignments, and overlap scheduling) should be used to gain important improvements in manufacturing performance.
- Manufacturing lead time estimating should be based on total work content.
- Due dates must be maintained to provide improvements in manufacturing performance.
- Due date filtering procedures should be used to diminish shop-floor nervousness.
- Increased use of cellular manufacturing systems and FMS are creating new opportunities and challenges for routine scheduling. Simple scheduling approaches should be considered first for these situations.

REFERENCES

Bagchi, U.; J. C. Hayya; and J. K. Ord. "Modeling Demand during Lead Time." *Decision Sciences* 15, no. 2 (Spring 1984).

Baker, Kenneth R., and David F. Pyke. "Solution Procedures for the Lot Streaming Problem." *Decision Sciences Journal* 21, no. 3 (Summer 1990), pp. 475–91.

————, and J. J. Kanet. "Job Shop Scheduling with Modified Due Dates." *Journal of Operations Management* 4, no. 1 (1983).

————, and G. B. McMahon. "Scheduling the General Job-Shop." *Management Science* 31, no. 5 (May 1985).

Berry, W. L., and V. Rao. "Critical Ratio Scheduling: An Experimental Analysis." *Management Science* 22, no. 2 (October 1975).

Bobrowski, Paul M., and Vincent A. Mabert. "Alternate Routing Strategies in Batch Manufacturing: An Evaluation." *Decision Sciences Journal* 19, no. 4 (Fall 1988) pp. 713–33.

Buzacott, J. A., and D. D. Yao. "Flexible Manufacturing Systems: A Review of Analytical Models." *Management Science* 32, no. 7 (July 1986).

Campbell, H. G.; R. A. Dudek; and M. L. Smith. "A Heuristic Algorithm for the n Job m Machine Sequencing Problem." *Management Science* 16, no. 10, June 1970.

Christy, David P., and John J. Kanet. "Open Order Rescheduling in Job Shops with Demand Uncertainty: A Simulation Study." *Decision Sciences Journal* 19, no. 4 (Fall 1988), pp. 801–18.

Conway, R. W.; W. L. Maxwell; and L. W. Miller. *Theory of Scheduling*. Reading, Mass.: Addison-Wesley, 1967.

Cruickshanks, A. B.; R. D. Drescher; and S. C. Graves. "A Study of Production Smoothing in a Job Shop Environment." *Management Science* 30, no. 3 (March 1984).

Denzler, D. R., and W. J. Boe. "Experimental Investigation of Flexible Manufacturing System Scheduling Rules." *International Journal of Production Research* 25, no. 7, 1987, pp. 979–94.

Elvers, D. A., and M. D. Treleven. "Job-Shop vs. Hybrid Flow—Shop Routing in a Dual Resource Constrained System." *Decision Sciences* 16, no. 2 (Spring 1985).

Fryer, J. S. "Labor Flexibility in Multiechelon Dual-Constraint Job Shops." *Management Science* 20, no. 7 (March 1974).

Graves, S. "A Review of Production Scheduling." *Operations Research* 29, no. 4 (July/August 1981).

Hausman, W. H., and G. D. Scudder. "Priority Scheduling Rules for Repairable Inventory Systems." *Management Science* 28, no. 11 (November 1982).

Hoffmann, T. R., and G. D. Scudder. "Priority Scheduling with Cost Considerations." *International Journal of Production Research* 21, no. 6 (1983).

Jacobs, F. R., and D. J. Bragg. "Repetitive Lots: Flow-Time Reductions through Sequencing and Dynamic Batch Sizing." *Decision Sciences Journal* 19, no. 2 (Spring 1988), pp. 281–94.

Jaikumar, R. "Postindustrial Manufacturing." *Harvard Business Review*, November–December 1986, pp. 69–76.

Johnson, S. M. "Optimal Two and Three-Stage Production Schedules with Setup Time Included." *Naval Research Logistics Quarterly* 1 (1954), pp. 61–68.

Luthi, H. J., and A. Polymeris. "Scheduling to Minimize Maximum Workload." *Management Science* 31, no. 11 (November 1985).

Mosier, C. T.; D. A. Elvers; and D. Kelly. "Analysis of Group Technology Scheduling Heuristics." *International Journal of Production Research* 22, no. 5 (1984), pp. 857–75.

Nanot, Y. R. "An Experimental Investigation and Comparative Evaluation of Priority Discipline in Job Shop Queueing Networks." Management Sciences Research Project, Research Report no. 87. University of California, Los Angeles, December 1963.

Nawoz, M. A.; M. Enscoci; and E. E. Ham. "A Heuristic Algorithm for the M Machine–N Job Flow Shop Sequencing Problem." *Omega* 41, no. 1, 1983.

Nelson, R. T. "Labor and Machine Limited Production Systems." *Management Science* 13, no. 9 (May 1967).

Penlesky, R. J. "Open Order Rescheduling Heuristics for MRP Systems in Manufacturing Firms." Ph.D. dissertation. Indiana University, 1982.

————; W. L. Berry; and U. Wemmerlöv. "Open Order Due Date Maintenance in MRP Systems." *Management Science* 35, no. 5 (May 1989), pp. 571–84.

————; U. Wemmerlöv; and W. L. Berry. "Filtering Heuristics for Rescheduling Open Orders in MRP Systems." *International Journal of Production Research,* forthcoming.

Posner, M. E. "A Sequencing Problem with Release Dates and Clustered Jobs." *Management Science* 32, no. 6 (June 1986).

Ragatz, G. L., and V. A. Mabert. "An Evaluation of Order Release Mechanisms in a Job-Shop Environment." *Decision Sciences Journal* 19, no. 1 (Winter 1988), pp. 167–89.

Raman, N.; F. B. Talbot; and R. V. Rachamadugu. "Due Date Based Scheduling in a General Flexible Manufacturing System." *Journal of Operations Management* 8, no. 2 (April 1989), pp. 115–32.

Russell, R. S., and B. W. Taylor III. "An Evaluation of Sequencing Rules for an Assembly Shop." *Decision Sciences* 16, no. 2 (Spring 1985).

Scudder, G. D. "Scheduling and Labor-Assignment Policies for a Dual-Constrained Repair Shop." *International Journal of Production Research* 24, no. 3 (May–June 1986).

Shanker, K., and Y. S. Tzen. "A Loading and Dispatching Problem in a Random Flexible Manufacturing System." *International Journal of Production Research* 23, no. 3 (1985), pp. 579–95.

Stecke, K. E., and J. J. Solberg. "Loading and Control Procedures for a Flexible Manufacturing System." *International Journal of Production Research* 19, no. 5 (1981), pp. 481–90.

Steele, D. C. "The Nervous MRP System: How to Do Battle." *Production and Inventory Management* 16 (4th quarter 1975).

Treleven, M. D., "The Timing of Labor Transfers in Dual Resource-Constrained Systems: Push v. Pull Rules." *Decision Sciences Journal* 18, no. 1 (Winter 1987), pp. 73–88.

————, and D. A. Elvers. "An Investigation of Labor Assignment Rules in a Dual-Constrained Job Shop." *Journal of Operations Management* 6, no. 1 (1986).

Wemmerlöv, U. *Production Planning and Control Procedures for Cellular Manufacturing Systems: Concepts and Practice.* Falls Church, Va.: American Production and Inventory Control Society, 1987.

————, and A. J. Vakharia. "Job and Family Scheduling a Flow Line Manufacturing Cell: A Simulation Study." *IIE Transactions,* December 1991.

DISCUSSION QUESTIONS

1. Why is rescheduling so important in production planning and control?
2. What kinds of performance measures would apply to "scheduling" at a college?
3. Are there college equivalents of the following concepts: flow shop structure, matched sets (common due dates), operation overlapping, and alternate routings?

4. Provide some examples of static and dynamic scheduling problems.
5. What sequencing rule do you use to do your homework?
6. If you were asked to audit lead times for a firm's MRP system, what would you look for?

PROBLEMS

1. The Pohl Pool Company has seven jobs waiting to be processed through its liner department. Each job's estimated processing times and due dates are as follows:

Job	Processing time (days)	Due date (days from now)
A	4	8
B	13	37
C	6	8
D	3	7
E	11	39
F	9	21
G	8	16

   a. Using the shortest processing time scheduling rule, in what order would the jobs be completed? Processing can start immediately.
   b. What is the average completion time (in days) of the sequence calculated in question a?
   c. What is the average job lateness (in days) of the sequence calculated in question a?

2. The Franklin Furniture Company has the following information on five jobs waiting to be processed in a work center. Processing can start immediately.

Job	Remaining processing time (days)	Due date (days from now)	Remaining number of operations
A	8	15	3
B	4	3	3
C	10	20	5
D	6	6	5
E	7	3	2

   a. Using the slack per operation scheduling rule, in what order would the jobs be started?
   b. What is the average job lateness (in days) of the sequence calculated in question a?

c. What is the average number of jobs in the system using the sequence in question a?

3. The Hyer-Than-Ever Kite Manufacturing Emporium must schedule the latest set of work orders through its frame-making department. Kites begin at frame making and then proceed through one, two, or three other departments, depending upon the ordered kite's sophistication. It is 8 A.M. Monday morning. Processing can start immediately. There is one operation per department. Shop scheduler Joan Weber faces the following set of orders listed in order of arrival (Assume no other jobs will arrive at the frame-making department until the five already there have been started.):

Kite order	Frame-making time (days)	Total processing time (days)	Total number of operations	Kite due date (days from now)
A	10	20	4	25
B	12	18	2	15
C	7	12	2	16
D	5	12	3	17
E	8	10	2	12

a. Using the order data, evaluate the first-come/first-served sequencing rule:
1. In what sequence should the frame making department process the jobs?
2. What will be the average completion time in frame making?
3. What will be the average number of jobs in the frame-making department?

b. Using the order data, evaluate the earliest due date sequencing rule:
1. In what sequence should the frame making department process the jobs?
2. Which specific jobs, if any, will be late in leaving frame making?
3. What will the average job lateness be in the frame-making department?

c. Using the order data, evaluate the shortest processing time sequencing rule:
1. In what sequence should the frame-making department process jobs?
2. On what days will order E be in process in frame making?
3. What will the average job lateness be in the frame-making department?

d. While Joan debated which rule to use, she received the following memo from Diane Britenbach, the president:

> Effective this morning, all departmental job sequencing should be performed using the slack time per operation rule. This should improve scheduling performance throughout the plant.

1. In what sequence should the frame-making department process the jobs if it uses the slack time per operation rule and considers due dates to be final due dates?
2. On what day will frame making complete order D?
3. What is the average completion time for this sequence in the frame-making department?

4. Prepare a two-machine schedule using the Johnson procedure for jobs A through D. The processing time per job (in days) is shown below:

	Machine	
Job	*I*	*II*
A	4	3
B	1	7
C	8	2
D	8	5

a. In what sequence should jobs be processed?
b. Construct a Gantt chart of the schedule for both machines.
c. Construct a Gantt chart, assuming there's no buffer storage between machines (e.g., machine I can't start a new job until machine II has started the old job).

5. Telly's Deli and Catering Company has five orders—each for a particular type of salad. Making each salad consists of two tasks: prepping and assembling. Telly can't start assembling an order until its prepping is complete. The salads must be complete in 24 hours.

Orders	Prepping (hours)	Assembling (hours)
Antipasto	6	4
Indonesian rice	3	4
Mixed vegetables	5	2
Marinated vegetables	8	6
German potato salad	2	1

a. What schedule produces the shortest time span in which these orders can be completed?
b. Can Telly complete all orders in 24 hours? How many hours does it take to complete the five orders?
c. For the schedule in question a, how many hours is the assembling task idle?
d. When will Telly finish the order for the mixed vegetable salad?
e. What is the average order completion time?

6. Flash Fasttrack (associate dean of janitorial services at Wombat University) must schedule the regular maintenance by his janitorial engineers. There are two crews: sweeping crew and waxing crew. Union rules prohibit sweepers from waxing and waxers from sweeping. A building's floors must first be swept before they can be waxed. Flash must schedule the order in which maintenance crews will visit each of Wombat U's six buildings. He wishes to minimize the sequence's total completion time. Times required to sweep and wax each building are as follows:

Building	Sweeping time (hours)	Waxing time (hours)
Astronomy	18	10
Biology	8	9
Chemistry	26	13
Drama	15	16
English	17	20
Foreign language	12	17

a. Schedule the crews through the buildings.

b. When will the waxing crew start and stop work in the chemistry building?

c. During the schedule, for how much time are sweepers and waxers available for other activities?

7. Currently, the marketing manager at Precision Parts, Inc. (a manufacturer of custom-made machined parts) promises six-week delivery time for all customer orders. The production manager is under pressure from top management to improve the firm's performance against quoted delivery dates. In his review of alternative procedures for making customer delivery date promises, he provides the following data on representative orders:

Order	Total processing time (in weeks)
1	2
2	4
3	6

a. Determine delivery time to be quoted on these orders, if the TWK (total work content) procedure is used. (Assume average time from order receipt to customer delivery is six weeks.)

b. Determine delivery time to be quoted on these orders if the SLK (slack) procedure is used. (Assume average time from order receipt to customer delivery is six weeks.)

c. Suppose the production manager knows the overall flow time for order #3 has averaged nine weeks in the past, and the flow time standard deviation for this order is one week. How would this information influence your recommendation of a delivery date setting procedure in this situation?

8. A single test operator staffs the SCM Corporation's three testing machines in its quality control lab. Only one machine is run at a time. SCM uses two scheduling rules: one for dispatching jobs at individual machines and another for assigning labor to machines. As of 8 this morning (on day 10), jobs waiting to be processed at each of the three machines are as follows:

		Test Machine A	
Job	*Due date**	*Processing time†*	*Arrival date**
1	12	1	9
5	18	6	8

		Test Machine B	
Job	*Due date**	*Processing time†*	*Arrival date**
2	14	3	6
3	16	4	7
6	14	6	6

		Test Machine C	
Job	*Due date**	*Processing time†*	*Arrival date**
4	13	4	5

* = Day number.
† = In days.

a. Assuming SCM uses the shortest processing time rule to sequence orders at each testing machine, determine to which machine the test operator should be assigned and which job will be processed when each of the following labor assignment decision rules are used (assume the operator has just completed his or her last job):
1. The shortest job.
2. The job that has been in the laboratory the longest.
3. The most jobs in the queue.
b. Suppose SCM uses the slack time dispatching rule to sequence jobs at each testing machine. To which machine should the test operator be assigned and which job will be processed when each of the following labor assignment decision rules are used? (Again it's 8 A.M. and the test operator is available for assignment.)
1. The shortest job.
2. The job that has been in the laboratory the longest.
3. The most jobs in the queue.
c. Would you recommend any other labor assignment decision rule under the conditions in question b above?
9. The materials manager at the Excello Grinding Wheel Company is concerned about the high volume of rescheduling exception messages his firm's MRP planners have to cope with each week. He's considering implementing one of several possible rescheduling filters (e.g., the Ability, Magnitude, or Horizon heuristic) and has provided the following example MRP record to illustrate such procedures' use:

Item A

Week number		1	2	3	4	5	6	7	8	9	10
Gross requirements		16	20	15	20	2	27	1	15	8	10
Scheduled receipts			30		30		30				
Projected available balance	12	-4	6	-9	1	-1	2	1	16	8	28
Planned order releases		30		30							

Q = 30; LT = 7; SS = 0

a. Assume total setup plus run times remaining on the three scheduled receipts in weeks 2, 4, and 6 are 2 weeks, 3.4 weeks (three weeks plus two days), and 4 weeks, respectively. What rescheduling exception messages would the Ability heuristic make?

b. Assume total setup plus run times remaining on the three scheduled receipts in weeks 2, 4, and 6, are 0.4 weeks (two days), 1 week, and 2 weeks, respectively. What rescheduling exception messages would the Ability heuristic make?

c. Assume $T_m = 1$. What rescheduling exception messages would the Magnitude heuristic make?

d. Assume $T_H = 2$. What rescheduling exception messages would the Horizon heuristic make?

e. Assume $T_m = 0$. What rescheduling exception messages would the Magnitude heuristic make?

f. Assume $T_m = \infty$. What rescheduling exception messages would the Magnitude heuristic make?

g. Suppose a rescheduling exception message calls for moving the 30 units in week 6 to week 8. Would this message be produced in $T_m = 1$?

10. The Medwitz Company has the following processing time data for machines 1, 2, and 3. Routing for each job begins at machine 1 and ends at machine 3:

	Processing time (in hours)		
Job	Machine 1	Machine 2	Machine 3
A	1	5	3
B	6	2	9
C	5	7	2
D	12	1	4

a. Prepare a schedule for a three-machine manufacturing cell using the Campbell, Dudek, and Smith (CDS) heuristic.

b. In what sequence should Medwitz process the jobs to minimize the make-span? Follow the same processing sequence at each machine in the cell.

c. Construct a Gantt chart of the best sequence schedule for all three machines.

11. The Philip Company has the following processing time data for machines 1, 2, 3, and 4. Routing for each job begins at machine 1 and ends at machine 4:

Job	Processing time (in hours)			
	Machine 1	Machine 2	Machine 3	Machine 4
A	6	2	4	2
B	1	8	7	5
C	5	7	2	3
D	4	4	1	6

a. Prepare a schedule for a four-machine manufacturing cell using the Campbell, Dudek, and Smith heuristic.

b. In what sequence should Philip process the jobs to minimize make-span? Follow the same processing sequence at each machine in the cell.

c. Construct a Gantt chart of the best sequence schedule for all four machines.

12. The Arnold Company has the following processing time data for the three-machine manufacturing cell. Each job's routing begins at machine 1 and ends at machine 3:

Job	Processing time (in hours)		
	Machine 1	Machine 2	Machine 3
A	1	7	3
B	4	2	8
C	2	9	2

a. A partial schedule, sequencing job A first and job B second, has already been prepared below. Use the NEH procedure that minimizes the makespan to determine the sequence through all three machines in the cell for jobs A, B, and C.

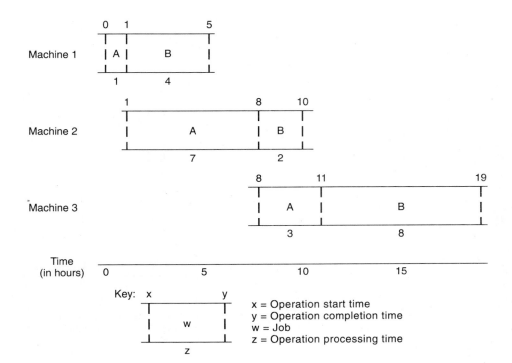

Time
(in hours)

Key:

x = Operation start time
y = Operation completion time
w = Job
z = Operation processing time

b. A partial schedule, sequencing job B first and job A second, has already been prepared below. Using the NEH procedure, determine the sequence through all three machines in the cell for jobs A, B, and C that minimizes the make-span. How does this schedule compare with that in part a of this problem?

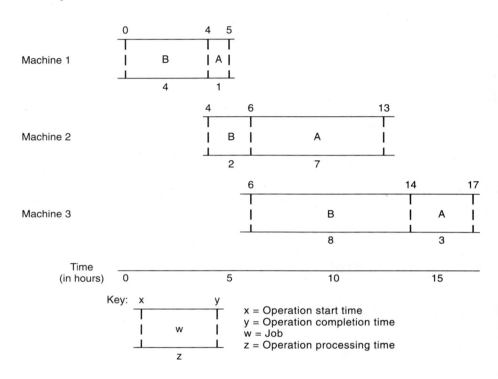

13. Prepare a schedule for a four-machine manufacturing cell using the Slack Time procedure and the following processing time data for the cell. Routing for each job begins at machine 1 and ends at machine 4.

		Processing time (in hours)				
Job	Family	Machine 1	Machine 2	Machine 3	Machine 4	Hours until due date
1	A	2	3	9	2	30
2	B	5	4	2	7	31
3	A	1	6	4	1	24
4	B	4	7	9	8	48

The due date for each order is set using the TWK rule; in this case TWK equals total processing plus setup time plus an allowance for queuing. Setup time is one hour when a change from one family to another is made and zero if parts of the same family are processed.

   a. Using the conventional version of the Slack Time procedure, determine the sequence for processing jobs 1 through 4 at machine 1 (and through the remaining machines).

   b. Using the Family version of the Slack Time procedure, determine the sequence for processing jobs 1 through 4 at machine 1 (and through the remaining machines).

14. A firm has installed a manufacturing cell containing five machines to process cylindrical parts in three different part families: A, B, and C. Though parts within each part family are similar in design, work can be routed through this cell in a variety of ways. Orders typically begin processing at either machine 1 or machine 2, and can be further processed at any or all of the remaining three machines. Therefore, a queue of orders exists at each of the five machines, and sequencing decisions are made at each machine in the cell.

    Two decisions are made in sequencing orders at each machine. At the time an order is completed at a machine, the foreman decides which part family to process next. Once she has selected the part family to process next, she uses the slack time dispatching rule to decide which order within the part family to process next. Exhibit A shows a dispatching report for machine 2 in this cell. This machine has just completed an order in part family A.

EXHIBIT A Machine 2 Dispatching Report

Shop order number	Item part number	Operation setup time*	Processing time†	Part family	Order due date‡	Total processing time remaining§
10–1234	5678	1	6.2	A	12	2
10–1240	1082	1	4.3	A	10	4
10–1241	1141	1	5.1	A	7	8
10–1231	1271	1	1.8	B	14	15
10–1229	4252	1	2.3	B	8	6
10–1215	8110	1	0.9	B	10	14
10–1251	1354	1	1.7	B	3	6
10–1249	1278	1	13.2	C	4	1
10–1225	7910	1	3.4	C	9	3
10–1242	6250	1	4.1	C	4	1
10–1260	5140	1	2.8	C	12	4
10–1261	6280	1	3.1	C	15	10
10–1042	1011	1	8.1	C	13	3

*In hours assuming a change in part family is required; otherwise setup time is zero.
†In hours.
‡Number of manufacturing days until the order due date.
§In manufacturing days.

---

    a. Which order should be processed next at machine 2 if the ECON procedure is used to select the part family to process next?

    b. Which order should be processed next at machine 2 if the WORK procedure is used to select the part family to process next?

    c. Which order should be processed next at machine 2 if the AVE procedure is used to select the part family to process next?

15. An FMS with 10 machines has been installed to process a variety of machined parts. These parts are automatically transferred from a 16-location loading station to each machine for processing. Upon completion at a machine, either a part is complete and therefore transferred from the FMS, or it waits for further processing in the FMS. In the latter case the part is stored in the loading station until it's dispatched to the next machine operation.

    The FMS dispatching report (Exhibit B) reflects the status of orders waiting to be processed at the FMS at 8 A.M. Monday. If the SPT/TOT dispatching rule is to be used in selecting the next order to process at each machine, what's the next order to be processed at each machine in the FMS?

EXHIBIT B   FMS Dispatching Report

Loading station number	Shop order number	Item part number	Operation number	Machine number	Operation processing time*	Total processing time remaining*	Order due date†
1	121–2	1234	10	2	0.5	4.4	2
2	100–10	4213	5	1	0.2	2.6	1
3	60–15	8819	10	2	0.4	8.1	1
4	151–41	1617	5	2	0.8	5.0	2
5	82–92	1002	5	6	0.3	3.4	8
6	130–14	4154	10	5	0.7	2.7	4
7	42–116	8213	10	10	0.6	9.8	4
8	44–210	1234	20	3	0.8	1.7	5
9	61–820	1617	10	3	0.2	4.2	2
10	81–419	1002	10	7	0.4	3.1	1
11	42–161	6150	15	6	0.9	5.6	1
12	45–1	1617	15	4	0.5	10.1	7
13	75–25	1234	30	6	0.8	0.9	1
14	18–191	8819	5	4	1.0	9.1	4
15	46–18	8213	20	2	0.2	6.1	2
16	53–114	1002	20	8	0.6	5.3	1

*In hours.
†Number of manufacturing days until the order due date.

Chapter 14

Advanced Concepts in Master Production Scheduling

This chapter addresses advanced concepts in master production scheduling. The techniques described are primarily oriented to the assemble-to-order (ATO) manufacturing environment. This environment is particularly challenging, because companies increasingly find themselves less able to predict the exact end-item configurations customers will order. Moreover, the number of these end items is rising in many firms. Using the advanced concepts described in this chapter, you can reduce the number of items to be master scheduled, improve the relationship between production and sales, and increase the MPS's flexibility. Implementation requires detailed design efforts and added complexities in maintaining bills of material—but the payoffs are significant.

The chapter is organized around four topics:

- Two-level master production scheduling: What is the structure of a two-level MPS and how does it work?
- The Hyster lift truck example: How does two-level master production scheduling work in a complex product environment?
- Additional techniques: What are some alternative methods for coping with the assemble-to-order manufacturing environment and how do they work?
- Methods for constructing planning bills of material: How do we construct the planning bills of material to support assemble-to-order manufacturing?

This chapter is closely related to Chapters 6, 7, and 8. Chapter 6 deals with the basics in master production scheduling, including time-phased record formats for master scheduling, available to promise logic, and planning bills of material. Chapter 7 describes linkages between the master production schedule and the production plan. Chapter 8 treats demand management and shows how some forms of planning bills lead to better ways to hedge against uncertainties in demand.

TWO–LEVEL MASTER PRODUCTION SCHEDULING

This section starts with a simplified example of two-level master production scheduling. Our intent is to show how the two-level MPS works, how records are processed as we book actual orders and update the record from period to period, and how actual orders consume forecasts of demand. The section ends by describing benefits of the approach; the net result is a much closer relationship between sales and manufacturing, where each function is better equipped to respond to customer needs.

Two-Level MPS Example

For our example product, let's use an over-bed table manufactured by the Hill-Rom Company. Figure 14.1 shows a simplified version of the table's **super bill of material**. A super bill creates an "average" unit—in this case, an average over-bed table. Customers have the option of ordering any of four models, in 10 different high-pressure laminate tops, with four different combinations of boots and casters. This results in 160 ($4 \times 10 \times 4 = 160$) end-item alternatives.

To illustrate two-level master production scheduling, let's focus attention on two of the four model types: the 622, which represents 50 percent of sales;

FIGURE 14.1   Hill-Rom Over-Bed Table Super Bill

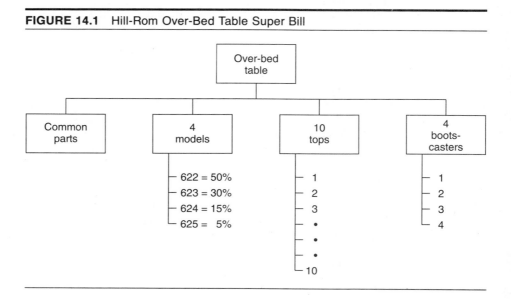

and the 623 model, which accounts for another 30 percent. Each "model" is only a collection number or "bag of parts." It represents the unique parts that (along with the common parts, a top and a boot-caster option) make a unique end item.

Figure 14.2 depicts MPS records for the over-bed table and the two model options, 622 and 623. The over-bed table is level 0 in the bill of materials, and the two model types are at level 1. The level 1 models and the level 1 common parts record explode directly into the MRP records, which provide detailed plans for manufacturing the parts. The collection of parts required to make a model 622, for example, is controlled by the level 1 MPS, which, in turn, is controlled by the MPS for the over-bed table line. The logic and records are consistent with time-phased MRP/MPS logic.

At the top of Figure 14.2, the over-bed table record is driven by the production plan. It shows a constant MPS rate of 100 units per week over the five weeks in our example. The "Orders" row shows the first two weeks of

---

FIGURE 14.2 Over-Bed Table Two-Level MPS Records

	Week				
	1	2	3	4	5
Over-bed table:					
Production plan	100	100	100	100	100
Orders	100	100			
Available					
Available to promise			100	100	100
MPS	100	100	100	100	100
On hand = 0; SS = 0					
622 model:					
Production forecast			50	50	50
Orders	50	50			
Available	10	10	10	10	10
Available to promise	10		50	50	50
MPS	50	50	50	50	50
On hand = 10; SS = 10					
623 model:					
Production forecast			30	30	30
Orders	30	30			
Available	15	15	15	15	15
Available to promise	15		30	30	30
MPS	30	30	30	30	30
On hand = 15; SS = 15					

the production plan and the MPS are completely sold. The "Available-to-promise" (ATP) row indicates no new customer orders can be promised earlier than week 3. There are ATP quantities of 100 units in weeks 3, 4, and 5.

For the second record, the 622 model, the "Production forecast" row is exploded from the ATP row of the over-bed table at level zero. The production forecasts of 50 units in weeks 3, 4, and 5 are what we expect to sell of the yet unbooked orders (i.e., 50 percent of the ATP amounts for over-bed tables). Similarly, the 30s in the third record come from the expectation that 30 percent of the over-bed table orders will be for the 623 model.

Explosion of the ATP quantities, rather than the MPS quantities, is because only the unsold items have uncertainty. There's no production forecast for the 622 and 623 models in weeks 1 and 2, because all 100 units to be built in those periods have been sold (consumed) and their *exact* component requirements have been allocated. For this reason, the production forecast is also called the "unconsumed forecast" in some systems.

The 622 model shows 10 units on hand and available to promise in week 1. This is a safety stock that protects against variability in this model's 50 percent usage rate estimate. The master scheduler determines when additional quantities should be scheduled to replenish the 622 model parts. Similar conventions apply to the 623 model.

Booking Customer Orders

Let's say a customer orders 40 units of the 623 model to be delivered as soon as possible. (We ignore the top and boot/caster options for this example.) The first check is: When are the over-bed units available to promise? Figure 14.2 shows week 3 is the first week in which sufficient ATP exists. The next question is whether a week 3 delivery of the 623 model is possible. Again, from Figure 14.2, the answer is yes: 30 units are available to promise in week 3, and an additional 15 units are ATP in week 1 (the safety stock).

Assuming the top and boot/caster options are also available to promise in week 3, the three records from Figure 14.2 would look like Figure 14.3 after the order is booked. Week 3's ATP of 100 over-bed tables now falls to 60. The former 30 units of ATP in week 3 for the 623 model have declined to zero, and week 1's ATP of 15 units is reduced to 5.

We now recalculate production forecasts for models 622 and 623. Since there are only 60 over-bed tables ATP in the level 0 record in week 3, expected sales volume for the 622 model is 30 units (50 percent) and 18 units are expected for the 623 model. These two models' records reflect this new set of expectations. One result is the 622 now seems to be overplanned (too much expected inventory), whereas the 623 is underplanned (an expected shortfall).

One immediate question is whether to revise the two MPS quantities for period 3 to compensate. If we revise the MPS quantity for the 622 to 30, the projected available balance returns to the safety stock level of 10. Similarly,

FIGURE 14.3 Over-Bed Table Two-Level MPS Records, after Booking the Order for 40 of the 623 Model in Week 3

	Week				
	1	2	3	4	5
Over-bed table:					
Production plan	100	100	100	100	100
Orders	100	100	40		
Available					
Available to promise			60	100	100
MPS	100	100	100	100	100
On hand = 0; SS = 0					
622 model:					
Production forecast			30	50	50
Orders	50	50			
Available	10	10	30	30	30
Available to promise	10		50	50	50
MPS	50	50	50	50	50
On hand = 10; SS = 10					
623 model:					
Production forecast			18	30	30
Orders	30	30	40		
Available	15	15	−13	−13	−13
Available to promise	5		0	30	30
MPS	30	30	30	30	30
On hand = 15; SS = 15					

the MPS for the 623 model in week 3 might be revised to 58 units to return its projected available balance to 15 units. These changes *won't* be made automatically. MPS quantities are firm planned orders; changing them requires consideration of the changes' overall impact. Moreover, there's the question of nervousness. Perhaps the next orders will naturally compensate for the order for 40 units of the 623 model. Note, even if no changes are made, five units of 623 are still available to promise during the three periods.

Managing with a Two-Level MPS

Two-level master production scheduling allows the company to focus attention separately on the overall rate of production (as determined by the production plan) and how various options are selling in the marketplace. When sales orders outstrip production, ATP dates for the 0-level item will stretch out. The opposite condition results in more near-term ATP and perhaps a decision about making certain items for inventory.

When the 0-level item has near-term ATP, but some level-1 items don't, it indicates the product mix isn't meeting expectations—or perhaps safety stocks are too small or in the wrong items. Combined marketing and MPS decision making is required.

Overall, the two-level MPS approach helps us align the MPC process closely to market forces. Products are planned and controlled in the way they're *sold,* as opposed to the way they're designed or manufactured. Flexibilities are defined and understood in both sales and manufacturing organizations.

THE HYSTER LIFT TRUCK EXAMPLE

Let's turn to a detailed example of the two-level MPS approach. Actual company records show how the MPS is designed and maintained. We'll also see how customer order promising is done in a complex product environment.

The Portland Plant

Hyster, Inc's Portland, Oregon, plant produces forklift trucks, straddle carriers, and towing winches (for mounting on the rear of crawler tractors). Straddle carriers are made to order, but both towing winches and lift trucks are assemble-to-order products with virtually unlimited end item possibilities. Lift trucks are divided into three model series with a super bill of materials for each; six super bills cover the towing winch models. Each super bill consists of various options with percentage usage factors.

Figure 14.4 indicates the end-item definition complexity. This figure represents the option choices for one model series of forklift truck. Option groups used in the super bill are closely related to their sales options; but they aren't identical since some options preclude others and dictate still other options. For example, the gasoline engine option dictates a particular frame and air cleaner. The powershift transmission *with* a gasoline engine dictates still another set of options, such as radiator, flywheel, and hydraulic hoses.

A 12-month production plan is established at corporate headquarters, stated in super bill terms. Each month, this plan is revised. (A new month is added.) It's stated in the number of units of each model series to have available to ship in each month. If no customer order exists at the time a commitment must be made to an exact end item (demand time fence), marketing issues a stock order for the model unit, which will become part of marketing's finished-goods inventory.

FIGURE 14.4 Product Option Matrix: Model Series A Lift Truck Order Entry

ORDER ENTRY

LIFT TRK 'A'			GAS O/C	GAS P/S	LPG O/C	LPG P/S	PERKINS O/C	PERKINS P/S	DETROIT O/C	DETROIT P/S
COMMON PARTS			001, 022, 100, 125, 133, 144, 001F, 144F, 125F, 133F, 103F	001, 023, 100, 125, 133, 144, 001F, 144F, 125F, 133F, 108F	001, 024, 100, 125, 133, 144, 001F, 144F, 125F, 133F, 108F	001, 025, 100, 125, 133, 144, 001F, 144F, 125F, 133F, 108F	001, 026, 100, 125, 133, 144, 001F, 144F, 125F, 133F, 108F	001, 027, 100, 125, 133, 144, 001F, 144F, 125F, 133F, 108F	001, 028, 100, 125, 133, 144, 001F, 144F, 125F, 133F, 108F	001, 029, 100, 125, 133, 144, 001F, 144F, 125F, 133F, 108F
ENGINE	GAS		002, 005, 006, 019, 030, 032, 035	002, 005, 006, 020, 031, 032, 036						
	LPG				003, 005, 019, 030, 032	003, 005, 020, 031, 033				
	PERKINS						008, 004, 006, 019, 032, 035	008, 004, 006, 020, 033, 036		
	DETROIT								007, 004, 006, 019, 035	007, 004, 006, 020, 036
AXLE	HYPOID	SOLID	012, 056, 134, 130	021, 037, 056, 134, 130	012, 056, 134, 130	021, 037, 056, 130, 130	012, 056, 134, 130	021, 037, 056, 134, 130	012, 056, 134, 130	021, 038, 056, 134, 130
		PNEUMATIC	012, 056, 134, 060, 130	021, 037, 056, 134, 060, 130	012, 056, 134, 060, 130	021, 037, 056, 060, 130	012, 056, 134, 060, 130	021, 037, 056, 134, 060, 130	012, 056, 134, 060, 130	021, 038, 056, 060, 130
	PLANETARY	SOLID	014, 057, 106, 116, 120	013, 057, 106, 116, 121	014, 057, 106, 116, 120	013, 057, 106, 116, 121	014, 057, 106, 116, 120	013, 057, 106, 116, 121	014, 057, 106, 116, 120	013, 057, 106, 116, 114
		PNEUMATIC	014, 057, 106, 116, 120, 061	013, 057, 106, 116, 121, 061	014, 057, 106, 116, 120, 061	013, 057, 106, 116, 121, 061	014, 057, 106, 116, 120, 061	013, 057, 106, 116, 121, 061	014, 057, 106, 116, 120, 061	013, 057, 106, 116, 114, 061
CAB			181, 619, 620, 626	181, 619, 621, 626	009, 181, 619, 620, 626	009, 181, 619, 621, 626	181, 619, 620, 626	181, 619, 621, 626	181, 619, 620, 626	181, 619, 621, 626
W/O CAB			182	182	010, 182	010, 182	182	182	182	182
HEATER & DEFROSTER			661	661	661	661	662	663	664	665
STANDARD VALVE			017, 080	018, 080	017, 080	018, 080	017, 080	018, 080	017, 080	018, 080
3-WAY VALVE	DIRECT		081, 015, 079, 086, 087	081, 016, 079, 086, 087	081, 015, 079, 086, 087	081, 016, 079, 086, 087	081, 015, 079, 086, 087	081, 016, 079, 086, 087	081, 015, 079, 086, 087	081, 016, 079, 086, 087
	PUSH		082, 015, 079, 086, 087	082, 016, 079, 086, 087	082, 015, 079, 086, 087	082, 016, 079, 086, 087	082, 015, 079, 086, 087	082, 016, 079, 086, 087	082, 015, 079, 086, 087	082, 016, 079, 086, 087
4-WAY VALVE			015, 079, 083, 085, 087	016, 079, 083, 085, 087	015, 079, 083, 085, 087	016, 079, 083, 085, 087	015, 079, 083, 085, 087	016, 079, 083, 085, 087	015, 079, 083, 085, 087	016, 079, 083, 085, 087
5-WAY VALVE			015, 079, 084, 085, 087	016, 079, 084, 085, 087	015, 079, 084, 085, 087	016, 079, 084, 085, 087	015, 079, 084, 085, 087	016, 079, 084, 085, 087	015, 079, 084, 085, 087	016, 079, 084, 085, 087
ATT PARTS – 3 / 4 / 5			680, 681, 682	680, 681, 682	680, 681, 682	680, 681, 682	680, 681, 682	680, 681, 682	680, 681, 682	680, 681, 682
STD A/C			044, 047, 050	044, 047, 050	044	044	044, 050	044, 050		
H D A/C			054	054			054	054	STD	STD

Note: Each box represents MPS option groups specified by salable option features for this model.

Source: W. L. Berry, T. E. Vollmann, and D. C. Whybark, *Master Production Scheduling: Principles and Practice* (Falls Church, Va.: American Production and Inventory Control Society, 1979), p. 140.

Order promising is based on available-to-promise logic. The first ATP checked is for the model common parts, then for each optional feature desired. The process is straightforward, but each order is so complex in terms of optional choices that, even with an interactive computer system, it takes an average of one hour to promise an order and peg each option to the customer order.

The MPS system at Hyster Portland was installed along with MRP. The MRP and MPS are run daily, using net change, and the MPS produces time-phased records for each of about 1,400 MPS super bills and options. The Portland factory had to make major changes in its approach to production planning/inventory control to adapt to the system, but results have been dramatic. Performance against factory schedules and delivery performance are much better, inventory levels have fallen, and marketing is kept fully informed as each order progresses through manufacturing's various stages.

Marketing is also extremely happy with its reduced lead time to issue a firm end-item stock order to the plant. Before the system was implemented, marketing had to specify the exact end-item configuration 45 working days before items were completed. This has been cut to 14 days, resulting in greater flexibility in responding to actual customer orders. Manufacturing is also happy because lift truck retrofitting has been greatly reduced.

The Two-Level MPS

Detailed records at Hyster illustrate consumption of the forecast by actual customer orders, time fencing, planning bills, and buffering. Before turning to these, we first consider a simplified version of the records to help in understanding.

Figure 14.5 is an overview of Hyster's two-level approach. Level 0 MPS decisions plan and control the model series (what we have been calling product family); those at level 1 deal with the common parts and options. The level 0 MPS record is for the super bill. Quantities are derived directly from the monthly production plan, reflecting the exact working days in each week.

FIGURE 14.5   Hyster Two-Level MPS

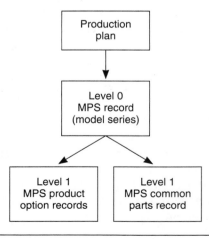

FIGURE 14.6 Hyster Simplified Example: Model A Series Lift Truck

Level 0 MPS record

Week Start day	8-21 178	8-28 183	9-5 188	9-11 192	9-18 197	9-25 202	10-2 207	10-9 212	10-16 217
Prod. forecast									
Sold									
Sched. rec.									
Proj. on hand									
MPS start	0	0	2	6	12	13	15	15	15

Level 1 common parts MPS record

Week Start day	8-21 178	8-28 183	9-5 188	9-11 192	9-18 197	9-25 202	10-2 207	10-9 212	10-16 217
Prod. forecast	0	0	2	6	12	13	15	15	15
Sold		13	10	8	2	2			
Sched. rec.		13							
Proj. on hand	0	0	0	0	0	0	0	0	0
MPS start	12	14	14	15	15	15	15	15	15

Lead time = 2 weeks.
Super bill percentage = 100%.

Level 1 oil clutch option parts MPS record

Week Start day	8-21 178	8-28 183	9-5 188	9-11 192	9-18 197	9-25 202	10-2 207	10-9 212	10-16 217
Prod. forecast	0	0	2	3	8	8	9	10	9
Sold		8	6	5	2	1			
Sched. rec.		9							
Proj. on hand	0	1	0	1	−1	0	0	−1	0
MPS start	7	9	8	10	9	9	10	9	9

Lead time = 2 weeks.
Super bill percentage = 62%.

The sum of the parts in the level 1 options, therefore, is closely coordinated with dictates of the production plan.

Figure 14.6 extracts data from the actual MPS records to illustrate the two-level MPS approach. It depicts the first nine weeks of Hyster's one-year planning horizon. The first MPS record in Figure 14.6 is a portion of the level 0

MPS record for the model series A (one of three lift truck families built at the Portland plant).

The Hyster records don't follow the same conventions as those we've used for MPS records. The Hyster records don't have an explicitly named available-to-promise row. They also combine some elements of an MRP record, including a lead time offset, which results in an MPS start date. The convention we've followed is to always show MPS quantities as of the date they'll be available for satisfying demand. Finally, the zero-level record doesn't contain the production plan at the forecast row, but it's netted against sold orders in the MPS start row. Nonetheless, the basic concepts are the same, and the Hyster two-level MPS accomplishes the same results. Several software packages support the two-level approach, but they have slightly different record formats.

The MPS quantities shown in the level 0 record in Figure 14.6 are derived from the production plan for this model series and are translated into weekly requirements that reflect the working days in each week. They are treated as firm planned orders. These production plan quantities are originally shown in the row labeled "MPS start," but are subsequently decremented by actual orders. The net result is that these "MPS start" quantities represent the unsold forecasts and are the available-to-promise amounts. Time buckets are printed in weekly increments for weeks 8/21 to 10/16. The production plan for this model series calls for production of 15 lift trucks per five-day week. (Some weeks have fewer than five working days and their production requirements are adjusted accordingly.)

The second record in Figure 14.6 is the level 1 record for the common parts. The row titled "Prod. forecast" is the same as the "MPS start" row from the level 0 record; that is, in the same way as in our earlier example, the level 0 record ATP is exploding to production forecasts for the level 1 records. For each model series A lift truck to be built, one set of common parts is required. The "Prod. forecast" row shows this explosion.

The common parts record in Figure 14.6 contains other useful information. For example, customer orders booked for model series A trucks are displayed in the "Sold" row. As an example, 13 model series A units have been booked for the week of 8/28. When Hyster books customer orders, both the "Prod. forecast" in the common parts MPS record at level 1 and the "MPS start" in the model series A MPS record at level 0 are decreased by the amount of the sales order. The resultant data are used to make available-to-promise calculations. For example, in booking an order for one unit for delivery in week 9/5, the value for that week in both the level 0 "MPS start" row and the common parts level 1 "Prod. forecast" row would decrease from 2 to 1. Likewise, the value of 10 in the level 1 common parts "Sold" row would increase to 11.

Firm planned orders for the level 1 common parts record are shown in the "MPS start" row. These orders are offset by a planned lead time of two weeks. As an example, the 12 units scheduled to be started in the week beginning 8/21 are to be completed at the start of week 9/5. This 12-unit firm planned order covers the requirements for 2 units of "Prod. forecast" in week 9/5 (exploded from the level 0 MPS record) and 10 units of sold orders in the same week.

The third record in Figure 14.6 is the level 1 MPS record for one of the model series A options, an oil clutch. For this record, the row labeled "Prod. forecast" is also derived from the "MPS start" row in the level 0 MPS record for model series A. The super bill percentage for this option is 62 percent. This indicates Hyster expects 62 percent of the model series A lift trucks to have an oil clutch specification. Thus, the "MPS start" row data in the level 0 MPS model series A record are multiplied by .62 to obtain the "Prod. forecast" data in the level 1 oil clutch MPS option record.

An exact multiplication of the "MPS start" data in the level 0 model series A MPS record by .62 won't yield the data in the "Prod. forecast" row of the level 1 oil clutch MPS option record. The system will always round up, but it keeps track of the cumulative round-up to an amount always less than one unit. For example, the "MPS start" row entry in the level 0 MPS record for 9/5 is 2. When multiplied by .62, the result is 1.24. The "Prod. forecast" row in the level 1 oil clutch MPS option record is rounded up to 2. In the following week, the level 0 record "MPS start" row entry is 6; when multiplied by .62, the result is 3.72. But since the earlier rounding was .76 (2 − 1.24), the "Prod. forecast" row value in the oil clutch MPS option record for week 9/11 is 3. There remains a cumulative roundup of .04 (.76 − .72).

Buffering

The safety stock aspects of the common and option parts MPS records in Figure 14.6 are contained in the projected on-hand balance ("Proj. on hand") row. The "Proj. on hand" row for these records equals the expected MPS firm planned order receipts, minus the "Prod. forecast," minus the "Sold" (orders), plus any "Proj. on hand" from the previous period. This projected on-hand balance number can be used to determine whether the MPS record contains any safety stock; that is, when the supply of parts exceeds the actual, plus expected usage for the option. Since the "Proj. on hand" row for the common parts in Figure 14.6 contains all zeros, there's no safety stock or overplanning for the common parts.

There's no overplanning for the common parts, since they're used one-for-one on each model series A truck. However, since some safety stock is often planned for product options, MPS records for these items often deliberately include a projected on-hand balance. Exception messages are generated to indicate those situations; that is, a positive balance in the projected on-hand row indicates the current material plan (supply) exceeds the projected usage (booked orders, plus production forecast), reflecting the existence of time-phased safety stock. Such an example can be seen for the oil clutch product option in Figure 14.6. Similarly, when the "Proj. on hand" row contains negative figures, the current material plan is inadequate to meet the projected usage for the item, and the exception message signals the master scheduler to review such an item. The negative balance probably doesn't indicate an inability to meet final assembly needs. It likely indicates a more limited ability to respond to customer orders than is indicated by the percentage usage of that option.

For example, in Figure 14.6's oil clutch option parts MPS record, the "Proj. on hand" is −1 in the week of 9/18. This comes about because eight units are in the "MPS start" row for 9/5, two units are sold for 9/18, and eight units are forecast to be sold that week. There is a "Proj. on hand" of one from the prior week (9/11). This one unit plus the eight-unit MPS is short of the expected total requirements (8 + 2) by one unit. But what this really means is that, if the "Prod. forecast" is perfect up to the week of 9/18, only seven more units can be provided for delivery in that week. In the following week, the expected deficiency is recovered.

The Hyster example also illustrates the safety stock's timing as well as its quantity. For example, the oil clutch option parts MPS record in Figure 14.6 shows the projected on hand to be zero for the week of 8/21 and one unit in 8/28. But for the week of 8/28, there's also no uncertainty! That is, the "Prod. forecast" is zero. The only requirement is eight actual customer orders. Thus, any safety stock would be redundant.

As actual customer orders come in, they'll specify the oil clutch option or they won't. The forecast for the oil clutch option will be consumed—albeit somewhat erratically. The 1 unit of projected on hand in week 9/11 means that, although 62 percent of the 6 units of model A left to be sold in week 9/11 are anticipated to be with an oil clutch (.62 × 6 = 3 with cumulative rounding), in fact, actual customer orders with an oil clutch can be promised for 3 + 1 = 4 units. The master production scheduler will take this into account when launching the next order for oil clutch options and in any necessary replanning of open shop orders.

What this means is safety stocks are only used to buffer uncertainty in actual customer orders. As order entry takes place, uncertainties are resolved. As soon as actual orders are known, there's no need to forecast them. Therefore, all safety stocks can be de-expedited in timing until a period where there's forecast demand. At this point, uncertainty exists about the option mix of customer orders.

The MPS Records

Figure 14.6's records illustrating operation of a two-level MPS system were drawn from Hyster's actual MPS system. Figure 14.7 is an example of a level 1 record. It is printed in weekly time buckets for weeks 8/21 to 1/15. For the remainder of the one-year MPS planning horizon, it is printed in four-week time buckets.

FIGURE 14.7 Model Series A: Level 1 Oil Clutch MPS Record

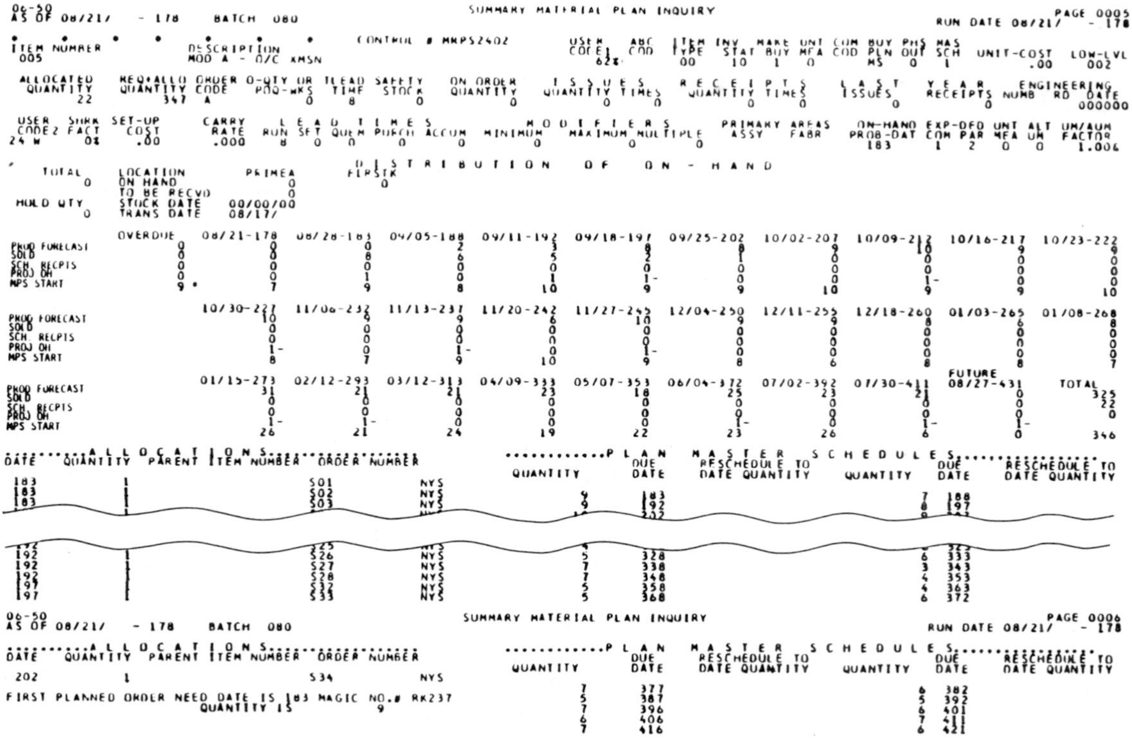

Note: Due to confidentiality, the above data is sample information only. Under normal operating circumstances, the "MPS start" would never be allowed to fall "Overdue."

Source: W. L. Berry, T. E. Vollmann, and D. C. Whybark, *Master Production Scheduling: Principles and Practice* (Falls Church, Va.: American Production and Inventory Control Society, 1979), p. 152.

Pegging information is shown in Figure 14.7's level 1 oil clutch MPS record. For example, customer orders, booked for the model series A trucks are displayed in the "sold" row and are listed along with the sales order number in the pegging data section in the lower left-hand corner of this record. For example, there are 8 model series A units booked for week 8/28; each individual order supporting these sales is shown in the sales order pegging data beginning on day 183. Likewise, firm planned orders in the "MPS start" row of this record are also listed (pegged) in Figure 14.7's lower right-hand corner. These data show due dates for the firm planned orders, instead of order release dates (e.g., 7 units are due in the week beginning with day 188).

The information in this record represents the type of information the master scheduler uses on a daily basis. While such information can be displayed in slightly different formats, the Hyster example illustrates use of a two-level master production scheduling system for coordinating the MPS and sales order promising. The impact on overall company operations has been impressive.

ADDITIONAL TECHNIQUES

In this section, we discuss some additional approaches to the techniques described for two-level master production scheduling. The first involves alternate ways to make available-to-promise calculations and the resultant implications. Thereafter, we treat the topics of consuming forecasts with actual demand, and the enhancing of the MPS with capacity planning that's linked to capacity consumption (based on actual orders).

Alternative ATP Explosion Conventions

In Figure 14.2 and 14.3, the production forecast (or unconsumed forecast) for level 1 items is derived from the ATP quantities at level 0; that is, for example, in Figure 14.3 after the customer order for 40 model 623s had been booked for week 3 delivery, the remaining ATP for the over-bed table (level 0) was 60 units. This quantity was exploded by the appropriate percentages to obtain new forecasts of 30 and 18 units, respectively, for the 622 and 623 models in week 3. This raises the question of whether we should revise the original forecasts of 50 percent and 30 percent, respectively, in light of one actual order. If the long-run percentages are indeed 50 and 30, then the master scheduler's actions based on the information in Figure 14.3 could induce undesirable nervousness into the MPC system because the MPS data drive the MRP system. Even though they're firm planned orders, the master scheduler can and should react to some changes. But the system itself may be indicating the need for MPS changes when, in fact, changes are unwarranted. This issue is illustrated by the question of how to deal with the potential problems in Figure 14.3.

The problem seen as an apparent shortage and surplus in Figure 14.3 will be particularly severe in a firm where demand is lumpy, such as when one customer infrequently buys relatively large amounts of one model. In some instances, demand coming from a distribution system can create the same effect. We once worked with a firm that sent a ship to the southern hemisphere once every four months. That shipment (of just a few models) induced a severe apparent "shortage."

One approach to dealing with this problem has been suggested by John Sari. In essence, he suggests using a longer horizon for allocating the unsold forecast at level 0 to the level 1 items. His approach starts with the sum of the expected sales for the two models from the data in Figure 14.2. The 100 units of actual orders and 150 units of forecast for the 622 model over the five weeks add up to 250 units. A similar calculation for the 623 model yields 150 units (60 + 90). After we book the order for 40 units and update the records in Figure 14.3, the resultant sum for the 622 model is 230 units (100 + 130); the sum for the 623 model is 178 (100 + 78).

Using our example in Figure 14.3, instead of exploding the remaining 60 units of ATP in week 3 by 50 percent and 30 percent, respectively, we might look on a cumulative basis further out into the unsold horizon. In our example at level zero for week 3 through 5 inclusive, 260 units are unsold out of an original forecast of 300 units. The original expectation was that 150 of these 300 units would be sold in the 622 model and 90 in the 623. One sales transaction has been recorded. It consumed 40 of the expected 90 units of model 623 and none of the 622. This infers the cumulative production forecast for weeks 3 through 5 for the 622 model should remain at 150 units (instead of the 130 in Figure 14.3), and the cumulative production forecast for 623 should be 50 units (instead of 78).

Achieving this result would require modifications to most MPS software. Even though this could provide more stability to the MPS records, there's a serious question regarding the desirability of doing so. The real issue is whether the firm really believes the long-run percentages are right; if so, a less responsive (more stable) approach to changing the MPS is indicated. Alternatively, the firm may well believe each sale indicates fundamental changes. If this is so, then actions indicated on the record in Figure 14.3 should be taken to best deal with the uncertainties inherent in the business. In each case, the master scheduler must manage the situation. To help in that process, the related issue of how best to buffer lumpy demand uncertainty must be addressed. Hedging techniques in demand management appear to be the best approach.

Consumption by Actual Orders

A problem similar to the issue of adjusting to the long-run or short-run product mix is the question of long-run or short-run consumption of the forecast by actual orders. To illustrate the problem, let's turn to Figure 14.8 (an extension of the over-bed table record in Figures 14.2 and 14.3). We show only the level 0 over-bed table status, assuming that we're just finishing week 3. Weeks 4 to 7 are also shown and have a planned production of 100 units per week.

Figure 14.8 raises the issue of what to do with the unconsumed forecast of 25 units if week 3 finishes with no new orders. Put another way, what should the record look like at the beginning of week 4, and what actions are required to create this condition? Figure 14.9 depicts the most obvious week 4 record, but we shall see it has problems.

Figure 14.9 assumes, since consumption of forecast by actual orders in week 3 is now complete, we should discard the unconsumed forecast from this week. (The 25 units of inventory will still be held in the level 1 records.)

FIGURE 14.8   Over-Bed Table Two-Level MPS Status at the End of Week 3

			Week		
	3	4	5	6	7
Over-bed table:					
Production plan	100	100	100	100	100
Orders	75				
Available					
Available to promise	25	100	100	100	100
MPS	100	100	100	100	100

On hand = 0; SS = 0.

FIGURE 14.9   Over-Bed Table Two-Level MPS Record at the Beginning of Week 4
(Discarding Unconsumed Forecast)

			Week		
	4	5	6	7	8
Over-bed table:					
Production plan	100	100	100	100	100
Orders					
Available					
Available to promise	100	100	100	100	100
MPS	100	100	100	100	100

On hand = 0; SS = 0.

However, discarding the forecast changes, in essence, the production plan. That plan dictated an output rate of 100 units per week; and if we discard 25 units of production plan or forecast in week 3, we'll also change the plans for component manufacturing and procurement. Timings for the reductions of different items in the bill of materials will vary, but the impact will be the same: some de-expedite messages and reductions in planned orders will occur.

Figure 14.10 shows an alternative formulation of the records as of week 4's start. In this case, the 25 units of unconsumed forecast is rolled over into the next week (4), thereby maintaining the production plan's overall dictates and sending the resultant set of signals down to the rest of the items in the bill of materials.

FIGURE 14.10 Over-Bed Table Two-Level MPS Record Alternative Formulation at the Beginning of Week 4 (Recognizing Unconsumed Forecast)

	Week				
	4	5	6	7	8
Over-bed table:					
Production plan	125	100	100	100	100
Orders					
Available	25				
Available to promise	125	100	100	100	100
MPS	100	100	100	100	100

On hand = 25; SS = 0.

The treatment of the record in Figure 14.10 allows for a quite uneven pattern of consumption of demand over time. The approach in Figure 14.9, on the other hand, may be based on the assumption a monthly forecast of 400 units is to be nearly evenly divided into 100 units each week. For many firms, this is an unreasonable assumption. There may be considerable week-to-week fluctuations due to random causes. In other cases, a pattern of demand exists where sales are higher near the end of the month or the end of the quarter. Clearly, if this is true, using the approach of Figure 14.9 could cause shortages at the end of the period and undermine the intent of the production plan made for the period.

The same problem exists if actual orders overconsume forecasts and the master scheduler changes the MPS. The net result will be to send signals down the explosion chain to replenish the overconsumption, without reducing any future forecasts to compensate. Richard Ling suggests one approach to this problem would be to carry another time bucket at the beginning of the planning horizon, which keeps track of the forecast underconsumption (or overconsumption). This "demand" could be monitored against some pre-specified limits; it also could be reviewed as one aspect of the production planning meeting.

A variant of this problem is described by Proud. He points out there will be two separate forecasts to be consumed for many items: production forecasts and service part forecasts. He suggests incorporating two forecast rows in the MPS record, as well as two actual demand rows and two ATP rows. We also need to keep track of how the available row is apportioned between the two sources of demand. The advantage of Proud's suggestion is, when actual demands are lumpy, "borrowing" between supply sources would be much more transparent. Moreover, over- and underconsumption issues might also be examined in terms of impacts on service part availability.

Capacity Planning

The final set of additional techniques for enhancing master production schedule practice deals with capacity planning and capacity consumption. The basic idea is that MPS record formats can be used to monitor capacity as the actual orders consume available capacity. Suppose, for example, every over-bed table took 2.5 hours of capacity in the welding department, and every bed model XYZ took 15 hours. If the over-bed table MPS was for 100 units per week and the XYZ bed was for 10, the resultant "production forecast" row in an MPS record for the welding department might look like Figure 14.11. The 400 hours required in week 1 would actually be for over-bed tables and beds to be sold in subsequent weeks since lead time for parts manufacture and subassembly must be taken into account.

Figure 14.11 shows an expected capacity requirement of 400 hours per week. This is determined by the resource profile method of capacity planning. For simplicity here, we'll use a constant lead time offset of two weeks for both beds and over-bed tables; that is, let's assume capacity requirements in week 1 for welding are to support an MPS in week 3. Let's say that actual orders for over-bed tables are 100, 0, and 4 units for weeks 3, 4, and 5, respectively, and orders for the XYZ bed are 10, 5, and 2 for the same weeks. The resultant "orders" for welding are 400, 75, and 40 hours, respectively. Figure 14.11 shows these values as well as a "master production schedule" of 450 hours per week. This 450 hours is, in fact, the work center's capacity. Treating it as a master production schedule allows us to generate ATP data for capacity and to monitor actual orders' consumption of capacity in key work centers.

This kind of capacity planning is particularly important in cases where many products consume capacity in a particular work center. It's also useful with potential bottlenecks. The approach lets us examine consequences of breakdowns or other problems. All that's required is to reduce the "MPS" or available capacity and to examine the impact on customer orders.

FIGURE 14.11   Welding Department Capacity Planning (in Welding Hours)

	Week				
	1	2	3	4	5
Production forecast	400	400	400	400	400
Orders	400	75	40		
Available	50	100	150	200	250
Available to promise	50	375	410	450	450
MPS	450	450	450	450	450

The approach can be extended to treatment of key vendors, and it lends itself to "scenario" generation. Most modern MPC systems allow us to create another "company" (in terms of the MPC data base), which is an identical copy of the present firm. This copy company can be used to examine a series of what-if questions without disturbing the data base that's actually being used to plan and control company operations.

METHODS FOR CONSTRUCTING PLANNING BILLS OF MATERIAL

Creating special planning bills of material to establish the master production schedule at an option level can provide major benefits in terms of reducing the number of items to be master scheduled, reducing resultant inventories, and increasing responsiveness to customer demands. However, the size and complexity of product structures for assemble-to-order products frequently make it hard to construct the necessary super bills of material. The major problem is separating the common parts from parts unique to particular options specified by the customer.

Managing the two-level approach and its impact is facilitated by a single part number for all common parts. For a part to be truly included in the common parts category, it has to remain uncommitted to an exact end item until final assembly. Devoting engineering talent to maximize the common parts category has significant payoffs in inventory, since no safety stocks are carried for common parts. We also achieve payoffs in responsiveness to customer requests, since parts aren't committed to the wrong items.

In firms whose product structures have thousands of individual components and vast numbers of optional product features, isolating common parts can be difficult. In this section, we describe two useful techniques for developing the super bills of material and for separating the common parts from the unique. The first technique, the **matrix bill of material,** is useful for problems of relatively small product complexity. The second approach, the **component commonality analysis system,** uses commercial software to deal with more complex problems.

The Matrix Bill of Material Approach

The matrix bill of material approach is described by Kneppelt. It involves constructing a table with columns for each end item and rows for each subassembly, component, and raw material part number. The entries in the table indicate usage of each component item in the row for each end item in the column. We can prepare the matrix manually or by a computer program. Figure 14.12 gives a partial example from Kneppelt based on bicycle components.

FIGURE 14.12 Matrix Bill of Material Structuring

Number	Component description	Unit	Model number used on			
			SS-53-RM-C-Y	SS-53-TM-C-Y	SS-53-RM-T	SS-53-TM-T
FR-53-S	Silver Frame (53CM)	EA	1			
FR-57-S	Silver Frame (57CM)	EA		1	1	
FR-60-S	Silver Frame (60CM)	EA				1
SA-T-M	Saddle Touring, Men	EA		1		1
SA-T-W	Saddle Touring, Women	EA				
SA-R-M	Saddle Racing, Men	EA	1		1	
SA-R-W	Saddle Racing, Women	EA				
SE-01	Seat Post (25.8MM)	EA	1	1	1	1
WH-C	Wheel Kit, Clincher	EA	1	1		
WH-T	Wheel Kit, Tubular	EA			1	1
CR-01	Crankset Kit	EA	1	1	1	1
PD-01	Pedals Kit	EA	1	1	1	1
De-01	Derailleur Kit	EA	1	1	1	1
BK-01	Brakes Kit	EA	1	1	1	1
HD-01	Handlebar Kit	EA	1	1	1	1
TC-01	Toe Clips/Straps	EA	1	1	1	1

Source: Leland R. Kneppelt, "Product Structuring Considerations for Master Production Scheduling," *Production and Inventory Management*, 1st quarter 1984.

We can use the matrix bill of material to identify common and unique parts in a particular product family and to create "families" based on commonality. Thereafter, it might be possible to group all the common parts for planning purposes. The common parts are those that are the same for all end items in the product family; they're also the same in usage quantity for each member of the family. By looking across the rows in Figure 14.12, we can determine which parts have constant usage in all end items in the product family. For example, the seat post is common to all four models listed in Figure 14.12, but the saddle depends on the particular model. To construct a super bill, the common parts are removed from the matrix and grouped under one part number; the remaining parts are grouped by option.

There can be, however, excessive complexity with this approach when a column is defined for each end item. To see this, consider the over-bed table example. We need 160 columns, since there are 160 unique end-item over-bed table possibilities. A simpler approach is to head the columns with the model or option choices available to the customer. In the case of the over-bed table, the result is 4 models + 10 top options + 4 boot-caster options = 18 columns. Each column comprises the set of parts associated with the particular model or option specified by the customer. Those parts common to all columns within a model or option choice (such as the four basic models) are removed and put into the part number that designates the common parts. The remaining parts in each column are the unique part numbers associated with a particular model or option choice. This approach bases the columns, not on how the product is made, but on how it's sold in the marketplace.

Commonality and Bill of Material Depth

Because the super bill is based on how a product is sold, we must understand some potential conflicts with the way products are manufactured. One such conflict is associated with determining common parts. For example, Figure 14.12 shows four different saddles for the bicycles. It well may be, however, that some saddle part, such as a spring, is common to all four saddles. The first question is how to "disentangle" the product structures and, next, how far to go down the structures to find common parts. A more fundamental question is whether to include the parts thus found in the list of common parts. The answer varies with circumstances.

For the over-bed table, the chipboard core for all 10 laminated tops is common. However, because of the lead time to make tops, the core must be committed to a particular color-laminated top before the final assembly begins. In this case, the core can't be a common part. Any part that's to be considered common for the purpose of master production scheduling with super bills has to remain uncommitted to a particular end item until final assembly.

In the case of the springs in the bicycle saddles, the question is whether saddles can be assembled as part of the final assembly process. If they're to be built ahead of final assembly, there must be overplanning in all saddle parts, including common springs, not just those unique to each particular saddle.

A different way to focus on this issue revolves around customer response time requirements. Kneppelt points out all end items in his bicycle example will have to be forecasted and carried in stock *if* competition forces the firm to ship immediately. If a one-week delay is possible, a final assembly process can be used where parts and assemblies are stocked and only be committed to exact end-item specifications after receipt of customer orders. If a two-week delay is possible, frames can be painted to order; if less than two weeks is permissible, frames must be held in specific colors.

We see, then, there can be important benefits of very small lead times for final assembly—and for a final assembly process that goes deeply into product structures. The deeper this becomes, the less the planning bills resemble the physical way products are built or the way products are considered from an engineering point of view. Product structures become increasingly "bags of parts," and other MPC techniques are required to dictate the products' actual assembly. These techniques include phantom bills, routings with operational entry (delivery) data, and just-in-time techniques.

A final concept relating to part commonality and product structures is design of the super bills to meet both marketing and manufacturing needs. Consider, for example, the Ford Taurus and Mercury Sable; each would be independently structured for sales purposes. On the other hand, from a manufacturing point of view, both cars would be built from a single MPS, based on a great deal of commonality. The same situation exists when part commonality *across* product families is used to define a new "family." Determining what is in fact a "family" is based on how the product is to be master scheduled and on how to maximize the dollar content of the common parts (which aren't overplanned). In some instances, larger product families can lead to fewer MPS items, but they may lead to a smaller set of common parts.

Component Commonality Analysis System

Another method for separating common parts from those unique to particular models and options uses relational data base software, which can be installed on many computer systems. This method is called Component Commonality Analysis System (CCAS) by Berry, Tallon, and Boe. CCAS provides a way to analyze computer-based bills of material using the sorting and summarizing features of relational data base software. It does the same work that's done manually in the matrix bill of material approach, but CCAS can analyze complex multiple product structures with high speed and accuracy. It's only limited by the storage capacity of the computer hardware on which it is run.

The CCAS approach includes four steps. First, we check data and record format descriptions to ensure the bill of material data is consistent with the

CCAS design. Second, we establish a CCAS data base for each product family, using bill of material records to identify part numbers that make up product options in the particular product family. Third, we process the CCAS data base for each family using the logic in Figure 14.13 to determine common and unique parts. Finally, we generate reports and make inquiries to the CCAS data base using various CCAS features.

Material Handling Equipment Manufacturer Example

An analysis of a material handling equipment manufacturer's bill of material records provides an example of applying CCAS. Common and unique parts are determined for the purpose of constructing super bills of material for master production scheduling. This analysis was performed for a company that produces material handling vehicles for a variety of specialized applications, including narrow-isle units for high-rise, high-density storage warehouses; order picking and selector equipment; and high-speed, low-lift transporters for truck and rail docks. Equipment is custom designed with many optional features to fit particular customer needs. Business volume has grown over the years, and manufacturing facilities have retained a batch manufacturing orientation involving fabrication and assembly operations.

CCAS analysis was performed using computer tapes supplied by the company for seven product families. In all, there were 70,651 bill of material records. For each item record, data include bill of material level, usage quantity, source (manufactured or purchased), unit cost, and description. Data cover approximately 20,000 different units sold out of roughly 240,000 possible end items. The CCAS analysis was used to determine the common parts for each of the seven product families. Figure 14.14 summarizes the analysis. More detailed results for product family A appear in Figure 14.15. Turning first to Figure 14.14, we see in section **A** sales volumes vary considerably between product lines. Given product diversity, it's difficult to forecast end-item demand in this kind of firm. Section **B** reinforces this idea; the number of end-item possibilities is huge.

Section **C** develops statistics for an average vehicle across all seven families. The most interesting result is the high percentage of common parts in an average product. Of the material in the average vehicle, 76 percent is made up of common parts. Section **D,** on the other hand, shows, to cover *all* of the options, it's necessary to carry a higher inventory of unique parts. However, the company should be able to achieve high inventory turnover on the common parts.

Section **E** breaks out the information in terms of purchased and manufactured parts. An obvious extension of this analysis is to price out the parts and to determine the monetary cost of various safety stock levels. Finally, section **F** shows the number of MPS super bill records required to plan and control each product family using super bills. The entire seven product fami-

FIGURE 14.13 CCAS Processing Logic

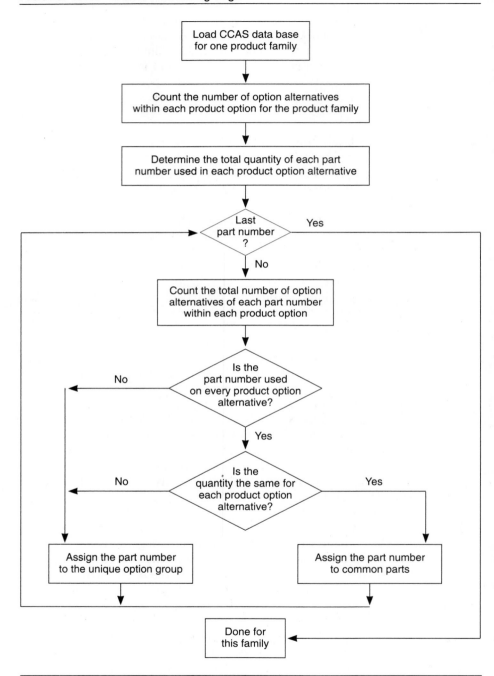

Source: W. L. Berry, W. J. Tallon, and W. J. Boe, "Product Structure Analysis for the Master Scheduling of Assemble-to-Order Products," working paper, University of North Carolina (Chapel Hill), 1988.

FIGURE 14.14   Material Handling Equipment Manufacturer: Summary Statistics

	Product family average							
	A	B	C	D	E	F	G	Total
A. Unit sales data:								
Percent sales volume	30%	15%	10%	30%	5%	5%	5%	
B. End-item configurations:								
As engineered	24192	99	129024	24	168	15552	69984	239043
As sold	1200	63	14080	24	54	768	3456	19645
C. Parts for an average vehicle:								
Common parts	1034	813	919	462	486	357	248	4319
Unique parts	287	93	288	49	106	173	283	1279
Total parts	1321	906	1207	511	592	530	531	5598
Percent common	78%	90%	76%	90%	82%	67%	47%	76%
D. Parts for the full family:								
Common parts	1034	813	919	462	486	357	248	4319
Unique parts	786	577	895	121	293	523	1048	4243
Total parts	1820	1390	1814	583	779	880	1296	8562
BOM records	9345	5968	11002	239	4591	16666	21840	70651
E. Purchased versus manufactured parts:								
Common purchased	651	499	598	280	292	237	158	2715
Common manufactured	383	314	321	182	194	120	90	1604
Unique purchased	455	329	453	57	194	284	663	2435
Unique manufactured	331	248	442	64	99	239	385	1808
F. Super bill MPS records:								
MPS records	18	11	27	10	10	22	26	124
Average vehicle unique purchased versus manufactured parts:								
Unique purchased	182	55	179	26	72	88	161	753
Unique manufactured	105	38	109	23	34	85	122	516

Source: W. L. Berry, W. J. Tallon, and W. J. Boe, "Product Structure Analysis for the Master Scheduling of Assemble-to-Order Products," working paper, University of North Carolina (Chapel Hill), 1988.

lies can be master scheduled with only 124 MPS records. These records in turn control the 4,243 unique parts and 4,139 common parts.

Figure 14.15 provides the resultant super bill for product family *A*. As can be seen, product family *A* has five optional features: options F through J. The super bill shows lead time and percentage uses for each option. The 17 MPS option records would be based on this super bill, along with the common parts record for this product family. The CCAS methodology was quite useful to the material handling equipment manufacturer in structuring its approach to master scheduling. After the analysis, the company was surprised at the extent of part commonality and how few MPS records it needed to manage the seven product families.

FIGURE 14.15 Material Handling Equipment Manufacturer: Super Bill of Material

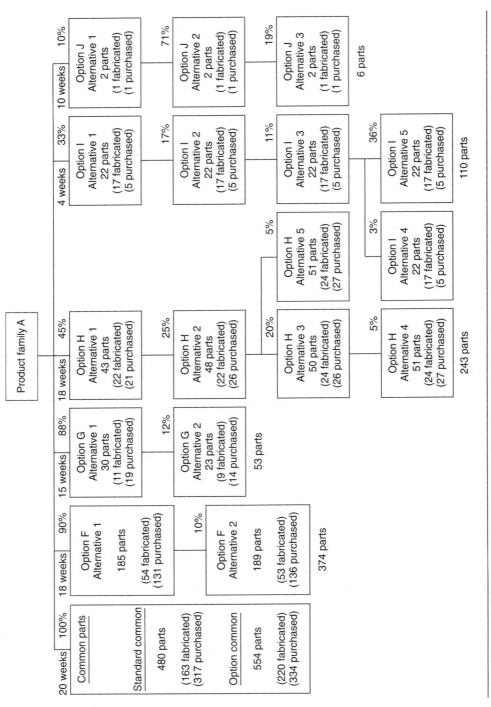

Source: W. L. Berry, W. J. Tallon, and W. J. Boe, "Product Structure Analysis for the Master Scheduling of Assemble-to-Order Products," working paper, University of North Carolina (Chapel Hill), 1988.

CONCLUDING PRINCIPLES

This chapter describes advanced concepts in master production scheduling, primarily techniques useful in the assemble-to-order manufacturing environment. Firms that implement these techniques can better respond to customer requests—and do so at lower costs of operation. We stress the following principles:

- Super bills and bill structuring techniques should be used to manage product mix in an assemble-to-order environment.
- Firms wanting to closely couple the production plan, MPS, and order entry should use two-level master production scheduling techniques.
- Very complex product environments can be master scheduled with relatively few MPS records, with the right product structuring.
- High costs of developing and maintaining planning bills for the MPS should be offset by benefits of closer coupling to the market where competition requires rapid response.
- Available-to-promise logic can provide a mechanism for managing consumption of capacity, as well as materials.
- Structuring bills of material to determine the common and unique parts must accommodate both how the products are sold and how they're manufactured.

REFERENCES

Berry, W. L.; T. E. Vollmann; and D. C. Whybark. *Master Production Scheduling: Principles and Practice.* Falls Church, Va.: American Production and Inventory Control Society, 1979.

————; Tallon; and W. J. Boe. "Product Structure Analysis for the Master Scheduling of Assemble-to-Order Products." Working paper, University of North Carolina (Chapel Hill), 1988.

Clay, Peter. "Advanced Available-to-Promise Concepts/Techniques." *1990 APICS International Conference Proceedings*, pp. 381–87.

Gessner, Robert A. *Master Production Schedule Planning.* New York: Wiley, 1986.

Karmarkar, Uday S. "Capacity Loading and Release Planning with Work-in-Progress (WIP) and Leadtimes." *Journal of Manufacturing and Operations Management* 2, no. 2 (1989), pp. 105–23.

Kneppelt, Leland R. "Product Structuring Considerations for Master Production Scheduling." *Production and Inventory Management* 25, no. 1 (1984), pp. 83–99.

McClelland, Marilyn K. "Order Promising and the Master Production Schedule." *Decision Sciences Journal* 19, no. 4 (Fall 1988), pp. 858–79.

Proud, John F. "Consuming the Master Production Schedule with Customer Orders." *APICS 26th Annual Conference Proceedings*, 1983, pp. 21–25.

Sari, John F. "The Planning Bill of Material—All It's Cracked Up to Be?" *APICS 25th Annual Conference Proceedings,* 1982, pp. 324–27.

Sridharan, Sri. V.; William L. Berry; and V. Udayabhanu. "Measuring Master Production Schedule Stability under Rolling Planning Horizons." *Decision Sciences Journal* 19, no. 1 (Winter 1988), pp. 147–66.

————, and William L. Berry. "Freezing the Master Production Schedule under Demand Uncertainty." *Decision Sciences Journal* 21, no. 1 (Winter 1990), pp. 97–120.

————; William L. Berry; and V. Udayabhanu. "Freezing the Master Production Schedule under Rolling Planning Horizons." *Management Science* 33, no. 9 (September 1987), pp. 1137–49.

————, and R. Lawrence LaForge. "An Analysis of Alternative Policies to Achieve Schedule Stability." *Journal of Manufacturing and Operations Management* 3, no. 1 (Spring 1990), pp. 53–73.

Sulser, Samuel S. "Advanced Concepts in Master Planning—A Matter of Systems Design." *APICS 27th Annual Conference Proceedings,* 1984.

Tallon, William J. "A Comparative Analysis of Master Production Scheduling Techniques for Assemble-to-Order Products." *Decision Sciences Journal* 20, no. 3 (Summer 1988), pp. 492–506.

Vollmann, Thomas E. *Master Planning Reprints.* Falls Church, Va.: American Production and Inventory Control Society, 1986.

Wemmerlöv, Urban. "Assemble-to-Order Manufacturing: Implications for Materials Management." *Journal of Operations Management* 4, no. 4 (1984), pp. 347–68.

DISCUSSION QUESTIONS

1. Why does two-level master production scheduling offer increasing benefits as the number of product options available to the customers goes up?
2. Why is the level 1 production forecast an explosion of the level 0 available to promise in a two-level MPS?
3. Why is safety stock not held in common parts?
4. Upstate University has majors in business administration, nursing, and physics. How might the university apply the super bill concept to the institution, considering each of the majors as a product family?
5. Upstate has just hired a new dean who believes in a liberal education. At her urging, the curriculum has been revised so every student takes a liberal arts core of courses during the first two years. How does this change the approach to Question 4?
6. Upstate University was using a two-level master production schedule for managing enrollments in the business school. Someone discovered accounting students apply two months later than other students. How might this influence the way Upstate consumes its forecasts of demand for classes?
7. Why is maximizing the monetary value of the common parts an important goal?

8. Jeff Miller, master scheduler for Ajax Widgets, has noticed one part in the circular widget family is used on every end item, but in one case two units are required. As a result, usage isn't the same and the part was not included in common. Jeff asks why he can't include one unit of the part in common and one more in the unique parts list for the circular widget that requires two units. Will this work?

9. Why is lead time for the final assembly an important issue?

PROBLEMS

1. Lefty Cardozo works at Wilbur's Wonderful Widgets as an order entry and master scheduling expert. Flaky Old Lefty (FOL) has come up with the following scheme: The company has four distinct products (A, B, C, and D) which come from two basic options. We start with a basic widget, but it can be equipped with either pink or red lining as well as automatic contour or no automatic contour. The breakdown is:

   A = Pink, automatic.
   B = Pink, no automatic.
   C = Red, automatic.
   D = Red, no automatic.

   FOL is having a hard time with the marketing group in trying to get specific forecast data for A, B, C, and D. However, FOL thinks a good guess is 3 to 1 for red to pink and 2 to 1 for automatic contours to no automatic contours. Design a super bill of materials for FOL.

2. FOL went to a seminar so he could change back to Smart Old Lefty (SOL). He learned another trick, which eliminates the need for super bills. To get the correct number of parts started for his master schedule of 100 per week, he plans the master schedule for 25 Bs (thereby accounting for all needed "pink" plus 25 of the 33 needed "no automatic"), 8 Ds (thereby finishing the "no automatic" and getting 8 of the 75 "red"), and 67 Cs (finishing the "red" and "automatic"). At the end of one week, SOL had the following sales data for a particular master schedule week:

15	A
25	C
30	D

   What's the implied product mix for the remaining 30 widgets?

3. A company makes products on a make-to-order basis instead of stocking finished-goods inventory. We find one particular item, product C, with a present inventory of zero and a total of 30 units sold. Assume the 30-unit present backlog position consists of 20 units promised in week 1 and 10 in week 2. Also assume an existing master schedule, which includes product C. The resultant master schedule quantities are shown in the weeks they're to be made. The following document has been produced.

Item #	Des-crip-tion	Assembly lead time	On hand	Lot size		Time fences	
						De-mand	Plan-ning
C		1	0	L–F–L		1	7
Week	1	2	3	4	5	6	7	8	9
Forecast	35	35	35	35	35	35	35	35	35
Orders	20	10							
Available									
Available to promise									
Master production schedule	87	63			12	88	50		

Calculate the "Available" and "Available to promise" rows of the preceding document.

4. Brent Gibson, master scheduler at Vincent Electric Company, explained the firm's approach to scheduling production: "Basically, it's fairly straightforward. We get an annual forecast from marketing, break it up into weeks, and set the lot size at two weeks' worth. Then, depending on the lead time, we can figure out when to order. In the case of the model 47 rotary arm, the forecast works out to 50 per week, which means lot size is 100. Since there's a four-week lead time, we place an order every other week for delivery four weeks later. Right now, we ordered last week, so orders are due in weeks 2 and 4. Next week we'll order the 100 units for week 6." When asked what happened when actual sales differed from the constant rate of 50 per week, Brent replied, "Well, first of all, you really have to learn how to handle those turkeys from marketing. What a bunch of cry babies. I usually just tell them I'd like to help them out, then ask where did they come in? By and large, we just hang in there, and the averages work out. If the overall forecast for the year is wrong, then we sometimes have to make an adjustment."

   a. Given the preceding information, the following set of booked orders and new orders to be promised, and the MPS techniques in this chapter, prepare an analysis based on the *current situation,* stating *all* the specific problems facing Brent. (Use an 11-week planning horizon in preparing your analysis. Assume assembly time is negligible.)

   b. Suppose assembling a #47 rotary arm requires a critical component, the #687 link, which is purchased from a supplier. Re-do your analysis in question a, considering the following information about the #687 link:

   One #47 rotary arm assembly requires three #687 links. Purchasing lead time for the #687 link equals 3 weeks. There's an open purchase order

Mode #47 Rotary Arm (On-Hand Inventory = 20)

Booked Orders		
Order number	Promise week	Quantity
1	1	10
2	1	5
3	1	2
4	2	60
5	3	60
6	4	50
7	4	60
8	6	20
9	7	10
New orders to be promised		
10	ASAP*	1
11	ASAP*	10
12	ASAP*	25

*As soon as possible.

(scheduled receipt) for 400 units scheduled to be received Monday morning of week 2 in the future. (The supplier has just advised you that 44 pieces of this order have been scrapped.)

Currently (at the start of week 1) there are 123 #687 links on hand in inventory. (Already 600 #687 links have been removed from inventory for use on the two MPS orders for the #47 rotary arm due to be completed in weeks 2 and 4.)

The current order quantity for the #687 link is 400 pieces. The supplier has made a commitment to ship a maximum of 100 pieces per week or 400 pieces in any four-week time interval.

   c. Now, suppose there's a $110 ordering cost and a $0.55 inventory carrying cost per piece per week for the #47 rotary arm. Recommend the best ordering policy and the time this ordering policy can be implemented. Indicate all factors you considered in this analysis.

   d. Identify all alternative courses of action Brent can take in the short run to solve the problems in questions a, b, and c. Give each alternative solution's pros and cons. Recommend one solution and justify your choice.

5. Tani produces several lines of specialty tanning equipment. The most successful is the Tawny Tanis line, whose most popular option by far (the Golden option) is requested in 80 percent of the orders for any Tawny Tanis product, on the average. Tani uses a planning fence at period 5 for the product line and period 4 for the options. Prepare the two-level MPS records for the Tawny Tanis line, the common parts, and the Golden option given the following information:

Production plan for Tawny Tanis:		500/period (1–6)
Master schedule for Tawny Tanis:		500/period (1–5)
Master schedule for common parts:		500/period (1–4)
Master schedule for Goldens:		400/period (1–4)

Actual orders for	*Tawny Tanis*	*Goldens*
Period 1	500	423
2	450	354
3	300	250
4	150	112

	Common parts	*Tawny Tanis*	*Goldens*
Inventory	0	0	45
Lot sizing	LFL	LFL	LFL
Safety stock	0	0	50

6. Underground Deli produces three sandwiches (jelly, peanut butter, and tunafish) on a roll, a bun, white bread, or dark bread. Each sandwich is carefully wrapped in yesterday's newspaper. It's a rough job managing the business, and the owners are thinking about installing an MPC system with a two-level MPS. They wonder about structuring a bill of material to aid this process. Help them by constructing the matrix bill of material for their 12 products. Use both the option and the end-item approach. The bills of material for white bread sandwiches appear below.

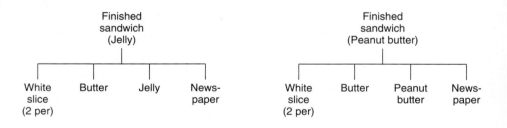

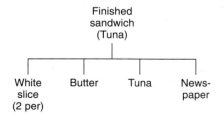

7. Neva's boutique manufactures and sells a number of special nightgowns. The Nevita model is available in several styles, with the most popular (the Flimsy) accounting for 60 percent of sales. Given the following records, update the two-level MPS as of period 2, scheduling the appropriate MPS quantities. At the very end of period 1, a new customer was granted a request for an order of five Nevitas in period 1, all in the Flimsy style, and five for period 5, of which three were Flimsys.

Planning fence at period 5

Nevitas	*1*	*2*	*3*	*4*	*5*	*6*
Prod. plan	100	100	100	100	100	100
Act. ords.	90	100	80	20		
Sch. rts.						
Avail.	0	0	0	0	0	−100
ATP	10		20	80	100	
MPS	100	100	100	100	100	

On hand = 0; lot-for-lot; SS = 0.

Planning fence at period 4

Filmsy	*1*	*2*	*3*	*4*	*5*
Prod. frcst.	6	0	12	48	60
Act. ords.	55	58	43	8	
Sch. rts.					
Avail.	4	6	11	15	−45
ATP	10	2	17	52	
MPS	60	60	60	60	

On hand = 5; lot-for-lot; SS = 5.

8. Jolly Joans are the latest fad among third and fourth graders. They come in three models, with Model A accounting for 50 percent of sales. The following records show the two-level MPS for this popular option. What would the records look like after booking an order to be shipped as soon as possible for six Jolly Joans, of which three were for Model A?

Planning fence at period 5

Jolly Joans	Period					
	1	2	3	4	5	6
Prod. plan	50	50	50	50	60	60
Act. ords.	50	50	40	20		
Sch. rts.						
Avail.	0	0	0	0	0	−60
ATP	0		10	30	60	
MPS	50	50	50	50	60	

On hand = 0; lot-for-lot; SS = 0.

Planning fence at period 4

Model A	Period				
	1	2	3	4	5
Prod. frcst.	0	0	5	15	30
Act. ords.	22	29	22	8	
Sch. rts.					
Avail.	8	4	2	4	−26
ATP	4		3	17	
MPS	25	25	25	25	

On hand = 5; lot-for-lot; SS = 5.

9. Develop a spreadsheet model of the two-level MPS records for Family A and Model I in the following situation. Family A consists of several models, of which Model I represents 50 percent of the demand. The company uses a planning fence at period 5 for the family and at period 4 for the models. What happens if the production plan is changed to eight per period for periods 3 through 6 (with no other changes)?

	Production plan for Family A:	10/period (1–6)
	Master schedule for Family A:	10/period (1–5)

	Actual Orders for		MPS for
	Family A	Model I	Model I
Period 1	10	6	0
2	8	4	5
3	6	5	6
4	2	1	5

	Family A	Model I
Sched. recpt.	0	5 (period 1)
Inventory	0	2
Lot sizing	LFL	LFL
Safety stock	0	3

10. Update the following MPS record after entering the four new customer orders. What actions, if any, should the master scheduler take?

	Week						
	1	2	3	4	5	6	7
Forecast	100	100	100	100	100	100	110
Orders	89	22	95	11	5	2	1
Available	12	112	12	112	12	112	12
ATP	23	83		184		197	
MPS		200		200		200	

On hand = 112
Assembly lead time = 3 weeks
Lot size = 200
Demand time fence = 3 weeks
Planning time fence = 6 weeks

New customer orders:

15 units, customer order number 6042 due in week 1.
19 units, customer order number 6044 due in week 1.
14 units, customer order number 6051 due in week 5.
25 units, customer order number 6056 due in week 3.

11. The master production schedule for the Carlson Company's product **family X** is prepared using a two-level MPS approach. This product family has two options (101 and 102). The product structure is:

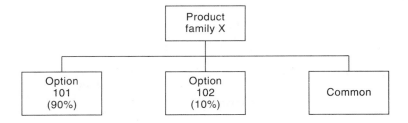

a. Given the following transactions and two-level MPS data, update the MPS. Roll the records ahead to the start of week 2.

Transactions:

The MPS quantity of 360 units of option 101 was completed.
A customer order for 15 units of option 101 in week 1 was canceled.
Ten units of scrap were reported on the MPS of 40 102's due in week 2.
An inventory adjustment of + 32 units was reported on option 101.

An inventory adjustment of -12 units was reported on option 102. No sales orders were booked for week 7.

Product family X	Week					
	1	*2*	*3*	*4*	*5*	*6*
Production plan	200	200	200	200	200	
Orders	200	200	140	190	50	10
Net	0	0	60	10	150	190

Option #101	Week					
	1	*2*	*3*	*4*	*5*	*6*
Forecast	0	0	54	9	135	171
Orders	155	140	91	117	30	6
Available	380	240	455	329	524	347
ATP	380	0	152	0	324	0
MPS (finish)	360	0	360	0	360	0
MPS (start)	360		360			

Forecast = 90%   Lead time = 2 weeks
Safety stock = 20   Lot size = 360
On hand = 175

Option #102	Week					
	1	*2*	*3*	*4*	*5*	*6*
Forecast	0	0	6	1	15	19
Orders	45	60	49	73	20	4
Available	20	0	-89	-124	-107	
ATP	20	-69	0	-53	0	36
MPS (finish)	0	40	0	40	0	40
MPS (start)	40		40			

Forecast = 10%   Lead time = 3 weeks
Safety stock = 50   Lot size = 40
On hand = 65

Common parts	Week					
	1	*2*	*3*	*4*	*5*	*6*
Forecast	0	0	60	10	150	190
Orders	200	200	140	190	50	10
Available	200	0	200	0	200	0
ATP	0	0	70	0	340	0
MPS (finish)	400	0	400	0	400	0
MPS (start)		400		400		

Forecast = 100%   Lead time = 1 week
Safety stock = 0   Lot size = 400
On hand = 0

b. On the basis of updated records from Problem 11a, what actions should the master scheduler take at the start of week 2?

12. The Carlson Company in Problem 11 has the following product structure when parts are added:

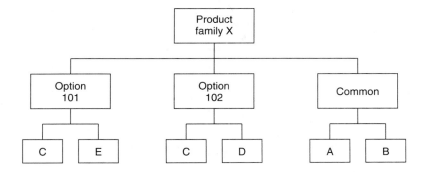

a. Prepare the MRP records for items A, B, C, D, and E, using data from Problem 11, the following records, and the associated lot size, safety stock, and lead time values given for each item. Assume one unit of each component is used in producing the parent items.

Item A	Week					
	1	2	3	4	5	6
Gross requirements						
Scheduled receipts						
Projected available balance						
Planned order release						

Q = LFL; SS = 10; LT = 1 week; On hand = 12

Item B	Week					
	1	2	3	4	5	6
Gross requirements						
Scheduled receipts						
Projected available balance						
Planned order release						

Q = LFL; SS = 10; LT = 1 week; On hand = 2

	Week					
Item C	1	2	3	4	5	6
Gross requirements						
Scheduled receipts						
Projected available balance						
Planned order release						

Q = LFL; SS = 10; LT = 1 week; On hand = 415

	Week					
Item D	1	2	3	4	5	6
Gross requirements						
Scheduled receipts						
Projected available balance						
Planned order release						

Q = LFL; SS = 50; LT = 1 week; On hand = 95

	Week					
Item E	1	2	3	4	5	6
Gross requirements						
Scheduled receipts						
Projected available balance						
Planned order release						

Q = LFL; SS = 15; LT = 1 week; On hand = 400

    b. How would the requirements for item C differ if item C was considered to be a common part? Why might it not be so considered?

13. A revision of the two-level MPS for product family X at the Carlson Company at the start of week 1 is shown below. This product family has two options. (See Problem 12 for a complete product structure.)

    a. The firm has just completed substantial work on a setup time reduction project as well as revisions for product family percentages. Therefore, lead times and lot sizes for options 101 and 102 have been reduced substantially, as seen in the following records. Given the revised MPS for product family X, and the associated lot size, safety stock, and lead time values given for each item, complete the MRP records for component items A, B, C, D, and E.

Assume one unit of each component is used in producing the parent items.

Product family X	Week					
	1	2	3	4	5	6
Production plan	200	200	200	200	200	200
Orders	200	200	140	190	50	10
Net	0	0	60	10	150	190

Option #101	Week					
	1	2	3	4	5	6
Forecast	0	0	36	6	90	114
Orders	155	140	91	117	30	6
Available	140	120	113	110	110	110
ATP	120	0	29	3	90	114
MPS (finish)	120	120	120	120	120	120
MPS (start)	120	120	120	120	120	

Forecast = 60%   Lead time = 1 week
Safety stock = 20   Lot size = 120
On hand = 175

Option #102	Week					
	1	2	3	4	5	6
Forecast	0	0	24	4	60	76
Orders	45	60	49	73	20	4
Available	100	120	127	130	130	130
ATP	100	20	31	7	60	76
MPS (finish)	80	80	80	80	80	80
MPS (start)	80	80	80	80	80	

Forecast = 40%   Lead time = 1 week
Safety stock = 50   Lot size = 80
On hand = 65

Common Parts		Week					
		1	2	3	4	5	6
Forecast		0	0	60	10	150	190
Orders		200	200	140	190	50	10
Available		0	0	0	0	0	0
ATP		0	0	60	10	150	190
MPS (finish)		200	200	200	200	200	200
MPS (start)		200	200	200	200	200	

Forecast = 100%   Lead time = 1 week
Safety stock = 0   Lot size = 200
On hand = 0

Item A		Week					
		1	2	3	4	5	6
Gross requirements							
Scheduled receipts							
Projected available balance							
Planned order releases							

Q = LFL; SS = 10; LT = 1 week; On hand = 212

Item B		Week					
		1	2	3	4	5	6
Gross requirements							
Scheduled receipts							
Projected available balance							
Planned order releases							

Q = LFL; SS = 10; LT = 1 week; On hand = 220

Item C		Week					
		1	2	3	4	5	6
Gross requirements							
Scheduled receipts							
Projected available balance							
Planned order releases							

Q = LFL; SS = 50; LT = 1 week; On hand = 415

Week

Item D		1	2	3	4	5	6
Gross requirements							
Scheduled receipts							
Projected available balance							
Planned order releases							

Q = LFL; SS = 50; LT = 1 week; On hand = 195

Week

Item E		1	2	3	4	5	6
Gross requirements							
Scheduled receipts							
Projected available balance							
Planned order releases							

Q = LFL; SS = 50; LT = 1 week; On hand = 400

b. How would the requirements for item C differ if item C was considered to be a common part? How (and why) is this answer different from that in Problem 12?

c. What would be necessary to use a kanban system to control items A, B, and C?

14. Hyster's manager of master production scheduling wants to adjust the MPS for the level 1 oil clutch option in Figure 14.7 so that a safety stock of 2 units is maintained in each period in the MPS. Please adjust the MPS to accomplish this in the following record. Why might safety stock be desirable on this item?

Week	8–21	8–28	9–5	9–11	9–18	9–25	10–2	10–9	10–16
Start day	178	183	188	192	197	202	207	212	217
Production forecast	0	0	2	3	8	8	9	10	9
Sold		8	6	5	2	1			
Scheduled receipts		9							
Projected on hand									
MPS start									

Lead Time = 2 weeks
Super Bill Percentage = 62%
Beginning On Hand Inventory = 0 units
Safety Stock = 2 units

15. The Fitzsimmons Pump Company produces industrial equipment on a make-to-order basis using a batch manufacturing process. The master production schedule is stated in terms of customer orders for each product family. The gear reducer product family's MPS is as follows. Standard manufacturing lead time quoted for component part production and final assembly is three weeks for this product family.

Gear reducer product family

	Week								
	1	2	3	4	5	6	7	8	9
Forecast	20	20	20	20	20	20	20	20	20
Orders	20	20	20	18	15	12	8	5	2
Available	0	0	0	0	0	0	0	0	0
ATP	0	0	0	2	5	8	12	15	18
MPS (complete)	20	20	20	20	20	20	20	20	20

Lead Time = 3 weeks   Lot Size = LFL
Safety Stock = 0   On Hand = 0

a. The shop is able to complete 20 units per week of this product family. Two work centers currently limit the output: the K&T machining center and the Essex gear hobber. The "typical" gear reducer requires an average of four standard hours per end item unit on the K&T machining center and two standard hours per unit of end product on the Essex gear hobber. The K&T machining center is operated on a two-shift 80-hour-per-week basis; the Essex gear hobber is run on a single shift 40-hour-per-week basis. Convert the MPS record for the gear reducer product family into standard hour-capacity equivalents.

b. Suppose the master production scheduler has just received a new order from sales for two gear reducers requiring three hours per unit on the K&T machining center and five hours per unit on the Essex gear hobber. When can this order be scheduled for delivery, assuming sufficient capacity exists at the other required work centers?

c. What assumptions are required in using this approach for monitoring shop capacity?

Chapter 15

Advanced Concepts in Production Planning

This chapter deals with modeling procedures for establishing an overall or aggregate production plan, and the disagggregation of the plan. The basic issue, given a set of product demands stated in some common denominator, is what levels of resources should be provided in each period? A long history of academic research has addressed aggregate production planning models. Theory has outstripped application by a wide margin. However, as firms implement MPC systems, there's a natural evolution toward questions of overall production planning which provide direction to the other MPC system modules. We're cautiously optimistic that theory and practice are converging. This chapter provides a basic understanding of how that convergence might occur. It's organized around five topics:

- Mathematical programming approaches: How can the aggregate production planning problem be formulated as a mathematical programming model?
- Other approaches: What other models have been developed and how do they compare?
- Disaggregation: What's the framework for disaggregating production plans into the master production schedule?
- Company example: What are the components of an actual advanced production planning and scheduling system, and what does it provide for the company?
- Applications potential: What are the roadblocks to implementing advanced production planning methods, and what's the prognosis for the future?

Chapter 15 has a close linkage with Chapter 7, which describes basic production planning approaches. There are also linkages to Chapters 6 and 14 on master production scheduling, and Chapter 4 on capacity planning. It's the aggregate production plan that constrains the master production schedule in a hierarchical view of manufacturing planning and control. Capacity planning procedures are used to determine the feasibility of an MPS.

MATHEMATICAL PROGRAMMING APPROACHES

In this section, we present an overview of some math programming models that have been suggested for the aggregate production planning problem. The academic literature has long been concerned with using formal decision models on this problem. We start by formulating the problem as a linear programming model. This approach is relatively straightforward, but is necessarily limited to cases where there are linear relationships in the input data. Thereafter, we describe a mixed integer programming approach for preparing aggregate production plans on a product line basis.

These approaches are already substantially more sophisticated than the practice found in most firms. More common are spreadsheet programs used to explore alternative production plans. Using forecasts of demand and factors relating to employment, productivity, overtime, and inventory levels, a series of what-if analyses helps to formulate production plans and evaluate alternative scenarios. "Level" and "chase" strategies are developed to bracket the options, and we can evaluate alternatives against the resultant bench marks. Although spreadsheet programs don't provide the optimal solutions reached by the models discussed in this chapter, they do help firms better understand the inherent trade-offs, and provide a focal point for important dialog among the functional areas of the firm.

Linear Programming

There are many linear programming formulations for the aggregate production planning problem. The objective is typically to find the lowest-cost plan, considering when to hire and fire, how much inventory to hold, when to use overtime and undertime, and so on, while always meeting the sales forecast. One formulation, based on measuring aggregate sales and inventories in terms of direct labor hours, follows:

Minimize:

$$\sum_{t=1}^{m} (C_H H_t + C_F F_t + C_R X_t + C_O O_t + C_I I_t + C_u U_t)$$

subject to:

1. Inventory constraint:

$$I_{t-1} + X_t + O_t - I_t = D_t$$
$$I_t \geq B_t$$

2. Regular time production constraint:

$$X_t - A_{1t}W_t + U_t = 0$$

3. Overtime production constraint:

$$O_t - A_{2t}W_t + S_t = 0$$

4. Work force level change constraint:

$$W_t - W_{t-1} - H_t + F_t = 0$$

5. Initializing constraints:

$$W_0 = A_3$$
$$I_0 = A_4$$
$$W_m = A_5$$

where:

C_H = The cost of hiring an employee.
C_F = The cost of firing an employee.
C_R = The cost per labor-hour of regular time production.
C_O = The cost per labor-hour of overtime production.
C_I = The cost per month of carrying one labor-hour of work.
C_u = The cost per labor-hour of idle regular time production.
H_t = The number of employees hired in month t.
F_t = The number of employees fired in month t.
X_t = The regular time production hours scheduled in month t.
O_t = The overtime production hours scheduled in month t.
I_t = The hours stored in inventory at the end of month t.
U_t = The number of idle time regular production hours in month t.
D_t = The hours of production to be sold in month t.
B_t = The minimum number of hours to be stored in inventory in month t.
A_{1t} = The maximum number of regular time hours to be worked per employee per month.
W_t = The number of people employed in month t.
A_{2t} = The maximum number of overtime hours to be worked per employee per month.
S_t = The number of unused overtime hours per month per employee.
A_3 = The initial employment level.
A_4 = The initial inventory level.
A_5 = The desired number of employees in month m (the last month in the planning horizon).
m = The number of months in the planning horizon.

Similar models have been successfully formulated for several variations of the production planning problem. In general, however, few real-world aggregate production planning decisions appear to be compatible with the linear assumptions. Some plans require discrete steps such as adding a second shift. For many companies, the unit cost of hiring or firing large numbers of employees is much larger than that associated with small labor force changes. Moreover, economies of scale aren't taken into account by linear programming formulations. Let's now turn to another approach, which partially overcomes the linear assumption limitations.

Mixed Integer Programming

The linear programming model provides a means of preparing low-cost aggregate plans for overall work force, production, and inventory levels. However, in some firms, aggregate plans are prepared on a product family basis. Product families are defined as groupings of products that share common manufacturing facilities and setup times. In this case, overall production, work force, and inventory plans for the company are essentially the summation of the plans for individual product lines. Mixed integer programming provides one method for determining the number of units to be produced in each product family. Chung and Krajewski describe a model for accomplishing this:

Minimize:

$$\sum_{i=1}^{n} \sum_{t=1}^{m} [C_{si} \, \sigma(X_{it}) + C_{mi}X_{it} + C_{Ii}I_{it}] + \sum_{t=1}^{m} [C_H H_t + C_F F_t + C_O O_t + A_{1t}C_R W_t]$$

subject to:

1. Inventory constraint:

$$I_{i,t-1} - I_{it} + X_{it} = D_{it} \qquad \text{(for } i = 1, \ldots, n \text{ and } t = 1, \ldots, m)$$

2. Production and setup time constraint:

$$A_{1t}W_t + O_t - \sum_{i=1}^{n} X_{it} - \sum_{i=1}^{n} \beta_i \sigma(X_{it}) \geq 0 \qquad \text{(for } t = 1, \ldots, m)$$

3. Work force level change constraint:

$$W_t - W_{t-1} - H_t + F_t = 0 \qquad \text{(for } t = 1, \ldots, m)$$

4. Overtime constraint:

$$O_t - A_{2t}W_t \leq 0 \qquad \text{(for } t = 1, \ldots, m)$$

5. Setup constraint:

$$-Q_i\sigma(X_{it}) + X_{it} \leq 0 \qquad \text{(for } t = 1, \ldots, m \text{ and } i = 1, \ldots, n)$$

6. Binary constraint for setups:

$$\sigma(X_{it}) = \begin{cases} 1 \text{ if } X_{it} > 0 \\ 0 \text{ if } X_{it} = 0 \end{cases}$$

7. Nonnegativity constraints:

$$X_{it}, I_{it}, H_t, F_t, O_t, W_t \geq 0$$

where:

X_{it} = Production in hours of product family i scheduled in month t.

I_{it} = The hours of product family i stored in inventory in month t.

D_{it} = The hours of product family i demanded in month t.

H_t = The number of employees hired in month t.

F_t = The number of employees fired in month t.

O_t = Overtime production hours in month t.

W_t = Number of people employed on regular time in month t.

$\sigma(X_{it})$ = Binary setup variable for product family i in month t.

C_{si} = Setup cost of product family i.

C_{Ii} = Inventory carrying cost per month of one labor-hour of work for product family i.

C_{mi} = Materials cost per hour of production of family i.

C_H = Hiring cost per employee.

C_F = Firing cost per employee.

C_O = Overtime cost per employee hour.

C_R = Regular time work force cost per employee hour.

A_{1t} = The maximum number of regular-time hours to be worked per employee in month t.

β_i = Setup time for product family i.

A_{2t} = Maximum number of overtime hours per employee in month t.

Q_i = A large number used to ensure the effects of binary setup variables; that is,

$$Q_i \geq \sum_{t=1}^{m} D_{it}.$$

n = Number of product families.

m = Number of months in the planning horizon.

The objective function and constraints in this model are similar to those in the linear programming model. The main difference is in the addition of product family setups in constraints 5 and 6. This model assumes all the setups for a product family occur in the month in which the end product is to be completed. Constraint 5 is a surrogate constraint for the binary variables used in constraint 6. This constraint forces $\sigma(X_{it})$ to be nonzero when $X_{it} > 0$ since Q_i is defined as at least the total demand for a product family over the planning horizon.

Additional constraints should be added to the model to specify the initial conditions at the start of the planning horizon; that is, constraints specifying beginning inventory for the product family, I_{io}, and work force level in the previous month, W_o, are required. Likewise, constraints specifying work force level at the end of the planning horizon, and minimum required closing inventory balance at the end of each month in the planning horizon, may be added.

OTHER APPROACHES

Several other models have been formulated for solving the aggregate production planning problem. In this section, we briefly describe three of them. The first is a classic academic work based on linear and quadratic cost assumptions. The other two approaches are enhancements of this model. The first, management coefficients, is based on managerial behavior as the way to develop general cost estimates. The second utilizes search approaches that allow for more general cost expressions.

The Linear Decision Rule

The linear decision rule model (LDR) for aggregate production planning was developed by Holt, Modigliani, Muth, and Simon in the 1950s. The primary application was in a paint producing company.

The major difference between the LDR model and linear programming models is the approach to cost input data. The four cost elements considered in LDR are regular payroll cost, hire/fire cost, overtime/undertime cost, and inventory/backlog cost.

The regular payroll cost is simply a linear function of the number of workers employed. For the other three cost elements, however, a quadratic cost function is used. For example, the hire/fire cost is defined as

$$64.3(W_t - W_{t-1})^2$$

where:

W_t = The work force to be established for the tth month.
W_{t-1} = The prior month's work force.
64.3 = Analytically derived coefficient for best fitting the squared differences in work force levels to actual operating cost results.

Figure 15.1 is an example quadratic hire/fire cost function, along with the presumed actual cost data. The presumed actual cost data only approximate the quadratic function. However, in some ways, the implication that each hire or fire decision results in ever-increasing unit costs is consistent with many managerial opinions.

FIGURE 15.1   LDR Quadratic Cost Function for Work Force Changes

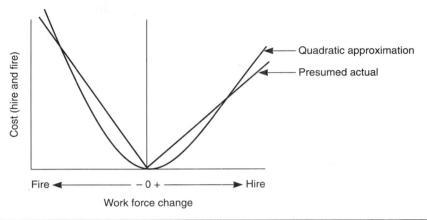

The total cost function for the paint company was made up of the four cost elements. The problem is to minimize the total cost function. Since cost data are linear and quadratic, we can derive the solution to the problem by calculus. The result is a set of two decision rules specifying the production output rate and the work force level in each month.

LDR was implemented at the paint company. Several years later, John Gordon visited the company and described the results:

> After considerable study and investigation it became apparent that although top management thought the rules were being used to determine aggregate production and work force, a more intuitive and long-standing system was in fact being used. The production control clerk whose responsibility it was to calculate the production and work force sizes, as well as convert these into item orders, was doing just that and posting the results in the form of job tickets on the production control board. When the foremen came into the production control office for a job ticket, they surveyed the available tickets for one that agreed with their intuitive feeling or judgment. If they found one they took it but if they did not they simply wrote out a ticket which corresponded with their feeling. Over the history of the use of the rules it turned out that about 50 percent of the tickets were used and the others ignored. Management, however, had the feeling that the rules were being used except in the odd case when judgment indicated that they should be overruled. At a later date the calculations associated with the rules were centralized with the installation of a data-processing center. The personnel in the center became concerned when their reports indicated that many of the production orders that they had issued were ignored. Consequently, and with the compliance of higher management, they instituted a reporting system, which fed back to the plant management, and the foremen, a cumulative listing of outstanding production orders. After a short delay the length of this cumulative list began to diminish until it all but vanished. But in the meantime the inventory of finished

goods associated with this plant rose steadily to alarming proportions, especially in some obsolete items. Further investigation revealed that although the rules were indicating the size of the work force, no action was ever taken to reduce the work force because it was against the policy of the company. This meant that the work force rule was indicating a reduction in the work force: the production rule, attempting to minimize costs given the present work force level but anticipating layoff, called for some production for the excess work force. The rules are interactive, but in this case the interaction had been eliminated.

The moral to the story seems clear: Never assume the real world matches the model without auditing. In fact, any system that's not readily understood by the users is more subject to overrides than one where the logic is transparent. This is a classic example of the informal system supplanting the formal system.

The Management Coefficients Model

A rather unique approach to aggregate production planning has been formulated by Bowman. He suggests the production rate for any period would be set by the following very general decision rule:

$$P_t = aW_{t-1} - bI_{t-1} + cF_{t+1} + K$$

where:

P_t = The production rate set for period t.
W_{t-1} = The work force in the previous period.
I_{t-1} = The ending inventory for the previous period.
F_{t+1} = The forecast of demand for the next period.
$a, b, c,$ and K = Constants.

Bowman's approach is to first gather historical data for P, W, I, and F. Thereafter, through regression analysis, the values of a, b, c, and K are estimated. The result is a decision rule based on past managerial behavior without any explicit cost functions. The assumption is managers know what's important, even if they can't readily state explicit costs. However, managers may either overreact to specific circumstances or delay in making adjustments. In both cases, a bias comes into their decision making. The regression analysis of the management coefficients model will average out this bias for future decisions.

Bowman compared the performance of the management coefficient model with LDR and actual company practice in four firms. In three out of four cases, the management coefficient model produced results superior to those made by the company; in two cases, results were also superior to LDR. Later research has tended to confirm the efficiency of Bowman's approach, lending credence to approaches that supplement the application of experienced judgment.

Search Decision Rules

As we've noted, linear programming models are limited by the linear cost assumptions. Similarly, LDR is restricted to linear and quadratic costs. The search decision rule (SDR) methodology helps overcome these restrictions. SDR approaches allow us to state cost data inputs in very general terms. The only requirement is that a computer program be constructed that will unambiguously evaluate any production plan's cost. The procedure then searches among alternative plans (in a guided fashion) for the plan of minimum cost. Unlike linear programming and LDR, there's no guarantee of mathematical optimality with SDR. However, the increased realism in input data provides the potential for solving a problem more in line with managerial perceptions.

Several researchers have worked with search procedures for the aggregate production planning problem. Taubert compared SDR with LDR for the paint company problem. He found SDR results were very close to those obtained by LDR. The technique has been applied in a number of companies. The versatility of the underlying approach provided makes it especially attractive for real-world applications.

DISAGGREGATION

Thus far, we've considered only establishing an overall plan of production. This plan must necessarily be disaggregated into specific products and detailed production actions. Moreover, the aggregate production planning problem, as formulated up to now, has been based on a single facility (although it's conceivable a facility subscript could be used in the linear programming model). For many firms, the problem of determining which facility will produce which products, in which quantities, is an important prerequisite to planning at each facility. In this section, we first consider an approach to disaggregation that closely parallels the managerial organization for making these decisions. We then turn to a mathematical programming model for determining the MPS within each product family (the product family planning having been done earlier with mixed integer programming).

The Disaggregation Problem

One issue receiving increased attention is converting overall aggregate production plans into detailed MPS plans; that is, managers must make day-to-day decisions on a product and unit basis rather than on the overall output level. The concept of disaggregation facilitates this process and avoids mismatches between plan and execution. In essence, disaggregation concerns overall production planning as well as consistent lower-level capacity decisions. It recognizes that aggregate decisions constrain the disaggregated actions. It therefore concerns how to break the total or aggregate plan into plans for subunits of product.

Disaggregation is an important topic. There has been some growth in both theory and practice, but the number of applications to date is limited. The disaggregation frame of reference is to maintain a match between the production plan and the master production schedule. The aggregate production plan must be the sum of the production called for by the detailed master production schedule (MPS). At issue is how to keep the two in concert. Some recent research offers potential help, but there's much to be done.

Hierarchical Production Planning

Hierarchical production planning is one approach to aggregate capacity analysis that's based upon disaggregation concepts and can accommodate multiple facilities. The approach incorporates a philosophy of matching product aggregations to decision-making levels in the organization. Thus, the approach isn't a single mathematical model but utilizes a series of models, where they can be formulated. Since the disaggregation follows organization lines, managerial input is possible at each stage. Figure 15.2 shows a schema of the approach.

A group of researchers (Bitran, Haas, Hax, Meal, and others) has been developing hierarchical production planning (HPP) over several years. Some of the work has involved mathematical contributions, while other work has increased the depth or breadth of application (incorporating distribution centers or levels of detail in a factory). All work, however, is based on some fundamental principles.

One principle has been mentioned already: disaggregation should follow organizational lines. Another principle is it's only necessary to provide information at the aggregation level appropriate to the decision. Thus, we don't need to use detailed part information for the plant assignment decisions. Finally, it's necessary to schedule only for the lead time needed to change decisions. That means detailed plans can be made for periods as short as the manufacturing lead times.

The process of planning follows Figure 15.2's schema. It first involves specifying which products to produce in which factories. Products are combined in logical family groupings to facilitate aggregation, assignment to factories, and modeling processes. Assignment to factories is based on minimizing capital investment cost, manufacturing cost, and transportation cost.

Once the assignment to factories has been done and managerial inputs incorporated, an aggregate production plan is made for each plant. The procedure for determining the aggregate production plan could be any of those discussed previously. The aggregate plan specifies production levels, inventory levels, overtime, and so on for the plant. This plan is constrained by the specific products and volumes assigned to the plant.

FIGURE 15.2   Hierarchical Planning Schema

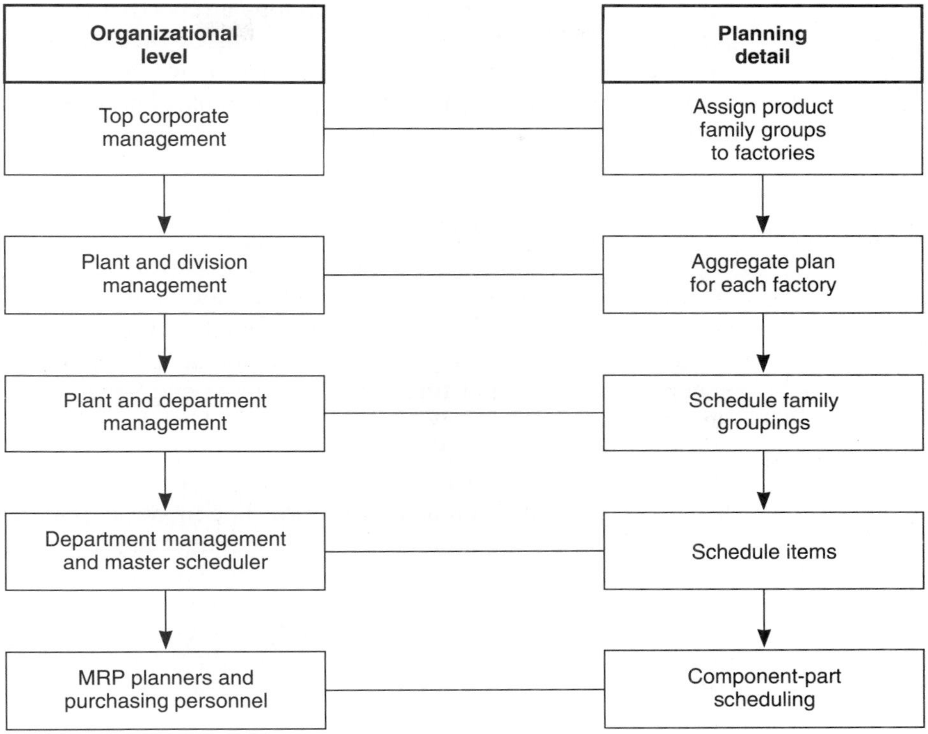

Organizational level		Planning detail
Top corporate management		Assign product family groups to factories
Plant and division management		Aggregate plan for each factory
Plant and department management		Schedule family groupings
Department management and master scheduler		Schedule items
MRP planners and purchasing personnel		Component-part scheduling

Source: G. D. Bitran, R. A. Haas, and A. C. Hax, "Hierarchical Production Planning: A Two-Stage System," *Operations Research,* March–April 1982, pp. 232–51.

The next step in the disaggregation calls for scheduling family groupings within the factory. The schedule is constrained by the aggregate production plan and takes into account any inventories that may exist for the group. The intention at this stage is to realize the economies of producing a family grouping together. Production lots (or share of the aggregate capacity) for the groups are determined and sequenced. If no major economies are achieved by scheduling the group as a unit, the procedure can move directly to the scheduling of individual items, the next stage in Figure 15.2.

Determining the individual item schedule is analogous to making a master production schedule. In the HPP schema, the MPS is constrained by the previously scheduled family groupings and may cover a shorter planning horizon. In some instances, we can use mathematical models to establish schedules. In all cases, items are scheduled within the capacity allocated for their family group. Detailed part and component scheduling can be done with MRP logic, order launching and inventory systems, or even mathematical modeling.

A recent extension to the basic HPP model is to use variable planning periods rather than the fixed planning period of 20 days used in the Bitran, Hass, and Hax approach. Oden develops a recursive algorithm to predict the length of the planning period that minimizes the annual sum of setup and inventory holding costs. Oden's model produces consistently lower production costs (overtime, setup, inventory holding) than those of the fixed period approach.

Disaggregation through Mathematical Programming

A disaggregation model by Chung and Krajewski illustrates how the aggregate production plan for each family, determined by their mixed integer programming model, can be disaggregated into a detailed master production schedule specifying lot size and timing for individual end products. Since family production and inventory levels have been established by the aggregate plan, the master production schedule must adhere to targets set by the aggregate plan. To do this the formulation includes resource requirements and limitations from the aggregate plan as constraints. The model uses information from the aggregate plan for each family, including setup time (β_i), setup status $(\sigma(X_{it}))$, production level (X_{it}), and inventory level (I_{it}). Also used are overtime (O_t), work force level (W_t), and regular time availability (A_{1t}) for each month. The formulation of the disaggregation model follows:

Minimize:

$$\sum_{i=1}^{n} \sum_{k \in Ki} \sum_{t' \in Nt} b_i^k B_{it'}^k + \sum_{t=1}^{m} (w^{3-} d_t^3 + w^{3+} d_t^{3+})$$

$$+ \sum_{i=1}^{n} \sum_{t=1}^{m} (w_i^{1-} d_{it}^{1-} + w_i^{1+} d_{it}^{1+} + w_i^{2-} d_{it}^{2-} + w_i^{2+} d_{it}^{2+})$$

subject to:

1. Inventory constraint:

$$I_{i,t'-1}^k - I_{it'}^k + X_{it'}^k + B_{it'}^k - B_{i,t'-1}^k = D_{it'}^k$$
$$\text{(for } i = 1, \ldots, n; \ k \in K_i; \ t' \in N_i; \text{ and } t = 1, \ldots, m)$$

2. Regular time and overtime production constraint:

$$\sum_{i=1}^{n} \sum_{k \in Ki} \sum_{m'=1}^{Li} \sum_{j=1}^{J} (r_{im'j}^k X_{i,t'+Li-m'}^k) + d_{t'}^{o-} - d_{t'}^{o+}$$

$$= \left(\frac{1}{4}\right) \left[A_{1t} W_t - \sum_{i=1}^{n} \beta_i \sigma(X_{it}) \right] \text{ (for } t' \in N_t \text{ and } t = 1, \ldots, m)$$

3. Overtime deviation constraint:

$$\sum_{t' \epsilon Nt} d_{t'}^{o+} + d_t^{3-} - d_t^{3+} = O_t \qquad \text{(for } t = 1, \ldots, m)$$

4. Regular time deviation constraint:

$$\sum_{k \epsilon K_i} \sum_{t' \epsilon N_t} X_{it'}^k + d_{it}^{1-} - d_{it}^{1+} = X_{it} \qquad \text{(for } i = 1, \ldots, n \text{ and } t = 1, \ldots, m)$$

5. Inventory deviation constraint:

$$\sum_{k \epsilon K_i} \sum_{t' \epsilon N_t} I_{it'}^k + d_{it}^{2-} - d_{it}^{2+} = I_{it}$$

$$\text{(for } i = 1, \ldots, n; t = 1, \ldots, m; \text{ and}$$
$$t' = 4(t-1) + 1, \ldots, 4(t-1) + 4)$$

6. Nonnegativity constraints:

$$X_{it'}^k, I_{it'}^k, B_{it'}^k, d_{t'}^{0-}, d_{t'}^{0+}, d_{it}^{2-}, d_{it}^{2+}, d_{it}^{3-}, d_{it}^{3+}, d_t^{1-}, d_t^{1+} \geq 0$$

where:

$X_{it'}^k$ = The production hours of end item k of product family i in week t'.

$I_{it'}^k$ = The hours of end item k of product family i in inventory at the end of week t'.

b_i^k = Cost to backorder one hour of production of end item k of product family i.

$B_{it'}^k$ = Backorder hours of end item k of product family i in week t'.

$D_{it'}^k$ = Hours of end item k of product family i demanded in week t'.

$d_{t'}^{0-}$ = Undertime in week t'; that is, the number of planned regular time hours not used in week t'.

$d_{t'}^{0+}$ = Overtime hours used in week t'.

d_t^{3-} = Negative deviation from the planned overtime level in month t.

d_t^{3+} = Positive deviation from the planned overtime level in month t.

d_{it}^{1-} = Negative deviation from the planned aggregate production level of product family i in month t.

d_{it}^{1+} = Positive deviation from the planned aggregate production level of product family i in month t.

d_{it}^{2-} = Negative deviation from the planned inventory of product family i at the end of month t.

d_{it}^{2+} = Positive deviation from the planned inventory of product family i at the end of month t.

$w_i^{1-}, w_i^{1+}, w_i^{2-}, w_i^{2+}, w^{3-}, w^{3+}$ = Weights (costs) assigned to the deviation variables.

$r_{im'j}^k$ = Proportion of total production labor-hours required for processing item k of product family i at operation or work center j in week m' (the week since production started on k) assuming at most one operation for each item at each work center j.

L_i = Production lead time for items in product family i.

n = Number of product families.

m = Production planning horizon length.

J = Number of work centers.

K_i = Set of end items within product family i.

N_t = Set of time-phased weeks, (t')'s, in month t. This example uses monthly time buckets for aggregate plans and uses weekly time buckets for the time-phased master production schedule. It's assumed there are 4 weeks in each month: that is, month 1 ($t = 1$) has $t' = 1, 2, 3, 4$; and month 2 ($t = 2$) has $t' = 5, 6, 7, 8$; and so on.

The first constraint represents production and inventory relationships from week to week with the backorder position for the individual product lines included. The second constraint defines the labor-hour requirements. The right-hand side value in this constraint is obtained from the aggregate plan solution, and it equals the regular labor-hours planned for the month, excluding hours consumed for setups for product families. The value of ¼ is used to translate the monthly (t) figures into weekly (timed-phased t') values, assuming there are four weeks in every month. Therefore, this constraint specifies the total employee hours at all work centers that produce items in the product families (including the under- and overtime adjustments) should equal overall employee capacity planned to be available each week (t').

The third, fourth, and fifth constraints use deviations to force the overtime, family production quantities, and closing inventory values to correspond to monthly goals set by the aggregate production plan; that is, O_t, X_{it}, and I_{it}, respectively. Weights placed on the deviations in the objective function control these deviations' magnitude and frequency.

The initial end item inventories and work force conditions must be included in the master production scheduling model. The beginning inventories of the end items within a product family must sum to the beginning inventory for the product family used in the aggregate planning model; that is:

$$\sum_{k \in K_i} I_{i,o}^k = I = I_{i,o} \, .$$

Also, the lead time (L_i) for each product family must be considered in solving the master production scheduling model. For example, if a four-week production lead time is used for a product family, including one week for the end item and three weeks for the components, then only the master schedule for week 4 and beyond can be changed. Components for weeks 1 to 3 must have been produced in the previous month. Therefore these items' resource requirements must be netted from resource capacities given for the first month in the second constraint. Likewise, demands for weeks 1 to 4 in the master production schedule must also be netted from the planned production in weeks 1 to 3.

Figure 15.3 presents a schematic that shows the relationship of the aggregate planning and the master production scheduling models to other MPC activities. This diagram indicates the sequential nature of the solution process. It begins with long-term capacity planning and demand forecasting. Then aggregate planning for each family is done. Factor values for X_{it}, I_{it}, $\sigma(X_{it})$, W_t, and O_t are then passed to the master production scheduling model,

FIGURE 15.3   A Schematic Diagram of a Sequential Production Planning Process

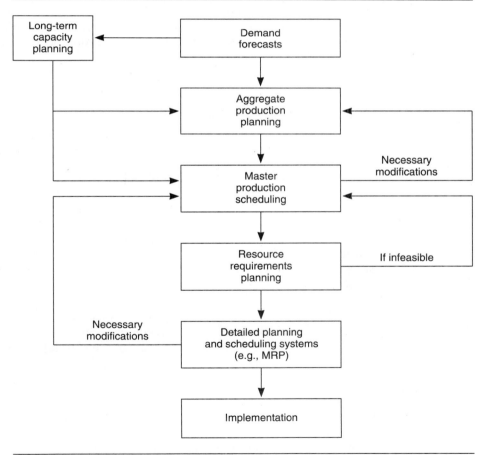

Source: C. Chung and Lee J. Krajewski, "Planning Horizons for Master Production Scheduling," *Journal of Operations Management,* August 1984.

which sometimes requires modifications to the aggregate production planning results. Thereafter, the master scheduling model's outputs are examined for feasibility in terms of resource requirements and then are passed to the company's detailed planning and scheduling systems. These include material requirements planning (MRP) and capacity planning, which are driven from the MPS. The detailed plans are integrated as well as possible, with the cycle repeated each month.

This approach represents one method for disaggregating overall production plans into detailed master production schedules.

COMPANY EXAMPLE

In this section, we present an application of advanced production planning models. We start with an overview of the problem, then present the aggregate production planning model, and end with a section detailing the disaggregation procedure.

Owens-Corning Fiberglas: Anderson, South Carolina, Plant

In 1982, Owens-Corning Fiberglas implemented a hierarchical approach for production planning and scheduling decisions at its Anderson, South Carolina, plant—one of the firm's largest plants. The problem structure and results are described by Oliff and Burch; the model details are explained in a subsequent paper by Oliff. The Anderson plant produces a fiberglass mat used in the marine industry for constructing boat hulls, as reinforcement in pipeline construction, and in bathtubs and showers. This mat is sold in a variety of widths and lengths, is treated with one of three process binders, and is frequently trimmed on one or both edges. Over 200 end products are produced, of which 28 represent 80 percent of the sales volume, the remainder being low-volume special-order products.

The fiberglass mat's production process consists of two high-volume batch production lines where demand exceeds capacity approximately six months of the year. Figure 15.4 gives an overview of the hierarchical planning system. This system determines:

- An aggregate plan reflecting relevant costs for inventory, production, and work force.
- Production lot sizes, line assignments, and inventory levels for each standard product.
- Specific production sequences and changeover costs for standard and special-order items.

FIGURE 15.4 Owens-Corning Fiberglas Hierarchical Planning System

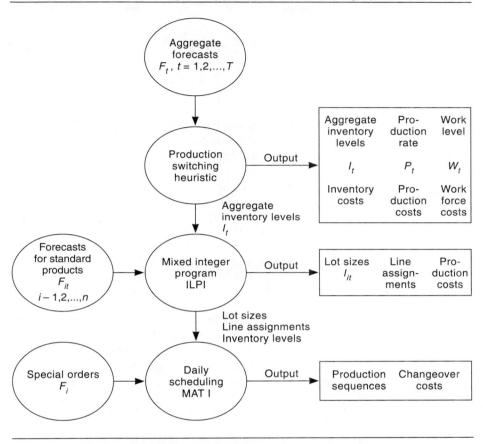

Source: M. D. Oliff and E. E. Burch, "Multiproduct Production Scheduling at Owens-Corning Fiberglas," *Interfaces,* September–October 1985, pp. 25–34.

While this system is an excellent example of applying advanced techniques for preparing aggregate plans, it also illustrates how these plans are disaggregated into the master production schedule for individual end products.

Aggregate Production Planning

The model used to prepare aggregate inventory level, work force, and production rate plans is a production switching heuristic (PSH). The objective function involves minimizing direct payroll, overtime, hiring and firing, and relevant inventory costs over a planning horizon ranging from 3 to 12 months, as shown in the following total cost equation:

Minimize:

$$\sum_i \sum_t C_{Ri} S_{it} + \sum_i \sum_t C_{Oi} O_{it} + \sum_i \sum_t C_{Hi} H_{it} + \sum_i \sum_t C_{Fi} F_{it} + \sum_t C(I_t)$$

$$(\text{for } i = 1, \ldots, k \text{ and } t = 1, \ldots, m)$$

where:

k = Number of production lines.
m = Number of production months.
C_{Ri} = Cost of variable wages per shift on production line i.
C_{Oi} = Cost of overtime per shift on production on line i.
C_{Hi} = Cost of increasing by one shift on production line i.
C_{Fi} = Cost of decreasing by one shift on production line i.
$C(I_t)$ = Cost of holding I_t pounds of inventory per month.
I_t = Pounds of inventory in month t.
S_{it} = Number of shifts on line i in month t.
O_{it} = Overtime shifts required on line i in month t.
H_{it} = Number of shifts added on line i from month $t - 1$ to month t.
F_{it} = Number of shifts reduced on line i from month $t - 1$ to month t.

The decision variables in the aggregate planning model are the monthly plant production rates (P_t) over the planning horizon (of m months), which can be adjusted by operating each of the two mat lines at the Anderson plant in one of several shift setups. Full production on line 1 involves using four 6-hour shifts per day with the plant running seven days per week on a 24-hour per day basis, while full production on line 2 is obtained by operating three 8-hour shifts per day on a five-day week, 24-hour per day schedule. A particular shift setting is denoted by (i,j) where i represents the number of shifts on line 1 and j is the number of shifts on line 2. The different shift settings (and monthly production rates) that are feasible because of process and organizational constraints at the Anderson plant are:

Decision variable	Shift setting (i,j)	Production in pounds per month
R_1	(4,2)	1,679,000
R_2	(4,1)	1,571,000
R_3	(3,3)	1,405,000
R_4	(3,2)	1,288,000
R_5	(3,1)	1,180,000

A production switching heuristic is used to fix the production rate for each month in the production plan. The particular heuristic used at the Anderson plant is shown in the following equation:

$$P_t = \begin{cases} R_1 \text{ if} & D_t - I_{t-1} + A > R_1 \\ R_5 \text{ if} & D_t - I_{t-1} + C < R_5 \\ R_2 \text{ if } R_1 > & D_t - I_{t-1} + A > R_2 \\ R_4 \text{ if } R_5 < & D_t - I_{t-1} + C > R_4 \\ R_3 \text{ otherwise} \end{cases}$$

where:

$$R_1 > R_2 > R_3 > R_4 > R_5.$$

D_t = Monthly forecast in pounds.
I_t = Inventory level in pounds.
P_t = Production level in pounds (determined by R_i).
S_{it} = Number of shifts on line i in month t (determined by R_i).
A = Minimum ending inventory target.
C = Maximum ending inventory target ($A \leq C$).

The switching heuristic works by setting the production rate for month t at R_1 if demand for month t, D_t, exceeds R_1 after deducting the starting inventory, I_{t-1}, and adding the minimum closing inventory target, A. Similarly, the production rate for month t is set at the minimum production rate R_5 if demand for month t, D_t, net of starting inventory, I_{t-1}, and after adding maximum closing inventory target, C, is less than the minimum production rate R_5. The monthly production rate can also be set at one of the intermediate production rates (R_2, R_3, or R_4) using the preceding equations.

Figure 15.5 gives an example of a production plan developed using the production switching heuristic. This plan was developed using actual sales, an initial inventory of 1,306,000 lbs., and a previous month production rate of $R_2(4,1)$. The minimum and maximum inventory targets (A and C) were both set to 1,100,000 in developing this plan.

The production switching heuristic is adaptable to a wide range of operating conditions. It's particularly applicable in production situations where capacity adjustments are made in large increments, such as by adding extra work shifts. Moreover, a variety of both regular and overtime production settings, inventory targets, and forecasts can be used in this model. In using the model, the production planner can restrict selection of decision variables and factors to be considered in examining a particular production plan's performance implications.

This flexibility in the production planning model means the control parameters—that is, the production rate factors (R_1 to R_5), the inventory targets (A and C), and the planning horizon length (m)—are context

FIGURE 15.5   Example Aggregate Production Plan Using Last Year's Sales as Forecast Data (No Overtime)

Month	Forecast (pounds)	Production (pounds)	Inventory (pounds)	Shift schedule (shifts)
0	0	0	1,306,000	4,1
1	1,190,000	1,180,000	1,296,000	3,1
2	1,280,000	1,180,000	1,196,000	3,1
3	1,408,000	1,288,000	1,076,000	3,2
4	1,714,000	1,679,000	1,041,000	4,2
5	1,184,000	1,288,000	1,145,000	3,2
6	1,546,000	1,405,000	1,004,000	3,3
7	1,216,000	1,288,000	1,076,000	3,2
8	1,167,000	1,288,000	1,197,000	3,2
9	1,404,000	1,288,000	1,081,000	3,2
10	1,204,000	1,288,000	1,165,000	3,2
11	1,154,000	1,180,000	1,191,000	3,1
12	1,114,000	1,180,000	1,231,000	3,1

Inventory cost	Payroll cost	Hire and fire cost	Overtime cost	Total cost
$87,673	$956,026	$65,806	0	$1,109,505

dependent. Owens-Corning Fiberglas determined the least cost values for these parameters using a simulation model, which reflected the firm's production planning situation and direct search methods. At the Anderson plant, production rate increments of 100,000 pounds (less than two days' production) were used in evaluating approximately 100,000 sets of parameter values. Starting conditions were provided to the simulation, and a total cost equation similar to the preceding objective function was used to determine the least cost set of control parameters. The control parameter values are reviewed periodically at the plant and the model parameters are updated by using the same procedures.

Disaggregating the Production Plan

Figure 15.4 shows the linkage between the production plan and the master production schedule. The monthly aggregate closing inventory position from the production switching heuristic, I_t, is passed to a mixed integer programming formulation that determines lot sizes, line assignments, and inventory levels for the 28 standard products (the MPS). The aggregate inventory balance is included as one constraint in the mixed integer program. These constraints include:

- Production balance equations to ensure individual product demand forecasts are satisfied while forcing inventory conservation.
- Inventory equations to adhere to aggregate inventory constraints, as well as to safety stock requirements.
- Noninterference equations to ensure feasible assignments of individual products to processing lines.
- Changeover equations to incorporate changeover costs from one product to another.

The mixed integer programming formulation is designed to minimize the combined costs of processing line changeovers and production in determining the 28 standard products' assignment to the two processing lines. There are three decision variables in the model:

X_{iltp} = 1 if line l's production capacity during month t and subperiod p is devoted to product i, and 0 otherwise.

σ_{ijltp} = 1 if a changeover occurs from product i to product j on line l during month t and at the beginning of subperiod p, and 0 otherwise.

I_{it} = the total pounds of product i in inventory at the end of the month t.

While the model determines production lot sizes, it's not primarily concerned with sequencing production lots at this level of decision making in Owens-Corning Fiberglas's manufacturing control system. Production lot sizes for the standard products can be determined from the mixed integer programming solution by simply multiplying the solution value for a particular X_{iltp} by the production rate for line l in month t and subperiod p. The objective function and constraints for this model are as follows:

Minimize:

$$\sum_i \sum_l \sum_t \sum_p (C_{pil} X_{iltp}) + \sum_i \sum_j \sum_l \sum_t \sum_p C_{sij} \sigma_{ijltp}$$

subject to:

Production balance:

$$I_{i,t-1} - I_{it} + \sum_l \sum_p r_{pil} \cdot X_{iltp} = D_{it} \text{ (for all } i \text{ and } t)$$

Inventory capacity:

$$\sum_i I_{it} = I_t \text{ (for all } t)$$
$$I_{it} \geq B_{it} \text{ (for all } i \text{ and } t)$$

Noninterference:

$$\sum_i X_{iltp} = 1 \text{ (for all } l, t, \text{ and } p)$$

Changeover (one per subperiod, p):

$$X_{ilt,p-1} - X_{iltp} = \sum_j \sigma_{ijltp} - \sum_i \sigma_{ijltp} \text{ (for all } l,t,p)$$

$$\sum_i \sum_j \sigma_{ijltp} = 1 \text{ (for all } l, t, p)$$

where:

C_{pil} = Production cost of product i on line l in subperiod p.

C_{sij} = Changeover cost from product i to j.

D_{it} = Demand for product i in month t.

r_{pil} = Production in subperiod p for product i in line l.

I_t = Desired aggregate inventory for month t.

B_{it} = Desired safety stock for product i in month t.

This mixed integer programming model results in a 10,000-row by 10,000-column matrix. Once a month a linear programming approximation generates feasible, near-optimal solutions to the mixed integer programming problem. The results provide Owen-Corning with monthly, quarterly, or yearly master production schedules. It yields inventory level information, lot sizes, and line assignments for coming months.

The disaggregation process is carried one step further by the daily scheduling part of the manufacturing control system in Figure 15.4. This phase produces a daily production schedule, using both customer orders for the 170 special ordered products as well as lot sizes and processing line assignments for standard products. It uses a sequencing heuristic that considers sequence-dependent setups, minimizing the cost of changing from one product to another. This heuristic produces a 7-day to 30-day schedule that includes sequences, complete product descriptions, expected run times, and expected changeover costs. Real-time responses to potential schedule changes as well as to new schedule requirements are obtained daily.

The integrated planning and scheduling system has provided Owens-Corning a number of major benefits. The average number of monthly changeovers decreased from 70 to fewer than 40 in just a year, providing annual savings of $100,000 or more. Operating savings have improved dramatically, and the system provides real-time cost estimates of actual and potential schedule changes. The system is an excellent example of using advanced techniques for planning and scheduling, as well as using a hierarchical process for disaggregating the production plan.

APPLICATIONS POTENTIAL

In this section, we consider the current situation in advanced production planning. We start by evaluating a rather disappointing history of the application of formal models. One limitation to applying modeling techniques is the data required, a topic we address as well. The section concludes with a prognosis for the future of advanced production planning concepts.

Application of Modeling Techniques

The application of quantitative models has been disappointing at best. The linear decision rule model (LDR) is more than 30 years old, but has seen very limited application. Linear programming models have been used extensively, but generally for firms with relatively homogenous output measures and simple product structures, such as oil refineries and feed mixing plants. Hierarchical production planning, which is relatively new and parallels the managerial organization, has had few applications.

Most interest in modeling techniques has been academic. Firms treating the aggregate production planning problem often combine approaches with long-range planning and budgeting cycles using simplified tabular or graphic methods. We see three key reasons for the lack of demonstrated applications of the theory.

The first reason is few firms actually make aggregate decisions in the way implied by the models. For most firms, the aggregate is merely the sum of several lower-level decisions. Among these decisions are output rates for particular factories, product lines, or even work centers. Management guidance on some overall or aggregate basis tends to be of a general nature, instead of providing a fixed set of constraints within which a process of disaggregation can unambiguously proceed. A related issue is the assumption of homogeneity. Some uniform measure of output makes more sense in a paint factory than in a multiproduct, multiplant firm. Also, there are great differences among workers. Analytical models treat them as equivalent.

A second reason for lack of application may be managerial understanding. It's difficult for many managers to understand quantitative models' analytic underpinnings. Our experience has shown the logic must be transparent to gain wide acceptance. This seems to be particularly true for aggregate planning models.

A final element inhibiting formal approaches' expanded application is the data requirements. Often real-world data don't correspond to model assumptions. In other cases, data required don't exist. This is an area where data bases like those developed for the manufacturing planning and control (MPC) system can help. We devote the next section to these important issues.

Data Issues

A high-quality data base for manufacturing, especially if it's linked to the cost accounting and financial analysis system, can help greatly in gathering data for aggregate capacity analysis. Even so, many problems remain. Looking just at the data requirements for the models raises several issues.

Although of limited applicability in some countries (or companies), one of the most difficult issues is how to estimate hiring and firing costs. Clearly, these costs' most important aspects aren't part of the accounting records. In determining hiring costs, the more easily estimated components are recruiting, interviewing, and training. More difficult estimations are length of time to become fully effective, time to reach required quality levels, and ability to assimilate into the firm's social environment. Firing costs can be extremely difficult to assess, particularly in terms of the influence on morale.

In most of the methods presented, we treated both hiring and firing costs as linear, with no constraints on the number of persons who could be hired or fired in a time period. In some examples, resulting labor force changes simply might not be possible. Moreover, costs of small adjustments are almost surely different from costs of large adjustments.

A somewhat similar problem relates to overtime/undertime costs. Overtime in modest amounts represents a useful means for dealing with short-term capacity problems. However, to go from a 40-hour workweek to a 60-hour workweek is quite severe. Experience indicates there's a reduction in hourly productivity when people work that many hours. The situation is aggravated when the number of weeks worked on overtime increases. Moreover, the ability to vary the workweek at will is clearly very limited. Undertime costs are difficult to measure, especially in terms of long-term influence on worker morale, turnover, and loss of skill.

Inventory carrying costs might be linear over fairly wide ranges, but could change as capital resources are strained or new opportunities are developed. The percentage rate used to represent inventory carrying costs is presumed to include more than the interest charges from bank loans and direct storage costs. Risks of obsolescence, physical deterioration, and having the wrong items in finished goods also have to be considered.

Again we raise the issue of finding a single aggregate output measure. Many firms rather cavalierly convert sales dollars to labor-hours to get such a measure. The assumptions are all sales dollars are equal, any labor-hour can be used to make any sales dollar, and any inventory is, in fact, useful to meet any marketplace demand. In fact, an overall single homogenous capacity measure often isn't a meaningful concept in many firms. For example, the company with several different product lines, particularly if one product is highly labor intensive and another is capital intensive, will have difficulty finding a uniform capacity measure.

The Future

We feel guarded optimism about use of advanced production planning methods. As more and more firms implement MPC systems and continue to improve them, the applicability of more formal production planning becomes clear. This is typically first manifested in basic production planning, where a monthly production plan is determined by a top-management group. As this process is refined, there's a natural tendency to consider formal models to support the effort. The trade-offs are better understood, as are costs to the firm.

Another reason for growing use of advanced production planning models comes from better managerial understanding of underlying methodologies. Most business school students now learn linear programming and other mathematical methods.

Finally, MPC systems in the years ahead will necessarily focus on fast response. High clerical costs of a detailed MPC system with many levels of planning will fall as more firms migrate toward just-in-time (JIT) systems. However, the front-end production planning for these JIT systems remains critically important. Concepts like level schedules, flow rates, and close matching of the market to manufacturing all need effective production planning.

In summary, we feel many firms will be increasingly interested in advanced production planning methods. However, implementing these systems will remain difficult. Lessons learned from application to application indicate we have a long way to go before well-established guidelines emerge.

CONCLUDING PRINCIPLES

This chapter has reviewed formal approaches to aggregate production planning. We stress the following principles:

- The match between the real world and the model should be as close as possible to make it easier to build the credibility necessary to use the model.
- Relatively homogeneous product lines or portions of lines should be used to have a close match between model and reality.
- Hierarchical approaches should be applied by management to match the production planning and disaggregation process to the appropriate organizational entities.
- Investing in training, enhancing data, improving basic MPC practices, and determining clear objectives must all be done before the full potential for using advanced techniques can be realized.
- Management must realize that significant efforts in model formulation, understanding, testing, and explanation all are important to successful applications.
- Advanced techniques must be built on a foundation of good basic practice. Modeling a mess doesn't make it better.

REFERENCES

Bechtold, Stephen E., and Larry W. Jacobs. "Subcontracting, Coordination, Flexibility, and Production Smoothing in Aggregate Planning." *Management Science* 36, no. 11 (November 1990), pp. 352–63.

Bitran, G. D.; E. A. Haas; and A. C. Hax. "Hierarchical Production Planning: A Two-Stage System." *Operations Research,* March–April 1982, pp. 232–51.

Bowman, E. H. "Consistency and Optimality in Managerial Decision Making." *Management Science,* January 1963.

Chung, C., and L. J. Krajewski. "Planning Horizons for Master Production Scheduling." *Journal of Operations Management,* August 1984, pp. 389–406.

Connell, B. C.; E. E. Adam, Jr.; and A. N. Moore. "Aggregate Planning in Health Care Foodservice Systems with Varying Technologies." *Journal of Operations Management* 5, no. 1, 1985.

Gelders, L. F., and L. N. Van Wassenhove. "Hierarchical Integration in Production Planning: Theory and Practice." *Journal of Operations Management* 3, no. 1, 1982.

Gordon, J. R. M. "A Multi-Model Analysis of an Aggregate Scheduling Decision." Ph.D. dissertation, Sloan School of Management, MIT, 1966 (published in Elwood S. Buffa, *Production-Inventory Systems: Planning and Control.* Homewood, Ill.: Richard D. Irwin, 1968, pp. 168–69).

Holt, C. C.; F. Modigliani; J. F. Muth; and H. A. Simon. *Planning Production, Inventories, and Workforce.* New York: Prentice-Hall, 1960, p. 16.

Holt, J. A. "A Heuristic Method for Aggregate Planning: Production Decision Framework." *Journal of Operations Management* 2, no. 1 (October 1981), pp. 43–51.

Lee, W. B.; E. Steinberg; and B. M. Khumawala. "Aggregate versus Disaggregate Production Planning: A Simulated Experiment Using LDR and MRP." *International Journal of Production Research* 21, no. 6 (1983), pp. 797–811.

Leong, G. Keong; Michael Oliff; and Robert E. Markland. "Improved Hierarchical Production Planning." *Journal of Operations Management* 8, no. 2 (April 1989), pp. 90–114.

Mackulak, G. T.; C. L. Moodie; and T. J. Williams. "Computerized Hierarchical Production Planning in Steel Manufacturing." *International Journal of Production Research* 18, no. 4 (1980), pp. 455–65.

Meal, H. L., and D. C. Whybark. "Material Requirements Planning in Hierarchical Planning Systems." *International Journal of Production and Operations Management* 25, no. 7 (1987), pp. 947–56.

Nelleman, David O. "Production Planning and Master Scheduling. Management's Game Plan." *APICS 22d Annual Conference Proceedings,* 1979. pp. 166–68.

Oden, Howard W. "Hierarchical Production Planning with Variable Planning Periods." DBA dissertation, Boston University, 1986.

Oliff, Michael D. "A Discrete Production Switching Rule for Aggregate Planning." *Decision Sciences* 18, no. 4 (Fall 1987), pp. 582–97.

Oliff, M. D., and E. E. Burch. "Multiproduct Production Scheduling at Owens-Corning Fiberglas." *Interfaces,* September–October 1985, pp. 25–34.

Taubert, W. H. "A Search Decision Rule for the Aggregate Scheduling Problem." *Management Science* 14, no. 6 (February 1968).

Tersine, R. J.; W. W. Fisher; and J. S. Morris. "Varying Lot Sizes as an Alternative to Undertime and Days Off in Aggregate Scheduling." *International Journal of Production Research* 24, no. 1 (1986), pp. 97–106.

Vickery, S. K., and R. E. Markland. "Integer Goal Programming for Multistage Lot Sizing: Experimentation and Implementation." *Journal of Operations Management* 5, no. 2, 1985.

Vollmann, T. E. "Capacity Planning: The Missing Link." *Production and Inventory Management Journal,* 1st quarter 1973.

DISCUSSION QUESTIONS

1. What is the least certain of the data inputs to aggregate production planning models? What can be done about this uncertainty?
2. What does a mixed integer programming formulation add over a linear programming approach?
3. What concerns should managers have about using aggregate production planning models?
4. How does the hierarchical approach match actual managerial practice?
5. How would you explain the Owens-Corning Fiberglas model to the workers?
6. Why have there been only limited documented successes of model applications to MPC activities?
7. In the disaggregation process, errors in the product mixes can occur. How do these creep in?
8. Why must we have effective MPC systems in place before we can anticipate adoption of advanced production planning concepts?

PROBLEMS

1. The Seymore Bikini Manufacturing Company of Boise, Idaho, has developed the following demand forecast for next year's bikinis:

Quarter	Sales
1	5,000
2	10,000
3	8,000
4	2,000

On January 1, there are 1,000 units in inventory. The firm has prepared the following data:

Hiring cost per employee = $200.
Firing cost per employee = $400.
Beginning work force = 60 employees.
Inventory carrying cost = $2 per unit per quarter of ending inventory.
Stockout cost = $5 per unit.
Regular payroll = $1,200 per employee per quarter.
Overtime cost = $2 per unit.
Each employee can produce 100 units per quarter. Demand not satisfied in any quarter is lost.

Using linear programming:
a. How much will Seymore produce during each quarter?
b. What's the total budget Seymore's plan will require for the next year?

2. Zebra Enterprises has collected the following information on one of its major products. (Use a spreadsheet to model a and b below.)

Most efficient production rate = 1,400 units per period.
Production change costs = $10 per unit of change (from 1,400 units/period).
Inventory costs = $3 per unit per period (on closing inventory balance).
Shortage costs = $7 per unit per period.
Beginning inventory = 200 units.

Period	Demand (units)
I	2,000
II	1,500
III	1,300
IV	1,800

a. Based on the preceding demand schedule, calculate a level production schedule that yields zero inventory at the end of period IV.
b. Calculate the total costs associated with the production schedule in part a.

3. The Kew Toy Company's production manager wonders whether to produce at a level production rate or a rate that matches sales each quarter. His analysis of company operations yields the following information:

Beginning employment level = 10 employees.
Beginning inventory = 0.
Hiring cost = $10 per employee.
Firing cost = $5 per employee.
Production per employee = 10 units per quarter.
Inventory carrying costs = $1 per unit per quarter (on ending inventory).

Target inventory at the end of the fourth quarter = 0.

Target employment level at the end of the fourth quarter equals the planned employment level specified by the plan under consideration.

Quarter	Sales forecast (units)
1	50
2	80
3	120
4	150

a. Which of the two policies would result in the lowest annual cost?

b. What's the total annual cost of the policy adopted in part a?

c. What production rate should he use each quarter?

d. Formulate this problem so it can be solved using linear programming.

4. The Happy Hatrack Company had been experimenting with aggregate production plans for several years. As the firm started to lay out plans for next year, management decided to use a two-month planning horizon to reduce computation cost. The firm's basic data were:

a. Employment level for this December = 10 people.

b. Demand forecast for January of next year = 100.

c. Demand forecast for February of next year = 120.

d. Inventory level planned for the end of this December = 0.

e. Backorders aren't allowed.

f. Desired February ending inventory = 0.

g. Regular time per month = $1,000 per worker-month.

h. Production = 10 units/person/month.

i. Overtime premium = 50 percent of regular time.

j. Overtime limit = 25 percent of regular time per person per month.

k. Inventory carrying cost = $100/month/unit (on the average per month).

l. Hiring (or firing) cost = $1,000 per person.

Use a spreadsheet model to compare the cost of a fixed employment production plan (i.e., 10 people) that uses overtime to meet demand to a level production plan that uses *no* overtime to meet production.

5. Consider the following information for the Big Red Express Company:

Production change cost: $15/unit of change (from a base of 1,300 units).

Most efficient production rate: 1,300 units per period.

Maximum amount of production capacity: 1,500 units per period.

Inventory costs: $2/unit per period on the ending inventory.

Shortage costs: $10/unit per period.

Subcontracting costs: $12/unit (maximum of 300 units per period).

Beginning inventory: 400 units.

Period	Demand (units)
1	2,000
2	1,500
3	2,000
4	1,700

a. As production manager of Big Red Express, you're in charge of meeting the preceding demand schedule for periods 1 through 4. President Rizzo specifies that you keep your production plan level for all four months. In addition, Rizzo limits the total amount of production per month to 1,500 units. Also, since the market Big Red Express competes in is very competitive, planned stockouts are *not* allowed. Therefore, you must decide if and when to subcontract. Maximum amount of subcontracting available in any one period is 300 units. Ending inventory at the end of period 4 should be zero. Use a spreadsheet to determine your plan's cost.

b. Assuming the same facts as in part a, except that *no* subcontracting is allowed, could you follow a pure chase strategy and still meet demand? Why or why not?

6. On December 31, the ABC Company forecast its next year's sales to be:

Quarter	Sales forecast
1st	9,000 units
2nd	12,000
3rd	16,000
4th	12,000
	49,000

Currently, the firm has 12 employees, each producing 1,000 units per quarter and earning $2,000 per quarter. The firm estimates its inventory carrying cost to be $2 per unit of ending inventory per quarter and its hiring or layoff costs to be $1,600 per employee. The firm could increase production by working overtime, but overtime work is limited to 25 percent of the regular production rate (or 250 units per employee per quarter). Overtime work is paid at the rate of 1.5 times the regular pay rate. Idle time costs the firm $4.16 per hour. (Assume there are 60 eight-hour days per quarter.) The company currently has an inventory of 1,000 units and wishes to have an ending inventory of 1,000 units at the end of the year. The company doesn't plan to incur inventory shortages.

a. Develop the total incremental cost expression for this problem.

b. Formulate this problem for solution, using linear programming.

c. What's the total incremental cost of a production plan that assumes a constant production rate (include both regular and overtime production) and level work force (assume 12 workers) each quarter?

   d. Compare the cost of the production plan in c to one based on a production rate equal to sales in each quarter. Which has the lowest cost?

7. The production manager at the Boston Paint Company is preparing production and inventory plans for next year. He has the following data concerning his firm:

Quarter	Sales forecast
1st	3,000 units
2nd	1,800
3rd	2,400
4th	3,500

Current inventory level = 300 units.
Current employment level = 600 people.
Production rate last quarter = 2,400 units (4 units/employee/quarter).
Inventory carrying cost = $20/unit/quarter (on ending inventory).
Hiring cost = $200/employee hired.
Layoff cost = $200/employee laid off.
Regular time production cost per unit = $320/unit.
Cost of overtime = $60/unit.
Desired closing inventory level = 100 units (minimum).

The production manager sees no equipment capacity limitations during the next two years. Employees are hired or laid off only at the beginning of each quarter.

   a. Formulate a linear programming production planning model for this company, in sufficient detail so the production manager can solve it to determine his production plan.

   b. After considering your recommendations, suppose the production manager revises his estimate of: (1) hiring and layoff costs and (2) overtime and idle time costs. The new costs are:

Cost of changing the employment level (y_1):

$$y_1 = \$200 \, (Y_n - Y_{n-1})^2$$

Cost of producing on overtime or permitting idle time (y_2):

$$y_2 = \$60(S_n - Ax_n)^2$$

where:

   x_n = The number of people employed in quarter n.
   S_n = The planned production rate in quarter n.
   A = The number of units produced per employee per quarter.

Given this new information, how would you change your recommendations to the production manager regarding formation of a production planning model?

8. The Old-n-Corny company produces great shower mats. Demand is seasonal, however. (Apparently people only get dirty at certain times of the year.) The company has a three-shift production capability and can staff 1, 2, or 3 shifts at the beginning of each quarter. The company can produce 10,000 mats a quarter for each shift staffed. At the moment, there are 2,000 mats in inventory and O-n-C would like to keep 2,000 as a buffer against uncertainty. The company is planning production for the next year and has compiled the following demand forecast. Determine production switching heuristics for the number of shifts to schedule each quarter.

	Quarter			
	1	2	3	4
Forecast	10,000	20,000	30,000	10,000

9. Apply the following production switching rules to the forecast given below. (You might want to use a spreadsheet.) What inventory do you end up with each quarter?

Current inventory = 1,250.
Max. inventory target = 1,000.
Min. inventory target = 1,000.
Rules: Production(i) = 1,000 if (Forecast(i) − Inventory(i − 1) + Min. inventory target) ≥ 850.
Production(i) = 500 if (Forecast(i) − Inventory(i − 1) + Max. inventory target) < 850.

	Quarter			
	1	2	3	4
Forecast	750	800	1,100	700

10. The Slick Switchers had developed a set of production switching rules for scheduling monthly production. They were quite anxious to try them out in a spreadsheet model. They had three shifts available, each of which could produce 20 units a month. Minimum target inventory was 20 units; maximum was 40. Current inventory was 35 units. What's the effect of the rules applied to the following forecast? What's the impact of reducing the maximum target inventory to 20 units?

Rules: Production(i) = 60 if (Forecast(i) − Inventory(i − 1) + **Min.** inventory target) > 60.

Production(i) = 20 if (Forecast(i) − Inventory(i − 1) + **Max.** inventory target) < 20.

Production(i) = 40 otherwise.

						Month						
	1	2	3	4	5	6	7	8	9	10	11	12
Forecast	25	30	45	55	40	30	40	50	65	60	30	20

11. The Pickwick Company's production director and marketing director are currently negotiating next year's production and sales plans. The marketing director confidently forecasts a major sales increase as follows:

	Quarter			
	1	2	3	4
Sales forecast (in units)	950	1,200	1,420	1,630

a. Assume that the beginning inventory is 200 units and the target ending inventory for the fourth quarter is 200 units. What production each quarter will provide a level production plan?
b. If average variable cost per unit for items produced by Pickwick is $500, what will be the monetary value of the finished-goods inventory at the end of the second quarter if it uses a level production plan?
c. If average variable labor cost per unit for the items produced by Pickwick is $200 and average wage rate is $5 per hour, what's the quarterly labor budget in hours and in dollars for a level production plan?

12. The Columbia Manufacturing Company's sales manager has prepared the following sales forecast for next year:

Sales forecast (in units)				
	Quarter			
Product family	1	2	3	4
A	250	350	500	125
B	150	225	360	75

The production manager has supplied the following data:

Product Family A:

Two direct labor-hours per unit of product.
Setup cost equals $700 per changeover from B.
Setup time equals eight direct labor-hours per changeover.
Inventory carrying cost per quarter equals $1.50 per one direct labor-hour of work left in inventory at the end of the quarter.
Materials cost equals $70 per direct labor-hour of production.
The beginning inventory equals zero.

Product Family B:

One direct labor-hour per unit of product.
Setup cost equals $600 per changeover from A.
Setup time equals 10 direct labor-hours per changeover.
Inventory carrying cost per quarter equals $1 per one direct labor-hour of work in inventory at the end of the quarter.
Materials cost equals $50 per direct labor-hour of production.
The beginning inventory equals zero.

In addition, these factors apply:

Hiring cost per employee equals $1,500.
Firing cost per employee equals $500.
Overtime cost per direct labor-hour equals $15.
Regular time cost per direct labor-hour equals $10.
The maximum number of regular time hours to be worked per employee per quarter equals 520.
Employees may not work more than 25 percent of the regular time hours on overtime each quarter.

a. Formulate this problem for solution using mixed integer programming.
b. Why is linear programming not appropriate for solving this production planning problem?

13. The Columbia Manufacturing Company's production planning manager (in Problem 12) has completed work on the firm's production plan for the coming year. He's now preparing the master production schedule for the next two weeks indicating the lot size and timing for the firm's individual products. Selected data concerning Product Family A's production plan for the next two weeks are as follows:

Item	Plan for next two weeks
Production level	38 direct labor-hours during the 2 weeks
Closing inventory	5 direct labor-hours
Overtime hours	0
Family setup made	Yes

Additional information:

Product Family A has two end products. (Seventy-five percent of the demand is for product 1; the remainder is for product 2.)

The plant consists of a single machining center and each end product is processed in one operation on the machining center.

Cost to backorder one hour of production of either end product is $100.

Production lead time for each end product is zero.

    a. Formulate this problem for solution using linear programming. Assume that a weight of 1.0 is assigned to each deviation variable.

    b. What are the advantages and limitations of solving the master production scheduling problem separately from the production planning problem?

14. The sales manager at the Universal Manufacturing Company has prepared the following sales forecast for next year:

Sales forecast (in units)				
			Quarter	
Product family	*1*	*2*	*3*	*4*
A	3,500	6,000	4,000	1,300
B	1,200	2,000	2,800	3,600

The production manager has supplied the following data:

Product Family A:

Three direct labor-hours are needed per unit of product.

Setup cost equals $3,000 per changeover from product family B.

Setup time equals 16 direct labor-hours per changeover.

Inventory carrying cost per quarter equals $.50 per direct labor-hour of work in inventory at the end of the quarter.

Materials cost equals $130 per direct labor-hour of production.

Beginning inventory equals zero.

Product Family B:

Two direct labor-hours are needed per unit of product.

Setup cost equals $1,800 per changeover from A.

Setup time equals 24 direct labor-hours per changeover.

Inventory carrying cost per quarter equals $.33 per direct labor-hour of work left in inventory at the end of the quarter.

Materials cost equals $105 per direct labor-hour of production.

Beginning inventory equals zero.

Additional information:

Hiring cost per employee equals $2,000.

Firing cost per employee equals $1,500.

Overtime cost per direct labor-hour equals $18.

Regular time cost per direct labor-hour equals $12.

The maximum number of regular time hours to be worked per employee per quarter equals 600.

Employees may not work more than 20 percent of the regular time hours on overtime each quarter.

a. Formulate this problem for solution using mixed integer programming.

b. Why is linear programming not appropriate for solving this production planning problem?

15. The production planning manager at the Universal Manufacturing Company (in Problem 14) has completed the firm's production plan for the coming year. He's now preparing the master production schedule for the next two weeks indicating lot size and timing for the firm's individual products. Selected data on the production plan for the next two weeks for product family B is as follows:

Item	Plan for next two weeks
Production level	370 direct labor-hours during the 2 weeks
Closing inventory	25 direct labor-hours
Overtime hours	0
Family setup made	yes

Additional information:

There are two end products in product family B. (Sixty percent of the demand is for product 1; the remainder is for product 2.)

The plant consists of a single machining center and each end product is processed in one operation on the machining center.

The cost to backorder one hour of production of either end product is $50.

Production lead time for each end product is zero.

a. Prepare a master production schedule for the next two weeks that conforms to the production plan.

b. What is the cost of labor, material, and inventory for this master production schedule?

Chapter 16

Short-Term Forecasting Systems

Forecasts of demand are one important input to manufacturing planning and control (MPC) systems. In this chapter, we treat short-term forecasting for individual items. Applying effective forecasting systems will result in low-cost routine forecasts and a set of monitors to indicate when forecasting problems are incurred. These forecasts of end items, spare parts, and other independent demand should be a part of the front-end modules of the MPC system. A key objective is to provide one, and only one, source for forecast data; this source is to be unbiased and usable by all areas in the firm.

Forecasts used for production and resource planning can be of many types, including subjective estimates, econometric models, and Delphi techniques. A detailed exposition of all these is a book in itself. Although many techniques could be applied to forecasting demand for individual end items, we focus here on short-term forecasts based on observations of past actual demand. The chapter is organized around seven topics:

- The forecasting problem: How is the forecasting problem defined for manufacturing planning and control purposes?
- Basic forecasting techniques: What are the basic techniques for forecasting short-term demand?
- Enhancing the basic exponential smoothing model: How can trend, seasonality, and other kinds of information be incorporated?
- Focus forecasting: What is the focus forecasting methodology and how does it produce forecasts?
- Comparisons of methods: Which forecasting techniques work best under which conditions? What are the lessons for managers?
- Using the forecasting system: How do we select initial forecasting parameter values and monitor forecast results?
- Forecasting in industry: How have these techniques been put into practice?

This chapter has very close linkages to Chapter 8 on demand management. Forecasting activities are accomplished in the demand management module of the MPC system. Chapter 17, on independent demand inventory management, and Chapter 19, on distribution requirements planning, both presume the existence of effective, routine forecasting procedures.

THE FORECASTING PROBLEM

In this chapter, we deal primarily with short-term forecasting techniques of the type most useful to routine decision making in manufacturing planning and control. However, other decision problems both within manufacturing and in other functional areas of the firm require different approaches to forecasting. First we briefly discuss these other situations before delving more deeply into developing short-term forecasting techniques for manufacturing. We also treat a vital forecasting question: How to evaluate a forecasting technique's performance.

Forecasting Perspectives

Managers need forecasts for a variety of decisions. Among these are long-run decisions involving such things as constructing a new plant, determining the type and size of aircraft for an airline fleet, extending a hotel's guest facilities, or changing the curriculum requirements in a university. Generally, these longer-run decisions require forecasts of aggregate levels of demand, utilizing such measures as annual sales volume, expected passenger volume, number of guest nights, or total number of students enrolled. In a sense, this is fortunate, since aggregate levels of an activity can usually be forecast more accurately than individual activities. As an example, a university administration probably has a pretty good estimate of how many students will enroll next term, even though the enrollment forecast for an elective course may be off by a considerable amount.

For aggregate forecasts, we may be able to use causal relationships and the statistical tools of regression and correlation. For example, household fixture sales are closely related to housing starts. The number of vacationers at resorts is related to the economy's net disposable income level. In such instances, the relationship may be statistically modeled, thereby providing the basis for a forecasting procedure. Managerial insight and judgment are also used extensively in developing aggregate forecasts of future activities for long-run decisions. Both statistical and qualitative forecasting methods can also be applied for medium-run decisions, such as the annual budgeting process. It's tempting to classify forecasting techniques as long-run or short-run, but this misses the point of developing and using techniques appropriate to the decision and situation.

Throughout this chapter we'll look at fairly mechanical procedures for making forecasts. Specifically, we'll look at models for "casting forward" historical information to make the "fore cast." Implicit in this process is a belief that past conditions that produced the historical data won't change. Although the procedures we'll develop are mechanical, we shouldn't draw from this the impression that managers always rely exclusively on past information to estimate future activity. In the first place, in certain instances, we simply have no past data. This occurs, for example, when a new product is introduced, a future sales promotion is planned, a new competitor appears, or new legislation affects our business. These circumstances all illustrate the need for managerial review and modification of the forecast where special knowledge should be taken into account. Don't lose sight of this as we move into this chapter's technical aspects.

We'll largely focus our attention on techniques for converting past information into forecasts. These are often statistical techniques. We'll also use statistical methods for evaluating forecasts' quality. The procedures are often called **statistical forecasting procedures.**

Forecast Evaluation

Ultimately, of course, the quality of any forecast is reflected in the quality of the decisions based on the forecast. This leads to suggesting the ideal comparison of forecasting procedures would be based on the costs of producing the forecast and the value of the forecast for the decision. From these data, the appropriate trade-off between the cost of developing and the cost of making decisions with forecasts of varying quality could be made. Unfortunately, neither cost is easily measured. In addition, such a scheme suggests that a different forecasting procedure might be required for each decision, an undesirably complex possibility. As a result of these complications, we rely on some direct measures of forecast quality.

One important criterion for any forecast procedure would be a low cost per forecast. For many manufacturing planning and control problems, we need to make forecasts for many thousands of items on a weekly or monthly basis; the result is the need for a simple, effective, low-cost procedure. Unlike the rare occasions when the decision is to add more factory capacity, routine short-term decisions are made frequently for many items, and can't require an expensive, time-consuming forecasting procedure. Moreover, since the resultant decisions are made frequently, any error in one forecast can be compensated for in the decision next time. However, a large expenditure for an aggregate long-term forecast may well be justified for making a factory capacity decision.

At one time, the vast data storage requirements and computer time needed for producing forecasts for several thousands or tens of thousands of items were major concerns. As computer time and storage costs decrease, this aspect of forecast procedure evaluation becomes less important. Nevertheless, we'll concentrate in this chapter on procedures that have the attribute of simplicity, are easy to use, and have low computer time/storage requirements.

For any forecasting procedure we develop, an important characteristic is honesty, or lack of **bias**; that is, the procedure should produce forecasts that are neither consistently high nor consistently low. Forecasts shouldn't be overly optimistic or pessimistic, but, rather, should tell it like it is. Since we're dealing with projecting past data, lack of bias means smoothing out past data's randomness so overforecasts are offset by underforecasts. To measure bias, we'll use the **mean error** as defined by Equation (16.1). In this equation, the forecast error in each period is actual demand in each period minus forecast demand for that period. Figure 16.1 shows an example calculation of bias:

$$\text{Mean error (bias)} = \frac{\sum_{i=1}^{n}(\text{Actual demand}_i - \text{Forecast demand}_i)}{n} \tag{16.1}$$

where:

i = Period number.
n = Number of periods of data.

As Figure 16.1 shows, when forecast errors tend to cancel one another out, the measure of bias tends to be low. Positive errors in some periods are offset by negative errors in others, which tends to produce an average error or bias near zero. In Figure 16.1, there's a bias and the demand was overforecast by an average of 25 units per period for the four periods.

Having unbiased forecasts is important in manufacturing planning and control, since the estimates, on average, are about right. But that's not enough.

FIGURE 16.1 Example Bias Calculation

		Period (i)		
	1	2	3	4
(1) Actual demand	1,500	1,400	1,700	1,200
(2) Forecast demand	1,600	1,600	1,400	1,300
Error (1) − (2)	−100	−200	300	−100

$$\text{Bias} = \sum_{i=1}^{4}\text{error}_i/4 = (-100 - 200 + 300 - 100)/4$$
$$= -100/4 = -25. \tag{16.1}$$

We still need to be concerned with the errors' magnitude. Note, for the example in Figure 16.1, we obtain the identical measure of bias if actual demand for the four periods had been 100, 100, 5,500, and 100, respectively. (This is shown as part of the calculations in Figure 16.2.) However, the individual errors are much larger, and this difference would have to be reflected in buffer inventories if we were to maintain a consistent level of customer service.

Let's now turn to a widely used measure of forecast error magnitude, the **mean absolute deviation (MAD)**. Figure 16.2 shows example calculations.

The formula is

$$\text{Mean absolute deviation (MAD)} = \frac{\sum_{i=1}^{n} |\text{Actual demand}_i - \text{Forecast demand}_i|}{n} \tag{16.2}$$

FIGURE 16.2   Sample MAD Calculations

	Period (i)			
	1	2	3	4
(1) Actual demand	1,500	1,400	1,700	1,200
(2) Forecast demand	1,600	1,600	1,400	1,300
Error (1) − (2)	−100	−200	300	−100

$$\text{MAD} = \sum_{i=1}^{4} |\text{error}_i|/4$$
$$= (|-100| + |-200| + |300| + |-100|)/4 = 175 \tag{16.2}$$

	Period (i)			
	1	2	3	4
(1) Actual demand	100	100	5,500	100
(2) Forecast demand	1,600	1,600	1,400	1,300
Error (1) − (2)	−1,500	−1,500	4,100	−1,200

$$\text{Bias} = \sum_{i=1}^{4} \text{error}_i/4 = (-1500 - 1500 + 4100 - 1200)/4$$
$$= -100/4 = -25 \tag{16.1}$$

$$\text{MAD} = \sum_{i=1}^{4} |\text{error}_i|/4$$
$$= (|-1,500| + |-1,500| + |4,100| + |-1,200|)/4$$
$$= 8,300/4 = 2,075 \tag{16.2}$$

where:

i = Period number.
n = Number of periods of data.
$|x|$ = Absolute value of x.

The mean absolute deviation expresses the size of the average error irrespective of whether it's positive or negative. It's the combination of bias and MAD that allows us to evaluate forecasting results. Bias is perhaps the most critical, since we can compensate for forecast errors through safety stocks, expediting, faster delivery means, and other kinds of responses. MAD indicates the expected compensation's size (e.g., required safety stock). However, if a forecast is consistently lower than demand, the entire material-flow pipeline will run dry; it will be necessary to start over again with raw materials. Other problems arise for a consistently high forecast. The great advantage of the techniques described in this chapter is they tend to be unbiased. Moreover, routine monitoring techniques identify bias when it's present. Judgmental forecasts, such as those made by marketing groups, are often biased because forecasting incorporates other goals (e.g., stimulate the sales force). The key is to clearly separate the *process* of forecasting from the *use* of forecasting. The process's goals are no bias and minimum MAD. What's *done* with the forecast is another issue.

Before turning to the techniques, one other relationship must be made. MAD measures error or deviation from an expected result (the forecast). The best-known measure of deviation or dispersion from statistics is the **standard deviation.** When errors are distributed normally, the standard deviation of the forecast errors is arithmetically related to MAD by Equation (16.3):

$$\text{Standard deviation of forecast errors} = 1.25 \text{ MAD}. \qquad (16.3)$$

BASIC FORECASTING TECHNIQUES

Now that we've identified the forecasting procedures' objectives, let's turn to some procedures that meet the objectives. In this section, we'll introduce some basic concepts that lie behind two very common short-term forecasting techniques: moving averages and exponential smoothing. Before we discuss these techniques, however, we present an example problem that allows us to continually relate the concepts and formulas to a real-world context.

Example Forecasting Situation

Enrique Martinez manages Panchos, a restaurant in a large hotel near Chicago's Loop. Panchos caters to both hotel guests and local street traffic. Its reputation has been growing. A few months ago, its capacity was expanded.

Enrique was studying ways to improve and routinize his decisions for managing the restaurant operations. He chose three situations to study in some detail: a new contract offer from his linen service, the trend in tequila-based drinks, and his twice weekly orders to the local wholesale grocery distributor. All these decisions depended on his ability to forecast demand. Although the decision in each situation involves placing orders, each presents quite a different forecasting problem. With the problem variety he'd chosen, Enrique felt he had a good basis for studying forecasting methods for Panchos.

The first situation involved a new contract proposal for tablecloths and napkins from the linen service supplying the restaurant. The linen supply firm's owner offered an attractive discount if Enrique would prespecify the quantity he wanted each week, rather than continuing the current practice whereby the linen supply firm had to bring enough clean linen to replace whatever quantity of dirty linen there might be each week. Enrique collected data on the number of tables served during the past nine weeks to determine how to forecast his linen needs. Figure 16.3 summarizes these data.

Before going on to the forecasting techniques, let's reiterate a point: We are *not* presently dealing with the decision of how to order tablecloths, how many to hold as safety stock, or any other decision. We're simply trying to forecast demand. The premise is that a *good* forecast will allow for better decisions, but we aren't combining the act of forecasting with any decision.

Orders for tequila-based drinks in the lounge were growing rapidly, creating a problem in determining how much tequila to order from the supplier each week. Enrique collected the number of tequila-based drinks served in each of the past nine weeks, as we see in Figure 16.4.

The final issue Enrique chose to review was how much food to order for the twice weekly delivery from the local wholesale grocery distributor. The distributor delivered on Tuesday (Panchos was closed on Mondays) and Fri-

---

FIGURE 16.3   Number of Tables Served during the Past Nine Weeks

	Week number								
	24	*25*	*26*	*27*	*28*	*29*	*30*	*31*	*32*
Tables served	1,600	1,500	1,700	900	1,100	1,500	1,400	1,700	1,200

FIGURE 16.4   Number of Tequila-Based Drinks Served during the Past Nine Weeks

	Week number								
	24	*25*	*26*	*27*	*28*	*29*	*30*	*31*	*32*
Number of drinks	16	71	40	85	196	351	254	261	364

day mornings. Deliveries consisted of canned goods, staples, condiments, and so on for use during the next three days. There was a substantially smaller volume of business during the first three days of the business week than during the last three. To get an idea of the pattern, Enrique tracked the usage of number 10 cans of vegetables for each of the six three-day periods as Figure 16.5 shows.

Figure 16.6 plots the number of tables served from the data in Figure 16.3. The number appears relatively stable (for the last nine weeks for which we have data) and seems to fluctuate randomly about some central value. If we were interested in using these past data to forecast the number of tables to

FIGURE 16.5   Number 10 Cans of Vegetables Used during the Past Six Three-Day Periods

	F–S–Sun	T–W–Th	F–S–Sun	T–W–Th	F–S–Sun	T–W–Th
Week number	29	30	30	31	31	32
Number of cans	48	35	47	30	51	37

FIGURE 16.6   Plot of Number of Tables Served at Panchos during the Past Nine Weeks

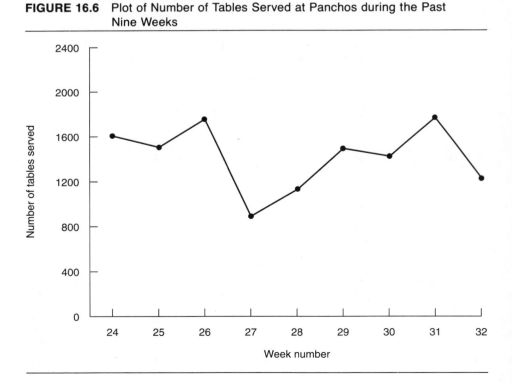

be served in future weeks, a tempting procedure would be to simply draw a line through the data points and use that line as our estimate for week 33 and subsequent weeks. This would estimate **average** or **expected demand** in future weeks; the process of drawing that line is an **averaging** or **smoothing** process. The process removes fluctuations around the line and focuses on the underlying average. It's this smoothing process that provides the basis for the techniques we now address.

Moving Averages

Rather than draw a line through the points in Figure 16.6 to find the average, we could simply calculate the arithmetic average of the nine historical observations. Since we're interested in averaged past data to project into the future, we could use an average of all past demand data available for forecasting purposes. There are several reasons, however, why this may not be a desirable way of smoothing. In the first place, there may be so many periods of past data that storing them all is an issue. Second, often the most recent history is most relevant in forecasting short-term demand in the near future. Recent data may reveal current conditions better than data several months or years old. For these reasons, many firms use the concept of a **moving average** for forecasting demand.

The moving average model for smoothing historical demand proceeds, as the name implies, by averaging a selected number of past periods of data. The average moves because a new average can be calculated whenever a period's demand is determined. Whenever a forecast is needed, the most recent past history of demand is used to do the averaging. Equation (16.4) shows the model for finding the moving average. The equation shows the moving average forecast always uses the most recent n periods of historical information available for developing the forecast. Notice the moving average is the forecast of demand for the next and subsequent periods. This timing convention needs to be clearly understood. We are at the end of period t; forecasts are made for periods $t + 1$, or $t + X$ periods into the future. Forecasts are *not* made for period t—that period's demand is known. Figure 16.7 shows sample calculations for the number of tables served.

$$\text{Moving average forecast (MAF)} \atop \text{at the end of period } t\text{: MAF}_t = \sum_{i=t-n+1}^{t} \text{Actual demand}_i/n \quad (16.4)$$

where:

i = Period number.

t = Current period (the period for which the most recent actual demand is known).

n = Number of periods in the moving average.

FIGURE 16.7   Example Moving Average Calculations

	Period					
	27	*28*	*29*	*30*	*31*	*32*
Actual demand	900	1,100	1,500	1,400	1,700	1,200

6-period MAF made at the end of period 32 $= \sum\limits_{27}^{32}$ Actual demand/6

$$= (900 + 1,100 + 1,500 + 1,400$$
$$+ 1,700 + 1,200)/6 = 1,300 \qquad (16.4)$$

3-period MAF made at the end of period 32 $= \sum\limits_{30}^{32}$ Actual demand/3

$$= (1,400 + 1,700 + 1,200)/3 = 1,433 \qquad (16.4)$$

The Basic Exponential Smoothing Model

You'll note the moving average model does smooth the historical data, but it does so with an equal weight on each piece of historical information. Thus, if we were at the end of period 29 with that demand (1,500) known, and demands for periods 30 and beyond unknown, the three-period moving average forecast for period 30 would be (900 + 1,100 + 1,500/3 = 1,167). At the end of period 30, the forecast for period 31 would be (1,100 + 1,500 + 1,400/3 = 1,333). If we look at a single period's demand, such as the 1,500 in period 29, it's used *only* for forecasts made at the end of periods 29, 30, and 31. In each case, the 1,500 has a weight in the forecast of one third. For forecasts made before the end of period 29 or after the end of period 32, this piece of demand data has no weight.

The **exponential smoothing** model for forecasting doesn't eliminate *any* past datum, but so adjusts the weights given to past data that older data get increasingly less weight (hence the name *exponential smoothing*). The basic idea is a fairly simple one and has a great deal of intuitive appeal. Each forecast is based on an average that's corrected each time there's a forecast error. For example, if we forecast 90 units of demand for an item in a particular period and that item's actual demand turns out to be 100 units, an appealing idea would be to increase our forecast by some portion of the 10-unit error in making the next period's forecast. In this way, if the error indicated demand was changing, we would begin to change the forecast. We may not want to incorporate the entire error (i.e., add 10 units), since the error may have just been due to random variations around the mean. The proportion of the error that will be incorporated into the forecast is called the *exponential smoothing constant* and is identified as α. The model for computing the new average appears in Equation (16.5) as we've just described it. Equation (16.6) gives the most common computational form of the exponentially smoothed average. The new exponentially smoothed average is again the forecast for the next and subsequent periods. The same timing convention is used; that is, the forecast is made at the end of period t for period $t + X$ in the future. Figure 16.8 shows example calculations for the number of tables served.

Exponential smoothing forecast (ESF) at the end of period t:

$$\text{ESF}_t = \text{ESF}_{t-1} + \alpha\,(\text{Actual demand}_t - \text{ESF}_{t-1}) \qquad (16.5)$$
$$= \alpha\,(\text{Actual demand}_t) + (1 - \alpha)\text{ESF}_{t-1} \qquad (16.6)$$

where:

α = The smoothing constant ($0 \le \alpha \le 1$).

t = Current period (the period for which the most recent actual demand is known).

ESF_{t-1} = Exponential smoothing forecast made one period previously (at the end of period $t - 1$).

Let's compare exponential smoothing and moving average forecasting procedures. Figure 16.9 compares a five-period MAF with an ESF with $\alpha = .3$. In preparing the forecast for period 28, the five-period MAF would apply a 20 percent weight to each of the five most recent actual demands. The ESF model (with $\alpha = .3$) would apply a 30 percent weight to the actual demand in period 27 as seen here:

$$\text{ESF}_{27} = \text{Period 28 forecast} = .3\,(\text{period 27 actual demand}) + .7\,(\text{ESF}_{26}).$$

By looking at the ESF for period 27, made at the end of period 26 (i.e., ESF_{26}), we see it was determined as

$$\text{ESF}_{26} = .3\,(\text{period 26 actual demand}) + .7\,(\text{ESF}_{25}).$$

FIGURE 16.8 Example Exponential Smoothing Calculations

	Period	
	27	28
Actual demand	900	1,100

Assume:
ESF_{26} = Exponential smoothing forecast made at the end of period 26 = 1,000,
$\qquad \alpha = .1$
ESF_{27} (made at the end of period 27 when actual demand for period 27 is known but actual demand in period 28 is not known) =

$1,000 + .1(900 - 1,000) = 990$	(16.5)
$.1(900) + (1 - .1)1,000 = 990$	(16.6)
$\text{ESF}_{28} = .1(1,100) + (1 - .1)990 = 1001$	(16.6)

FIGURE 16.9   Relative Weights Given to Past Demand by a Moving Average and Exponential Smoothing Model

	Period								
	20	21	22	23	24	25	26	27*	28
5-period MAF weights	0%	0%	0%	20%	20%	20%	20%	20%	—
ESF weights (α = .3)	2%	4%	5%	7%	10%	15%	21%	30%	—

*Forecast made at the end of period 27.

By substitution, ESF_{27} can be shown to be

$$ESF_{27} = .3 \text{ (period 27 actual demand)}$$
$$+ .7 \, [.3 \text{ (period 26 actual demand)} + .7 \, (ESF_{25})].$$

This results in a weight of .21 (.7 × .3) being applied to the actual demand in period 26 when the forecast for periods 28 and beyond is made at the end of period 27. By similar substitution, we can derive the entire line for the ESF weights in Figure 16.9.

Figure 16.9 shows, for the forecast made at the end of period 27, 30 percent of the weight is attached to actual demand in period 27, 21 percent for period 26, and 15 percent for period 25. The sum of these weights, 66 percent, is the weight placed on the last three periods of demand. The sum of all the weights given for the ESF model in Figure 16.9 is 94 percent. If we continue to find the weights for periods 19, 18, and so on, the sum for all weights is 1.0, as intuition would tell us. If the smoothing constant were .1 instead of .3, a table like Figure 16.9 would have values of .1, .09, and .081 for the weights of periods 27, 26, and 25, respectively. The sum of these three (27 percent) is the weight placed on the last three periods. Moreover, (1 − 27% = 73%) is the weight given to all actual data *more than* three periods old.

This result shows larger values of α give more weight to recent demands and utilize older demand data less than is the case for smaller values of α; that is, larger values of α provide more responsive forecasts, and smaller values produce more stable forecasts. The same argument can be made for the number of periods in an MAF model. This points out the basic trade-off in determining what smoothing constant (or length of moving average) to use in a forecasting procedure. The higher the smoothing constant or the shorter the moving average, the more responsive forecasts are to changes in underlying demand, but the more "nervous" they are in the presence of randomness. Similarly, smaller smoothing constants or longer moving averages provide stability in the face of randomness but slow reactions to changes in the underlying demand.

ENHANCING THE BASIC EXPONENTIAL SMOOTHING MODEL

Thus far, the exponential smoothing model we've described assumes demand is essentially constant, with only random variations around an average. However, if we have any indication of a systematic pattern underlying the randomness of demand, we must take it into account to improve our forecast accuracy. Two broad categories of factors might explain nonrandom patterns. The first category of factors includes characteristics of the marketplace, such as cycles and seasonal or trend patterns in demand. Note these need not be natural but could be induced by events like new model introductions or special shows. The second broad category of nonrandom patterns are results of plans we may have for future events that will influence demand. These events could consist of special product promotions, timing of a major customer's orders, special ordering requirements for customers, or new product announcements (either ours or our competitors'). In this section, we'll first look at the changes that can be made to an exponential smoothing model to incorporate market factors; we'll close with methods for taking into account plans or knowledge of events that will affect future demand.

Trend Enhancement

The first market characteristic we'll take into account is a **trend** in the product's demand. This is represented by the increase in tequila-based drinks Enrique Martinez faces. A cursory glance at the demand for tequila-based drinks (see Figure 16.4) indicates substantial growth (trend) in demand. To take the trend into account explicitly, we need a method to estimate the trend (i.e., the change in basic demand from period to period). Thinking again of drawing a line through the data to make a smoothed estimate of the trend leads us to consider using smoothing procedures to make the estimate. As a first step in the process of developing the trend estimate, we use exponential smoothing to smooth the random fluctuations and create a new *base value*. As was true with the basic model, we perform the calculations as soon as we find out what actual demand was during the period; that is, at the end of the period. The method for computing the new base value is

$$\text{Base value}_t = \alpha \, (\text{Actual demand}_t)$$
$$+ \, (1 - \alpha)(\text{Base value}_{t-1} + \text{Trend}_{t-1}) \qquad (16.7)$$

where:

α = Base value smoothing constant $(0 \leqslant \alpha \leqslant 1)$.

t = Current period (the period for which the most recent actual demand is known).

Base value$_{t-1}$ = Base value computed one period previously (at the end of period $t - 1$).

Trend$_{t-1}$ = Trend value computed one period previously (at the end of period $t - 1$).

Once the new base value for the current period is determined, we apply exponential smoothing to develop the most up-to-date estimate of the trend value. A second smoothing constant, β, is introduced to do this and is applied as shown in Equation (16.8).

$$\text{Trend}_t = \beta \, (\text{Base value}_t - \text{Base value}_{t-1}) + (1 - \beta)(\text{Trend}_{t-1}) \quad (16.8)$$

where:

β = Trend smoothing constant $(0 \leq \beta \leq 1)$.

Base value$_t$ = Base value computed at the end of period t.

Base value $_{t-1}$ = Base value computed one period previously (at the end of period $t - 1$).

Trend $_{t-1}$ = Trend value computed one period previously (at the end of period $t - 1$).

Once we've determined most current trend value and the new base value, we can make **trend-enhanced forecasts (TEF)** of future demand. Equation (16.9) gives the method. Notice this equation can be used to forecast for more than one period in the future by multiplying the trend value by the number of periods in the future. Figure 16.10 provides sample calculations.

FIGURE 16.10   Example Trend-Enhanced Forecast Calculations

Data: $\alpha = .2$, $\beta = .1$, Trend$_{t-1}$ = 5, Base value$_{t-1}$ = 100

TEF$_t$ (that is, the trend-enhanced forecast made at the end of period $t - 1$ before actual demand for period t was known) =

$$\text{TEF}_t = \text{Base value}_{t-1} + \text{Trend}_{t-1} = 100 + 5 = 105$$

At the end of period t, demand for period t becomes known:

$$\text{Actual demand}_t = 107$$

We now produce forecasts using Equations (16.7), (16.8), and (16.9):

$$\text{Base value}_t = .2(107) + (1 - .2)(100 + 5) = 105.4 \quad (16.7)$$
$$\text{Trend}_t = .1(105.4 - 100) + (1 - .1)(5) = 5.04 \quad (16.8)$$
$$\text{Forecast for next period (i.e., TEF}_{t+1}) = 105.4 + (1)5.04 = 110.44 \quad (16.9)$$
Forecast for four periods from now
$$\text{(i.e., TEF}_{t+4}) = 105.4 + (4)5.04 = 125.56 \quad (16.9)$$

The trend-enhanced forecast for X periods in the future at the end of period t is measured as follows:

$$\text{TEF}_{t+X} = \text{Base value}_t + X\,(\text{Trend}_t) \qquad (16.9)$$

where:

$$X = \text{Number of periods beyond period } t \text{ for which the forecast is desired.}$$
$$t = \text{Current period for which actual demand is known.}$$
$$\text{Base value}_t = \text{Exponentially smoothed base value computed at the end of period } t.$$
$$\text{Trend}_t = \text{Exponentially smoothed estimate of trend per period computed at the end of period } t.$$

An alternative to the TEF model might be to use the ESF with a high smoothing constant, since we noted how high values of the smoothing constant were responsive to changes in demand. In fact, this alternative won't produce results that are as good as the TEF when there is a definitive trend component in the demand. This can be seen in Figure 16.11, which compares an ESF with $\alpha = .8$ and a TEF ($\alpha = .2$, $\beta = .2$) for the tequila-based drink data. As noted, the TEF approach produces better measures of both bias and MAD. By looking at the graph, you'll see the ESF lags actual demand, which leads to a much higher bias.

Figure 16.11 clearly demonstrates the advantage of explicitly including information about patterns in demand when such patterns do, in fact, exist. However, the opposite statement is also true: If a pattern in demand doesn't clearly exist, then using a model that assumes such a pattern will lead to poor forecasts. For example, if we applied the TEF model to the data for tables served at Panchos (Figure 16.6), the result could be larger values for both bias and MAD than those obtained with an appropriate ESF.

Seasonal Enhancement

Many products have a **seasonal** demand pattern. We can all think of examples like baseballs, skis, lawn furniture, antifreeze, and holiday greeting cards. Perhaps less obvious are seasons for a product or service during a day, week, or month—like increased bank deposits on paydays, travel on the weekend (or mornings and evenings), mail deposits on Friday afternoons, or long-distance phone calls during the day. A useful indication of the degree of seasonal variation for a product is the *seasonal index*. This index estimates how much demand during a season will be above or below the product's average demand. For example, a bathing suit line may sell 100 units a month on the average, but 150 suits a month in peak season and 75 in the off-season. The

FIGURE 16.11   Comparison of Trend-Enhanced and Basic Exponential Smoothing Models

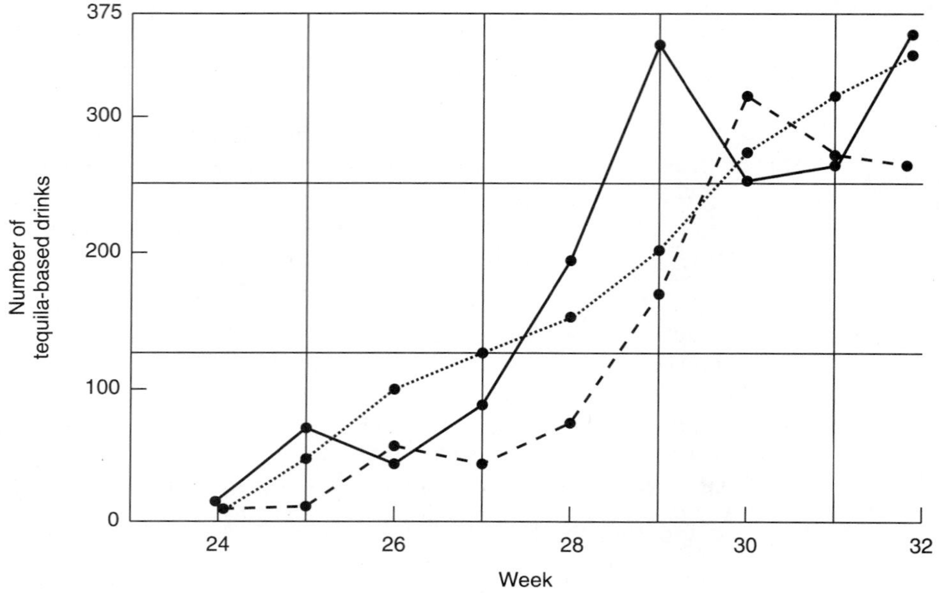

Symbol	Forecast	Bias	MAD
●———●	Actual demand		
●············●	Trend enhanced		
	$\alpha = .2, \beta = .2$	7.26	45.81
●– – – ●	Basic model		
	$\alpha = .8$	45.50	65.08

index for peak season sales would be 1.50 and 0.75 for the off-season for bathing suits. The seasonal index is used to adjust forecasts for seasonal patterns.

Much the same kind of logic we used for adjusting for the trend is applied in adjusting for a product's seasonality. As before, we'll use exponential smoothing to estimate seasonal indexes of demand for a product. A seasonal pattern during the week is quite clear in the number 10 cans of vegetables used during three-day periods at Panchos (Figure 16.5). Demand is much less in the early part of the week than in the latter part of the week. Enrique's forecasting model should take this into account. Computing **seasonally enhanced forecasts (SEF)** again involves updating some new base value at the end of some period t. Next, there's an updating of the seasonal index that applies to the period just past. Finally, there's a forecast of demand one or more periods in the future based on the latest base value and the appropriate seasonal index for the period to be forecast. Equations (16.10) and (16.11) give the formulas for updating the base value and seasonal index. The first equation produces the new base by exponentially smoothing the "deseasonalized" demand data (i.e., actual demand divided by the seasonal index); the second equation uses another smoothing constant to produce an updated estimate of the seasonal index for the period just past:

$$\text{Base value}_t = \alpha \left(\frac{\text{Actual demand}_t}{\text{Old index}_s} \right) + (1 - \alpha)(\text{Base value}_{t-1}) \quad (16.10)$$

$$\text{New index}_s = \gamma \left(\frac{\text{Actual demand}_t}{\text{Base value}_t} \right) + (1 - \gamma)(\text{Old index}_s) \quad (16.11)$$

where:

α = Smoothing constant for base value.
γ = Smoothing constant for seasonal indexes.
s = Season indicator.
Base value_{t-1} = Base value computed one period previously (at the end of period$_{t-1}$).
Old index_s = Index value for season s, calculated one full cycle ago.

Updating the base at the end of period t is similar to updating the TEF model, except actual demand is first deseasonalized. A seasonal index is maintained and updated for however many periods are included in a cycle of demand. For the number 10 cans of vegetables, there would be two seasonal indexes: one for the three-day period of F-S-Sun, and the other for T-W-Th. For a monthly series in an annual cycle, there would be 12 seasonal indexes, each updated when that month's actual demand becomes available and is used to forecast that month in the future whenever the forecast is desired.

The method for forecasting demand is to multiply the updated base value obtained using Equation (16.10) by the appropriate seasonal index. This is shown as Equation (16.12); Figure 16.12 gives examples. The seasonal enhanced forecast (SEF) for season s in the future at the end of period t is

$$\text{SEF}_s = \text{Base value}_t (\text{Index}_s) \quad (16.12)$$

FIGURE 16.12 Example Seasonal-Enhanced Forecast Calculations

Data: α = .2, γ = .3, old index$_{\text{F-S-Sun}}$ = 1.22, old index$_{\text{T-W-Th}}$ = .77
   Base value$_4$ = 40.2, Actual demand$_5$ = 51 (this was a F–S–Sun)
   Base value$_5$ = .2(51/1.22) + (1 − .2)(40.2) = 40.5   (16.10)
   New index$_{\text{F-S-Sun}}$ = .3(51/40.5) + (1 − .3) 1.22 = 1.23   (16.11)
   SEF$_6$ = 40.5(.77) = 31.2 (this was a T–W–Th)   (16.12)

---

where:

Base value$_t$ = Most recent deseasonalized base value at the end of period t.
Index$_s$ = Seasonal index associated with the season s.

If Panchos wanted to forecast demand for number 10 cans of vegetables for the next F-S-Sun period now (i.e., at the end of period 5 before actual demand for period 6 is known), using Figure 16.12's data,

$$SEF_7 = 40.5(1.23) = 49.8.$$

Trend and Seasonal Enhancement

A logical extension of either trend enhancement or seasonal enhancement is a model that incorporates both patterns in demand. Without going into a detailed explanation, the approach is to first apply an expanded version of Equation (16.10) to update the base value; this is shown as Equation (16.13). Next, we compute the seasonal index with Equation (16.11) as before, and the updated trend component with Equation (16.8). Finally, the forecast Equation (16.14) incorporates the trend enhancement from Equation (16.9) and the seasonal enhancement from Equation (16.12). The entire sequence is

$$\text{Base value}_t = \alpha \left(\frac{\text{Actual demand}_t}{\text{Old index}_s} \right) + \qquad (16.13)$$
$$(1 - \alpha)(\text{Base value}_{t-1} + \text{Trend}_{t-1})$$

$$\text{New index}_s = \gamma \left(\frac{\text{Actual demand}_t}{\text{Base value}_t} \right) + (1 - \gamma)(\text{Old index}_s) \qquad (16.11)$$

$$\text{Trend}_t = \beta (\text{Base value}_t - \text{Base value}_{t-1})$$
$$+ (1 - \beta)(\text{Trend}_{t-1}). \qquad (16.8)$$

The **trend seasonal enhanced forecast (TSEF)** for X periods in the future at the end of period t is

$$TSEF_{t+x} = (\text{Base value}_t + X \text{ trend}_t) \text{ Index}_{t+X-m} \qquad (16.14)$$

where:

Base value$_t$ = Most recent deseasonalized base value at the end of period t.
Index$_s$ = Seasonal index for the season s.
Trend$_t$ = Trend estimate per period calculated at the end of period t.
X = Number of periods beyond period t for which the forecast is desired.
m = Number of periods in the seasonal cycle.

The subscript on the index in Equation (16.14) deserves an additional note. The forecast is made at the end of period t, for X periods into the future; hence, $t + X$. Subtracting m moves the index subscript back to the seasonal value computed at the end of that last season in the cycle. For example, if, at the end of month 5 (May), we wish to forecast for August, the index subscript's 5 + 3 − 12, or the seasonal index computed nine months previously at the end of the preceding August.

Other Enhancements

We've illustrated use of multiplicative seasonal factors and additive trend factors. Other approaches have used additive seasonality and/or multiplicative trend. Still other approaches allow for acceleration, which would be applicable to fad products; others utilize the fitting of mathematical functions, such as sine waves to past data. The point is, *if* a genuine pattern in the underlying demand exists, then a model incorporating this pattern can produce better forecasts. We saw this in Figure 16.11, comparing a trend model and the basic exponential smoothing model for the tequila-based drink data. A similar comparison could be made for the number 10 cans of vegetables, using models with and without seasonal enhancement. However, once again we caution against use of these enhanced models if the underlying data don't clearly support their application. We know of a firm that has a definitive seasonal sales pattern with growth—when we talk about overall sales in dollars. However, the same demand pattern doesn't exist for many individual items in this company.

FOCUS FORECASTING

Two observations lie behind developing the **focus forecasting** technique. First, the forecasting approach that worked best last time may work best this time. Second, persons experienced in making short-term forecasts tend to use simple models. Bernard Smith developed focus forecasting by putting these two concepts together in a forecasting system.

Focus forecasting uses the *one* forecasting model that would have performed the best in the recent past to make the next forecast. The procedure starts by simulating past periods' forecasts for a variety of simple forecasting models. Next, the performance that would have been achieved in preceding periods by each of these models is reviewed. Finally, the model that would have had the best performance in the past is used for forecasting the next period.

The focus forecasting procedure works with rolling quarterly data. The **mean absolute percentage error (MAPE)** is the criterion used to choose the forecasting model for making the next forecast. We define MAPE as the absolute error (as was used in Equation (16.2) to calculate MAD) divided by actual demand. The model with the lowest value of MAPE for the last quarter is used to make the forecast for the next quarter.

The models used in the focus forecasting system are ones that have actually been used successfully in practice. They should be understood by the personnel responsible for producing the forecasts. The models are, therefore, quite straightforward approaches to forecasting. Typical of the models used in the system are:

1. The forecast for the next quarter is the actual demand for the same three-month period last year.
2. The forecast for the next quarter is 110 percent of the actual demand for the same three-month period last year.
3. The forecast for the next quarter is half the total actual demand for the past six months (a two-quarter moving average).
4. The forecast for the next quarter is the actual demand for the previous three-month period.
5. The forecast for the next quarter is the actual demand for the same three-month period last year multiplied by the growth or decline since last year as measured by the ratio of the demand for the previous quarter to the demand for the same quarter last year.
6. If demand in the past six months is less than 40 percent of the demand for the six months preceding that, the forecast for the next quarter is 110 percent of the demand for the same three-month period last year (i.e., we're coming into the upturn of a seasonal swing).
7. If demand in the past six months is more than 2.5 times the demand for the six months preceding that, the forecast for the next quarter is the same as the demand for the same three-month period last year (i.e., we're starting into the downside of a seasonal swing).

COMPARISONS OF METHODS

At one time we intended to devote a whole chapter to advanced forecasting techniques. We abandoned that idea after reviewing the work of Spyros Makridakis and his colleagues. Their research and some later work evaluating focus forecasting will be overviewed here. The results contain a key message for practice: simple models usually outperform more complex procedures, especially for short-term forecasting.

The Forecasting Competition

A variety of forecasting techniques have been developed and more are being created all the time. They range from very simple to mathematically complex, from aggregate-business–oriented to stockkeeping-unit–oriented, and from very costly to inexpensive. Among the techniques at the business planning level are those involving expert opinion and consensus, causal or regression approaches that link activities in one sector with those in another sector, and economic or business analysis approaches. For the more operations-oriented forecasts, techniques range from attempts to characterize past data by using mathematical approaches (e.g., spectral analysis, Box-Jenkins, or trigonometric patterns) to simple projections of past performance using moving averages or exponential smoothing.

Spyros Makridakis organized a forecasting competition in which seven experts evaluated 21 forecasting models. The competition was based on 1,001 different actual time series. Some of these were yearly, some quarterly, and still others monthly. Some of the series were microdata (e.g., for business firms, divisions, or subdivisions); others were for macrodata (e.g., GNP or its major components). Some series were comprised of seasonal data; others weren't. Expert proponents of a variety of forecasting models analyzed the data, determined appropriate model parameters, and made forecasts of the series. The forecasting horizon's length varied from 1 to 18 periods into the future. Forecasting accuracy was determined with five different measures.

There was no one model that consistently outperformed all the others for all series, all measures, or all forecasting horizons. Some models do better than others on macrodata, while others are better for microdata. Similarly, some models were better for monthly data than for quarterly or yearly data, and still others were good for longer forecasting horizons. Therefore, one conclusion that comes out of this work is that a forecast user can improve forecast accuracy by choosing a model that fits the criterion and the environment in which he or she is interested (e.g., microdata versus macrodata, short versus long horizon, and measure of accuracy).

Since we're concerned here with short-term horizons, the general conclusion that simple methods do better than the more sophisticated models, especially over short horizons for microdata, is important. Such techniques as simple exponential smoothing tend to outperform sophisticated methods, such as Box-Jenkins or econometric models.

Figure 16.13 summarizes the rankings for some of the procedures (for one-period forecasting horizons). For most of the criteria shown, exponential smoothing models do quite well. Figure 16.13 is for all the 1,001 data series; the best techniques do even better for just the microdata. One of the research's surprises is the combination technique's performance. It indicates the focused forecasting idea of selecting the *one* forecasting technique with the lowest error to make the forecast was partially right. It might be even better to average the forecasts from the several models used each period.

The Focus Forecasting Comparison

To further test the idea that averaging might be better than choosing a single technique, Flores and Whybark performed an experiment involving focus forecasting and an average of all the models' forecasts. Since the focus fore-

FIGURE 16.13   Performance Rank for Forecasting Techniques among 21
Methods for a One-Period Planning Horizon

*Criterion**

Method (all adjusted for seasonality)	MAPE (mean average percent error)	MSE (mean squared error)	Average ranking relative to all other techniques	Median APE (median value of percentage error)
Naive (Forecast = Current actual)	7	17	8	8
Moving average	15	20	10	11
Simple exponential smoothing	3	13	7	7
Exponential smoothing with trend	4	7	2	4
Exponential smoothing with trend and seasonal factors	4	7	2	2
Combination (an average of the forecasts from six methods)	1	10	1	1

*The best performance on the criterion is 1, the worst is 21.

Source: Makridakis et al., "The Accuracy of Extrapolation (Time Series) Methods: Results of a Forecasting Competition," *Journal of Forecasting* 1, no. 2 (1982).

casting approach requires that several forecasting models be in place anyway, averaging forecasts from all the models was a simple extension of the technique. Averaging could lead to better results—and is also consistent with the desire for simplicity and understandability.

A focus forecasting system comprised of the seven models listed earlier in the chapter was used as the basis for the experiment. The focus forecasting results were compared to the results of averaging all the forecasts, and a basic exponential smoothing model was used as a basis for comparison. Both simulated and actual demand data were used to test the approaches.

MAD and MAPE were used as criteria to evaluate the three procedures' forecasting performance. The results were the same for both criteria. For the simulated demand data, there were significant differences between all three procedures—going from averaging (best) to exponential smoothing (worst). The rankings were changed and the level of significance reduced when actual data were used, however. Exponential smoothing performed best, but focus forecasting and averaging weren't statistically different.

These experiments' pragmatic implications are clear. Forecasting actual demand is difficult. Unfortunately, the results don't provide a consistently superior choice of forecasting technique. The results, though, support the use of simple forecasting models.

The important conclusion for practitioners is that more sophisticated and expensive models aren't necessarily better. It means those who advocate using complex forecasting models need to justify their choice. They need to clearly demonstrate that they can provide better forecasts than the simpler procedures, and that the error measures are more consistent with the needs of the decision makers. This "show-me" attitude becomes even more important when we consider the preparation cost for using many of the sophisticated models. In addition to computer and other costs, we should also add the cost to the organization of using a procedure that's difficult for non-experts to understand.

USING THE FORECASTING SYSTEM

Using the forecasting system requires a heavy dose of common sense, as well as application of techniques. In this section, we'll look at some methods for incorporating external information into the forecasting system. We'll also look at the problem of establishing the forecasting model parameters and of monitoring the forecasting model results. In exponential smoothing, it's not enough to determine demand patterns and to select the forecasting model that appears to provide minimum bias and MAD. Before we can start making forecasts, it's necessary to choose the smoothing constants and to establish the initial base value, trend value, and seasonal indexes. Once forecasting has started, of course, these initial values are recalculated with each new piece of demand information.

But we still aren't done! We must continue evaluating forecasts' quality to make sure the model chosen is still appropriate, to determine whether market conditions have changed, and to learn quickly when something has gone awry. We first turn our attention to the topics of external information, getting started, and monitoring. Thereafter, we briefly raise some strategic issues relating to forecasting.

Incorporating External Information

Many kinds of information can and should be used to make good forecasts. For example, in a college town on the day of a football game, traffic around the stadium is a mess. An intelligent forecaster adjusts travel plans on game days to avoid the stadium traffic, if possible. He or she modifies the forecast due to knowing the game's impact on traffic. An exponential smoothing model based on observations during the week would probably forecast very little traffic around the stadium. We certainly wouldn't use the exponential smoothing forecast without adjusting it for game day. That simple principle is applicable to business forecasting as well, but it's surprising how often people fail to make these adjustments.

Examples of activities that will influence demand and perhaps invalidate the routine forecasting model are special promotions, product changes, competitors' actions, and economic changes. We have two primary ways to incorporate information about such future activities into the forecast. The first is to change the forecast directly; the second is to change the forecasting model. We might use the first method if we knew, for example, there was to be a promotion of a product in the future, or we were going to open more retail outlets, or we were going to introduce a competing product. In these instances, we could adjust the forecast directly to account for the activities, just as we do for the game day. By recognizing explicitly future conditions won't reflect past conditions, we can modify the forecast directly to reflect our assessment of the future.

The second method for dealing with future activities would be to change the model itself. This might work best when we're unsure of what these activities' effect will be. If, for example, we know one of our competitors is going to introduce a new product, we suspect the market will change, but we may not be sure of the change's direction or magnitude. If the product is expensive, we may gain sales; if it's novel, we may lose sales. All we know is there may be a change. In this instance, we could increase the smoothing constant, making the model more responsive to changes in the marketplace. In this way, we can incorporate changes into our forecasts more quickly. If we know something of what may happen, we could change both the forecast and the smoothing constant. Both methods help to incorporate our information about the future into the forecasts before using the forecasts to make decisions.

Getting Started

When historical demand data are available, there's nothing like a plot of those data for getting started. If there's a pattern to the demand, we can easily plot it. The plots also help us set the initial values for doing the forecasting in a way consistent with the historical data. If, for example, plots reveal seasonal factors, we can estimate the base value by taking the average for at least one seasonal cycle.

We can find seasonal indexes by averaging the indexes calculated for each period in the cycle. Similarly, a plot of the values for trend data would enable us to draw in a trend line (or we could average the period-to-period changes) to estimate the trend value. The plot would also help determine the base value to use for forecasts. In every instance (constant data, trend data, or seasonal data), plots will help us determine whether it's desirable to use the more recent data in setting starting values.

Once starting values have been determined, we can use relatively high smoothing constants for the early forecasts to quickly overcome any errors in the starting values. It's also desirable to make simulated forecasts of the past few periods of historical data as test data for the model. By using, say, 75 percent of the historical data to estimate initial values, and then simulating forecasts for the remaining 25 percent of the data, values for starting the initial forecasts would already have been smoothed by the forecasting model.

The choice of smoothing constants for use in the models for forecasting is a matter of balancing responsiveness with stability. This isn't an easy balance, however, and practice has provided some guidance. For smoothing the average or base value, an α of about .1 to .2 has been found useful in practice. The β value is generally held to less than the α value, about .05 to .1. The value for γ depends on how frequently the seasonal index is recalculated. If often, such as every few weeks, a low γ (.1) is acceptable. If less frequently (yearly), $\gamma = .3$ to .4 might be used. In practice, some simulation with past data can be useful. However, we feel this approach is of limited value, since the objective is to forecast well in the *future*. The issue always comes down to the stability-responsiveness trade-off, based on how stable the future environment is judged to be.

Demand Filter Monitoring

The smoothing models in this chapter all incorporate actual demand data into the forecasts as soon as the information is available. Therefore, actual demand data must be correct. One way to help ensure this is through demand filtering (i.e., checking actual demand against a range of reasonable values). An effective approach is to screen actual demand values against some limit before updating the forecasts, and to have some thinking person (not a computer) determine whether exceptions are correct. A common screening limit is four MADs in either direction of the forecast demand for the period. Since 4 MADs correspond to 3.2 standard deviations, this limit provides a probability of less than .001 of the demand value being a random occurrence for normally distributed forecast errors. If an actual demand falls outside this limit, a manual review is applied.

Once the filter catches a value outside the limits, the review might consist of checking to make sure there wasn't a clerical error in recording demand, there wasn't some explainable cause for the big change, or conditions really have changed and demand will be changed significantly. If conditions are changing, the situation may call for techniques for modifying the forecasts discussed earlier.

The limits to use for filtering individual actual demand observations depend on a manual review's cost compared to an error's cost. The probabilities of exceeding the limits can be determined from a normal table using the relationship between the number of standard deviations and MAD, given in Equation (16.3). This provides insight into setting limits on the observations.

Demand filtering can be very important in actual practice. We've seen many examples where average demand for some product such as a particular chair at Ethan Allen might be, say, 20 units per month. All of a sudden an order comes along for 300 chairs! Someone opened a restaurant. Demand filtering will pick up this situation, first asking if a data entry error has occurred. The thinking analyst well might not allow this order to influence the average or forecast. At Ethan Allen it would be treated as a "contract sale," which is only forecast in overall dollars. It's too difficult to forecast the exact timings and actual items of contract sales.

Tracking Signal Monitoring

The approach of exponential smoothing can also be used to compute a useful statistic called the **tracking signal.** The tracking signal helps in monitoring the forecast's quality. We use the methods of exponential smoothing to make a smoothed average of the bias and MAD. Equations for doing this, (16.15) and (16.16), follow. These equations simply smooth the same error measures we introduced in Equations (16.1) and (16.2) early in this chapter. By using exponential smoothing, the measures incorporate and weight most heavily the recent demand information. The smoothing constant, δ, is between 0 and 1 and has the same properties as the smoothing constant in the exponential smoothing forecasting model. The larger the δ, the more heavily weighted or responsive the tracking signal is to the most recent forecast error. Figure 16.14 shows some sample calculations:

$$\text{Smoothed bias,} = \delta(\text{Actual demand}_t - \text{Forecast}_t)$$
$$+ (1 - \delta)(\text{Smoothed bias}_{t-1}) \qquad (16.15)$$
$$\text{Smoothed MAD}_t = \delta|\text{Actual demand}_t - \text{Forecast}_t|$$
$$+ (1 - \delta)(\text{Smoothed MAD}_{t-1}) \qquad (16.16)$$

FIGURE 16.14   Example Smoothed Bias and MAD Calculations

Data: $\text{Forecast}_t = 100$, actual $\text{demand}_t = 90$, $\delta = .1$.

    Smoothed $\text{bias}_{t-1} = -1$, smoothed $\text{MAD}_{t-1} = 5$

    Smoothed $\text{bias}_t = .1(90 - 100) + (1 - .1)(-1) = -1.9$ (16.15)

    Smoothed $\text{MAD}_t = .1|90 - 100| + (1 - .1)(5) = 5.5$ (16.16)

    Tracking signal $= -1.9/5.5 = -.345$

---

where:

$$0 \leq \delta \leq 1.$$

$$|\text{Actual demand} - \text{Forecast}_t| = \text{Absolute value of the forecast error observed during period } t.$$

We combine the smoothed bias and smoothed MAD to calculate the tracking signal as shown in Equation (16.17). Note the smoothed MAD provides an estimate of the expected error (i.e., the average error) and the bias shows consistent over- or underforecasting. The tracking signal varies between -1 and $+1$. Either of these extreme values indicates all the forecasts are of the same sign. If the forecast is unbiased, the tracking signal will be near zero, irrespective of the MAD's value. The tracking signal allows us to compute a measure of bias that's independent of MAD, one that will have the same numerical meaning for every item forecast. As the tracking signal deviates from zero in any significant way, manual review of the particular item is called for.

$$\text{Tracking signal}_t = \frac{\text{Smoothed bias}_t}{\text{Smoothed MAD}_t} \qquad (16.17)$$

where:

$$-1 \leq \text{Tracking signal} \leq +1.$$

The tracking signal is an indicator of forecast bias that's consistent for all observations. Its use is essentially the same as that described for demand filtering; that is, by isolating those items for which the tracking signal is deviating significantly from the nominal value of zero, we can take corrective actions. For example, if an item were forecast with the basic exponential smoothing model (ESF), and an underlying trend existed in the data, the tracking signal would move away from zero.

The issue of what tracking signal value to use for initiating a review is essentially the same as that for demand filtering. The closer the limit is to zero, the sooner poor forecasts are discovered. On the other hand, with small limits, the number of times a review will be necessary rises. Also, the chance for reaching an erroneous conclusion from the review rises. The appropriate value is also not independent of δ. Small values of the smoothing constant for MAD and bias result in more stability and less responsiveness in these measures. Stability means that it will take longer for the tracking signal to respond to an underlying change in conditions.

Strategic Issues

There are a number of strategic and managerial questions about forecasting that we passed over rather rapidly or haven't discussed. Certainly we haven't had space to discuss all possible forecasting models, and it wouldn't be fair to leave this discussion without indicating there are several approaches to short-term forecasting we haven't mentioned here.

Although it wasn't indicated for any of the Pancho's restaurant problems, it's often necessary to make longer-term decisions for which the item-level, short-term forecasts simply aren't adequate. Among these decisions are capital expansion projects, proposals to develop a new product line, and merger or acquisition opportunities. For these long-term decisions, forecasts based on causal or econometric models (or simply based on managerial insight and judgment) can often produce improved results. Causal models are those that relate the firm's business to indicators that are more easily forecast or are available as general information. Early in this chapter, we used household fixture sales and their relation to housing starts as an indication of a causal relationship. Substantial managerial judgment is required in reviewing forecasts that form the basis for making long-term decisions. The general principle indicated here is the nature of the forecast must be matched with the nature of the decision. The level of aggregation, the amount of management review, the cost, and the quality of the forecast needed really depend on the nature of the decision being made. Many short-term operating decisions don't warrant use of expensive forecasting techniques, which has been one reason for focusing on short-term projection techniques. For strategic decisions with more at risk than two extra bottles of tequila, the investment in more expensive procedures (more management involvement) is needed. Figure 16.15 presents a general schema.

In the ongoing management of forecasts, strategic questions can also come about from a review triggered by forecast monitoring. For example, the forecasting model might be appropriate, but there are insufficient adjustments to account for known actions in the marketplace. Forecasting procedure must be managed to make sure special knowledge is included in the forecasts.

A review might indicate the model is no longer appropriate. There may be trend or seasonal effects that should now be included or dropped, or perhaps a compound model that has both trend and seasonal enhancements should be developed. In such cases, the model needs to be adjusted accordingly.

Yet another instance, where the model may not be appropriate, is where demand depends on other decisions in the firm. For example, demand for tires in an auto factory depends on the number of cars being produced. That's quite a different forecasting problem from trying to determine how many cars the public wants to buy. A dependent demand relationship should always be looked for.

It's apparent forecasting is a pervasive, central activity in managing operations. To be effective, the forecasting system must be linked closely to a number of other systems. Certainly, those decisions requiring forecast information must be linked directly to the forecasting system's output. Since all forecasting models presented in this chapter require demand data, there must be close linkage between the order entry system and the forecasting system. Many firms will use sales data or shipment data instead of demand for adjusting their forecasts. In cases where demand information isn't available, this may be warranted; but there's a difference between sales, shipments, and demand.

FIGURE 16.15   Applicability of Various Forecast Attributes to Decision Attributes*

Decision attributes

Level	Frequency	Money	Time
Mission	Rare	Much	Long run
Strategic	Occasional	Some	Medium run
Tactical	Often	Little	Short run

Forecast attribute	Increasing aggregation	Item level	Product family	Total sales or output
	Cost/ forecast	Low	Medium	High
	Degree of management involvement	Low	Medium	High
	Nature of forecast model	Projection technique	Econometric casual	Management judgment

*The darker the area, the greater the applicability.

Since it's demand we're interested in forecasting, the link with the order entry system should be capable of picking up demand information. If we don't have the stock available to make the sale or shipment, this will affect our customer service—but not the fact there was a demand.

FORECASTING IN INDUSTRY

We come now to the last of our seven topics in forecasting. In this section, we briefly describe the approach used by one firm, the Ethan Allen Furniture Company. The firm utilizes an exponential-smoothing–based forecasting system to forecast its products' demand. Forecasting models are part of an overall managerial system that provides for monitoring demand, developing forecasts, reviewing and modifying forecasts, aggregating information, producing sales history data, and developing a variety of other management reports. Figure 16.16 provides one example of the type of report that can be produced by the forecasting system. This particular report can be produced on request for any product management might wish to scrutinize. The forecasting model used to produce the forecasts in Figure 16.16 was a seasonally enhanced model using a smoothing constant of .2. The report shows monthly seasonal factors along with forecasts, actual demand, errors, and percent errors. Note also manual adjustments can be made, and MAD and the tracking signal can be reported.

Figure 16.17 is one of the monitoring reports produced by the system whenever a manual review is indicated by the system. The first product in Figure 16.17, a governor's chair, has triggered a review because the error exceeds 50 percent of the forecast. The limit of tolerance is shown at the top of the report. This triggered the inclusion of this particular governor's chair on the sales screening report, which suggests possible manual correction. The report also includes data on the last three forecasts, actual demand, MAD, and other review data. Adjustments are made manually, if needed, and will appear in subsequent runs of the report if actual demand continues to fall outside the limits for review. The next two items in Figure 16.17 are included in the report because one individual customer order was larger than the stated percentage of the total forecast. The report shows any information on past changes to the forecast as well. This keeps the entire process explicit to the reviewer. The sales screening process ensures the ultimate responsibility for forecasting rests with management.

CONCLUDING PRINCIPLES

Forecasts provide an important input to manufacturing planning and control systems. Although many kinds of forecasts are possible, this chapter has focused on short-term forecasts based on past data. We have shown how exponential smoothing models can be used to make these short-term forecasts and how routine forecast monitoring can be achieved.

We have tried to emphasize that forecasting is too important to leave to a forecasting model. Firms that use forecasting models wisely employ them to support, not to supplant, managerial judgment. The importance of taking external information into account is one example. Another is the necessary judgment required in a review resulting from forecast monitoring. For example, a tracking signal can indicate a review. It takes a thinking person to decide precisely how to do the review, how (or whether) to change the model, and how to modify the forecasting model data.

We stress the following basic concepts or principles:

- Evaluative criteria must be chosen for the short-term forecasting system. The choices implied in this chapter are minimum bias, minimum MAD, low cost, and simplicity.

FIGURE 16.16 Ethan Allen, Inc., Sales and Forecasts

MIRROR

FOR ITEM 11-9008- 225

	JAN	FEB	MAR	APR	MAY	JUN	JUL	AUG	SEP	OCT	NOV	DEC
SEASONAL FACTORS	.74	1.12	1.39	.63	.72	.85	.79	1.17	1.73	1.01	.79	1.06

AVG SALES 44.5

ADJUSTMENTS TO AMOUNT FOR

ADJUSTMENTS TO AMOUNT FOR MAD

TRACK SGNL

NUMBER OF UNITS FORECAST AND SOLD
0...20...40...60...80...100..120..140..160..180..200..220..240

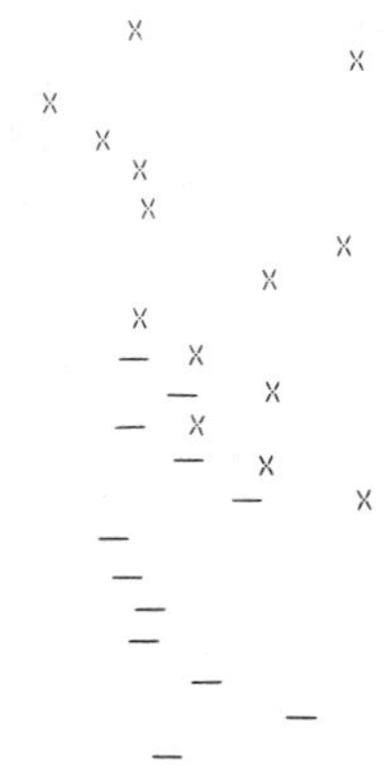

SALES	TOTAL FCST	ERROR	PCT ERROR	DATE
27				FEB
70				MAR
16				APR
20				MAY
28				JUN
29				JUL
66				AUG
53				SEPT
28				OCT
38	26	+12	+46%	NOV
53	37	+16	+43%	DEC
38	25	+13	+52%	JAN
52	38	+14	+36%	FEB
71	48	+23	+47%	MAR
	22			APR
	24			MAY
	29			JUN
	27			JUL
	40			AUG
	59			SEPT
	34			OCT

— TOTAL FORECAST
X SALES

FIGURE 16.17 Ethan Allen, Inc., Sales Screening Exception Report

FOR MAY

UPPER LIMIT PERCENT = 50%   LOWER LIMIT PERCENT = 50%   NUMBER OF MADS = 2.5

PERCENT/MAD LIMITS EXCEEDED

ADJUSTED FORECAST	ACTUAL SALES	ERROR	PERCENT ERROR	MAD ERROR	CHR GOV SALES RANGE FROM - TO	AV SLS	SEAS	FORECAST	ADJUSTMENT	REASON	MAD	MAD/AV	TRACK SGNL
268	84	-184	-68%	2.2	134 402	372.4	0.72	268			81.7	21%	+5.5
232	282	+50	+21%		TWO MONTHS AGO			232					
534	379	-155	-29%		THREE MONTHS AGO			534					
434	236	-198	-45%		FOUR MONTHS AGO			434					

30-6050-A   218 R   CHR GOV CRVR

LARGE INDIVIDUAL ORDER   CUST ACCT NO 17-4870-0   ORDER DATE 5/21   QUANTITY 12   CONSOLIDATION NO. 30-6050-A   218   FACTORY 018
AVERAGE SALES   97.9   ORDER % OF AV SLS   12%

PERCENTAGE LIMITS EXCEEDED

ADJUSTED FORECAST	ACTUAL SALES	ERROR	PERCENT ERROR	MAD ERROR	SALES RANGE FROM - TO	AV SLS	SEAS	FORECAST	ADJUSTMENT	REASON	MAD	MAD/AV	MAD LIM	CUMUL ERROR	TRACK SGNL
70	34	-36	-51%	0.8	35 105	97.9	0.72	70			40.8	41%	104%	+225	
68	44	-24	-35%		TWO MONTHS AGO			68							
143	171	+28	+19%		THREE MONTHS AGO			143							
123	60	-63	-51%		FOUR MONTHS AGO			123							

30-6052-   218 R   CHR CPTN

LARGE INDIVIDUAL ORDER   CUST ACCT NO 35-3595-0   ORDER DATE 5/01   QUANTITY 12   AVERAGE SALES 61.3   ORDER % OF AV SLS 19%
LARGE INDIVIDUAL ORDER   CUST ACCT NO 13-544B-0   ORDER DATE 5/24   QUANTITY 24   AVERAGE SALES 61.3   ORDER % OF AV SLS 39%
CONSOLIDATION NO. 30-6052-   218   FACTORY 018

ADJUSTED FORECAST	ACTUAL SALES	ERROR	PERCENT ERROR	MAD ERROR	SALES RANGE FROM - TO	AV SLS	SEAS	FORECAST	ADJUSTMENT	REASON	MAD	MAD/AV	MAD LIM	CUMUL ERROR	TRACK SGNL
44	42	-2	-4%	0.0	0 0	61.3	0.72	44			28.4	46%	116%	+22	+0.7
46	12	-34	-73%		TWO MONTHS AGO			46							
95	123	+28	+29%		THREE MONTHS AGO			95							
82	40	-42	-51%		FOUR MONTHS AGO			82							

30-6055-   218 R   DRY SINK

CONSOLIDATION NO. 30-6055-   218   FACTORY 022

PERCENTAGE LIMITS EXCEEDED

ADJUSTED FORECAST	ACTUAL SALES	ERROR	PERCENT ERROR	MAD ERROR	SALES RANGE FROM - TO	AV SLS	SEAS	FORECAST	ADJUSTMENT	REASON	MAD	MAD/AV	MAD LIM	CUMUL ERROR	TRACK SGNL
30	9	-21	-70%	1.3	15 45	42.3	0.72	30			16.0	37%	95%	+109	+6.8
28	23	-5	-17%		TWO MONTHS AGO			28							
58	77	+19	+32%		THREE MONTHS AGO			58							
52	20	-32	-61%		FOUR MONTHS AGO			52							

- The most critical problem is for management to control bias. It's often easier to live with larger errors (larger MAD) if that's what it takes to reduce bias.
- Use of short-term forecasts must be separated from the act of forecasting.
- Methods for monitoring forecasts over time must be installed.
- Forecasting needs to be embedded in a management structure.
- Forecasting is not a computer program, and the result shouldn't be monitored by the computer department.
- Simple forecasting methods seem to work better than sophisticated procedures for short-term forecasts of microdata.

REFERENCES

Armstrong, J. S. "The Ombudsman: Research on Forecasting: A Quarter Century Review, 1960–1984." *Interfaces,* January–February 1986.

Box, G. E. P., and G. M. Jenkins. *Time Series Analysis: Forecasting and Control.* New York: Holden-Day, 1970.

Brown, R. G. *Smoothing, Forecasting and Prediction of Discrete Time Series.* Englewood Cliffs, N.J.: Prentice-Hall, 1962.

Chambers, J. C.; S. K. Mullick; and D. D. Smith. "How to Choose the Right Forecasting Technique." *Harvard Business Review,* July–August 1971, pp. 45–74.

Flores, B. E., and D. C. Whybark. "A Comparison of Focus Forecasting with Averaging and Exponential Smoothing Strategies." *Production and Inventory Management,* 3rd quarter 1986.

————. "Forecasting 'Laws' for Management." *Business Horizons,* July/August 1985.

Flowers, A. D. "A Simulation Study of Smoothing Constant Limits for an Adaptive Forecasting System." *Journal of Operations Management* 1, no. 2, 1980.

Gardner, Everette S., Jr., and E. McKenzie. "Seasonal Exponential Smoothing with Damped Trends." *Management Science* (Note) 35, no. 3 (March 1989), pp. 372–75.

Georgoff, D. M., and R. G. Murdick. "Manager's Guide to Forecasting." *Harvard Business Review,* January–February 1986, pp. 110–20.

Gupta, S., and P. C. Wilton. "Combination of Forecasts: An Extension." *Management Science* 33, no. 3 (March 1987).

Lawrence, M. J.; R. H. Edmundson; and M. J. O'Connor. "The Accuracy of Combining Judgmental and Statistical Forecasts." *Management Science* 32, no. 12 (December 1986).

Lee, T. S., and E. E. Adam, Jr. "Forecasting Error Evaluation in Material Requirements Planning Production-Inventory Systems." *Management Science* 32, no. 9 (September 1986).

————; Everett E. Adam; and Ronald J. Ebert. "An Evaluation of Forecast Error in Master Production Scheduling for Material Requirements Planning Systems." *Decision Sciences* 18, no. 2 (Spring 1987), pp. 292–307.

Mabert, V. A. "An Introduction to Short-Term Forecasting Using the Box Jenkins Methodology." Atlanta: AIIE Monograph, 1975.

Makridakis, S.; A. Andersen; R. Carbone; R. Fildes; M. Hibon; R. Lewandowski; J. Newton; E. Parzen; and R. Winkler. "The Accuracy of Extrapolation (Time Series) Methods: Results of a Forecasting Competition." *Journal of Forecasting* I (1982), pp. 111–53.

_____; and R. L. Winkler, "Averages of Forecasts: Some Empirical Results." *Management Science* 29, no. 9 (September 1983).

McLain, F. O. "Restarting a Forecasting System When Demand Suddenly Changes." *Journal of Operations Management,* October 1981, pp. 53–61.

Smith, B. T. *Focus Forecasting Computer Techniques for Inventory Control.* Boston: CBI Publishing, 1978.

Trigg, D. W., and A. G. Leach. "Exponential Smoothing with an Adaptive Response Rate." *Operations Research Quarterly,* March 1967, pp. 53–59.

Wheelwright, S. C., and S. Makridakis. *Forecasting Methods for Management.* 5th ed. New York: Wiley, 1989.

Winters, P. R. "Forecasting Sales by Exponentially Weighted Moving Averages." *Management Science,* April 1960, pp. 324–42.

DISCUSSION QUESTIONS

1. Provide some examples of both short- and long-term forecast needs. What are some of the "special" types of information that should be taken into account, in addition to past history, in making these forecasts?
2. Some experts have argued it's more important to have low bias than to have low forecast error. Why would they argue this way?
3. What concerns would you have with using the data in Figure 16.4 to project future Tequila-based drink demand?
4. Would you use a seasonal-enhanced model for forecasting individual book sales in a book store?
5. A friend has just installed a focus forecasting system and wants you to add a new model just developed by the statistics department of a major university. What would you advise him?
6. How does a high-smoothing constant for the first few forecasts help in starting a forecasting model?
7. If you had a tracking signal in place and were using it to routinely monitor forecasts, what actions would you take if the tracking signal exceeded your limit and called for a review?

PROBLEMS

1. The Acme Machine Company's master production scheduler has been analyzing options on orders received for the firm's industrial valve product line. All customer orders for industrial valves require either option A or option B. During

the past six months the following customer order data have been collected. Each customer order is for one valve.

Month	Total number of industrial valve orders received	Number of orders requiring option A	Number of orders requiring option B
February	30	6	24
March	35	10	25
April	40	12	28
May	50	18	32
June	55	25	30
July	65	28	37

a. Develop a basic exponential smoothing model to forecast the percentage of customer orders requiring the use of option A for August. Assume the initial forecast of orders with option A is 15 percent, and Acme wants to use a smoothing constant of $\alpha = .4$.

b. What alternative forecasting models might be applied to forecast the percentage of customer orders requiring option A?

2. The Granger Transmission Company's manager of shop operations is concerned with forecasting weekly output in terms of standard labor-hours for the BD Chucker work center. He has collected the following data concerning the standard labor-hour output of this work center for the past four weeks:

Week number	525	526	527	528
Standard labor-hour output	549	579	581	564

a. Using a basic exponential smoothing model, a beginning average of 550 as of the end of week number 524, and a smoothing constant of $\alpha = .2$, prepare a forecast of the standard labor-hour output of this work center for week 529 as of the end of week 528.

b. What factors will influence this work center's actual standard labor-hour output? What forecasting model alternatives would you suggest for this situation?

3. Delta Electronics' warehouse manager would like to develop a forecasting model for the firm's data entry product line. The following data cover the past five months for this product line.

Month	Sales (in units)
1	1,800
2	1,860
3	1,920
4	2,050
5	2,120

a. Using a basic exponential smoothing model, a beginning average at the start of month 1 of 1,780 units, and a smoothing constant value of α = .2, prepare a forecast for month 6 as of the end of month 5.

b. What alternative forecasting models might be considered for the data entry product line?

4. Delta Electronics' master production scheduler would like to develop forecasting models for two of the end products in the firm's data entry product line. The scheduler collected unit sales data for two end products: the J401 and H212 models.

Month	J401 sales (in units)	H212 sales (in units)
July	500	170
August	510	180
September	480	490
October	530	230
November	640	590

a. Develop a basic exponential smoothing model for the J401 model using a beginning average as of the end of June of 480 and a smoothing constant value of α = .2. Forecast December sales as of the end of November.

b. Develop a basic exponential smoothing model for the H212 model using a beginning average of 200 as of the end of June and a smoothing constant value of α = .2. Prepare a forecast of December sales as of the end of November.

c. What alternative forecasting models might be considered for these products?

5. Talbot Publishing Company's production planning manager has provided the following historical sales data for its leading textbook on forecasting:

		Year		
	4	5	6	7
Sales*	21	18	20	17

*In 1,000 units

The firm is considering using a basic exponential smoothing model with $\alpha = .2$ to forecast this item's sales.

a. Assuming sales have averaged 20,000 units through year 3, prepare forecasts for years 5 through 7 as of the end of year 4.

b. Calculate the MAD value for the forecasts using the actual sales data provided.

6. Repeat Problem 5, updating the forecasts for years 6 and 7 at the end of years 5 and 6, respectively.

7. The ACME Company has recorded the following data for one of its new products over a six-month period. (The company assumes no trend or seasonal effects.)

Month	Demand (in units of product)
January	40
February	70
March	60
April	120
May	100
June	90

a. What would the February forecast have been if made at the end of January, using exponential smoothing with $\alpha = .2$ and a forecast for January of 30 units?

b. What would the May forecast have been if made at the end of April, using a four-month moving average?

c. What would be the mean absolute deviation (MAD) of the forecast errors for May and June, given forecasts for these two months were 105 and 95 units, respectively?

8. Sue Sayer is employed as a forecasting analyst for the Barry M. Stiff Casket Corporation. Ms. Sayer has collected the following sales data on Stiff's best-selling casket, Model 12–A:

Period	Sales
1	28
2	32
3	39
4	40
5	38
6	47
7	50
8	59
9	56

a. What's the three-period moving average forecast for period 8 made at the end of period 7?

b. If the forecast for period 5 were 35, what would be the forecast for period 6 made at the end of period 5, using basic exponential smoothing and a smoothing constant of $\alpha = .2$?

c. If the base at the end of period 7 were 54 and the trend at the end of period 7 were 4 ($\alpha = .2$, $\beta = .4$), what would be the forecast for period 9 made at the end of period 8?

d. Given the model described in c, what would be the forecast for period 10 made at the end of period 9?

e. Given the model described in c and d, what would be the forecast for period 11 made at the end of period 9?

f. What would be the mean absolute deviation of forecast errors, given forecasts for periods 1, 2, 3, and 4 were 30, 31, 35, and 38, respectively?

9. Edsel Muffler, Inc., showed the following sales figures for its stainless steel muffler, Rusty, over the past six months:

Month	Sales
July	125
August	84
September	60
October	44
November	36
December	44

Assume the base value at the end of November was 40 units and the trend value was -23 units. What would be the forecast for the following February made at the end of December ($\alpha = .3$, $\beta = .4$)?

10. The Tidy Corporation's sales manager has given you the following information regarding the sales history of one of its products, the widget:

Quarter	Year 1	Year 2
1	50	75
2	100	125
3	25	50
4	75	100

a. Plot the quarterly sales data for each year.

b. What seasonal index for each quarter could be used to forecast this sales product for year 3?

11. The Angel Wing Company's production manager is using a trend-enhanced exponential smoothing model to forecast demand for an end product called

Sparkles. At the start of the second quarter, the forecasting model had the following parameter values:

$$\alpha = .5.$$
$$\beta = .5.$$
$$\text{Base}_1 = 50 \text{ units per quarter.}$$
$$\text{Trend}_1 = 5 \text{ units per quarter.}$$
$$\text{Smoothed bias}_1 = -5 \text{ units per quarter.}$$
$$\text{Smoothed MAD}_1 = 10 \text{ units per quarter.}$$

a. What's the forecast for the demand for Sparkles for the third quarter made at the end of the first quarter?
b. If demand for Sparkles is 60 units during the second quarter, what are the values of the base and trend at the end of the second quarter?
c. Assuming demand for Sparkles is 60 units during the second quarter, what's the forecast of demand for Sparkles for the fourth quarter made at the end of the second quarter?
d. What's the value of the tracking signal for Sparkles as of the end of the first quarter?
e. What information does a tracking signal convey to the production manager?

12. The Alpha Corporation has a product with seasonal differences in sales between the halves of the year. Sales in the first half of the year are generally less than for the second half of the year. For this reason they've split sales from the two previous years (7 and 8) into two halves and would like to use these two parts to predict sales in the coming year (9). They also feel there's an upward trend in sales. Sales for the past two years were:

	Year 7	Year 8
First half (F)	100	105
Second half (S)	110	130

Company executives estimated the trend at the end of year 6 was 5 units per half year (T_{6S}) = 5. The base at the end of year 6 (B_{6S}) = 95. Seasonal factors were 1.05 and 0.95 for the first and second half of the year, respectively. (Note these factors are set purposely to the opposite of what they should be to demonstrate how they'll be corrected.)

$$\alpha = .3.$$
$$\beta = .5.$$
$$\gamma = .5.$$

a. Using a trend and seasonally enhanced model, forecast each of the four half years, sequentially updating the model at the end of each half year.
b. At the end of year 8 prepare a forecast for each half year in year 9.
c. Graph values of the updated seasonal factor for each half year.

13. For the first five months of the year, demand for Focii has been 14, 23, 12, 17, and 18. Farquart Focus has a focus forecasting system (naturally) using just

two forecasting techniques. The first is a two-period moving average and the second is simply that this period's demand will equal last period's. Farquart uses the MAD for the past three months as the criterion for choosing which model will make the forecast for the next month.

a. What will the forecast be for June and which model will be used?

b. Would it make any difference if demand for March had been 30 instead of 12?

14. Use a spreadsheet program to compare a three-period moving average forecasting model with a basic (ESF) exponential smoothing model. Five periods of past data exist (27, 26, 32, 41, and 28); the five future periods to be forecast have demands of 35, 43, 47, 28, and 38. Develop MAD values for each technique (forecasting one period ahead) for the five periods.

a. Using the average of the five periods of history to start the exponential smoothing model, what smoothing constant produces the MAD value closest to that of the moving average approach? Which model gives the lower MAD?

b. What changes when you use the average of the past three periods to start the exponential smoothing model?

15. The following two demand sets are to be used to test two different basic exponential smoothing models. The first model uses $\alpha = .1$, and the second uses $\alpha = .5$. In both cases, the model should be initialized with a beginning forecast value of 50; that is, the ESF forecast for period 1 made at the end of period 0 is 50 units. In each of the four cases (two models on two demand sets), compute the average forecast error and MAD. Use a spreadsheet program to do this analysis. What do the results mean?

Demand set I		Demand set II	
Period	Demand	Period	Demand
1	51	1	77
2	46	2	83
3	49	3	90
4	55	4	22
5	52	5	10
6	47	6	80
7	51	7	16
8	48	8	19
9	56	9	27
10	51	10	79
11	45	11	73
12	52	12	88
13	49	13	15
14	48	14	21
15	43	15	85
16	46	16	22
17	55	17	88
18	53	18	75
19	54	19	14
20	49	20	16

Chapter 17

Independent Demand Inventory Management

This chapter concerns managing inventory items that aren't used directly to manufacture products. These items include finished goods in factory or field warehouses, spare-parts inventories, office and factory supplies, and maintenance materials. Techniques in this chapter are used to determine appropriate order quantities and timing for individual items, and to manage multi-item systems. If we perform these functions well, we can provide appropriate levels of customer service without excess levels of inventory or management costs.

The chapter is organized around nine topics:

- Basic concepts: What is independent demand and what are the functions of independent demand inventory?
- Management issues: How can routine inventory decisions be implemented and how is performance measured?
- Inventory-related costs: How are costs of the inventory system measured and used?
- Economic order quantity model: What techniques are used to determine the quantity to order?
- Quantity discount model: How can discounts for purchasing large quantities be considered in determining order quantities?
- Order timing decisions: How can we determine timing of orders and set the level of safety stock?
- Order quantity and reorder point interactions: How can the quantity and timing of orders be jointly determined?
- Multi-item management: What techniques are available for focusing management attention on the important items?
- Multiple items from a single source: How can ordering several items from one source be coordinated?

Chapter 17 is related to material requirements planning in Chapters 2 and 11. Both these MRP chapters discuss lot sizing (order quantity) and order timing decisions. Principles of independent demand inventory management apply, even though MRP decisions are based on dependent demand. Chapters 3 and 12 present JIT concepts that consider inventory management when the order cost is very low and velocity is the key criterion. The forecasting material in Chapter 16 provides a key input into the inventory management system. Systems for distribution inventory management appear in Chapter 18.

BASIC CONCEPTS

The investment in inventory typically represents one of the largest single uses of capital in a business, often over 25 percent of total assets. In this section, we discuss different types of inventory, distinguishing between independent and dependent demand inventories. We also describe functions of different types of inventories (transit, cycle, safety, and anticipation stock).

Independent versus Dependent Demand Items

This chapter concerns managing independent demand inventories. The demand for items contained in independent demand inventories (such as those stocked in the field warehouses in Figure 17.1) is primarily influenced by factors outside of company decisions. These external factors induce random variation in demand for such items. As a result, demand forecasts for these items are typically projections of historical demand patterns. These forecasts estimate the average usage rate and pattern of random variation.

Demand for the items in the manufacturing inventories in Figure 17.1 (e.g., the raw material and component items) is directly dependent on internal factors well within the firm's control, such as the final assembly schedule (FAS) or master production schedule (MPS); that is, demand for raw materials and component items is a derived demand, which we can calculate exactly once we have the FAS or MPS. Therefore, demand for end product items is called *independent demand,* while demand for items contained in manufacturing inventories is called *dependent demand.*

Figure 17.1 gives other examples of independent demand inventories. Items subject to random use such as spare parts for production equipment, office supplies, or production supplies used to support the process all have independent demands. The techniques described in this chapter are suitable for all such items. Demand for these items can't be calculated from a production schedule or other direct management program.

FIGURE 17.1 Dependent and Independent Demand Inventories

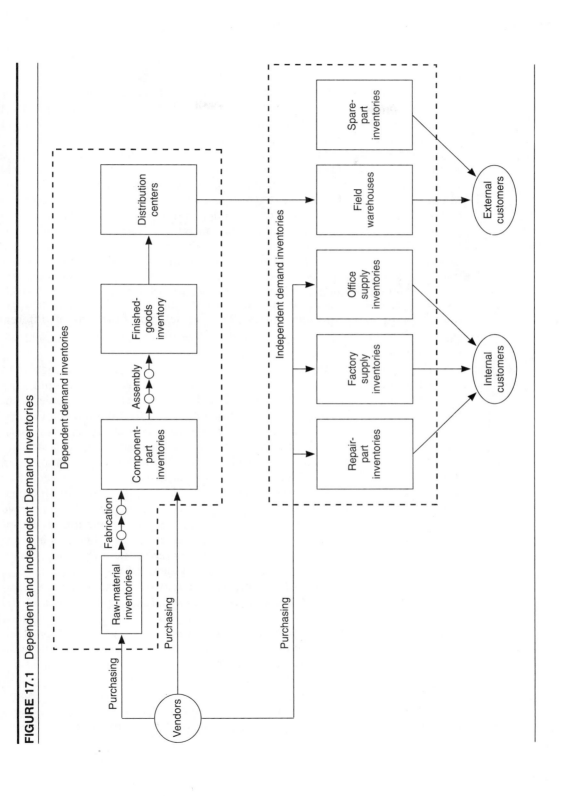

Functions of Inventory

An investment in inventory enables us to decouple successive operations or anticipate changes in demand. Inventory also enables us to produce goods at some distance from the actual consumer. This section describes four types of inventories that perform these functions.

Transit stock depends on the time to transport goods from one location to another. These inventories (along with those in distribution centers, field warehouses, and customers' locations) are also called *pipeline inventories*. Management can influence the magnitude of the transit stock by changing the distribution system's design. For example, in-transit inventory between the raw material vendor and factory can be cut by: (1) changing the transportation method (e.g., switching from rail to air freight) or (2) switching to a supplier closer to the factory to reduce transit time. These choices, however, involve cost and service trade-offs, which need to be considered carefully. For example, shipping raw material by air freight instead of by rail may cut transit time in half and therefore reduce average pipeline inventory by 50 percent, but it might increase unit cost due to higher transportation costs. Therefore, the consequences of changing suppliers or transport modes should be weighed against investing in more (or less) inventory.

Cycle stock exists whenever orders are made in larger quantities than needed to satisfy immediate requirements. For example, a warehouse may sell two units of a given end item weekly. However, because of scale economies with larger shipping quantities, it might choose to order a batch of eight units once each month. By investing in cycle stock, it can satisfy many periods of demand, rather than immediate need, and keep shipping costs down.

Safety stock provides protection against irregularities or uncertainties in an item's demand or supply. That is, when demand exceeds what's forecast or when resupply time is longer than anticipated. Safety stock ensures that customer demand can be satisfied immediately, and that customers won't have to wait while their orders are backlogged. For example, a portion of the inventory held at distribution centers may be safety stock. Suppose average demand for a given product in a distribution center is 100 units a week with a restocking lead time of one week and weekly demand might be as large as 150 units with replenishment lead time as long as two weeks. To ensure meeting the maximum demand requirements in this situation, a safety stock of 100 units might be created.

An important management question concerns the amount of safety stock actually required; that is, how much protection is desirable? This question represents an inventory-investment trade-off between protection against demand and supply uncertainties and costs of investing in safety stock.

Anticipation stock is needed for products with seasonal patterns of demand and uniform supply. Manufacturers of children's toys, air-conditioners, and calendars all face peak demand conditions where the production facility is frequently unable to meet peak seasonal demand. Therefore, anticipation stocks are built up in advance and depleted during the peak demand periods. Again, trade-offs must be considered. An investment in additional factory capacity could reduce the need for anticipation stocks.

MANAGEMENT ISSUES

Several issues surround the management of independent demand inventories. In this section we look at three: making routine inventory decisions, determining inventory system performance, and timing implementation.

Routine Inventory Decisions

Basically only two decisions need to be made in managing independent demand inventories: *how much to order (size)* and *when to order (timing)*. These two decisions can be made routinely using any one of the four inventory control *decision rules* in Figure 17.2. The decision rules involve placing orders for either a fixed or a variable order quantity, with either a fixed or a variable time between successive orders. For example, under the commonly used order point (Q,R) rule, an order for a fixed quantity (Q) is placed whenever the stock level reaches a reorder point (R). Likewise, under the S,T rule, an order is placed once every T periods for an amount equaling the difference between current on-hand balance and a desired inventory level (S) upon receipt of the replenishment order.

Effective use of any of these decision rules involves properly determining decision rule parameter values (e.g., Q, R, S, and T). This chapter details procedures for determining order quantity (Q) and reorder point (R) parameters for the order point rule, and it gives references covering determination of parameter values for the other decision rules in Figure 17.2.

FIGURE 17.2   Inventory Decision Rules

	Order quantity	
Order frequency	*Fixed (Q)**	*Variable (S)†*
Variable (R)‡	Q,R	S,R
Fixed (T)§	Q,T	S,T

*Q = Order a fixed quantity (Q).
†S = Order up to a fixed expected opening inventory quantity (S).
‡R = Place an order when the inventory balance drops to (R).
§T = Place an order every (T) periods.

Determining Inventory System Performance

A key management issue is determining the inventory control system's performance. We've already mentioned how large the investment in inventory can be. That investment's size makes it a visible performance measure. Because of this, some managers simply specify inventory reduction targets as the performance measure. Unfortunately this is usually too simplistic. It doesn't reflect trade-offs between the inventory investment and other benefits or activities in the company.

A common measure of inventory performance, **inventory turnover,** relates inventory levels to the product's sales volume. Inventory turnover is computed as annual sales volume divided by average inventory investment. Thus, a product with annual sales volume of $200,000 and average inventory investment of $50,000 has inventory turnover of 4. That is, the inventory was replaced (turned) four times during the year.

Turnover is often used to compare an individual firm's performance with others in the same industry or to monitor the effects of a change in inventory decision rules. High inventory turnover suggests a high rate of return on inventory investment. Nevertheless, though it does relate inventory level to sales activity, it doesn't reflect benefits of having the inventory.

To incorporate a major benefit of inventory, some firms use customer service to assess their inventory system performance. One common measure of customer service is the **fill rate** (the percentage of units immediately available when requested by customers). Thus, a 98 percent fill rate means only 2 percent of the units requested weren't on the shelf when a customer asked for them. A 98 percent fill rate sounds good. On the other hand, a 2 percent rate of unsatisfied customers doesn't. Some firms now use a dissatisfaction measure to focus attention on continuous improvement of customer service.

Other measures of inventory-related customer service can be used, but all attempt to formalize trade-offs in costs and benefits. Among the alternatives, we find percentage of the different items ordered that were available, number of times any shortage occurred in a time period, length of time before the item was made available, and percentage of customers who suffered a lack of availability. The correct measure or measures depend upon the reason for having the inventory, the item's importance, the nature of the business, and the firm's objectives.

Timing the Implementation

After analysis of the appropriate decision rules and performance measures, the critical management task is making the changes to improve inventory performance. Appropriate timing of these changes is important. Informal procedures may be quite effective for managing inventories in a small-scale warehouse; as the number of products and sales volumes increases, more formal inventory control methods are needed to assure continued growth. Further improvements might be warranted as the business grows and as inventory management technology improves.

Some of this chapter's concepts require new mind-sets, such as the distinction between dependent and independent demand. Other concepts require new organizational objectives and role changes throughout the company. Both these issues must be explicitly considered in timing implementation. One final caveat in implementation, especially for highly automated computer systems: the basic systems must be in place first. If inventory accuracy is poor, computerizing only means that mistakes can be made at the speed of light! If the warehouse currently runs on informal knowledge of what's where and how much is available, or if some inventory is held back by salespersons for "their" customers, a formal system won't help. Basic disciplines and understandings must be in place before formal decision rules are developed.

INVENTORY–RELATED COSTS

Investment in inventory isn't the only cost associated with managing inventories, even though it may be the most visible. This section treats three other cost elements: cost of preparing an order for more inventory, cost of keeping that inventory on hand until a customer requests it, and cost implied when there is a shortage of inventory. We'll also discuss incremental costs in the context of inventory management.

Order Preparation Costs

Order preparation costs are incurred each time an inventory replenishment order is placed. Included are the variable clerical costs associated with issuing the paperwork, plus any one-time costs involved in transporting goods between plants and warehouses. Work measurement techniques, such as time study, can be used to measure the labor content of order preparation. Determining other order preparation costs is sometimes more subtle. For instance, the inventory balance might need to be verified before ordering. Sometimes there may be a fixed cost for filling out a form and a variable cost for each item ordered. Companies frequently bear large costs of maintaining files, controlling quality, and verifying accurate receipts, as well as other hidden costs.

Inventory Carrying Costs

Inventory commits management to certain costs that are related to inventory quantity, items' value, and length of time the inventory is carried. By committing capital to inventory, a firm forgoes use of these funds for other purposes (e.g., to acquire new equipment, to develop new products, or to invest in short-term securities). Therefore, a cost of capital, which is expressed as an annual interest rate, is incurred on the inventory investment.

The cost of capital may be based on the cost of obtaining bank loans to finance the inventory investment (e.g., 10 to 20 percent), the interest rate on short-term securities the firm could earn if funds weren't invested in inventory (e.g., 5 to 15 percent), or the rate of return on capital investment projects that can't be undertaken because funds must be committed to inventory. For example, the cost of capital for inventory investment might be 25 percent in the case where a new machine would yield a 25 percent return on investment. In any case, capital cost for inventory might be determined by alternative uses for funds. Cost of capital typically varies from 5 to 35 percent, but climbs substantially higher in some cases.

The cost of capital is only one part of inventory holding cost. Others are the variable costs of taxes and insurance on inventories, costs of inventory obsolescence or product shelf life limitations, and operating costs involved in storing inventory—for example, rental of public warehousing space, or costs of owning and operating warehouse facilities (such as heat, light, and labor).

As an example, if capital cost is 10 percent, and combined costs of renting warehouse space, product obsolescence, taxes, and insurance come to an additional 10 percent of the average value of the inventory investment, total cost of carrying inventory is 20 percent of the cost of an inventory item. In this example, an inventory item costing $1 per unit would have an inventory carrying cost of $.20/unit/year.

Shortage and Customer Service Costs

A final set of inventory-related costs are those incurred when demand exceeds the available inventory for an item. This cost is more difficult to measure than the order preparation or inventory carrying costs.

In some cases, shortage costs may equal the product's contribution margin when the customer can purchase the item from competing firms. In other cases, it may only involve the paperwork required to keep track of a backorder until a product becomes available. However, this cost may be very substantial in cases where significant customer goodwill is lost. The major emphasis placed on meeting delivery requirements in many firms suggests while shortage and customer service costs are difficult to measure, they're critical in measuring inventory performance.

Customer service measures are frequently used as surrogate measures for inventory shortage cost—for example, the fill rate achieved in meeting product demand (e.g., the percentage of demand supplied directly from inventory upon demand). If the annual demand for an item is 1,000 units and 950 units are supplied directly from inventory, a 95 percent fill rate is achieved.

The level of customer service can be measured in several ways—examples include the fill rate, average length of time required to satisfy backorders, or percentage of replenishment order cycles in which one or more units are back ordered. Level of customer service can also be translated into level of inventory investment required to achieve a given level of customer service. As an example, a safety stock of 1,000 units may be required to achieve an 85 percent customer service level, while 2,000 units of safety stock may be required to achieve a 98 percent customer service level. Translating customer service level objectives into the inventory investment required often is useful in determining customer service level/inventory trade-offs.

Incremental Inventory Costs

Two criteria are useful in determining which costs are relevant to a particular inventory management decision: (1) Does the cost represent an actual out-of-pocket expenditure or a forgone profit? (2) Does the cost actually vary with the decision being made? Determining the item cost used in calculating inventory carrying cost is a good illustration of applying these criteria. The item's cost should represent the actual out-of-pocket cost of purchasing or producing the item and placing it in inventory (i.e., an item's variable material, labor, and overhead costs). An element of the overhead cost, such as a cost allocation for general administrative expenses, isn't an actual out-of-pocket expenditure.

Another example involves measuring clerical costs incurred in preparing replenishment orders. If clerical staff size remains constant throughout the year, regardless of the number of replenishment orders placed, this cost is not relevant to the decision being made (i.e., the replenishment order quantity). These examples are not meant to be exhaustive, but rather illustrative of the careful analysis required in determining costs to be considered in evaluating inventory management performance.

Example Cost Trade-Offs

Order quantity decisions primarily affect the amount of inventory held in cycle stocks at the various stocking points in Figure 17.1. Large order quantities mean orders are placed infrequently and lead to low annual costs of preparing replenishment orders, but they also increase cycle stock inventories and annual costs of carrying inventory. Determining replenishment order quantities focuses on the question of what quantity provides the most economic trade-off between order preparation and inventory carrying costs. An example item stocked in a field warehouse is used to illustrate this trade-off.

The Model 100 movie camera is sold to several hundred retail stores from a field warehouse. To avoid excessive inventories, stores place orders frequently and in small quantities. The demand for the movie camera at a typical field warehouse was obtained from past sales records. It averages 5 units per weekday (or 1,250 units per year). The movie camera can be obtained within a one-day lead time from the distribution center (DC) serving the field warehouse. This requires preparing an order and faxing it to the DC. The variable cost of preparing a replenishment order is estimated to be $6.25. The firm's cost of carrying inventory is estimated at 25 percent of the item cost per year, including variable costs of capital, insurance, taxes, and obsolescence. Since the camera's unit cost is $100, inventory carrying cost is $25/unit/year.

Currently, the field warehouse orders the Model 100 movie camera on a daily basis in lots of five units. The solid line in Figure 17.3 plots the inventory level versus time for this decision rule. This plot assumes demand is

---

FIGURE 17.3   Inventory Level versus Time for Model 100 Movie Camera

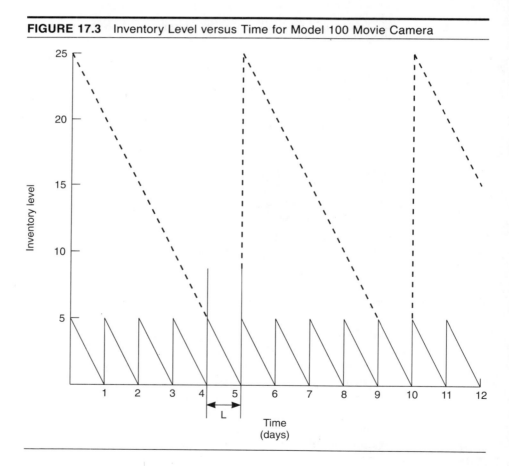

constant at 5 units per day, and the resulting average inventory level is 2.5 units. Since orders are placed daily, 250 orders are placed per year, costing a total of $1,562.50/year ($6.25 × 250). The average inventory of 2.5 units represents an annual inventory carrying cost of $62.50 a year (2.5 × $25), yielding an overall combined cost of $1,625/year for placing orders and carrying inventory.

The dashed line in Figure 17.3 shows the inventory level plot for an alternative order quantity of 25 units, or placing orders weekly. In this case, average inventory is 12.5 units and 50 orders are placed annually. The larger order quantity in this case provides important savings in ordering cost ($312.50 versus the previous $1,562.50) with an increase in annual inventory cost ($312.50 versus the previous $62.50). Overall, a shift to a larger order quantity produces a favorable trade-off between ordering and inventory carrying costs, which cuts total cost to $625 per year.

A number of order quantities should be evaluated to determine the best trade-off between ordering and inventory carrying costs. The economic order quantity model enables us to determine the lowest-cost order quantity directly.

ECONOMIC ORDER QUANTITY MODEL

The order quantity decision is formally stated in the **economic order quantity (EOQ)** model. This equation describes the relationship between costs of placing orders, costs of carrying inventory, and the order quantity. This model makes several simplifying assumptions: the demand rate is constant, costs remain fixed, and production and inventory capacity are unlimited. Despite these seemingly restrictive assumptions, the EOQ model provides useful guidelines for ordering decisions—even in operating situations that depart substantially from these assumptions.

The total annual cost equation for the economic order quantity is

$$\text{TAC} = (A/Q)\, C_P + (Q/2)\, C_H. \tag{17.1}$$

This equation contains two terms. The first term, $(A/Q)\, C_P$, represents annual ordering cost, where A is annual demand for the item, Q is order quantity, and C_P is cost of order preparation. Therefore, the total ordering cost per year is proportional to the number of orders placed annually (A/Q).

The second term, $(Q/2)\, C_H$, represents annual inventory carrying cost, where average inventory is assumed to be half the order quantity (Q), and C_H is the inventory carrying cost per unit per year; that is, item cost (v) times the annual inventory carrying cost rate (C_r).

Combined costs of ordering and carrying inventory are expressed as a function of the order quantity (Q) in Equation (17.1), enabling us to evaluate the total cost of any given order quantity.

Solving the EOQ Model

One method of determining the lowest-cost ordering quantity is to graph the total cost equation for various order quantities. Figure 17.4 plots the total cost equation for the Model 100 movie camera for several different order quantities based on the following data:

$$A = 1,250.$$
$$C_P = 6.25.$$
$$C_H = 25.$$
$$\text{TAC} = (1,250/Q)6.25 + (Q/2)25.$$

Minimum total cost can be found graphically to equal 25 (i.e., placing orders weekly). Both terms of the total-cost equation are also plotted.

FIGURE 17.4   Cost versus Order Quantity for Model 100 Movie Camera

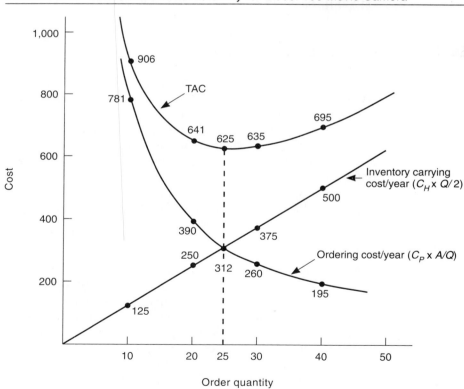

We should note several facts in these graphs. First, inventory carrying costs increase in a straight line as order quantity is increased, while ordering cost diminishes rapidly at first and then at a slower rate as the ordering cost is allocated over an increasing number of units. Second, in the EOQ model, the minimum cost solution exists where the ordering cost per year equals annual inventory carrying cost. (This observation is used in developing lot-sizing decision rules for dependent demand items.) Finally, total cost is relatively flat around the minimum cost solution ($Q = 25$ in this case), indicating inventory management performance is relatively insensitive to small changes in order quantity around the minimum-cost solution.

A second and more direct method of solving for the minimum-cost order quantity is by using the EOQ formula in Equation (17.2):

$$\text{EOQ} = \sqrt{2C_P A/C_H}. \tag{17.2}$$

This formula is derived from the total cost equation (17.1) using calculus. That is, Equation (17.1) is differentiated with respect to the decision variable Q and solved by setting the resulting equation equal to zero, as Equations (17.3) through (17.6) show:

$$dTAC/dQ = -C_P(A/Q^2) + C_H/2 \tag{17.3}$$

$$C_P A/Q^2 = C_H/2 \tag{17.4}$$

$$Q^2 = 2 C_P A/C_H \tag{17.5}$$

$$\text{EOQ} = \sqrt{2C_P A/C_H} \tag{17.6}$$

where:

EOQ = The optimal value of Q.

Using the EOQ formula for the Model 100 movie camera produces a lot size of 25; that is, $\sqrt{[(2)(6.25)(1250)]/25}$. In using this expression, we must make sure both demand and inventory carrying cost are measured in the same units (1,250 units/year and \$25/unit/year, in this case).

In addition to its use in determining order quantities, the EOQ formula can also be used to develop another important measure in the control of inventories—the **economic time between orders (TBO).** The formula to calculate TBO in weeks is

$$\text{TBO} = \text{EOQ}/\overline{D} \tag{17.7}$$

where:

$\overline{D}$ = Average weekly usage rate.

For the Model 100 movie camera, the TBO equals one week (25 units/order)/(25 units/week). This measure can be used to determine an economic ordering frequency or time between inventory reviews. In the case of the movie camera, we might consider using a Q,T decision rule; that is, order an economic lot size weekly.

QUANTITY DISCOUNT MODEL

One assumption underlying the EOQ model is unit cost remains fixed over the range of order quantities considered. This frequently isn't the case for purchased items where price discounts and transportation rate breaks are quoted when these items are ordered in large quantities. When discounts are possible, trade-offs reflected in the decision model become more complex, and solution procedures require more computations. The trade-offs involve a reduction in both ordering cost and item cost for an increase in inventory carrying cost, when the order quantity rises (and vice versa).

The total annual cost expression (TAC) for the quantity discount model includes three terms: annual purchase cost (vA), annual ordering cost ($A/Q)C_P$, and annual inventory carrying cost ($Q/2)C_H$. The quantity discount TAC equation, (17.8), differs in two respects from Equation (17.1). First, total annual cost of purchasing the item (CA) is a function of the item unit cost (v); second, inventory carrying cost (C_H) also depends on the item unit cost ($C_H = vC_r$, where C_r is the annual inventory carrying cost rate). The item unit cost, in turn, depends on order quantity.

$$\text{TAC} = (v)A + (A/Q)C_P + (Q/2)C_H \qquad (17.8)$$

For example, the variable transportation cost for shipping the Model 100 movie camera from the distribution center to the field warehouse is $10 per unit for lot sizes of less than 40 units, and $5 per unit for lot sizes of 40 or more units. The quantity 40 is called the **break point** for getting the lower cost. If we assume the movie camera's $100 unit cost was delivered cost, based on daily shipments of 5 units, this means the item value is $100/unit ($v_1$) for order quantities below 40 units, and $95/unit ($v_2$) for larger order quantities.

Since item cost isn't a continuous function of the order quantity, the quantity discount model can't be solved for an exact solution using calculus, and the computational procedure involves several steps. Magee and Boodman suggest the following five-step method for direct calculation of the minimum-cost order quantity:

1. Calculate the economic order quantity using minimum unit cost. If this quantity falls within the range for which the cost is correct, it's a *valid* economic order quantity and will result in the minimum cost for the particular item.

2. If the EOQ calculated in Step 1 isn't valid, find total annual cost for each break point quantity at which item cost changes.
3. Calculate an EOQ for each item cost.
4. Calculate total annual cost for each valid EOQ determined in Step 3.
5. The minimum-cost order quantity is that associated with the lowest cost found in either Step 2 or Step 4.

We illustrate this procedure using the Model 100 movie camera and $v_1 =$ $100, $v_2 =$ $95, and a minimum order quantity (b) of 40 for the $95 cost. The calculations at each step are:

1. EOQ $= \sqrt{2\ (1{,}250)\ (6.25)/(.25)\ 95} = 26$.
   (This EOQ is *invalid,* since it's less than the minimum order quantity of 40 required for the minimum cost, $95.)
2. $TAC_b = (95)\ (1{,}250) + (6.25)(1250/40) + (.25)(95)(40/2) = \$119{,}420$.
3. EOQ $= \sqrt{[(2)\ (1{,}250)\ (6.25)]/(.25)\ (100)} = 25$.
4. $TAC_{EOQ} = (100)(1{,}250) + (6.25)\ (1250/25) + (.25)(100)(25/2) = \$125{,}625$.
5. Minimum-cost order quantity is therefore the break point, 40.

Figure 17.5 illustrates each step in this process. The convention used for indicating the curves in Figure 17.5 is: TAC_1 is the total annual cost curve when the transportation cost is $100; TAC_2 is for a cost of $95; TAC_b is the cost at the break point of 40 units; TAC_{EOQ} is the cost at the EOQ. Note EOQ of 26 on TAC_2 isn't a feasible solution, since it lies below the break point of 40 units; and TAC_b is less than TAC_1.

ORDER TIMING DECISIONS

In this section we describe timing of replenishment orders under the order point rule Q,R from Figure 17.2. This means calculating the reorder point (R). The inventory level is assumed to be under continuous monitoring (review), and, when the stock level reaches the reorder point, a replenishment order for a fixed quantity (Q) is issued. Setting the reorder point is influenced by four factors: demand rate, lead time required to replenish inventory, amount of uncertainty in the demand rate and in the replenishment lead time, and management policy regarding the acceptable level of customer service.

When there's no uncertainty in an item's demand rate or lead time, safety stock isn't required, and determination of the reorder point is straightforward. For example, if the Model 100 movie camera's demand rate is assumed to be exactly five units per day, and replenishment lead time is exactly one day, a reorder point of five units provides sufficient inventory to cover demand until the replenishment order is received.

FIGURE 17.5   Purchase Discount Total Annual Cost Curves

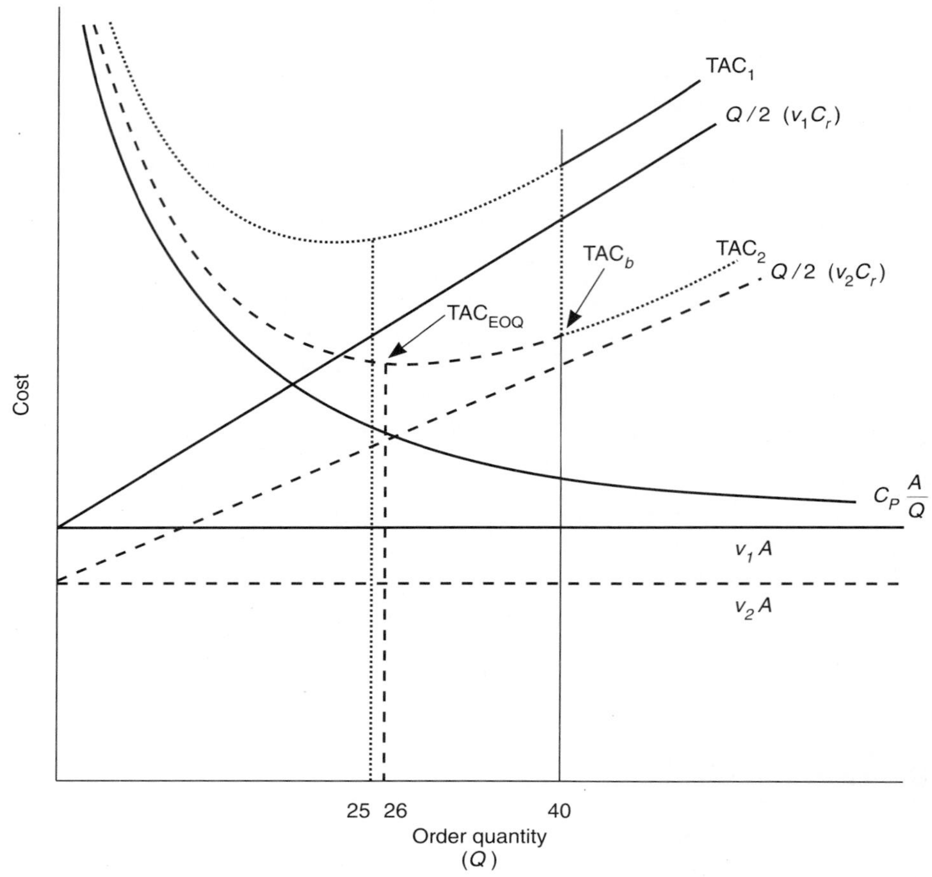

Sources of Demand and Supply Uncertainty

The assumptions of fixed demand rate and constant replenishment lead time are rarely justified in actual operations. Random fluctuations in demand for individual products occur because of variations in the timing of consumers' purchases of the product. Likewise, the replenishment lead time often varies because of machine breakdowns, employee absenteeism, material shortages, or transportation delays in the factory and distribution operations.

The Model 100 movie camera illustrates the amount of uncertainty usually experienced in demand for end product items. Analysis of this item's warehouse sales and inventory records indicates replenishment lead time is quite

FIGURE 17.6 Model 100 Movie Camera Daily Demand

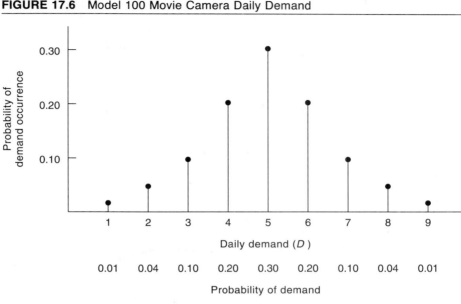

stable, requiring a one-day transit time from the distribution center to the field warehouse. However, daily demand (D) varies considerably for the camera. While daily demand averages five units, demands of from one to nine units have been experienced, as Figure 17.6 shows.

If the reorder point is set at five units to cover average demand during the one-day replenishment lead time, inventory shortages of one to four units can result when daily demand exceeds the average of five units; that is, when demand equals six, seven, eight, or nine units. Therefore, if we're to protect against inventory shortages when there's uncertainty in demand, the reorder point must be greater than average demand during the replenishment lead time. The difference between the average demand during lead time and the reorder point is called **safety stock (S).** Increasing the reorder point to nine units would provide a safety stock of four units, for example. It would also prevent any stock-outs from occurring if the Model 100 movie camera's historical pattern of demand does not change.

The Introduction of Safety Stock

Figure 17.7 illustrates introducing safety stock into the reorder point setting. The reorder point (R) in this diagram has two components: safety stock level (S), and level of inventory ($R - S$) required to satisfy average demand ($\bar{d}$) during the average replenishment lead time (L). The reorder point is the sum

FIGURE 17.7   Safety Stock as a Buffer against Demand Variability

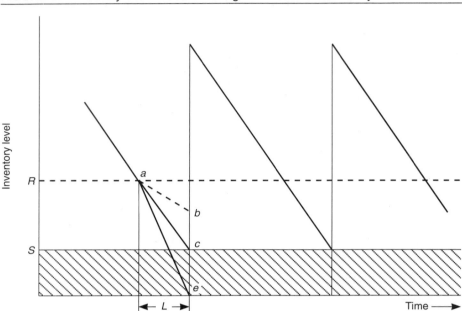

of these two: $R = \overline{d} + S$. To simplify this explanation, lead time in Figure 17.7 is assumed to be constant while demand rate varies.

When a replenishment order is issued (at point a), demand variations during the replenishment lead time mean the inventory level can drop to a point between b and e. In the movie camera's case, inventory level may drop by one to nine units (points b and e, respectively) before a replenishment order is received. When demand equals the average rate of five units or less, the inventory level reaches a point between b and c, and the safety stock isn't needed. However, when the demand rate exceeds the five-unit average and inventory level drops to a point between c and e, a stock-out will occur unless safety stock is available. (We can construct a similar diagram when both demand rate and lead time vary.)

Determining the Safety Stock.   Before deciding the safety stock level, we must establish a criterion for determining how much protection against inventory shortages is warranted. One of two different criteria is often used: the probability of stocking out in any given replenishment order cycle, or the desired level of customer service in satisfying product demand immediately out of inventory (the fill rate). We illustrate both criteria using the demand distribution for the Model 100 movie camera in Figure 17.6.

Stock-Out Probability.   One method for determining the required level of safety stock is to specify an acceptable risk of stocking out during any given replenishment order cycle. Figure 17.6 provides demand distribution data for this analysis for the Model 100 movie camera. There is a .05 probability of demand exceeding seven units (i.e., a demand of either eight or nine units occurring). A safety stock level of two units (meaning a reorder point of seven units) would provide a risk of stocking out in 5 percent (1 out of 20) of the replenishment order cycles. This safety stock level provides a .95 probability of meeting demand during any given replenishment order cycle. Note this means there is a .05 probability of stocking out by *either* one or two units when demand exceeds seven units.

We can reduce the risk of stocking out by investing more in safety stock; that is, with safety stock of three units, the probability of stocking out can be cut to .01, and with four units of safety stock the risk of stocking out is 0, assuming the demand distribution doesn't change. Thus, one method of determining the required level of safety stock is to specify an acceptable trade-off between the probability of stocking out during a replenishment order cycle and investment of funds in inventory.

Customer Service Level.   A second method for determining the required level of safety stock is to specify an acceptable fill rate. For doing this, we define the customer service level as the percentage of demand, measured in units, that can be supplied directly out of inventory. Figure 17.8 provides data for calculations for the Model 100 movie camera. It shows a safety stock of 1 unit, which enables 95.8 percent of the annual demand of 1,250 units for this item to be supplied directly out of inventory to the customer. We compute the service level as follows:

FIGURE 17.8   Safety Stock Determination

Reorder point (R)	Safety stock (S)	Demand probability (P(d) = R)	Probability of stocking out (P(d) > R)	Average number of shortages per replenishment order cycle*	Service† level (SL)
5	0	.30	.35	.56	88.8%
6	1	.20	.15	.21	95.8
7	2	.10	.05	.06	98.8
8	3	.04	.01	.01	99.8
9	4	.01	.00	0	100.0

*This is calculated by:

$$\sum_{d=R+1}^{d_{MAX}} P(d)(d - R)$$

†Assuming the replenishment order quantity is five units.

$$SL = 100 - (100/Q) \sum_{d=R+1}^{d_{MAX}} P(d)(d-R) \qquad (17.9)$$

where:

Q = The order quantity.

R = The reorder point.

$P(d)$ = The probability of a demand of d units during the replenishment lead time.

d_{MAX} = The maximum demand during the replenishment lead time.

For example, when the safety stock is set at one unit in Figure 17.8, we compute the service level as

$$SL = 95.8 = 100 - (100/5)[(.01)(3) + (.04)(2) + (0.1)(1)]. \qquad (17.10)$$

A service level of 95.8 percent means 4.2 percent of the annual demand, or $(.042)(1,250) = 52.5$ units, can't be supplied directly out of inventory. Since the current lot size (Q) is 5 units and the item is ordered 250 times per year, the average number of stock-outs per reorder cycle is .21 (i.e., 52.5/250), as shown in Figure 17.8.

Figure 17.8 shows the impact of increasing the safety stock level on both the service level and the average number of shortages per replenishment order cycle. The service level can be raised to 100 percent by increasing safety stock to four units. Again, as in the case of the stock-out probability method described previously, choice of the required safety stock level depends on determining an acceptable trade-off between customer service level and inventory investment.

So far, the safety stock and order quantity parameters for an order point system have been determined separately. These two parameters are, however, interdependent in their effect on customer service level performance. We can see this interactive effect in Equation (17.9), since both safety stock level and order quantity size affect the level of customer service. A more comprehensive example later in this chapter will address this interaction.

Continuous Distributions

Two different criteria for determining the required safety stock level and the reorder point have been described (i.e., use of a stock-out probability and a desired level of customer service). In discussing both criteria, we used a discrete distribution to describe the uncertainty in demand during the replenishment lead time (order cycle). It's frequently convenient to approximate a discrete distribution with a continuous distribution to simplify the safety stock and reorder point calculations. One distribution that often provides a

---

FIGURE 17.9 Normal Approximation to the Empirical Demand Distribution*

Midpoint X	Discrete distribution probability	Interval	Normal distribution probability	Probability of demand exceeding (X + 0.5)	Expected number of stock-outs when reorder point = X[†]
1	.01	.5–1.5	.0085	.9902	4.0068
2	.04	1.5–2.5	.0380	.9522	3.0128
3	.11	2.5–3.5	.1109	.8413	2.0591
4	.20	3.5–4.5	.2108	.6305	1.2303
5	.30	4.5–5.5	.2610	.3695	.5983
6	.20	5.5–6.5	.2108	.1587	.2255
7	.10	6.5–7.5	.1109	.0478	.0641
8	.04	7.5–8.5	.0380	.0098	.0127
9	.01	8.5–9.5	.0085	.0013	.0018

*A χ^2 test indicates that these two distributions are not significantly different. ($\chi^2 = 8.75$ versus 20.09 at the 0.01 level of significance.)
†This is σ_d E(Z) based on the E(Z) values from R. G. Brown, *Decision Rules for Inventory Management* (New York: Holt, Rinehart & Winston, 1967) pp. 95–103.

---

close approximation to empirical data is the normal distribution. In this section, we indicate the changes required in the calculations when the normal distribution is used to describe uncertainty in demand during the replenishment lead time.

Figure 17.9's data enable us to compare the empirically derived probability values for the Model 100 movie camera demand in Figure 17.6, with similar values derived from the normal distribution. The comparison shows the normal distribution closely approximates the empirical observations and can be used to determine safety stock and reorder point levels.

Probability of Stocking-Out Criterion

When the probability of stocking out is used as the safety stock criterion, the required level of safety stock and the reorder point values are easily computed using the normal distribution. First, we determine the mean and standard deviation for the distribution of demand during the replenishment lead time. These values have been calculated using the empirical distribution data for the Model 100 movie camera in Figure 17.6 and are shown in Figure 17.10 along with examples of the area (probability) under the normal distribution.

Next, we can calculate the safety stock (or reorder point) value using a table of normal probability values. For example, suppose sufficient safety stock is desired for the Model 100 movie camera that the probability of stock-

FIGURE 17.10 Daily Demand Distribution

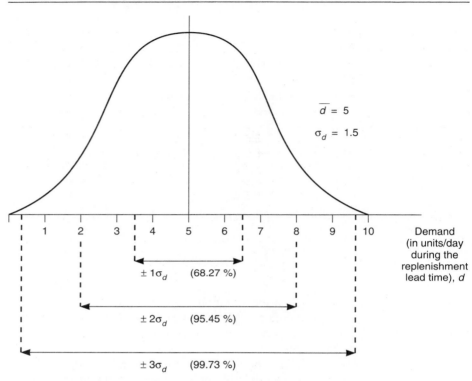

ing out in any given replenishment order cycle is .05. We determine the safety stock level and the reorder point as follows:

Safety stock $= Z\sigma_d$ (17.11)

Reorder point $=$ Mean demand during the replenishment
lead time $+ Z\sigma_d$ (17.12)

where:

$Z =$ The appropriate value from a table of standard normal distribution probabilities.

$\sigma_d =$ Demand during the replenishment lead time standard deviation.

The Z value for a .05 probability of stocking out is 1.645 (from a table of standard normal distribution probabilities). The required level of safety stock, therefore, is 2.5 units—that is, (1.645)(1.5)—and the reorder point is 7.5(8) units. The reorder point can also be determined directly from the data in Figure 17.9, where the probability of demand exceeding 7.5 is .0478.

Customer Service Criterion

When the customer service level is used as the safety stock criterion, we can also determine the desired level of safety stock using the normal distribution approximation. For this case, we need the average number of stock-outs per replenishment order cycle. To get this, the quantity $\sum_{d=R+1}^{d_{MAX}} P(d)(d - R)$ shown in Equations (17.9) and (17.10), is replaced by $\sigma_d E(Z)$. The σ_d still equals the standard deviation of the normal distribution being used to approximate demand during replenishment lead time. The $E(Z)$ value is the partial expectation of the normal distribution called the service function. It's the expected *number* of stock-outs when Z units of safety stock are held in the standard normal curve. A graph of the service function, $E(Z)$, is plotted in Figure 17.11. Note when Z is less than -1, the service function, $E(Z)$, is approximately linear.

The safety stock and reorder point calculations are similar to those shown earlier in Equations (17.9) and (17.10). As an illustration, suppose we want a 95 percent service level for the Model 100 movie camera, and we go back to using an order quantity of five units. The required value for $E(Z)$ is computed using Equation (17.14), which we derive from Equation (17.13):

$$SL = 100 - (100/Q)(\sigma_d E(Z)) \qquad (17.13)$$

or

$$E(Z) = [(100 - SL)Q]/100 \, \sigma_d. \qquad (17.14)$$

In this case, the service function value, $E(Z)$, equals .167; that is,

$$E(Z) = [(100 - 95)(5)]/[(100)(1.5)] = .167$$

and

$$\sigma_d E(Z) = 0.25.$$

From the service function table in Figure 17.11, we find an $E(Z)$ *of* .167 represents a Z value of approximately $+ .6\sigma_d$. The safety stock level therefore is .9 = (.6)(1.5). The reorder point would be 5.9. Alternatively, from Figure 17.9, we find $R = 6$ when $\sigma_d E(Z) = .2255$. Note this is the same result we got using the empirical discrete distribution earlier.

Time Period Correction Factor

In the preceding examples, the demand data were expressed as units per day and the lead time was one day. Sometimes the demand data are provided in a different number of time units than the lead time. For example, we might

FIGURE 17.11   Service Function

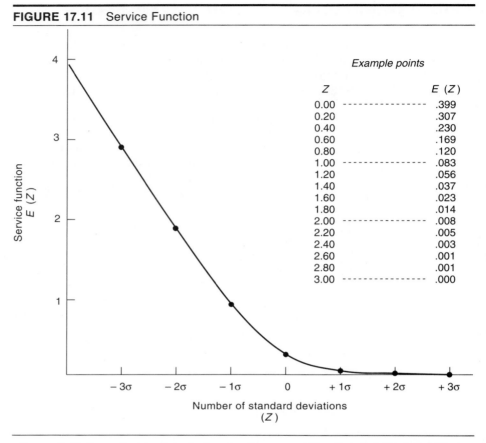

Z		E (Z)
0.00	- - - - - - - - - - - - -	.399
0.20		.307
0.40		.230
0.60		.169
0.80		.120
1.00	- - - - - - - - - - - - -	.083
1.20		.056
1.40		.037
1.60		.023
1.80		.014
2.00	- - - - - - - - - - - - -	.008
2.20		.005
2.40		.003
2.60		.001
2.80		.001
3.00	- - - - - - - - - - - - -	.000

Source: R. G. Brown, *Decision Rules for Inventory Management* (New York: Holt, Rinehart & Winston, 1967), pp. 95–103.

have weekly demand and a two-week lead time. In such case, an adjustment must be made as Equation (17.15) shows:

$$\text{Safety stock} = Z\,(\sigma_D \sqrt{m}) \tag{17.15}$$

where:

m = The lead time expressed as a multiple of the time period used for the demand distribution.

σ_D = Standard deviation of the demand per period.

If lead time for the Model 100 movie camera were three days instead of one day, required safety stock would be 4.3 units; that is, $(1.645)(1.5)\sqrt{3}$, and the reorder point would be 19.3 units: (3 days)(5 units/day) + 4.3 units. Since lead time in this example is three times the demand interval of one day, the

$\sqrt{3}$ factor has been included in calculating required safety stock. The resulting safety stock level increases for the three-day lead time to allow for the possible increase in variation in demand over the additional two days.

Forecast Error Distribution

In many inventory management software packages, demand values for the economic order quantity and reorder point calculations are forecast using statistical techniques such as exponential smoothing. When these forecasting techniques are used, the required safety stock level depends on the forecasting model's accuracy—how much variation there is around the forecast. Very little safety stock is required when forecast errors are small, and vice versa, for a fixed level of customer service. One commonly used measure of forecasting model accuracy is the mean absolute deviation (MAD) of the forecast errors.

The methods for determining the safety stock and reorder point levels described earlier in this chapter are relevant when product demand is forecast and a MAD value is maintained for the forecasting model. As an illustration, suppose an exponential smoothing model is used to forecast demand for the Model 100 movie camera, a .05 probability of stocking out during a reorder cycle is specified, and the forecast errors are normally distributed, as Figure 17.12 shows. We calculate the safety stock as follows:

$$\text{Safety stock} = Z\sigma_E = Z\,(1.25\ \text{MAD}) \tag{17.16}$$

where:

Z = The appropriate value from a table of standard normal distribution probabilities.

σ_E = Forecast error distribution standard deviation. (The value of σ_E can be approximated by 1.25 MAD when the forecast errors are normally distributed.)

Since the Z value is 1.645 for a .05 probability of stocking out and the MAD value equals 1.2 from Figure 17.12, the required level of safety stock is 2.5 units; that is, (1.645) (1.25) (1.2). The reorder point would be 7.5 units, as we found before. In this example, the forecast interval is the same as replenishment lead time—one day. When this isn't the case, an adjustment must be made to the MAD value in a manner analogous to Equation (17.13). The adjustment is:

$$\text{Safety stock} = Z\,(1.25\ \text{MAD})\,\sqrt{m} \tag{17.17}$$

where:

m = The lead time expressed as a multiple of the forecast interval.

FIGURE 17.12   Model 100 Movie Camera Forecast Error Distribution

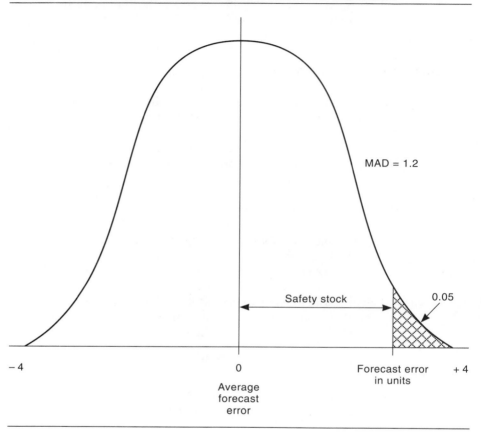

Safety stock can also be determined when customer service level is used as the safety stock criterion and product demand is forecast. In this case, Equation (17.15) is modified slightly:

$$E(Z) = [(100 - SL)\, Q] / [100\, (1.25\, MAD)]. \qquad (17.18)$$

After making this change, the safety stock and the reorder point calculations are similar to those involving $E(Z)$ shown earlier in this chapter.

ORDER QUANTITY AND REORDER POINT INTERACTIONS

We saw earlier that there is an interaction between the reorder point and the order quantity in terms of their impact on the customer service level. In this section we use an example problem to explore this interaction. The problem allows us to determine the demand during lead time distribution using discrete probability distributions for lead time and demand. We also introduce the shortage cost concept, which we use to expand the total cost equation. We then present two ways of jointly determining the reorder point and order quantity using total cost as well as service level criteria: a search procedure and an iterative procedure.

Service Levels and Order Quantities

Figure 17.13 provides data for a single inventory item that will serve as our example throughout this section. Note both lead time and weekly demand are uncertain. This means the process of determining distribution of demand during lead time must account for both uncertainties. We use a tree diagram to illustrate all possible combinations of demand and lead time that can occur. These are then summarized to give the distribution of demand during lead time.

The lead time and weekly demand distribution data in Figure 17.13 have been used to develop the three diagram in Figure 17.14. As an example of the calculations, we'll use the branch at the bottom of Figure 17.14. The 4-unit demand occurs during the replenishment lead time only when the lead time is 2 weeks and 2 units are demanded each week. The probability of these events occurring is: $(.2)(.2)(.2) = .008$. Figure 17.14 enumerates the remaining combinations of daily demand and lead time values; the results are summarized in the demand during lead time distribution in Figure 17.15.

Figure 17.15 also shows the expected number of units short, $E\{s\}$, for specified reorder points. A shortage can only occur when demand exceeds the

---

FIGURE 17.13   Data for Interaction Example

Notation

Item cost	ν	$500 per unit
Annual demand	A	35 units per year
Fixed ordering cost	C_P	$45 per order
Shortage cost	C_s	$60 per unit short
Inventory carrying cost rate	C_r	25% of item cost per year
Economic order quantity	EOQ	5 units
Replenishment lead time	L	1 week, probability = .8
		2 weeks, probability = .2
Average lead time	$\overline{L}$	1.2 weeks
Weekly demand	D	0 units, probability = .5
		1 unit, probability = .3
		2 units, probability = .2
Average weekly demand	$\overline{D}$	.7 units

FIGURE 17.14 Tree Diagram of Possible Occurrences of Lead Time and Demand for Example Problem

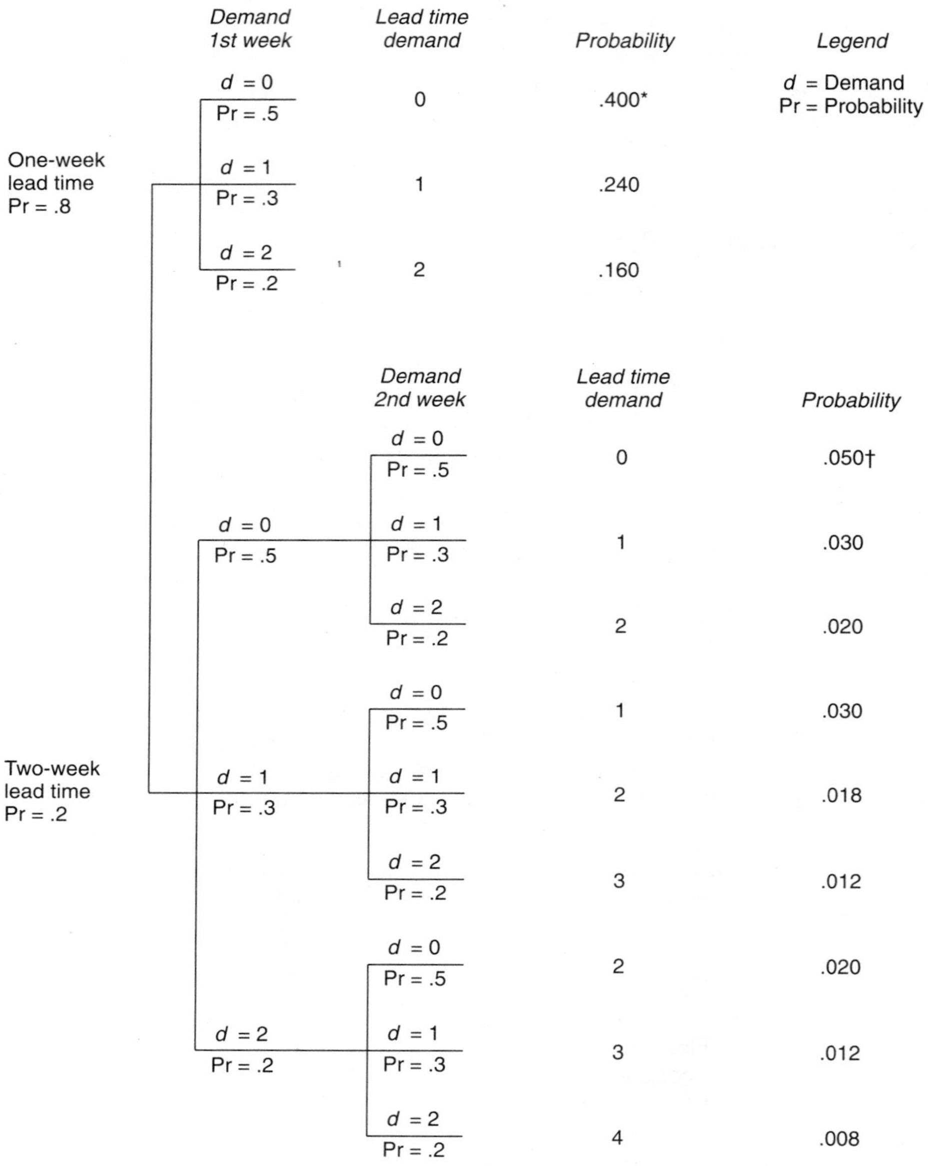

	Demand 1st week	Lead time demand	Probability	Legend
	d = 0 Pr = .5	0	.400*	d = Demand Pr = Probability
One-week lead time Pr = .8	d = 1 Pr = .3	1	.240	
	d = 2 Pr = .2	2	.160	

	Demand 2nd week	Lead time demand	Probability
	d = 0 Pr = .5	0	.050†
d = 0 Pr = .5	d = 1 Pr = .3	1	.030
	d = 2 Pr = .2	2	.020
	d = 0 Pr = .5	1	.030
Two-week lead time Pr = .2	d = 1 Pr = .3	d = 1 Pr = .3	2	.018
	d = 2 Pr = .2	3	.012
	d = 0 Pr = .5	2	.020
d = 2 Pr = .2	d = 1 Pr = .3	3	.012
	d = 2 Pr = .2	4	.008

*L = 1 week (Pr = .8) and d = 0 for that week (Pr = .5). Probability of both occurring is .4 = (.8 x .5).

†L = 2 weeks (Pr = .2) and d = 0 (Pr = .5) for each week. Probability of all three occurring is .05 = (.2 x .5 x .5).

FIGURE 17.15 Demand during Lead Time Distribution

Demand (d)	Probability of demand = d	Probability of demand > d	Reorder point (R)	Expected number of units short (E{s}) when reorder point = R
0	.450	.550	0	.840 units
1	.300	.250	1	.290 units
2	.218	.032	2	.040 units
3	.024	.008	3	.008 units
4	.008	0	4	0 units
Expected demand during lead time = .840				

reorder point. So, for example, when the reorder point is 4, no shortages can occur. If the reorder point is 2, there's a 1-unit shortage if the demand is 3 and a 2-unit shortage if demand is 4. The probabilities of these demands occurring are .024 and .008, respectively. This means the expected number of stock-outs are $(.024 \times 1) + (.008 \times 2) = .040$ when the reorder point is 2.

To show the interaction between reorder point and order quantity, we use the data in Figure 17.15. Suppose the item was currently ordered about five times per year in quantities of 7. If the reorder point was set to 1 unit, the expected number of units short *per reorder cycle* would be .29. This would mean, for the 5 cycles per year, about 1.5 units would be out of stock in a year. This corresponds to a service level of about 95 percent $[(35 - 1.5)/35]$.

If the order quantity is changed to 35, only 1 reorder cycle per year would occur. There would be an expected .29 units short in the cycle if the reorder point was 1, but that's now the expected number short for the year, as well. This corresponds to a service level of 99 percent $[(35 - .29)/35]$. Even a reorder point of 0 would provide a level of service of about 97 percent $[(35 - .84)/35]$ when 35 units are ordered at a time.

The order quantity of 35 requires more cycle stock than the order quantity of 7. For the larger order, the exposure to stock-out is only once a year, as opposed to 5 times per year when the order quantity is 7. The increased cycle stock when the order quantity is 35 protects against demand fluctuations—acting much like safety stock.

So far we have seen that the order quantity and reorder point both affect service levels and inventory costs. One way of providing data for management consideration is to develop tables that make explicit the trade-off between inventory costs (carrying, ordering, and safety stock costs) and customer service levels for various order quantities and reorder points. Figure 17.16 shows such a table using the example data.

To illustrate the calculations for Figure 17.16, consider a reorder point of 2 and an order quantity of 4. Annual ordering cost would be $(35/4)(45) =$

FIGURE 17.16   Inventory Costs* and Service Levels

	Order quantity		
Reorder point	*4*	*5*	*6*
0	$643.75 79%	$627.50 83%	$637.50 86%
1	663.75 93	647.50 94	657.50 95
2	788.75 99	772.50 99	782.50 99

*Ordering costs plus cycle and safety stock carrying costs.

$393.75. Cycle stock carrying cost would be (4/2)(.25)(500) = $250, while cost of safety stock would be (2 − .84)(.25)(500) = $145. This totals $788.75.

Total Cost Criterion

So far, our example has used fill rate (percentage of units ordered met from stock) as the measure of customer service. In some cases, costs of not having the units in stock (e.g., lost profits, penalty costs, and loss of customer good-will) can be quantified. Use of this cost, the *shortage cost(C_s)*, permits a more comprehensive examination of inventory decisions since the ordering, carry-ing, and shortage costs can all be considered in determining inventory parameters.

The equation for total annual cost (TAC) (17.19) contains terms for the costs of placing orders, carrying inventory, and incurring inventory short-ages. This equation requires estimates for all the costs, probably the most difficult of which is the shortage cost. Once cost estimates are made, the expression can be used to find the lowest total cost set of order quantity (*Q*) and reorder point (*R*) parameter values. The equation is

$$\text{TAC} = A/Q \left[C_P + C_s \left(\sum_{d=R+1}^{d_{max}} (d - R) P(d) \right) \right] + C_H[Q/2 + (R - \bar{d}] \quad (17.19)$$

where:

A = Annual demand.
Q = Order quantity.
C_P = Fixed ordering cost.
C_H = Inventory carrying cost per unit per period = vC_r.
v = Item unit cost.
C_r = Annual inventory carrying cost rate.
d = Demand during the replenishment lead time.
$\bar{d}$ = Average demand during the replenishment lead time.
$P(d)$ = Probability of demand during lead time equaling d.
R = Reorder point.

$(R - \bar{d})$ = Safety stock level.

$\quad C_s$ = Shortage cost per unit.

The first part of Equation 17.19 includes the ordering cost and the stock-out cost. The number of reorder cycles per period is A/Q, which can be used to convert costs per cycle to period costs. The expression $(A/Q)C_P$ is the cost per period of placing orders. The expected number of units short, $(E\{s\})$, for a reorder point of R is

$$\left[\sum_{d=R+1}^{d_{max}} (d - R) P(d) \right].$$

Multiplying this by $(A/Q)C_s$ gives cost per period of inventory shortages. Cost per period of carrying cycle stock is $(Q/2)C_H$, and cost per period of carrying safety stock inventory is $C_H(R - \bar{d})$.

Any particular solution to Equation 17.19 provides total cost per period for a given setting of the order quantity (Q) and the reorder point (R). If the unit cost of acquiring an item depends on quantity ordered, this model requires additional terms. This can occur when volume or transportation discounts are available.

We'll present two methods for using Equation 17.19 to determine the least cost order point/order quantity values: a grid search approach and an iterative approach. We'll use the example problem to illustrate the approaches.

Grid Search Procedure

Using the example problem of Figure 17.13 and Equation (17.19), we get an expression for total annual cost as a function of the order quantity Q and reorder point R:

$$\text{TAC} = (35/Q)[45 + 60\ E\{s\}] + 125[(Q/2) + (R - .84)].$$

Evaluating this expression for several values of Q and R provides the results in Figure 17.17, which serve as the basis for a grid search.

Our strategy for performing the grid search is to start with Q equal to the economic order quantity (5 units). Next we search on the reorder point (starting at the maximum of $R = 4$) until the costs reach a minimum and start to increase. This occurs at $R = 1$ when $Q = 5$.

The next step is to vary Q around $Q = 5$, when $R = 0, 1,$ and 2, to see if costs increase. Since they decrease for $Q = 6$, when $R = 0$ and 1, we continue on to $Q = 7$. At this point only the value at $R = 0$ is still decreasing, so we go on to $Q = 8$, where we finish. (We've underlined the values that need *not* be completed.) We've now identified the point of minimum total cost: $R = 1, Q = 6$.

Several observations can be made from Figure 17.17. The solution suggested is $Q = 6, R = 1$. It trades off some exposure to stock-outs by increas-

---

FIGURE 17.17   Total Costs for Several Reorder Points and Order Quantities

Reorder point	Order quantity				
	4	5	6	7	8
0	979.75	875.30	826.50	809.50	812.38
1	816.00	769.30	759.00	769.50	793.00
2	790.85	789.30	796.50	808.70	852.38
3	917.95	899.18	910.30	934.90	967.93
4	1,038.75	1,022.50	1,032.50	1,057.50	1,091.88

---

ing the order quantity over the economic order quantity. In some instances, the solution reduces the reorder point as well. That is the reason for checking the costs at $R = 0$ and $Q = 8$. The economic order quantity (EOQ) provides a reasonable starting point for the search, although the solution will be further from the EOQ the larger the stock-out cost. We'll use some of these observations in the iterative procedure.

The Iterative (Q, R) Procedure

Figure 17.18 summarizes the iterative procedure. It starts with the EOQ, as we did with the grid search. The value of $P(d > R)$ at Step 2 is found by equating the extra annual inventory carrying cost incurred by increasing the reorder point by 1 unit, C_H, to the savings in shortage costs that can be attributed to the additional unit of inventory; that is, $C_H = (A/Q)C_s(E\{s\}_R - E\{s\}_{R+1})$. Since $(E\{s\}_R - E\{s\}_{R-1}) = P(d > R)$, $C_H = (A/Q)C_s \cdot P(d > R)$. The calculation of Q at Step 4 is obtained by differentiating Equation (17.19) with respect to Q, setting the resulting expression equal to 0, and solving for Q:

$$Q = \sqrt{\frac{2A[C_P + C_s E\{s\}_R]}{C_H}}.$$

To illustrate the procedure, we'll use the example problem. Since the procedure iterates from calculating Q to calculating R, it's sometimes called the Q, R procedure.

Step 1: $Q = \sqrt{(2)(35)(45)/(.25)(500)} = 5$.

Step 2: $P(d > R) = (5)(.25)(500)/(35)(60) = .30$.

   The closest value in Figure 17.15 is .250 at $R = 1$.

Step 3: $E\{s\} = .290$ (from Figure 17.15 when $R = 1$).

Step 4: $Q = \sqrt{2(35)[45 + (.29)(60)]/(.25)(500)} = 5.91 \approx 6$.

Step 5: $P(d > R) = (6)(.25)(500)/(35)(60) = .36$. $R = 1$.

FIGURE 17.18   The Iterative Procedure for Finding Q and R

1. Compute the EOQ $= \sqrt{2AC_P/C_H}$.

2. Compute $P(d > R) = QC_H/AC_s$ and determine the value of R by comparing the value of $P(d > R)$ with the cumulative demand during lead time distribution values.

3. Determine $E\{s\}$, the expected inventory shortages, using the value of R from Step 2.

4. Compute $Q = \sqrt{2A[C_P + C_sE\{s\}_R]/C_H}$.

5. Repeat Steps 2 through 4 until convergence occurs; i.e., until sequential values for Q at Step 4 and R at Step 2 are equal.

Source: R. B. Felter and W. C. Dalleck, *Decision Models for Inventory Management* (Homewood, Ill.: Richard D. Irwin, 1961).

The five-step procedure converged quickly on the same solution values for Q and R that were indicated in Figure 17.17. Since the procedure considers the expected shortage cost in determining Q in Step 4, and since the computation effort is minimal, it's often a useful approach for determining the order quantity and reorder point values. In cases where the magnitude of the shortage cost, C_s, is large, this procedure takes it into account and adjusts the order quantity and reorder point accordingly. This may mean an increase in the order quantity over the EOQ and a reduction in reorder point are required to reduce total cost. The five-step procedure explicitly accounts for the interaction between inventory shortages and ordering costs in solving for the minimum-cost reorder point/order quantity values.

MULTI–ITEM MANAGEMENT

In this section we consider the management of multiple items in inventory. In particular we look at a method for categorizing items so the most important will receive management attention. The technique is called **ABC analysis.** It's first discussed with a single criterion for classification and later with multiple criteria.

Single-Criterion ABC Analysis

A single-criterion ABC analysis consists of separating the inventory items into three groupings according to their annual cost volume usage (unit cost × annual usage). These groups are: A items having a high dollar usage, B items having an intermediate dollar usage, and C items having a low dollar usage.

Figure 17.19 shows the results of a typical ABC analysis. For this inventory, 20 percent of the items are A items, which account for 65 percent of the annual dollar usage. The B category comprises 30 percent of the items and

FIGURE 17.19   ABC Analysis

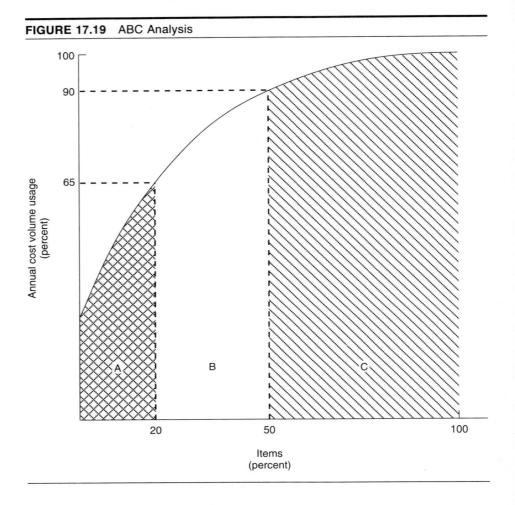

25 percent of the dollar usage, while the remaining 50 percent of the items are C items accounting for only 10 percent of the annual dollar usage. While percentages may vary from firm to firm, it's common to find a small percentage of the items accounting for a large percentage of the annual cost volume usage.

ABC analysis provides a tool for identifying those items that will make the largest impact on the firm's overall inventory cost performance when improved inventory control procedures are implemented. A perpetual inventory system, improvements in forecasting procedures, or a careful analysis of the order quantity and timing decisions for A items will provide a larger improvement in inventory cost performance than will similar efforts on the C items. Therefore, ABC analysis is often a useful first step improving inventory performance.

ABC analysis helps focus management attention on what's really important. Managers concentrate on the "significant few" (the A items) and spend less time on the "trivial many" (the C items). Unfortunately, classifying items into A, B, and C categories based on just one criterion may overlook other important criteria.

Multiple-Criteria ABC Analysis

Several noncost criteria are important in inventory management. Among them are lead time, obsolescence, availability, substitutability, and criticality. Flores and Whybark looked into the use of noncost criteria in managing maintenance inventories. Criticality seemed to sum up managers' feelings about most noncost aspects of the maintenance items. It takes into account such factors as severity of the impact of running out, how quickly the item could be purchased, whether a substitute was available, and even political consequences of being out. Some of these criticality notions may even weigh more heavily than dollar usage in managing the item—much like the proverbial cobbler's nail.

That isn't to say managers shouldn't still be concerned about dollar usage implications of maintenance inventory. To have separate ABC categories for dollar usage and criticality, however, could lead to a large number of combinations, each of which could require a different management policy. The potentially large number of different policies violates the principle of simplicity (a recurring theme in this book). To keep the number of inventory management policies to a workable few, the number of combinations of criteria needs to be kept small. This means combining criteria somehow (e.g., combining high-cost noncritical items with low-cost critical items).

The procedure for doing this consists of several steps. First we produce the dollar usage distribution and associated ABC categories. The second step involves establishing the "ABC" categories of criticality. To keep the confusion level down, we use I, II, and III to designate the criticality categories. The criteria to establish these categories are more implicit and intuitive. Category I, for example, might include items that would bring the plant to a stop and for which there's no easy substitute, alternative supply, or quick fix. The III items, on the other hand, are the ones for which there would be little if any impact if there were a shortage. The II items are the ones left over. Figure 17.20 shows the distributions of dollar usage and criticality for a sample of maintenance inventory items at a consumer durable manufacturing plant.

There's substantially less dollar usage in category I than in category A. This should not be surprising, given that the criteria for I included things like impact of outage and ease of replacement. Figure 17.21 presents a matrix of the dollar usage and criticality classifications. There's an entry for every combination. That means both low dollar usage and high dollar usage items can

FIGURE 17.20 Distributions of Dollar Usage and Criticality for a Sample of Maintenance Inventory Items

	Dollar usage				Criticality		
Category	Number of items	Percentage of items	Percentage of dollar usage	Category	Number of items	Percentage of items	Percentage of dollar usage
A	15	11%	84%	I	5	4%	40%
B	25	15	15	II	48	39	56
C	88	74	1	III	75	57	4
Total	128	100	100		128	100	100

Source: B. E. Flores and D. C. Whybark, "Implementing Multiple Criteria ABC Analysis," *Journal of Operations Management* 7, no. 1 (Fall 1987).

FIGURE 17.21 Number of Items Classified by Dollar Usage and Criticality

	Criticality			
Dollar usage	I	II	III	Total
A	2	12	1	15
B	1	19	5	25
C	2	17	69	88
Total	5	48	75	128

Source: B. E. Flores and D. C. Whybark, "Implementing Multiple Criteria ABC Analysis," *Journal of Operations Management* 7, no. 1 (Fall 1987).

have high criticality (or low criticality). It also means the problem of combining still remains.

Multiple-Criteria ABC Management Policies. There are nine possible combinations in Figure 17.21 that could each require a different management policy. The next step is to reduce the number, although R. G. Brown argues this isn't necessary when the computer can keep track of any number of policies. We, however, are concerned about having a number with which people can cope. A simple mechanical procedure is used to combine classifications to provide three initial categories of items. These categories, AA, BB, and CC, provide a starting point for management to reassess the item's classifications. The procedure simply assigns every item in A-I (see Figure 17.21), A-II, and B-I to AA; every item in A-III, C-I, and B-II to BB; and every item in B-III, C-II, and C-III to CC. This results in 15 AA items, 22 BB items, and 91 CC items.

FIGURE 17.22 Multiple-Criteria Distributions

Combined category	Number of items	Percentage of items	Percentage of dollar usage
AA	14	11%	78%
BB	16	13	12
CC	98	76	10
Total	128	100	100

Source: B. E. Flores and D. C. Whybark, "Implementing Multiple Criteria ABC Analysis," *Journal of Operations Management* 7, no. 1 (Fall 1987).

The next step is to ask management to review each item's classification. The final step is to define specific policies for managing each category. In fact, it's helpful to develop tentative policies first. These can be a guideline in reviewing each item's classification. With the policies in mind, the question to ask when reviewing each item is: Should it be managed with the procedures that apply to its classification? Figure 17.22 shows the manager's reclassification of the items. There were changes from the mechanical assignments in each category. Managers even created a fourth category, although we haven't shown it. In evaluating the items, they found nearly half shouldn't be carried in inventory at all. This demonstrated our observation that it's hard to enhance a system that doesn't have sound basics.

Specific inventory management policies are needed for each category to bring meaning to phrases like "closer management" or "more management attention." Policies are developed to cover four areas: inventory record verification, order quantity, safety stock, and classification of the item itself. The first area, verification, is to prevent the unpleasant surprises that often occur when the computer record doesn't agree with the physical count. To improve accuracy, more frequent counts should be made. This implies a higher frequency for the AA items than for the BB or CC items. Order quantity and safety stock levels are established for each item depending on both the economics and criticality. Finally, since it's a changing world, a specific period for reconsidering the item's classification is established.

Figure 17.23 shows the specific values chosen for each area. The frequency of counting was established using an average counting rate of 40 items per labor-hour and taking into account past difficulties with the inventory records and transaction reporting system. The order quantities were roughly based on the EOQ values, while safety stock was based on the item's criticality. For both order quantity and safety stock, each part was considered individually. Finally, in order not to leave the impression the item's category was "frozen," a specific frequency of review of each item's classification was established.

FIGURE 17.23   Inventory Management Policy Parameters for Multiple-Criteria ABC Items

	Category		
	AA	*BB*	*CC*
Counting frequency	Monthly	Every six months	Yearly
Order quantity	Small for costly items	Medium: EOQ-based	Large quantities
Safety stock	Large for critical items	Large for critical items	Low or none
Reclassify review	Every six months	Every six months	Yearly

Source: B. E. Flores and D. C. Whybark, "Implementing Multiple Criteria ABC Analysis," *Journal of Operations Management* 7, no. 1 (Fall 1987).

The multiple-criteria ABC categories take into account many factors not normally considered in classifying inventory items for management purposes. When combined with clear, specific policies for each category, they can substantially improve the use of scarce talent in managing the inventories.

Example.   Figure 17.24 illustrates software to support multiple-criteria ABC analysis. The block labeled "Value class rules" contains the different criteria reported here: lead time, unit cost, and annual value (annual usage value). In this example, the various criteria haven't been combined into a single category, but are kept separate by their individual A, B, C, and D classifications. Thus, for example, the first part, AA-05, is classified as B in terms of lead time, C in unit cost, and A in annual usage value. Note that it represents almost 25 percent of the annual usage value ("Perct total value").

MULTIPLE ITEMS FROM A SINGLE SOURCE

In this section, we consider the economies of jointly ordering several items from a single source. In many independent demand situations, the stocking point may receive different items from the same source. Examples include most wholesale inventories, inventories of spare parts held in a central facility, and indirect supply items being purchased from a single vendor. Placing orders for several of these items at the same time can result in very significant inventory cost savings. Moreover, by timing the orders to different vendors so receipt of shipments is smoothed out, we can reduce warehouse costs for restocking, shelf space, and demurrage.

FIGURE 17.24 Multicriterion ABC Analysis Software

```
XYZ COMPANY
REPORT NO. IC-480A
REQUESTED BY - YOUR NAME
```

A B C   I N V E N T O R Y   C L A S S I F I C A T I O N

VALUE CLASS	------VALUE CLASS RULES------				USAGE WEIGHT FACTORS	POST VALUE CLASS
	LEAD TIME	UNIT COST	ANNUAL VALUE	PERCT VALUE		
A	20	100.00	100,000	25	YTD 50	N
B	15	50.00	50,000	50	GROSS 50	
C	10	5.00	5,000	80		
D	0	.00	0	100		

PART NO/ DESC	CODES TY AC	PART COUNT	PERCT TOTAL COUNT	ANNUAL USAGE	LEAD TIME DAYS	CURRENT UNIT COST	ANNUAL USAGE VALUE	CUMM USAGE VALUE	PERCT TOTAL VALUE	VALUE CLASS CURR	PREV
AA-05 CENTER MEMBER	1 3	1	14.3	11,000	15 B	10.00 C	110,000 A	110,000	24.5 A	A	A
AA-09 RAW MATERIAL	3 4	2	28.6	-12,500	30 A	7.50 C	93,750 B	203,750	45.4 B	A	A
AA-11 KNOB & LOCK	1 3	3	42.8	20,000	15 B	4.50 D	90,000 B	293,750	65.5 C	B	B
AA-10 GLUE	3 4	4	57.1	15,000	10 C	5.00 C	75,000 B	368,750	82.2 D	B	C
AA-13 HINGE	1 3	5	71.4	240,000	10 C	.25 D	60,000 B	428,750	95.6 D	B	B
AA-12 LOCK CATCH	1 3	6	85.7	30,000	10 C	.50 D	15,000 C	443,750	98.9 D	C	C
AA-14 SCREW	1 3	7	100.0	480,000	5 D	.01 D	4,800 D	448,550	100.0 D	D	D

1. User specified ABC parameters for leadtime, unit cost, annual dollar value, and per cent total value determine value class rules
2. Annual usage can be weighted by year-to-date and/or planned usage percentages for more effective ranking
3. Value class ranking based on highest value in accordance with specified parameters

Source: MAC-PAC Manufacturing Planning and Control System General Description Manual (Chicago: Arthur Andersen & Co., 1980), p. 11.

We treat joint ordering by successively increasing the interdependency among the items considered. The first approaches presented are based on triggering a joint order's release from individual item reorder points. Several methods for determining how much to order in total and how much of each item to order are presented. The next approaches base the joint order's release on a group reorder point. We conclude the section with some experimental evidence of the savings possible from applying joint ordering techniques.

The joint ordering circumstance is so common that many companies have developed software packages for managing single supplier items. Approaches differ from package to package. These differences can lead to different service levels and inventory cost performances. The techniques in this section illustrate the differences. The last technique in this section describes an approach used in a commercial software package.

Methods Based on Individual Item Reorder Points

Figure 17.25 presents an example problem for five items from the same source. In the example, price and weight per unit, forecasts, and order quantities and reorder points are calculated for each item independently. If inventory were to be managed item by item, the only item to order (based on the current status) would be item A, since it's the only item below its reorder point. The annual inventory carrying costs and ordering costs for managing these five items independently are about $330 (applying Equation (17.19) and ignoring the stock-out and safety stock costs).

One approach to joint ordering is to place a single order whenever any *one* item drops below its reorder point. How much of each item to order could be

---

FIGURE 17.25   Example: Five Items from Same Source

Item	Price	Weight (lbs.)	Monthly usage forecast Units	Monthly usage forecast $	Monthly usage forecast Lbs.	EOQ	ROP	Current on-hand plus on-order
A	$1.10	5	100	110	500	330	79	50
B	.40	1	80	32	80	490	67	200
C	3.80	10	200	760	2,000	250	168	400
D	.25	.5	25	6	13	350	16	75
E	.05	.1	70	4	7	1,300	58	60
			475	912	2,600			

Individual order cost = $10.
Joint order cost = $10 + $2/item.
Inventory carrying rate = 20%/year.
Full truck load quantity = 10,000 lbs.

determined from the individual economic order quantities. These quantities would then be used to construct one consolidated order. The choice of which items to order could be determined by how close each is to its reorder point. After including any items below reorder point, others could be added using the ratio of the current on-hand plus on-order to the reorder point. This assures ordering those items most likely to run out first. Individual items could be added to the joint order until some minimum dollar amount or weight is attained.

Applying this logic to the example problem would mean adding 330 units of item A first, 1,300 units of item E next, item C next, and so on until the joint ordering criterion was met. Note not all items need to be ordered with this approach although all could be ordered. Using this approach doesn't necessarily result in regular timing of orders from the vendor. The next order is placed when the next item reaches the reorder point, which depends on how many items are ordered this time, variations in usage rates, and so on.

A method that does produce a regular cycle of orders to the vendor is based on developing a joint economic order quantity. This can be done by treating all the items from a vendor as a whole and determining an economic time between orders or an economic dollar value order. In general, the expression for the economic dollar value order is

$$Q = \sqrt{\frac{2(\sum_{i} v_i A_i)C_P}{C_r}} \tag{17.20}$$

where:

$\sum v_i A_i$ = The total dollar volume per year for all items from the vendor. (v_i is the value of item i and A_i is the demand per year for item i.)

C_P = The cost of placing an order (including any individual item order costs).

C_r = The annual inventory carrying cost rate.

The economic time between orders is

$$TBO = Q/\sum_{i} v_i A_i. \tag{17.21}$$

Applying Equation (17.20) to the data in Figure 17.25 gives

$$Q = \sqrt{2(912 \times 12)20/.2} = \$1,479.$$

Note the dollar forecast has been annualized and C_P is $20, the joint order cost of $10 plus $2 for each of the five items. Applying Equation (17.21) to this result gives

$$TBO = 1,479/912 = 1.62 \text{ months.}$$

The individual order quantities now must be established. Using the **economic dollar value order** method, we could order each item in 1.62-month supply quantities. The results of applying this logic appear in the first section of Figure 17.26 for each of the five sample items.

Ordering each item in a 1.62-month supply leads to different quantities than using the EOQ for each item, but suffers the same limitation. It ignores present inventory positions. An alternative would be to order individual quantities and bring each item up to a 1.62-month supply above its individual reorder point. This takes into account current inventory position, and it would mean all items would reach their individual reorder points at the same time **(simultaneous reorder point** method) if the forecast was perfect.

To illustrate the concept, item A's current inventory position of 50 is 29 units below its reorder point of 79. Thus, the order quantity would be 162 + 29 = 191 units. The middle section of Figure 17.26 shows the present inventory position of each item in terms of the number of months supply in excess of the order point. Items B and D don't need to be ordered, because they have more than a 1.62-month supply over their order points. Order quantities are shown for those items that should be ordered. Note this procedure will result in a regular pattern of orders to a supplier if the forecasts are generally correct.

A third alternative that uses individual reorder points is to base the combined order on some quantity discount or low-cost transportation alternative, the **full truckload** method. For our example, a truckload quantity is 10,000 pounds. When an order is made, the question is how to allocate the 10,000 pounds appropriately to the individual items. The analysis starts by noting a 10,000-pound order represents a 3.85-month supply using the combined de-

FIGURE 17.26   Three Joint Ordering Alternatives for Example Using Individual Item Reorder Points

	Economic dollar value order		Simultaneous reorder point		Full truck load		
Item	Dollar Value of Order	Item order quantity*	Months supply over reorder point	Item order quantity†	Pounds over reorder point	Pounds to be ordered	Item order quantity‡
A	$ 178	162	−.29	191	−145.0	2,520	504
B	53	130	1.66	—	133.0	247	247
C	1,232	324	1.16	92	2,320.0	7,180	718
D	10	41	2.36	—	29.5	30	60
E	6	113	.03	111	.2	33	330
	$1,479				2,337.7	10,010	

*1.62 months' supply (1.62 × monthly forecast).
†For items with less than a 1.62 months' supply over reorder point, order = 1.62 months' supply − (On-hand + On-order).
‡4.75 months' supply − (On-hand + On-order) all values in pounds.

mand of 2,600 pounds per month for all items. When the individual item current on-hand plus on-order values are converted to pounds, there's an "excess" of 2,337.7 pounds above the reorder point (as shown in the last section of Figure 17.26). This represents approximately a 0.9-month supply. Thus, if a 10,000-pound order is to be placed now, it must be so placed to bring the total inventory to 4.75 (3.85 + 0.9) months above the reorder point quantities.

Allocating the 10,000-pound order to the individual items can be illustrated with item A. The present inventory is 29 units, or 145 pounds below the reorder point value for this item. A 4.75-month supply in pounds is 2,375 pounds. To account for the 145 pounds below reorder point, we must order 2,520 pounds or 504 units of item A. The last section of Figure 17.26 shows the pounds and quantities of each item to be ordered. Note the small rounding error in the number of pounds to be ordered.

One weakness of all procedures based on individual reorder points concerns service levels attained relative to service levels expected. In each procedure, only *one* item, say item A, needs to be below the reorder point for the combined order to be placed. This means other items might be above their reorder points, thus providing *higher* service than expected for them. On the next order, it might be another item that's below reorder point, raising the service level for item A as well as for others.

This provision of "over service" and higher than necessary inventory is difficult to overcome with methods where the joint order is triggered by individual reorder points. One "fix" is to specify individual items' reorder points so they represent the *minimum* service level for each item. Note only the item below reorder point will be exposed to this low service possibility. In the next section, we look at treating the group as a whole for deciding when to order.

Methods Based on Group Reorder Points

In using individual reorder points to trigger joint orders, we've argued individual item and group service levels would be higher than those associated with independent reorder points. One way to overcome this problem is to create a group reorder point providing a service level for the group in total. This can be done by basing the joint reorder point on the individual ones. Figure 17.27 summarizes several ways of doing this for the example problem of Figure 17.25. It shows reorder points that use units, dollars, and months of supply on hand as the basis for reordering.

The first group reorder point in Figure 17.27 is simply the sum of the individual reorder points. A joint order would be triggered whenever the combined inventory positions of items A through E fell below 388 units. This approach can combine "apples and bananas" in summing reorder points. The second approach converts item reorder points into dollars and sums these to find a group reorder point based on the joint inventory's dollar value. The third alternative in Figure 17.27 is based on the reorder point for each item

FIGURE 17.27   Group Reorder Point Alternatives

	Item					Group reorder point
	A	B	C	D	E	
Reorder point units	79	67	168	16	58	388
Reorder point ($)	$86.90	$26.80	$638.40	$4.00	$2.90	$759.00
Average reorder point month's supply	.79	.84	.84	.64	.83	.79
Weighted month's supply						388/475 = .82

expressed in months of supply (dividing the reorder point by the forecast). The group reorder point is the arithmetic average of the individual reorder points. The final approach is similar but weights the group reorder point by the individual reorder points. This is done by dividing the sum of the reorder points by the total item forecast to calculate the months of supply reorder point for the group. These same approaches can be used to set group service levels directly by using the demand during lead time distribution measured in the same units.

To determine when to place a joint order, we compare the group reorder point value to the inventory position for all items in the group. When units, dollars, or months of supply *in total* fall below the group reorder point, a group order is placed. In each case, other tests can be added, which prevent a single item from accumulating too many stock-outs, falling below some minimum point, or creating unjustified orders for everything in the group.

Once a group order is triggered, we can use any of the approaches in Figure 17.26 to determine individual order quantities. For the group order point policies, however, only the last two methods in Figure 17.26 provide relative balancing between items.

An alternative approach to joint ordering is to place orders on a single vendor on a periodic basis, where the period is based on the economic order frequency, Equation (17.21). Quantities ordered could bring inventory levels for each item up to the level necessary to provide the service desired until the next order. Orders for different vendors could be spread to smooth warehouse labor requirements.

A Group Service Level Method

Many commercial software packages contain joint ordering logic. A version of this logic uses a group reorder point, based on desired group service levels, to determine when to place an order. The logic used for determining individual order quantities is simultaneous reorder point.

Periodically, the inventory is reviewed to see if an order should be placed. The decision is based on calculating the expected number of stock-outs that would occur for each item if no order was placed. This involves: (1) determining the variance of the demand during lead and review time to see how many standard deviations of each item are currently on hand and (2) calculating the expected number of stock-outs if no order is placed.

The total expected number of stock-outs is compared to the group level required to meet desired service levels. The allowable level of stock-outs is derived from the economic time between orders and the desired service level. If total projected stock-outs exceed the allowed number, an order is initiated.

The order is made at least equal to the economic dollar value order. It's allocated so each item has an inventory that will provide its service level and will reach the group reorder point at the same time as the other items given current forecasts. Not every item needs to be ordered in each cycle, and labor smoothing is done by scheduling other vendors' shipments at other times.

Simulation Experiments

Kleijnen and Rens performed an extensive set of simulation experiments to evaluate the group ordering procedure just described. They studied this procedure in a wide variety of operating situations by varying the following factors:

The number of items per group.
The length of the review period.
The ratio of the order interval to the review period.
The ratio of the fixed to the variable ordering cost.
The ratio of the optimum to the actual joint order size.
The lead time variance.
The desired customer service level values in a group.

Figure 17.28 summarizes these experiments' results for four measures of performance (customer service, inventory carrying cost, ordering cost, and total cost). The results indicate the joint ordering procedure can have a major effect on inventory system performance. The joint ordering procedure reduced total cost by nearly 50 percent compared to independent ordering. Much of this improvement came from lower ordering costs, reflecting the joint ordering procedure's objective. However, this procedure also dominates independent ordering in terms of inventory carrying cost due to lower order quantities.

The joint ordering procedures provide actual customer service levels close to those obtained by independent ordering. Across all 16 experiments, joint ordering averaged within .5 percent of the customer service level for independent ordering. Also, joint ordering is more effective than independent ordering when actual customer service level is compared with desired customer

FIGURE 17.28   Experimental Results

Experiment number	Customer service level ratio*	Inventory carrying cost ratio†	Ordering cost ratio‡	Total cost ratio§
1	.999	.489	.664	.555
2	1.041	.787	1.251	.896
3	.989	.452	1.649	.629
4	1.001	.337	.672	.423
5	1.015	.411	.441	.424
6	1.017	.400	.874	.579
7	.988	.599	.614	.605
8	.945	.269	.463	.339
9	.983	.558	.792	.630
10	.981	.361	.528	.430
11	.946	.351	.381	.364
12	.987	.525	.680	.592
13	1.012	.461	1.319	.686
14	1.105	.469	.651	.522
15	1.008	.179	.692	.348
16	1.000	.446	.864	.607
Average	.995	.443	.783	.539

Notes:
*Average customer service level (group ordering)/average customer service level (independent ordering).
†Average inventory carrying cost (group ordering)/average inventory carrying cost (independent ordering).
‡Average ordering cost (group ordering)/average ordering cost (independent ordering).
§Average total cost (group ordering)/average total cost (independent ordering).

Source: J. P. C. Kleijnen and P. J. Rens, "IMPACT Revisited: A Critical Analysis of IBM's Inventory Package—IMPACT," *Production and Inventory Management,* first quarter 1978.

service level. Actual customer service level under joint ordering was less than desired customer service level in half the experiments, while under independent ordering this occurred in 10 of 16 experiments.

CONCLUDING PRINCIPLES

This chapter presents considerable theory on independent demand inventory management. Despite the material's technical nature, several management principles emerge:

- The difference between dependent and independent demand must serve as the first basis for determining appropriate inventory management procedures.

- Organizational criteria must be clearly established before we set safety stock levels and measure performance.
- A sound basic independent demand system must be in place before we attempt to implement some of the advanced techniques presented here.
- Savings in inventory-related costs can be achieved by a joint determination of the order point and order quantity parameters.
- Combined ordering of several inventory items obtained from a single source can cut inventory-related costs.
- All criteria should be taken into account in classifying inventory items for management priorities.
- The policies developed for each ABC classification should be used to guide the classification of each item as well as to manage its inventories.
- Management must be sure the organization is prepared to take on advanced systems before attempting implementation.

REFERENCES

Baker, K. R.; M. J. Magazine; and H. L. W. Nuttle. "The Effect of Commonality on Safety Stock in a Simple Inventory Model." *Management Science* 32, no. 8 (August 1986).

Ballou, Ronald H. "Estimating and Auditing Aggregate Inventory Levels at Multiple Stocking Points." *Journal of Operations Management* 1, no. 3 (February 1981), pp. 143–54.

Billington, P. J. "The Classic Economic Production Quantity Model with Setup Cost as a Function of Capital Expenditure." *Decision Sciences* 18, no. 1 (Winter 1987).

Brown, R. G. *Decision Rules for Inventory Management*. New York: Holt, Rinehart & Winston, 1967.

Carlson, J. G., and C. S. Gopal. "The Numbering and Taxonomy of Inventoried Items." *International Journal of Operations and Production Management* 3, no. 1, 1983, pp. 10–19.

Chikan, Attila. "Heuristic Modeling of a Multi-Echelon Production-Inventory System." *Production Emerging Trends and Issues*. Netherlands: Elsevier, 1985.

Constable, G. C., and D. C. Whybark. "The Interaction of Transportation and Inventory Decisions." *Decision Sciences* 9, no. 4 (October 1978), pp. 688–99.

Davis, E. W., and D. C. Whybark. "Inventory Management." *Small Business Bibliography 75*. Washington, D.C.: Small Business Administration, 1980.

Flores, B. E., and D. C. Whybark. "Multiple Criteria ABC Analysis." *International Journal of Operations and Production Management* 6, no. 3 (Fall 1986).

————. "Implementing Multiple Criteria ABC Analysis." *Journal of Operations Management* 7, no. 2 (September 1987).

Greene, J. H. *Production and Inventory Control Handbook*. 2nd ed. New York: McGraw-Hill, 1986.

Inventory Management Reprints. Falls Church, Va.: American Production and Inventory Control Society, 1989.

Jayaraman, R., and M. T. Tabucanon. "Co-Ordinated versus Independent Replenishment Inventory Control." *International Journal of Operations and Production Management* 4, no. 1, 1984, pp. 61–69.

Kleijnen, J. P. C., and P. J. Rens. "Impact Revisited: A Critical Analysis of IBM's Inventory Package—IMPACT." *Production and Inventory Management,* first quarter, 1978.

Krupp, James A. "Inventory Turn Rates as a Management Control Tool." *Inventories and Production* 6, no. 4 (September–October 1989).

MAC-PAC Manufacturing Planning and Control System General Description Manual. Chicago: Arthur Andersen & Co., 1980.

Magee, J. F., and D. M. Boodman. *Production Planning and Inventory Control*. New York: McGraw-Hill, 1967.

Porteus, E. L. "Investing in Reduced Setups in the EOQ Model." *Management Science* 31, no. 8 (August 1985).

Rosenblatt, M. J. "Fixed Cycle, Basic Cycle, and EOQ Approaches to Multi-Item Single Supplier Inventory System." *International Journal of Production Research* 22, no. 6, 1984, pp. 1131–39.

Service Parts Management Reprints. Falls Church, Va.: American Production and Inventory Control Society, 1982.

Silver, E. A., and R. Peterson. *Decision Systems for Inventory Management and Production Planning*. 2nd ed. New York: Wiley, 1985.

Vandemark, Robert L. "The Case for the Combination Systems." *APICS 23rd Annual Conference Proceedings,* 1980, pp. 125–27.

DISCUSSION QUESTIONS

1. The concepts of independent and dependent demand are important in inventory management, but sometimes they aren't clearly distinguishable. Can you find the elements of dependency and independency in the following situations: selling snacks at a football game, producing bumpers for automobiles, and selling greeting cards at a shopping center.
2. What would you expect to find as the predominant type of inventory in the following businesses: a ski manufacturer, a make-to-order tugboat manufacturer, and a printer?
3. Which of the inventory costs (carrying, shortage, or preparation) are most difficult to measure? How would you determine if you needed more precision in the estimate of the cost?
4. Why might the EOQ model not be appropriate for a dependent demand item?
5. A friend of yours wants to leap-frog the basic independent demand inventory ideas and go directly to implementing some of the more advanced concepts. What arguments would you raise against this strategy? What counterarguments might persuade you the leap-frog strategy would be OK?

6. How many pairs of socks do you own? How many dress suits? How does this relate to the ABC concept?

7. How would you classify the three following items using multiple criteria ABC analysis to decide between AA, BB, or CC: a car, a calculator, and writing paper?

8. Use the analogy of writing home for money to explain the concept of demand during lead time. Be sure to account for the fact that both lead time and demand are variables.

9. How can Figure 17.16 be used to estimate management's estimated (implied) cost of stock-outs?

10. A manufacturing firm uses a color coding scheme for placing joint orders on a single source of supply. Each vendor is coded with an individual color in the Kardex file. When any one of the items from a vendor reaches reorder point, an economic order quantity is ordered for *all* items from the same vendor, which are easily found by the color coding. Comment on this procedure.

PROBLEMS

1. An end product has an average usage of 8 units/period and an ordering cost of $40/order. The item's inventory carrying cost is $0.10/unit/period. No safety stock is carried for this item.
   a. Calculate the economic order quantity.
   b. Calculate the average cycle stock for this item using the order quantity in question a.
   c. Assuming there are 13 periods per year, calculate total cost per year.

2. Barnes Trucking Company's purchasing manager received the following quote from a supplier of a maintenance item:

Purchase quantity/order	Price
Less than 1,000 units	$100/unit
1,000 units or more	85/unit

If ordering cost equals $50, annual volume equals 2,000 units, and cost of holding inventory is 20 percent of item value per unit per year, what's the total annual cost of the minimum-cost order quantity?

3. Develop a general spreadsheet for finding order quantities when there are quantity discounts. Use Problem 2 as a test problem. When you have a program that takes different values of the costs and the price/quantity schedules, find the least-cost quantity to purchase for the following set of quantity discount data (order cost = $50, inventory holding cost = 20 percent of item value, and annual demand = 400):

Quantity ordered	Price
Fewer than 50 units	$27.00
50–99 units	26.00
100–199 units	25.50
200–499 units	25.00
More than 499 units	24.75

4. Demand during lead time for Fuzzies is distributed as follows:

Probability	.1	.1	.3	.2	.1	.1	.1
Demand	14	15	16	17	18	19	20

    a. Use a spreadsheet program to evaluate the expected number of units short per reorder cycle for reorder points of 14 to 20. What's the expected shortage cost per reorder cycle when the reorder point is 14 and cost per unit short is $10?

    b. What happens to the expected shortage cost (when $R = 14$ and $C_s = 10) if the demand distribution shifts as follows?

Probability	.2	.4	.2	.1	.1	0	0
Demand	14	15	16	17	18	19	20

5. The ICU Optical Clinic has recently introduced a new line of eyeglasses that incorporates a highly fashionable frame and special lenses that darken in sunlight and lighten indoors, thus eliminating the need to purchase a separate pair of sunglasses. ICU purchases frames for the glasses from an outside vendor. The lenses are manufactured on site. The clinic has recently noted the following demand distribution for its new glasses:

Demand/month	Probability
12	.10
13	.15
14	.15
15	.20
16	.20
17	.10
18	.05
19	.05

Purchase price to the ICU Optical Clinic is $30 per frame. The clinic has an ordering cost of $25. Cost of carrying inventory is 25 percent of the item value per unit per year. Order lead time is constant at one month. The work year consists of 12 months.

a. Compute the economic order quantity for the frames ICU purchases.

b. Compute the reorder point and buffer stock level for a 99 percent customer service level. Assume order quantity and reorder point can be computed independently.

c. What's the total annual cost of carrying the buffer stock computed in question c?

6. The corner grocery store orders a two-week supply of flour whenever on-hand inventory falls below the reorder point. Lead time for flour is one day. On Tuesday, September 28, the clerk found 20 bags of flour in stock. The sales are normally distributed with a mean of 18 bags per day and a standard deviation of 10 bags per day. Desired customer service level is 99 percent.

a. What reorder point for flour provides a 99 percent customer service level?

b. How many bags of flour should be ordered on September 28 if the store is open five days per week?

7. The Seldom Seen Ranch in Muckinfut, Texas, is in the process of developing an inventory control system for purchasing hay that will cope with Texas-size uncertainty. Seldom Seen foreman Horace Cints prepared the following information on Seldom Seen's hay use:

Average demand during lead time = 1,000 bales.
Lead time = 1 month.
Economic order quantity = 2,500 bales.
Forecast interval = 1 month.
Mean absolute deviation of forecast error = 40 bales.
Desired probability of stocking out = 0.10.

a. How much safety stock will be required?

b. What's the reorder point?

c. What's the customer service level for this policy?

d. What decision rule should Horace Cints use in ordering hay, assuming constant hay usage throughout the year?

8. Consider the following demand and lead time data for a company's major product:

Demand per day	Probability	Manufacturing lead time (in days)	Probability
1	.3	2	.4
2	.7	3	.6

a. Determine the distribution for demand during lead time and calculate average demand during lead time.

b. For each reorder point from 2 to 6 units (assume the item is ordered 10 times per year and stock-out cost is $10 per unit), what are the expected stockouts and shortage costs per year?

9. A company has an item in stock whose annual inventory carrying cost is $10 per unit per year, stock-out cost is $4 per unit, ordering cost is $5 per order, and the annual requirement is 100 units. Demand during replenishment lead time has the following probability distribution:

Demand during lead time (in units)	Probability
0	.14
1	.27
2	.27
3	.17
4	.09
5	.04
6	.02

Determine the minimum-cost lot size and reorder point for this item.

10. Solihull Distributors stocks a product having ordering cost of $216, inventory carrying cost of $5/unit/year, stock-out cost of $100 per unit short, an annual requirement of 60 units, and one-month production lead time. Demand during replenishment lead time has the following probability distribution:

Demand during lead time (in units)	Probability
0	.01
1	.04
2	.08
3	.12
4	.15
5	.20
6	.15
7	.12
8	.08
9	.04
10	.01

What's the minimum-cost lot size and reorder point for this item?

11. The Anderson Company produces a spare part with ordering cost of $101, inventory carrying cost of $20/unit/year, stock-out cost of $100 per unit short, an annual requirement of 988 units, and one-week replenishment lead time. Demand during replenishment lead time has the following probability distribution:

Demand during lead time (in units)	Probability
15	.02
16	.05
17	.09
18	.15
19	.38
20	.15
21	.09
22	.05
23	.02

Find the minimum-cost lot size and reorder point for this item.

12. After the merger, Old Framkranz found himself the only inventory clerk left at Allied Breakwater Company (ABC). The merger had left ABC with only 10 parts in inventory, and he'd been told to manage them as effectively as possible. Each part was to be reviewed once a week. (ABC worked a very exhausting five-day week.) Framkranz knew inventory carrying cost was 20 percent per dollar of cost per year. He also knew, to really manage a part's inventory well, he would have to spend a full day on the review, but he could do a reasonably good job in half a day. Even the most cursory review would require one quarter day per part, however. Files for the 10 parts contain the following information. How should Framkranz schedule the review of the parts?

Part number	Vendor	Unit cost ($)	Shipping cost/unit	Cost per order ($)	Annual usage	Reorder point	Order quantity
1	A	.20	0	10	1000	50	70
2	B	1.00	.10	20	10	1	45
3	B	.25	.10	15	12	2	95
4	C	3.00	.25	15	100	20	71
5	A	10.00	0	10	300	15	55
6	B	7.00	.10	15	2	1	7
7	C	.50	.25	20	10	2	13
8	A	5.00	0	10	400	20	39
9	C	20.00	.30	20	2	1	4
10	A	2.00	0	10	200	10	100

13. Here's a sample of items from the Soaring Eagle Hang Glider Company's maintenance inventory.
    a. Rank the items in descending dollar usage order using a spreadsheet. How many items does it take to represent 50 percent of the total? How many does it take to represent the last 10 percent?

					Item					
	a	b	c	d	e	f	g	h	i	j
Cost ($)	83	68	23	45	10	2	94	51	87	24
Usage	14	47	105	24	75	43	56	5	48	81

b. Management made A, B, and C as well as I, II, and III assignments for dollar usage and criticality, respectively, as shown below. Using the mechanical procedure to develop AA, BB, and CC categories, what are the classifications for each item? What's the dollar usage distribution for AA to CC?

					Item					
	a	b	c	d	e	f	g	h	i	j
A–C	C	B	B	C	C	C	A	C	A	B
I–III	III	II	I	III	I	III	I	II	III	III

14. The Regis Book and Stationery store is negotiating with a new vendor that can supply several grades of paper. Before a final decision can be reached on whether to purchase from this firm, management wants to know whether to order the various paper grades separately or to place combined (joint) orders for the items. The following data were collected on a sample of four grades of paper to aid the inventory planner's analysis:

	Paper grade			
	#1	#2	#3	#4
Daily sales requirements*	2,000 lb.	960 lb.	400 lb.	1,440 lb.
Inventory holding cost (per thousand pounds per day)	$1.00	$ .80	$ .40	$ .60
Purchase cost (per pound)	1.00	.80	.40	.60
Individual order cost	8.00	8.00	8.00	8.00
Joint order cost	$8 plus $1 per item ordered			
Inventory carrying cost rate	25% of item cost per year			

*Sales rate based on 250 days per year.

Should Regis Book and Stationery negotiate for individual orders or not? State any assumptions you made regarding your recommendation.

15. The Vickers Automotive Components Company currently purchases two fabricated assemblies (items 1234 and 1235) from a supplier. The transportation cost to ship these items from the supplier to Vickers is $250 per order. The trucking firm Vickers currently uses has offered to reduce transportation cost to $450 per order if Vickers orders the two items jointly. Using the following data, determine the minimum-cost independent order quantity for both items and evaluate whether Vickers should order the two items jointly.

	Item 1234	*Item 1235*
Annual demand	1,000 units	1,500 units
Item value	$250/unit	$75/unit
Inventory carrying cost	40% of item value/year	40% of item value/year

Chapter 18

Distribution Requirements Planning

Distribution requirements planning (DRP) provides the basis for tying the physical distribution system to the manufacturing planning and control (MPC) system. The set of DRP techniques described in this chapter can help firms that maintain distribution inventories improve their ability to link marketplace requirements with manufacturing activities. DRP relates current inventory positions and forecasts of field demand to manufacturing's master production scheduling and material planning modules. A well-developed DRP system helps management anticipate future requirements in the field, closely match material supply to demand, effectively deploy inventory to meet customer service requirements, and rapidly adjust to the vagaries of the marketplace. In addition, the system encourages significant logistics savings through better planning of aggregate transportation capacity needs and dispatching of shipments. This chapter will show how the techniques work, how they tie into the MPC system, and how they can be used to realize the potential savings.

The chapter is organized around four topics:

- DRP in manufacturing planning and control systems: How does DRP fit into the MPC system?
- DRP techniques: How does DRP work and how is it used to manage the demand and supply of field inventories?
- Management issues with DRP: What organizational questions must be addressed to fully realize the system's potential?
- Company example: How does DRP work in an actual firm?

This chapter closely relates to Chapter 16 on forecasting, Chapter 8 on demand management, and Chapters 6 and 14 on master production scheduling. The DRP system is driven by forecasts, ties into demand management activities, and provides a significant input to the master production scheduling process. Record processing in DRP is consistent with that described for material requirements planning (MRP) in Chapter 2. The theory on independent demand inventories is found in Chapter 17.

DRP IN MANUFACTURING PLANNING AND CONTROL SYSTEMS

Distribution requirements planning (DRP) is best conceived as one part of demand management. Figure 18.1 shows DRP's general relationship to some manufacturing planning and control system modules. DRP is a link between the marketplace, demand management, and master production scheduling. The link is effected through time-phased information on inventories and through material and shipping plans that coordinate activities in these modules.

Finished-goods inventories are often positioned in a complicated physical system, consisting of field warehouses, intermediate distribution centers, and a central supply. In such systems, a key task is effectively managing the required flow of goods and inventories between the firm and the market. In performing this task, DRP has a central coordinating role similar to material requirements planning's role in coordinating materials in manufacturing. DRP's role is to provide the necessary data for matching customer demand with the supply of products at various stages in the physical distribution system and products being produced by manufacturing.

Key elements of these data are the planned timings and quantities for replenishing inventories throughout the physical distribution system. These data take into account currently available field inventories and forecasts. Planners use these data to evaluate the quality of the current match between supply and demand and to make adjustments as required.

FIGURE 18.1 Distribution Requirements Planning in the MPC System

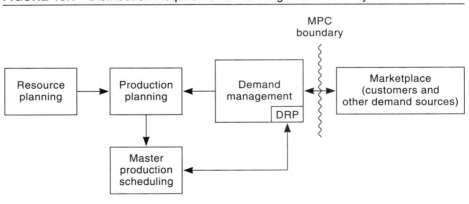

Distribution requirements planning provides information to the master production scheduler in a format consistent with the MRP records. By using standard MRP software approaches for DRP, the full range of MRP techniques (such as firm planned orders, pegging, and exception messages) are available to manage distribution inventories. This also provides the basis for integrating the data base throughout the MPC system—from purchasing through distribution. Evaluation of alternative plans, with the integrated data base, provides a complete view of the material planning implications. This is particularly valuable in master production scheduling.

DRP data provide the basis for adjusting the master production schedule (MPS) to reflect changes in demand or in the product mix. If manufacturing and distribution system priorities can't be adjusted to respond to these requirements, the implications can be evaluated and communicated to customers in a timely fashion. Common records and system integration mean there's complete visibility to see how best to use available inventories and to adjust future schedules. DRP provides a solid base of information to make these decisions, instead of relying on political negotiations between field and factory.

DRP and the Marketplace

DRP starts in the marketplace—or as close to it as possible. In most instances, this means in a warehouse or other distribution facility. In some instances, however, it could actually be at a customer location. Some firms gather information on inventory levels and on product usage directly from key customers. This offers them a major strategic advantage in providing service and advice to these customers.

DRP records start at the independent demand interface; that is, they are derived from forecasts. Since customers make their own ordering decisions (except in the instance just described, where knowledge of the customer's inventories and usage may allow the firm to advise the customer on ordering), demand is *independent* of the company's decisions. From that point on, however, decisions are under the company's control. Timing and sizes of replenishment shipments, manufacturing batch sizes, and purchase order policies are all under management control.

The DRP approach allows us to pick up all the detailed local information for managing physical distribution and for coordinating with the factory. Since customer demand is independent, each warehouse needs detailed forecasts of end-item demand. However, careful attention to actual customer demand patterns may be useful in tailoring forecasts to local conditions. We know of one instance, for example, where the local warehouse manager was able to identify several products purchased late in the month by some large firms. This produced a different demand pattern in the forecast than the constant weekly demand throughout the month resulting from a standard forecasting software package. The modified forecasts produced important inventory savings by more closely matching demand with supply at this location.

Two types of demand data may be available locally that can help us manage field inventories. Information on future special orders can help us provide service to other customers while satisfying special orders. Planned inventory adjustments by customers can also be reflected in the system, again providing data for more closely managing the distribution process. In each of these cases, the system allows the company to respond to advance notice of conditions, rather than treating them as surprises when they occur.

All management decisions for controlling inventories are reflected in the plans for resupplying warehouses. Planned shipment information provides valuable data for managing the local facility. Personnel required for unloading incoming material and stocking shelves can be planned. If there are problems in satisfying local demands, realistic promises can be made to waiting customers. Also, the amount of capital tied up in local inventory can be more realistically estimated for funds management.

In summary, DRP serves two purposes at the warehouse level. First, DRP enables us to capture data, including local demand conditions, for modifying the forecast and to report current inventory positions. The second purpose is to provide data for managing the local facility and to provide the data base for consistent communications with customers and the rest of the company.

DRP and Demand Management

The demand management module is the gateway between the manufacturing facility and the marketplace. In some systems with field inventories, it's where information on demand is taken in and where product for the field warehouses is sent out. This process requires detailed matching of supply to demand in every location—and requires providing supply to meet all sources of demand. DRP is a method for managing the resultant large volume of dynamic information and for generating the set of plans for manufacturing and replenishment.

Plans derived from the DRP information and from the resultant shipping requirements are the basis for managing the logistics system. Figure 18.2 shows the relationship between DRP and these activities. Vehicle capacity planning is the process of planning the vehicle availability for the set of future shipments as generated by DRP. Shipping requirements also are used to determine vehicle loads, vehicle dispatching, and warehouse receipts planning.

By planning future replenishment needs, DRP establishes the basis for more effective vehicle dispatching decisions. These decisions are continually adjusted to reflect current conditions. Long-term plans help to determine the necessary transportation capacity. Warehouses' near-term needs are used to

---

FIGURE 18.2 Distribution Requirements Planning and the Logistics System

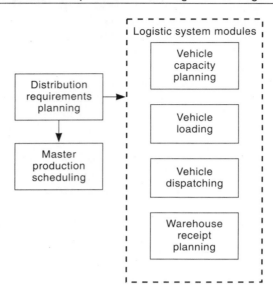

---

efficiently load vehicles without compromising customer service levels. Data on planned resupply of the warehouses can be used for scheduling the labor force in the warehouses.

As actual field demands vary around the forecasts, adjustments to plans are required. DRP continually makes these adjustments, sending the inventories from the central warehouse to those distribution centers where they're most needed. In circumstances where insufficient total inventory exists, DRP provides the basis for deciding on allocations. The planning information facilitates ideas, such as providing stock sufficient to last the same amount of time at each location, or favoring the "best" customers, or even being able to accurately say when availability will be improved and delivery can be expected.

DRP and Master Production Scheduling

Perhaps DRP's greatest payoff to master production scheduling is from integrating records and information. Since the formats of DRP and MRP records are compatible and all MPC modules are linked, DRP allows us to extend MPC visibility into the distribution system. This, however, has political implications. One company we know decided not to integrate the records, and it established a committee to resolve issues concerning inventories' size and composition. A sister company installed an integrated system using DRP. When it became evident the integrated system was superior, the political cost of dismantling the committee was high. All committee members had "permanent" jobs.

DRP collects detailed information in the field and summarizes it so MPS decisions can respond to overall company needs. DRP permits evaluation of current conditions to determine if manufacturing priorities should be revised. It provides insights into how they should be changed and into implications to the field if they aren't. Thus, more reasoned trade-offs can be made in the use of limited capacity or materials.

Shipping plans in DRP provide the master scheduler better information to match manufacturing output with shipping needs. Requirements based on shipments to the distribution centers can be quite different from demand in the field. Manufacturing should be closely coordinated with the former. For example, firms matching shipment timings and sizes with manufacturing batches can achieve substantial inventory savings. We turn next to DRP's technical details.

DRP TECHNIQUES

In this section, we develop the logic of DRP. We start by introducing the basic record and how DRP information is processed. Then we turn to the time-phased order point, how to link several warehouse records, ways to manage day-to-day variations from plans, and how to use safety stocks in a DRP-based system.

The Basic DRP Record

The DRP system's basic data elements are detailed records for individual products at locations as close to the final customer as possible. Records are maintained centrally as a part of the MPC system data base, but continually updated information on inventory and demand are passed between the central location and the field sites either on some periodic basis or on-line. For illustrative purposes, we'll consider the record for a single **stock-keeping unit (SKU)** at a field warehouse.

To integrate DRP into the overall MPC system, we expand the bill of materials beyond its usual context. The zero level in the bill of material is defined as the SKU in a field warehouse. Thus, an item isn't seen as completed simply when the raw materials have been transformed into a finished product, but only after it has been delivered to the location where it satisfies a customer demand. This extension of the bill of materials into field locations allows us to use standard MRP explosion techniques to link the field with all other MPC systems.

Regardless of the physical location (central inventory at the plant, a distribution center, a field warehouse, or even a customer's shelf), the item's ultimate demand comes from the customer. The warehouse is where the company's internal world of dependent demand must deal with the customer's independent demand. The customer, within wide ranges, decides how much and when to order; these decisions are usually independent of the company's decisions. Planners in the company, on the other hand, decide when and how much product to make. They also decide when and how much to send to field locations. To link company decisions with the customers' we must start with a forecast. This is recorded in the first row of the basic DRP record in Figure 18.3.

The record looks like an MRP record, but there are some subtle differences other than the use of forecast data in the requirements row. For example, since it's for a specific location, it not only provides time-phased data on how much and when, but also tells us where. It's not the differences, however, that are important. It's the consistency of format and processing logic that provide many of DRP's benefits. To explain this, let's go through the record in some detail.

Figure 18.3 shows a change in the forecast in period 5. This could be due to a revision by someone at the warehouse who has information on local demand, or due to a sales promotion. The fact that these variations can be incorporated into the system at this level provides one of the advantages of using DRP for managing field inventories.

The second row shows shipments in transit to the warehouse. In Figure 18.3, one shipment is scheduled to arrive in time for use in period 2. Thus, time for unloading and shelving the products must be accounted for in setting lead time to show the order available for use in period 2. The equivalent row in a manufacturing MRP record is called "scheduled receipts" (open orders). However, more than the name of the in-transit row is different between man-

---

FIGURE 18.3 Field Warehouse DRP Record

		Period						
		1	*2*	*3*	*4*	*5*	*6*	*7*
Forecast requirements		20	20	20	20	30	30	30
In transit			60					
Projected available balance	45	25	65	45	25	55	25	55
Planned shipments				60		60		

Safety stock = 20; shipping quantity = 60; lead time = 2.

ufacturing and distribution. In manufacturing, there is some flexibility in the timing of open orders. They can be speeded up or slowed down to a certain extent, by changing priorities in the shop-floor system. This is more difficult with goods in transit. Once a shipment is on a vehicle bound for a particular location, there's little opportunity to change the arrival time.

The projected inventory balance row contains the current inventory balance (45 for the example in Figure 18.3) and projections of available inventory for each period in the planning horizon (7 periods here). A safety stock value of 20 has been determined as sufficient to provide the customer service level desired for the item. The economics of transportation or packaging indicate a normal shipment of this product to this location is 60 units. Finally, it takes two periods to load, ship, unload, and store the product.

The projected available balance is generated by using the forecast requirements. The process is identical to that used for processing MRP records. In Figure 18.3, the available balance for the end of period 1 is determined by subtracting the forecast requirement of 20 from the initial inventory of 45. The 25 units at the end of period 1, plus the in-transit quantity of 60 to be received in period 2, minus the forecast of 20 for period 2, give the closing balance of 65 for period 2.

Planned shipments are indicated for those periods in which a shipment would have to be made to avoid a projected balance having less than the safety stock. The projected balance for period 4, for example, is 25 units. The forecast for period 5 is 30 units. Therefore, a shipment of product that will be available in period 5 is needed. Since lead time is two periods, a planned shipment of 60 units (the shipping quantity) is shown for period 3. Similarly, the planned shipment in period 5 is needed to cover the forecast of 30 in period 7, since there is only a 25-unit projected available balance at the end of period 6.

The result of these calculations for each product at each location is a plan for future shipments needed to provide the customer service levels the company desires. These plans depend on forecasts, but they incorporate management decisions for shipping quantities and safety stocks in planning resupply schedules. These plans provide the visibility planners need to match supply and demand.

Time-Phased Order Point (TPOP)

Many companies use economic order or shipping quantity/reorder point (Q,R) procedures based on demand forecasts for managing their field inventories. This means decisions for resupply are made independently at the field location, with no integrated forward planning; that is, when the on-hand quantity at a location reaches the reorder point, the shipping quantity is ordered with no thought given to any other items ordered, to the situation at the factory, or to warehouses—or to when the next order might be needed. Also, Q,R assumes a constant usage. Whenever forecast information is used as the requirements and a time-phased MRP approach is used to develop planned shipments, it's called **time-phased order point (TPOP).** Time-phased order point can be used for constant usage situations and even when the usage forecast varies from period to period. We use Figure 18.4 to illustrate the approach.

If a Q,R system were used, the reorder point for the situation in Figure 18.4 would be 20 units, comprised of the safety stock (5) plus the demand during lead time (15), assuming continuous review of inventory balances. Simulating the Q,R rules for Figure 18.4's data would lead to orders in periods 2, 4, and 7. If we use DRP logic, the planned shipments are in periods 1, 3, and 6. Thus, the timing of the orders in the TPOP record in Figure 18.4 doesn't exactly match the expected timing of orders using Q,R.

The results are, however, very close. The differences would largely disappear if the periods were made small (e.g., days instead of weeks) since they're primarily due to the fact that Q,R assumes continuous review. The TPOP approach is based on the MRP logic of planning a shipment that prevents the ending balance in the period from falling below the safety stock level.

One advantage of TPOP over Q,R is the TPOP record shows *planned* shipment data. These aren't part of the Q,R approach. In addition, TPOP isn't limited to use of constant requirement assumptions. When forecast usages vary, differences between TPOP and Q,R can be much larger than those in Figure 18.4.

Not only is it important to have planned shipment data, it's also critical to capture *all* demand information. Forecast sales requirements are only one source of demand input. DRP can use TPOP plus actual order data plus service part requirements plus interplant demands. *All* these demand sources can be integrated into the demand data driving DRP.

FIGURE 18.4   Example Time-Phased Order Point (TPOP) Record

		Period						
		1	*2*	*3*	*4*	*5*	*6*	*7*
Forecast requirements		15	15	15	15	15	15	15
In transit								
Projected available balance	22	7	32	17	42	27	12	37
Planned shipments		40		40			40	

Safety stock = 5; shipping quantity = 40; lead time = 1.

Linking Several Warehouse Records

Once DRP records are established for the field warehouses, information on planned shipments is passed through the distribution centers (if any) to the central facility. This process is sometimes referred to as **implosion.** The concept indicates we're gathering information from a number of field locations and aggregating it at the manufacturing facility. This is different from the explosion notion in manufacturing (where a finished product is broken into its components), but the process is the same, and in both cases it's based on bills of material.

The record in Figure 18.5 is for the central warehouse inventory. The gross requirements correspond to planned shipments to the two warehouses in Figures 18.3 and 18.4. This relationship reflects the logic that, if there were a shipment of 40 units of product to warehouse number 2 in period 1, there

FIGURE 18.5 Field Warehouse to Central Warehouse Records for DRP

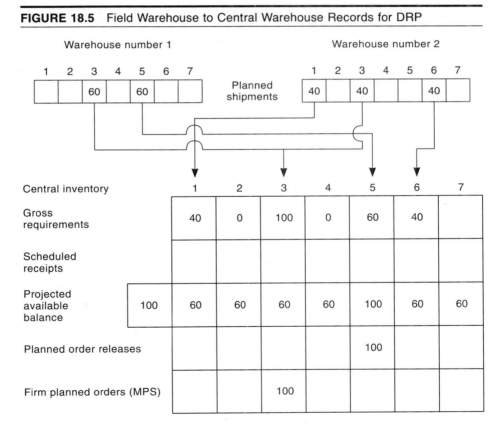

Safety stock = 50; order quantity = 100; lead time = 0

would be a "demand" on the central warehouse for 40 units in period 1. The "demand," however, is dependent, having come from the company's shipping department. It is, therefore, a gross requirement and not a forecast requirement. We have crossed over from the independent demand world of the customers to the dependent demand world of the company.

The central warehouse's gross requirement is shown in the same period as the planned shipment, since lead time to ship, unload, and put away the product has already been accounted for in the field warehouse record. For example, the shipment of 60 units of product planned for period 3 in warehouse number 1 allows for the lead time before it's needed to meet forecast demand.

The logic for imploding planned shipment information holds also for much more complicated distribution systems. If there were intermediate distribution centers, they would have gross requirements derived from the warehouses they served. At distribution centers, gross requirements are established for any period in which replenishment shipments are planned to warehouses.

A primary task at the central facility is to create the master production schedule. The central inventory record in Figure 18.5 can be used for this purpose. The record shows projected available balances for the central inventory and the planned order releases, which provide the quantities needed to maintain the 50 units of safety stock. The master production schedule is created by using zero lead time and firm planned orders. Firm planned orders are created by the planners and aren't under system control; that is, they aren't automatically replanned as conditions change but are maintained in the periods and quantities designated by the planners.

The MPS states when manufacturing is to have the product completed and available for shipment to the field warehouses. In the example in Figure 18.5, the MPS quantity (firm planned order) has replaced a planned order, although this need not be the case.

Our example implies creating an MPS for each end item. This may not be desirable in firms that assemble or package a large variety of end items from common modules, subassemblies, or bulk materials. If the MPS is stated in subassemblies or bulk materials, then a final assembly schedule (FAS) needs to be created for managing the conversion to assemblies or packed products. The resultant records usually don't need to be frozen or firm planned over extensive planning horizons, since they only deal with conversion of, say, bulk material into some specific packaged products. Figure 18.6 shows how this conversion can be managed. This packaged product's gross requirements are exploded from the shipments planned to go to all the field locations. Packaging this specific product from the bulk material takes one period. Firm planned orders, up to the planning fence at period 3, are used to schedule the packaging operation. The record also shows 10 units of packaged product held at the central facility to provide flexibility in shipping to the field warehouses.

Figure 18.7 shows how the various sizes of packaged products can be combined into a bulk inventory record for creating the factory's MPS. In the example, two package sizes consume the bulk inventory. The packages are in

FIGURE 18.6   FAS Record for Packaging Bulk Materials

Packaged product		Period						
		1	2	3	4	5	6	7
Gross requirements		40	80	100	0	60	20	100
Scheduled receipts		40						
Projected available balance	10	10	10	10	10	10	10	10
Planned order releases					60	20	100	
Firm planned orders		80	100					

Q = lot-for-lot; lead time = 1; SS = 10; planning fence: period 3.

grams, while the bulk item is in kilograms. The explosion process works from packaged item to bulk, but the grams have been converted to kilograms to get the gross requirements for the bulk material (e.g., for period 1: 100 units × 200 grams = 20,000 grams = 20 kilograms). The firm planned orders for the bulk material are the factory MPS, stating when the bulk inventory must be replenished to meet the packaging schedules.

Managing Day-to-Day Variations from Plan

On a daily basis, disbursals for actual customer demand, receipts of inventory, and other transactions are processed. These transactions are used to periodically update the DRP records. If forecasts and execution of plans were perfect, we wouldn't need to do anything but add the new period of information at the end of the planning horizon each time the records were processed. Unfortunately, we haven't found a company where such conditions hold. Figure 18.8 shows a more likely set of circumstances.

In this example, actual sales vary from 16 to 24 around the forecast of 20 units. The actual sales of 18 in period 1 have no impact on planned shipments, while actual sales of 24 units in period 2 change the plan. The additional sales in period 2 increase net requirements, which leads to planning a shipment in period 3 rather than period 4. Sales in period 3 were lower than expected, so net requirements are less and the planned shipment in period 5 is changed to period 6. Thus, the gross to net logic results in modifying shipping plans to keep them matched to the current market situation.

One negative aspect of the logic is clear from Figure 18.8's example. Actual sales' deviations around the forecast were reflected in changed shipping plans. These changes could have a destabilizing impact on the master schedule and

FIGURE 18.7   Bulk Material Record and MPS

200-gram product	Period						
	1	2	3	4	5	6	7
Planned orders					60	40	
Firm planned orders	100		100				

500-gram product	Period						
	1	2	3	4	5	6	7
Planned orders				10	10		20
Firm planned orders		20					

Bulk material—kilograms		Period						
		1	2	3	4	5	6	7
Gross requirements		20	10	20	5	17	8	10
Scheduled receipts								
Projected available balance	5	25	15	35	30	13	5	35
Planned orders								40
Firm planned orders (MPS)		40		40				

Q = 40; SS = 0; lead time = 0; planning fence: period 5.

shop. Two techniques for stabilizing the information flow are firm planned orders and error addback.

Figure 18.9 applies the firm planned order (shipment) concept to the warehouse example. By using firm planned shipments, the record shows the what-if results of maintaining the present order pattern. By using DRP records to display this pattern, we generate standard exception messages. For example, if a present plan violates a stated safety stock objective, exception messages highlight it.

In the Figure 18.9 record for period 3, the firm planned shipment of 40 in period 4 isn't rescheduled to period 3, even though the projected available inventory balance for period 4 is less than the safety stock. Thus, the master scheduler can review the implications of *not* changing before deciding whether

FIGURE 18.8 Records for a Single SKU at One Warehouse over Four Periods

Period		1	2	3	4	5
Forecast requirements		20	20	20	20	20
In transit		40				
Projected available balance	6	26	6	26	6	26
Planned shipments			40		40	
Actual demand for period 1 = 18.						

Period		2	3	4	5	6
Forecast requirements		20	20	20	20	20
In transit						
Projected available balance	28	8	28	8	28	8
Planned shipments		40		40		
Actual demand for period 2 = 24.						

Period		3	4	5	6	7
Forecast requirements		20	20	20	20	20
In transit		40				
Projected available balance	4	24	44	24	44	24
Planned shipments		40		40		
Actual demand for period 3 = 16.						

Period		4	5	6	7	8
Forecast requirements		20	20	20	20	20
In transit		40				
Projected available balance	28	48	28	8	28	8
Planned shipments				40		

Shipping Q = 40; SS = 6; lead time = 1.

FIGURE 18.9   Record for a Single SKU at One Warehouse with Firm Planned Order (Shipments) Logic

Period		1	2	3	4	5
Forecast requirements		20	20	20	20	20
In transit		40				
Projected available balance	6	26	6	26	6	26
Firm planned shipments			40		40	

Actual demand for period 1 = 18.

Period		2	3	4	5	6
Forecast requirements		20	20	20	20	20
In transit						
Projected available balance	28	8	28	8	28	8
Firm planned shipments		40		40		40

Actual demand for period 2 = 24.

Period		3	4	5	6	7
Forecast requirements		20	20	20	20	20
In transit		40				
Projected available balance	4	24	4	24	4	24
Firm planned shipments			40		40	

Actual demand for period 3 = 16.

Period		4	5	6	7	8
Forecast requirements		20	20	20	20	20
In transit						
Projected available balance	28	8	28	8	28	8
Firm planned shipments		40		40		40

Shipping Q = 40; SS = 6; lead time = 1.

changes should be made. In this case, the decision might be to opt for consistency in the information, knowing there still is some projected safety stock and the next order is due to arrive in period 5.

An alternative for stabilizing the information is the error addback method. This approach assumes forecasts are unbiased or accurate on the average. This means any unsold forecast in one period will be made up for in a subsequent period, or any sales exceeding forecast now will reduce sales in a subsequent period. Using this method, errors are added (or subtracted) from future requirements to reflect the expected impact of actual sales on projected sales. Figure 18.10 applies this concept to the warehouse example. Note the planned shipments are under system control; that is, firm planned orders aren't used.

This example's records show the planned orders in exactly the same periods as in the firm planned shipment case. Adjustments to the forecast requirements ensure stability in the information. It's apparent this technique's effectiveness diminishes if the forecast isn't unbiased. For example, if the reduced demand in periods 3 and 4 is part of a continuing trend, the procedure will break down. DRP will continue to build inventory as though the reduced demand will be made up in the future. This means forecasts must be carefully monitored and changed when necessary so the procedure can be started again. One convenient measure for evaluating forecast accuracy is the cumulative forecast error. If this exceeds a specified quantity, the item forecast should be reviewed. For example, in period 4, cumulative error has reached a + 7 (a value exceeding the safety stock); this might indicate the need to review the forecast for this item.

Safety Stock in DRP

Distribution requirements planning provides the means for carrying inventories and safety stocks at any location in the system. In the examples of Figures 18.3 through 18.5, we show safety stock in field locations and at the central facility.

With DRP, it's possible to use safety lead time as well. In those circumstances where the uncertainty is more likely to be in terms of timing (as in delivering product to the field), it may be better to use safety lead time. In the case of uncertainty in quantity (as with variable yields in manufacturing), safety stock is more typically used.

Where and how much safety stock (or safety lead time) to carry is still an open issue. Research and company experience are just now beginning to provide answers. In terms of quantity, the theory of relating safety stock to the uncertainty in our demand forecasts is clearly valid. The choice would be made on the basis of trade-offs between customer service levels and inventory required.

FIGURE 18.10   Record for a Single SKU at One Warehouse with Error Addback

Period		1	2	3	4	5
Forecast requirements		20	20	20	20	20
In transit		40				
Projected available balance	6	26	6	26	6	26
Planned shipments			40		40	
Period 1 demand = 18; cumulative error = +2.						

Period		2	3	4	5	6
Forecast requirements		22	20	20	20	20
In transit						
Projected available balance	28	6	26	6	26	6
Planned shipments		40		40		
Period 2 demand = 24; cumulative error = −2.						

Period		3	4	5	6	7
Forecast requirements		18	20	20	20	20
In transit		40				
Projected available balance	4	26	6	26	6	26
Planned shipments			40		40	
Period 3 demand = 16; cumulative error = +2.						

Period		4	5	6	7	8
Forecast requirements		22	20	20	20	20
In transit						
Projected available balance	28	6	26	6	26	6
Planned shipments		40		40		
Period 4 demand = 15; cumulative error = +7.						

Shipping Q = 40; SS = 6; lead time = 1.

On the other hand, in distribution, we're not concerned just about how much uncertainty there is but where it is. Less is known about where to put safety stocks. One principle is to carry safety stocks where there's uncertainty. This implies the location closest to the customer and, perhaps, to intermediate points, where there's some element of independent demand. The argument would imply no safety stock where there's dependent demand.

If the uncertainty from several field locations could be aggregated, it should require less safety stock than having stock at each field location. This argument has led to the concept of a "national level" safety stock popularized by Robert G. Brown in his work on materials management. The idea is to have some central stock that can be sent to field locations as conditions warrant, or to permit transshipments between field warehouses. This added flexibility should provide higher levels of service than maintaining multiple field safety stocks. The issue is clouded, however, by the fact that stock in the central facility isn't where customers are.

Figure 18.11 shows simulation results for evaluating whether national-level safety stocks should be carried. The study's results indicate it's more effi-

FIGURE 18.11　Service Levels as a Function of Amount and Location of Safety Stock

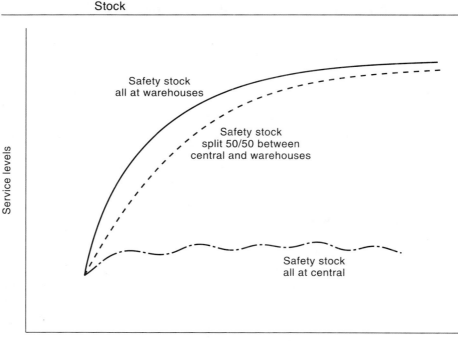

Source: W. B. Allen, "A Comparative Simulation of Central Inventory Control Policies for Positioning Safety Stock in a Multi-Echelon Distribution System," Ph.D. dissertation, Indiana University, 1983, p. 86.

cient to carry the stock in the field than to divide it between field and central. In all of William Allen's runs, the results were the same. Similar conclusions were reached for some Phillips TV parts distributed throughout Europe. In a study at Eindhoven University in the Netherlands, van Donselaar found the parts inventory should be sent to the field instead of being split between field and central facility.

Still another example of the benefits of pushing safety stocks into the field is seen in a large food company. In one division, the primary factory has little warehouse space and end items have considerable bulk. The result is a need to ship all production to distribution centers within hours of manufacture. This, in turn, forces the factory to keep in touch with actual demand and to continually adjust to actual circumstances. The company uses DRP to link distribution centers to all its factories. The products provided by this factory consistently have higher fill rates and fewer stock-outs than other products sold through the same distribution centers. Also, inventory turnover on immediately shipped products is higher than for other products. Other plants have their own warehouses where stocks are held. These stocks reduce inventory turns and don't provide immediate service to customers.

MANAGEMENT ISSUES WITH DRP

With an operational DRP system integrated with other MPC systems, management has the ability to rationalize material flows from purchasing through distribution. Achieving this desired state, however, raises several critical management questions. We've already discussed some of the DRP issues, such as planning parameters, safety stock, stability, and the form of the master scheduling interface. More fundamental issues relate to assuring the system has appropriate data entry procedures, has an organizational setup facilitating an integrated MPC approach, and uses DRP to solve specific distribution problems. We now turn to each of these managerial topics.

Data Integrity and Completeness

Let's start by considering the record for an item at a location as close to the customer as possible. We've called this location the field warehouse, but it could be at a distribution center, at the customer location itself, or even at the central facility. The issue concerns *where* the forecast is to be input. Since this is the source of demand data for planning throughout the system, it must be correctly determined and maintained. For this record, there are two key data items on which all plans are based: forecast requirements and inventory balance (including any in transit). The axiom of garbage-in, garbage-out holds in DRP as elsewhere. To have confidence in the forecast data, we must assign responsibility for both forecast preparation and adjustments.

A key issue in forecast data integrity for DRP systems is use of aggregate forecasts, which are thereafter broken down into detailed forecasts. An example is a pharmaceutical firm that forecasts annual U.S. insulin sales in total ounces, based on the number of diabetics in the country. This total is multiplied by the company market share, which is in turn broken down into package sizes, weeks, and locations as the basis for field forecasts. As the total is broken down, relative errors increase, but the MPS is based on the totals as brought through the DRP system. This is the summation of the detailed forecasts, after field modifications, so the errors should tend to cancel out.

It's imperative, however, that adjustments of the detailed forecasts don't result in a systematic bias that doesn't balance out. DRP systems are designed to respond to forecast *errors,* but forecast bias is a problem we must avoid.

Once the basic forecast has been generated, people in the field can be given some authority to modify it according to local information and needs. This should be constrained by some rule, like "plus or minus 20 percent adjustment" or "only the timing can be changed but the monthly totals must remain the same." Some mechanism must be put in place for picking up this kind of local intelligence for adjusting forecasts, but it's important to define the limits for proper control.

Also, management programs should be established to monitor this process. Monitoring is more complex when records for items are at customer locations (e.g., a large hospital). In all cases, standard forecast monitoring techniques need to be applied, particularly to discover bias introduced through the adjustment process.

Inventory accuracy depends on transaction processing routines and discipline. Procedures for quick and accurate reporting of shipments to customers, allocations to customers, returns, adjustments, receipts, and the like must all be in place. Another source of errors is incorrect balances of material in transit; these will affect all calculations in the subsequent record processing. Computer auditing can help find outliers in the data for all these cases, but tight procedural controls are clearly needed.

Organizational Support

Figure 18.12 shows conflicting functional objectives and their impact on inventory, customer service, and total costs. This figure illustrates some inherent conflicts that need resolution in an integrated MPC system. These conflicts are particularly real when DRP is part of the overall MPC system. In many firms, minimization of transportation costs, for example, is a transportation department's objective. The resultant impact on other parts of the organization is often not clearly understood.

In a comprehensive MPC system with DRP, linkages across functional boundaries are encouraged; but organizational support and evaluation mea-

FIGURE 18.12   Conflicting Functional Objectives

Functional objectives	Impact of objectives on. . .		
	Inventory	Customer service	Total costs
● High customer service	△	▲	△
● Low transportion costs	△	▽	▼
● Low warehousing costs	▼	▽	▼
● Reduce inventories	▼	▽	▼
● Fast deliveries	△	▲	△
● Reduced labour costs	△	▽	▼
● Desired results	▼	▲	▼

Source: T. C. Jones and D. W. Riley, "Using Inventory for Competitive Advantage through Supply Chain ·Management," *International Journal of Physical Distribution and Materials Management,* 1985, p. 16.

sures need to be established that will minimize suboptimization of overall enterprise goals. Many firms have implemented a **materials management** form of organization to help align responsibilities to the material flow needs. Materials management organizations are responsible for all aspects of materials, from purchasing to final distribution to the customers. Their responsibilities include determining what to make and when, when to take delivery of raw materials, how much to allocate to field locations, and how to relieve short-term materials problems.

As firms improve MPC systems either through more comprehensive approaches (e.g., DRP) or through such enhancements as JIT, emphasis shifts from material control to material velocity. Basic discipline and data integrity aren't abandoned—they're assumed. Time and responsiveness become the most important objectives. Implied is a need to reduce the organizational fragmentation that has built-in time delays. New organizational structures will increasingly be required, ones with overarching authority to dictate actions that provide rapid response to customer needs.

Integration of DRP into a comprehensive MPC system in many cases provides the integrative framework for such an organization. Eli Lilly has recognized this in its materials management organization. Robert Dille, director of purchasing, is quoted by Tom Feare, managing editor of *CPI Purchasing,* as saying, "Our buyers are really requirements analysts. In working with the system, they have seen the advantages of moving to smaller and smaller buckets. Having gone from monthly to weekly, they are now talking about daily buckets."

A concept first popularized in Europe is called **supply chain management.** Figure 18.13 illustrates the concept, showing the many organizational entities that need to be coordinated. The ideas parallel those of materials management, but they focus on the process of building the products. Coordinating the chain in Figure 18.13 requires integrated information in the DRP form and an organizational form, such as materials management.

Key to implementing an MPC system with DRP, regardless of the organizational form, is developing planners. The titles can vary in different organizations. Planners establish firm planned orders, evaluate alternative means to solve short-term problems, coordinate problems that cross functional boundaries, and help evaluate trends. They are also responsible for checking feasibility of changes, monitoring data integrity, and assessing the impact of new situations.

Problem Solving

We have already discussed changing conditions due to uncertainty in demand and the techniques that help to deal with them. Other problems come from changing market conditions, product lines, or marketing plans. Examples are product substitutions, promotions, changes in warehouse locations or customer assignments, and controlling the age of stock. DRP records help us solve these problems. We'll describe three of these as illustrative: a sales promotion, closing a warehouse, and monitoring stock aging.

Figure 18.14 presents a sales promotion served by a warehouse. For simplicity, we consider only one product at a single warehouse and the product's packaging line. The process starts with modifying the demand forecast at the warehouse. In the example, the promotion is planned for weeks 5 though 8. The impact is estimated to double sales (from 20 to 40) during the first two weeks and to have a reduced impact during the next two weeks. Note the promotion "steals" from demand in weeks 9 and 10.

The safety stock has been left at five units for the promotion period, although it might have been increased. Stock for the sales promotion is planned for delivery in the weeks in which needed, although it also might have been planned for earlier delivery if there were a need to set up some special display area. The shipping quantity is in multiples of 20, representing a shipping carton or pallet load. Planned shipments are exploded to the packaging record. Using the firm planned orders, the planner has scheduled product packaging at a constant rate of output for the next five weeks (just under the capacity of the packaging line), building inventory in anticipation of the promotion.

The anticipation inventories are shown as remaining at the packaging facility, but they might be sent to the field if space is available or if a truck is on its way with available cargo space. The pattern of inventory buildup could be different, depending on the trade-off between level production and varying the production levels. The DRP records facilitate both planning the buildup

FIGURE 18.13 Supply Chain Management

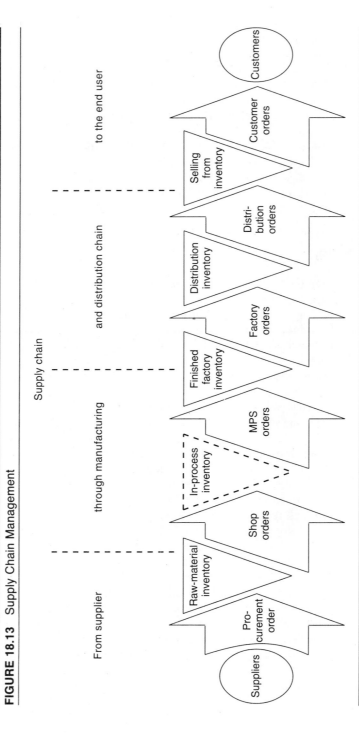

Source: T. C. Jones and D. W. Riley, "Using Inventory for Competitive Advantage through Supply Chain Management," *International Journal of Physical Distribution and Materials Management* 15, no. 1 (1985), p. 17.

FIGURE 18.14 A Sales Promotion

Warehouse		Period									
		1	2	3	4	5	6	7	8	9	10
Forecast requirements		20	20	20	20	40	40	30	30	10	10
In transit		20									
Projected available balance	27	27	7	7	7	7	7	17	7	17	7
Planned shipments			20	20	40	40	40	20	20		

Q = 20; lead time = 1; SS = 5.

Packaging		Period									
		1	2	3	4	5	6	7	8	9	10
Gross requirements		0	20	20	40	40	40	20	20		
Scheduled receipts											
Projected available balance	0	0	12	24	16	8	0	0	0		
Planned order releases							20	20			
Firm planned orders		32	32	32	32	32					

Q = lot-for-lot; lead time = 1; SS = 0; planning fence: .3.
Note: Packaging capacity = 35 units/period.

and analyzing alternative ways to meet the need (or even arguing for a postponement if capacity is a problem).

Our second example deals with a warehouse closure. Firms often change distribution systems, so closing warehouses and changing shipping patterns are ongoing activities. In Figure 18.15, warehouse 1 is scheduled to close at the end of four weeks, and warehouse 2 is to start supplying the customers. Again, the process of managing the cutover starts with the forecasts. The requirements at warehouse 1 stop at the end of week 4 and are picked up by warehouse 2, reflecting the timing and quantities of the closedown and transfer.

In the example, warehouse 1 would normally have had a planned shipment for 60 units in week 3. The planner has overridden the planned order with a firm planned order for the exact need, reducing safety stock to zero. Note the system plans an order to restore the safety stock and, eventually, the

---

FIGURE 18.15   Warehouse Closure

		Period						
Warehouse 1		*1*	*2*	*3*	*4*	*5*	*6*	*7*
Forecast requirements		30	30	30	30	0	0	0
In transit		60						
Projected available balance	43	73	43	13	60	60	60	60
Planned shipments				60				
Firm planned shipments				17				

Q = 60; lead time = 1; SS = 10.

		Period						
Warehouse 2		*1*	*2*	*3*	*4*	*5*	*6*	*7*
Forecast requirements		100	100	100	100	130	130	130
In transit								
Projected available balance	207	107	207	107	207	77	147	217
Planned shipments						200	200	
Firm planned shipments		200		200				

Q = 200; lead time = 1; SS = 20.

---

planner will need to change the safety stock parameter to zero in order not to send the wrong signal to the master production scheduler.

As time goes on, the actual quantity needed for warehouse 1 may well be different, but there are two weeks before the exact determination must be made. The planner may decide to send even less to warehouse 1 because it might be easier to pick up some unsatisfied demand from warehouse 2 than to deal with the remnant inventories at warehouse 1. The point is DRP systems provide visibility for solving these problems. Even though the exact quantity to be sent to warehouse 1 won't be determined until week 3, the amount produced and made available to send to either warehouse will be correct, since planned shipments from both are exploded to the requirements at the central facility. The main questions are how to divide the available stock between the two facilities and how to manage the transition.

Our final example deals with controlling inventory of products whose shelf life is a concern. Certainly one aspect of this problem is to use strict first-in/first-out physical movement and to make this part of the training for warehouse personnel. The second way to deal effectively with shelf life problems is to identify those products that may be headed for a problem before it's too late. One way to do this is to use the DRP records and build in exception messages to flag potential shelf life problems.

If, for example, demand for a product at some location is dropping for some reason, the forecast should be reduced. This means any available inventory or in transit stock will cover a longer time period; as a result, the first planned shipment will be several periods in the future. This condition can be detected, calling for a review of the particular product's shelf life at this location. Perhaps it should be immediately shipped to another warehouse. This feature is incorporated in the actual system, to which we now turn our attention.

COMPANY EXAMPLE

Abbott Laboratories, Ltd., of Canada produces health-care products. Three lines (pharmaceuticals, hospital products, and infant nutrition products) are produced in three plants. About 750 end items are distributed through DCs (distribution centers) throughout Canada.

Abbott Laboratories uses DRP to manage field inventories. Detailed forecast data are entered into warehouse DRP records. These records are represented as the zero level of the bill of materials (BOM). Figure 18.16 illustrates these records for two locations: Vancouver and Montreal. Forecast data are entered as the requirements rows. The first 20 weeks are displayed in weekly time periods (buckets). Thereafter, monthly buckets are used for a total planning horizon of two years.

In the Montreal record, an entry of 120 in the week of 8/7 appears in a row labeled "Customer orders." This row allows the inclusion of specific advance order information. The information is not added to the gross requirements, however. It's there for detailed planning of shipments and to recognize advanced special orders.

Lead time in the Vancouver record is 35 days (five weeks) including safety lead time. DRP time-phased records are produced using these parameters. For example, projected on-hand balance at the end of week 9/4 for Vancouver is insufficient to meet the gross requirements of 9/11. The result is a planned order for a quantity of 24, five weeks earlier in the week of 8/7.

Figure 18.17 shows the DRP records for the central warehouse and for the bulk item used to make the end item product B. Gross requirements for cen-

FIGURE 18.16 Abbott DRP Records for Vancouver and Montreal

Description	Size	um	Std bactr	FC	BY	PL	IT	C	Scrap	Total landed costs	OP	Life	O-LT	P-LT	O-LT	QA-LT	T-LT	On hand	OA inventory	Allocated	Safety stock
Product-B	200	BL	1872	A2		08	IT	C	1.0			1095	14	0	0	0	14	2520.0	0.0	0.0	0.0

21-MONTREAL

	PAST DUE	7/24	7/31	08/07	08/14	08/21	08/28	09/04	09/11	09/18	09/25	10/02	10/09	10/16
CUSTOMER ORDERS				120										
GROSS REQUIREMENTS		601	601	601	601	576	556	556	556	578	633	633	633	633
SCHEDULED RECEIPTS														
ON HAND	2520.0	1919	1318	717	116	1412	856	300	1616	1038	405	1644	1011	378
FIRM PLANNED ORDERS														
PLANNED ORDERS				1872			1872			1872			1872	

MONTHLY

	10/23	10/30	11/06	11/13	11/20	11/27	12/04	01/01	01/29	02/26	03/26	04/23	05/21	06/18
CUSTOMER ORDERS														
GROSS REQUIREMENTS	777	801	801	801	633	507	2100	2386	2744	2167	2356	2404	2479	2212
SCHEDULED RECEIPTS														
ON HAND	1473	672	1743	942	309	1674	1446	932	60	1637	1153	621	14	1546
FIRM PLANNED ORDERS														
PLANNED ORDERS	1872		1872			1872	1872	1872	1872	3744	1872	1872	1872	3744

	07/16	08/13	09/10	10/08	11/05	12/03	12/31	01/28	02/25	03/24	04/21	05/19	06/16	TOTAL
CUSTOMER ORDERS	2418													
GROSS REQUIREMENTS	2418	2289	2400	2844	2742	2100	2323	2480	2285	2356	2280	2348	2944	62735
SCHEDULED RECEIPTS	1000													
ON HAND	1000	583	55	955	85	1729	1278	670	257	1645	1237	761	1561	
FIRM PLANNED ORDERS														
PLANNED ORDERS	3744	1872	1872	3744	1872	3744	1872	1872	1872	3744	1872	1872	3744	61776

NOTES & COMMENTS

Comm list no.
List xyz

03-VANCOUVER

Description	Size	um	Std bactr	FC	BY	PL	IT	C	Scrap	Total landed costs	OP	Life	O-LT	P-LT	O-LT	QA-LT	T-LT	On hand	OA inventory	Allocated	Safety stock
Product-B	200		24			08	F	C	1.00	0.000		1095	35		0	0	35	36.0	0.0	0.0	0.0

	PAST DUE	7/24	7/31	08/07	08/14	08/21	08/28	09/04	09/11	09/18	09/25	10/02	10/09	10/16
CUSTOMER ORDERS															
GROSS REQUIREMENTS		5	5	5	5	5	5	5	5	5	5	5	5	5
SCHEDULED RECEIPTS															
ON HAND	36	31	26	21	16	11	6	1	20	15	10	5	5	19
FIRM PLANNED ORDERS															
PLANNED ORDERS									24					24

MONTHLY

	10/23	10/30	11/06	11/13	11/20	11/27	12/04	01/01	01/29	02/26	03/26	04/23	05/21	06/18
CUSTOMER ORDERS															
GROSS REQUIREMENTS	7	7	7	7	5	4	17	21	23	20	20	20	20	20
SCHEDULED RECEIPTS															
ON HAND	12	5	22	15	10	6	13	16	17	21	1	5	9	13
FIRM PLANNED ORDERS															
PLANNED ORDERS			24				24	24	24	24		24	24	24

	07/16	08/13	09/10	10/08	11/05	12/03	12/31	01/28	02/25	03/24	04/21	05/19	06/16	TOTAL
CUSTOMER ORDERS														
GROSS REQUIREMENTS	20	20	20	24	23	17	24	20	20	20	20	20	26	533
SCHEDULED RECEIPTS														
ON HAND	17	21	1	1	2	9	9	13	17	21	1	5	3	
FIRM PLANNED ORDERS														
PLANNED ORDERS	24	24		24	24	24	24	24	24	24		24	24	504

NOTES & COMMENTS

Comm list no.

List xyz

Source: W. L. Berry, T. E. Vollmann, and D. C. Whybark, *Master Production Scheduling: Principles and Practice* (Falls Church, Va.: American Production and Inventory Control Society, 1979), p. 87.

FIGURE 18.17 Abbott DRP Records for Central and Bulk

Description	Size	um	Std bactr	FC	BY	PL	IT	C	Scrap	Total landed costs	OP	Life	O-LT	P-LT	O-LT	QA-LT	T-LT	On hand	OA inventory	Allocated	Safety stock
Product-B	200	BL	7619	A2	00	03	C	0	1.07			1095		12	0	11	23	5220.0	0.0	144.0	1000.0

01-CENTRAL

	PAST DUE	7/24	7/31	08/07	08/14	08/21	08/28	09/04	09/11	09/18	09/25	10/02	10/09	10/16
CUSTOMER ORDERS	0	0	0	0	0	0	0	0	0	0	0	0	0	0
GROSS REQUIREMENTS	204	12	12	1908	12	12	2916	156	36	1884	12	1104	1980	12
SCHEDULED RECEIPTS	0	0	0	7619	0	0	0	0	0	0	0	0	0	0
ON HAND	4872	4860	4848	10078	10066	10054	7138	6982	6946	5062	5050	3946	9104	9092
FIRM PLANNED ORDERS	0	0	0	0	0	0	0	0	7619	0	0	0	0	0
PLANNED ORDERS	0	0	0	0	0	0	0	0	0	0	0	0	0	0

MONTHLY

	10/23	10/30	11/06	11/13	11/20	11/27	12/04	01/01	01/29	02/26	03/26	04/23	05/21	06/18
CUSTOMER ORDERS	0	0	0	0	0	0	0	0	0	0	0	0	0	0
GROSS REQUIREMENTS	2952	12	36	1884	96	1092	1944	4968	3108	3012	4968	3108	2076	3108
SCHEDULED RECEIPTS	0	0	0	0	0	0	0	0	0	0	0	0	0	0
ON HAND	6140	6128	6092	4208	4112	3020	8214	10384	7276	4264	6434	3326	8388	5280
FIRM PLANNED ORDERS	0	0	0	0	7619	0	7619	0	0	7619	0	7619	0	7619
PLANNED ORDERS	0	0	0	0	0	0	0	0	0	0	0	0	0	0

	07/16	08/13	09/10	10/08	11/05	12/03	12/31	01/28	02/25	03/24	04/21	05/19	06/16	TOTAL
CUSTOMER ORDERS	0	0	0	0	0	0	0	0	0	0	0	0	0	0
GROSS REQUIREMENTS	4740	3120	3108	4980	2052	3108	3012	4980	2964	2052	3108	4956	2808	87612
SCHEDULED RECEIPTS	0	0	0	0	0	0	0	0	0	0	0	0	0	7619
ON HAND	7678	4558	8588	10746	8694	5586	2574	4732	1768	6854	3746	5928	4956	0
FIRM PLANNED ORDERS	0	7619	7619	0	0	0	7619	0	7619	0	7619	0	0	60952
PLANNED ORDERS	0	0	0	0	0	0	0	0	0	0	0	0	0	22857

NOTES & COMMENTS

Comm list no.
List xyz

Description	Size	um	Std bactr	FC	BY	PL	IT	C	Scrap	Total landed costs	OP	Life	O-LT	P-LT	O-LT	QA-LT	T-LT	On hand	OA inventory	Allocated	Safety stock
Product-B	200	L	4000			00	B	C	1.00		N						0	0	0	0	0

84-Bulk

	PAST DUE	7/24	7/31	08/07	08/14	08/21	08/28	09/04	09/11	09/18	09/25	10/02	10/09	10/16
CUSTOMER ORDERS														
GROSS REQUIREMENTS														
SCHEDULED RECEIPTS									4000					
ON HAND														
FIRM PLANNED ORDERS									4000					
PLANNED ORDERS														

MONTHLY

	10/23	10/30	11/06	11/13	11/20	11/27	12/04	01/01	01/29	02/26	03/26	04/23	05/21	06/18
CUSTOMER ORDERS														
GROSS REQUIREMENTS					4000		4000			4000		4000		4000
SCHEDULED RECEIPTS														
ON HAND														
FIRM PLANNED ORDERS														
PLANNED ORDERS				4000			4000		4000			4000		4000

	07/16	08/13	09/10	10/08	11/05	12/03	12/31	01/28	02/25	03/24	04/21	05/19	06/16	TOTAL
CUSTOMER ORDERS														
GROSS REQUIREMENTS			4000				4000		4000		4000			44000
SCHEDULED RECEIPTS														
ON HAND		4000	4000				4000		4000	4000	4000			
FIRM PLANNED ORDERS														
PLANNED ORDERS		4000	4000				4000		4000	4000	4000			44000

NOTES & COMMENTS

Comm list no.
List xyz

Source: W. L. Berry, T. E. Vollmann, and D. C. Whybark, *Master Production Scheduling: Principles and Practice* (Falls Church, Va.: American Production and Inventory Control Society, 1979), p. 88.

tral are based on planned orders from all the DCs. For example, in the week of 8/7, the gross requirement of 1,908 consists of 24 from Vancouver, 1,872 from Montreal, and 12 from some other DC.

The batch size shown for the central warehouse in Figure 18.17 is 7,619—the number of product B units yielded from a batch of 4,000 in the unit of measure for the bulk product; that is, the 7,619 shown as a firm planned order in 9/11 becomes a gross requirement of 4,000 for bulk in that week.

The master production scheduler works with the DRP record for central. This person's job is to convert planned orders into firm planned orders, and to manage timing of the firm planned orders. For example, all orders at central are firm planned in Figure 18.17 until 12/31 in the last row. The last three orders (12/31, 2/25, and 4/21) are only planned orders. DRP logic can replan these as needed. Only the master production scheduler can move the firm planned orders that appear early in the record. The result is a stable MPS for the bulk production.

Figure 18.18 illustrates the information available for actual shipment planning. The requirements due to be shipped this week or in the next two weeks (also any past-due shipments) are summarized by distribution center and product type. This enables the planner to look at the current requirements or future planned shipments in making up carloads destined for a distribution center. Since the information is available in terms of cube, weight, and pallet load, the planner can use the most limited resource in making a shipment decision. This flexibility enables planners to efficiently use transportation resources to meet product needs.

The application of DRP at Abbott Laboratories resulted in benefits in many areas of the company. The product obsolesence costs were reduced, inventory turnover improved, and customer service increased. The logistics information also led to reductions in transportation and warehouse costs.

CONCLUDING PRINCIPLES

This chapter has presented a technique for integrating field inventories, distribution centers, and warehouses into the firm's manufacturing planning and control system. The technique, distribution requirements planning (DRP), utilizes record formats and processing logic consistent with MRP. To effectively use DRP, we see the following general principles:

- The top-level records for a DRP system should cover items in a location as close to the customer as possible (or even at the customer, if feasible).
- Local information on demand patterns should be incorporated into the DRP record.
- Data and performance measurement systems should be put in place to monitor forecast adjustments in the field.

FIGURE 18.18 Abbott Short-Term Shipping Information

MPDR0030 DISTRIBUTION REQUIREMENTS PLAN PAGE 206

DC 21 – MONTREAL

DIVISION 2

LIST/SIZE	QUANTITY	PALLET	WEIGHT	CUBE	QUANTITY	PALLET	WEIGHT	CUBE
	PAST DUE				WEEK 1			
XYZ-200 PRODUCT B	0	0.0	0.0	0.0	0	0.0	0.0	0.0
	WEEK 2				WEEK 3			
	0	0.0	0.0	0.0	1872	2.0	3744.0	112.3

PRIORITY 96: 5076 AVAILABLE IN CENTRAL

DC 03 – VANCOUVER

LIST/SIZE	QUANTITY	PALLET	WEIGHT	CUBE	QUANTITY	PALLET	WEIGHT	CUBE
	PAST DUE				WEEK 1			
XYZ-200 PRODUCT B	0	0.0	0.0	0.0	0	0.0	0.0	0.0
	WEEK 2				WEEK 3			
	0	0.0	0.0	0.0	24	0.0	48.0	1.4

PRIORITY 96: 5076 AVAILABLE IN CENTRAL

Source: W. L. Berry, T. E. Vollmann, and D. C. Whybark, *Master Production Scheduling: Principles and Practice* (Falls Church, Va.: American Production and Inventory Control Society, 1979), p. 90.

- Matching supply to demand requires close control of supply as well as data on demand.
- Projections of future requirements should be used to decide inventory allocation in periods of short supply.
- Transparent records and consistent processing logic should be used to integrate the system.
- What-if analysis should be based on integrated records of the system.
- Uncertainty filters, like firm planned orders or error addback, should be available to the master production scheduler.
- The organization form should be consistent with the supply chain being managed.

REFERENCES

Allen, W. B. "A Comparative Simulation of Central Inventory Control Policies in Positioning Safety Stock in a Multi-Echelon Distribution System." DBA dissertation, Indiana University, 1984.

Ballou, R. H. "Estimating and Auditing Aggregate Inventory Levels at Multiple Stocking Points." *Journal of Operations Management* 1, no. 3 (1981).

Berry, W. L.; T. E. Vollmann; and D. C. Whybark. *Master Production and Scheduling—Principles and Practice.* Falls Church, Va.: American Production and Inventory Control Society, 1979.

Bregman, R. L. "Enhanced Distribution Requirements Planning." *The Journal of Business Logistics* 11, no. 1 (1990), pp. 49–68.

Brown, R. G. *Materials Management Systems.* New York: Wiley Interscience, 1977.

Dube, W. R. "Closed Loop Planning for Manufacturing and Distribution." *International Journal of Physical Distribution and Materials Management* 16, no. 1 (1986), pp. 5–13.

Feare, T. "Lilly—How They Buy." *CPI Purchasing,* February 1985, pp. 26–34.

Ford, Q. "Distribution Requirements Planning and MRP." *APICS 24th Annual Conference Proceedings (1981),* pp. 275–78.

Friedman, W. F. "Physical Distribution: The Concept of Shared Services." *Harvard Business Review,* March–April 1983.

Glover, F.; G. Jones; D. Karney; and J. Mote. "An Integrated Production Distribution and Inventory Planning System." *Interfaces,* November 1979, pp. 21–35.

Jones, T. C., and D. W. Riley. "Using Inventory for Competitive Advantage through Supply Chain Management." *International Journal of Physical Distribution and Materials Management* 15, no. 1 (1985), pp. 16–26.

Ling, R. C. "Demand Management: Let's Put More Emphasis on This Term." *APICS 26th Annual Conference Proceedings* (1983), pp. 11–12.

Martin, A. "DRP: Another Resource Planning System." *Production and Inventory Management Review,* December 1982.

———— *Distribution Resource Planning.* Essex Junction, Vt.: Oliver Wight Publications, 1983.

Perry, W. "The Principles of Distribution Resource Planning (DRP)." *Production and Inventory Management,* December 1982.

Schwarz, L. B. "Physical Distribution: The Analysis of Inventory and Location." *AIIE Transactions* 13, no. 2 (June 1981).

Smith, B. "DRP Improves Productivity, Profit, and Service Levels." *Modern Materials Handling,* July 1985, pp. 63–65.

Stenger, A. J., and J. L. Cavinato. "Adapting MRP to the Outbound Side—Distribution Requirements Planning." *Production and Inventory Management,* 4th quarter 1979, pp. 1–14.

Turner, J. R. "DRP: Theory and Reality." *APICS 1990 Annual Conference Proceedings.*

van Donselaar, K. "Commonality and Safety Stocks." *Pre-Prints of the 4th International Working Seminar on Production Economics,* 1986, pp. 446–80.

Vaughn, O.; T. Perez; and R. Stemwedel. Short Cycle Replenishment at 3M." *APICS 1990 Annual Conference Proceedings.*

Wemmerlöv, U. "A Time-Phased Order Point System in Environments with and without Demand Uncertainty: A Comparative Analysis of Nonmonetary Performance Variables." *International Journal of Production Research* 24, no. 2, 1986, pp. 343–58.

Zmolek, J. "Global DRP: Using the MPS as a Coordinating Mechanism." *APICS 1990 Annual Conference Proceedings.*

DISCUSSION QUESTIONS

1. What is meant by the statement, "The real task of managing materials is matching supply to demand"?
2. Describe how DRP helps bridge the gap between the market and the factory.
3. What are some of the benefits a warehouse manager derives from improved resupply information?
4. Discuss risks and benefits of local warehouse personnel modifying forecast data for field warehouses.
5. A consumer goods manufacturer uses a periodic review system to manage field inventories. This means, once a period, field inventory clerks check to see if inventory is below reorder point; if so they order an amount equal to the EOQ. The materials manager argues this is the same as using TPOP records. What's your opinion?
6. How can you monitor forecast data modifications by field personnel to ensure changes result in improvements over original computer forecasts?
7. What are the differences between the planned shipments, as indicated in Figure 18.3 and planned orders from MRP records?
8. How are the supply chain management concept and materials management similar? How do they differ?

9. The Abbott records in Figure 18.16 show a customer order for Montreal on 8/07. Why do you think this customer order isn't added to the forecast requirements ("Gross requirements" in their terminology) in the record?

PROBLEMS

1. For the warehouse of Figure 18.8, suppose actual demands were 24, 16, and 18 in periods 1, 2, and 3, respectively. The initial DRP record is:

		Period				
		1	2	3	4	5
Forecast requirements		20	20	20	20	20
Scheduled receipts		40				
Projected available balance	6	26	6	26	6	26
Planned shipments			40		40	

Q = 40; lead time = 1; SS = 6.

a. Create the other three DRP records as done in Figure 18.8. What problem is created in period 2? What can be done?
b. Does error addback work for this set of circumstances?

2. The Hazy Company maintains a West Coast distribution center (DC), which is supplied from the plant warehouse in the Midwest. It takes exactly one week to ship to the distribution center from the Midwest. One of its products has an ordering cost of $10 per order, inventory carrying cost of $1 per unit per week, and average weekly demand at the DC of 5 units (although it has fluctuated uniformly between 0 and 10 units per week in actuality). Over the years, the product's safety stock level has varied. There are currently (early Monday morning) nine units in inventory at the DC. The company is willing to risk a probability of stocking out of 0.10 in any order cycle.
a. If the company uses an economic order quantity, reorder point system, what should the order quantity and reorder point be?
b. If the company adapted DRP logic to this DC supply situation and decided to ship only on Mondays (to consolidate shipments), what would you suggest for the planned shipping pattern over the next 10 weeks? (Use a safety stock level of two units.)
c. If actual demand for the upcoming week were six units, what would the new shipping pattern be? (Assume any planned shipments in the current week are released.)

3. The Hercules Mining Company's distribution manager has supplied the following information pertaining to its product, the H208 oscillator, which is stocked at the firm's field warehouse:

Average weekly demand = 25 units.
Current on-hand balance = 15 units.
Open order (due next week) = 60 units.
Economic order quantity = 60 units.
Shipping time = 1 week.
Safety stock = 0 units.

Complete the record indicating the planned shipment schedule for the next eight weeks:

	Week							
	1	2	3	4	5	6	7	8
Forecast requirements								
In transit								
Projected available balance								
Planned shipments								

4. The XYZ Company's finance manager is currently studying projected inventory investment for product 101 at the central (plant) warehouse. Quantities of product 101 are shipped from the central warehouse to warehouses A and B, based on individual warehouse requirements. This product costs $100 per unit.

   Using Exhibit A's time-phased order point and inventory information for product 101, determine the planned order releases at the central warehouse. Construct a graph indicating projected inventory investment at warehouse A for this product at the beginning and end of each week over the next six-week period. Assume this product's demand is equally divided among the days of each week.

EXHIBIT A   Product 101

Product 101 Warehouse A	Week					
	1	2	3	4	5	6
Forecast requirements	20	20	20	20	20	20
In transit	15					
Projected available balance (10)						
Planned shipments						

Q = 40; LT = 1; SS = 0.

	Week					
Product 101 Warehouse B	*1*	*2*	*3*	*4*	*5*	*6*
Forecast requirements	10	10	10	10	10	10
In transit						
Projected available balance	15					
Planned shipments						

Q = 20; LT = 1; SS = 0.

	Week					
Product 101 Warehouse C	*1*	*2*	*3*	*4*	*5*	*6*
Gross requirements						
Scheduled receipts						
Projected available balance	60					
Planned order releases						

Q = lot-for-lot; LT = 1; SS = 0.

5. The distribution of Dyna-Pep is from the plant in Dunham to two warehouses and then to the customers. The distribution pattern is as follows:

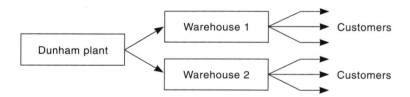

For the purpose of analysis, assume that the warehouse will sell *exactly* 20 cases of Dyna Pep per week. Inventory carried either at the plant or at the warehouse locations costs 10 cents per week per case (based on Friday night inventory). The ordering cost is $9 per order at each facility, and it takes exactly one week to transport the product from the plant to either of the two warehouses or to produce the product. The beginning inventory at each warehouse is 20 units, and the beginning inventory at the Dunham plant is 120 units.
   a. Determine the inventory carrying cost plus setup cost per week at Dunham and both warehouses managing the system with EOQ at all locations.
   b. Determine the inventory carrying cost and setup cost per week using DRP to manage the system.

6. The MVA Pet Food Company distributes one of its products, Gro-Pup, through two warehouses in Seattle and Portland. A central warehouse at the St. Louis

plant distributes Gro-Pup to these two warehouses in serving the Northwest regional market.

a. Develop a distribution schedule for the two warehouses and a production schedule for the plant for the next eight-week period, using the DRP worksheet in Exhibit B. (Note the sales forecast for this product is 20 units per week at the Seattle warehouse, and 43 units of the product are currently on hand. The shipment order quantity is 60 units, planned shipment lead time is one week, and there's no safety stock requirement. Similar information is included in Exhibit B for the other two facilities.) The product is packed in lots of 50 units at St. Louis. The packaging process takes two weeks.

b. Each Gro-Pup package requires one unit of packaging material. Develop a purchasing schedule for the packaging material assuming:
   1. There are currently 25 units on hand.
   2. An open order for 100 units is due to be received from the vendor next week.
   3. The purchase order quantity equals 100 units.
   4. Purchasing lead time is four weeks.

c. Update the distribution and production schedules as of the start of week 2, assuming the following transactions occurred during week 1:

	Receipts	Order releases	Sales (disbursements)	Inventory adjustment	Open order scrap
Gro-Pup/Seattle	0	0	19	−3	0
Gro-Pup/Portland	15	15	20	+3	0
Gro-Pup/Plant	0	50	15	0	0
Gro-Pup/Packaging	90	100	50	+4	10

EXHIBIT B   Distribution Requirements Worksheet

			Week							
Gro-Pup Seattle Warehouse		1	2	3	4	5	6	7	8
Forecast requirements		20	20	20	20	20	20	20	20
In transit									
Projected available balance	43								
Planned shipments									

Q = 60; LT = 1; SS = 0.

Gro-Pup Portland Warehouse		Week							
		1	2	3	4	5	6	7	8
Forecast requirements		10	10	10	10	10	10	10	10
In transit		15							
Projected available balance	2								
Planned shipments									

Q = 15; LT = 1; SS = 0.

Gro-Pup Plant Warehouse		Week							
		1	2	3	4	5	6	7	8
Gross requirements									
Scheduled receipts			50						
Projected available balance	30								
Planned order releases									

Q = 50; LT = 2; SS = 0.

Gro-Pup Packaging Material		Week							
		1	2	3	4	5	6	7	8
Gross requirements									
Scheduled receipts		100							
Projected available balance	25								
Planned order releases									

Q = 100; LT = 4; SS = 0.

7. The Drip Producers are planning a promotion of one of its products, the Dead Drop. Dead Drops are distributed through only one of the warehouses. The promotion is to begin in period 4 and run through period 6. The sales forecast is normally 30 per period, but during the promotion the company expects sales to be 60 per period. This is shown below with other data.

	Period					
	1	*2*	*3*	*4*	*5*	*6*
Sales forecast	30	30	30	60	60	60

	Warehouse	*Central*
Ship/ord. quantity	40 units	Lot-for-lot
Current inventory	12 units	0 units
Lead time	1 period	1 period
Safety stock	10 units	0 units
Sched. rec./ship.	40 units in period 1	40 units in period 1

a. Use a spreadsheet to develop the warehouse and central facility's records.

b. Suppose the central facility's capacity was limited to 50 units per period. How would you provide the material for the promotion in the field?

8. Develop a spreadsheet for the Cranstable Company's two warehouses and one central facility system using the following data. What happens if the central order quantity changes to 200?

	Warehouse 1	*Warehouse 2*	*Central*
Forecast requirements	20/period	30/period	
Ship/ord. quantity	48 units	60 units	100 units
Current inventory	23 units	12 units	0 units
Lead time	1 period	1 period	1 period
Safety stock	10 units	10 units	0 units
Sched. rec./ship.	48 units in period 1	60 units in period 1	None

9. The Cranstable Company was just getting into the swing of DRP when old Barnstable moved away from warehouse 1's region never to be seen again by anyone at the company. (Barnstable demand was lost to a competitor.) Now this wouldn't normally cause problems, but for Cranstable it meant sales were halved at warehouse 1. Using Problem 8's data, what's the impact of Barnstable's departure?

10. Demand for Drips averages five per week at a particular warehouse. The Water Company ships Drips in lots of 10 and maintains a safety stock level of two Drips at the warehouse. The company currently uses a periodic review system to replenish the warehouse. Each Monday morning, the warehouse inventory clerk checks the Drips inventory. If it's less than the reorder point of 12 Drips (demand of 10 during lead and review time, plus safety stock of 2 Drips), an order for 10 is placed on the Water Company. It takes one week for the order

to reach the warehouse. If demand isn't met, sales are lost (i.e., there are no backorders).

a. Use a spreadsheet to set up the DRP record for the warehouse for the next 10 periods and generate the planned shipments. Use an opening inventory of eight units and no in-transit shipments.

b. Develop a spreadsheet that can "simulate" the current practice. Show the Monday morning balance; test for an order; and show demand for the week, the Friday balance, and the following Monday balance. The following format may be helpful:

	Week			
	1	2	3	—
Monday balance				
Demand				
Friday balance				
Order				

c. Compare the orders from the simulation with the planned orders under DRP when demand is exactly five units per week for all 10 weeks. What happens in the simulation when demand is 7, 3, 5, 1, 4, 6, 5, 9, 6, and 4 for the 10 weeks? Is the order pattern different from the DRP plan?

11. The XYZ Chemical Company performs sales forecasting on a national basis. Each month it prepares a sales forecast of monthly sales for the coming year for each of the firm's end products. At each of the firm's three distribution warehouses, time-phased order point records are maintained for the individual products. The national sales forecast for product A is 2,500 units per four-week period. The sales of this product at each of the three warehouses is split as follows: 50 percent from warehouse A, 30 percent from warehouse B, and 20 percent from warehouse C. An analysis of the sales history of product A at warehouse C indicates weekly sales of this product are distributed as follows: 10 percent in week 1, 30 percent in week 2, 50 percent in week 3, and 10 percent in week 4 in a four-week period. Complete the following time-phased order point record for Product A.

		Week							
Product A / Warehouse C		1	2	3	4	5	6	7	8
Forecast requirements									
In transit			160	278					
Projected available balance	65								
Planned shipments									

Q = LFL; SS = 0; LT = 3 weeks.

12. The Excello Corporation's distribution planner is concerned about the variability of transit times between the plant in Madison, Indiana, and the firm's distribution center in Atlanta. It's been decided to incorporate a one-week safety lead time into the time-phased order point records for items stocked in the Atlanta distribution center. Please complete the record for product X in the Atlanta distribution center.

Forecast: 200 units per week.
In transit: 410 units scheduled for receipt in week 1.
On-hand inventory: 15 units.
Order quantity: 600 units.
Lead time: 1 week.
Safety stock = 0.
Safety lead time = 1 week.

Product X		Week							
		1	2	3	4	5	6	7	8
Forecast requirements									
In transit									
Projected available balance									
Planned shipments									

13. The Allied Product Company's sales director has decided to close the firm's San Diego warehouse and open a new warehouse in Los Angeles. The San Diego warehouse is scheduled to close the end of week 4; the new warehouse in Los Angeles is scheduled to open for business at the start of week 5. Product B is currently stocked in the San Diego warehouse and will be stocked in the new Los Angeles warehouse. Sales forecast for product B is 100 units per week. Please complete the following time-phased order point records for product B, reflecting the change in the company's warehousing plans for the item and the implications for the plant.

Product B / San Diego Warehouse		Week							
		1	2	3	4	5	6	7	8
Forecast requirements									
In transit									
Projected available balance	205								
Planned shipments									

Q = LFL; SS = 10 units; LT = 2 weeks.

Product B / Los Angeles Warehouse

		Week 1	2	3	4	5	6	7	8
Forecast requirements									
In transit									
Projected available balance									
Planned shipments									

Q = LFL: SS = 10 units; LT = 1 week.

Product B / Plant

		Week 1	2	3	4	5	6	7	8
Gross requirements									
Scheduled receipts									
Projected available balance	300								
Planned order releases									

Q = 200 units; SS = 15 units; LT = 3 weeks.

14. The Stasik Pharmaceutical Company's distribution planning manager is concerned about the rising cost of outdated inventory in the firm's distribution warehouses. Shelf life for the company's products shouldn't exceed four weeks. Government regulations require products in stock longer than four weeks to be scrapped. Please devise an exception notice test that can be applied to the time-phased order point records in the firm's distribution warehouses to direct the inventory planner's attention to items where excess inventory may exist. Apply this procedure to the time-phased order point record below to determine whether out-of-date inventory is likely to occur for product W.

Product W

		Week 1	2	3	4	5	6	7	8
Forecast requirements		300	300	300	300	300	300	300	300
In transit			900						
Projected available balance	350	50	650	350	950	650	1250	950	1550
Firm planned orders			900		900		900		
Planned shipment									

Q = 900; SS = 50 units; LT = 2 weeks.

15. Product D is stocked only at the AMC Chemical Company's Dallas warehouse and at the company's plant warehouse in Akron. The sales director has forecast product sales from the Dallas warehouse to be 40 units per week. Product D is manufactured at the firm's Akron plant using 2 units of ingredient X per unit of product D.

a. Please complete the time-phased order point records for product D at the Dallas warehouse and the plant warehouse as well as the MRP record for ingredient X using the following information. (Assume ingredient X is only used in product D.)

Product D / Dallas Warehouse		Week							
		1	2	3	4	5	6	7	8
Forecast requirements									
In transit									
Projected available balance	85								
Planned shipments									

Q = LFL: SS = 5 units; LT = 2 weeks.

Product D / Plant Warehouse		Week							
		1	2	3	4	5	6	7	8
Gross requirements									
Scheduled receipts									
Projected available balance	42								
Planned order releases									

Q = LFL: SS = 2 units; LT = 2 weeks.

Ingredient X		Week							
		1	2	3	4	5	6	7	8
Gross requirements									
Scheduled receipts		320							
Projected available balance	4								
Planned order releases									

Q = 320 units; SS = 2 units; LT = 4 weeks.

b. Actual sales for product D at the Dallas warehouse in week 1 were 48 units. Prepare time-phased order point and MRP records as of the beginning of week 2. What actions should the master production scheduler take on the basis of this information?

Product D / Dallas Warehouse	Week							
	2	3	4	5	6	7	8	9
Forecast requirements								
In transit								
Projected available balance								
Planned shipments								

Q = LFL; SS = 5 units; LT = 2 weeks.

Product D / Plant Warehouse	Week							
	2	3	4	5	6	7	8	9
Gross requirements								
Scheduled receipts								
Projected available balance								
Planned order releases								

Q = LFL; SS = 2 units; LT = 2 weeks.

Ingredient X	Week							
	2	3	4	5	6	7	8	9
Gross requirements								
Scheduled receipts								
Projected available balance								
Planned order releases								

Q = 320 units; SS = 2 units; LT = 4 weeks.

c. Suppose product D's actual sales at the Dallas warehouse in week 2 were 31 units. What action would the master scheduler take on the basis of this information? To avoid this situation, what modifications to the master production schedule might be taken?

	Week							
Product D / Dallas Warehouse	*3*	*4*	*5*	*6*	*7*	*8*	*9*	*10*
Forecast requirements								
In transit								
Projected available balance 85								
Planned shipments								

Q = LFL; SS = 5 units; LT = 2 weeks.

	Week							
Product D / Plant Warehouse	*3*	*4*	*5*	*6*	*7*	*8*	*9*	*10*
Gross requirements								
Scheduled receipts								
Projected available balance 42								
Planned order releases								

Q = LFL; SS = 2 units; LT = 2 weeks.

	Week							
Ingredient X	*3*	*4*	*5*	*6*	*7*	*8*	*9*	*10*
Gross requirements								
Scheduled receipts								
Projected available balance 4								
Planned order releases								

Q = 320 units; SS = 2 units; LT = 4 weeks.

Chapter 19

MPC Frontiers

In this final chapter of the book, we identify some new developments and future directions for manufacturing planning and control (MPC) systems. Manufacturing planning and control is a dynamic field that will continue to evolve. MRP was only possible with high-speed random access computing. User-friendly languages and dedicated computers enable use of on-line systems for detailed shop scheduling based on finite loading. Artificial intelligence languages are increasingly being applied to MPC problems.

Not all MPC frontiers are being pushed by computer technology. The just-in-time (JIT) approaches and resultant drive for simplicity in many cases results in *less* computerization. Emphasis on hidden factory costs, simplicity, and fast response to customer needs is profoundly affecting MPC design and operation.

This chapter starts by using the general MPC schematic to evaluate new concepts. We then deal with three related new and different approaches to manufacturing planning and control, followed by discussion of some resultant challenges for evolution in MPC practice.

The chapter centers around five topics:

- The MPC system schematic: How can our general model of a manufacturing planning and control system be used to assess new systems?
- Optimized production technology (OPT): What is this approach and where is it heading?
- The periodic control system at Kumera Oy: How did a European firm quickly improve manufacturing performance?
- MPC for the process industries (PRISM): What changes are needed for process technology firms and how are they handled?
- Observations on MPC evolution: What generalizations and conclusions can be drawn for future MPC practice?

Chapter 19 relates to many other chapters. The comparison and evaluation of new MPC approaches utilizes ideas in Chapter 1. The OPT system uses finite loading concepts described in Chapter 5. Related scheduling issues are described in Chapter 13. Just-in-time concepts are discussed in Chapters 3 and 12. Finally, Chapter 10 addresses the issues encountered in implementing MPC systems.

THE MPC SYSTEM SCHEMATIC

In this section, we relate our general model for manufacturing planning and control systems to other systems and approaches. The MPC system schematic is presented as Figure 19.1. We first discuss the concept of a standard for MPC systems and then turn to using the schematic.

FIGURE 19.1   Manufacturing Planning and Control System

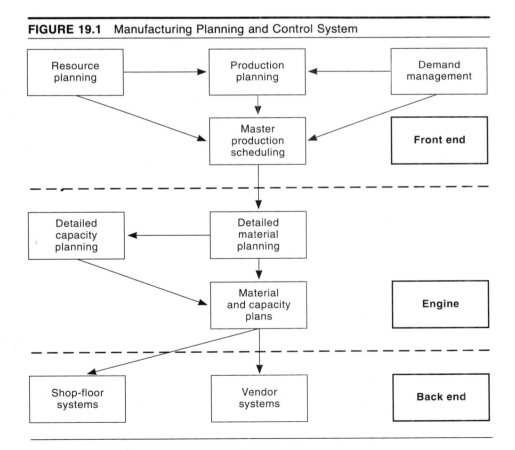

The Standard for MPC Systems

It's tempting to say the MPC system shown in Figure 19.1 is the standard for evaluating and comparing alternative systems. This is accurate in a limited but important sense. The schematic presents a set of functions that must be performed. The standard is that each of these functions must be accomplished in any firm; that is, there must be front-end, engine, and back-end activities.

What Figure 19.1 doesn't capture is the emphasis or importance each module will have in a particular company. For example, firms with only a limited number of suppliers would have very different vendor systems than a company with many. Figure 19.1 also doesn't detail *costs* associated with maintaining satisfactory performance in each of the three areas. Many firms are reducing transaction costs associated with tracking in the back end.

The differences between firms aren't only manifested in different emphases; different systems or techniques will be used, as well. Moreover, all this needs to be viewed from an evolutionary point of view. *No* system is incapable of improvement. Once a fairly complete integrated (and transparent) system is in place, it's easier to evaluate and implement new items.

In summary, we see the MPC system schematic as the standard for the general functions to be performed. The specific procedures applied to accomplish these functions must be evaluated on an ongoing basis, both in terms of alternatives and in terms of what's currently viewed as important. There must be systems that perform these functions—but it's not important that they be *specific* types of systems. Moreover the systems need to be continuously improved.

Beyond the Schematic

Figure 19.1 can be used as a checklist for auditing an MPC system. The crucial questions are whether each activity is performed, how well they're integrated, and how well each subsystem works. As noted previously, subsystems can always be improved.

An example involving no real change in fundamental theory is the increasing use of on-line systems. Many firms have well-functioning MPC systems based on a weekly cycle for MRP planning. Going to an on-line system allows the company to move to a daily cycle, with resultant reductions in inventory. In many cases, the on-line system's operating costs also are less.

For many of the newer approaches to MPC, an upgrade of existing methodologies isn't enough. It's also necessary to consider the broad management and sociotechnical environments in which MPC systems are embedded. Sometimes it's necessary to make major changes in managerial practice and in the evaluation systems for manufacturing performance to realize improvement. This is particularly true for firms considering the kinds of approaches we'll discuss next.

OPTIMIZED PRODUCTION TECHNOLOGY (OPT)

A great deal of attention has been focused on the theory of constraints (TOC) and a proprietary system called OPT. (This acronym originally stood for optimized production timetable, but now stands for optimized production technology.) We'll discuss OPT first, then turn to TOC. OPT is trademarked in two ways. The first trademark is "The OPT Concepts" (the philosophy); the second is "OPT," which refers to the software package (OPT/SERVE). Many questions about OPT have arisen. What does it do? What's the experience of OPT users? Layered on top of the professionals' questions are broad issues raised by OPT's creators themselves: What's important in scheduling a factory? What are the impacts of modern-day cost accounting systems on MPC? In addition, those who created and sell OPT have evolved their own thinking and beliefs, so their views have changed somewhat over time.

We believe the best way to understand OPT is to view its role in basic MPC systems. This view allows those who understand MPC systems to see where OPT fits. We also believe, when viewed from this vantage point, OPT (at least when OPT is seen as a software product rather than as a philosophy) can be added to most state-of-the-art MPC software packages as an enhancement. Viewing OPT in terms of software *and* in philosophical terms will permit us to see some important objectives for future enhancements to manufacturing planning and control.

Basic Concepts of OPT

When we first studied OPT, the conclusion was OPT fits into the "back-end" section of Figure 19.1; that is, it was only a sophisticated shop-floor control system based on finite loading procedures that concentrates on a subset of work centers (the bottlenecks). It uses an algorithm developed by Eliyahu Goldratt (published details of which are still unavailable) to do the finite loading (scheduling) very quickly.

This view doesn't recognize all OPT's contributions. OPT begins its process by combining data in the bill of material file with data in the routing file. The result is a network, or extended tree diagram, where each part in the product structure also has its operational data attached directly. These data are then combined with the MPS to form the "product network." Figure 19.2 symbolically represents an OPT product network. Customer orders are linked to the final assembly process, which, in turn, is linked to the completed components, detailed fabrication steps, and raw materials. Additional data typically included in the OPT files are capacities, maximum inventories, minimum

---

FIGURE 19.2 Sample Product Network

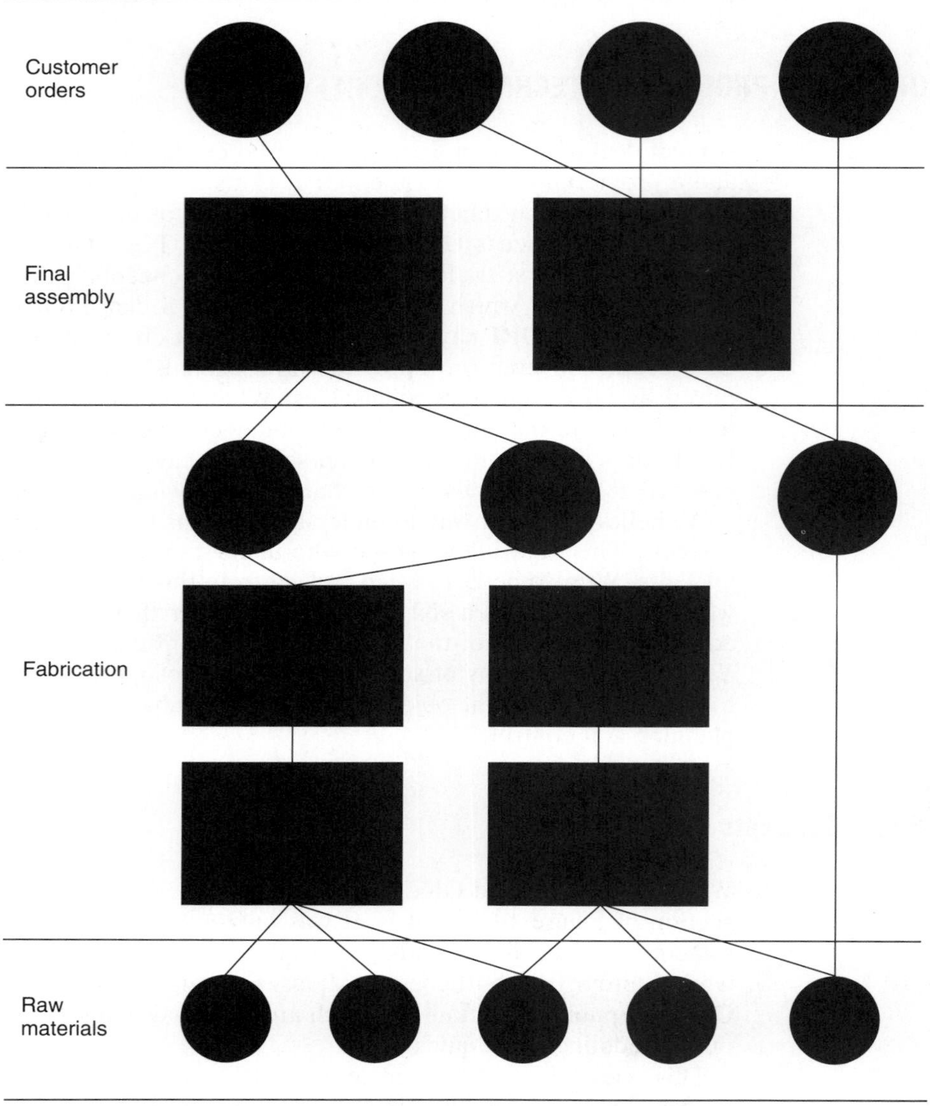

Source: F. Robert Jacobs, "OPT Uncovered: Many Production Planning and Scheduling Concepts Can Be Applied with or without the Software," *Industrial Engineering*, October 1984.

---

batch quantities, order quantities, due dates, alternate machine routings, labor constraints, and other data typically used in finite loading models. In Figure 19.3, these data would be part of the "resource description." Product network and resource descriptions are then fed into a set of routines called

FIGURE 19.3 OPT/SERVE Information Flow

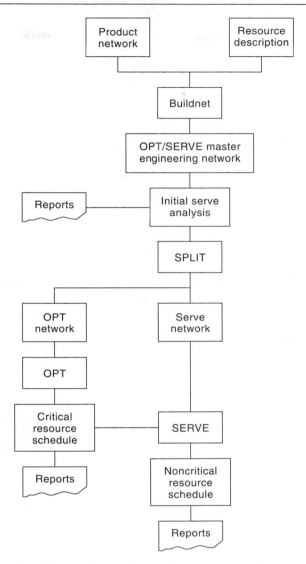

Source: F. Robert Jacobs, "OPT Uncovered: Many Production Planning and Scheduling Concepts Can Be Applied with or without the Software," *Industrial Engineering,* October 1984.

BUILDNET and SERVE that identify the bottleneck resources. The BUILDNET routine combines the product network and resource information to form an OPT network of the bottleneck resources. The SERVE routine uses the information on nonbottleneck resources to backward schedule from the order due dates, using logic similar to MRP, assuming infinite capacity for the resources. The initial analysis provides reports indicating bottleneck resources.

The system incorporates a rough-cut capacity planning routine that provides much of the information of other capacity planning procedures. Since the OPT product network includes both the parts and their routings, a pass through this network can result in an estimate of the capacity required at each work center. Moreover, a gross-to-net calculation can be made at each step to improve the capacity requirement estimates. Lot sizes at this rough-cut stage are based on lot-for-lot rules. The resultant capacity needs, when divided by the number of weeks in the planning horizon, are the average capacity requirements for each resource. When divided by the resource capacities, the result is average expected loads.

The average loads on machine centers are sorted in descending order, and the most heavily loaded are studied by analysts. Typical questions include: Are the data correct? Are the time standards accurate? Can we easily increase capacity? Can we use alternate routing for some items? Any changes based on these questions result in another run to see if the bottleneck resources change.

At this point, the OPT product network is split into two portions using a routine called SPLIT. Figure 19.4 shows a symbolic example of SPLIT's results. The lower section (the "SERVE network") includes all operations that precede the bottleneck resources. The upper portion (the "OPT network") incorporates all bottleneck resources and all succeeding operations, including market demand for end products with parts that have processing on the bottleneck resources. The OPT network is forward finite loaded, using Goldratt's algorithm.

The "SERVE network" (see Figure 19.3) encompasses all the nonbottleneck part operations (the noncritical resource schedule), which are back scheduled using MRP logic. In the initial scheduling pass using SERVE, due dates are offset from customer order due dates for all part operations. In the second pass, however, due dates for any part operations that feed bottlenecks are based on those established by the OPT finite loading of bottlenecks. Schedules for those part operations are so set that material will be available in time for the first operation in the OPT network.

One advantage of the OPT-MRP split is we can readily see where attention should be focused. Not only is bottleneck capacity utilized more intensively by finite loading of this small subset of work centers, but identifying bottlenecks allows us to target efforts in quality and production improvements on these resources.

Now we can identify clearly one of OPT's primary contributions. When finite loading through bottleneck resources is complete, the result is a *doable master production schedule*. For this reason, OPT is sometimes considered to

FIGURE 19.4 Product-Network: Critical, Noncritical SPLIT

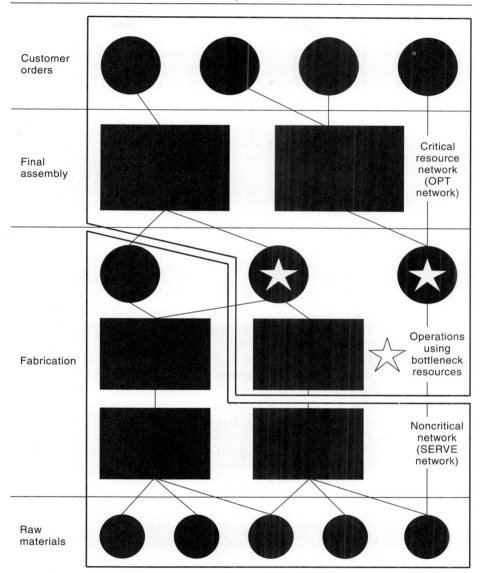

Customer orders

Final assembly

Fabrication

Raw materials

Critical resource network (OPT network)

Operations using bottleneck resources

Noncritical network (SERVE network)

Source: F. Robert Jacobs, "OPT Uncovered: Many Production Planning and Scheduling Concepts Can Be Applied with or without the Software," *Industrial Engineering,* October 1984.

be a "front-end" system (i.e., a master production scheduling technique). We see it less as an MPS technique than as an enhancement to the MPS. OPT conceivably can take any MPS as input and determine the extent to which it is doable.

This means OPT makes an explicit computer-based analysis of the feedback from the engine (and back end) to the front end—an important enhancement to MRP systems. It means a valid MPS is generated—one the firm has a strong chance of achieving—based on the capacity parameters used in the scheduling.

A secondary contribution at this stage comes from the way OPT schedules the nonbottleneck resources. The easiest way to see this is to assume (as is often the case in practice) there are *no* bottlenecks. In that case, OPT schedules are based on MRP logic. The difference is that OPT in this case will change batch sizes (reducing them) to the point where some resources almost become bottlenecks. The result is less WIP: reduced lead time, greater material velocity, and a move toward "zero inventory" manufacturing. OPT does much of this by overlapping schedules using unequal batch sizes for transferring and processing.

OPT's third important contribution is to virtually eliminate the fundamental issue of conflicting priorities between MRP and finite loading. By only finite loading a small fraction of the work centers, priority conflict issues should largely disappear. Moreover, computational time required to do finite loading should be dramatically cut by only dealing with a subset of orders and work centers/resources.

In operation as a shop-floor control technique, OPT has a few other differences from usual practice. A fundamental tenet in OPT is an hour lost in a bottleneck resource is an hour lost to the entire factory's output, while an hour lost in a nonbottleneck resource has no real cost. This means capacity utilization of bottleneck resources is important. OPT increases utilization by using WIP buffers in front of bottlenecks, and where output from a bottleneck joins with some other parts. OPT also runs large batch sizes at bottleneck operations, thereby reducing relative time spent in setup downtime.

In practice, the variable lot size issue has two major implications. First, lead times should be shorter: smaller batches will move faster through nonbottleneck work centers. Second, (and less felicitous), procedures have to be developed to split/join batches as they go through production.

OPT and the MPC Framework

When OPT was first introduced, some people saw it as a replacement for an integrated MPC system. In fact, this isn't correct. OPT is better seen as encompassing much of what's in the engine and back end of Figure 19.1—but combining them in a way that allows us to plan both materials and capacity at the same time; that is, OPT accomplishes many functions in the MPC framework, but not all.

A fundamental principle of OPT is only the bottleneck operations (resources) are of critical scheduling concern. The argument is production output is limited by the bottleneck operations. Increased throughput can only come via better capacity utilization of the bottleneck facilities, and increased batch sizes are one way to increase utilization.

OPT calculates different batch sizes throughout the plant, depending upon whether a work center is a bottleneck. This has several MPC implications. In Figure 19.1, lot sizes are produced as part of the MRP explosion process. In typical finite loading procedures, the batch size is fixed. Such isn't the case with OPT. It also follows that a batch size for one operation on a part could be different for other operations on the same part. This implies special treatment will be required for any paperwork that travels with shop orders. In fact, OPT is designed to do order splitting. In usual practice, order splitting is done on backlogged (bottleneck) machines; in this situation OPT would do the opposite.

The key to lot sizing in OPT is distinguishing between a *transfer* batch (that quantity that moves from operation to operation) and a *process* batch (the total lot size released to the shop). Any differences are held as work-in-process inventories in the shop. In essence, no operation can start until at least a transfer batch is built up behind it. Also, whatever the buildup behind the work center, it *only* produces a transfer batch unless the finite scheduling routine calls for multiple batches.

The lot-sizing issue is closely related to a scheduling approach in OPT called **drum-buffer-rope.** The name comes from the bottleneck defining the schedule (i.e., it's the drum), "pull" scheduling in nonbottleneck operations (the rope), and buffers at both the bottleneck and finished goods (but not at nonbottlenecks). The basic concept is to move material as quickly as possible through nonbottleneck work centers until it reaches the bottleneck. There, work is scheduled for maximum efficiency (large batches). Thereafter, work again moves at maximum speed to finished goods. What this means for lot sizing is very small transfer batches to and from the bottleneck, with a large process batch at the bottleneck. In fact, JIT operating conditions are used everywhere except at the bottleneck.

It's been argued OPT doesn't have the same needs for data accuracy that MRP scheduling has. This is partially correct, if you feel less accuracy is required for nonbottleneck parts and work centers. But going into the process of using OPT, you may not realize very well what these bottleneck operations are. Both OPT and MRP require detailed knowledge of product structures and processes. Data bases, accurate transaction processing, and the right managerial commitment are required for both as well.

Returning to Figure 19.1, we see OPT can't uniquely be put in the front end, the engine, or the back end. It works in all three areas, and does some things quite differently than when scheduling is done by other approaches. However, OPT uses most of the same data as other MPC applications. We still need a front end, engine, and back end (both shop-floor and vendor systems). For the firm with an operating MPC system, the basic data base and closed-loop understanding exist. Implementing OPT as an enhancement seems to be a logical extension. OPT is another example of separating the vital few from the trivial many, and thereafter providing a mechanism to exploit this knowledge for better manufacturing planning/control. It allows a firm to simultaneously plan materials and capacities and to integrate important concepts from finite loading into the MPC system.

Philosophical Underpinnings

Figure 19.5 summarizes several principles of OPT. Two of these were identified previously. First, utilization of a bottleneck is critical. Second, reduced utilization of nonbottleneck resources costs nothing. On the other hand, in the traditional cost accounting view, people should always be working. But if these people are at nonbottleneck resources, the net result of their work will increase WIP and cause confusion for scheduling at other work centers. Under OPT, as with JIT, it's quite all right to not work if there's no work to do. In fact, working (by the usual definition) in this situation will *cause* problems.

The primary OPT objective is stated as making money. This is achieved by maximizing throughput. Throughput is limited by the bottleneck resources, so all efforts are devoted to maximizing capacity utilization in these work centers. Capacity can never be totally balanced. It's *flow* (that can be sold), not overall capacity utilization, that's important.

FIGURE 19.5 OPT Principles

1. Balance flow, not capacity.
2. The level of utilization of a nonbottleneck is determined not by its own potential but by some other constraint in the system.
3. Utilization and activation of a resource are not synonymous.
4. An hour lost at a bottleneck is an hour lost forever.
5. An hour saved at a nonbottleneck is just a mirage.
6. Bottlenecks govern both throughput and inventory in the system.
7. The transfer batch may not, and many times should not, be equal to the process batch.
8. The process batch should be variable, not fixed.
9. Schedules should be established by looking at all constraints simultaneously. Lead times are the result of a schedule and can't be predetermined.

The best way to utilize labor effectively is to work on cross-training so unique skills are less of a constraint. Workers who aren't at bottleneck operations shouldn't be paced by a 100 percent work load. They should utilize extra time in other activities, such as quality improvement, industrial engineering, and skill enhancement.

These philosophical underpinnings of OPT are important in their own right. To some extent, many of the basic arguments have been made before, but OPT carries these arguments to a more operational level. The link is clearly made between the OPT position on cost accounting and the resultant scheduling and shop practices. The result is a set of manufacturing practices that are sometimes counterintuitive—and difficult to implement without a belief at the top in the OPT philosophy.

OPT philosophical underpinnings are important to achieving benefits in many manufacturing improvement programs. The net result is a need for education throughout the company, a change in mores for many firms and top management's commitment to the basic concept, philosophy, and resultant actions required (e.g., let nonbottleneck people do some nondirect production work or even be idle). As an example of changed philosophies, an OPT firm's response to the recession was to reduce batch sizes, not work force!

Theory of Constraints

The OPT philosophy has evolved. To more clearly separate OPT's philosophical concepts from the computer software, Eliyahu Goldratt and his associates have coined the term **theory of constraints (TOC)** to represent their ideology. The concept of a bottleneck has been generalized into "constraint," which includes marketplace constraints. In fact, they argue one goal is to have company output constrained by the marketplace, not by constraints over which the firm has more control.

The theory of constraints has many of the same underpinnings as linear programming; the notion of shadow prices, for example, is directly applicable. But it adds some more operational concepts for dealing with constraining situations. Constraints are explicitly identified, and following the drum-buffer-rope concepts, they're buffered with inventory. Also, the constraint's importance is made clear to the entire factory. For example, bottleneck work centers are operated over coffee breaks and lunch, and are worked a maximum of overtime hours. Moreover, jobs are closely examined to find any that can be alternatively routed—even if the result is "excess cost" for the work so routed. The goal is always to break a constraint or bottleneck condition —and thereafter identify the *next* constraint. Continuous improvement is an integral part of the theory of constraints philosophy. Moreover, the *path* for the improvement is directed by the theory—always following the constraints.

Implementation Issues

OPT presents several difficulties in implementation. To those who have been through a major MPC system implementation, such as JIT or MRP, there are both similarities and important differences. In general, OPT is not for the novice. Companies need to understand basic finite scheduling concepts. They also need sound basic systems, education, top-management support, and a willingness to unlearn some ingrained habits.

Some of these relate directly to the philosophy. The procedures go hand in hand with the philosophical arguments. We know of one firm that has been working for several years to implement OPT, without great success, because it has strong pressures to fully utilize all direct labor-hours. The cost accounting tenets of OPT have just not been accepted.

A related issue is the unpublished algorithm used for scheduling the plant. We have argued throughout this book for "system transparency." OPT is anything *but* transparent. It's difficult to understand, and it's even more difficult to understand why some schedules have been produced. Many OPT results are quite counterintuitive; it's often hard to see why they are as they are.

There's always a difficulty of implementation if the basis for the schedule isn't clear to the shop-floor people responsible for its execution. This is aggravated when the shop-floor people's performance evaluation isn't directly related to schedule execution. Some evidence of these problems was expressed by members of an OPT users group who talked about the time required to get schedule adherence among foremen on the floor.

Another problem of OPT is the certainty assumptions used in processing. To the extent that data are incorrect on capacities, batch time requirements, and so on, the system will produce imperfect results. Control techniques, such as input/output, help in this regard.

OPT buffers the schedules for critical operations at bottleneck operations by using both safety stock and safety lead time. In scheduling a sequence of jobs on the same machine, safety lead time can be introduced between subsequent batches. This provides a cushion against variations adversely affecting the flow of jobs through this same operation.

To protect against having these variations affect subsequent operations on the same job, safety lead time is again employed. In this case, the start of the next operation on the same job isn't scheduled immediately after the current operation is completed. A delay is introduced to perform the buffering here. Note there can be another job in process during the delay; its completion will affect the actual start date for the arriving job. Each of these allowances means actual conditions will vary from the OPT schedule. The question for the supervisors at some point could easily be *which* job to run next.

To ensure there's always work at the bottleneck operation (to provide maximum output), there are safety stocks in front of these work centers. Thus, whenever one job is completed, another is ready to go on the bottleneck machine.

To protect the assembly schedule against shortages that could severely cut output, a safety stock of completed parts from the bottleneck operations are held before assembly. The idea is that disrupting the bottleneck operation to produce a part needed by final assembly will reduce output. Part shortages that can be made up by going through operations that aren't bottlenecks won't cut capacity.

Management factors also enter the OPT scheduling system. These include making realistic schedules that meet management criteria. Factors involving the levels of work-in-process inventory, the capacity utilization attainable, degree of schedule protection, and batch size controls can all be applied to the OPT procedure. These help take into account the company culture as the procedure is implemented.

The Repetitive Lot Concept

The OPT use of different batch sizes, depending on whether a work center is a bottleneck, has led to related research efforts on the scheduling frontiers of MPC systems. Jacobs and Bragg combine shop scheduling decisions with lot sizes in their work on the **repetitive lot concept.** They present simulation results indicating major improvements in the average flow time for manufactured lots and work-in-process inventory using conventional priority scheduling procedures and transfer batches for job shop production.

Jacobs and Bragg permit the *original order quantities* released to the shop for manufacturing to be split into smaller *transfer batches* that can flow immediately to the next operation prior to the operation's completion at its current work center. The transfer batches are predetermined integral fractions of the original order quantity. They provide a work center with the flexibility to start producing an order before it is completed at the previous work center. Such flexibility, frequently referred to as "lot-splitting" and "overlap or line scheduling," reduces order flow times, improves machine utilization, cuts setup times, and smoothes work flow in the shop to yield better use of capacity. This flexibility also means the number of units produced during a given work center setup, *operation batch size*, can vary between a transfer batch and the original order quantity.

Figure 19.6 illustrates the repetitive lot concept and its effect on order flow time. Using fixed operation batch sizes of 1,000 (equal to the original order quantity) in Part A, the order is completed at hour 2,250. In Part B, while the original order quantity is used at operation 1, a transfer batch size of 100 is used to permit processing the order simultaneously at operations 2 and 3 for completion by hour 1,125. In this case, the operation batch sizes for work centers 2 and 3 are 200 and 500, respectively. Although Figure 19.6 doesn't consider the fact that other jobs may be competing for the resources each operation uses, the simulation took this into account when assessing the potential benefit of the repetitive lot concept.

FIGURE 19.6   A Comparison of Fixed versus Variable Operation and Transfer Batch Sizes for a Single Job

	Operation	Time per part (minutes)
Original order quantity is 1,000 units. Three operations are required to produce the part. Processing time per part is given. There is no setup time required.	1	1.00
	2	0.50
	3	0.75

A. Fixed operation batch size = 1,000
Transfer batch size = 1,000

Operation	Schedule	Completion time of last unit
1		1,000
2		1,500
3		2,250

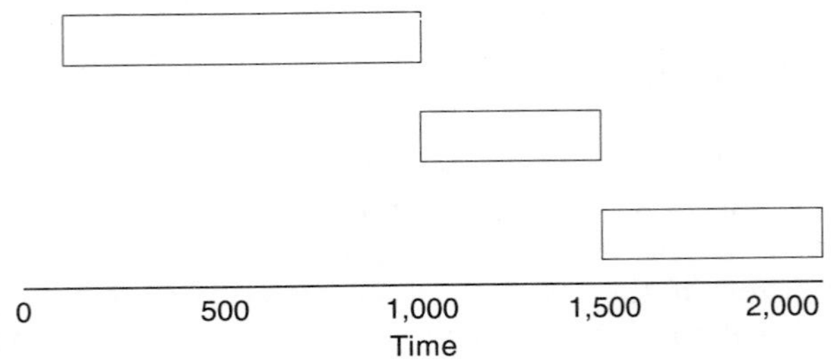

0   500   1,000   1,500   2,000
Time

B. Variable operation batch size
Transfer batch size = 100

Operation	Schedule	Operation batch size	Completion time of last unit
1		1,000	1,000
2		200	1,050
3		500	1,125

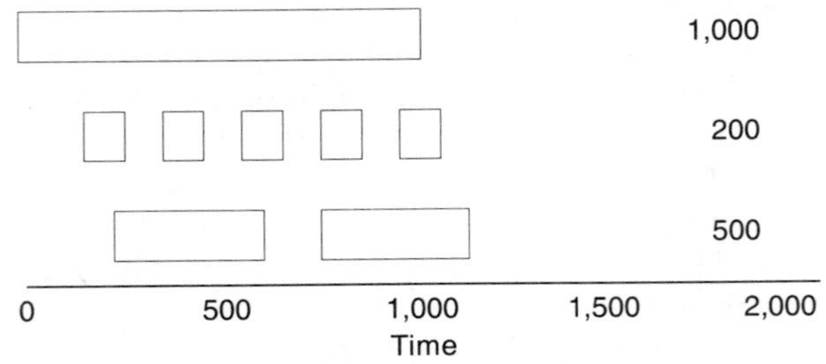

0   500   1,000   1,500   2,000
Time

Source: F. Robert Jacobs and Daniel J. Bragg, "Repetitive Lots: Job Flow Considerations in Production Sequencing and Batch Sizing," *Decision Sciences* 19, no. 2 (1989), pp. 281–94.

The repetitive lot concept can be applied by using any standard priority scheduling method (e.g., shortest processing time, critical ratio). When an order is completed under traditional priority scheduling rules, the highest-priority order in the queue is selected for processing next. Under the repetitive lot concept, a work center may contain transfer batches coming from many released orders. In this case, the queue is searched for transfer batches of the same item that has just been completed at the work center. If such an item is available, it's processed next, regardless of priority; otherwise, the highest-priority transfer batch in the queue is selected and a new setup is made at the work center. If the queue contains no transfer batches, the next batch to arrive at the work center is processed.

Jacobs and Bragg report simulation results in which the repetitive lot size concept is tested, using a model of a shop with 10 work centers. The original order quantity for released orders was varied from 120 to 400 in these experiments, and two different transfer batch sizes (50 and 10) were used. A 38 percent average improvement in the mean order flow time was observed when a transfer batch size of 50 was used; a 44 percent average improvement was obtained with a transfer batch size of 10. Total setup time at the work centers fell 23 to 27 percent when transfer batches were used in conjunction with small original order quantities (120 to 200) for the released orders. However, total setup time rose 13 to 16 percent when larger original order quantities (250 to 400) were used.

This study indicates joint scheduling and batch sizing procedures developed by using repetitive lot concepts can provide the benefits of small lot production without requiring capacity increases or investment in reducing setup times. While the repetitive lot concept may raise material handling costs and make tracking orders in a shop more complex, it appears to be a promising new integrative approach for improving manufacturing performance. High-volume manufacturers with limited product lines having numerous operations appear to benefit most from the reduced order flow times, lower levels of work-in-process inventory, and potential gains in customer service provided by using the repetitive lot concept.

THE PERIODIC CONTROL SYSTEM AT KUMERA OY

Kumera Oy (headquartered about 40 miles north of Helsinki in Riihimäki, Finland) produces a broad range of gear-driven, speed-reducing power transmissions. It engages in all phases of manufacturing from engineering to parts fabrication and assembly. Kumera employs about 450 people and has been using the periodic control system for several years.

A typical product consists of a housing, mounting devices, shafts, gears, assembly hardware, and perhaps a motor. Some 50 different part numbers are usual, with 100 or more individual pieces. Of the 50 part numbers, one third to one half are manufactured (generally the large items, such as gears, shafts, and housings) and the rest purchased. Lead times for some items exceed two months, though most are within five weeks. The large number of end items possible means virtually no finished-goods inventory can be carried, but competitive pressures require short delivery times to customers. The periodic control system has helped resolve this basic conflict between sales and production.

The System

To describe the periodic control system, we follow the MPC framework in Figure 19.1. The system's description starts with the front-end activities, passes to the engine, and concludes with the back end. The production planning part of the front end is an intimate part of company game planning. The resultant commitment to an integrated plan provides direction to the specific production planning and control activities.

The first step in installing the periodic control system was designating product groups. The planning and scheduling activities could not be based on specific end items, since the number is too large, and the company must quote delivery times less than total product lead times. Five product groups have been formed on the basis of production process similarities. Within each group, divisions exist for product options. The groupings don't exactly conform to the catalogue product families, but it's easy to translate from customer orders to the groups for order entry purposes.

The groupings facilitate front-end activities. Forecasts are made at the main option level and are aggregated by product group as the basis for demand management, budgeting, profit and cash planning, and production planning. Production and resource planning specify the overall production rate and any capacity expansion the company will undertake. The resulting plan (which reflects sales, finance, and engineering objectives as well) has a one-year horizon, by week, in units, for each of the five product groups.

The product groups provide the basis for master production scheduling. The basic process will be illustrated using an example with the five groups, shown in Figure 19.7. To start, rough-cut capacity planning ensures capacity at key machines will be available to produce the forecast product group mix before going forward with master production scheduling.

The master production schedule is based on the period length of the periodic control system. It's determined by dividing the year into equal increments. Each group of products will be scheduled *once* during a period. Kumera calls the specific products to be produced a *production set*. Actual customer orders are assigned to the production set in a period using available-to-promise concepts; that is, actual orders replace the period quantities used for planning. Shop orders for production of all components necessary to produce actual customer orders in particular production sets are released at one time. In general, the components are *only* for those specific items in the production set. The sequence of specific products within a production set is based on manufacturing efficiencies in changing from one group to another.

The choice of a repetitive cycle of production helps production control and coordinates production orders' release dates. The length of a period determines how many units of each group will be made in a production set. The quantity effects manufacturing costs. The length of period also affects how much of the production can be made strictly to order and how much to forecast. Figure 19.8 shows these and other factors influencing period length. The

FIGURE 19.7 Group Forecasts, Capacity Allocation, and Periodic Quantities

Product group	*Annual forecast*	*Share of available labor (percent)*	*Capacity of key machine (percent)*	*Period quantities*
A	500	15%	25%	50
B	100	20	5	10
C	1,000	10	30	100
D	2,000	25	20	200
E	1,500	30	10	150

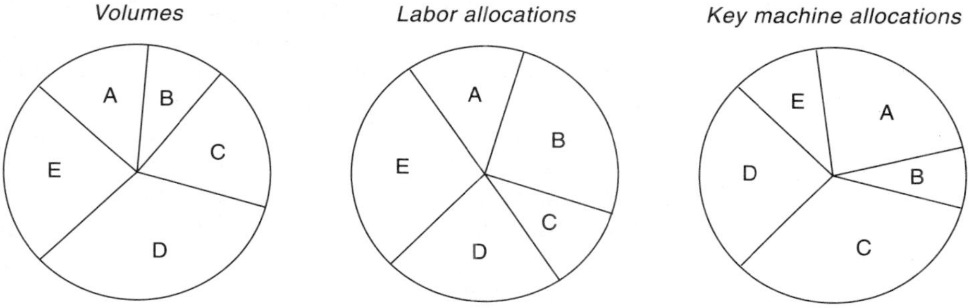

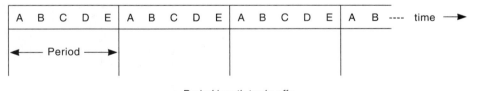

Volumes
A
B
C
D
E
Labor allocations
A
B
C
D
E
Key machine allocations
E
A
B
C
D

Source: D. C. Whybark, "Production Planning and Control at Kumera Oy," *Production and Inventory Management,* first quarter 1984.

Figure 19.8 Sequence for Releasing Production Orders for Groups

A B C D E
A B C D E
A B C D E
A B ---- time
Period

Period length trade-offs

Longer period	*Shorter period*
Manufacturing efficiency	Quicker customer response
Larger purchase quantities	Lower inventories (cycle stock)
Fewer setups	Higher potential market share

Source: D. C. Whybark, "Production Planning and Control at Kumera Oy," *Production and Inventory Management,* first quarter 1984.

final decision requires management to assess the trade-offs between all factors.

For the example in Figure 19.7, a five-week period was chosen. This means each group will be scheduled 10 times per year (using a 50-week year). Once the period length is chosen, forecasts are used to calculate the number of products to be produced in each production set each period. Figure 19.9 shows the master production schedule for the example. The manufacturing objective is to have completed, by the fifth week of the period, all materials required for assembly of the actual products in each production set. This means, as the next group A is launched, the previous group A is being scheduled into assembly. (See the top of the schedule.) Kumera thinks in terms of the launch date for each production set. (See the bottom of the schedule.) It widely distributes timing table data to customers and vendors as well as within the company.

To specify material requirements, the exact customer orders to be produced in a production set must be determined. This is the job of order entry in demand management. An attempt is made to assign each order to the production schedule that will meet the delivery date requested by the customer. For example, a request for an item from the B product group for delivery in week 14 would be assigned to B_8, if possible. (See Figure 19.9.)

Order promising is done in accordance with the commitment between marketing and manufacturing. No more orders may be entered into a production set than the period quantity (e.g., 10 units for B in each production set). Once the period quantity is reached, the next customer order must be delayed until a later production set, or the order must be exchanged for one in an earlier

FIGURE 19.9   Master Production Schedule (the Timing Table)

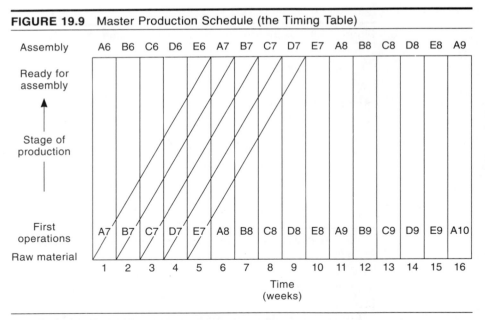

Source: D. C. Whybark, "Production Planning and Control at Kumera Oy," *Production and Inventory Management,* first quarter 1984.

set (if possible). Kumera looks at the timing table as a train schedule. When the launch date arrives, production orders for the set are released. (The train leaves.) The sales department can make any allocation of schedule dates among customers until the train leaves for the factory. After that, no more changes; they have to wait for the next train. No release is made for any order that doesn't have the engineering completed, thereby avoiding potential mix-ups in manufacturing. If the launch date arrives and the "train is not full," sales can decide to make some orders in anticipation (i.e., against a forecast), but this is very dangerous to this firm since most orders are custom-made. The alternative is to pull up an order from a production set scheduled for a later release. At the launch date, the train *leaves*—sometimes with empty seats! That is, if the production set isn't up to the period quantity, the company only makes to order. As one Kumera manager says, " If there are empty beds in the hospital, you don't go out and make people sick just to fill the beds."

When the release date has arrived, a composite bill of material is prepared for the actual set of customer orders in the production set. The composite bill is built up from the separate bills for each product in the set. Actual inventories and allocations are checked against the computer records (Kumera runs well over 95 percent accuracy) before calculating net requirements and release to the shop.

Lot sizing is performed for selected items. The bill of material processor can look ahead to the actual orders in the next and subsequent production sets for end items using the same parts. Subject to parameters on how far ahead to look, some aggregation of order demands can be used to justify producing more than the net requirements.

Once quantities have been determined, items are released to the factory floor and the shop-floor system carries them through to completion. Purchased items are scheduled to be received as needed on the week before the staging week (providing one week of safety lead time). They're collected into an assembly storage area where they're joined by the manufactured parts.

Purchase and shop orders are released at the same time. Shop orders have an implicit due date at assembly staging time (five weeks hence in the example). Purchased raw materials are due prior to the assembly staging week. Kumera numbers the production sets in release date sequence. The sequence numbers are displayed conspicuously on all shop papers associated with the production set. They provide a very simple shop-floor priority rule: "Always work next on the item with lowest-numbered production set."

The production set sequence numbers are also used for other circumstances. For example, emergency production of pieces not required in assembly (e.g., a part needed because a customer's machine is down) can be accomplished by assigning a low number. The scheme can also be used for producing low-priority items (e.g., service parts not immediately required) as capacity is available. The assembly foreman schedules the assembly sequence to make an efficient progression through the products in the production set.

If any products are still awaiting parts, their assembly can be delayed until the latter part of the week. The late parts will have the lowest number (highest priority) on the shop floor, and so will go through as quickly as possible.

Not all purchase orders can be delayed until the release date for the production set because of long lead times. In these cases, purchase quantities are calculated to provide just enough inventory to produce the period quantities scheduled for future production. Kumera works at keeping the parts and raw materials ordered this way highly common. The commonality and gross to net logic used in calculating the production quantities help keep the residual inventories low. To help key vendors plan their production, Kumera gives them timing table data. Vendor follow-up is done from this information or the staged inventories for assembly.

Using the System

The periodic control system's simplicity enabled Kumera Oy to do the implementation in six months. The totally manual initial system was converted to the computer two years later. The system required certain disciplines and induced others. Some were necessary to use the system effectively, while others facilitated later improvements.

Inventory accuracy is essential at the time of net requirement calculation. The physical check at the time the bill of material is processed ensures correct counts. At Kumera, the required accuracy is maintained even though not all inventory is locked up. The short period helps maintain accuracy, since there's less time for shrinkage; short periods also result in smaller production lots and lower inventory quantities.

The marketing disciplines were the most difficult—no more maximization of sales or arbitrary delivery promises. There were to be no more orders in a given production set than the period quantities. Customers were to be told honestly when delivery could be expected and, if the date wasn't good enough, exchanges with previously scheduled customers would have to be worked out or the order lost. There would be no more "we'll try" answers.

At the management level, adherence to the game plan and willingness to back the order entry function when difficult decisions were being made was required. Management integrated the budget and profit-planning activities with the periodic control concept. That way, all management activities were keyed to the same basic information and control reports.

In production, the required disciplines weren't as difficult, but some major changes were required. The new priorities had to be followed and reliance on "hot" or shortage lists had to stop. The quantities indicated on the shop orders would be the quantities produced. (No more running a "few extra" since the machine is set up.) The short period meant more setups and lower inventories.

One reaction to Kumera's improved delivery reliability was a shift in customer behavior. As they recognized promised delivery dates were honest, customers began to rely on them. Distributing timing table data to customers shifted their ordering patterns to correspond more closely with Kumera's schedules.

Much the same thing happened with vendors. As vendors realized the timing table really was used for purchasing, they found they could make better plans. Distributing the data to key vendors reinforced this. As Kumera reduced "panic" buying, the schedule's utility for the vendors was further reinforced.

Internal changes evolved too. Engineering was added to the order entry checks. No customer order is released to the shop without complete engineering. This means order entry must check before promising a delivery date (assignment to a production set) for any product with substantial engineering to be completed. The same is true for orders needing long lead-time items that aren't part of the common items purchased prior to release. Purchasing is consulted before a promise date is given.

Payoffs

There have been many tangible benefits from installing the periodic control system and some intangible ones as well. Tangible benefits include lower inventories, improved margins and customer service, and fewer expeditors, as Figure 19.10 shows.

Developing production groupings has greatly facilitated the forecasting task and helped focus marketing on meeting its sales objectives for the groups. The scheduling and priority systems have routinized management of the production and assembly activities. The problems are very visible. This focuses

FIGURE 19.10   Some Payoffs from Periodic Control

	Before periodic control	Six months after implementation	Two years after implementation
Inventory turns	2.5	9.2	10.1
Late deliveries	50%	10%	0%
Gross margin	10%	27%	30%
People: production control	4	1	1
People: expediting	2	1	0

Source: D. C. Whybark, "Production Planning and Control at Kumera Oy," *Production and Inventory Management,* first quarter 1984.

management attention on them quickly, much as the just-in-time approach does.

The system's simplicity makes it easy to work with. The system is transparent to everyone in the company, so staff can make well-informed decisions. No one guesses at priorities, changes instructions on work-in-process, or shifts the schedule around. This has improved all working relationships and enhances the attitude of working together to solve problems as they appear.

Engineering changes are much easier to implement and manage. They are always tied to a production set in a way similar to that done by some advanced JIT users. With low inventories in the shop and actual customer orders in every production set, the effectivity dates for engineering changes are much easier to determine and control.

One strong test of the system and its benefits was demonstrated during a recession in Finland. Kumera managed to expand business during the period, largely due to its customer-service capability. The firm also has vertically integrated. The periodic control system was installed in most new acquisitions as a first step in improving the operations.

Kumera's periodic control system is more than a technique for planning and controlling production. It's a management tool and state of mind. Sound management principles at the system's front end are key to successfully applying the engine and back-end techniques.

MPC FOR THE PROCESS INDUSTRIES (PRISM)

Some years ago, MPC authorities attempted to apply systems and techniques from discrete manufacturing to the process industries. These efforts met with limited success. Moreover, with the growing use of just-in-time (JIT) approaches to MPC, there's now some question as to whether the flow of ideas should be reversed: increasingly, discrete manufacture is becoming a continuous flow. For instance, PRISM is a software package designed explicitly for manufacturing planning and control in the process industries. It is of special interest to firms with continuous flow processes but provides insights into batch processes.

Process industries share many features with discrete industries, but there are also situations and problems calling for unique approaches. A catsup manufacturer can buy tomatoes with 10 to 20 percent water content, 12 to 17 percent sugar content, and differing numbers of blemishes. All are within acceptable specifications, but each different lot purchased needs to have the differences identified, be processed accordingly, and have more or fewer other ingredients added. If tomatoes have a high water content, catsup manufacturing takes longer; if sugar content is low, more sugar is required; if blemishes are numerous, it may take more tomatoes per bottle of catsup. Things get even more complex when multiple products are made. Whole tomatoes can only be packed from top-quality input raw material, whereas catsup and to-mato paste can be made from lesser grades. Moreover, tomatoes and most food materials deteriorate rapidly; their "specifications" are constantly changing, and this information needs to be reflected in the "on-hand balance" data used for manufacturing planning and control.

Another unique feature of the process industries is the frequent occurrence of by-products, coproducts, different grades of output, materials that are reused in the process, and waste products. It's sometimes difficult to predict the amounts of these elements at the outset. Many companies in the process industries want to reflect these considerations in their MPC systems and monitor actual results against expectations.

Furthermore, for many process industry companies, several classes of resources go beyond "materials" that need to be planned and controlled at a detailed level. Included are utilities, equipment capacities, crews, and special items such as vessels for transporting product or processing units that must follow a set pattern of use and cleanout.

Finally, company management wants to evaluate the performance of manufacturing! A cynical accountant at one food manufacturer said when things went well, it was chalked up to excellent manufacturing; when things went poorly, it was a bad growing season. The MPC system needs to identify the particular ingredients, their associated properties, and many other conditions of manufacturing to determine if performance is superior or mediocre.

Production Resources

The heart of the PRISM system is based on the concept of **production resources.** A resource is defined as any element that is required for production or results from production. Examples include machine-hours, labor-hours, materials, utilities, tools, completed products, waste products, and effluent. Production resources are inputs and outputs from the various tasks in the production process that management wants to control. Each resource is tied to a "production model" in that the resource is explicitly defined as an input to or output from a production task.

All resources are specified and controlled at the level of "class," "subclass," and "resource." For example, within the class of utilities, there might be subclasses of electricity, natural gas, coal, and steam. Defined resources for a subclass might be meters 1, 2, and 3 which monitor the specific amounts of natural gas used in three different furnaces in the plant. Readings before and after a particular production batch is produced provide data on specific resource usages and outputs.

Resources can be defined at whatever level of detail management wishes. The lower the level of detail, the more transaction costs will be incurred, and the greater the detail about consumption of resources. The process industries, to a much greater degree than batch manufacturing, need to be concerned with many resources other than material resources. Moreover, the output resources are produced at many more points in the process than the typical *finished-goods* units produced in discrete batch manufacturing. A petroleum refinery might have inputs such as distillates or additives coming in at various steps or tasks in the process, and outputs such as sulfur or heating gases arising at other steps.

PRISM distinguishes between *balance* and *nonbalance* resources. Utilities, labor-hours, and machine-hours are nonbalance resources, since we don't carry them in inventory from period to period. On the other hand, raw materials and additives are carried over from one period to the next (and perhaps from one task to another) and require inventory accounting for proper planning and control.

PRISM also allows inventories to have a multitude of classifications, such as "hold for quality check," restricted use, quarantine, potency, shelf life, or any other attribute that's important for manufacturing. This capability provides great flexibility. For example, goods coming from a "certified" vendor don't have to go through the same set of quality checks as those for other vendors; finished goods (resources) for a particular customer can be earmarked for special transportation or packing.

This resource classification capability provides for *lot tracking and control.* This is critical in food and pharmaceutical production, but it's becoming increasingly important in other MPC environments as well. Lot control is found in many software packages, but the ability to use it on *any* resource is particularly important for environmental monitoring and other situations when more than end products need to be tracked and controlled. Lot tracking in PRISM can be turned off and on as resources go through the system. During some phases it's either impossible or meaningless to maintain lot control. When lot control is in place during these phases, extra cost is incurred without any benefit.

Production Models

PRISM utilizes a **production model** as the fundamental way to organize MPC, instead of the traditional bill of material and routing file as used in discrete manufacturing. The production model combines features of bills of materials and routings with the required resource inputs (whatever they may be) and the expected resource outputs (whatever they may be and wherever they may occur). Production model building is essentially process analysis, where a unique flow diagram is designed for each product to be manufactured and for each way this manufacturing can take place. For example, if a product might be manufactured with two alternative processes, there will be two production models to describe how manufacturing can be done. Figure 19.11 depicts the production model concept.

FIGURE 19.11 The PRISM Production Model Concept

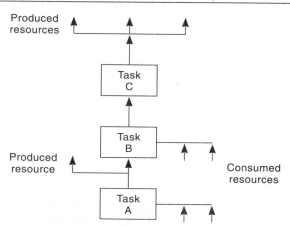

Each production model is divided into tasks, which represent the lowest level at which resource consumption, resource creation, and yield information can be determined. This definition means any process can be defined in great detail or in highly aggregated terms depending on management needs. If a processing unit does not require detailed control, it could be considered as a task; if more detailed analysis is required for monitoring the unit's performance, then many subtasks could be explicitly identified.

Transaction Processing

In PRISM, tasks are defined as either reportable or nonreportable based on the desired level of subsequent analysis. PRISM recognizes the trade-off between the costs of transaction processing and having more detail for analysis with a feature called *spotlighting*. Nonreportable transactions can use a theoretical calculation; for example, the standard quantity of a resource such as labor can be assumed to have been consumed (like backflushing, as done in JIT). This relieves the system from paying the transaction cost of picking up the precise amount of labor consumed in that task. Whenever the assumption for some nonreportable transaction needs to be tested, it can be reset as a reportable transaction, (i.e., it can be spotlighted).

Nonreportable transactions can also be used for resource outputs, such as the amount of steam generated in a chemical process's product. Whether you want to report the actual transaction is a function of both predictability and importance. Some transactions are more critical than others to overall performance.

PRISM can also perform "aggregate" reporting. In this case, a user makes only one transaction for a resource consumed in several tasks. After that, the overall amount is prorated to the particular tasks and production plans on some predetermined basis such as output volumes. Another example might be to allocate the overall labor cost for a packaging line to the number of each product produced during the period.

Expected yield data are part of the production model. For each task, the expected yields are specified and actual yields collected. Actual yields can be quite different than expected if there are significant differences in raw materials.

Performance Measurement

One of PRISM's best features is its flexibility in terms of performance measurement. Figure 19.12 shows the reporting schema for collecting measurement data. *Any* resource can be monitored at the class, subclass, or resource level. This means measurements can evolve as needed with minor adjustments to the system. If a transaction against a resource is being reported and now the firm believes the resultant measurements are no longer critical, those transactions' status can be easily changed to nonreported. Similarly, the new measurements may require some nonreported transactions to now be reported. This open-ended measurement feature allows MPC to take on the meaning it truly deserves—*manufacturing* planning and control.

By having an open-ended ability to change measurements, PRISM allows a company to concentrate on what's important *now,* including more than planning and control of materials. If quality is poor in a particular task, detailed metrics can be used to aid the detective work necessary to fix the problem.

FIGURE 19.12 Reporting Detail, Accuracy, and Costs

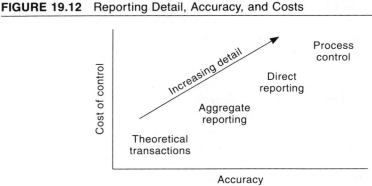

When it's fixed, a new problem can be identified, and new metrics formulated. The overall result is an MPC system consistent with continuous improvement.

Aggregate performance measurement is less accurate and expensive than performance based on detailed transaction processing, but may suffice. For example, daily consumption of natural gas can be determined from a meter, and thereafter allocated to the products produced that day on the basis of the theoretical amount each requires. In many instances, aggregate performance is good enough—it indicates details are indeed proceeding as planned.

Direct reporting of performance measurement requires that direct issues and other transactions be posted to the particular production models to which they're assigned. In some cases this is the only way to determine if a process is under control. On the other hand, direct measurement is expensive; a desired goal in MPC should be to reduce the cost of control.

The most accurate and most expensive performance measure in Figure 19.12 is for real time process control. Clearly this is possible and desirable in some instances. Many users have found coupling process controls to MPC software to be an excellent way to provide timely information to managers.

The fundamental concept imbedded in PRISM's spotlighting is to only implement more detailed transaction reporting when it's warranted. If an actual overall result is consistent with the sum of detailed theoretical transactions, then we'd probably conclude more expensive controls are unwarranted. Periodic sampling based on more accurate reporting can be used to test the assumptions' efficacy.

Implications for MPC Practice

PRISM has many common features with other MPC systems, but several features are unique. The major difference is being able to plan and control many more resources than with MRP-based systems. The critical resources may not be material or capacity; they might be energy, waste management limitations, labor, or some other resource. In its planning phase, PRISM has an innovation somewhat similar to OPT; in essence material and capacity plans are made simultaneously, rather than sequentially. This means all resources, not just material, are planned at the same time. The master production schedule is processed through each planning model to determine the resources required. Resources, in turn, are categorized as being either infinite, limited, or finite. The infinite assumption is similar to that used for typical MRP processing. Limited resources are those that *can* be expanded, but at extra cost and to a limited degree, such as direct labor with overtime. Finite resources are fixed and must be scheduled with finite loading techniques. The overall result is similar to that used in OPT, where finite constraints dictate the achievable MPS, and other resources are scheduled with greater flexibility. The major addition is that PRISM does the planning for *all* resources specified in the production model.

PRISM also allows for lead times to be either input as fixed times or calculated internally. In the latter case, run times can be linear (Setup time + Time per unit) or batch (fixed time irrespective of number of units). Clearly the choice depends on the particular conditions of the process. Cleanup times are also explicitly included since they're often required in the process industry. Also, when feasible, PRISM can do overlap scheduling. For example, if the first task in a production model takes longer than the second, PRISM can start the first soon enough to build sufficient inventory in front of task 2 so the two tasks finish the batch at the same time.

OBSERVATIONS ON MPC EVOLUTION

In this chapter, we have examined three quite different approaches to manufacturing planning and control. Each represents an evolution in MPC systems. TOC and OPT are important because they make the treatment of constrained resources explicit. Thus the feedbacks from the back end to the front end assure the MPS is a *doable* MPS. Separating work into what crosses bottlenecks and what doesn't allows a critical focus on what's really important. The philosophy also makes clear what's *not* important: utilization of nonbottleneck resources.

The Kumera Oy system pushes the frontiers because it's a time-driven system, rather than a material-driven system; that is, the Kumera Oy system assigns orders to time periods in the same way passengers get on a train. If you miss a train, you wait for the next one. The objective is to have high-speed trains and frequent schedules. Factory execution of all material flow dictates is based on time assignments. By slotting a particular customer into a unique time bucket, all parts of the order are coordinated in a relatively simple way. The series of these time buckets makes up a "train schedule." This schedule is broadcast to both customers and suppliers. The result is an integrative force for better interfirm cooperation.

The PRISM system, designed explicitly for process industries, expands the criteria as well as the issues addressed in MPC systems. The concept of "material" is widened to include all resources that flow into and out of process "tasks." The system also supports variability in resources, such as quality or potency.

One of PRISM's most interesting features is the evolutionary posture taken to performance measurement and data collection. Increasing costs of detailed transaction processing can be traded off against the attendant benefits. As problems (such as a particular yield situation) are solved, new ones are identified, and the data collection system is modified to reflect this change in managerial attention.

CONCLUDING PRINCIPLES

As we review these new approaches and learn of others, we're reminded that MPC systems and philosophies continue to evolve. From these frontiers, however, some principles do emerge:

- There is no "ultimate weapon" in MPC systems. Firms should continually evaluate improvements.
- The MPC framework is useful in assessing where a particular approach or system fits within the system.
- A working MPC system should be in place before you undertake improvements.
- You should concentrate on bottleneck resources to focus material planning and capacity planning on the vital few.
- Variable lot sizes should be used to reduce manufacturing lead time and reduce work-in-process.
- Material-based scheduling systems should be applied to process industries only where appropriate. Rate-based systems might be better suited.
- A broad range of resources and strategic objectives can be simultaneously planned with some of the emerging MPC systems concepts.
- Clear communication of master schedule information should be used to improve customer and vendor coordination.
- Management should use direct reporting or process control transaction detail only when the accuracy is cost justified.

REFERENCES

Billington, P. J.; J. O. McClain; and L. J. Thomas. "Heuristics for Multilevel Lot-Sizing with a Bottleneck." *Management Science* 32, no. 8 (August 1986).

Bolander, S. F.; R. C. Heard; S. M. Seward; and S. G. Taylor. *Manufacturing Planning and Control in the Process Industries*. Falls Church, Va.: American Production and Inventory Control Society, 1981.

Carlson, J. G. "Microcomputers for Demonstrating G/T, JIT, Kanban and MRP." *Production Emerging Trends and Issues*. Netherlands: Elsevier, 1985.

Cooper, R. and R. S. Kaplan. "Profit Priorities from Activity-Based Costing." *Harvard Business Review,* May–June 1991.

Dixon, J. Robb; A. J. Nanni; and T. E. Vollmann. *The New Performance Challenge*. Homewood, Ill.: Dow Jones–Irwin, 1990.

Ebner, M. L., and T. E. Vollmann. "Manufacturing Systems for the 1990's." In *Intelligent Manufacturing: Proceedings from the First International Conference on Expert Systems and the Leading Edge in Production Planning and Control,* ed. M. Oliff. Menlo Park, Calif.: Benjamin/Cummings, 1987.

Everdell, Romeyn. "MRPII, JIT, and OPT: Not a Choice But a Synergy." *1984 APICS Conference Proceedings.*

Fogarty, D. W.; J. H. Blackstone; and T. R. Hoffmann. *Production and Inventory Management,* 2nd ed. Cincinnati: Southwestern, 1991, chapter 19.

Fox, Robert E. "MRP, Kanban or OPT, What's Best?" *Inventories and Production,* January–February 1982.

_____. "OPT: An Answer for America, Part II." *Inventories and Production,* November–December 1982.

_____. "OPT: An Answer for America, Part III." *Inventories and Production,* January–February 1983.

_____. "OPT: An Answer for America, Part IV." *Inventories and Production,* March–April 1983.

Goldratt, Eliyahu. "The Unbalanced Plant." *APICS 1981 International Conference Proceedings,* pp. 195–99.

_____, and J. Cox. *The Goal.* Norwich, Conn.: North River Press, 1984.

_____. *The Theory of Constraints Journal* 1, 1989.

_____. *The Haystack Syndrome.* Norwich, Conn.: North River Press, 1990.

Jacobs, F. Robert. "The OPT Scheduling System: A Review of a New Production Scheduling System." *Production and Inventory Management,* 3rd quarter 1983.

_____. "OPT Uncovered: Many Production Planning and Scheduling Concepts Can Be Applied with or without the Software." *Industrial Engineering,* October 1984, pp. 89–95.

_____, and D. J. Bragg. "Repetitive Lots: Job Flow Considerations in Production Sequencing and Batch Sizing." *Decision Sciences* 19, no. 2 (1989), pp. 281–94.

Kumpulainen, Vesa. *Periodic Production Control.* 1100 Riihimaki 10, Finland: Kumera Oy, 1983.

Lundrigan, R. "What Is This Thing Called OPT?" *Production and Inventory Management,* 2nd quarter 1986, pp. 2–12.

Macchietto, S. "The Integration of Planning, Scheduling, and Control Systems in the Process Industry." *Proceedings of the "Beyond MRP" Workshop,* The Hague, November 1990.

Mattila, Veli-Pekka. *Periodic Control System.* 1100 Riihimaki 10, Finland: Kumera Oy, 1983.

Meleton, M. P., Jr. "OPT—Fantasy or Breakthrough?" *Production and Inventory Management,* 2nd quarter 1986, pp. 13–21.

Moily, J. P. "Optimal and Heuristic Procedures for Component Lot-Splitting in Multi-Stage Manufacturing Systems." *Management Science* 32, no. 1 (January 1980).

Nakane, J., and R. W. Hall. "Management Specs for Stockless Production." *Harvard Business Review,* May/June 1983, pp. 84–91.

Suresh, N. C., and J. R. Meredith. "Achieving Factory Automation through Group Technology Principles." *Journal of Operations Management* 5, no. 2 (1985).

Swann, D. "Using MRP for Optimized Schedules (Emulating OPT)." *Production and Inventory Management,* 2nd quarter 1986, pp. 30–37.

Taylor, S. G., and S. F. Bolander. "Process Flow Scheduling Principles." *Production and Inventory Management* 2, no. 31, 1st quarter 1991, pp. 67–71.

Thompson, Olin W., and S. J. Connor. "Upgrading Accounting and Costing Systems." *Manufacturing Systems,* May 1986.

————. "Computer-Integrated Manufacturing in the Process Industry." *P&IM Review*, February 1991.

Vollmann, Thomas E. "OPT as an Enhancement to MRPII." *Production and Inventory Management*, 2nd quarter 1986, pp. 38–47.

Zijlstra, P., and J. Wijngaard. "An Application of MRP in Batch Process Industry." *Proceedings of the "Beyond MRP" Workshop*, The Hague, November 1990.

DISCUSSION QUESTIONS

1. How would you respond to the question, "Which is better: MPC or MRP?"
2. How can the MPC system of Figure 19.1 be used to evaluate newly developed software?
3. A recent promotion implied a progression from EOQ to MRP to JIT and finally to OPT. How would you react to this claim?
4. What do you feel about the transparency of a system? Must people know how a procedure works to be able to use it?
5. The priority rule at Kumera is "always work on the items from the lowest-numbered production set next." What shop-floor dispatching rule is that equivalent to? Could it be improved upon?
6. It's tempting to conclude it's unimportant what the system's technical choice is, given the importance of management. Can you provide examples of where the technical choice is important?
7. Why does the text suggest the repetitive lot concept would be most beneficial to high-volume manufacturers with limited product lines having numerous operations?
8. What concerns would you have about jobs in the queue when the repetitive lots concept is used?
9. It's easy to imagine the co- and by-products PRISM helps manage in process industries. Examples include gas and petroleum, chemical by-products, and slag from steel. Some of these could require actions that could limit the facility's capacity. Are there such problems in discrete industries?

PROBLEMS

1. Marucheck's makeshift manufacturing facility had three departments: shaping, pickling, and packing. Marucheck's orders averaged 100 pieces each. Each of the three shaping machines required one hour setup, but could run a piece in one minute. The pickling department lowered baskets of pieces into brine tanks and subjected them to low-voltage current, a heating and cooling, and a rinse. The whole process took four hours for any number of baskets or pieces. The only brine tank could hold four baskets, each of which could contain 50 pieces. (Baskets were loaded while another load was in the tank.) Each piece was inspected and wrapped in bubble pack in the packing department. Each of the four people in the department could do this at the rate of 25 pieces per hour.

Marucheck had heard of OPT and wanted to identify the bottleneck department. Which is it?

2. The Optima Shop has two work centers: Big Mess and No Problem. The Monday list of orders to be filled this week shows the total setup and total run time requirements (in hours) at the two centers.

	Big Mess	*No Problem*
Customer order 1	Part A: setup 5, run 10	Part C: setup 1, run 2
	Part B: setup 2, run 5	
Customer order 2	Part A: setup 5, run 3	Part D: setup 2, run 2
Customer order 3	Part B: setup 2, run 5	Part C: setup 1, run 3

Joe Biggs, the scheduler of Optima, said his first criterion is to minimize setups, and the second is to prioritize customer orders in numerical order.
   a. How should he schedule part production in the two centers? (Please illustrate using a Gantt chart.)
   b. Can customer delivery promises be met without overtime if capacity is 40 hours in each work center?
   c. How does the answer to part b change if only one person is assigned to do the setup in *both* work centers?
   d. Is the schedule consistent with the OPT philosophy?

3. Consider a work center with a 40-hour-per-week capacity. MRP planned orders for that work center show the following requirements for the next several weeks.

Job	*A*	*B*	*C*	*D*	*E*	*F*	*G*	*H*
Hours required	5	20	15	30	45	15	20	10
Week release planned	1	1	2	2	3	4	4	4

   a. Plot the weekly load against capacity.
   b. How would you adjust the schedule to meet capacity?

4. Consider the following chart showing the capacity load by week for a department with 40 hours per week of available capacity.

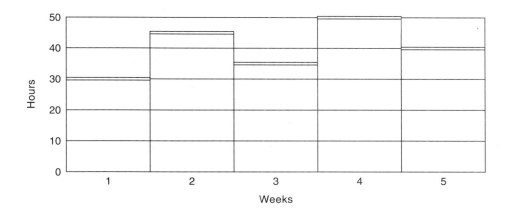

a. How would you adjust the load (assuming jobs could be started in one week and be finished the next)?
b. What "decision rule" would you use to do the adjustment in a dynamic situation?

5. In an effort to make better use of the painting shop, Farshad Raffi decided to take capacity into account as he developed his schedule for the shop. Painting had an eight-hour day, had no overtime, and didn't allow jobs to carry over into the next day. On a Monday morning Farshad had the following orders to be finished in the current five-day week:

Job	1	2	3	4	5	6	7	8	9	10	11	12	13	14	15
Hours	1	3	2	2	1	6	2	3	1	2	5	2	1	1	2
Color	W	W	B	B	W	B	W	W	B	W	B	W	W	B	B
Due	M	M	W	Tu	Tu	W	W	Th	Th	F	W	F	Th	F	F

It takes no time to go from running white (W) to running black (B), but it takes an hour to clean the system when going from black to white. Is there a schedule that will allow him to meet all due dates? How would you describe your process to somebody else? (Assume the shop last ran white.)

6. The Ace Tool Company is considering implementing the repetitive lot concept in scheduling the firm's fabrication shop. The production manager selected an example order to use in evaluating benefits and potential costs of this scheduling approach. A transfer batch size of 100 units was suggested for this item. The example order is for a quantity of 1,000 units and has the following routing data:

Operation	Work center	Setup time	Run time/unit
1	1	40 minutes	2.4 minutes/unit
2	2	20 minutes	1.44 minutes/unit

a. Assuming a single-shift, eight-hour day, five-day week for work centers 1 and 2, prepare a Gantt chart showing the earliest start- and finish-time schedule for this order when the repetitive lot concept is used, and under a conventional scheduling approach when all items in the order are processed at one time. What are the earliest start and finish times for each transfer batch at work center 2, assuming none of the transfer batches are processed together to save setup time?

b. What's the difference in the order-completion times under the two scheduling approaches in part a above?

c. What are the benefits and potential costs of this scheduling approach?

7. Recently, the Universal Machine Tool Company introduced the repetitive lot concept in scheduling its fabrication shop. The following dispatching report lists the orders ready for processing at the K&T machining center at 8 A.M. Tuesday. The K&T machining center has just completed an order for 20 units of part number 6633 on shop order number WE 433. If the earliest due date dispatching rule is used to schedule orders at this work center, determine the sequence in which orders should be processed (assuming no new orders enter the work center).

Shop order number	Part number	Order quantity	Operation quantity	Setup time	Run time	Order due date
XX 234	8965	80	10	0.5	3.0	951
RT 435	6123	50	50	0.1	6.5	918
GI 209	6754	160	16	0.4	5.8	941
TV 244	9087	54	54	0.9	8.2	925
WE 433	6633	100	20	0.8	4.4	944
US 899	7831	210	210	0.7	5.0	923
XX 234	8965	80	10	0.5	3.0	951
WE 433	6633	100	20	0.8	4.4	944
GI 209	6754	160	16	0.4	5.8	941
WE 433	6633	100	20	0.8	4.4	944

8. Calculate completion times for part B of Figure 19.6. Assume there are no other jobs in the shop and each operation will know when a transfer batch will reach it. There doesn't need to be a complete operation batch available to start work at any of the three operations but, once started, the operation batch must

be completely processed without waiting for additional transfer batches (i.e., no idle time during an operation batch's run).

9. Consider the following data for three jobs processed in machine center 48Z and the queue information for machine center 48Z.

Job	Setup time (minutes)	Run time/unit(minutes)	Batch size
A	10	0.1	100
B	30	0.05	200
C	5	0.2	100

Queue data for machine center 48Z:

Job	Arrival time	Job	Arrival time	Job	Arrival time
A	9:41	C	10:05	C	10:35
A	9:43	B	10:15	C	10:40
B	9:47	B	10:20	A	10:41
C	10:01	B	10:22	B	10:50
A	10:02	C	10:23	A	10:55
C	10:03	A	10:30	C	10:59

a. If machine center 48Z used a first-come/first-served rule to process jobs, how long would it take to process the queue? (Assume no other jobs arrive, all jobs in queue are for one batch each, and a job C has just been completed.)
b. How long would it take to process all jobs in the queue using a repetitive lot logic?

10. A firm considering using the periodic control system has divided its product line into four groups with the following annual forecast in units (assuming a 50-week year):

Group	Forecast
I	400
II	200
III	500
IV	100

a. The company has decided on an eight-week period. Calculate the period quantities.

b. If group I takes 40 direct labor-hours per unit, group II takes 20, group III takes 5, and group IV takes 50, what is the allocation of labor capacity to the groups?

c. Plot the "timing table" for the items.

11. After several months of operation, the order board (future production sets) for a company with periodic control and five product groups looks like this:

Product group	Period quantity	Production set number							
		1	2	3	4	5	6	7	8
A	100	50	20	0	0	10	0	0	0
B	50	50	50	50	50	50	50	40	30
C	200	200	200	120	100	50	0	20	0
D	10	10	10	8	5	2	0	1	0
E	40	20	2	0	0	0	0	0	0

a. What do you make of the current situation?

b. What actions would you recommend?

12. What is the composite bill of material for a periodic control group that has 10 items (3 of part 2, 2 of part 7, and 5 of part 10) with the following individual bills?

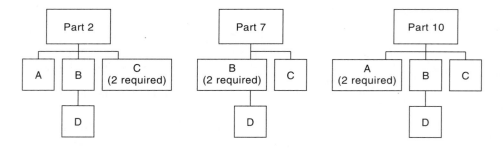

What quantities should be ordered if the inventories are:

Part	Amount
2	0
7	0
10	0
A	10
B	0
C	20
D	100

13. A small Norwegian firm produced three types of decorative wooden dolls: trolls, gnomes, and elves. Three manufacturing steps were used to produce the dolls: roughing out the shape, finishing the shaping, and painting. All dolls went through all three departments in the same order, from rough through finish to paint. Total labor-hours required per doll type and the number of workers in each department are as follows:

| | Work departments | | | Period |
	Rough	Finish	Paint	quantities
Trolls	5	3	7	15
Gnomes	3	2	6	20
Elves	8	5	11	10
Number of workers	2	1	3	

For several months the company has been trying the periodic control system. It has worked out that a production set of each type can be released each week on a rotating basis. Thus, each three weeks, one doll type will be started into production and will be completed in three weeks. This means each department has one week to finish its work on each set. The company has been able to do quite well with the system by using period quantities of 15 for trolls, 20 for gnomes, and 10 for elves. The company works a 40-hour week. A little overtime has been required in the finishing department in order to stick to the schedule.

a. Use a spreadsheet to calculate the labor-hour load by department by week. How well is it balanced?

b. After a few more months product mix was clearly changing. Elves' popularity was increasing at gnomes' expense. Sales felt overall demand for dolls was still about the same, so they suggested a simple change in the period quantities. They reasoned that a switch to 10 for the gnomes and 20 for the elves would allow them to get back to a reasonable delivery promise time. They also felt this would make no difference to the factory since the quantities would remain the same. What will be the impact of this suggestion?

c. Using the data from part b, what suggestions would you make about employment and staffing for the three departments?

14. Develop a PRISM production model for the following brewing process. Brewing consumes yeast, hops, grains, and water while producing raw beer, CO_2, and garbage. The filtering step consumes filters and energy, while producing finished beer. The bottling step consumes bottles, energy, and caps, while producing finished beer and garbage.

15. A process for depositing a thin layer of a special protective alloy on ceramic plates for use in space research has been developed. It requires dipping the plate into a special solution, providing a very intense short electrical charge, removing the plate (which is covered with the alloy and a film of contamination), and rinsing the plate in a cleansing solution from which the

contaminants must be removed in another process. The PRISM production model is shown in the figure.

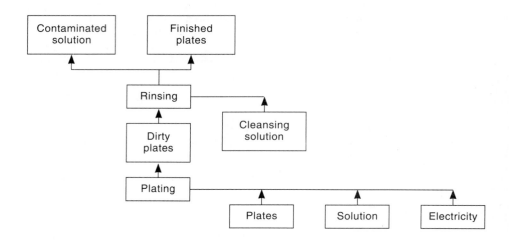

A new aerospace project requires 100,000 coated ceramic plates. Uncoated plates can be supplied at the rate of 1,000 per day; it takes 44 pounds of solution to produce 1,000 coated plates. Electricity can be provided in any quantity required, but the special solution can only be obtained at the rate of 200 pounds/week. The cleansing solution can be purchased in any quantity, but decontamination of the solution after use can only be done at the rate of 200 gallons per day. Each plate needs an average of one half gallon of cleansing solution to remove the contaminated film. If the firm works a seven-day week, how many days will it take to complete the 100,000 plates required?

Index